9/93   H3152351   COSO99 mt   58.95

ADVANCED COMPUTER ARCHITECTURE:
Parallelism, Scalability, Programmability

McGraw-Hill Series in Computer Science

Senior Consulting Editor

C. L. Liu, *University of Illinois at Urbana-Champaign*

Consulting Editor

Allen B. Tucker, *Bowdoin College*

Computer Organization and Architecture

Bartee: *Computer Architecture and Logic Design*
Bell and Newell: *Computer Structures, Readings and Examples*
Cavanagh: *Digital Computer Arithmetic, Design and Implementation*
Gear: *Computer Organization and Programming*
Hamacher, Vranesic, and Zaky: *Computer Organization*
Hayes: *Computer Architecture and Organization*
Hwang: *Advanced Computer Architecture: Parallelism, Scalability, Programmability*
Kogge: *The Architecture of Symbolic Computing*
Scragg: *Computer Organization: A Top-Down Approach*
Sieworek, Bell, and Newell: *Computer Structures: Principles and Examples*
Ward and Halstead: *Computation Structures**

* Co-published by the MIT Press and McGraw-Hill, Inc.

Networks, Parallel and Distributed Computing

Ahuja: *Design and Analysis of Computer Communication Networks*
Filman and Friedman: *Coordinated Computing: Tools and Techniques for Distributed Software*
Hwang: *Advanced Computer Architecture: Parallelism, Scalability, Programmability*
Hwang and DeGroot (eds.): *Parallel Processing for Supercomputers and Artificial Intelligence*
Keiser: *Local Area Networks*
Kershenbaum: *Telecommunications Network Design Algorithms*
Lakshmivarahan and Dhall: *Analysis and Design of Parallel Algorithms*
Quinn: *Designing Efficient Algorithms for Parallel Computers*
Siegel: *Interconnection Networks for Large-Scale Parallel Processing*

McGraw-Hill Series in Computer Engineering

Senior Consulting Editors

Stephen W. Director, *Carnegie Mellon University*

C. L. Liu, *University of Illinois at Urbana-Champaign*

Bartee: *Computer Architecture and Logic Design*
Bell and Newell: *Computer Structure: Readings and Examples*
Garland: *Introduction to Microprocessor System Design*
Hamacher, Vranesic, and Zaky: *Computer Organization*
Hayes: *Computer Architecture and Organization*
Horvath: *Introduction to Microprocessors using the MC 6809 or the MC 68000*
Hwang: *Advanced Computer Architecture: Parallelism, Scalability, Programmability*
Kohavi: *Switching and Finite Automata Theory*
Lawrence-Mauch: *Real-Time Microcomputer System Design: An Introduction*
Levine: *Vision in Man and Machine*
Navabi: *VHDL: Analysis and Modeling of Digital Systems*
Peatman: *Design of Digital Systems*
Peatman: *Design with Microcontrollers*
Peatman: *Digital Hardware Design*
Ritterman: *Computer Circuit Concepts*
Rosen: *Discrete Mathematics and Its Applications*
Sandige: *Modern Digital Design*
Sze: *VLSI Technology*
Taub: *Digital Circuits and Microprocessors*
Wear, Pinkert, Wear, and Lane: *Computers: An Introduction to Hardware and Software Design*
Wiatrowski and House: *Logic Circuits and Microcomputer Systems*

Trademark Notice

ALLCACHE and KSR-1 are trademarks of Kendall Square Research.

Alewife, J-Machine, and *T are trademarks of Massachusetts Institute of Technology.

CDC 6600, CDC 7600, Cyber 205, and Cyber 2000 are trademarks of Control Data Corporation.

CM-2, CM-5, C*, *Lisp, Paris, CM Fortran, and CMOST are trademarks of of Thinking Machines Corporation.

Convex C3800 is a trademark of Convex Computer Corporation.

Cray 1, Cray 2, Cray X-MP, Cray Y-MP, C90, Cray/MPP, CFT77, T3D, UNICOS, COS, and CAL are trademarks of Cray Research, Inc.

DAP600 and DAP610 are trademarks of Advanced Memory Technology.

Dash and TeraDash are trademarks of Stanford University.

Futurebus, SCI, and Futurebus+ are trademarks of the Institute of Electrical and Electronic Engineers, Inc.

IBM 360/91, System/360, System/370, 701, 801, 7030, 3090, GF11, RP3, RS/6000, POWER, ES/9000, and System/390 are trademarks of International Business Machines.

Illiac IV, Cedar, and Parafrase are trademarks of University of Illinois.

Intel 960 CA, 80486, i860, Multibus II, iPSC, iPSC/860, Paragon XP/S are trademarks of Intel Corporation.

MC68000, 68040, 88100, 88110, and 88200 are trademarks of Motorola Corporation.

MIMDizer, FORGE, and VAST are trademarks of Pacific Sierra.

MIPS R2000, R3000, and R4000 are trademarks of MIPS Computer Systems.

Mach/OS, Warp, C.mmp, and Cm* are trademarks of Carnegie-Mellon University.

MasPar MP-1 is a trademark of MasPar Computer Corporation.

Multiflow is a trademark of Multiflow Corporation.

Multimax and Ultramax are trademarks of Encore Computer Corporation.

NYU Ultracomputer is a trademark of New York University.

Ncube 3200, 6400, Ncube/10, Ncube/2 are trademarks of Ncube Corporation.

OSF/1 is a trademark of Open Software Foundation, Inc.

PFC and ParaScope are trademarks of Rice University.

SX/2, SX/3, and SX-X are trademarks of NEC Information Systems.

Stardent 3000 is a trademark of Stardent Computer, Inc.

Symmetry is a trademark of Sequent Computers.

TC-2000 is a trademark of BBN Advanced Computers.

TI ASC is trademark of Texas Instruments.

TSO, PSO, SPARC, and SunView are trademarks of Sun Microsystems.

Tera is a trademark of Tera Computer Systems.

The Cosmic Cube and Mosaic C are trademarks of California Institute of Technology.

UNIX 4.3 BSD and 4.4 BSD are trademarks of the University of California at Berkeley.

UNIX and UNIX F77 are trademarks of AT&T Bell laboratories.

VAX 11/780, VAX 8600, VAX 9000, VMS, ULTRIX, and Alpha AXP are trademarks of Digital Equipment Corporation.

VP2000 and VPP500 are trademarks of Fujitsu Ltd.

ADVANCED COMPUTER ARCHITECTURE:

Parallelism, Scalability, Programmability

Kai Hwang

**Professor of Electrical Engineering
and Computer Science
University of Southern California**

McGraw-Hill, Inc.

New York St. Louis San Francisco Auckland Bogotá
Caracas Lisbon London Madrid Mexico
Milan Montreal New Delhi Paris
San Juan Singapore Sydney Tokyo Toronto

ADVANCED COMPUTER ARCHITECTURE:
Parallelism, Scalability, Programmability

2 3 4 5 6 7 8 9 0 DOH DOH 9 0 9 8 7 6 5 4 3

ISBN 0-07-031622-8

This book was set in Times Roman using LaTeXand FrameMaker
at the University of Southern California.
The book was produced with camera-ready master supplied by
the author.
The editor was Eric M. Munson; the copyeditor was Carol J. Dean;
the production supervisor was Denise L. Puryear.
R. R. Donnelley & Sons Company was printer and binder.

Library of Congress Cataloging-in-Publication Data

Hwang, Kai, 1943–

    Advanced computer architecture: parallelism, scalability, programmability/
Kai Hwang.

    p.   cm. — (McGraw-Hill computer science series. Computer organization and architecture. Networks, parallel and distributed computing. McGraw-Hill computer engineering series.)
Includes bibliographical references (p.) and index.
ISBN 0-07-031622-8

    1. Computer architecture.   I. Title.   II. Series.

QA76.9.A73H87   1993
004′.35—dc20   92-44944

About the Author

Kai Hwang is a Professor of Electrical Engineering and Computer Science at the University of Southern California. Prior to joining USC, he was on the faculty of Purdue University for 10 years. He received his undergraduate education at the National Taiwan University in China. He earned his Ph.D. degree from the University of California at Berkeley. Dr. Hwang has engaged in research and teaching on digital computer architectures and parallel processing for over 20 years. He has authored or coauthored five books and 120 journal and conference papers in the computer science and engineering areas. He is a founding Coeditor-in-Chief of the *Journal of Parallel and Distributed Computing.*

He has served as the founding Director of the USC Computer Research Institute and a Distinguished Visitor of the IEEE Computer Society. He has chaired several international computer conferences and lectured worldwide on advanced computer topics. His research has been supported by NSF, IBM, AT&T, AFOSR, ONR, DOT, Alliant, and Intel. He has been a consultant for IBM, JPL, Fujitsu, Japan's ETL, GMD in Germany, and ITRI and Academia Sinica in China. He is presently on the advisory boards of several international journals and research organizations.

The Institute of Electrical and Electronics Engineers elected him an IEEE Fellow in 1986 for contributions in computer architectures, digital arithmetic, and parallel processing. He was the holder of the Distinguished CDC Visiting Chair Professorship in Computer Science at the University of Minnesota during the spring quarter of 1989. He has produced over a dozen Ph.D. students from Purdue and USC. His present research work focuses on scalable multiprocessor systems, parallel programming tools, and supercomputing applications.

This book is dedicated to those who are eager to learn in a rapidly changing world, to those who teach and share knowledge without discrimination, and to those who are determined to make a contribution through creative work.

K.H.

Contents

PART II   HARDWARE TECHNOLOGIES   **155**

PART IV   SOFTWARE FOR PARALLEL PROGRAMMING   545

Foreword

Kai Hwang has introduced the issues in designing and using high performance parallel computers at a time when a plethora of scalable computers utilizing commodity microprocessors offer higher peak performance than traditional vector supercomputers. These new machines, their operating environments including the operating system and languages, and the programs to effectively utilize them are introducing more rapid changes for researchers, builders, and users than at any time in the history of computer structures.

For the first time since the introduction of Cray 1 vector processor in 1975, it may again be necessary to change and evolve the programming paradigm — provided that massively parallel computers can be shown to be useful outside of research on massive parallelism. Vector processors required modest data parallelism and these operations have been reflected either explicitly in Fortran programs or implicitly with the need to evolve Fortran (e.g., Fortran 90) to build in vector operations.

So far, the main line of supercomputing as measured by the usage (hours, jobs, number of programs, program portability) has been the shared memory, vector multi-processor as pioneered by Cray Research. Fujitsu, IBM, Hitachi, and NEC all produce computers of this type. In 1993, the Cray C90 supercomputer delivers a peak of 16 billion floating-point operations per second (a Gigaflops) with 16 processors and costs about $30 million, providing roughly 500 floating-point operations per second per dollar.

In contrast, massively parallel computers introduced in the early 1990s are nearly all based on utilizing the same powerful, RISC-based, CMOS microprocessors that are used in workstations. These scalar processors provide a peak of $\approx$ 100 million floating-point operations per second and cost $20 thousand, providing an order of magnitude more peak per dollar (5000 flops per dollar). Unfortunately, to obtain peak power requires large-scale problems that can require $O(n^3)$ operations over supers, and this significantly increases the running time when peak power is the goal.

The multicomputer approach interconnects computers built from microprocessors through high-bandwidth switches that introduce latency. Programs are written in either an evolved parallel data model utilizing Fortran or as independent programs that communicate by passing messages. The book describes a variety of multicomputers including Thinking Machines' CM5, the first computer announced that could reach a teraflops using 8K independent computer nodes, each of which can deliver 128 Mflops utilizing four 32-Mflops floating-point units.

The architecture research trend is toward scalable, shared-memory multiprocessors in order to handle general workloads ranging from technical to commercial tasks and workloads, negate the need to explicitly pass messages for communication, and provide memory addressed accessing. KSR's scalable multiprocessor and Stanford's Dash prototype have proven that such machines are possible.

The author starts by positing a framework based on evolution that outlines the main approaches to designing computer structures. He covers both the scaling of computers and workloads, various multiprocessors, vector processing, multicomputers, and emerging scalable or multithreaded multiprocessors. The final three chapters describe parallel programming techniques and discuss the host operating environment necessary to utilize these new computers.

The book provides case studies of both industrial and research computers, including the Illinois Cedar, Intel Paragon, TMC CM-2, MasPar M1, TMC CM-5, Cray Y-MP, C-90, and Cray MPP, Fujitsu VP2000 and VPP500, NEC SX, Stanford Dash, KSR-1, MIT J-Machine, MIT *T, ETL EM-4, Caltech Mosaic C, and Tera Computer.

The book presents a balanced treatment of the theory, technology, architecture, and software of advanced computer systems. The emphasis on parallelism, scalability, and programmability makes this book rather unique and educational.

I highly recommend Dr. Hwang's timely book. I believe it will benefit many readers and be a fine reference.

C. Gordon Bell

Preface

The Aims

This book provides a comprehensive study of scalable and parallel computer architectures for achieving a proportional increase in performance with increasing system resources. System resources are scaled by the number of processors used, the memory capacity enlarged, the access latency tolerated, the I/O bandwidth required, the performance level desired, etc.

Scalable architectures delivering a sustained performance are desired in both sequential and parallel computers. Parallel architecture has a higher potential to deliver scalable performance. The scalability varies with different architecture-algorithm combinations. Both hardware and software issues need to be studied in building scalable computer systems.

It is my intent to put the reader in a position to design scalable computer systems. Scalability is defined in a broader sense to reflect the interplay among architectures, algorithms, software, and environments. The integration between hardware and software is emphasized for building cost-effective computers.

We should explore cutting-edge technologies in scalable parallel computing. Systems architecture is thus studied with generality, scalability, programmability, and performability in mind.

Since high technology changes so rapidly, I have presented the material in a generic manner, unbiased toward particular machine implementations. Representative processors and systems are presented only if they contain important features which may last into the future.

Every author faces the same dilemma in writing a technology-dependent book which may become obsolete quickly. To cope with the problem, frequent updates with newer editions become a necessity, and I plan to make revisions every few years in the future.

The Contents

This book consists of twelve chapters divided into four parts covering *theory, technology, architecture,* and *software* aspects of parallel and vector computers as shown in the flowchart:

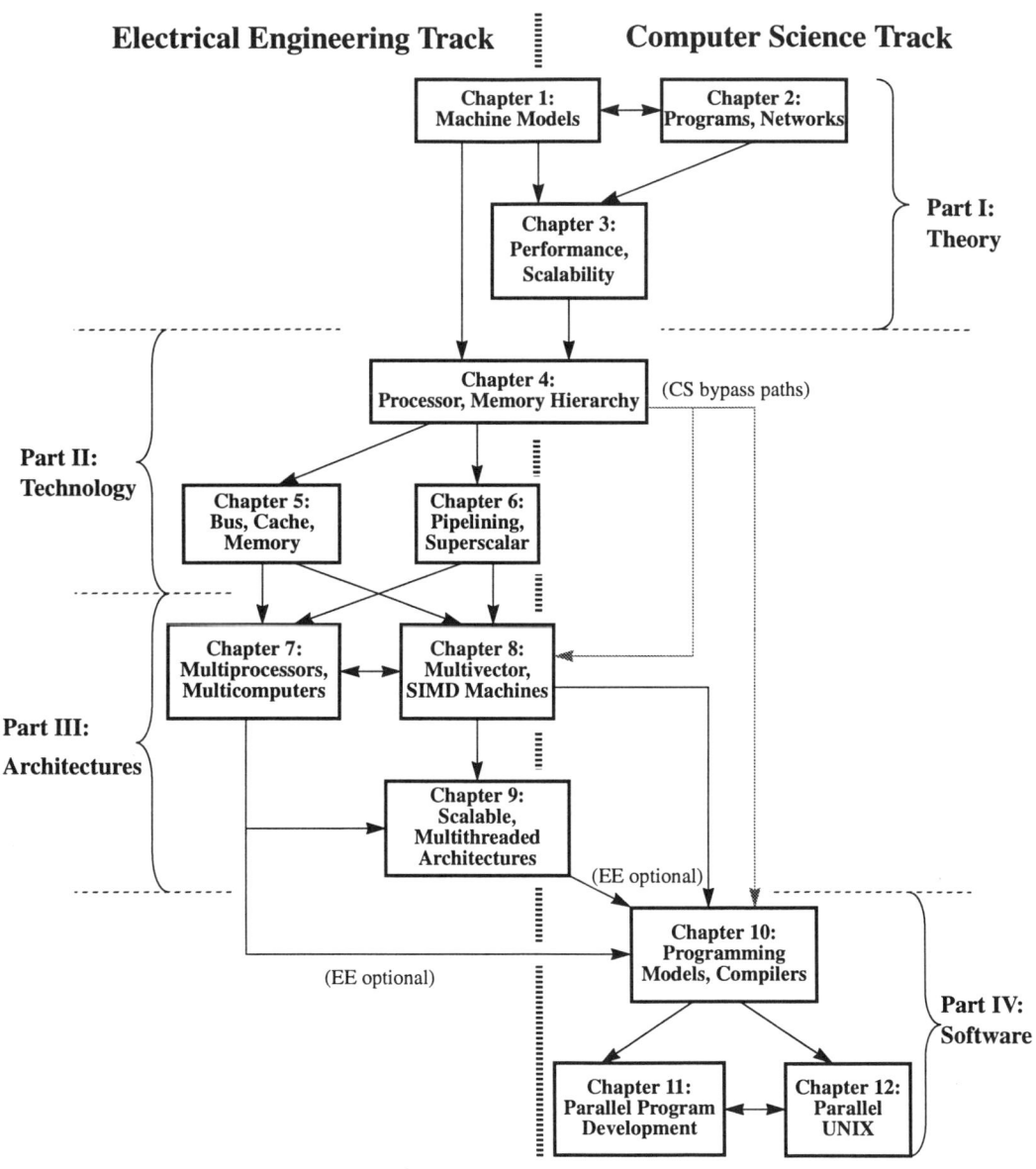

Readers' Guide

Part I presents principles of parallel processing in three chapters. These include parallel computer models, scalability analysis, theory of parallelism, data dependences, program flow mechanisms, network topologies, benchmark measures, performance laws, and program behaviors. These chapters lay the necessary foundations for readers to study hardware and software in subsequent chapters.

In Part II, three chapters are devoted to studying advanced processors, cache and memory technology, and pipelining techniques. Technological bases touched include RISC, CISC, superscalar, superpipelining, and VLIW architectures. Shared memory, consistency models, cache architecture, and coherence protocols are studied.

Pipelining is extensively applied in memory-access, instruction execution, scalar, superscalar, and vector arithmetic operations. Instruction prefetch, data forwarding, software interlocking, scoreboarding, branch handling, and out-of-order issue and completion are studied for designing advanced processors.

In Part III, three chapters are provided to cover shared-memory multiprocessors, vector and SIMD supercomputers, message-passing multicomputers, and scalable or multithreaded architectures. Since we emphasize scalable architectures, special treatment is given to the IEEE Futurebus+ standards, multistage networks, cache coherence, latency tolerance, fast synchronization, and hierarchical and multidimensional structures for building shared-memory systems.

Massive parallelism is addressed in message-passing systems as well as in synchronous SIMD computers. Shared virtual memory and multithreaded architectures are the important topics, in addition to compound vector processing on pipelined supercomputers and coordinated data parallelism on the CM-5.

Part IV consists of three chapters dealing with parallel programming models, multiprocessor UNIX, software environments, and compiler development for parallel/vector computers. Both shared variables and message-passing schemes are studied for interprocessor communications. Languages, compilers, and software tools for program and benchmark development and performance monitoring are studied.

Among various UNIX extensions, we discuss master-slave, floating-executive, and multithreaded kernels for resource management in a network of heterogeneous computer systems. The Mach/OS and OSF/1 are studied as example systems.

This book has been completely newly written based on recent material. The contents are rather different from my earlier book coauthored with Dr. Faye Briggs in 1983. The two books, separated by 10 years, have very little in common.

The Audience

The material included in this text is an outgrowth of two graduate-level courses: Computer Systems Architecture (EE 557) and Parallel Processing (EE 657) that I have taught at the University of Southern California, the University of Minnesota (Spring 1989), and National Taiwan University (Fall 1991) during the last eight years.

The book can be adopted as a textbook in senior- or graduate-level courses offered by Computer Science (CS), Computer Engineering (CE), Electrical Engineering (EE), or Computational Science programs. The flowchart guides the students and instructors in reading this book.

The first four chapters should be taught to all disciplines. The three technology chapters are necessary for EE and CE students. The three architecture chapters can be selectively taught to CE and CS students, depending on the instructor's interest and the computing facilities available to teach the course. The three software chapters are written for CS students and are optional to EE students.

Five course outlines are suggested below for different audiences. The first three outlines are for 45-hour, one-semester courses. The last two outlines are for two-quarter courses in a sequence.

(1) For a Computer Science course on *Parallel Computers and Programming*, the minimum coverage should include Chapters 1–4, 7, and 9–12.

(2) For an exclusive Electrical Engineering course on *Advanced Computer Architecture*, the minimum coverage should include Chapters 1–9.

(3) For a joint CS and EE course on *Parallel Processing Computer Systems*, the minimum coverage should include Chapters 1–4 and 7–12.

(4) Chapters 1 through 6 can be taught to a senior or first-year graduate course under the title *Computer Architecture* in one quarter (10 weeks / 30 hours).

(5) Chapters 7 through 12 can be taught to a graduate course on *Parallel Computer Architecture and Programming* in a one-quarter course with course (4) as the prerequisite.

Instructors may wish to include some advanced research topics treated in Sections 1.4, 2.3, 3.4, 5.4, 6.2, 6.5, 7.2, 7.3, 8.3, 10.4, 11.1, 12.5, and selected sections from Chapter 9 in each of the above course options. The architecture chapters present four different families of commercially available computers. Instructors may choose to teach a subset of these machine families based on the accessibility of corresponding machines on campus or via a public network. Students are encouraged to learn through hands-on programming experience on parallel computers.

A *Solutions Manual* is available to instructors only from the Computer Science Editor, College Division, McGraw-Hill, Inc., 1221 Avenue of the Americas, New York, NY 10020. Answers to a few selected exercise problems are given at the end of the book.

The Prerequisites

This is an advanced text on computer architecture and parallel programming. The reader should have been exposed to some basic computer organization and programming courses at the undergraduate level. Some of the required background material can be found in *Computer Architecture: A Quantitative Approach* by John Hennessy and David Patterson (Morgan Kaufman, 1990) or in *Machine and Assembly Language Programming* by Arthur Gill (Prentice-Hall, 1978).

Students should have some knowledge and experience in logic design, computer hardware, system operations, assembly languages, and Fortran or C programming. Because of the emphasis on scalable architectures and the exploitation of parallelism in practical applications, readers will find it useful to have some background in probability, discrete mathematics, matrix algebra, and optimization theory.

Acknowledgments

I have tried to identify all sources of information in the bibliographic notes. As the subject area evolves rapidly, omissions are almost unavoidable. I apologize to those whose valuable work has not been included in this edition. I am responsible for all omissions and for any errors found in the book. Readers are encouraged to contact me directly regarding error correction or suggestions for future editions.

The writing of this book was inspired, taught, or assisted by numerous scholars or specialists working in the area. I would like to thank each of them for intellectual exchanges, valuable suggestions, critical reviews, and technical assistance.

First of all, I want to thank a number of my former and current Ph.D. students. Hwang-Cheng Wang has assisted me in producing the entire manuscript in LaTeX. Besides, he has coauthored the Solutions Manual with Jung-Gen Wu, who visited USC during 1992. Weihua Mao has drawn almost all the figure illustrations using FrameMaker, based on my original sketches. I want to thank D.K. Panda, Joydeep Ghosh, Ahmed Louri, Dongseung Kim, Zhi-Wei Xu, Sugih Jamin, Chien-Ming Cheng, Santosh Rao, Shisheng Shang, Jih-Cheng Liu, Scott Toborg, Stanley Wang, and Myungho Lee for their assistance in collecting material, proofreading, and contributing some of the homework problems. The errata from Teerapon Jungwiwattanaporn were also useful. The Index was compiled by H.C. Wang and J.G. Wu jointly.

I want to thank Gordon Bell for sharing his insights on supercomputing with me and for writing the Foreword to motivate my readers. John Hennessy and Anoop Gupta provided the Dash multiprocessor-related results from Stanford University. Charles Seitz has taught me through his work on Cosmic Cube, Mosaic, and multicomputers. From MIT, I received valuable inputs from the works of Charles Leiserson, William Dally, Anant Agarwal, and Rishiyur Nikhil. From University of Illinois, I received the Cedar and Perfect benchmark information from Pen Yew.

Jack Dongarra of the University of Tennessee provided me the Linpack benchmark results. James Smith of Cray Research provided up-to-date information on the C-90 clusters and on the Cray/MPP. Ken Miura provided the information on Fujitsu VPP500. Lionel Ni of Michigan State University helped me in the areas of performance laws and adaptive wormhole routing. Justin Rattner provided information on the Intel Delta and Paragon systems. Burton Smith provided information on the Tera computer development.

Harold Stone and John Hayes suggested corrections and ways to improve the presentation. H.C. Torng of Cornell University, Andrew Chien of University of Illinois, and Daniel Tabak of George-Mason University made useful suggestions. Among my colleagues at the University of Southern California, Jean-Luc Gaudiot, Michel Dubois, Rafael Saavedra, Monte Ung, and Viktor Prasanna have made concrete suggestions to improve the manuscript. I appreciate the careful proofreading of an earlier version of the manuscript by D.K. Panda of the Ohio State University. The inputs from Vipin Kumar of the University of Minnesota, Xian-He Sun of NASA Langley Research Center, and Alok Choudhary of Syracuse University are also appreciated.

In addition to the above individuals, my understanding on computer architecture and parallel processing has been influenced by the works of David Kuck, Ken Kennedy,

Jack Dennis, Michael Flynn, Arvind, T.C. Chen, Wolfgang Giloi, Harry Jordan, H.T. Kung, John Rice, H.J. Siegel, Allan Gottlieb, Philips Treleaven, Faye Briggs, Peter Kogge, Steve Chen, Ben Wah, Edward Davidson, Alvin Despain, James Goodman, Robert Keller, Duncan Lawrie, C.V. Ramamoorthy, Sartaj Sahni, Jean-Loup Baer, Milos Ercegovac, Doug DeGroot, Janak Patel, Dharma Agrawal, Lenart Johnsson, John Gustafson, Tse-Yun Feng, Herbert Schewetman, and Ken Batcher. I want to thank all of them for sharing their vast knowledge with me.

I want to acknowledge the research support I have received from the National Science Foundation, the Office of Naval Research, the Air Force Office of Scientific Research, International Business Machines, Intel Corporation, Alliant Computer Systems, and American Telephone and Telegraph Laboratories.

Technical exchanges with the Electrotechnical Laboratory (ETL) in Japan, the German National Center for Computer Research (GMD) in Germany, and the Industrial Technology Research Institute (ITRI) in Taiwan are always rewarding experiences to the author.

I appreciate the staff and facility support provided by Purdue University, the University of Southern California, the University of Minnesota, the University of Tokyo, National Taiwan University, and Academia Sinica during the past twenty years. In particular, I appreciate the encouragement and professional advices received from Henry Yang, Lofti Zadeh, Richard Karp, George Bekey, Authur Gill, Ben Coates, Melvin Breuer, Jerry Mendel, Len Silverman, Solomon Golomb, and Irving Reed over the years.

Excellent work by the McGraw-Hill editorial and production staff has greatly improved the readability of this book. In particular, I want to thank Eric Munson for his continuous sponsorship of my book projects. I appreciate Joe Murphy and his coworkers for excellent copy editing and production jobs. Suggestions from reviewers listed below have greatly helped improve the contents and presentation.

The book was completed at the expense of cutting back many aspects of my life, spending many long hours, evenings, and weekends in seclusion during the last several years. I appreciate the patience and understanding of my friends, my students, and my family members during those tense periods. Finally, the book has been completed and I hope you enjoy reading it.

Kai Hwang

Los Angeles, California

Reviewers:

Andrew A. Chien, *University of Illinois*;
David Culler, *University of California, Berkeley*;
Ratan K. Guha, *University of Central Florida*;
John P. Hayes, *University of Michigan*;
John Hennessy, *Stanford University*;
Dhamir Mannai, *Northeastern University*;
Michael Quinn, *Oregon State University*;
H. J. Siegel, *Purdue University*;
Daniel Tabak, *George-Mason University*.

ADVANCED COMPUTER ARCHITECTURE:
Parallelism, Scalability, Programmability

Part I

Theory of Parallelism

> ## Chapter 1
> ## Parallel Computer Models
>
> ## Chapter 2
> ## Program and Network Properties
>
> ## Chapter 3
> ## Principles of Scalable Performance

Summary

This theoretical part presents computer models, program behavior, architectural choices, scalability, programmability, and performance issues related to parallel processing. These topics form the foundations for designing high-performance computers and for the development of supporting software and applications.

Physical computers modeled include shared-memory multiprocessors, message-passing multicomputers, vector supercomputers, synchronous processor arrays, and massively parallel processors. The theoretical parallel random-access machine (PRAM) model is also presented. Differences between the PRAM model and physical architectural models are discussed. The VLSI complexity model is presented for implementing parallel algorithms directly in integrated circuits.

Network design principles and parallel program characteristics are introduced. These include dependence theory, computing granularity, communication latency, program flow mechanisms, network properties, performance laws, and scalability studies. This evolving theory of parallelism consolidates our understanding of parallel computers, from abstract models to hardware machines, software systems, and performance evaluation.

1

Chapter 1

Parallel Computer Models

Parallel processing has emerged as a key enabling technology in modern computers, driven by the ever-increasing demand for higher performance, lower costs, and sustained productivity in real-life applications. Concurrent events are taking place in today's high-performance computers due to the common practice of multiprogramming, multiprocessing, or multicomputing.

Parallelism appears in various forms, such as lookahead, pipelining, vectorization, concurrency, simultaneity, data parallelism, partitioning, interleaving, overlapping, multiplicity, replication, time sharing, space sharing, multitasking, multiprogramming, multithreading, and distributed computing at different processing levels.

In this chapter, we model physical architectures of parallel computers, vector supercomputers, multiprocessors, multicomputers, and massively parallel processors. Theoretical machine models are also presented, including the *parallel random-access machines* (PRAMs) and the complexity model of VLSI (*very large-scale integration*) circuits. Architectural development tracks are identified with case studies in the book. Hardware and software subsystems are introduced to pave the way for detailed studies in subsequent chapters.

1.1 The State of Computing

Modern computers are equipped with powerful hardware facilities driven by extensive software packages. To assess state-of-the-art computing, we first review historical milestones in the development of computers. Then we take a grand tour of the crucial hardware and software elements built into modern computer systems. We then examine the evolutional relations in milestone architectural development. Basic hardware and software factors are identified in analyzing the performance of computers.

1.1.1 Computer Development Milestones

Computers have gone through two major stages of development: mechanical and electronic. Prior to 1945, computers were made with mechanical or electromechanical parts. The earliest mechanical computer can be traced back to 500 BC in the form of the abacus used in China. The abacus is manually operated to perform decimal arithmetic with carry propagation digit by digit.

Blaise Pascal built a mechanical adder/subtractor in France in 1642. Charles Babbage designed a difference engine in England for polynomial evaluation in 1827. Konrad Zuse built the first binary mechanical computer in Germany in 1941. Howard Aiken proposed the very first electromechanical decimal computer, which was built as the Harvard Mark I by IBM in 1944. Both Zuse's and Aiken's machines were designed for general-purpose computations.

Obviously, the fact that computing and communication were carried out with moving mechanical parts greatly limited the computing speed and reliability of mechanical computers. Modern computers were marked by the introduction of electronic components. The moving parts in mechanical computers were replaced by high-mobility electrons in electronic computers. Information transmission by mechanical gears or levers was replaced by electric signals traveling almost at the speed of light.

Computer Generations   Over the past five decades, electronic computers have gone through five generations of development. Table 1.1 provides a summary of the five generations of electronic computer development. Each of the first three generations lasted about 10 years. The fourth generation covered a time span of 15 years. We have just entered the fifth generation with the use of processors and memory devices with more than 1 million transistors on a single silicon chip.

The division of generations is marked primarily by sharp changes in hardware and software technologies. The entries in Table 1.1 indicate the new hardware and software features introduced with each generation. Most features introduced in earlier generations have been passed to later generations. In other words, the latest generation computers have inherited all the nice features and eliminated all the bad ones found in previous generations.

Progress in Hardware   As far as hardware technology is concerned, the first generation (1945–1954) used vacuum tubes and relay memories interconnected by insulated wires. The second generation (1955–1964) was marked by the use of discrete transistors, diodes, and magnetic ferrite cores, interconnected by printed circuits.

The third generation (1965–1974) began to use *integrated circuits* (ICs) for both logic and memory in *small-scale* or *medium-scale integration* (SSI or MSI) and multi-layered printed circuits. The fourth generation (1974–1991) used *large-scale* or *very-large-scale integration* (LSI or VLSI). Semiconductor memory replaced core memory as computers moved from the third to the fourth generation.

The fifth generation (1991–present) is highlighted by the use of high-density and high-speed processor and memory chips based on even more improved VLSI technology. For example, 64-bit 150-MHz microprocessors are now available on a single chip with over one million transistors. Four-megabit dynamic *random-access memory* (RAM) and 256K-bit static RAM are now in widespread use in today's high-performance computers.

It has been projected that four microprocessors will be built on a single CMOS chip with more than 50 million transistors, and 64M-bit dynamic RAM will become

Table 1.1 Five Generations of Electronic Computers

Generation	Technology and Architecture	Software and Applications	Representative Systems
First (1945–54)	Vacuum tubes and relay memories, CPU driven by PC and accumulator, fixed-point arithmetic.	Machine/assembly languages, single user, no subroutine linkage, programmed I/O using CPU.	ENIAC, Princeton IAS, IBM 701.
Second (1955–64)	Discrete transistors and core memories, floating-point arithmetic, I/O processors, multiplexed memory access.	HLL used with compilers, subroutine libraries, batch processing monitor.	IBM 7090, CDC 1604, Univac LARC.
Third (1965–74)	Integrated circuits (SSI/-MSI), microprogramming, pipelining, cache, and lookahead processors.	Multiprogramming and time-sharing OS, multiuser applications.	IBM 360/370, CDC 6600, TI-ASC, PDP-8.
Fourth (1975–90)	LSI/VLSI and semiconductor memory, multiprocessors, vector supercomputers, multicomputers.	Multiprocessor OS, languages, compilers, and environments for parallel processing.	VAX 9000, Cray X-MP, IBM 3090, BBN TC2000.
Fifth (1991–present)	ULSI/VHSIC processors, memory, and switches, high-density packaging, scalable architectures.	Massively parallel processing, grand challenge applications, heterogeneous processing.	Fujitsu VPP500, Cray/MPP, TMC/CM-5, Intel Paragon.

available in large quantities within the next decade.

The First Generation   From the architectural and software points of view, first-generation computers were built with a single *central processing unit* (CPU) which performed serial fixed-point arithmetic using a program counter, branch instructions, and an accumulator. The CPU must be involved in all memory access and *input/output* (I/O) operations. Machine or assembly languages were used.

Representative systems include the ENIAC (Electronic Numerical Integrator and Calculator) built at the Moore School of the University of Pennsylvania in 1950; the IAS (Institute for Advanced Studies) computer based on a design proposed by John von Neumann, Arthur Burks, and Herman Goldstine at Princeton in 1946; and the IBM 701, the first electronic stored-program commercial computer built by IBM in 1953. Subroutine linkage was not implemented in early computers.

The Second Generation   Index registers, floating-point arithmetic, multiplexed memory, and I/O processors were introduced with second-generation computers. *High-level languages* (HLLs), such as Fortran, Algol, and Cobol, were introduced along with compilers, subroutine libraries, and batch processing monitors. Register transfer language was developed by Irving Reed (1957) for systematic design of digital computers.

Representative systems include the IBM 7030 (the Stretch computer) featuring instruction lookahead and error-correcting memories built in 1962, the Univac LARC (Livermore Atomic Research Computer) built in 1959, and the CDC 1604 built in the 1960s.

The Third Generation   The third generation was represented by the IBM/360–370 Series, the CDC 6600/7600 Series, Texas Instruments ASC (Advanced Scientific Computer), and Digital Equipment's PDP-8 Series from the mid-1960s to the mid-1970s.

Microprogrammed control became popular with this generation. Pipelining and cache memory were introduced to close up the speed gap between the CPU and main memory. The idea of multiprogramming was implemented to interleave CPU and I/O activities across multiple user programs. This led to the development of time-sharing *operating systems* (OS) using virtual memory with greater sharing or multiplexing of resources.

The Fourth Generation   Parallel computers in various architectures appeared in the fourth generation of computers using shared or distributed memory or optional vector hardware. Multiprocessing OS, special languages, and compilers were developed for parallelism. Software tools and environments were created for parallel processing or distributed computing.

Representative systems include the VAX 9000, Cray X-MP, IBM/3090 VF, BBN TC-2000, etc. During these 15 years (1975–1990), the technology of parallel processing gradually became mature and entered the production mainstream.

The Fifth Generation   Fifth-generation computers have just begun to appear. These machines emphasize *massively parallel processing* (MPP). Scalable and latency-tolerant architectures are being adopted in MPP systems using VLSI silicon, GaAs technologies, high-density packaging, and optical technologies.

Fifth-generation computers are targeted to achieve Teraflops (10^{12} floating-point operations per second) performance by the mid-1990s. *Heterogeneous processing* is emerging to solve large-scale problems using a network of heterogeneous computers with shared virtual memories. The fifth-generation MPP systems are represented by several recently announced projects at Fujitsu (VPP500), Cray Research (MPP), Thinking Machines Corporation (the CM-5), and Intel Supercomputer Systems (the Paragon).

1.1.2   Elements of Modern Computers

Hardware, software, and programming elements of a modern computer system are briefly introduced below in the context of parallel processing.

Computing Problems   It has been long recognized that the concept of computer architecture is no longer restricted to the structure of the bare machine hardware. A modern computer is an integrated system consisting of machine hardware, an instruction set, system software, application programs, and user interfaces. These system elements are depicted in Fig. 1.1. The use of a computer is driven by real-life problems demanding fast and accurate solutions. Depending on the nature of the problems, the solutions may require different computing resources.

For numerical problems in science and technology, the solutions demand complex mathematical formulations and tedious integer or floating-point computations. For alphanumerical problems in business and government, the solutions demand accurate transactions, large database management, and information retrieval operations.

For artificial intelligence (AI) problems, the solutions demand logic inferences and symbolic manipulations. These computing problems have been labeled *numerical computing*, *transaction processing*, and *logical reasoning*. Some complex problems may demand a combination of these processing modes.

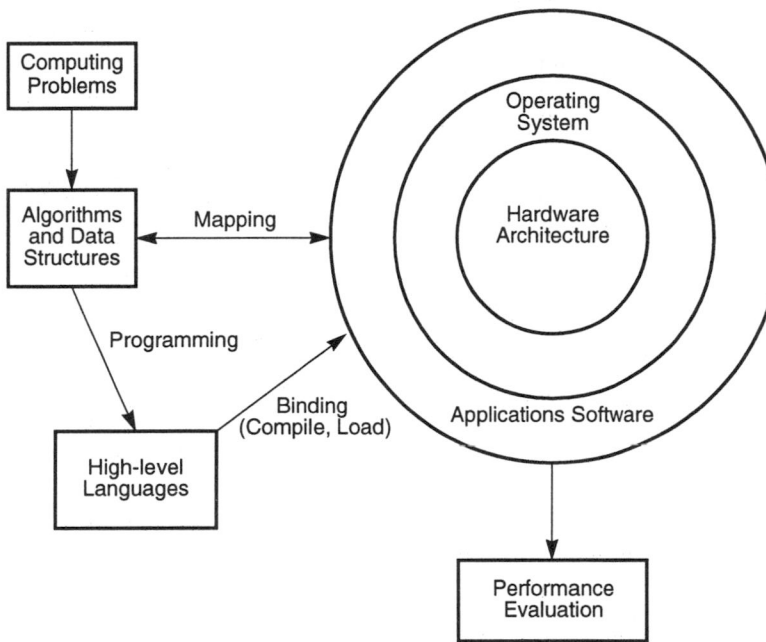

Figure 1.1 Elements of a modern computer system.

Algorithms and Data Structures Special algorithms and data structures are needed to specify the computations and communications involved in computing problems. Most numerical algorithms are deterministic, using regularly structured data. Symbolic processing may use heuristics or nondeterministic searches over large knowledge bases.

Problem formulation and the development of parallel algorithms often require interdisciplinary interactions among theoreticians, experimentalists, and computer programmers. There are many books dealing with the design and mapping of algorithms or heuristics onto parallel computers. In this book, we are more concerned about the resources mapping problem than about the design and analysis of parallel algorithms.

Hardware Resources   The system architecture of a computer is represented by three nested circles on the right in Fig. 1.1. A modern computer system demonstrates its power through coordinated efforts by hardware resources, an operating system, and application software. Processors, memory, and peripheral devices form the hardware core of a computer system. We will study instruction-set processors, memory organization, multiprocessors, supercomputers, multicomputers, and massively parallel computers.

Special hardware interfaces are often built into I/O devices, such as terminals, workstations, optical page scanners, magnetic ink character recognizers, modems, file servers, voice data entry, printers, and plotters. These peripherals are connected to mainframe computers directly or through local or wide-area networks.

In addition, software interface programs are needed. These software interfaces include file transfer systems, editors, word processors, device drivers, interrupt handlers, network communication programs, etc. These programs greatly facilitate the portability of user programs on different machine architectures.

Operating System   An effective operating system manages the allocation and deallocation of resources during the execution of user programs. We will study UNIX extensions for multiprocessors and multicomputers in Chapter 12. Mach/OS kernel and OSF/1 will be specially studied for multithreaded kernel functions, virtual memory management, file subsystems, and network communication services. Beyond the OS, application software must be developed to benefit the users. Standard benchmark programs are needed for performance evaluation.

Mapping is a bidirectional process matching algorithmic structure with hardware architecture, and vice versa. Efficient mapping will benefit the programmer and produce better source codes. The mapping of algorithmic and data structures onto the machine architecture includes processor scheduling, memory maps, interprocessor communications, etc. These activities are usually architecture-dependent.

Optimal mappings are sought for various computer architectures. The implementation of these mappings relies on efficient compiler and operating system support. Parallelism can be exploited at algorithm design time, at program time, at compile time, and at run time. Techniques for exploiting parallelism at these levels form the core of parallel processing technology.

System Software Support   Software support is needed for the development of efficient programs in high-level languages. The source code written in a HLL must be first translated into object code by an optimizing compiler. The *compiler* assigns variables to registers or to memory words and reserves functional units for operators. An *assembler* is used to translate the compiled object code into machine code which can be recognized by the machine hardware. A *loader* is used to initiate the program execution through the OS kernel.

Resource binding demands the use of the compiler, assembler, loader, and OS kernel to commit physical machine resources to program execution. The effectiveness of this process determines the efficiency of hardware utilization and the programmability of the computer. Today, programming parallelism is still very difficult for most programmers due to the fact that existing languages were originally developed for sequential computers. Programmers are often forced to program hardware-dependent features instead of programming parallelism in a generic and portable way. Ideally, we need to develop a parallel programming environment with architecture-independent languages, compilers, and software tools.

To develop a parallel language, we aim for efficiency in its implementation, portability across different machines, compatibility with existing sequential languages, expressiveness of parallelism, and ease of programming. One can attempt a new language approach or try to extend existing sequential languages gradually. A new language approach has the advantage of using explicit high-level constructs for specifying parallelism. However, new languages are often incompatible with existing languages and require new compilers or new passes to existing compilers. Most systems choose the language extension approach.

Compiler Support   There are three compiler upgrade approaches: *preprocessor*, *precompiler*, and *parallelizing compiler*. A preprocessor uses a sequential compiler and a low-level library of the target computer to implement high-level parallel constructs. The precompiler approach requires some program flow analysis, dependence checking, and limited optimizations toward parallelism detection. The third approach demands a fully developed parallelizing or vectorizing compiler which can automatically detect parallelism in source code and transform sequential codes into parallel constructs. These approaches will be studied in Chapter 10.

The efficiency of the binding process depends on the effectiveness of the preprocessor, the precompiler, the parallelizing compiler, the loader, and the OS support. Due to unpredictable program behavior, none of the existing compilers can be considered fully automatic or fully intelligent in detecting all types of parallelism. Very often *compiler directives* are inserted into the source code to help the compiler do a better job. Users may interact with the compiler to restructure the programs. This has been proven useful in enhancing the performance of parallel computers.

1.1.3   Evolution of Computer Architecture

The study of computer architecture involves both hardware organization and programming/software requirements. As seen by an assembly language programmer, computer architecture is abstracted by its instruction set, which includes opcode (operation codes), addressing modes, registers, virtual memory, etc.

From the hardware implementation point of view, the abstract machine is organized with CPUs, caches, buses, microcode, pipelines, physical memory, etc. Therefore, the study of architecture covers both instruction-set architectures and machine implementation organizations.

Over the past four decades, computer architecture has gone through evolutional rather than revolutionary changes. Sustaining features are those that were proven performance deliverers. As depicted in Fig. 1.2, we started with the von Neumann architecture built as a sequential machine executing scalar data. The sequential computer was improved from bit-serial to word-parallel operations, and from fixed-point to floating-point operations. The von Neumann architecture is slow due to sequential execution of instructions in programs.

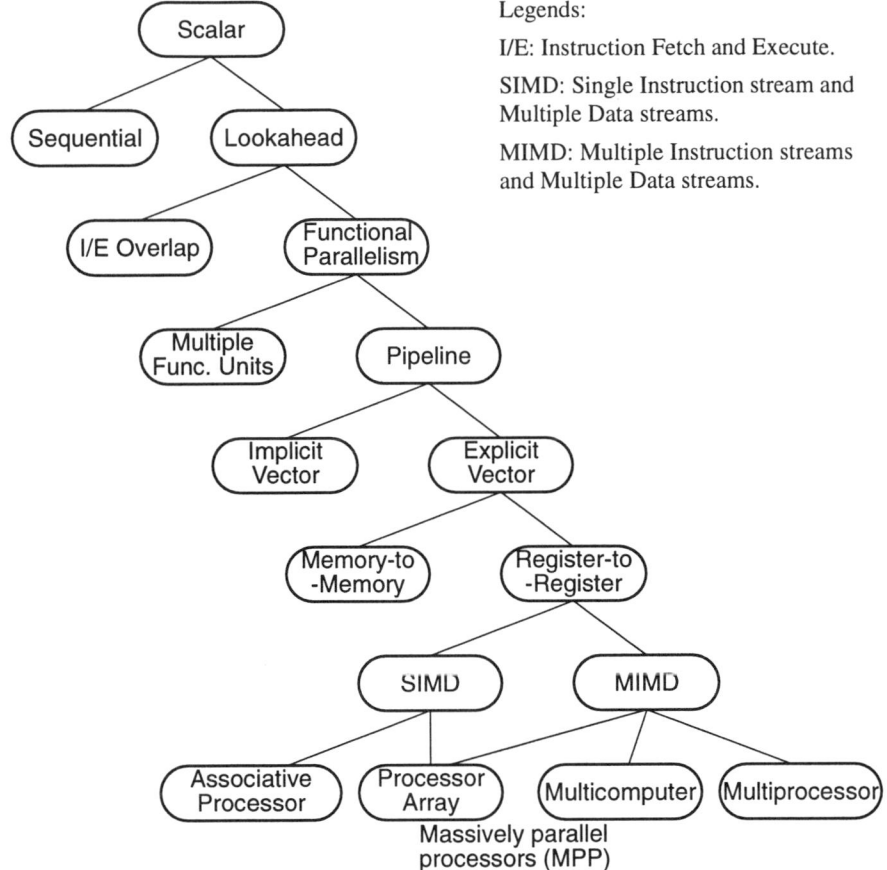

Legends:

I/E: Instruction Fetch and Execute.

SIMD: Single Instruction stream and Multiple Data streams.

MIMD: Multiple Instruction streams and Multiple Data streams.

Figure 1.2 Tree showing architectural evolution from sequential scalar computers to vector processors and parallel computers.

Lookahead, Parallelism, and Pipelining   Lookahead techniques were introduced to prefetch instructions in order to overlap I/E (instruction fetch/decode and execution) operations and to enable functional parallelism. Functional parallelism was supported by two approaches: One is to use multiple functional units simultaneously, and the other is to practice pipelining at various processing levels.

The latter includes pipelined instruction execution, pipelined arithmetic computations, and memory-access operations. Pipelining has proven especially attractive in performing identical operations repeatedly over vector data strings. Vector operations were originally carried out implicitly by software-controlled looping using scalar pipeline processors.

Flynn's Classification   Michael Flynn (1972) introduced a classification of various computer architectures based on notions of instruction and data streams.   As illustrated in Fig. 1.3a, conventional sequential machines are called SISD (*single instruction stream over a single data stream*) computers. Vector computers are equipped with scalar and vector hardware or appear as SIMD (*single instruction stream over multiple data streams*) machines (Fig. 1.3b).   Parallel computers are reserved for MIMD (*multiple instruction streams over multiple data streams*) machines.

An MISD (*multiple instruction streams and a single data stream*) machines are modeled in Fig. 1.3d. The same data stream flows through a linear array of processors executing different instruction streams.   This architecture is also known as *systolic arrays* (Kung and Leiserson, 1978) for pipelined execution of specific algorithms.

Of the four machine models, most parallel computers built in the past assumed the MIMD model for general-purpose computations. The SIMD and MISD models are more suitable for special-purpose computations. For this reason, MIMD is the most popular model, SIMD next, and MISD the least popular model being applied in commercial machines.

Parallel/Vector Computers   Intrinsic parallel computers are those that execute programs in MIMD mode. There are two major classes of parallel computers, namely, *shared-memory multiprocessors* and *message-passing multicomputers*.   The major distinction between multiprocessors and multicomputers lies in memory sharing and the mechanisms used for interprocessor communication.

The processors in a multiprocessor system communicate with each other through *shared variables* in a common memory. Each computer node in a multicomputer system has a local memory, unshared with other nodes. Interprocessor communication is done through *message passing* among the nodes.

Explicit vector instructions were introduced with the appearance of *vector processors*. A vector processor is equipped with multiple vector pipelines that can be concurrently used under hardware or firmware control. There are two families of pipelined vector processors:

Memory-to-memory architecture supports the pipelined flow of vector operands directly from the memory to pipelines and then back to the memory.   *Register-to-register* architecture uses vector registers to interface between the memory and functional pipelines. Vector processor architectures will be studied in Chapter 8.

Another important branch of the architecture tree consists of the SIMD computers for synchronized vector processing.   An SIMD computer exploits *spatial parallelism* rather than *temporal parallelism* as in a pipelined computer.   SIMD computing is achieved through the use of an array of *processing elements* (PEs) synchronized by the same controller. Associative memory can be used to build SIMD associative processors. SIMD machines will be treated in Chapter 8 along with pipelined vector computers.

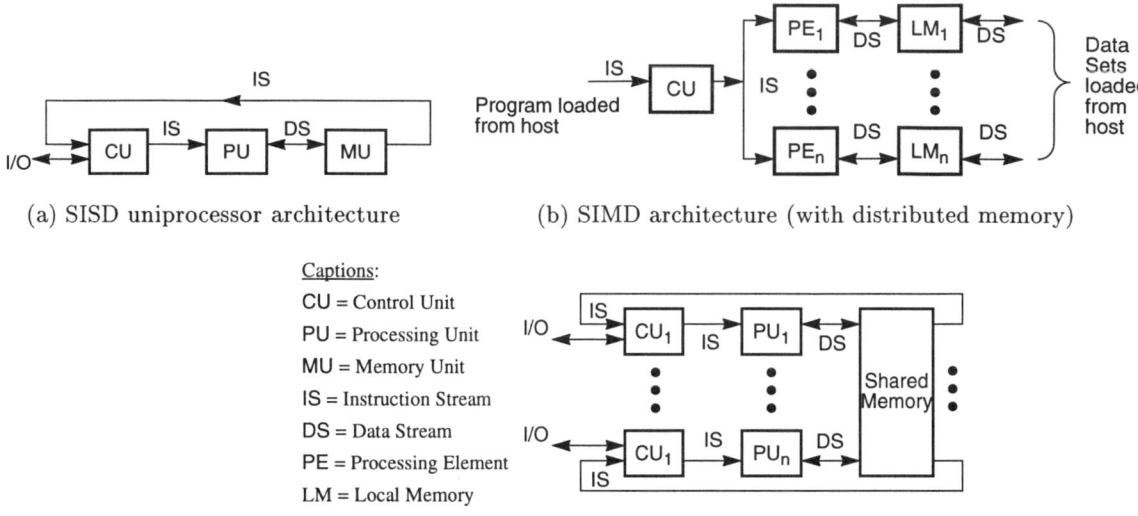

(a) SISD uniprocessor architecture   (b) SIMD architecture (with distributed memory)

Captions:

CU = Control Unit
PU = Processing Unit
MU = Memory Unit
IS = Instruction Stream
DS = Data Stream
PE = Processing Element
LM = Local Memory

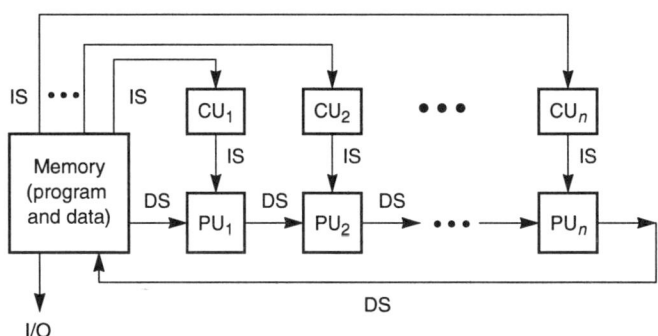

(c) MIMD architecture (with shared memory)

(d) MISD architecture (the systolic array)

Figure 1.3 Flynn's classification of computer architectures. (Derived from Michael Flynn, 1972)

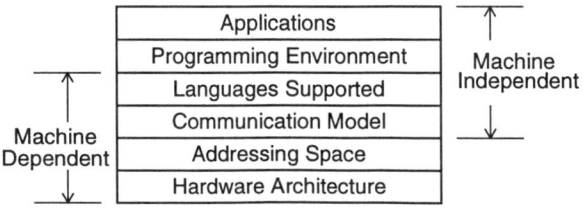

Figure 1.4 Six layers for computer system development. (Courtesy of Lionel Ni, 1990)

Development Layers   A layered development of parallel computers is illustrated in Fig. 1.4, based on a recent classification by Lionel Ni (1990). Hardware configurations differ from machine to machine, even those of the same model. The address space of a processor in a computer system varies among different architectures. It depends on the memory organization, which is machine-dependent. These features are up to the designer and should match the target application domains.

On the other hand, we want to develop application programs and programming environments which are machine-independent. Independent of machine architecture, the user programs can be ported to many computers with minimum conversion costs. High-level languages and communication models depend on the architectural choices made in a computer system. From a programmer's viewpoint, these two layers should be architecture-transparent.

At present, Fortran, C, Pascal, Ada, and Lisp are supported by most computers. However, the communication models, shared variables versus message passing, are mostly machine-dependent. The Linda approach using *tuple spaces* offers an architecture-transparent communication model for parallel computers. These language features will be studied in Chapter 10.

Application programmers prefer more architectural transparency. However, kernel programmers have to explore the opportunities supported by hardware. As a good computer architect, one has to approach the problem from both ends. The compilers and OS support should be designed to remove as many architectural constraints as possible from the programmer.

New Challenges   The technology of parallel processing is the outgrowth of four decades of research and industrial advances in microelectronics, printed circuits, high-density packaging, advanced processors, memory systems, peripheral devices, communication channels, language evolution, compiler sophistication, operating systems, programming environments, and application challenges.

The rapid progress made in hardware technology has significantly increased the economical feasibility of building a new generation of computers adopting parallel processing. However, the major barrier preventing parallel processing from entering the production mainstream is on the software and application side.

To date, it is still very difficult and painful to program parallel and vector computers. We need to strive for major progress in the software area in order to create a user-friendly environment for high-power computers. A whole new generation of programmers need to be trained to program parallelism effectively. High-performance computers provide fast and accurate solutions to scientific, engineering, business, social, and defense problems.

Representative real-life problems include weather forecast modeling, computer-aided design of VLSI circuits, large-scale database management, artificial intelligence, crime control, and strategic defense initiatives, just to name a few. The application domains of parallel processing computers are expanding steadily. With a good understanding of scalable computer architectures and mastery of parallel programming techniques, the reader will be better prepared to face future computing challenges.

1.1.4   System Attributes to Performance

The ideal performance of a computer system demands a perfect match between machine capability and program behavior. Machine capability can be enhanced with better hardware technology, innovative architectural features, and efficient resources management. However, program behavior is difficult to predict due to its heavy dependence on application and run-time conditions.

There are also many other factors affecting program behavior, including algorithm design, data structures, language efficiency, programmer skill, and compiler technology. It is impossible to achieve a perfect match between hardware and software by merely improving only a few factors without touching other factors.

Besides, machine performance may vary from program to program. This makes *peak performance* an impossible target to achieve in real-life applications. On the other hand, a machine cannot be said to have an average performance either. All performance indices or benchmarking results must be tied to a program mix. For this reason, the performance should be described as a range or as a harmonic distribution.

We introduce below fundamental factors for projecting the performance of a computer. These performance indicators are by no means conclusive in all applications. However, they can be used to guide system architects in designing better machines or to educate programmers or compiler writers in optimizing the codes for more efficient execution by the hardware.

Consider the execution of a given program on a given computer. The simplest measure of program performance is the *turnaround time*, which includes disk and memory accesses, input and output activities, compilation time, OS overhead, and CPU time. In order to shorten the turnaround time, one must reduce all these time factors.

In a multiprogrammed computer, the I/O and system overheads of a given program may overlap with the CPU times required in other programs. Therefore, it is fair to compare just the total CPU time needed for program execution. The CPU is used to execute both system programs and user programs. It is the user CPU time that concerns the user most.

Clock Rate and CPI   The CPU (or simply the *processor*) of today's digital computer is driven by a clock with a constant *cycle time* (τ in nanoseconds). The inverse of the cycle time is the *clock rate* ($f = 1/\tau$ in megahertz). The size of a program is determined by its *instruction count* (I_c), in terms of the number of machine instructions to be executed in the program. Different machine instructions may require different numbers of clock cycles to execute. Therefore, the *cycles per instruction* (CPI) becomes an important parameter for measuring the time needed to execute each instruction.

For a given instruction set, we can calculate an *average* CPI over all instruction types, provided we know their frequencies of appearance in the program. An accurate estimate of the average CPI requires a large amount of program code to be traced over a long period of time. Unless specifically focusing on a single instruction type, we simply use the term CPI to mean the average value with respect to a given instruction set and a given program mix.

Performance Factors   Let I_c be the number of instructions in a given program, or the instruction count. The CPU time (T in seconds/program) needed to execute the program is estimated by finding the product of three contributing factors:

$$T = I_c \times \text{CPI} \times \tau \qquad (1.1)$$

The execution of an instruction requires going through a cycle of events involving the instruction fetch, decode, operand(s) fetch, execution, and store results. In this cycle, only the instruction decode and execution phases are carried out in the CPU. The remaining three operations may be required to access the memory. We define a *memory cycle* as the time needed to complete one memory reference. Usually, a memory cycle is k times the processor cycle τ. The value of k depends on the speed of the memory technology and processor-memory interconnection scheme used.

The CPI of an instruction type can be divided into two component terms corresponding to the total processor cycles and memory cycles needed to complete the execution of the instruction. Depending on the instruction type, the complete instruction cycle may involve one to four memory references (one for instruction fetch, two for operand fetch, and one for store results). Therefore we can rewrite Eq. 1.1 as follows:

$$T = I_c \times (p + m \times k) \times \tau \qquad (1.2)$$

where p is the number of processor cycles needed for the instruction decode and execution, m is the number of memory references needed, k is the ratio between memory cycle and processor cycle, I_c is the instruction count, and τ is the processor cycle time. Equation 1.2 can be further refined once the CPI components (p, m, k) are weighted over the entire instruction set.

System Attributes   The above five performance factors (I_c, p, m, k, τ) are influenced by four system attributes: instruction-set architecture, compiler technology, CPU implementation and control, and cache and memory hierarchy, as specified in Table 1.2.

The instruction-set architecture affects the program length (I_c) and processor cycle needed (p). The compiler technology affects the values of I_c, p, and the memory reference count (m). The CPU implementation and control determine the total processor time $(p \cdot \tau)$ needed. Finally, the memory technology and hierarchy design affect the memory access latency $(k \cdot \tau)$. The above CPU time can be used as a basis in estimating the execution rate of a processor.

MIPS Rate   Let C be the total number of clock cycles needed to execute a given program. Then the CPU time in Eq. 1.2 can be estimated as $T = C \times \tau = C/f$. Furthermore, CPI $= C/I_c$ and $T = I_c \times \text{CPI} \times \tau = I_c \times \text{CPI}/f$. The processor speed is often measured in terms of *million instructions per second* (MIPS). We simply call it the MIPS rate of a given processor. It should be emphasized that the MIPS rate varies with respect to a number of factors, including the clock rate (f), the instruction count (I_c), and the CPI of a given machine, as defined below:

$$\text{MIPS rate} \;=\; \frac{I_c}{T \times 10^6} \;=\; \frac{f}{\text{CPI} \times 10^6} \;=\; \frac{f \times I_c}{C \times 10^6} \qquad (1.3)$$

Table 1.2 Performance Factors Versus System Attributes

System Attributes	Instr. Count, I_c	Average Cycles per Instruction, CPI			Processor Cycle Time, τ
		Processor Cycles per Instruction, p	Memory References per Instruction, m	Memory-Access Latency, k	
Instruction-set Architecture	X	X			
Compiler Technology	X	X	X		
Processor Implementation and Control		X			X
Cache and Memory Hierarchy				X	X

Based on Eq. 1.3, the CPU time in Eq. 1.2 can also be written as $T = I_c \times 10^{-6}/\text{MIPS}$. Based on the system attributes identified in Table 1.2 and the above derived expressions, we conclude by indicating the fact that the MIPS rate of a given computer is directly proportional to the clock rate and inversely proportional to the CPI. All four system attributes, instruction set, compiler, processor, and memory technologies, affect the MIPS rate, which varies also from program to program.

Throughput Rate Another important concept is related to how many programs a system can execute per unit time, called the *system throughput* W_s (in programs/second). In a multiprogrammed system, the system throughput is often lower than the *CPU throughput* W_p defined by:

$$W_p = \frac{f}{I_c \times \text{CPI}} \tag{1.4}$$

Note that $W_p = (\text{MIPS}) \times 10^6/I_c$ from Eq. 1.3. The unit for W_p is programs/second. The CPU throughput is a measure of how many programs can be executed per second, based on the MIPS rate and average program length (I_c). The reason why $W_s < W_p$ is due to the additional system overheads caused by the I/O, compiler, and OS when multiple programs are interleaved for CPU execution by multiprogramming or time-sharing operations. If the CPU is kept busy in a perfect program-interleaving fashion, then $W_s = W_p$. This will probably never happen, since the system overhead often causes an extra delay and the CPU may be left idle for some cycles.

Example 1.1 MIPS ratings and performance measurement

Consider the use of a VAX/780 and an IBM RS/6000-based workstation to execute a hypothetical benchmark program. Machine characteristics and claimed performance are given below:

Machine	Clock	Performance	CPU Time
VAX 11/780	5 MHz	1 MIPS	$12x$ seconds
IBM RS/6000	25 MHz	18 MIPS	x seconds

These data indicate that the measured CPU time on the VAX 11/780 is 12 times longer than that measured on the RS/6000. The object codes running on the two machines have different lengths due to the differences in the machines and compilers used. All other overhead times are ignored.

Based on Eq. 1.3, the instruction count of the object code running on the RS/6000 is 1.5 times longer than that of the code running on the VAX machine. Furthermore, the average CPI on the VAX/780 is assumed to be 5, while that on the RS/6000 is 1.39 executing the same benchmark program.

The VAX 11/780 has a typical CISC (*complex instruction set computing*) architecture, while the IBM machine has a typical RISC (*reduced instruction set computing*) architecture to be characterized in Chapter 4. This example offers a simple comparison between the two types of computers based on a single program run. When a different program is run, the conclusion may not be the same.

We cannot calculate the CPU throughput W_p unless we know the program length and the average CPI of each code. The system throughput W_s should be measured across a large number of programs over a long observation period. The message being conveyed is that one should not draw a sweeping conclusion about the performance of a machine based on one or a few program runs. ■

Programming Environments　　The programmability of a computer depends on the programming environment provided to the users. Most computer environments are not user-friendly. In fact, the marketability of any new computer system depends on the creation of a user-friendly environment in which programming becomes a joyful undertaking rather than a nuisance. We briefly introduce below the environmental features desired in modern computers.

Conventional uniprocessor computers are programmed in a *sequential environment* in which instructions are executed one after another in a sequential manner. In fact, the original UNIX/OS kernel was designed to respond to one system call from the user process at a time. Successive system calls must be serialized through the kernel.

Most existing compilers are designed to generate sequential object codes to run on a sequential computer. In other words, conventional computers are being used in a sequential programming environment using languages, compilers, and operating systems all developed for a uniprocessor computer.

When using a parallel computer, one desires a *parallel environment* where parallelism is automatically exploited. Language extensions or new constructs must be developed to specify parallelism or to facilitate easy detection of parallelism at various granularity levels by more intelligent compilers.

Besides parallel languages and compilers, the operating systems must be also extended to support parallel activities. The OS must be able to manage the resources behind parallelism. Important issues include parallel scheduling of concurrent events, shared memory allocation, and shared peripheral and communication links.

Implicit Parallelism   An implicit approach uses a conventional language, such as C, Fortran, Lisp, or Pascal, to write the source program. The sequentially coded source program is translated into parallel object code by a parallelizing compiler. As illustrated in Fig. 1.5a, this compiler must be able to detect parallelism and assign target machine resources. This compiler approach has been applied in programming shared-memory multiprocessors.

With parallelism being implicit, success relies heavily on the "intelligence" of a parallelizing compiler. This approach requires less effort on the part of the programmer. David Kuck of the University of Illinois and Ken Kennedy of Rice University and their associates have adopted this implicit-parallelism approach.

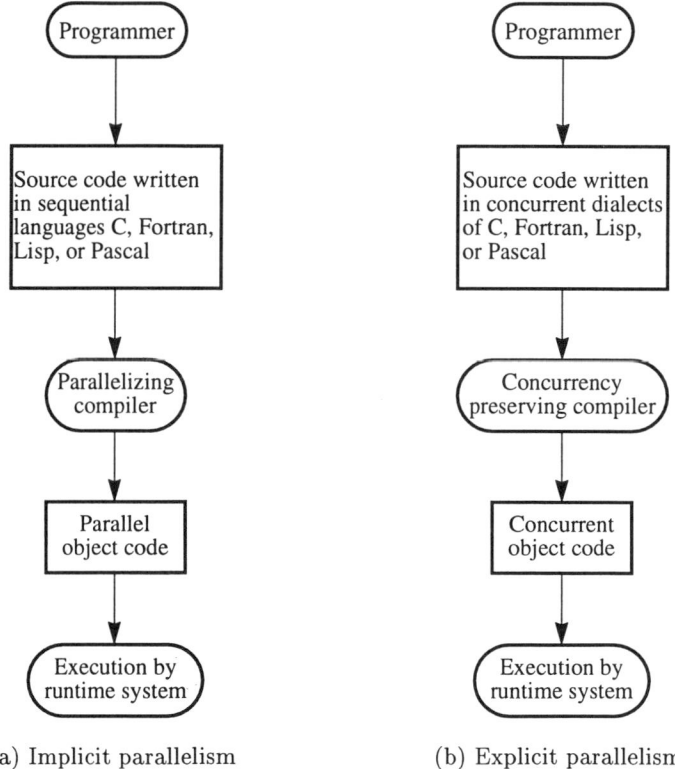

(a) Implicit parallelism   (b) Explicit parallelism

Figure 1.5 Two approaches to parallel programming.   (Courtesy of Charles Seitz; reprinted with permission from "Concurrent Architectures", p. 51 and p. 53, *VLSI and Parallel Computation*, edited by Suaya and Birtwistle, Morgan Kaufmann Publishers, 1990)

Explicit Parallelism The second approach (Fig. 1.5b) requires more effort by the programmer to develop a source program using parallel dialects of C, Fortran, Lisp, or Pascal. Parallelism is explicitly specified in the user programs. This will significantly reduce the burden on the compiler to detect parallelism. Instead, the compiler needs to preserve parallelism and, where possible, assigns target machine resources. Charles Seitz of California Institute of Technology and William Dally of Massachusetts Institute of Technology adopted this explicit approach in multicomputer development.

Special software tools are needed to make an environment more friendly to user groups. Some of the tools are parallel extensions of conventional high-level languages. Others are integrated environments which include tools providing different levels of program abstraction, validation, testing, debugging, and tuning; performance prediction and monitoring; and visualization support to aid program development, performance measurement, and graphics display and animation of computational results.

1.2 Multiprocessors and Multicomputers

Two categories of parallel computers are architecturally modeled below. These physical models are distinguished by having a shared common memory or unshared distributed memories. Only architectural organization models are described in Sections 1.2 and 1.3. Theoretical and complexity models for parallel computers are presented in Section 1.4.

1.2.1 Shared-Memory Multiprocessors

We describe below three shared-memory multiprocessor models: the *uniform-memory access* (UMA) model, the *nonuniform-memory-access* (NUMA) model, and the *cache-only memory architecture* (COMA) model. These models differ in how the memory and peripheral resources are shared or distributed.

The UMA Model In a UMA multiprocessor model (Fig. 1.6), the physical memory is uniformly shared by all the processors. All processors have equal access time to all memory words, which is why it is called uniform memory access. Each processor may use a private cache. Peripherals are also shared in some fashion.

Multiprocessors are called *tightly coupled systems* due to the high degree of resource sharing. The system interconnect takes the form of a common bus, a crossbar switch, or a multistage network to be studied in Chapter 7.

Most computer manufacturers have *multiprocessor* (MP) extensions of their *uniprocessor* (UP) product line. The UMA model is suitable for general-purpose and time-sharing applications by multiple users. It can be used to speed up the execution of a single large program in time-critical applications. To coordinate parallel events, synchronization and communication among processors are done through using shared variables in the common memory.

When all processors have equal access to all peripheral devices, the system is called a *symmetric* multiprocessor. In this case, all the processors are equally capable of running the executive programs, such as the OS kernel and I/O service routines.

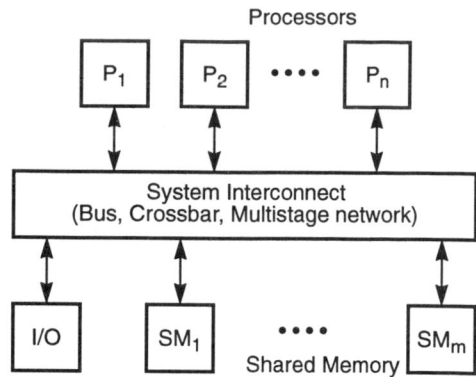

Figure 1.6 The UMA multiprocessor model (e.g., the Sequent Symmetry S-81).

In an *asymmetric* multiprocessor, only one or a subset of processors are executive-capable. An executive or a master processor can execute the operating system and handle I/O. The remaining processors have no I/O capability and thus are called *attached processors* (APs). Attached processors execute user codes under the supervision of the master processor. In both MP and AP configurations, memory sharing among master and attached processors is still in place.

Example 1.2 Approximated performance of a multiprocessor

This example exposes the reader to parallel program execution on a shared-memory multiprocessor system. Consider the following Fortran program written for sequential execution on a uniprocessor system. All the arrays, A(I), B(I), and C(I), are assumed to have N elements.

L1:		**Do** 10 I = 1, N
L2:		A(I) = B(I) + C(I)
L3:	10	**Continue**
L4:		SUM = 0
L5:		**Do** 20 J = 1, N
L6:		SUM = SUM + A(J)
L7:	20	**Continue**

Suppose each line of code L2, L4, and L6 takes 1 machine cycle to execute. The time required to execute the program control statements L1, L3, L5, and L7 is ignored to simplify the analysis. Assume that k cycles are needed for each interprocessor communication operation via the shared memory.

Initially, all arrays are assumed already loaded in the main memory and the short program fragment already loaded in the instruction cache. In other words, instruction fetch and data loading overhead is ignored. Also, we ignore bus contention or memory access conflicts problems. In this way, we can concentrate on the analysis of CPU demand.

The above program can be executed on a sequential machine in $2N$ cycles under the above assumptions. N cycles are needed to execute the N independent iterations in the I loop. Similarly, N cycles are needed for the J loop, which contains N recursive iterations.

To execute the program on an M-processor system, we partition the looping operations into M sections with $L = N/M$ elements per section. In the following parallel code, **Doall** declares that all M sections be executed by M processors in parallel:

```
Doall K=1,M
        Do 10 I = L(K−1) + 1, KL
            A(I) = B(I) + C(I)
10      Continue
        SUM(K) = 0
        Do 20 J = 1, L
            SUM(K) = SUM(K) + A(L(K−1) + J)
20      Continue
    Endall
```

For M-way parallel execution, the sectioned I loop can be done in L cycles. The sectioned J loop produces M partial sums in L cycles. Thus $2L$ cycles are consumed to produce all M partial sums. Still, we need to merge these M partial sums to produce the final sum of N elements.

The addition of each pair of partial sums requires k cycles through the shared memory. An l-level binary adder tree can be constructed to merge all the partial sums, where $l = \log_2 M$. The adder tree takes $l(k + 1)$ cycles to merge the M partial sums sequentially from the leaves to the root of the tree. Therefore, the multiprocessor requires $2L + l(k + 1) = 2N/M + (k + 1)\log_2 M$ cycles to produce the final sum.

Suppose $N = 2^{20}$ elements in the array. Sequential execution of the original program takes $2N = 2^{21}$ machine cycles. Assume that each IPC synchronization overhead has an average value of $k = 200$ cycles. Parallel execution on $M = 256$ processors requires $2^{13} + 1608 = 9800$ machine cycles.

Comparing the above timing results, the multiprocessor shows a speedup factor of 214 out of the maximum value of 256. Therefore, an efficiency of $214/256 = 83.6\%$ has been achieved. We will study the speedup and efficiency issues in Chapter 3.

The above result was obtained under favorable assumptions about overhead. In reality, the resulting speedup might be lower after considering all software overhead and potential resource conflicts. Nevertheless, the example shows the promising side of parallel processing if the interprocessor communication overhead can be reduced to a sufficiently low level. ∎

The NUMA Model   A NUMA multiprocessor is a shared-memory system in which the access time varies with the location of the memory word. Two NUMA machine models are depicted in Fig. 1.7. The shared memory is physically distributed to all processors, called *local memories*. The collection of all local memories forms a global address space accessible by all processors.

It is faster to access a local memory with a local processor. The access of remote memory attached to other processors takes longer due to the added delay through the interconnection network. The BBN TC-2000 Butterfly multiprocessor assumes the configuration shown in Fig. 1.7a.

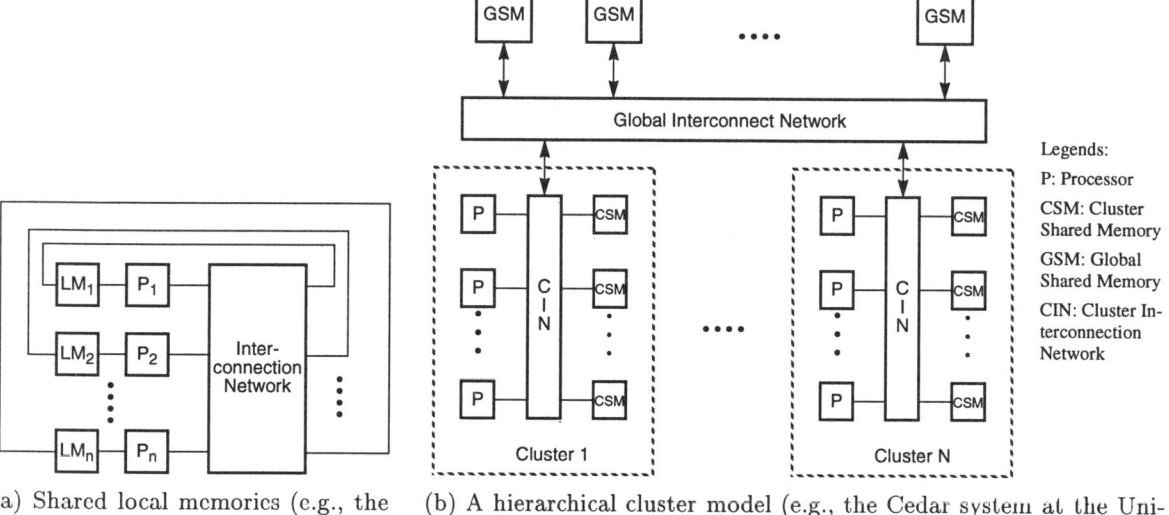

(a) Shared local memories (e.g., the BBN Butterfly)

(b) A hierarchical cluster model (e.g., the Cedar system at the University of Illinois)

Figure 1.7 Two NUMA models for multiprocessor systems.

Besides distributed memories, globally shared memory can be added to a multiprocessor system. In this case, there are three memory-access patterns: The fastest is local memory access. The next is global memory access. The slowest is access of remote memory as illustrated in Fig. 1.7b. As a matter of fact, the models shown in Figs. 1.6 and 1.7 can be easily modified to allow a mixture of shared memory and private memory with prespecified access rights.

A hierarchically structured multiprocessor is modeled in Fig. 1.7b. The processors are divided into several *clusters*. Each cluster is itself an UMA or a NUMA multiprocessor. The clusters are connected to *global shared-memory* modules. The entire system is considered a NUMA multiprocessor. All processors belonging to the same cluster are allowed to uniformly access the *cluster shared-memory* modules.

All clusters have equal access to the global memory. However, the access time to the cluster memory is shorter than that to the global memory. One can specify the access right among intercluster memories in various ways. The Cedar multiprocessor, built at the University of Illinois, assumes such a structure in which each cluster is an Alliant FX/80 multiprocessor.

Table 1.3 Representative Multiprocessor Systems

Company and Model	Hardware and Architecture	Software and Applications	Remarks
Sequent Symmetry S-81	Bus-connected with 30 i386 processors, IPC via SLIC bus; Weitek floating-point accelerator.	DYNIX/OS, KAP/Sequent preprocessor, transaction multiprocessing.	i486-based multiprocessor available 1991. i586-based systems to appear.
IBM ES/9000 Model 900/VF	Model 900/VF has 6 ES/9000 processors with vector facilities, crossbar connected to I/O channels and shared memory.	OS support: MVS, VM KMS, AIX/370, parallel Fortran, VSF V2.5 compiler.	Fiber optic channels, integrated cryptographic architecture.
BBN TC-2000	512 M88100 processors with local memory connected by a Butterfly switch, a NUMA machine.	Ported Mach/OS with multiclustering, use parallel Fortran, time-critical applications.	Shooting for higher performance using faster processors in future models.

The COMA Model   A multiprocessor using cache-only memory assumes the COMA model. Examples of COMA machines include the Swedish Institute of Computer Science's Data Diffusion Machine (DDM, Hagersten et al., 1990) and Kendall Square Research's KSR-1 machine (Burkhardt et al., 1992). The COMA model is depicted in Fig. 1.8. Details of KSR-1 are given in Chapter 9.

The COMA model is a special case of a NUMA machine, in which the distributed main memories are converted to caches. There is no memory hierarchy at each processor node. All the caches form a global address space. Remote cache access is assisted by the distributed cache directories (D in Fig. 1.8). Depending on the interconnection network used, sometimes hierarchical directories may be used to help locate copies of cache blocks. Initial data placement is not critical because data will eventually migrate to where it will be used.

Besides the UMA, NUMA, and COMA models specified above, other variations exist for multiprocessors. For example, a *cache-coherent non-uniform memory access* (CC-NUMA) model can be specified with distributed shared memory and cache directories. Examples of the CC-NUMA model include the Stanford Dash (Lenoski et al., 1990) and the MIT Alewife (Agarwal et al., 1990) to be studied in Chapter 9. One can also insist on a cache-coherent COMA machine in which all cache copies must be kept consistent.

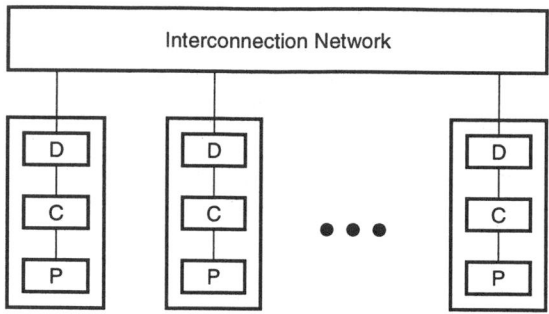

Figure 1.8 The COMA model of a multiprocessor. (P: Processor, C: Cache, D: Directory; e.g., the KSR-1)

Representative Multiprocessors   Several commercially available multiprocessors are summarized in Table 1.3. They represent four classes of multiprocessors. The Sequent Symmetry S81 belongs to a class called minisupercomputers. The IBM System/390 models are high-end mainframes, sometimes called near-supercomputers. The BBN TC-2000 represents the MPP class.

The S-81 is a transaction processing multiprocessor consisting of 30 i386/i486 microprocessors tied to a common backplane bus. The IBM ES/9000 models are the latest IBM mainframes having up to 6 processors with attached vector facilities. The TC-2000 can be configured to have 512 M88100 processors interconnected by a multistage Butterfly network. This is designed as a NUMA machine for real-time or time-critical applications.

Multiprocessor systems are suitable for general-purpose multiuser applications where programmability is the major concern. A major shortcoming of multiprocessors is the lack of scalability. It is rather difficult to build MPP machines using centralized shared-memory model. Latency tolerance for remote memory access is also a major limitation.

Packaging and cooling impose additional constraints on scalability. We will study scalability and programmability in subsequent chapters. In building MPP systems, distributed-memory multicomputers are more scalable but less programmable due to added communication protocols.

1.2.2   Distributed-Memory Multicomputers

A distributed-memory multicomputer system is modeled in Fig. 1.9. The system consists of multiple computers, often called *nodes*, interconnected by a message-passing network. Each node is an autonomous computer consisting of a processor, local memory, and sometimes attached disks or I/O peripherals.

The message-passing network provides point-to-point static connections among the nodes. All local memories are private and are accessible only by local processors. For this reason, traditional multicomputers have been called *no-remote-memory-access* (NORMA) machines. However, this restriction will gradually be removed in future multicomputers with distributed shared memories. Internode communication is carried out by passing messages through the static connection network.

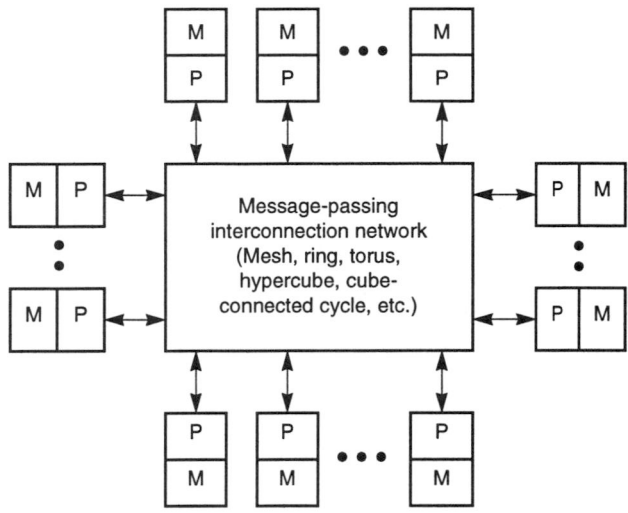

Figure 1.9 Generic model of a message-passing multicomputer.

Multicomputer Generations   Modern multicomputers use hardware routers to pass messages. A computer node is attached to each router. The boundary router may be connected to I/O and peripheral devices. Message passing between any two nodes involves a sequence of routers and channels. Mixed types of nodes are allowed in a heterogeneous multicomputer. The internode communications in a heterogeneous multicomputer are achieved through compatible data representations and message-passing protocols.

Message-passing multicomputers have gone through two generations of development, and a new generation is emerging.

The *first generation* (1983–1987) was based on processor board technology using hypercube architecture and software-controlled message switching. The Caltech Cosmic and Intel iPSC/1 represented the first-generation development.

The *second generation* (1988–1992) was implemented with mesh-connected architecture, hardware message routing, and a software environment for medium-grain distributed computing, as represented by the Intel Paragon and the Parsys SuperNode 1000.

The emerging *third generation* (1993–1997) is expected to be fine-grain multicomputers, like the MIT J-Machine and Caltech Mosaic, implemented with both processor and communication gears on the same VLSI chip.

In Section 2.4, we will study various static network topologies used to construct multicomputers. Famous topologies include the *ring, tree, mesh, torus, hypercube, cube-connected cycle*, etc. Various communication patterns are demanded among the nodes, such as one-to-one, broadcasting, permutations, and multicast patterns.

Important issues for multicomputers include message-routing schemes, network flow control strategies, deadlock avoidance, virtual channels, message-passing primitives, and program decomposition techniques. In Part IV, we will study the programming issues of three generations of multicomputers.

Representative Multicomputers   Three message-passing multicomputers are summarized in Table 1.4. With distributed processor/memory nodes, these machines are better in achieving a scalable performance. However, message passing imposes a hardship on programmers to distribute the computations and data sets over the nodes or to establish efficient communication among nodes. Until intelligent compilers and efficient distributed OSs become available, multicomputers will continue to lack programmability.

Table 1.4 Representative Multicomputer Systems

System Features	Intel Paragon XP/S	nCUBE/2 6480	Parsys Ltd. SuperNode1000
Node Types and Memory	50 MHz i860 XP computing nodes with 16–128 Mbytes per node, special I/O service nodes.	Each node contains a CISC 64-bit CPU, with FPU, 14 DMA ports, with 1–64 Mbytes/node.	EC-funded Esprit supernode built with multiple T-800 Transputers per node.
Network and I/O	2-D mesh with SCSI, HIPPI, VME, Ethernet, and custom I/O.	13-dimensional hypercube of 8192 nodes, 512-Gbyte memory, 64 I/O boards.	Reconfigurable interconnect, expandable to have 1024 processors.
OS and Software task parallelism Support	OSF conformance with 4.3 BSD, visualization and programming support.	Vertex/OS or UNIX supporting message passing using wormhole routing.	IDRIS/OS is UNIX-compatible, supported.
Application Drivers	General sparse matrix methods, parallel data manipulation, strategic computing.	Scientific number crunching with scalar nodes, database processing.	Scientific and academic applications.
Performance Remarks	5–300 Gflops peak 64-bit results, 2.8–160 GIPS peak integer performance.	27 Gflops peak, 36 Gbytes/s I/O, challenge to build even larger machine.	200 MIPS to 13 GIPS peak, largest supernode in use contains 256 Transputers.

The Paragon system assumes a mesh architecture, and the nCUBE/2 has a hypercube architecture. The Intel i860s and some custom-designed VLSI processors are used as building blocks in these machines. All three OSs are UNIX-compatible with extended functions to support message passing.

Most multicomputers are being upgraded to yield a higher degree of parallelism with enhanced processors. We will study various massively parallel systems in Part III where the tradeoffs between scalability and programmability are analyzed.

1.2.3 A Taxonomy of MIMD Computers

Parallel computers appear as either SIMD or MIMD configurations. The SIMDs appeal more to special-purpose applications. It is clear that SIMDs are not size-scalable, but unclear whether large SIMDs are generation-scalable. The fact that CM-5 has an MIMD architecture, away from the SIMD architecture in CM-2, may shed some light on the architectural trend. Furthermore, the boundary between multiprocessors and multicomputers has become blurred in recent years, Eventually, the distinctions may vanish.

The architectural trend for future general-purpose computers is in favor of MIMD configurations with distributed memories having a globally shared virtual address space. For this reason, Gordon Bell (1992) has provided a taxonomy of MIMD machines, reprinted in Fig. 1.10. He considers shared-memory multiprocessors as having a single address space. Scalable multiprocessors or multicomputers must use distributed shared memory. Unscalable multiprocessors use centrally shared memory.

Multicomputers use distributed memories with multiple address spaces. They are scalable with distributed memory. Centralized multicomputers are yet to appear. Many of the identified example systems will be treated in subsequent chapters. The evolution of fast LAN (*local area network*)-connected workstations will create "commodity supercomputing". Bell advocates high-speed workstation clusters interconnected by high-speed switches in lieu of special-purpose multicomputers. The CM-5 development has already moved in this direction.

The scalability of MIMD computers will be further studied in Section 3.4 and Chapter 9. In Part III, we will study distributed-memory multiprocessors (KSR-1, SCI, etc.); central-memory multiprocessors (Cray, IBM, DEC, Fujitsu, Encore, etc.); multicomputers by Intel, TMC, and nCUBE; fast LAN-based workstation clusters; and other exploratory research systems.

1.3 Multivector and SIMD Computers

In this section, we introduce supercomputers and parallel processors for vector processing and data parallelism. We classify supercomputers either as pipelined vector machines using a few powerful processors equipped with vector hardware, or as SIMD computers emphasizing massive data parallelism.

1.3.1 Vector Supercomputers

A vector computer is often built on top of a scalar processor. As shown in Fig. 1.11, the vector processor is attached to the scalar processor as an optional feature. Program

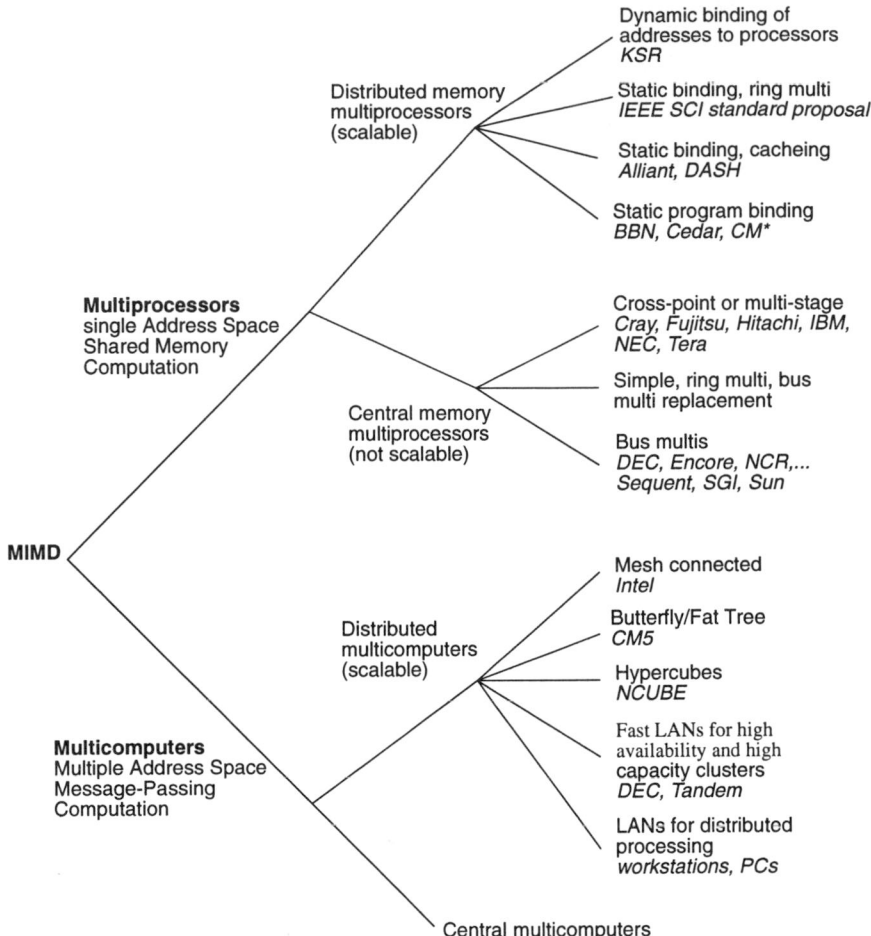

Figure 1.10 Bell's taxonomy of MIMD computers. (Courtesy of Gordon Bell; reprinted with permission from the *Communications of ACM*, August 1992)

and data are first loaded into the main memory through a host computer. All instructions are first decoded by the scalar control unit. If the decoded instruction is a scalar operation or a program control operation, it will be directly executed by the scalar processor using the scalar functional pipelines.

If the instruction is decoded as a vector operation, it will be sent to the vector control unit. This control unit will supervise the flow of vector data between the main memory and vector functional pipelines. The vector data flow is coordinated by the control unit. A number of vector functional pipelines may be built into a vector processor. Two pipeline vector supercomputer models are described below.

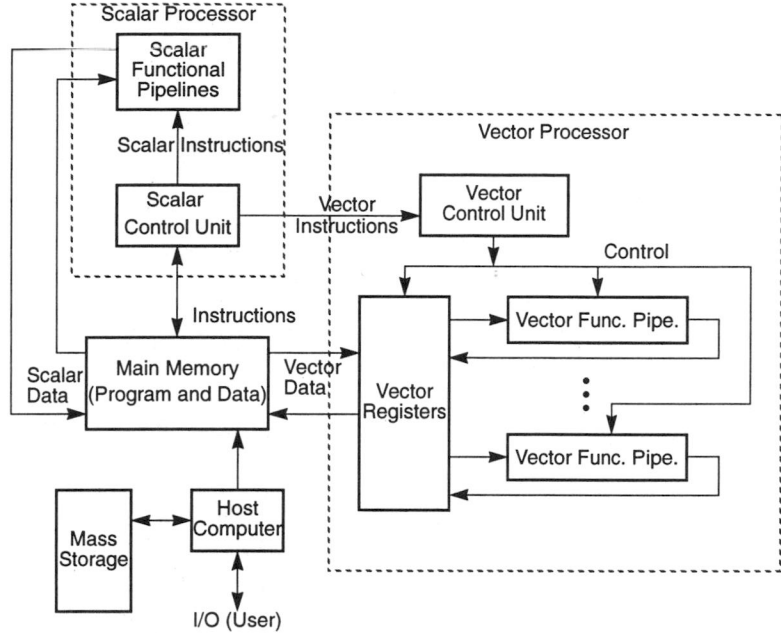

Figure 1.11 The architecture of a vector supercomputer.

Vector Processor Models   Figure 1.11 shows a *register-to-register* architecture. Vector registers are used to hold the vector operands, intermediate and final vector results. The vector functional pipelines retrieve operands from and put results into the vector registers. All vector registers are programmable in user instructions. Each vector register is equipped with a component counter which keeps track of the component registers used in successive pipeline cycles.

The length of each vector register is usually fixed, say, sixty-four 64-bit component registers in a vector register in a Cray Series supercomputer. Other machines, like the Fujitsu VP2000 Series, use reconfigurable vector registers to dynamically match the register length with that of the vector operands.

In general, there are fixed numbers of vector registers and functional pipelines in a vector processor. Therefore, both resources must be reserved in advance to avoid resource conflicts between different vector operations. Several vector-register based supercomputers are summarized in Table 1.5.

A *memory-to-memory* architecture differs from a register-to-register architecture in the use of a vector stream unit to replace the vector registers. Vector operands and results are directly retrieved from the main memory in superwords, say, 512 bits as in the Cyber 205.

Pipelined vector supercomputers started with uniprocessor models such as the Cray 1 in 1976. Recent supercomputer systems offer both uniprocessor and multiprocessor models such as the Cray Y-MP Series. Most high-end mainframes offer multiprocessor models with add-on vector hardware, such as the VAX 9000 and IBM 390/VF models.

Representative Supercomputers Over a dozen pipelined vector computers have been manufactured, ranging from workstations to mini- and supercomputers. Notable ones include the Stardent 3000 multiprocessor equipped with vector pipelines, the Convex C3 Series, the DEC VAX 9000, the IBM 390/VF, the Cray Research Y-MP family, the NEC SX Series, the Fujitsu VP2000, and the Hitachi S-810/20.

Table 1.5 Representative Vector Supercomputers

System Model	Vector Hardware Architecture and Capabilities	Compiler and Software Support
Convex C3800 family	GaAs-based multiprocessor with 8 processors and 500-Mbyte/s access port. 4 Gbytes main memory. 2 Gflops peak performance with concurrent scalar/vector operations.	Advanced C, Fortran, and Ada vectorizing and parallelizing compilers. Also support inter-procedural optimization, POSIX 1003.1/OS plus I/O interfaces and visualization system
Digital VAX 9000 System	Integrated vector processing in the VAX environment, 125–500 Mflops peak performance. 63 vector instructions. 16 x 64 x 64 vector registers. Pipeline chaining possible.	MS or ULTRIX/OS, VAX Fortran and VAX Vector Instruction Emulator (VVIEF) for vectorized program debugging.
Cray Research Y-MP and C-90	Y-MP runs with 2, 4, or 8 processors, 2.67 Gflop peak with Y-MP8256. C-90 has 2 vector pipes/CPU built with 10K gate ECL with 16 Gflops peak performance.	CF77 compiler for automatic vectorization, scalar optimization, and parallel processing. UNICOS improved from UNIX/V and Berkeley BSD/OS.

The Convex C1 and C2 Series were made with ECL/CMOS technologies. The latest C3 Series is based on GaAs technology.

The DEC VAX 9000 is Digital's largest mainframe system providing concurrent scalar/vector and multiprocessing capabilities. The VAX 9000 processors use a hybrid architecture. The vector unit is an optional feature attached to the VAX 9000 CPU. The Cray Y-MP family offers both vector and multiprocessing capabilities.

1.3.2 SIMD Supercomputers

In Fig. 1.3b, we have shown an abstract model of SIMD computers having a single instruction stream over multiple data streams. An operational model of SIMD computers is presented below (Fig. 1.12) based on the work of H. J. Siegel (1979). Implementation models and case studies of SIMD machines are given in Chapter 8.

SIMD Machine Model   An operational model of an SIMD computer is specified by a 5-tuple:

$$M = \langle N, C, I, M, R \rangle \qquad\qquad (1.5)$$

where

(1) N is the number of *processing elements* (PEs) in the machine. For example, the Illiac IV has 64 PEs and the Connection Machine CM-2 uses 65,536 PEs.

(2) C is the set of instructions directly executed by the *control unit* (CU), including scalar and program flow control instructions.

(3) I is the set of instructions broadcast by the CU to all PEs for parallel execution. These include arithmetic, logic, data routing, masking, and other local operations executed by each active PE over data within that PE.

(4) M is the set of masking schemes, where each mask partitions the set of PEs into enabled and disabled subsets.

(5) R is the set of data-routing functions, specifying various patterns to be set up in the interconnection network for inter-PE communications.

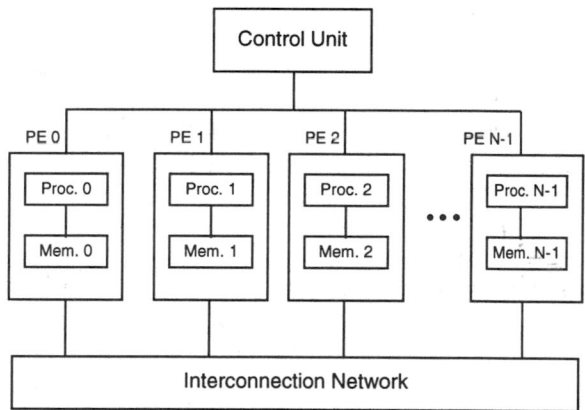

Figure 1.12 Operational model of SIMD computers.

One can describe a particular SIMD machine architecture by specifying the 5-tuple. An example SIMD machine is partially specified below:

Example 1.3 Operational specification of the MasPar MP-1 computer

We will study the detailed architecture of the MasPar MP-1 in Chapter 7. Listed below is a partial specification of the 5-tuple for this machine:

(1) The MP-1 is an SIMD machine with $N = 1024$ to 16,384 PEs, depending on which configuration is considered.

(2) The CU executes scalar instructions, broadcasts decoded vector instructions to the PE array, and controls the inter-PE communications.

(3) Each PE is a register-based load/store RISC processor capable of executing integer operations over various data sizes and standard floating-point operations. The PEs receive instructions from the CU.

(4) The masking scheme is built within each PE and continuously monitored by the CU which can set and reset the status of each PE dynamically at run time.

(5) The MP-1 has an X-Net mesh network plus a global multistage crossbar router for inter-CU-PE, X-Net nearest 8-neighbor, and global router communications.

∎

Representative SIMD Computers   Three SIMD supercomputers are summarized in Table 1.6. The number of PEs in these systems ranges from 4096 in the DAP610 to 16,384 in the MasPar MP-1 and 65,536 in the CM-2. Both the CM-2 and DAP610 are fine-grain, bit-slice SIMD computers with attached floating-point accelerators for blocks of PEs.

Each PE of the MP-1 is equipped with a 1-bit logic unit, 4-bit integer ALU, 64-bit mantissa unit, and 16-bit exponent unit. Therefore, the MP-1 is a medium-grain SIMD machine. Multiple PEs can be built on a single chip due to the simplicity of each PE. The MP-1 implements 32 PEs per chip with forty 32-bit registers per PE. The 32 PEs are interconnected by an *X-Net mesh*, which is a 4-neighbor mesh augmented with diagonal dual-stage links.

The CM-2 implements 16 PEs as a mesh on a single chip. Each 16-PE mesh chip is placed at one vertex of a 12-dimensional hypercube. Thus $16 \times 2^{12} = 2^{16} = 65,536$ PEs form the entire SIMD array.

The DAP610 implements 64 PEs as a mesh on a chip. Globally, a large mesh (64×64) is formed by interconnecting these small meshes on chips. Fortran 90 and modified versions of C, Lisp, and other synchronous programming languages have been developed to program SIMD machines.

1.4   PRAM and VLSI Models

Theoretical models of parallel computers are abstracted from the physical models studied in previous sections. These models are often used by algorithm designers and VLSI device/chip developers. The ideal models provide a convenient framework for developing parallel algorithms without worry about the implementation details or physical constraints.

The models can be applied to obtain theoretical performance bounds on parallel computers or to estimate VLSI complexity on chip area and execution time before the

Table 1.6 Representative SIMD Supercomputers

System Model	SIMD Machine Architecture and Capabilities	Languages, Compilers and Software Support
MasPar Computer Corporation MP-1 Family	Available in configurations from 1024 to 16,384 processors with 26,000 MIPS or 1.3 Gflops. Each PE is a RISC processor, with 16 Kbytes local memory. An X-Net mesh plus a multistage crossbar interconnect.	Fortran 77, MasPar Fortran (MPF), and MasPar Parallel Application Language; UNIX/OS with X-window, symbolic debugger, visualizers and animators.
Thinking Machines Corporation, CM-2	A bit-slice array of up to 65,536 PEs arranged as a 10-dimensional hypercube with 4×4 mesh on each vertex, up to 1M bits of memory per PE, with optional FPU shared between blocks of 32 PEs. 28 Gflops peak and 5.6 Gflops sustained.	Driven by a host of VAX, Sun, or Symbolics 3600, Lisp compiler, Fortran 90, C*, and *Lisp supported by PARIS
Active Memory Technology DAP600 Family	A fine-grain, bit-slice SIMD array of up to 4096 PEs interconnected by a square mesh with 1K bits per PE, orthogonal and 4-neighbor links, 20 GIPS and 560 Mflops peak performance.	Provided by host VAX/VMS or UNIX Fortran-plus or APAL on DAP, Fortran 77 or C on host. Fortran-plus affected Fortran 90 standards.

chip is fabricated. The abstract models are also useful in scalability and programmability analysis, when real machines are compared with an idealized parallel machine without worrying about communication overhead among processing nodes.

1.4.1 Parallel Random-Access Machines

Theoretical models of parallel computers are presented below. We define first the time and space complexities. Computational tractability is reviewed for solving difficult problems on computers. Then we introduce the *random-access machine* (RAM), *parallel random-access machine* (PRAM), and variants of PRAMs. These complexity models facilitate the study of asymptotic behavior of algorithms implementable on parallel computers.

Time and Space Complexities   The complexity of an algorithm for solving a problem of size s on a computer is determined by the execution time and the storage space required. The *time complexity* is a function of the problem size. The time complexity function in order notation is the *asymptotic time complexity* of the algorithm. Usually, the worst-case time complexity is considered. For example, a time complexity $g(s)$ is said to be $O(f(s))$, read "order $f(s)$", if there exist positive constants c and s_0 such that $g(s) \leq cf(s)$ for all nonnegative values of $s > s_0$.

The *space complexity* can be similarly defined as a function of the problem size *s*. The *asymptotic space complexity* refers to the data storage of large problems. Note that the program (code) storage requirement and the storage for input data are not considered in this.

The time complexity of a serial algorithm is simply called *serial complexity*. The time complexity of a parallel algorithm is called *parallel complexity*. Intuitively, the parallel complexity should be lower than the serial complexity, at least asymptotically. We consider only *deterministic algorithms*, in which every operational step is uniquely defined in agreement with the way programs are executed on real computers.

A *nondeterministic algorithm* contains operations resulting in one outcome in a set of possible outcomes. There exist no real computers that can execute nondeterministic algorithms. Therefore, all algorithms (or machines) considered in this book are deterministic, unless otherwise noted.

NP-Completeness   An algorithm has a *polynomial complexity* if there exists a polynomial $p(s)$ such that the time complexity is $O(p(s))$ for any problem size *s*. The set of problems having polynomial-complexity algorithms is called *P-class* (for polynomial class). The set of problems solvable by nondeterministic algorithms in polynomial time is called *NP-class* (for nondeterministic polynomial class).

Since deterministic algorithms are special cases of the nondeterministic ones, we know that P $\subset$ NP. The P-class problems are computationally *tractable*, while the NP $-$ P-class problems are *intractable*. But we do not know whether P $=$ NP or P $\neq$ NP. This is still an open problem in computer science.

To simulate a nondeterministic algorithm with a deterministic algorithm may require exponential time. Therefore, intractable NP-class problems are also said to have exponential-time complexity.

Example 1.4 Polynomial- and exponential-complexity algorithms

Polynomial-complexity algorithms are known for sorting *n* numbers in $O(n \log n)$ time and for multiplication of two $n \times n$ matrices in $O(n^3)$ time. Therefore, both problems belong to the P-class.

Nonpolynomial algorithms have been developed for the traveling salesperson problem with complexity $O(n^2 2^n)$ and for the knapsack problem with complexity $O(2^{n/2})$. These complexities are *exponential*, greater than the polynomial complexities. So far, deterministic polynomial algorithms have not been found for these problems. Therefore, exponential-complexity problems belong to the NP-class.
∎

Most computer scientists believe that P $\neq$ NP. This leads to the conjecture that there exists a subclass, called *NP-complete* (NPC) problems, such that NPC $\subset$ NP but NPC $\cap$ P $= \emptyset$ (Fig. 1.13). In fact, it has been proved that if any NP-complete problem is polynomial-time solvable, then one can conclude P $=$ NP. Thus NP-complete problems are considered the hardest ones to solve. Only approximation algorithms were derived for solving some of the NP-complete problems.

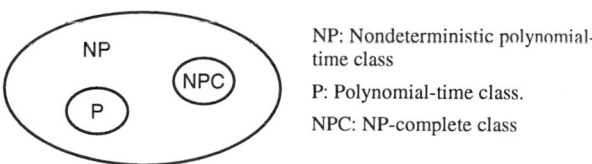

NP: Nondeterministic polynomial-
time class

P: Polynomial-time class.

NPC: NP-complete class

Figure 1.13 The relationships conjectured among the NP, P, and NPC classes of computational problems.

PRAM Models   Conventional uniprocessor computers have been modeled as *random-access machines* (RAM) by Sheperdson and Sturgis (1963). A *parallel random-access machine* (PRAM) model has been developed by Fortune and Wyllie (1978) for modeling idealized parallel computers with zero synchronization or memory access overhead. This PRAM model will be used for parallel algorithm development and for scalability and complexity analysis.

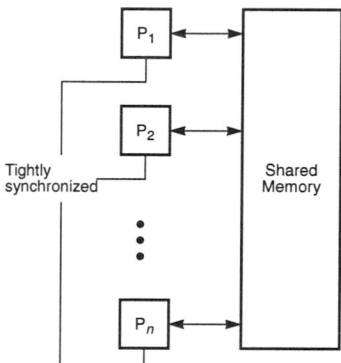

Figure 1.14 PRAM model of a multiprocessor system with shared memory, on which all n processors operate in lockstep in memory access and program execution operations. Each processor can access any memory location in unit time.

An *n*-processor PRAM (Fig. 1.14) has a globally addressable memory. The shared memory can be distributed among the processors or centralized in one place. The *n* processors [also called *processing elements* (PEs) by other authors] operate on a synchronized read-memory, compute, and write-memory cycle. With shared memory, the model must specify how concurrent read and concurrent write of memory are handled. Four memory-update options are possible:

- *Exclusive read* (ER) — This allows at most one processor to read from any memory location in each cycle, a rather restrictive policy.

- *Exclusive write* (EW) — This allows at most one processor to write into a memory location at a time.
- *Concurrent read* (CR) — This allows multiple processors to read the same information from the same memory cell in the same cycle.
- *Concurrent write* (CW) — This allows simultaneous writes to the same memory location. In order to avoid confusion, some policy must be set up to resolve the write conflicts.

Various combinations of the above options lead to several variants of the PRAM model as specified below. Since CR does not create a conflict problem, variants differ mainly in how they handle the CW conflicts.

PRAM Variants   Described below are four variants of the PRAM model, depending on how the memory reads and writes are handled.

(1) The *EREW-PRAM model* — This model forbids more than one processor from reading or writing the same memory cell simultaneously (Snir, 1982; Karp and Ramachandran, 1988). This is the most restrictive PRAM model proposed.

(2) The *CREW-PRAM model* — The write conflicts are avoided by mutual exclusion. Concurrent reads to the same memory location are allowed.

(3) The *ERCW-PRAM model* — This allows exclusive read or concurrent writes to the same memory location.

(4) The *CRCW-PRAM model* — This model allows either concurrent reads or concurrent writes at the same time. The conflicting writes are resolved by one of the following four policies (Fortune and Wyllie, 1978):

  - *Common* — All simultaneous writes store the same value to the hot-spot memory location.
  - *Arbitrary* — Any one of the values written may remain; the others are ignored.
  - *Minimum* — The value written by the processor with the minimum index will remain.
  - *Priority* — The values being written are combined using some associative functions, such as summation or maximum.

Example 1.5 Multiplication of two $n \times n$ matrices in $O(\log n)$ time on a PRAM with $n^3/\log n$ processors (Viktor Prasanna, 1992)

Let A and B be the input matrices. Assume n^3 PEs are available initially. We later reduce the number of PEs to $n^3/\log n$. To visualize the algorithm, assume the memory is organized as a three-dimensional array with inputs A and B stored in two planes. Also, for sake of explanation, assume a three-dimensional indexing of the PEs. PE(i, j, k), $0 \le k \le n-1$ are used for computing the (i, j)th entry of the output matrix, $0 \le i, j \le n-1$, and n is a power of 2.

In step 1, n product terms corresponding to each output are computed using n PEs in $O(1)$ time. In step 2, these are added to produce an output in $O(\log n)$ time.

The total number of PEs used is n^3. The result is available in $C(i, j, 0), 0 \leq i, j \leq n-1$. Listed below are programs for each PE(i, j, k) to execute. All n^3 PEs operate in parallel for n^3 multiplications. But at most $n^3/2$ PEs are busy for $(n^3 - n^2)$ additions. Also, the PRAM is assumed to be CREW.

Step 1:
1. Read A(i, k)
2. Read B(k, j)
3. Compute A(i, k) × B(k, j)
4. Store in C(i, j, k)

Step 2:
1. $\ell \leftarrow n$
2. **Repeat**
   $\ell \leftarrow \ell/2$
   if $(k < \ell)$ then
     begin
       Read C(i, j, k)
       Read C(i, j, k + ℓ)
       Compute C(i, j, k) + C(i, j, k + ℓ)
       Store in C(i, j, k)
     end
   until $(\ell = 1)$

To reduce the number of PEs to $n^3/\log n$, use a PE array of size $n \times n \times n/\log n$. Each PE is responsible for computing $\log n$ product terms and summing them up. Step 1 can be easily modified to produce $n/\log n$ partial sums, each consisting of $\log n$ multiplications and $(\log n - 1)$ additions. Now we have an array C(i, j, k), $0 \leq i, j \leq n - 1, 0 \leq k \leq n/\log n - 1$, which can be summed up in $\log(n/\log n)$ time. Combining the time spent in step 1 and step 2, we have a total execution time $2\log n - 1 + \log(n/\log n) \approx O(\log n)$ for large n. ∎

Discrepancy with Physical Models   PRAM models idealized parallel computers, in which all memory references and program executions by multiple processors are synchronized without extra cost. In reality, such parallel machines do not exist. An SIMD machine with shared memory is the closest architecture modeled by PRAM. However, PRAM allows different instructions to be executed on different processors simultaneously. Therefore, PRAM really operates in synchronized MIMD mode with a shared memory.

Among the four PRAM variants, the EREW and CRCW are the most popular models used. In fact, every CRCW algorithm can be simulated by an EREW algorithm. The CRCW algorithm runs faster than an equivalent EREW algorithm. It has been proved that the best n-processor EREW algorithm can be no more than $O(\log n)$ times slower than any n-processor CRCW algorithm.

The CREW model has received more attention in the literature than the ERCW model. The CREW models MISD machines, which have attracted little attention. For our purposes, we will use the CRCW-PRAM model unless otherwise stated. This particular model will be used in defining scalability in Chapter 3.

For complexity analysis or performance comparison, various PRAM variants offer an ideal model of parallel computers. Therefore, computer scientists use the PRAM model more often than computer engineers. In this book, we design parallel/vector computers using physical architectural models rather than PRAM models.

The PRAM model will be used for scalability and performance studies in Chapter 3 as a theoretical reference machine. For regularly structured parallelism, the PRAM can model much better than practical machine models. Therefore, sometimes PRAMs can indicate an upper bound on the performance of real parallel computers.

1.4.2   VLSI Complexity Model

Parallel computers rely on the use of VLSI chips to fabricate the major components such as processor arrays, memory arrays, and large-scale switching networks. An AT^2 model for two-dimensional VLSI chips is presented below, based on the work of Clark Thompson (1980). Three lower bounds on VLSI circuits are interpreted by Jeffrey Ullman (1984). The bounds are obtained by setting limits on memory, I/O, and communication for implementing parallel algorithms with VLSI chips.

The AT^2 Model   Let A be the chip area and T be the latency for completing a given computation using a VLSI circuit chip. Let s by the problem size involved in the computation. Thompson stated in his doctoral thesis that for certain computations, there exists a lower bound $f(s)$ such that

$$A \times T^2 \geq O(f(s)) \tag{1.6}$$

The chip area A is a measure of the chip's complexity. The latency T is the time required from when inputs are applied until all outputs are produced for a single problem instance. Figure 1.15 shows how to interpret the AT^2 complexity results in VLSI chip development. The chip is represented by the base area in the two horizontal dimensions. The vertical dimension corresponds to time. Therefore, the three-dimensional solid represents the history of the computation performed by the chip.

Memory Bound on Chip Area A   There are many computations which are memory-bound, due to the need to process large data sets. To implement this type of computation in silicon, one is limited by how densely information (bit cells) can be placed on the chip. As depicted in Fig. 1.15a, the memory requirement of a computation sets a lower bound on the chip area A.

The amount of information processed by the chip can be visualized as information flow upward across the chip area. Each bit can flow through a unit area of the horizontal chip slice. Thus, the chip area bounds the amount of memory bits stored on the chip.

I/O Bound on Volume AT   The volume of the rectangular cube is represented by

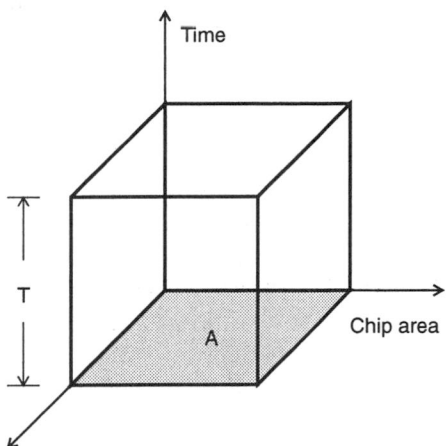

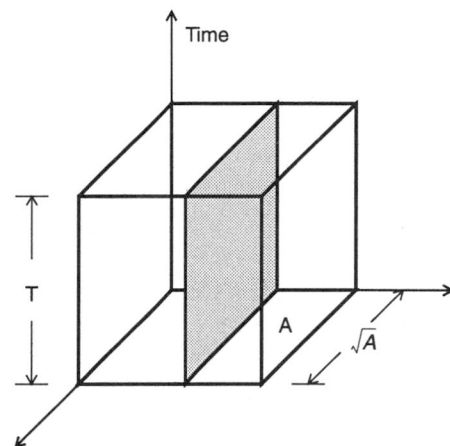

(a) Memory-limited bound on chip area
A and I/O-limited bound on chip history
represented by the volume AT

(b) Communication-limited bound on the
bisection $\sqrt{A}T$

Figure 1.15 The AT^2 complexity model of two-dimensional VLSI chips.

the product AT. As information flows through the chip for a period of time T, the
number of input bits cannot exceed the volume. This provides an I/O-limited lower
bound on the product AT, as demonstrated in Fig. 1.15a.

The area A corresponds to data into and out of the entire surface of the silicon
chip. This areal measure sets the maximum I/O limit rather than using the peripheral
I/O pads as seen in conventional chips. The height T of the volume can be visualized as
a number of snapshots on the chip, as computing time elapses. The volume represents
the amount of information flowing through the chip during the entire course of the
computation.

Bisection Communication Bound, $\sqrt{A}T$ Figure 1.15b depicts a communication
limited lower bound on the bisection area $\sqrt{A}T$. The bisection is represented by the
vertical slice cutting across the shorter dimension of the chip area. The distance of this
dimension is at most $\sqrt{A}$ for a square chip. The height of the cross section is T.

The bisection area represents the maximum amount of information exchange between the two halves of the chip circuit during the time period T. The cross-section
area $\sqrt{A}T$ limits the communication bandwidth of a computation. VLSI complexity
theoreticians have used the square of this measure, AT^2, as the lower bound.

Charles Seitz (1990) has given another interpretation of the AT^2 result. He considers the area-time product AT the cost of a computation, which can be expected to vary
as $1/T$. This implies that the cost of computation for a two-dimensional chip decreases
with the execution time allowed.

When three-dimensional (multilayer) silicon chips are used, Seitz asserted that the cost of computation, as limited by volume-time product, would vary as $1/\sqrt{T}$. This is due to the fact that the bisection will vary as $(AT)^{2/3}$ for 3-D chips instead of as $\sqrt{AT}$ for 2-D chips.

Example 1.6 VLSI chip implementation of a matrix multiplication algorithm (Viktor Prasanna, 1992)

This example shows how to estimate the chip area A and compute time T for $n \times n$ matrix multiplication C = A $\times$ B on a mesh of processing elements (PEs) with a broadcast bus on each row and each column. The 2-D mesh architecture is shown in Fig. 1.16. Inter-PE communicaiton is done through the broadcast buses. We want to prove the bound $AT^2 = O(n^4)$ by developing a parallel matrix multiplication algorithm with time $T = O(n)$ in using the mesh with broadcast buses. Therefore, we need to prove that the chip area is bounded by $A = O(n^2)$.

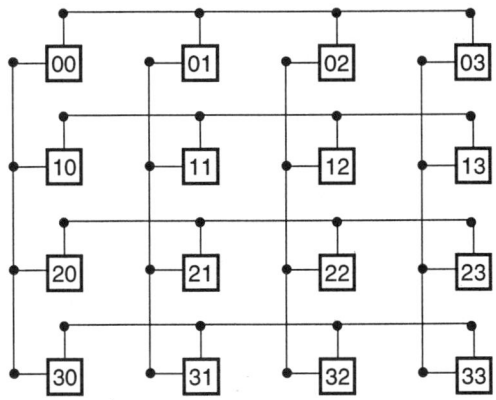

Figure 1.16 A 4×4 mesh of processing elements (PEs) with broadcast buses on each row and on each column. (Courtesy of Prasanna Kumar and Raghavendra; reprinted from *Journal of Parallel and Distributed Computing*, April 1987)

Each PE occupies a unit area, and the broadcast buses require $O(n^2)$ wire area. Thus the total chip area needed is $O(n^2)$ for an $n \times n$ mesh with broadcast buses. We show next that the $n \times n$ matrix multiplication can be performed on this mesh chip in $T = O(n)$ time. Denote the PEs as PE(i,j), $0 \leq i, j \leq n-1$.

Initially the input matrix elements A(i,j) and B(i,j) are stored in PE(i,j) with no duplicated data. The memory is distributed among all the PEs. Each PE can access only its own local memory. The following parallel algorithm shows how to perform the dot-product operations in generating all the output elements C(i,j) = $\sum_{k=0}^{n-1}$ A(i,k) $\times$ B(k,j) for $0 \leq i, j \leq n-1$.

        Doall 10 for $0 \leq i, j \leq n-1$
10   PE(i,j) sets C(i,j) to 0 /Initialization/

> **Do** 50 for $0 \leq k \leq n - 1$
>   **Doall** 20 for $0 \leq i \leq n - 1$
> 20   PE(i,k) broadcasts A(i,k) along its row bus
>   **Doall** 30 for $0 \leq j \leq n - 1$
> 30   PE(k,j) broadcasts B(k,j) along its column bus
>   /PE(i,j) now has A(i, k) and B(k, j), $0 \leq i, j \leq n - 1$/
>   **Doall** 40 for $0 \leq i, j \leq n - 1$
> 40   PE(i, j) computes C(i, j) $\leftarrow$ C(i, j) + A(i, k) $\times$ B(k, j)
> 50 **Continue**

The above algorithm has a sequential loop along the dimension indexed by k. It takes n time units (iterations) in this k-loop. Thus, we have $T = O(n)$. Therefore, $AT^2 = O(n^2) \cdot (O(n))^2 = O(n^4)$.

■

1.5   Architectural Development Tracks

The architectures of most existing computers follow certain development tracks. Understanding features of various tracks provides insights for new architectural development. We look into six tracks to be studied in later chapters. These tracks are distinguished by similarity in computational models and technological bases. We review below a few representative systems in each track.

1.5.1 Multiple-Processor Tracks

Generally speaking, a multiple-processor system can be either a shared-memory multiprocessor or a distributed-memory multicomputer as modeled in Section 1.2. Bell listed these machines at the leaf nodes of the taxonomy tree (Fig. 1.10). Instead of a horizontal listing, we show a historical development along each important track of the taxonomy.

Shared-Memory Track   Figure. 1.17a shows a track of multiprocessor development employing a single address space in the entire system. The track started with the C.mmp system developed at Carnegie-Mellon University (Wulf and Bell, 1972). The C.mmp was an UMA multiprocessor. Sixteen PDP 11/40 processors are interconnected to 16 shared-memory modules via a crossbar switch. A special interprocessor interrupt bus is provided for fast interprocess communication, besides the shared memory. The C.mmp project pioneered shared-memory multiprocessor development, not only in the crossbar architecture but also in the multiprocessor operating system (Hydra) development.

Both the NYU Ultracomputer project (Gottlieb et al., 1983) and the Illinois Cedar project (Kuck et al., 1987) were developed with a single address space. Both systems used multistage networks as a system interconnect. The major achievements in the Cedar project are in parallel compilers and performance benchmarking experiments. The Ultracomputer developed the combining network for fast synchronization among multiple processors to be studied in Chapter 7.

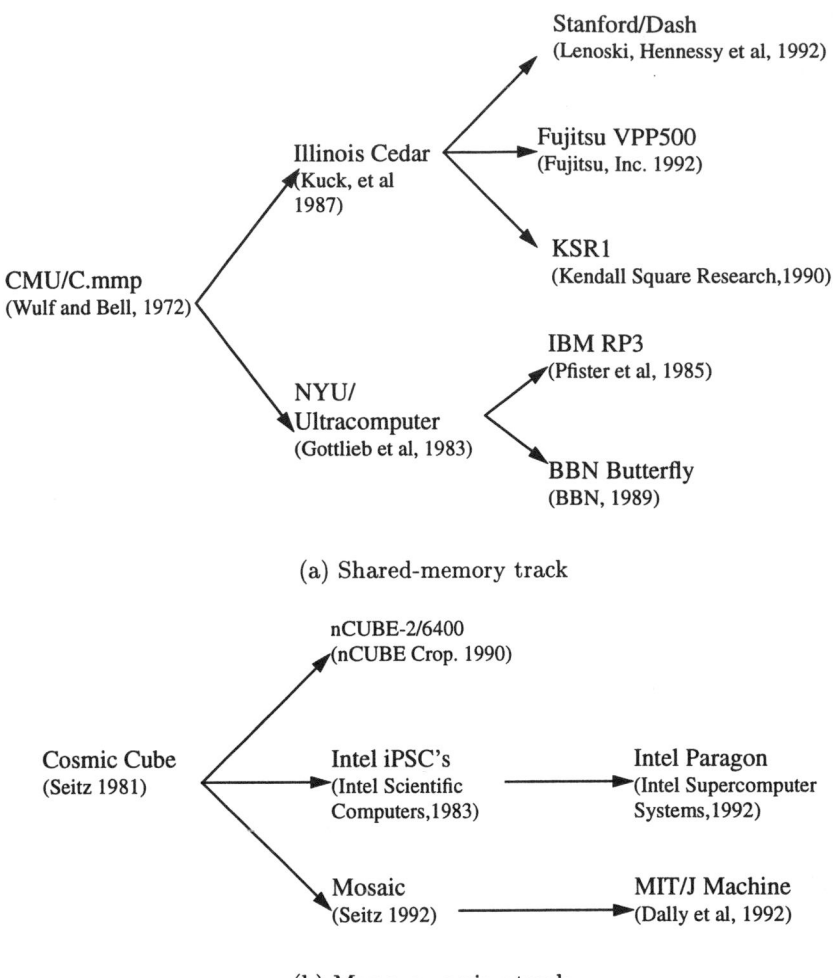

(a) Shared-memory track

(b) Message-passing track

Figure 1.17 Two multiple-processor tracks with and without shared memory.

The Stanford Dash (Lenoski, Hennessy et al., 1992) is a NUMA multiprocessor with distributed memories forming a global address space. Cache coherence is enforced with distributed directories. The KSR-1 is a typical COMA model. The Fujitsu VPP 500 is a 222-processor system with a crossbar interconnect. The shared memories are distributed to all processor nodes. We will study the Dash and the KSR-1 in Chapter 9 and the VPP500 in Chapter 8.

Following the Ultracomputer are two large-scale multiprocessors, both using multistage networks but with different interstage connections to be studied in Chapters 2 and 7. Among the systems listed in Fig. 1.17a, only the KSR-1, VPP500, and BBN Butterfly (BBN Advanced Computers, 1989) are commercial products. The rest are research systems; only prototypes have been built in laboratories.

Message-Passing Track   The Cosmic Cube (Seitz et al., 1981) pioneered the development of message-passing multicomputers (Fig. 1.17b). Since then, Intel has produced a series of medium-grain hypercube computers (the iPSCs). The nCUBE 2 also assumes a hypercube configuration. The latest Intel system is the Paragon (1992) to be studied in Chapter 7. On the research track, the Mosaic C (Seitz, 1992) and the MIT J-Machine (Dally et al., 1992) are two fine-grain multicomputers to be studied in Chapter 9.

1.5.2   Multivector and SIMD Tracks

The multivector track is shown in Fig. 1.18a, and the SIMD track in Fig. 1.18b.

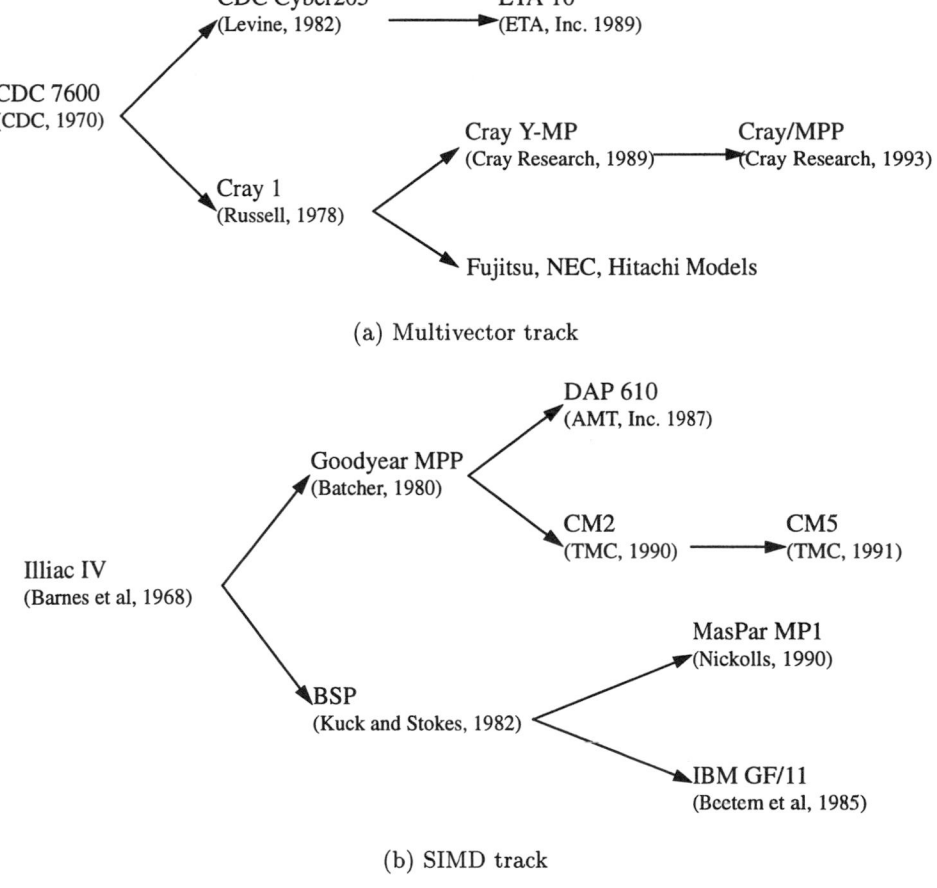

(a) Multivector track

(b) SIMD track

Figure 1.18 Multivector and SIMD tracks.

Both tracks are used for concurrent scalar/vector processing. Detailed studies can be found in Chapter 8.

Multivector Track   These are traditional vector supercomputers. The CDC 7600 was the first vector dual-processor system. Two subtracks were derived from the CDC 7600. The Cray and Japanese supercomputers all followed the register-to-register architecture. Cray 1 pioneered the multivector development in 1978. The latest Cray/MPP is a massively parallel system with distributed shared memory. It is supposed to work as a back-end accelerator engine compatible with the existing Cray Y-MP Series.

The other subtrack used memory-to-memory architecture in building vector supercomputers. We have identified only the CDC Cyber 205 and its successor the ETA10 here. Since the production of both machines has been discontinued now, we list them here simply for completeness in tracking different supercomputer architectures.

The SIMD Track   The Illiac IV pioneered the construction of SIMD computers, even the array processor concept can be traced back far earlier to the 1960s. The subtrack, consisting of the Goodyear MPP, the AMT/DAP610, and the TMC/CM-2, are all SIMD machines built with bit-slice PEs. The CM-5 is a synchronized MIMD machine executing in a multiple-SIMD mode.

The other subtrack corresponds to medium-grain SIMD computers using word-wide PEs. The BSP (Kuck and Stokes, 1982) was a shared-memory SIMD machine built with 16 processors updating a group of 17 memory modules synchronously. The GF11 (Beetem et al., 1985) was developed at the IBM Watson Laboratory for scientific simulation research use. The MasPar MP1 is the only medium-grain SIMD computer currently in production use. We will describe the CM-2, MasPar MP1, and CM-5 in Chapter 8.

1.5.3   Multithreaded and Dataflow Tracks

These are two research tracks (Fig. 1.19) that have been mainly experimented with in laboratories. Both tracks will be studied in Chapter 9. The following introduction covers only basic definitions and milestone systems built today.

The conventional von Neumann machines are built with processors that execute a single context by each processor at a time. In other words, each processor maintains a single thread of control with limited hardware resources. In a multithreaded architecture, each processor can execute multiple contexts at the same time. The term *multithreading* implies that there are multiple threads of control in each processor. Multithreading offers an effective mechanism for hiding long latency in building large-scale multiprocessors.

As shown in Fig. 1.19a, the multithreading idea was pioneered by Burton Smith (1978) in the HEP system which extended the concept of scoreboarding of multiple functional units in the CDC 6400. The latest multithreaded multiprocessor projects are the Tera computer (Alverson, Smith et al., 1990) and the MIT Alewife (Agarwal et al., 1989) to be studied in Section 9.4. Until then, all multiprocessors studied use single-threaded processors as building blocks.

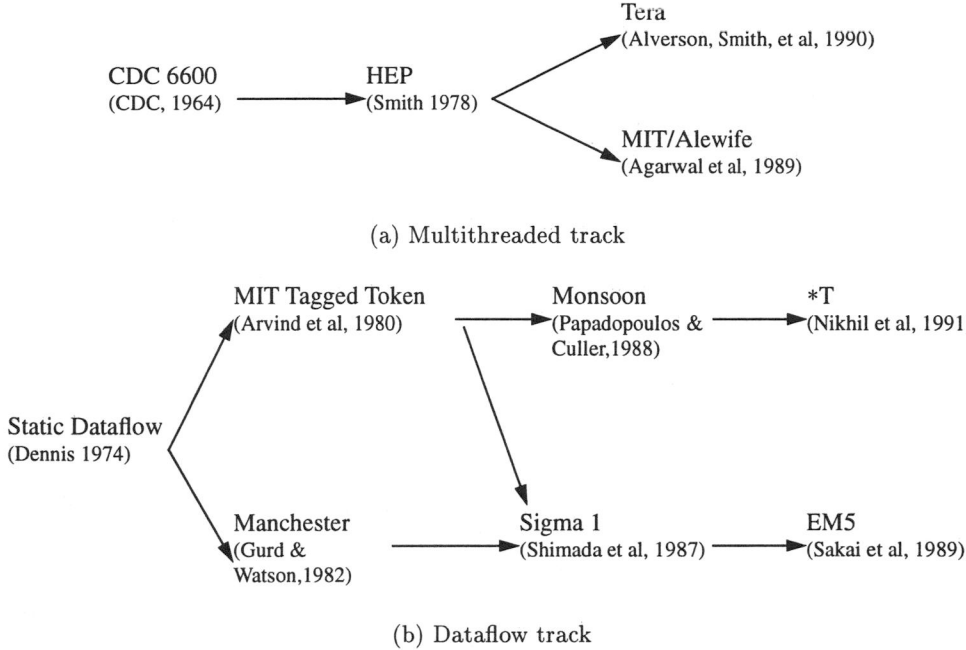

(a) Multithreaded track

(b) Dataflow track

Figure 1.19 Multithreaded and dataflow tracks.

The Dataflow Track   We will introduce the basic concepts of dataflow computers in Section 2.3. Some experimental dataflow systems are described in Section 9.5. The key idea is to use a dataflow mechanism, instead of a control-flow mechanism as in von Neumann machines, to direct the program flow. Fine-grain, instruction-level parallelism is exploited in dataflow computers.

As listed in Fig. 1.19b, the dataflow concept was pioneered by Jack Dennis (1974) with a "static" architecture. The concept later inspired the development of "dynamic" dataflow computers. A series of tagged-token architectures was developed at MIT by Arvind and coworkers. We will describe the tagged-token architecture in Section 2.3.1 and then the *T prototype (Nikhil et al., 1991) in Section 9.5.3.

Anther important subtrack of dynamic dataflow computer is represented by the Manchester machine (Gurd and Watson, 1982). The ETL Sigma 1 (Shimada et al., 1987) and EM5 have evolved from the MIT and Manchester machines. We will study the EM5 (Sakai et al., 1989) in Section 9.5.2. These dataflow machines are still in the research stage.

1.6 Bibliographic Notes and Exercises

Various machine architectures were classified by [Flynn72]. A recent collection of papers on architectural alternatives for exploiting parallelism can be found in the tutorial [Lilja92]. [Almasi89] and Gottlieb provided a thorough survey of research done on parallel computing up to 1989. [Hwang89a] and DeGroot presented critical reviews on parallel processing techniques developed for supercomputers and artificial intelligence.

[Hennessy90] and Patterson provided an excellent treatment of uniprocessor computer design. For a treatment of earlier parallel processing computers, readers are referred to [Hwang84] and Briggs. The layered classification of parallel computers was proposed in [Ni91]. [Bell92] introduced the MIMD taxonomy. Systolic array was introduced by [Kung78] and Leiserson. [Prasanna Kumar87] and Raghavendra proposed the mesh architecture with broadcast buses for parallel computing.

Multiprocessor issues were characterized by [Gajski85] and Pier and by [Dubois88], Scheurich, and Briggs. Multicomputer technology was assessed by [Athas88] and Seitz. An introduction to existing parallel computers can be found in [Trew91] and Wilson. Key references of various computer systems, listed in Bell's taxonomy (Fig. 1.10) and in different architectural development tracks (Figs. 1.17 through 1.19), are identified in the figures as well as in the bibliography. Additional references to these case-study machines can be found in later chapters. A collection of reviews, a bibliography, and indexes of resources in parallel systems can be found in [ACM91] with an introduction by Charles Seitz.

A comparative study of NUMA and COMA multiprocessor models can be found in [Stenström92], Joe, and Gupta. SIMD machines were modeled by [Siegel79]. The RAM model was proposed by [Sheperdson63] and Sturgis. The PRAM model and variations were studied in [Fortune78] and Wyllie, [Snir82], and [Karp88]. The theory of NP-completeness is covered in the book by [Cormen90], Leiserson, and Rivest. The VLSI complexity model was introduced in [Thompson80] and interpreted by [Ullman84] and by [Seitz90] subsequently.

Readers are referred to the following journals and conference records for information on recent developments:

- *Journal of Parallel and Distributed Computing* (Academic Press, since 1983).
- *Journal of Parallel Computing* (North Holland, Amsterdam, since 1984).
- *IEEE Transactions on Parallel and Distributed Systems* (IEEE Computer Society, since 1990).
- *International Conference on Parallel Processing* (Pennsylvania State University, since 1972).
- *International Symposium on Computer Architecture* (IEEE Computer Society, since 1972).
- *Symposium on the Frontiers of Massively Parallel Computation* (IEEE Computer Society, since 1986).
- *International Conference on Supercomputing* (ACM, since 1987).
- *Symposium on Architectural Support for Programming Languages and Operating Systems* (ACM, since 1975).
- *Symposium on Parallel Algorithms and Architectures* (ACM, since 1989).
- *International Parallel Processing Symposium* (IEEE Computer Society, since 1986).

- *IEEE Symposium on Parallel and Distributed Processing* (IEEE Computer Society, since 1989).

Exercises

Problem 1.1 A 40-MHz processor was used to execute a benchmark program with the following instruction mix and clock cycle counts:

Instruction type	Instruction count	Clock cycle count
Integer arithmetic	45000	1
Data transfer	32000	2
Floating point	15000	2
Control transfer	8000	2

Determine the effective CPI, MIPS rate, and execution time for this program.

Problem 1.2 Explain how instruction set, compiler technology, CPU implementation and control, and cache and memory hierarchy affect the CPU performance and justify the effects in terms of program length, clock rate, and effective CPI.

Problem 1.3 A workstation uses a 15-MHz processor with a claimed 10-MIPS rating to execute a given program mix. Assume a one-cycle delay for each memory access.

(a) What is the effective CPI of this computer?

(b) Suppose the processor is being upgraded with a 30-MHz clock. However, the speed of the memory subsystem remains unchanged, and consequently two clock cycles are needed per memory access. If 30% of the instructions require one memory access and another 5% require two memory accesses per instruction, what is the performance of the upgraded processor with a compatible instruction set and equal instruction counts in the given program mix?

Problem 1.4 Consider the execution of an object code with 200,000 instructions on a 40-MHz processor. The program consists of four major types of instructions. The instruction mix and the number of cycles (CPI) needed for each instruction type are given below based on the result of a program trace experiment:

Instruction type	CPI	Instruction mix
Arithmetic and logic	1	60%
Load/store with cache hit	2	18%
Branch	4	12%
Memory reference with cache miss	8	10%

(a) Calculate the average CPI when the program is executed on a uniprocessor with the above trace results.

(b) Calculate the corresponding MIPS rate based on the CPI obtained in part (a).

Problem 1.5   Indicate whether each of the following statements is *true* or *false* and justify your answer with reasoning and supportive or counter examples:

(a) The CPU computations and I/O operations cannot be overlapped in a multiprogrammed computer.

(b) Synchronization of all PEs in an SIMD computer is done by hardware rather than by software as is often done in most MIMD computers.

(c) As far as programmability is concerned, shared-memory multiprocessors offer simpler interprocessor communication support than that offered by a message-passing multicomputer.

(d) In an MIMD computer, all processors must execute the same instruction at the same time synchronously.

(e) As far as scalability is concerned, multicomputers with distributed memory are more scalable than shared-memory multiprocessors.

Problem 1.6   The execution times (in seconds) of four programs on three computers are given below:

Program	Execution Time (in seconds)		
	Computer A	Computer B	Computer C
Program 1	1	10	20
Program 2	1000	100	20
Program 3	500	1000	50
Program 4	100	800	100

Assume that 100,000,000 instructions were executed in each of the four programs. Calculate the MIPS rating of each program on each of the three machines. Based on these ratings, can you draw a clear conclusion regarding the relative performance of the three computers? Give reasons if you find a way to rank them statistically.

Problem 1.7   Characterize the architectural operations of SIMD and MIMD computers. Distinguish between multiprocessors and multicomputers based on their structures, resource sharing, and interprocessor communications. Also, explain the differences among UMA, NUMA, and COMA, and NORMA computers.

Problem 1.8   The following code segment, consisting of six instructions, needs to be executed 64 times for the evaluation of vector arithmetic expression: $D(I) = A(I) + B(I) \times C(I)$ for $0 \le I \le 63$.

            Load R1, B(I)   /R1 ← Memory $(\alpha + I)$/

Load R2, C(I)	/R2 ← Memory $(\beta + I)$/
Multiply R1, R2	/R1 ← (R1) × (R2)/
Load R3, A(I)	/R3 ← Memory $(\gamma + I)$/
Add R3, R1	/R3 ← (R3) + (R1)/
Store D(I), R3	/Memory $(\theta + I)$ ← (R3)/

where R1, R2, and R3 are CPU registers, (R1) is the content of R1, α, β, γ, and θ are the starting memory addresses of arrays B(I), C(I), A(I), and D(I), respectively. Assume four clock cycles for each Load or Store, two cycles for the Add, and eight cycles for the Multiply on either a uniprocessor or a single PE in an SIMD machine.

(a) Calculate the total number of CPU cycles needed to execute the above code segment repeatedly 64 times on an SISD uniprocessor computer sequentially, ignoring all other time delays.

(b) Consider the use of an SIMD computer with 64 PEs to execute the above vector operations in six synchronized vector instructions over 64-component vector data and both driven by the same-speed clock. Calculate the total execution time on the SIMD machine, ignoring instruction broadcast and other delays.

(c) What is the speedup gain of the SIMD computer over the SISD computer?

Problem 1.9 Prove that the best parallel algorithm written for an n-processor EREW-PRAM model can be no more than $O(\log n)$ times slower than any algorithm for a CRCW model of PRAM having the same number of processors.

Problem 1.10 Consider the multiplication of two n-bit binary integers using a 1.2-μm CMOS multiplier chip. Prove the lower bound $AT^2 > kn^2$, where A is the chip area, T is the execution time, n is the word length, and k is a technology-dependent constant.

Problem 1.11 Compare the PRAM models with physical models of real parallel computers in each of the following categories:

(a) Which PRAM variant can best model SIMD machines and how?

(b) Repeat the question in part (a) for shared-memory MIMD machines.

Problem 1.12 Answer the following questions related to the architectural development tracks presented in Section 1.5:

(a) For the shared-memory track (Fig. 1.17), explain the trend in physical memory organizations from the earlier system (C.mmp) to more recent systems (such as Dash, etc.).

(b) Distinguish between medium-grain and fine-grain multicomputers in their architectures and programming requirements.

(c) Distinguish between register-to-register and memory-to-memory architectures for building conventional multivector supercomputers.

(d) Distinguish between single-threaded and multithreaded processor architectures.

Problem 1.13 Design an algorithm to find the maximum of n numbers in $O(\log n)$ time on an EREW-PRAM model. Assume that initially each location holds one input value. Explain how you would make the algorithm processor time optimal.

Problem 1.14 Develop two algorithms for fast multiplication of two $n \times n$ matrices with a system of p processors, where $1 \leq p \leq n^3/\log n$. Choose an appropriate PRAM machine model to prove that the matrix multiplication can be done in $T = O(n^3/p)$ time.

(a) Prove that $T = O(n^2)$ if $p = n$. The corresponding algorithm must be shown, similar to that in Example 1.5.

(b) Show the parallel algorithm with $T = O(n)$ if $p = n^2$.

Problem 1.15 Match each of the following eight computer systems: KSR-1, RP3, Paragon, Dash, CM-2, VPP500, EM-5, and Tera, with one of the best descriptions listed below. The mapping is a one-to-one correspondence.

(a) A massively parallel system built with multiple-context processors and a 3-D torus architecture.

(b) A data-parallel computer built with bit-slice PEs interconnected by a hypercube/mesh network.

(c) A ring-connected multiprocessor using a cache-only memory architecture.

(d) An experimental multiprocessor built with a dynamic dataflow architecture.

(e) A crossbar-connected multiprocessor built with distributed processor/memory nodes forming a single address space.

(f) A multicomputer built with commercial microprocessors with multiple address spaces.

(g) A scalable multiprocessor built with distributed shared memory and coherent caches.

(h) An MIMD computer built with a large multistage switching network.

Chapter 2

Program and Network Properties

This chapter covers fundamental properties of program behavior and introduces major classes of interconnection networks. We begin with a study of computational granularity, conditions for program partitioning, matching software with hardware, program flow mechanisms, and compilation support for parallelism. Interconnection architectures introduced include static and dynamic networks. Network complexity, communication bandwidth, and data-routing capabilities are discussed.

2.1 Conditions of Parallelism

The exploitation of parallelism has created a new dimension in computer science. In order to move parallel processing into the mainstream of computing, H.T. Kung (1991) has identified the need to make significant progress in three key areas: *computation models* for parallel computing, *interprocessor communication* in parallel architectures, and *system integration* for incorporating parallel systems into general computing environments.

A theoretical treatment of parallelism is thus needed to build a basis for the above challenges. In practice, parallelism appears in various forms in a computing environment. All forms can be attributed to levels of parallelism, computational granularity, time and space complexities, communication latencies, scheduling policies, and load balancing. Very often, tradeoffs exist among time, space, performance, and cost factors.

2.1.1 Data and Resource Dependences

The ability to execute several program segments in parallel requires each segment to be independent of the other segments. The independence comes in various forms as defined below separately. For simplicity, to illustrate the idea, we consider the dependence relations among instructions in a program. In general, each code segment may contain one or more statements.

We use a *dependence graph* to describe the relations. The nodes of a dependence graph correspond to the program statements (instructions), and the directed edges with different labels show the ordered relations among the statements. The analysis of dependence graphs shows where opportunity exists for parallelization and vectorization.

Data Dependence   The ordering relationship between statements is indicated by the data dependence. Five types of data dependence are defined below:

(1) *Flow dependence*: A statement S2 is *flow-dependent* on statement S1 if an execution path exists from S1 to S2 and if at least one output (variables assigned) of S1 feeds in as input (operands to be used) to S2. Flow dependence is denoted as S1 $\longrightarrow$ S2.

(2) *Antidependence*: Statement S2 is *antidependent* on statement S1 if S2 follows S1 in program order and if the output of S2 overlaps the input to S1. A direct arrow crossed with a bar as in S1 $\longmapsto$ S2 indicates antidependence from S1 to S2.

(3) *Output dependence*: Two statements are *output-dependent* if they produce (write) the same output variable. S1 $\circ\!\!\longrightarrow$ S2 indicates output dependence from S1 to S2.

(4) *I/O dependence*: Read and write are I/O statements. I/O dependence occurs not because the same variable is involved but because the same file is referenced by both I/O statements.

(5) *Unknown dependence*: The dependence relation between two statements cannot be determined in the following situations:

- The subscript of a variable is itself subscribed (indirect addressing).
- The subscript does not contain the loop index variable.
- A variable appears more than once with subscripts having different coefficients of the loop variable.
- The subscript is nonlinear in the loop index variable.

When one or more of these conditions exist, a conservative assumption is to claim unknown dependence among the statements involved.

Example 2.1   Data dependence in programs

Consider the following code fragment of four instructions:

S1:	Load R1, A	/R1 $\leftarrow$ Memory(A)/
S2:	Add R2, R1	/R2 $\leftarrow$ (R1) + (R2) /
S3:	Move R1, R3	/R1 $\leftarrow$ (R3)/
S4:	Store B, R1	/Memory(B) $\leftarrow$ (R1)/

As illustrated in Fig. 2.1a, S2 is flow-dependent on S1 because the variable A is passed via the register R1. S3 is antidependent on S2 because of potential conflicts in register content in R1. S3 is output-dependent on S1 because they both modify the same register R1. Other data dependence relationships can be similarly revealed on a pairwise basis. Note that dependence is a partial ordering relation; that is, the members of not every pair of statements are related. For example, the statements S2 and S4 in the above program are totally *independent*.

Next, we consider a code fragment involving I/O operations:

S1:   Read (4), A(I)   /Read array A from tape unit 4/
S2:   Rewind(4)   /Rewind tape unit 4 /
S3:   Write (4), B(I)   /Write array B into tape unit 4/
S4:   Rewind (4)   /Rewind tape unit 4/

As shown in Fig. 2.1b, the read/write statements, S1 and S3, are I/O-dependent on each other because they both access the same file from tape unit 4. The above data dependence relations should not be arbitrarily violated during program execution. Otherwise, erroneous results may be produced with changed program order. The order in which statements are executed in a program is often well defined. Repetitive runs should produce identical results. On a multiprocessor system, the program order may or may not be preserved, depending on the memory model used. Determinism yielding predictable results can be controlled by a programmer as well as by constrained modification of writable data in a shared memory.

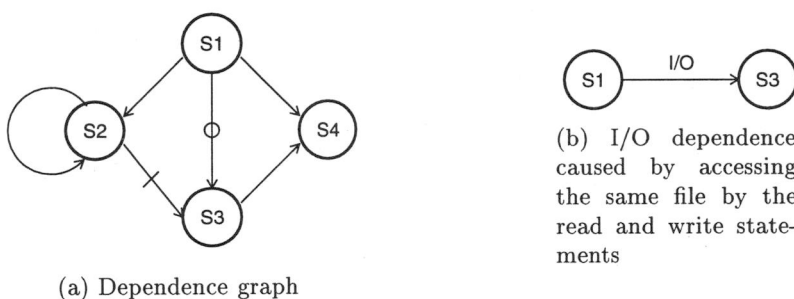

(a) Dependence graph

(b) I/O dependence caused by accessing the same file by the read and write statements

Figure 2.1 Data and I/O dependences in the program of Example 2.1.

Control Dependence   This refers to the situation where the order of execution of statements cannot be determined before run time. For example, conditional statements (IF in Fortran) will not be resolved until run time. Different paths taken after a conditional branch may introduce or eliminate data dependence among instructions. Dependence may also exist between operations performed in successive iterations of a looping procedure. In the following, we show one loop example with and another without control-dependent iterations. The successive iterations of the following loop are *control-independent*:

        Do 20 I = 1, N
            A(I) = C(I)
            IF (A(I) .LT. 0) A(I) = 1
    20   **Continue**

The following loop has *control-dependent* iterations:

        Do 10 I = 1, N
          IF (A(I − 1) .EQ. 0) A(I) = 0
   10 **Continue**

Control dependence often prohibits parallelism from being exploited. Compiler techniques are needed to get around the control dependence in order to exploit more parallelism.

Resource Dependence This is different from data or control dependence, which demands the independence of the work to be done. *Resource dependence* is concerned with the conflicts in using shared resources, such as integer units, floating-point units, registers, and memory areas, among parallel events. When the conflicting resource is an ALU, we call it *ALU dependence*.

If the conflicts involve workplace storage, we call it *storage dependence*. In the case of storage dependence, each task must work on independent storage locations or use protected access (such as locks or monitors to be described in Chapter 11) to shared writable data.

The transformation of a sequentially coded program into a parallel executable form can be done manually by the programmer using explicit parallelism, or by a compiler detecting implicit parallelism automatically. In both approaches, the decomposition of programs is the primary objective.

Program partitioning determines whether a given program can be partitioned or split into pieces that can execute in parallel or follow a certain prespecified order of execution. Some programs are inherently sequential in nature and thus cannot be decomposed into parallel branches. The detection of parallelism in programs requires a check of the various dependence relations.

Bernstein's Conditions In 1966, Bernstein revealed a set of conditions based on which two processes can execute in parallel. A *process* is a software entity corresponding to the abstraction of a program fragment defined at various processing levels. We define the *input set* I_i of a process P_i as the set of all input variables needed to execute the process.

Similarly, the *output set* O_i consists of all output variables generated after execution of the process P_i. Input variables are essentially operands which can be fetched from memory or registers, and output variables are the results to be stored in working registers or memory locations.

Now, consider two processes P_1 and P_2 with their input sets I_1 and I_2 and output sets O_1 and O_2, respectively. These two processes can execute in parallel and are denoted $P_1 \parallel P_2$ if they are independent and do not create confusing results.

Formally, these conditions are stated as follows:

$$\left. \begin{array}{ccc} I_1 \cap O_2 &=& \emptyset \\ I_2 \cap O_1 &=& \emptyset \\ O_1 \cap O_2 &=& \emptyset \end{array} \right\} \qquad (2.1)$$

These three equations are known as *Bernstein's conditions*. The input set I_i is also called the *read set* or the *domain* of P_i by other authors. Similarly, the output set O_i has been called the *write set* or the *range* of a process P_i. In terms of data dependences, Bernstein's conditions simply imply that two processes can execute in parallel if they are flow-independent, antiindependent, and output-independent.

The parallel execution of two processes produces the same results regardless of whether they are executed sequentially in any order or in parallel. This is possible only if the output of one process will not be used as input to the other process. Furthermore, the two processes will not modify (write) the same set of variables, either in memory or in the registers.

In general, a set of processes, $P_1, P_2, ..., P_k$, can execute in parallel if Bernstein's conditions are satisfied on a pairwise basis; that is, $P_1 \parallel P_2 \parallel P_3 \parallel \cdots \parallel P_k$ if and only if $P_i \parallel P_j$ for all $i \neq j$. This is exemplified by the following program illustrated in Fig. 2.2.

Example 2.2   Detection of parallelism in a program using Bernstein's conditions

Consider the simple case in which each process is a single HLL statement. We want to detect the parallelism embedded in the following five instructions labeled P_1, P_2, P_3, P_4, and P_5 in program order.

$$\left.\begin{array}{llll} P_1: & C & = & D \times E \\ P_2: & M & = & G + C \\ P_3: & A & = & B + C \\ P_4: & C & = & L + M \\ P_5: & F & = & G \div E \end{array}\right\} \tag{2.2}$$

Assume that each statement requires one step to execute. No pipelining is considered here. The dependence graph shown in Fig. 2.2a demonstrates flow dependence as well as resource dependence. In sequential execution, five steps are needed (Fig. 2.2b).

If two adders are available simultaneously, the parallel execution requires only three steps in Fig. 2.2c. Pairwise, there are 10 pairs of statements to check against Bernstein's conditions. Only 5 pairs, $P_1 \parallel P_5, P_2 \parallel P_3, P_2 \parallel P_5, P_5 \parallel P_3$, and $P_4 \parallel P_5$, can execute in parallel as revealed in Fig. 2.2a if there are no resource conflicts. Collectively, only $P_2 \parallel P_3 \parallel P_5$ is possible (Fig. 2.2c) because $P_2 \parallel P_3$, $P_3 \parallel P_5$, and $P_5 \parallel P_2$ are all possible. ∎

In general, the parallelism relation $\parallel$ is commutative; i.e., $P_i \parallel P_j$ implies $P_j \parallel P_i$. But the relation is not transitive; i.e., $P_i \parallel P_j$ and $P_j \parallel P_k$ do not necessarily guarantee $P_i \parallel P_k$. For example, we have $P_1 \parallel P_5$ and $P_5 \parallel P_2$, but $P_1 \nparallel P_2$, where $\nparallel$ means P_1 and P_2 cannot execute in parallel. In other words, the order in which P_1 and P_2 are executed will make a difference in the computational results.

Therefore, $\parallel$ is not an equivalence relation. However, $P_i \parallel P_j \parallel P_k$ implies associativity; i.e., $(P_i \parallel P_j) \parallel P_k = P_i \parallel (P_j \parallel P_k)$, since the order in which the parallel

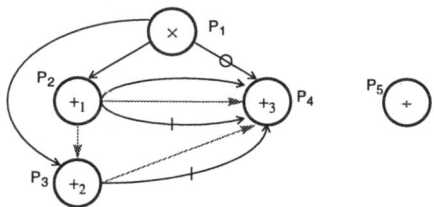

(a) A dependence graph showing both data dependence (solid arrows) and resource dependence (dashed arrows)

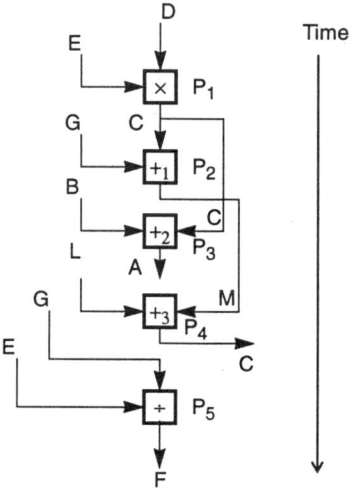

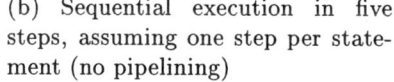

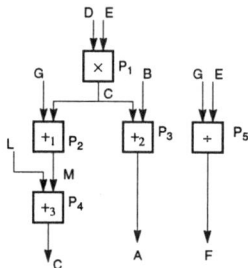

(b) Sequential execution in five steps, assuming one step per statement (no pipelining)

(c) Parallel execution in three steps, assuming two adders are available per step

Figure 2.2 Detection of parallelism in the program of Example 2.2.

executable processes are executed should not make any difference in the output sets. It should be noted that the condition $I_i \cap I_j \neq \emptyset$ does not prevent parallelism between P_i and P_j.

Violations of any one or more of the three conditions in Eq. 2.1 prohibits parallelism between two processes. In general, violation of any one or more of the $3n(n-1)/2$ Bernstein's conditions among n processes prohibits parallelism collectively or partially. Many program constructs may prohibit parallelism.

Any statements or processes which depend on run-time conditions are not transformed to parallel form. These include IF statements or conditional branches. Recursive computations in successive iterations also prohibit parallelism. In general, data dependence, control dependence, and resource dependence all prevent parallelism from being exploitable.

The statement-level dependence can be generalized to higher levels, such as code segment, subroutine, process, task, and program levels. The dependence of two higher-level objects can be inferred from the dependence of statements in the corresponding objects. The goals of analyzing the data dependence, control dependence, and resource dependence in a code are to identify opportunities for parallelization or vectorization.

Very often program restructuring or code transformations need to be performed before such opportunities can be revealed. The dependence relations are used in instruction issue and pipeline scheduling operations described in Chapter 6. We wish to build intelligent compilers to detect parallelism automatically. Further discussion of compiler techniques is given in Chapter 10.

2.1.2   Hardware and Software Parallelism

For implementation of parallelism, we need special hardware and software support. In this section, we address these support issues. We first distinguish between hardware and software parallelism. The mismatch problem between hardware and software is discussed. Then we describe the fundamental concept of compilation support needed to close the gap between hardware and software.

Details of special hardware functions and software support for parallelism will be treated in the remaining chapters. The key idea being conveyed is that parallelism cannot be achieved free. Besides theoretical conditioning, joint efforts between hardware designers and software programmers are needed to exploit parallelism in upgrading computer performance.

Hardware Parallelism   This refers to the type of parallelism defined by the machine architecture and hardware multiplicity. Hardware parallelism is often a function of cost and performance tradeoffs. It displays the resource utilization patterns of simultaneously executable operations. It can also indicate the peak performance of the processor resources.

One way to characterize the parallelism in a processor is by the number of instruction issues per machine cycle. If a processor issues k instructions per machine cycle, then it is called a *k-issue* processor.

A conventional processor takes one or more machine cycles to issue a single instruction. These types of processors are called *one-issue* machines, with a single instruction pipeline in the processor. In a modern processor, two or more instructions can be issued per machine cycle.

For example, the Intel i960CA is a three-issue processor with one arithmetic, one memory access, and one branch instruction issued per cycle. The IBM RISC/System 6000 is a four-issue processor capable of issuing one arithmetic, one memory access, one floating-point, and one branch operation per cycle.

A multiprocessor system built with n k-issue processors should be able to handle a maximum number of nk threads of instructions simultaneously.

Software Parallelism   This type of parallelism is defined by the control and data dependence of programs. The degree of parallelism is revealed in the program profile or in the program flow graph. Software parallelism is a function of algorithm, programming style, and compiler optimization. The program flow graph displays the patterns of simultaneously executable operations. Parallelism in a program varies during the execution period. It often limits the sustained performance of the processor.

Example 2.3 Mismatch between software parallelism and hardware parallelism (Wen-Mei Hwu, 1991)

Consider the example program graph in Fig. 2.3a. There are eight instructions (four *loads* and four *arithmetic* operations) to be executed in three consecutive machine cycles. Four *load* operations are performed in the first cycle, followed by two *multiply* operations in the second cycle and two *add/subtract* operations in the third cycle. Therefore, the parallelism varies from 4 to 2 in three cycles. The average software parallelism is equal to $8/3 = 2.67$ instructions per cycle in this example program.

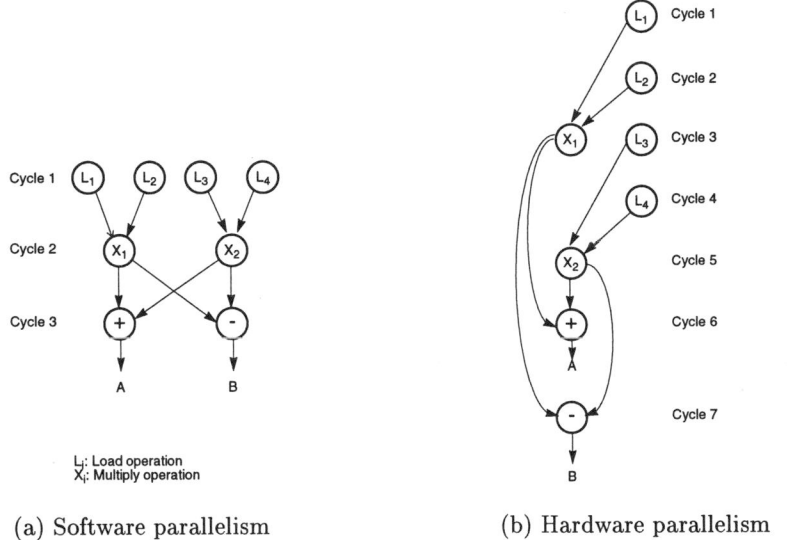

(a) Software parallelism   (b) Hardware parallelism

Figure 2.3 Executing an example program by a two-issue superscalar processor.

Now consider execution of the same program by a two-issue processor which can execute one memory access (*load* or *write*) and one arithmetic (*add*, *subtract*, *multiply*, etc.) operation simultaneously. With this hardware restriction, the program must execute in seven machine cycles as shown in Fig. 2.3b. Therefore, the *hardware parallelism* displays an average value of $8/7 = 1.14$ instructions executed per cycle. This demonstrates a mismatch between the software parallelism and the hardware parallelism.

Let us try to match the software parallelism shown in Fig. 2.3a in a hardware platform of a dual-processor system, where single-issue processors are used.

The achievable hardware parallelism is shown in Fig. 2.4, where L/S stands for *load/store* operations. Note that six processor cycles are needed to execute the 12 instructions by two processors. S_1 and S_2 are two inserted *store* operations, and l_5 and l_6 are two inserted *load* operations. These added instructions are needed for interprocessor communication through the shared memory.

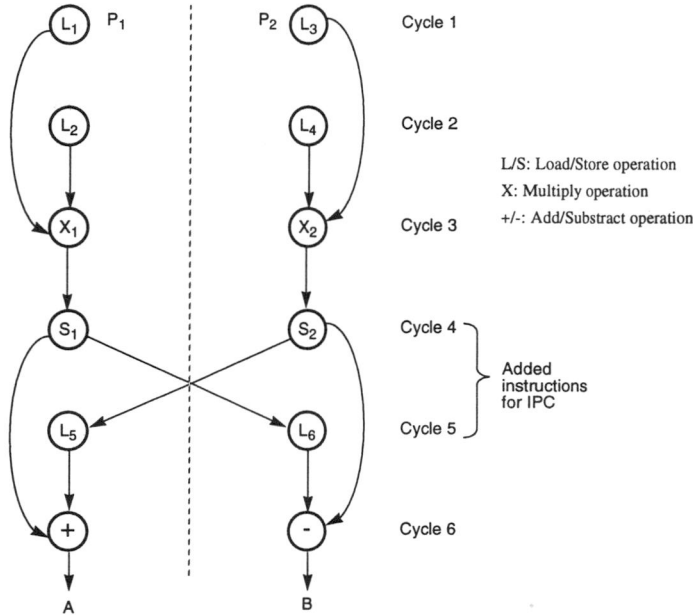

Figure 2.4 Dual-processor execution of the program in Fig. 2.3a.

Of the many types of software parallelism, two are most frequently cited as important to parallel programming: The first is *control parallelism*, which allows two or more operations to be performed simultaneously. The second type has been called *data parallelism*, in which almost the same operation is performed over many data elements by many processors simultaneously.

Control parallelism, appearing in the form of pipelining or multiple functional units, is limited by the pipeline length and by the multiplicity of functional units. Both pipelining and functional parallelism are handled by the hardware; programmers need take no special actions to invoke them.

Data parallelism offers the highest potential for concurrency. It is practiced in both SIMD and MIMD modes on MPP systems. Data parallel code is easier to write and to debug than control parallel code. Synchronization in SIMD data parallelism is handled by the hardware. Data parallelism exploits parallelism in proportion to the quantity of data involved. Thus data parallel computations appeal to scaled problems, in which the performance of a MPP does not drop sharply with small sequential fraction in the program.

To solve the mismatch problem between software parallelism and hardware parallelism, one approach is to develop compilation support, and the other is through hardware redesign for more efficient exploitation by an intelligent compiler. These two approaches must cooperate with each other to produce the best result.

Hardware processors can be better designed to exploit parallelism by an optimizing compiler. Pioneer work in processor technology with this objective can be found in the IBM 801, Stanford MIPS, and Berkeley RISC. Most processors use a large register file and sustained instruction pipelining to execute nearly one instruction per cycle. The large register file supports fast access to temporary values generated by an optimizing compiler. The registers are exploited by the code optimizer and global register allocator in such a compiler.

The instruction scheduler exploits the pipeline hardware by filling *branch* and *load* delay slots. In superscalar and superpipelining, hardware and software branch prediction, multiple instruction issue, speculative execution, high bandwidth instruction cache, and support for dynamic scheduling are needed to facilitate the detection of parallelism opportunities. The architecture must be designed interactively with the compiler.

2.1.3   The Role of Compilers

Compiler techniques are used to exploit hardware features to improve performance. The pioneer work on the IBM PL.8 and Stanford MIPS compilers has aimed for this goal. Other optimizing compilers for exploiting parallelism include the CDC STACKLIB, Cray CFT, Illinois Parafrase, Rice PFC, Yale Bulldog, and Illinois IMPACT.

In Chapter 10, we will study loop transformation, software pipelining, and features developed in existing optimizing compilers for supporting parallelism. Interaction between compiler and architecture design is a necessity in modern computer development. Most existing processors issue one instruction per cycle and provide a few registers. This may cause excessive spilling of temporary results from the available registers. Therefore, more software parallelism may not improve performance in conventional scalar processors.

There exists a vicious cycle of limited hardware support and the use of a naive compiler. To break the cycle, one must design the compiler and the hardware jointly at the same time. Interaction between the two can lead to a better solution to the mismatch problem between software and hardware parallelism.

The general guideline is to increase the flexibility in hardware parallelism and to exploit software parallelism in control-intensive programs. Hardware and software design tradeoffs also exist in terms of cost, complexity, expandability, compatibility, and performance. Compiling for multiprocessors is much more involved than for uniprocessors. Both granularity and communication latency play important roles in the code optimization and scheduling process.

2.2   Program Partitioning and Scheduling

This section introduces the basic definitions of computational granularity or level of parallelism in programs. Communication latency and scheduling issues are illustrated with programming examples.

2.2.1   Grain Sizes and Latency

Grain size or *granularity* is a measure of the amount of computation involved in a software process. The simplest measure is to count the number of instructions in a grain (program segment). Grain size determines the basic program segment chosen for parallel processing. Grain sizes are commonly described as *fine, medium*, or *coarse*, depending on the processing levels involved.

Latency is a time measure of the communication overhead incurred between machine subsystems. For example, the *memory latency* is the time required by a processor to access the memory. The time required for two processes to synchronize with each other is called the *synchronization latency*. Computational granularity and communication latency are closely related. We reveal their relationship below.

Parallelism has been exploited at various processing levels. As illustrated in Fig. 2.5, five levels of program execution represent different computational grain sizes and changing communication and control requirements. The lower the level, the finer the granularity of the software processes.

In general, the execution of a program may involve a combination of these levels. The actual combination depends on the application, formulation, algorithm, language, program, compilation support, and hardware limitations. We characterize below the parallelism levels and review their implementation issues from the viewpoints of a programmer and of a compiler writer.

Instruction Level   At instruction or statement level, a typical grain contains less than 20 instructions, called *fine grain* in Fig. 2.5. Depending on individual programs, fine-grain parallelism at this level may range from two to thousands. Butler et al. (1991) has shown that single-instruction-stream parallelism is greater than two. Wall (1991) finds that the average parallelism at instruction level is around five, rarely exceeding seven, in an ordinary program. For scientific applications, Kumar (1988) has measured the average parallelism in the range of 500 to 3000 Fortran statements executing concurrently in an idealized environment.

The advantage of fine-grain computation lies in the abundance of parallelism. The exploitation of fine-grain parallelism can be assisted by an optimizing compiler which should be able to automatically detect parallelism and translate the source code to a parallel form which can be recognized by the run-time system. Instruction-level parallelism is rather tedious for an ordinary programmer to detect in a source code.

Loop Level   This corresponds to the iterative loop operations. A typical loop contains less than 500 instructions. Some loop operations, if independent in successive iterations, can be vectorized for pipelined execution or for lock-step execution on SIMD machines.

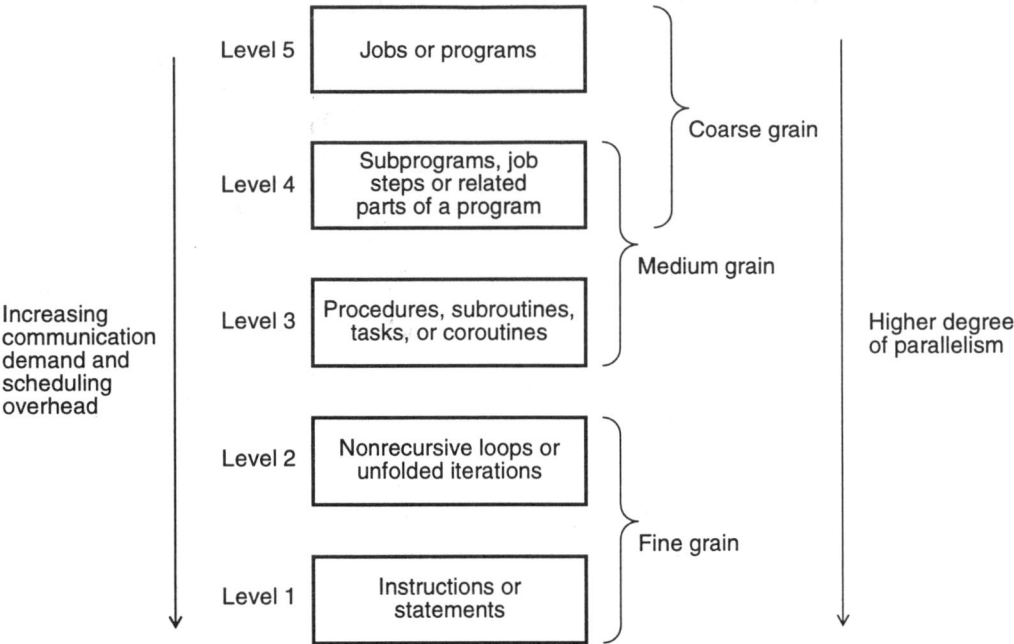

Figure 2.5 Levels of parallelism in program execution on modern computers. (Reprinted from Hwang, *Proc. IEEE*, October 1987)

Some loop operations can be self-scheduled for parallel execution on MIMD machines.

Loop-level parallelism is the most optimized program construct to execute on a parallel or vector computer. However, recursive loops are rather difficult to parallelize. Vector processing is mostly exploited at the loop level (level 2 in Fig. 2.5) by a vectorizing compiler. The loop level is still considered a fine grain of computation.

Procedure Level   This level corresponds to medium-grain size at the task, procedural, subroutine, and coroutine levels. A typical grain at this level contains less than 2000 instructions. Detection of parallelism at this level is much more difficult than at the finer-grain levels. Interprocedural dependence analysis is much more involved and history-sensitive.

The communication requirement is often less compared with that required in MIMD execution mode. SPMD execution mode is a special case at this level. Multitasking also belongs in this category. Significant efforts by programmers may be needed to restructure a program at this level, and some compiler assistance is also needed.

Subprogram Level   This corresponds to the level of job steps and related subprograms. The grain size may typically contain thousands of instructions. Job steps can overlap across different jobs. Subprograms can be scheduled for different processors in

at&t

SAEED ALAHMARI
20411 COLONIAL HILL DR UNIT 206
TAMPA, FL 33647-3689

Confirmation number
52R7MRCIT08/031
2010

Past Due Amount: $353.94
Total Amount Due: $353.94
Account Number: 52326733 1048
Date: May 14, 2016

Wireless Number(s): 571-303-8961, 937-554-3510

SAEED ALAHMARI:

Regretfully, we have canceled your wireless service because your account remains unpaid. Our records reflect an unpaid balance of $353.94. If you do not pay the balance owed, AT&T may refer your account(s) to a third party collector. This may result in a negative reference on your credit report.

If you have already made your payment, please disregard this reminder. If not, please remit payment immediately using the remittance slip and envelope. For your convenience, you may also pay by electronic check, debit card, or major credit card by calling 1-800-947-5096.

If you have any questions about your account, please call us at 1-800-947-5096 and an AT&T Representative will be glad to

Thank you for your prompt attention to this matter.

TLFNL1

AMOUNT DUE: $353.94

Account Number **52326733I048**

Please include account number on your check.

Make checks payable to:

SAEED ALAHMARI
20411 COLONIAL HILL DR UNIT 206
TAMPA, FL 33647-3689

AT&T MOBILITY
PO BOX 536216
ATLANTA GA 30353-6216

923005232673310480000000000035394000000035394001

SPMD or MPMD mode, often on message-passing multicomputers.

Multiprogramming on a uniprocessor or on a multiprocessor is conducted at this level. In the past, parallelism at this level has been exploited by algorithm designers or programmers, rather than by compilers. We do not have good compilers for exploiting medium- or coarse-grain parallelism at present.

Job (Program) Level This corresponds to the parallel execution of essentially independent jobs (programs) on a parallel computer. The grain size can be as high as tens of thousands of instructions in a single program. For supercomputers with a small number of very powerful processors, such coarse-grain parallelism is practical. Job-level parallelism is handled by the program loader and by the operating system in general. Time-sharing or space-sharing multiprocessors explore this level of parallelism. In fact, both time and space sharing are extensions of multiprogramming.

To summarize, fine-grain parallelism is often exploited at instruction or loop levels, preferably assisted by a parallelizing or vectorizing compiler. Medium-grain parallelism at the task or job step demands significant roles for the programmer as well as compilers. Coarse-grain parallelism at the program level relies heavily on an effective OS and on the efficiency of the algorithm used. Shared-variable communication is often used to support fine-grain and medium-grain computations.

Message-passing multicomputers have been used for medium- and coarse-grain computations. In general, the finer the grain size, the higher the potential for parallelism and the higher the communication and scheduling overhead. Fine grain provides a higher degree of parallelism, but heavier communication overhead, as compared with coarse-grain computations. Massive parallelism is often explored at the fine-grain level, such as data parallelism on SIMD or MIMD computers.

Communication Latency By balancing granularity and latency, one can achieve better performance of a computer system. Various latencies are attributed to machine architecture, implementing technology, and communication patterns involved. The architecture and technology affect the design choices for latency tolerance between subsystems. In fact, latency imposes a limiting factor on the scalability of the machine size. For example, memory latency increases with respect to memory capacity. Thus memory cannot be increased indefinitely without exceeding the tolerance level of the access latency. Various latency hiding or tolerating techniques will be studied in Chapter 9.

The latency incurred with interprocessor communication is another important parameter for a system designer to minimize. Besides signal delays in the data path, IPC latency is also affected by the communication patterns involved. In general, n tasks communicating with each other may require $n(n - 1)/2$ communication links among them. Thus the complexity grows quadratically. This leads to a communication bound which limits the number of processors allowed in a large computer system.

Communication patterns are determined by the algorithms used as well as by the architectural support provided. Frequently encountered patterns include *permutations* and *broadcast*, *multicast*, and *conference* (many-to-many) communications. The communication demand may limit the granularity or parallelism. Very often tradeoffs do exist between the two.

The communication issue thus involves the reduction of latency or complexity, the prevention of deadlock, minimizing blocking in communication patterns, and the trade-off between parallelism and communication overhead. We will study techniques that minimize communication latency, prevent deadlock, and optimize grain size throughout the book.

2.2.2   Grain Packing and Scheduling

Two fundamental questions to ask in parallel programming are: (i) How can we partition a program into parallel branches, program modules, microtasks, or grains to yield the shortest possible execution time? and (ii) What is the optimal size of concurrent grains in a computation?

This grain-size problem demands determination of both the number and the size of grains (or microtasks) in a parallel program. Of course, the solution is both problem-dependent and machine-dependent. The goal is to produce a short schedule for fast execution of subdivided program modules.

There exists a tradeoff between parallelism and scheduling/synchronization overhead. The time complexity involves both computation and communication overheads . The program partitioning involves the algorithm designer, programmer, compiler, operating system support, etc. We describe below a *grain packing* approach introduced by Kruatrachue and Lewis (1988) for parallel programming applications.

Example 2.4 Program graph before and after grain packing (Kruatrachue and Lewis, 1988)

The basic concept of program partitioning is introduced below. In Fig. 2.6, we show an example *program graph* in two different grain sizes. A program graph shows the structure of a program. It is very similar to the dependence graph introduced in Section 2.1.1. Each node in the program graph corresponds to a computational unit in the program. The *grain size* is measured by the number of basic machine cycles (including both processor and memory cycles) needed to execute all the operations within the node.

We denote each node in Fig. 2.6 by a pair (n, s), where n is the *node name* (id) and s is the grain size of the node. Thus grain size reflects the number of computations involved in a program segment. Fine-grain nodes have a smaller grain size, and coarse-grain nodes have a larger grain size.

The edge label (v, d) between two end nodes specifies the output variable v from the source node or the input variable to the destination node, and the communication delay d between them. This delay includes all the path delays and memory latency involved.

There are 17 nodes in the fine-grain program graph (Fig. 2.6a) and 5 in the coarse-grain program graph (Fig. 2.6b). The coarse-grain node is obtained by combining (grouping) multiple fine-grain nodes. The fine grain corresponds to the following program:

Var $a, b, c, d, e, f, g, h, i, j, k, l, m, n, o, p, q$

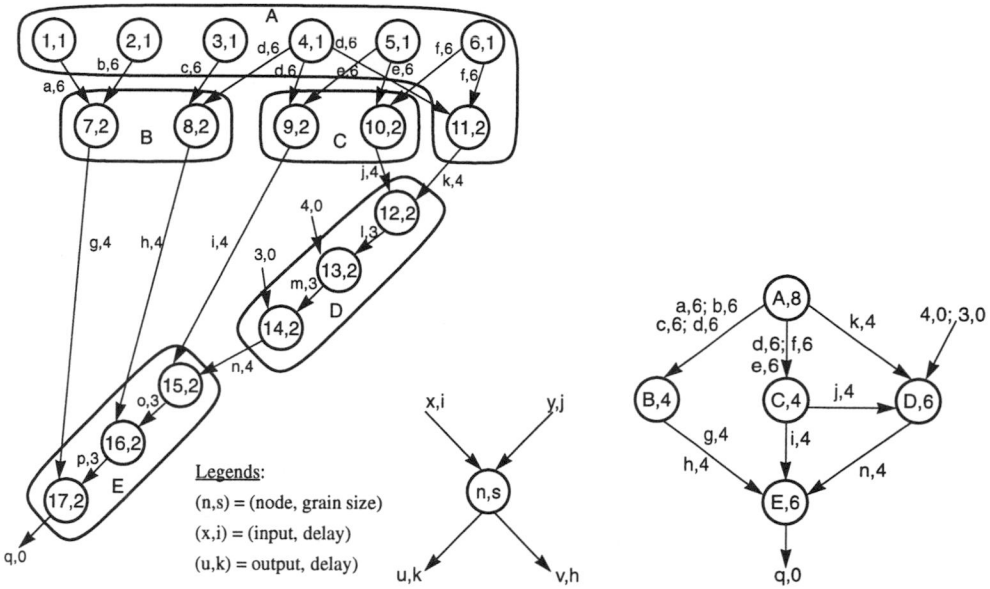

(a) Fine-grain program graph before packing

(b) Coarse-grain program graph after packing

Figure 2.6 A program graph before and after grain packing in Example 2.4. (Modified from Kruatrachue and Lewis, *IEEE Software*, Jan. 1988)

Begin

1.	$a := 1$	10.	$j := e \times f$
2.	$b := 2$	11.	$k := d \times f$
3.	$c := 3$	12.	$l := j \times k$
4.	$d := 4$	13.	$m := 4 \times l$
5.	$e := 5$	14.	$n := 3 \times m$
6.	$f := 6$	15.	$o := n \times i$
7.	$g := a \times b$	16.	$p := o \times h$
8.	$h := c \times d$	17.	$q := p \times q$
9.	$i := d \times e$		

End

Nodes 1, 2, 3, 4, 5, and 6 are memory reference (data fetch) operations. Each takes one cycle to address and six cycles to fetch from memory. All remaining nodes (7 to 17) are CPU operations, each requiring two cycles to complete. After packing, the coarse-grain nodes have larger grain sizes ranging from 4 to 8 as shown.

The node (A,8) in Fig. 2.6b is obtained by combining the nodes (1,1), (2,1), (3,1), (4,1), (5,1), (6,1), and (11,2) in Fig. 2.6a. The grain size, 8, of node A is the summation of all grain sizes $(1 + 1 + 1 + 1 + 1 + 1 + 2 = 8)$ being combined.

■

The idea of grain packing is to apply fine grain first in order to achieve a higher degree of parallelism. Then one combines (packs) multiple fine-grain nodes into a coarse-grain node if it can eliminate unnecessary communications delays or reduce the overall scheduling overhead.

Usually, all fine-grain operations within a single coarse-grain node are assigned to the same processor for execution. Fine-grain partition of a program often demands more interprocessor communication than that required in a coarse-grain partition. Thus grain packing offers a tradeoff between parallelism and scheduling/communication overhead.

Internal delays among fine-grain operations within the same coarse-grain node are negligible because the communication delay is contributed mainly by interprocessor delays rather than by delays within the same processor. The choice of the optimal grain size is meant to achieve the shortest schedule for the nodes on a parallel computer system.

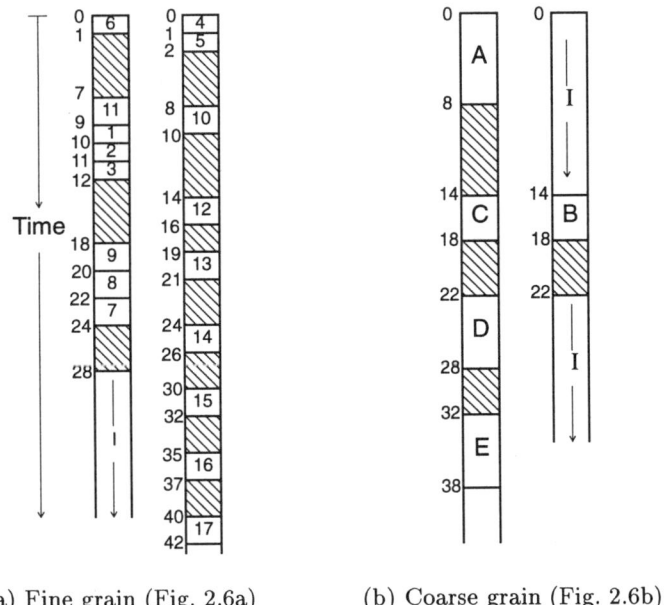

(a) Fine grain (Fig. 2.6a)   (b) Coarse grain (Fig. 2.6b)

Figure 2.7 Scheduling of the fine-grain and coarse-grain programs. (I: idle time; shaded areas: communication delays)

With respect to the fine-grain versus coarse-grain program graphs in Fig. 2.6, two multiprocessor schedules are shown in Fig. 2.7. The fine-grain schedule is longer (42 time units) because more communication delays were included as shown by the shaded area. The coarse-grain schedule is shorter (38 time units) because communication delays among nodes 12, 13, and 14 within the same node D (and also the delays among 15, 16, and 17 within the node E) are eliminated after grain packing.

2.2.3   Static Multiprocessor Scheduling

Grain packing may not always produce a shorter schedule. In general, dynamic multiprocessor scheduling is an NP-hard problem. Very often heuristics are used to yield suboptimal solutions. We introduce below the basic concepts behind multiprocessor scheduling using static schemes.

Node Duplication   In order to eliminate the idle time and to further reduce the communication delays among processors, one can duplicate some of the nodes in more than one processor.

Figure 2.8a shows a schedule without duplicating any of the five nodes. This schedule contains idle time as well as long interprocessor delays (8 units) between P1 and P2. In Fig. 2.8b, node A is duplicated into A′ and assigned to P2 besides retaining the original copy A in P1. Similarly, a duplicated node C′ is copied into P1 besides the original node C in P2. The new schedule shown in Fig. 2.8b is almost 50% shorter than that in Fig. 2.8a. The reduction in schedule time is caused by elimination of the (a, 8) and (c, 8) delays between the two processors.

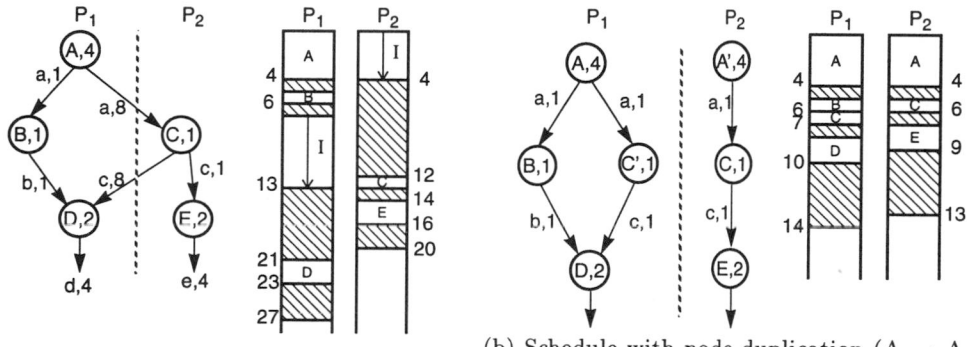

(a) Schedule without node duplication

(b) Schedule with node duplication (A → A and A′; C → C and C′)

Figure 2.8 Node-duplication scheduling to eliminate communication delays between processors. (I: idle time; shaded areas: communication delays)

Grain packing and node duplication are often used jointly to determine the best grain size and corresponding schedule. Four major steps are involved in the grain determination and the process of scheduling optimization:

Step 1. Construct a fine-grain program graph.
Step 2. Schedule the fine-grain computation.
Step 3. Grain packing to produce the coarse grains.
Step 4. Generate a parallel schedule based on the packed graph.

The purpose of multiprocessor scheduling is to obtain a minimal time schedule for the computations involved. The following example clarifies this concept.

Example 2.5 Program decomposition for static multiprocessor scheduling (Kruatrachue and Lewis, 1988)

Figure 2.9 shows an example of how to calculate the grain size and communication latency. In this example, two 2×2 matrices A and B are multiplied to compute the sum of the four elements in the resulting product matrix $C = A \times B$. There are eight multiplications and seven additions to be performed in this program, as written below:

$$\left[\begin{array}{cc} A_{11} & A_{12} \\ A_{21} & A_{22} \end{array} \right] \times \left[\begin{array}{cc} B_{11} & B_{12} \\ B_{21} & B_{22} \end{array} \right] = \left[\begin{array}{cc} C_{11} & C_{12} \\ C_{21} & C_{22} \end{array} \right]$$

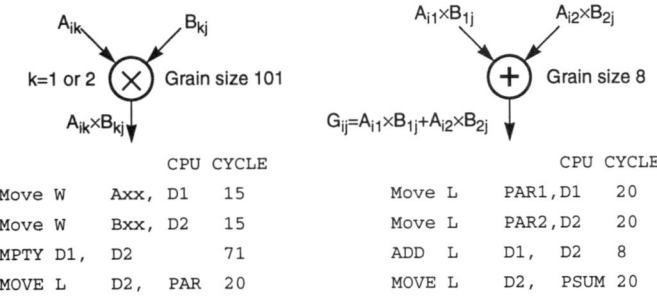

(a) Grain size calculation in M68000 assembly code at 20-MHz cycle

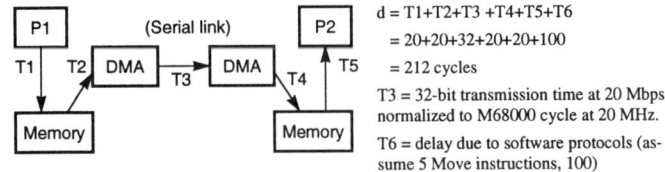

(b) Calculation of communication delay d

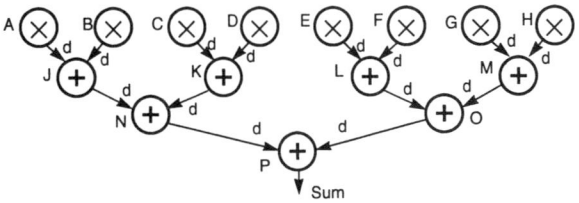

(c) Fine-grain program graph

Figure 2.9 Calculation of grain size and communication delay for the program graph in Example 2.5. (Courtesy of Kruatrachue and Lewis; reprinted with permission from *IEEE Software*, 1988)

$$
\begin{aligned}
C_{11} &= A_{11} \times B_{11} + A_{12} \times B_{21} \\
C_{12} &= A_{11} \times B_{12} + A_{12} \times B_{22} \\
C_{21} &= A_{21} \times B_{11} + A_{22} \times B_{21} \\
C_{22} &= A_{21} \times B_{12} + A_{22} \times B_{22} \\
\text{Sum} &= C_{11} + C_{12} + C_{21} + C_{22}
\end{aligned}
$$

As shown in Fig. 2.9a, the eight multiplications are performed in eight $\otimes$ nodes, each of which has a grain size of 101 CPU cycles. The remaining seven additions are performed in a 3-level binary tree consisting of seven $\oplus$ nodes. Each additional node requires 8 CPU cycles.

The interprocessor communication latency along all edges in the program graph is eliminated as $d = 212$ cycles by adding all path delays between two communicating processors (Fig. 2.9b).

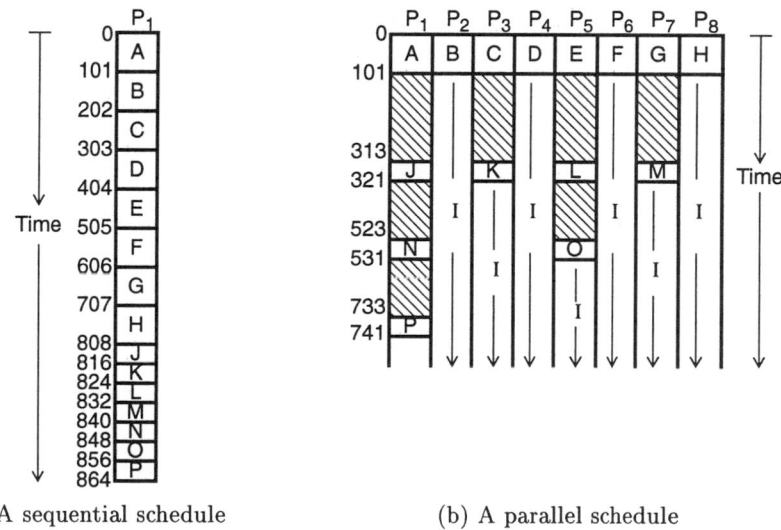

(a) A sequential schedule   (b) A parallel schedule

Figure 2.10 Sequential versus parallel scheduling in Example 2.5.

A fine-grain program graph is thus obtained in Fig. 2.9c. Note that the grain size and communication delay may vary with the different processors and communication links used in the system.

Figure 2.10 shows scheduling of the fine-grain program first on a sequential uniprocessor (P1) and then on an eight-processor (P1 to P8) system (Step 2). Based on the fine-grain graph (Fig. 2.9c), the sequential execution requires 864 cycles to complete without incurring any communication delay.

Figure 2.10b shows the reduced schedule of 741 cycles needed to execute the 15 nodes on 8 processors with incurred communication delays (shaded areas). Note that the communication delays have slowed down the parallel execution significantly, resulting in many processors idling (indicated by I), except for P1 which produces the final sum. A speedup factor of $864/741 = 1.16$ is observed.

Next we show how to use grain packing (Step 3) to reduce the communication overhead. As shown in Fig. 2.11, we group the nodes in the top two levels into four coarse-grain nodes labeled V, W, X, and Y. The remaining three nodes (N, O, P) then form the fifth node Z. Note that there is only one level of interprocessor communication required as marked by d in Fig. 2.11a.

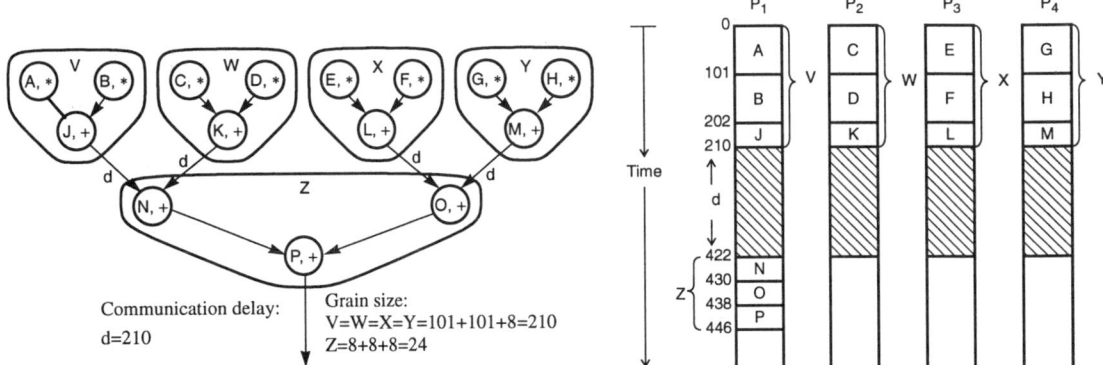

(a) Grain packing of 15 small nodes into 5 bigger nodes   (b) Parallel schedule for the packed program

Figure 2.11 Parallel scheduling for Example 2.5 after grain packing to reduce communication delays.

Since the maximum degree of parallelism is now reduced to 4 in the program graph, we use only four processors to execute this coarse-grain program. A parallel schedule is worked out (Fig. 2.11) for this program in 446 cycles, resulting in an improved speedup of $864/446 = 1.94$.

2.3 Program Flow Mechanisms

Conventional computers are based on a control flow mechanism by which the order of program execution is explicitly stated in the user programs. Dataflow computers are based on a data-driven mechanism which allows the execution of any instruction to be driven by data (operand) availability. Dataflow computers emphasize a high degree of parallelism at the fine-grain instructional level. Reduction computers are based on a demand-driven mechanism which initiates an operation based on the demand for its results by other computations.

2.3.1 Control Flow Versus Data Flow

Conventional von Neumann computers use a *program counter* (PC) to sequence the execution of instructions in a program. The PC is sequenced by instruction flow in a program. This sequential execution style has been called *control-driven*, as program flow is explicitly controlled by programmers.

Control-flow computers use shared memory to hold program instructions and data objects. Variables in the shared memory are updated by many instructions. The execution of one instruction may produce side effects on other instructions since memory is shared. In many cases, the side effects prevent parallel processing from taking place.

In fact, a uniprocessor computer is inherently sequential, due to use of the control-driven mechanism. However, control flow can be made parallel by using parallel language constructs or parallel compilers. In this book, we study primarily parallel control-flow computers and their programming techniques. Until the data-driven or demand-driven mechanism is proven to be cost-effective, the control-flow approach may continue to dominate the computer industry.

In a *dataflow computer*, the execution of an instruction is driven by data availability instead of being guided by a program counter. In theory, any instruction should be ready for execution whenever operands become available. The instructions in a data-driven program are not ordered in any way. Instead of being stored in a shared memory, data are directly held inside instructions.

Computational results (*data tokens*) are passed directly between instructions. The data generated by an instruction will be duplicated into many copies and forwarded directly to all needy instructions. Data tokens, once consumed by an instruction, will no longer be available for reuse by other instructions.

This data-driven scheme requires no shared memory, no program counter, and no control sequencer. However, it requires special mechanisms to detect data availability, to match data tokens with needy instructions, and to enable the chain reaction of asynchronous instruction executions. No memory sharing results in no side effects.

Asynchrony implies the need for handshaking or token-matching operations. A pure dataflow computer exploits fine-grain parallelism at the instruction level. Massive parallelism would be possible if the data-driven mechanism could be cost-effectively implemented with low instruction execution overhead.

A Dataflow Architecture There are quite a few experimental dataflow computer projects. Arvind and his associates at MIT have developed a tagged-token architecture for building dataflow computers. As shown in Fig. 2.12, the global architecture consists of n processing elements (PEs) interconnected by an $n \times n$ routing network. The entire system supports pipelined dataflow operations in all n PEs. Inter-PE communications are done through the pipelined routing network.

Within each PE, the machine provides a low-level *token-matching* mechanism which dispatches only those instructions whose input data (tokens) are already available. Each datum is tagged with the address of the instruction to which it belongs and the context in which the instruction is being executed. Instructions are stored in the program memory. Tagged tokens enter the PE through a local path. The tokens can also be passed to

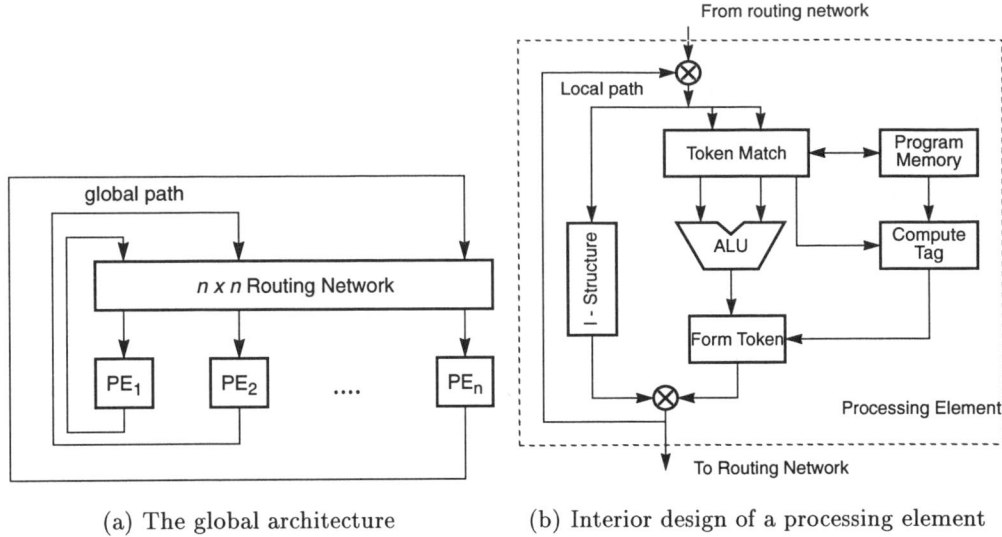

(a) The global architecture   (b) Interior design of a processing element

Figure 2.12 The MIT tagged-token dataflow computer. (Adapted from Arvind and
        Iannucci, 1986 with permission)

other PEs through the routing network. All internal token circulation operations are
pipelined without blocking.

One can think of the instruction address in a dataflow computer as replacing the
program counter, and the context identifier replacing the frame base register in a control
flow computer. It is the machine's job to match up data with the same tag to needy
instructions. In so doing, new data will be produced with a new tag indicating the
successor instruction(s). Thus, each instruction represents a synchronization operation.
New tokens are formed and circulated along the PE pipeline for reuse or to other PEs
through the global path, which is also pipelined.

Another synchronization mechanism, called the *I-structure*, is provided within each
PE. The I-structure is a tagged memory unit for overlapped usage of a data structure
by both the producer and consumer processes. Each word of I-structure uses a 2-bit
tag indicating whether the word is *empty*, is *full*, or has *pending read* requests. The use
of I-structure is a retreat from the pure dataflow approach. The purpose is to reduce
excessive copying of large data structures in dataflow operations.

Example 2.6 Comparison of dataflow and control-flow computers (Gajski,
Padua, Kuck, and Kuhn, 1982)

The dataflow graph in Fig. 2.13a shows that 24 instructions are to be executed (8
divides, 8 *multiplies*, and 8 *adds*). A dataflow graph is similar to a dependence graph
or program graph. The only difference is that data tokens are passed around the
edges in a dataflow graph. Assume that each *add*, *multiply*, and *divide* requires 1, 2,

and 3 cycles to complete, respectively. Sequential execution of the 24 instructions on a control flow uniprocessor takes 48 cycles to complete, as shown in Fig. 2.13b.

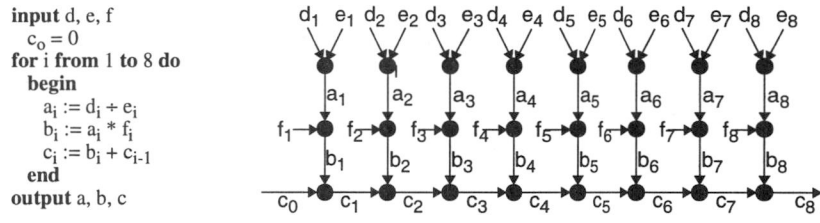

```
input d, e, f
  c_0 = 0
for i from 1 to 8 do
  begin
    a_i := d_i + e_i
    b_i := a_i * f_i
    c_i := b_i + c_{i-1}
  end
output a, b, c
```

(a) A sample program and its dataflow graph

(b) Sequential execution on a uniprocessor in 48 cycles

(c) Data-driven execution on a 4-processor dataflow computer in 14 cycles

$s_1=b_2+b_1,\ t_1=b_3+s_1,\ c_1=b_1+c_0,\ c_5=b_5+c_4$

$s_2=b_4+b_3,\ t_2=s_1+s_2,\ c_2=s_1+c_0,\ c_6=s_3+c_4$

$s_3=b_6+b_5,\ t_3=b_7+s_3,\ c_3=t_1+c_0,\ c_7=t_3+c_4$

$s_4=b_8+b_7,\ t_4=s_4+s_3,\ c_4=t_2+c_0,\ c_8=t_4+c_4$

(d) Parallel execution on a shared-memory 4-processor system in 14 cycles

Figure 2.13 Comparison between dataflow and control-flow computers. (Adapted from Gajski, Padua, Kuck, and Kuhn, 1982; reprinted with permission from *IEEE Computer*, Feb. 1982)

On the other hand, a dataflow multiprocessor completes the execution in 14 cycles in Fig. 2.13c. Assume that all the external inputs (d_i, c_i, f_i for $i = 1, 2, ..., 8$ and c_0) are available before entering the Do loop. With four processors, instructions a_1, a_2, a_3, and a_4 are all ready for execution in the first three cycles. The results produced then trigger the execution of a_5, b_1, a_6, and a_7 starting from cycle 4. The data-driven chain reactions are shown in Fig. 2.13c. The output c_8 is the last one to produce, due to its dependence on all the previous c_i's.

Figure 2.13d shows the execution of the same set of computations on a conventional multiprocessor using shared memory to hold the intermediate results (s_i and t_i for $i = 1, 2, 3, 4$). Note that no shared memory is used in the dataflow implementation. The example does not show any time advantage of dataflow execution over control flow execution.

The theoretical minimum time is 13 cycles along the critical path $a_1 b_1 c_1 c_2...c_8$. The chain reaction control in dataflow is more difficult to implement and may result in longer overhead, as compared with the uniform operations performed by all the processors in Fig. 2.13d.

■

One advantage of tagging each datum is that data from different contexts can be mixed freely in the instruction execution pipeline. Thus, instruction-level parallelism of dataflow graphs can absorb the communication latency and minimize the losses due to synchronization waits. Besides token matching and I-structure, compiler technology is also needed to generate dataflow graphs for tagged-token dataflow computers. The dataflow architecture offers an ideal model for massively parallel computations because all far-reaching side effects are removed. Side effects refer to the modification of some shared variables by unrelated operations.

2.3.2 Demand-Driven Mechanisms

In a *reduction machine*, the computation is triggered by the demand for an operation's result. Consider the evaluation of a nested arithmetic expression $a = ((b + 1) \times c - (d \div e))$. The data-driven computation chooses a bottom-up approach, starting from the innermost operations $b + 1$ and $d \div e$, then proceeding to the $\times$ operation, and finally to the outermost operation $-$. Such a computation has been called *eager evaluation* because operations are carried out immediately after all their operands become available.

A *demand-driven* computation chooses a top-down approach by first demanding the value of a, which triggers the demand for evaluating the next-level expressions $(b+1) \times c$ and $d \div e$, which in turn triggers the demand for evaluating $b + 1$ at the innermost level. The results are then returned to the nested demander in the reverse order before a is evaluated.

A demand-driven computation corresponds to *lazy evaluation*, because operations are executed only when their results are required by another instruction. The demand-driven approach matches naturally with the functional programming concept. The removal of side effects in functional programming makes programs easier to parallelize. There are two types of reduction machine models, both having a recursive control mechanism as characterized below.

Reduction Machine Models   In a *string reduction* model, each demander gets a separate copy of the expression for its own evaluation. A long string expression is reduced to a single value in a recursive fashion. Each reduction step has an operator followed by an embedded reference to demand the corresponding input operands. The operator is suspended while its input arguments are being evaluated. An expression is said to be fully reduced when all the arguments have been replaced by literal values.

In a *graph reduction* model, the expression is represented as a directed graph. The graph is reduced by evaluation of branches or subgraphs. Different parts of a graph or subgraphs can be reduced or evaluated in parallel upon demand. Each demander is given a pointer to the result of the reduction. The demander manipulates all references to that graph.

Graph manipulation is based on sharing the arguments using pointers. This traversal of the graph and reversal of the references are continued until constant arguments are encountered. This proceeds until the value of a is determined and a copy is returned to the original demanding instruction.

2.3.3 Comparison of Flow Mechanisms

Control-flow, dataflow, and reduction computer architectures are compared in Table 2.1. The degree of explicit control decreases from control-driven to demand-driven to data-driven. Highlighted in the table are the differences between *eager evaluation* and *lazy evaluation* in data-driven and demand-driven computers, respectively.

Furthermore, control tokens are used in control-flow computers and reduction machines, respectively. The listed advantages and disadvantages of the dataflow and reduction machine models are based on research findings rather than on extensive operational experience.

Even though conventional von Neumann model has many disadvantages, the industry is still building computers following the control-flow model. The choice was based on cost-effectiveness, marketability, and the narrow windows of competition used by the industry. Program flow mechanisms dictate architectural choices. Both dataflow and reduction models, despite a higher potential for parallelism, are still in the research stage. Control-flow machines still dominate the market.

In this book, we study mostly control-flow parallel computers. But dataflow and multithreaded architectures will be further studied in Chapter 9. Dataflow or hybrid von Neumann and dataflow machines offer design alternatives.

As far as leading-edge computer architecture is concerned the dataflow or hybrid models cannot be ignored. Both the Electrotechnical Laboratory (ETL) in Japan and the Massachusetts Institute of Technology have paid great attention to these approaches. The book edited by Gaudiot and Bic (1991) provides details of some recent development on dataflow computers.

2.4 System Interconnect Architectures

Static and dynamic networks for interconnecting computer subsystems or for constructing multiprocessors or multicomputers are introduced below. We study first the distinctions between direct networks for static connections and indirect networks for dynamic connections. These networks can be used for internal connections among pro-

Table 2.1 Control-Flow, Dataflow, and Reduction Computers

Machine Model	Control Flow (control-driven)	Dataflow (data-driven)	Reduction (demand-driven)
Basic Definition	Conventional computation; token of control indicates when a statement should be executed	Eager evaluation; statements are executed when all of their operands are available	Lazy evaluation statements are executed only when their result is required for another computation
Advantages	Full control	Very high potential for parallelism	Only required instructions are executed
	Complex data and control structures are easily implemented	High throughput	High degree of parallelism
		Free from side effects	Easy manipulation of data structures
Disadvantages	Less efficient	Time lost waiting for unneeded arguments	Does not support sharing of objects with changing local state
	Difficult in programming	High control overhead	Time needed to propagate demand tokens
	Difficult in preventing run-time error	Difficult in manipulating data structures	

(Courtesy of Wah, Lowrie, and Li; reprinted with permission from *Computers for Artificial Intelligence Processing* edited by Wah and Ramamoorthy, Wiley and Sons, Inc., 1990)

cessors, memory modules, and I/O disk arrays in a centralized system, or for distributed networking of multicomputer nodes.

Various topologies for building networks are specified below. Then we focus on the communication properties of interconnection networks. These include latency analysis, bisection bandwidth, and data-routing functions. Finally, we analyze the scalability of parallel architecture in solving scaled problems.

The communication efficiency of the underlying network is critical to the performance of a parallel computer. What we hope to achieve is a low-latency network with a high data transfer rate and thus a wide communication bandwidth. These network properties help make design choices for machine architecture.

2.4.1 Network Properties and Routing

The topology of an interconnection network can be either static or dynamic. *Static networks* are formed of point-to-point direct connections which will not change during program execution. *Dynamic networks* are implemented with switched channels, which are dynamically configured to match the communication demand in user programs.

Static networks are used for fixed connections among subsystems of a centralized system or multiple computing nodes of a distributed system. Dynamic networks include buses, crossbar switches, and multistage networks, which are often used in shared-memory multiprocessors. Both types of networks have also been implemented for inter-

PE data routing in SIMD computers.

Before we analyze various network topologies, let us define several parameters often used to estimate the complexity, communication efficiency, and cost of a network. In general, a network is represented by the graph of a finite number of nodes linked by directed or undirected edges. The number of nodes in the graph is called the *network size*.

Node Degree and Network Diameter The number of edges (links or channels) incident on a node is called the *node degree d*. In the case of unidirectional channels, the number of channels into a node is the *in degree*, and that out of a node is the *out degree*. Then the node degree is the sum of the two. The node degree reflects the number of I/O ports required per node, and thus the cost of a node. Therefore, the node degree should be kept a constant, as small as possible in order to reduce cost. A constant node degree is very much desired to achieve modularity in building blocks for scalable systems.

The *diameter D* of a network is the maximum shortest path between any two nodes. The path length is measured by the number of links traversed. The network diameter indicates the maximum number of distinct hops between any two nodes, thus providing a figure of communication merit for the network. Therefore, the network diameter should be as small as possible from a communication point of view.

Bisection Width When a given network is cut into two equal halves, the minimum number of edges (channels) along the cut is called the *channel bisection width b*. In the case of a communication network, each edge corresponds to a *channel* with w bit wires. Then the *wire bisection width* is $B = bw$. This parameter B reflects the wiring density of a network. When B is fixed, the *channel width* (in bits) $w = B/b$. Thus the bisection width provides a good indicator of the maximum communication bandwidth along the bisection of a network. All other cross sections should be bounded by the bisection width.

Another quantitative parameter is the *wire length* (or channel length) between nodes. This may affect the signal latency, clock skewing, or power requirements. We label a network *symmetric* if the topology is the same looking from any node. Symmetric networks are easier to implement or to program. Whether the nodes are homogeneous, the channels are buffered, or some of the nodes are switches are other useful properties for characterizing the structure of a network.

Data-Routing Functions A data-routing network is used for inter-PE data exchange. This routing network can be static, such as the hypercube routing network used in the TMC/CM-2, or dynamic such as the multistage network used in the IBM GF11. In the case of a multicomputer network, the data routing is achieved through message passing. Hardware routers are used to route messages among multiple computer nodes.

We specify below some primitive data-routing functions implementable on an inter-PE routing network. The versatility of a routing network will reduce the time needed for data exchange and thus can significantly improve the system performance.

Commonly seen data-routing functions among the PEs include *shifting*, *rotation*, *permutation* (one-to-one), *broadcast* (one-to-all), *multicast* (many-to-many), *personalized communication* (one-to-many), *shuffle, exchange,* etc. These routing functions can be implemented on ring, mesh, hypercube, or multistage networks.

Permutations   For n objects, there are $n!$ permutations by which the n objects can be reordered. The set of all permutations form a *permutation group* with respect to a composition operation. One can use cycle notation to specify a permutation function.

For example, the permutation $\pi = (a, b, c)(d, e)$ stands for the bijection mapping: $a \rightarrow b, b \rightarrow c, c \rightarrow a, d \rightarrow e$, and $e \rightarrow d$ in a circular fashion. The cycle (a, b, c) has a period of 3, and the cycle (d, e) a period of 2. Combining the two cycles, the permutation π has a period of $2 \times 3 = 6$. If one applies the permutation π six times, the identity mapping $I = (a), (b), (c), (d), (e)$ is obtained.

One can use a crossbar switch to implement the permutation. Multistage networks can implement some of the permutations in one or multiple passes through the network. Permutations can also be implemented with shifting or broadcast operations. The permutation capability of a network is often used to indicate the data routing capability. When n is large, the permutation speed often dominates the performance of a data-routing network.

Perfect Shuffle and Exchange   Perfect shuffle is a special permutation function suggested by Harold Stone (1971) for parallel processing applications. The mapping corresponding to a perfect shuffle is shown in Fig. 2.14a. Its inverse is shown on the right-hand side (Fig. 2.14b).

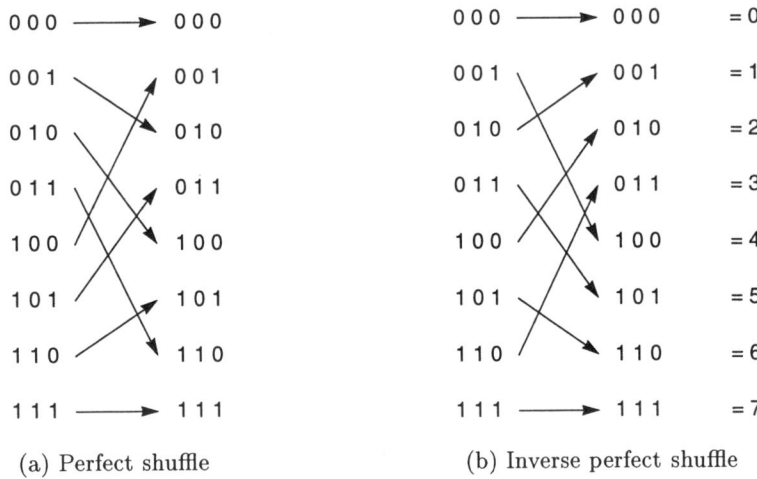

(a) Perfect shuffle   (b) Inverse perfect shuffle

Figure 2.14 Perfect shuffle and its inverse mapping over eight objects. (Courtesy of H. Stone; reprinted with permission from *IEEE Trans. Computers*, 1971)

In general, to shuffle $n = 2^k$ objects evenly, one can express each object in the domain by a k-bit binary number $x = (x_{k-1}, \ldots, x_1, x_0)$. The perfect shuffle maps x to y, where $y = (x_{k-2}, \ldots, x_1, x_0, x_{k-1})$ is obtained from x by shifting 1 bit to the left and wrapping around the most significant to the least significant position.

Hypercube Routing Functions   A three-dimensional binary cube network is shown in Fig. 2.15. Three routing functions are defined by three bits in the node address. For example, one can exchange the data between adjacent nodes which differ in the least significant bit C_0, as shown in Fig. 2.15b.

Similarly, two other routing patterns can be obtained by checking the middle bit C_1 (Fig. 2.15c) and the most significant bit C_2 (Fig. 2.15d), respectively. In general, an n-dimensional hypercube has n routing functions, defined by each bit of the n-bit address. These data exchange functions can be used in routing messages in a hypercube multicomputer.

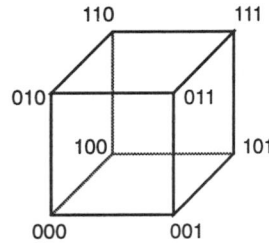

(a) A 3-cube with nodes denoted as $C_2 C_1 C_0$ in binary

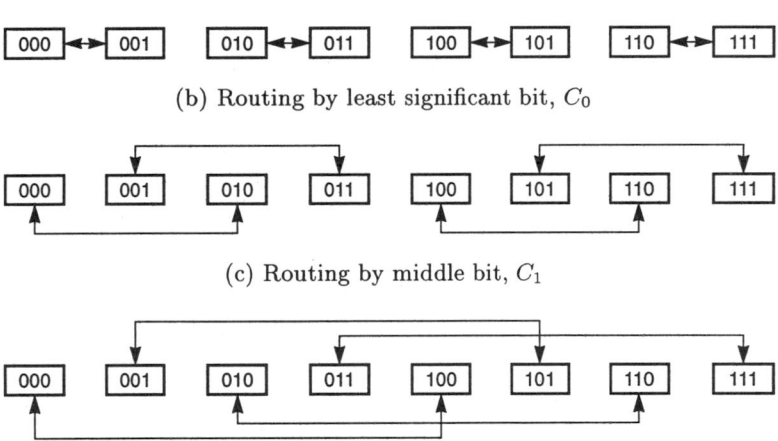

(b) Routing by least significant bit, C_0

(c) Routing by middle bit, C_1

(d) Routing by most significant bit, C_2

Figure 2.15 Three routing functions defined by a binary 3-cube.

Broadcast and Multicast   *Broadcast* is a one-to-all mapping. This can be easily achieved in an SIMD computer using a broadcast bus extending from the array controller to all PEs. A message-passing multicomputer also has mechanisms to broadcast messages. *Multicast* corresponds to a mapping from one subset to another (many to many).

Personalized broadcast sends personalized messages to only selected receivers. Broadcast is often treated as a global operation in a multicomputer. Personalized broadcast may have to be implemented with matching of destination codes in the network.

Network Performance   To summarize the above discussions, the performance of an interconnection network is affected by the following factors:

(1) *Functionality* — This refers to how the network supports data routing, interrupt handling, synchronization, request/message combining, and coherence.

(2) *Network latency* — This refers to the worst-case time delay for a unit message to be transferred through the network.

(3) *Bandwidth* — This refers to the maximum data transfer rate, in terms of Mbytes/s transmitted through the network.

(4) *Hardware complexity* — This refers to implementation costs such as those for wires, switches, connectors, arbitration, and interface logic.

(5) *Scalability* — This refers to the ability of a network to be modularly expandable with a scalable performance with increasing machine resources.

2.4.2   Static Connection Networks

Static networks use direct links which are fixed once built. This type of network is more suitable for building computers where the communication patterns are predictable or implementable with static connections. We describe their topologies below in terms of network parameters and comment on their relative merits in relation to communication and scalability.

Linear Array   This is a one-dimensional network in which N nodes are connected by $N-1$ links in a line (Fig. 2.16a). Internal nodes have degree 2, and the terminal nodes have degree 1. The diameter is $N-1$, which is rather long for large N. The bisection width $b = 1$. *Linear arrays* are the simplest connection topology. The structure is not symmetric and poses a communication inefficiency when N becomes very large.

For very small N, say $N = 2$, it is rather economic to implement a linear array. As the diameter increases linearly with respect to N, it should not be used for large N. It should be noted that a linear array is very different from a *bus* which is time-shared through switching among the many nodes attached to it. A linear array allows concurrent use of different sections (channels) of the structure by different source and destination pairs.

Ring and Chordal Ring   A *ring* is obtained by connecting the two terminal nodes of a linear array with one extra link (Fig. 2.16b). A ring can be unidirectional or

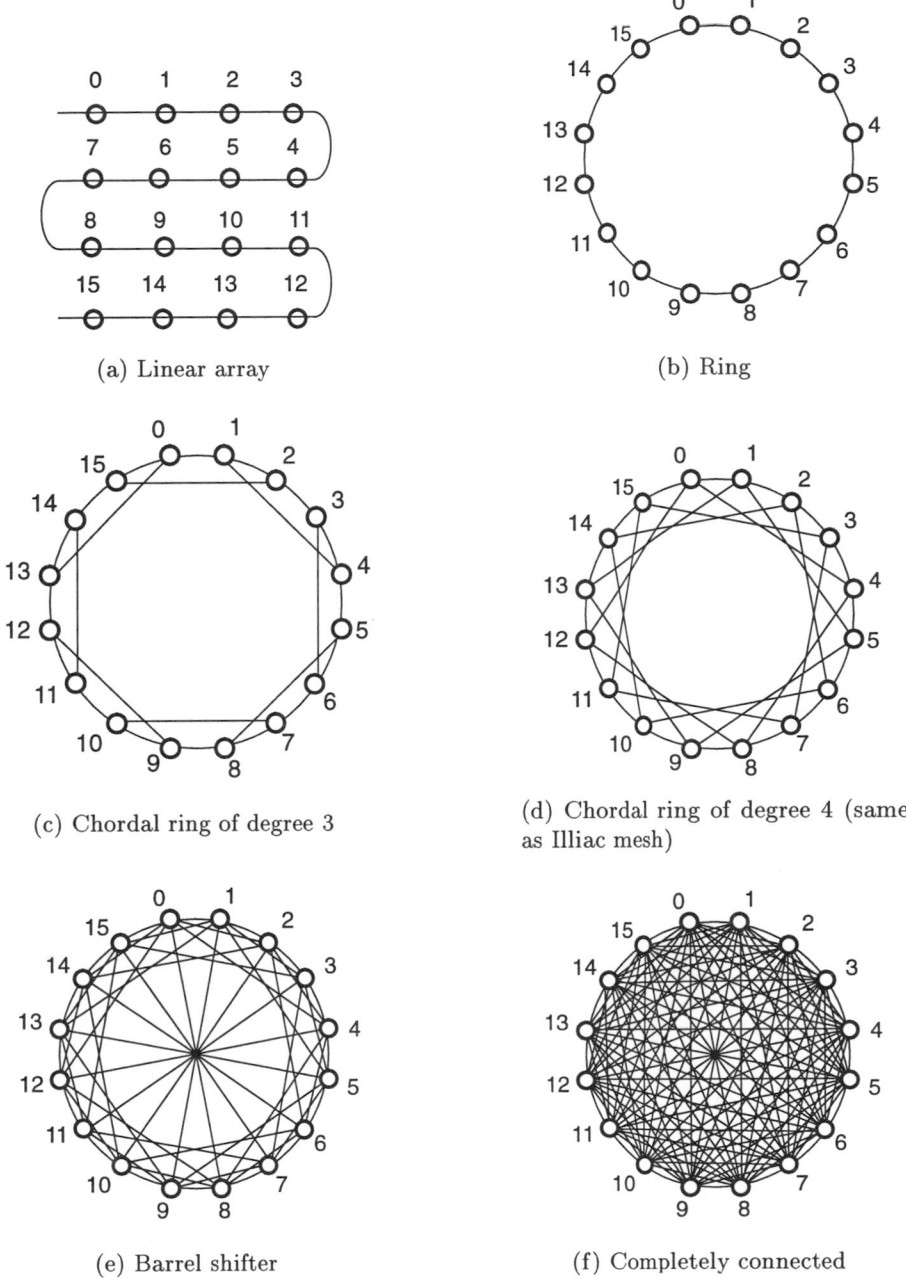

Figure 2.16 Linear array, ring, chordal rings of degrees 3 and 4, barrel shifter, and completely connected network.

bidirectional. It is symmetric with a constant node degree of 2. The diameter is $\lfloor N/2 \rfloor$ for a bidirectional ring, and N for unidirectional ring.

The IBM *token ring* has this topology, in which messages circulate along the ring until they reach the destination with a matching token. Pipelined or packet-switched, rings have been implemented in the CDC Cyberplus multiprocessor (1985) and in the KSR-1 computer system (1992) for interprocessor communications.

By increasing the node degree from 2 to 3 or 4, we obtain two *chordal rings* as shown in Figs. 2.16c and 2.16d, respectively. One and two extra links are added to produce the two chordal rings, respectively. In general, the more links added, the higher the node degree and the shorter the network diameter.

Comparing the 16-node ring (Fig. 2.16b) with the two chordal rings (Figs. 2.16c and 2.16d), the network diameter drops from 8 to 5 and to 3, respectively. In the extreme, the *completely connected network* in Fig. 2.16f has a node degree of 15 with the shortest possible diameter of 1.

Barrel Shifter   As shown in Fig. 2.16e for a network of $N = 16$ nodes, the *barrel shifter* is obtained from the ring by adding extra links from each node to those nodes having a distance equal to an integer power of 2. This implies that node i is connected to node j if $|j - i| = 2^r$ for some $r = 0, 1, 2, \cdots, n - 1$ and the network size is $N = 2^n$. Such a barrel shifter has a node degree of $d = 2n - 1$ and a diameter $D = n/2$.

Obviously, the connectivity in the barrel shifter is increased over that of any chordal ring of lower node degree. For $N = 16$, the barrel shifter has a node degree of 7 with a diameter of 2. But the barrel shifter complexity is still much lower than that of the completely connected network (Fig. 2.16f).

Tree and Star   A *binary tree* of 31 nodes in five levels is shown in Fig. 2.17a. In general, a k-level, completely balanced binary tree should have $N = 2^k - 1$ nodes. The maximum node degree is 3 and the diameter is $2(k - 1)$. With a constant node degree, the binary tree is a scalable architecture. However, the diameter is rather long.

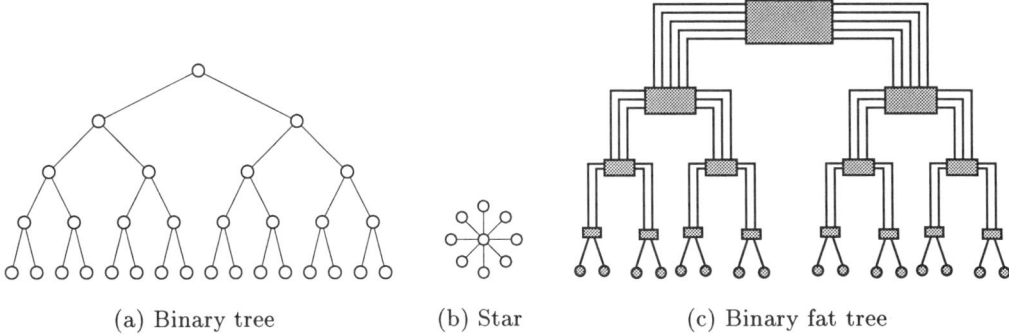

(a) Binary tree   (b) Star   (c) Binary fat tree

Figure 2.17 Tree, star, and fat tree.

The *star* is a two-level tree with a high node degree of $d = N - 1$ (Fig. 2.17b) and a small constant diameter of 2. A DADO multiprocessor was built at Columbia University (1987) with a 10-level binary tree of 1023 nodes. The star architecture has been used in systems with a centralized supervisor node.

Fat Tree The conventional tree structure used in computer science can be modified to become the *fat tree*, as introduced by Leiserson in 1985. A binary fat tree is shown in Fig. 2.17c. The channel width of a fat tree increases as we ascend from leaves to the root. The fat tree is more like a real tree in that branches get thicker toward the root.

One of the major problems in using the conventional binary tree is the bottleneck problem toward the root, since the traffic toward the root becomes heavier. The fat tree was proposed to alleviate the problem. The idea of a fat tree has been applied in the Connection Machine CM-5 to be studied in Chapter 8. The idea of binary fat trees can also be extended to multiway fat trees.

Mesh and Torus A 3×3 mesh network is shown in Fig. 2.18a. This is a popular architecture which has been implemented in the Illiac IV, MPP, DAP, CM-2, and Intel Paragon with variations.

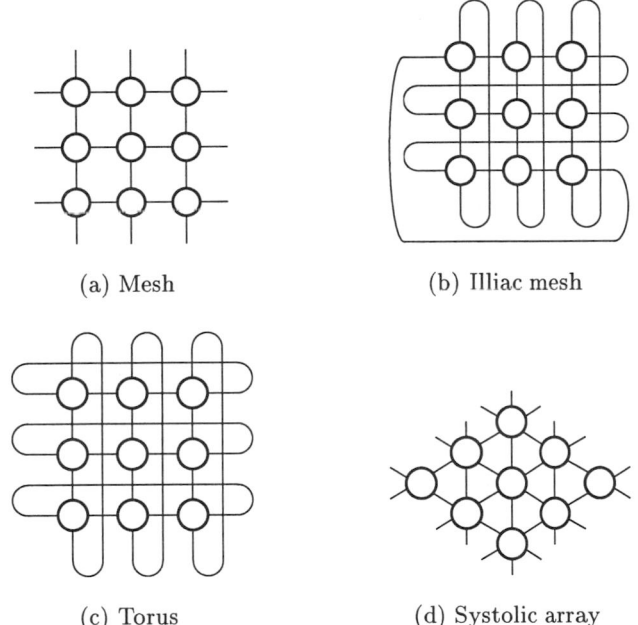

(a) Mesh

(b) Illiac mesh

(c) Torus

(d) Systolic array

Figure 2.18 Mesh, Illiac mesh, torus, and systolic array.

In general, a k-dimensional mesh with $N = n^k$ nodes has an interior node degree of

$2k$ and the network diameter is $k(n-1)$. Note that the pure mesh as shown in Fig. 2.17c is not symmetric. The node degrees at the boundary and corner nodes are 3 or 2.

Figure 2.18b shows a variation of the mesh by allowing wraparound connections. The Illiac IV assumed an 8×8 Illiac mesh with a constant node degree of 4 and a diameter of 7. The Illiac mesh is topologically equivalent to a chordal ring of degree 4 as shown in Fig. 2.16d for an $N = 9 = 3 \times 3$ configuration.

In general, an $n \times n$ Illiac mesh should have a diameter of $d = n - 1$, which is only half of the diameter for a pure mesh. The *torus* shown in Fig. 2.18c can be viewed as another variant of the mesh with an even shorter diameter. This topology combines the ring and mesh and extends to higher dimensions.

The torus has ring connections along each row and along each column of the array. In general, an $n \times n$ binary torus has a node degree of 4 and a diameter of $2\lfloor n/2 \rfloor$. The torus is a symmetric topology. All added wraparound connections help reduce the diameter by one-half from that of the mesh.

Systolic Arrays   This is a class of multidimensional pipelined array architectures designed for implementing fixed algorithms. What is shown in Fig. 2.18d is a systolic array specially designed for performing matrix-matrix multiplication. The interior node degree is 6 in this example.

In general, static systolic arrays are pipelined with multidirectional flow of data streams. The commercial machine Intel iWarp system (Anaratone et al., 1986) was designed with a systolic architecture. The systolic array has become a popular research area ever since its introduction by Kung and Leiserson in 1978.

With fixed interconnection and synchronous operation, a systolic array matches the communication structure of the algorithm. For special applications like signal/image processing, systolic arrays may offer a better performance/cost ratio. However, the structure has limited applicability and can be very difficult to program. Since this book emphasizes general-purpose computing, we will not study systolic arrays further. Interested readers may refer to the book by S.Y. Kung (1988) for using systolic and wavefront architectures in building VLSI array processors.

Hypercubes   This is a binary n-cube architecture which has been implemented in the iPSC, nCUBE, and CM-2 systems. In general, an n-cube consists of $N = 2^n$ nodes spanning along n dimensions, with two nodes per dimension. A 3-cube with 8 nodes is shown in Fig. 2.19a.

A 4-cube can be formed by interconnecting the corresponding nodes of two 3-cubes, as illustrated in Fig. 2.19b. The node degree of an n-cube equals n and so does the network diameter. In fact, the node degree increases linearly with respect to the dimension, making it difficult to consider the hypercube a scalable architecture.

Binary hypercube has been a very popular architecture for research and development in the 1980s. Both Intel iPSC/1, iPSC/2, and nCUBE machines were built with the hypercube architecture. The architecture has dense connections. Many other architectures, such as binary trees, meshes, etc., can be embedded in the hypercube.

With poor scalability and difficulty in packaging higher-dimensional hypercubes, the hypercube architecture is gradually being replaced by other architectures. For

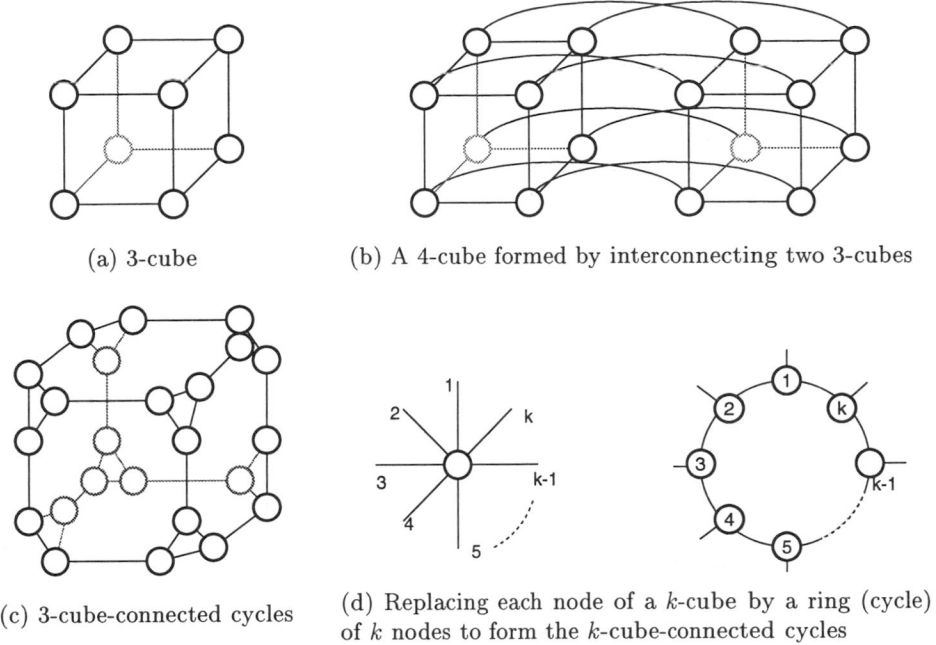

(a) 3-cube   (b) A 4-cube formed by interconnecting two 3-cubes

(c) 3-cube-connected cycles   (d) Replacing each node of a k-cube by a ring (cycle) of k nodes to form the k-cube-connected cycles

Figure 2.19 Hypercubes and cube-connected cycles.

example, the CM-5 chooses the fat tree over the hypercube implemented in the CM-2. The Intel Paragon chooses a two-dimensional mesh over its hypercube predecessors. Topological equivalence has been established among a number of network architectures. The bottom line for an architecture to survive in future systems is packaging efficiency and scalability to allow modular growth.

Cube-Connected Cycles This architecture is modified from the hypercube. As illustrated in Fig. 2.19c, a 3-cube is modified to form *3-cube-connected cycles* (CCC). The idea is to cut off the corner nodes (vertices) of the 3-cube and replace each by a ring (cycle) of 3 nodes.

In general, one can construct *k-cube-connected cycles* from a k-cube with $n = 2^k$ cycles nodes as illustrated in Fig. 2.19d. The idea is to replace each vertex of the k-dimensional hypercube by a ring of k nodes. A k-cube can be thus transformed to a k-CCC with $k \times 2^k$ nodes.

The 3-CCC shown in Fig. 2.19b has a diameter of 6, twice that of the original 3-cube. In general, the network diameter of a k-CCC equals $2k$. The major improvement of a CCC lies in its constant node degree of 3, which is independent of the dimension of the underlying hypercube.

Consider a hypercube with $N = 2^n$ nodes. A CCC with an equal number of N nodes must be built from a lower-dimension k-cube such that $2^n = k \cdot 2^k$ for some $k < n$.

For example, a 64-node CCC can be formed by replacing the corner nodes of a 4-cube with cycles of four nodes, corresponding to the case $n = 6$ and $k = 4$. The CCC has a diameter of $2k = 8$, longer than 6 in a 6-cube. But the CCC has a node degree of 3, smaller than the node degree of 6 in a 6-cube. In this sense, the CCC is a better architecture for building scalable systems if latency can be tolerated in some way.

k-ary n-Cube Networks   Rings, meshes, tori, binary n-cubes (hypercubes), and Omega networks are topologically isomorphic to a family of *k-ary n-cube* networks. Figure 2.20 shows a 4-ary 3-cube network.

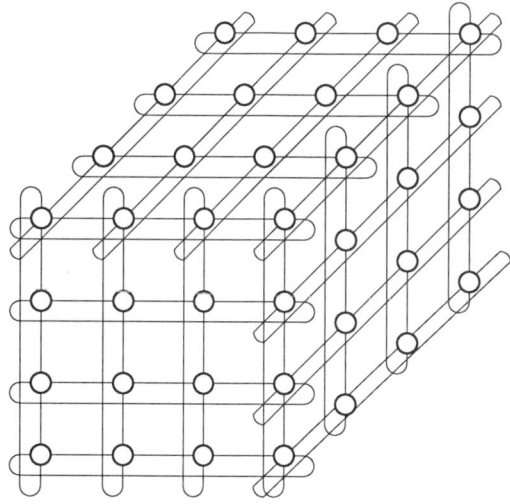

Figure 2.20 The k-ary n-cube network shown with k = 4 and n = 3; hidden nodes or connections are not shown.

The parameter n is the dimension of the cube and k is the *radix*, or the number of nodes (multiplicity) along each dimension. These two numbers are related to the number of nodes, N, in the network by:

$$N = k^n, \quad (k = \sqrt[n]{N}, \quad n = \log_k N) \tag{2.3}$$

A node in the k-ary n-cube can be identified by an n-digit radix-k address $A = a_0 a_1 a_2 \cdots a_n$, where a_i represents the node's position in the ith dimension. For simplicity, all links are assumed bidirectional. Each line in the network represents two communication channels, one in each direction. In Fig. 2.20, the lines between nodes are bidirectional links.

Traditionally, low-dimensional k-ary n-cubes are called *tori*, and high-dimensional binary n-cubes are called *hypercubes*. The long end-around connections in a torus can be avoided by folding the network as shown in Fig. 2.21. In this case, all links along the ring in each dimension have equal wire length when the multidimensional network is embedded in a plane.

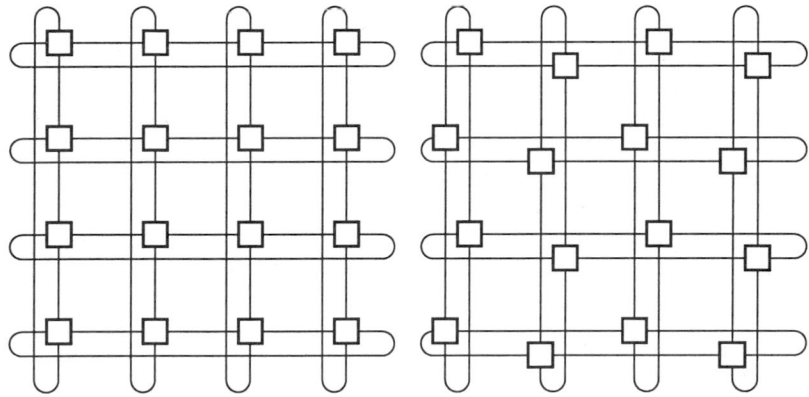

(a) Traditional torus (a 4-ary 2-cube)   (b) A torus with folded connections

Figure 2.21 Folded connections to equalize the wire length in a torus network.
(Courtesy of W. Dally; reprinted with permission from *IEEE Trans. Computers*,
June 1990)

William Dally (1990) has revealed a number of interesting properties of k-ary n-cube networks. The cost of such a network is dominated by the amount of wire, rather by the number of switches required. Under the assumption of constant wire bisection, low-dimensional networks with wide channels provide lower latency, less contention, and higher hot-spot throughput than higher-dimensional networks with narrow channels.

Network Throughput The *network throughput* is defined as the total number of messages the network can handle per unit time. One method of estimating throughput is to calculate the capacity of a network, the total number of messages that can be in the network at once. Typically, the maximum throughput of a network is some fraction of its capacity.

A *hot spot* is a pair of nodes that accounts for a disproportionately large portion of the total network traffic. Hot-spot traffic can degrade performance of the entire network by causing congestion. The *hot-spot throughput* of a network is the maximum rate at which messages can be sent from one specific node P_i to another specific node P_j.

Low-dimensional networks operate better under nonuniform loads because they have more resource sharing. In a high-dimensional network, wires are assigned to particular dimensions and cannot be shared between dimensions. For example, in a binary n-cube, it is possible for a wire to be saturated while a physically adjacent wire assigned to a different dimension remains idle. In a torus, all physically adjacent wires are combined into a single channel which is shared by all messages.

Minimum network latency is achieved when the network radix k and dimension n

are chosen to make the components of communication latency due to distance D (the number of hops between nodes) and the message aspect ratio L/W (message length L normalized to the channel width W) approximately equal. The minimum latency occurs at a very low dimension, 2 for up to 1024 nodes.

Low-dimensional networks reduce contention because having a few high-bandwidth channels results in more resource sharing and thus a better queueing performance than having many low-bandwidth channels. While network capacity and worst-case blocking latency are independent of dimension, low-dimensional networks have a higher maximum throughput and lower average block latency than do high-dimensional networks.

Both fat tree networks and k-ary n-cube networks are considered universal in the sense that they can efficiently simulate any other network of the same volume. Dally claimed that any point-to-point network can be embedded in a 3-D mesh with no more than a constant increase in wiring length.

Table 2.2 Summary of Static Network Characteristics

Network type	Node degree, d	Network diameter, D	No. of links, l	Bisection width, B	Symmetry	Remarks on network size
Linear Array	2	$N-1$	$N-1$	1	No	N nodes
Ring	2	$\lfloor N/2 \rfloor$	N	2	Yes	N nodes
Completely Connected	$N-1$	1	$N(N-1)/2$	$(N/2)^2$	Yes	N nodes
Binary Tree	3	$2(h-1)$	$N-1$	1	No	Tree height $h = \lceil \log_2 N \rceil$
Star	$N-1$	2	$N-1$	$\lfloor N/2 \rfloor$	No	N nodes
2D-Mesh	4	$2(r-1)$	$2N-2r$	r	No	$r \times r$ mesh where $r = \sqrt{N}$
Illiac Mesh	4	$r-1$	$2N$	$2r$	No	Equivalent to a chordal ring of $r = \sqrt{N}$
2D-Torus	4	$2\lfloor r/2 \rfloor$	$2N$	$2r$	Yes	$r \times r$ torus where $r = \sqrt{N}$
Hypercube	n	n	$nN/2$	$N/2$	Yes	N nodes, $n = \log_2 N$ (dimension)
CCC	3	$2k - 1 + \lfloor k/2 \rfloor$	$3N/2$	$N/(2k)$	Yes	$N = k \times 2^k$ nodes with a cycle length $k \geq 3$
k-ary n-cube	$2n$	$n\lfloor k/2 \rfloor$	nN	$2k^{n-1}$	Yes	$N = k^n$ nodes

Summary of Static Networks   In Table 2.2, we summarize the important characteristics of static connection networks. The node degrees of most networks are less than 4, which is rather desirable. For example, the INMOS Transputer chip is a compute-communication microprocessor with four ports for communication.

With a constant node degree of 4 or less, a Transputer (such as the T800) becomes applicable as a building block. The node degrees for the completely connected and star networks are both bad. The hypercube node degree increases with $\log_2 N$ and is also bad when the value of N becomes large.

Network diameters vary over a wide range. With the invention of hardware routing (wormhole routing), the diameter has become less critical an issue because the communication delay between any two nodes becomes almost a constant with a high degree of pipelining. The number of links affects the network cost. The bisection width affects the network bandwidth.

The property of symmetry affects scalability and routing efficiency. It is fair to say that the total network cost increases with d and l. A smaller diameter is still a virtue. But the average distance between nodes may be a better measure. The bisection width can be enhanced by a wider channel width. Based on the above analysis, ring, mesh, torus, k-ary n-cube, and CCC all have some desirable features for building future MPP systems.

2.4.3   Dynamic Connection Networks

For multipurpose or general-purpose applications, we need to use dynamic connections which can implement all communication patterns based on program demands. Instead of using fixed connections, switches or arbiters must be used along the connecting paths to provide the dynamic connectivity. In increasing order of cost and performance, dynamic connection networks include *bus systems*, *multistage interconnection networks* (MIN), and *crossbar switch networks*.

The price tags of these networks are attributed to the cost of the wires, switches, arbiters, and connectors required. The performance is indicated by the network bandwidth, data transfer rate, network latency, and communication patterns supported. A brief introduction to dynamic connection networks is given below. Details can be found in subsequent chapters.

Digital Buses   A *bus system* is essentially a collection of wires and connectors for data transactions among processors, memory modules, and peripheral devices attached to the bus. The bus is used for only one transaction at a time between a source (master) and a destination (slave). In case of multiple requests, the bus arbitration logic must be able to allocate or deallocate the bus servicing the requests one at a time.

For this reason, the digital bus has been called *contention bus* or a *time-sharing bus* among multiple functional modules. A bus system has a lower cost and provides a limited bandwidth compared to the other two dynamic connection networks. Many industrial and IEEE bus standards are available.

Figure 2.22 shows a bus-connected multiprocessor system. The system bus provides a common communication path between the processors or I/O subsystem and the memory modules or secondary storage devices (disks, tape units, etc.). The system bus is often implemented on a backplane of a printed circuit board. Other boards for processors, memories, or device interfaces are plugged into the backplane board via connectors or cables.

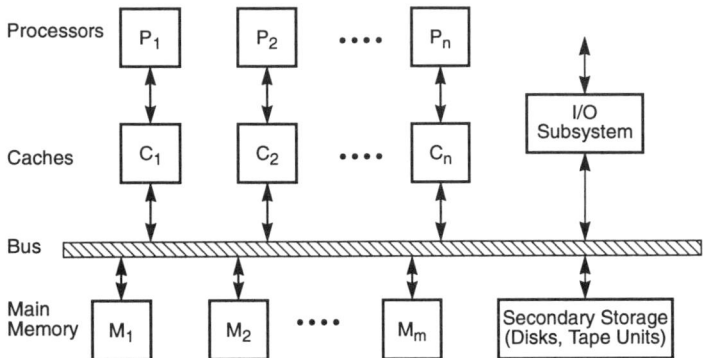

Figure 2.22 A bus-connected multiprocessor system, such as the Sequent Symmetry S1.

The active or master devices (processors or I/O subsystem) generate requests to address the memory. The passive or slave devices (memories or peripherals) respond to the requests. The common bus is used on a time-sharing basis, and important busing issues include the bus arbitration, interrupts handling, coherence protocols, and transaction processing. We will study various bus systems, such as the VME bus and the IEEE Futurebus+, in Chapter 5. Hierarchical bus structures for building larger multiprocessor systems are studied in Chapter 7.

Switch Modules   A *a × b switch module* has *a* inputs and *b* outputs. A *binary switch* corresponds to a 2 × 2 switch module in which $a = b = 2$. In theory, *a* and *b* do not have to be equal. However, in practice, *a* and *b* are often chosen as integer powers of 2; that is, $a = b = 2^k$ for some $k \geq 1$.

Table 2.3 list several commonly used switch module sizes: 2 × 2, 4 × 4, and 8 × 8. Each input can be connected to one or more of the outputs. However, conflicts must be avoided at the output terminals. In other words, one-to-one and one-to-many mappings are allowed; but many-to-one mappings are not allowed due to conflicts at the output terminal.

When only one-to-one mappings (permutations) are allowed, we call the module an $n \times n$ crossbar switch. For example, a 2 × 2 crossbar switch can connect two possible patterns: *straight* or *crossover*. In general, an $n \times n$ crossbar can achieve $n!$ permutations. The numbers of legitimate connection patterns for switch modules of various sizes are listed in Table 2.3.

Table 2.3 Switch Modules and Legitimate States

Module Size	Legitimate States	Permutation Connections
2×2	4	2
4×4	256	24
8×8	16,777,216	40,320
$n \times n$	n^n	$n!$

Multistage Networks MINs have been used in both MIMD and SIMD computers. A generalized multistage network is illustrated in Fig. 2.23. A number of $a \times b$ switches are used in each stage. Fixed interstage connections are used between the switches in adjacent stages. The switches can be dynamically set to establish the desired connections between the inputs and outputs.

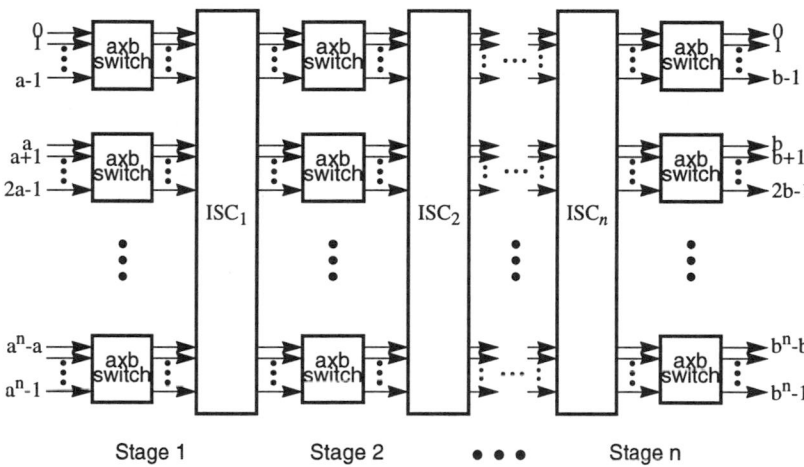

Figure 2.23 A generalized structure of a multistage interconnection network (MIN) built with a × b switch modules and interstage connection patterns $ISC_1, ISC_2, ..., ISC_n$.

Different classes of MINs differ in the switch modules used and in the kind of *interstage connection* (ISC) patterns used. The simplest switch module would be the 2×2 switches ($a = b = 2$ in Fig. 2.23). The ISC patterns often used include *perfect shuffle, butterfly, multiway shuffle, crossbar, cube connection*, etc. Some of these ISC patterns are shown below with examples.

Omega Network Figures 2.24a to 2.24d show four possible connections of 2×2 switches used in constructing the Omega network. A 16×16 Omega network is shown in Fig. 2.24e. Four stages of 2×2 switches are needed. There are 16 inputs on the left and 16 outputs on the right. The ISC pattern is the perfect shuffle over 16 objects.

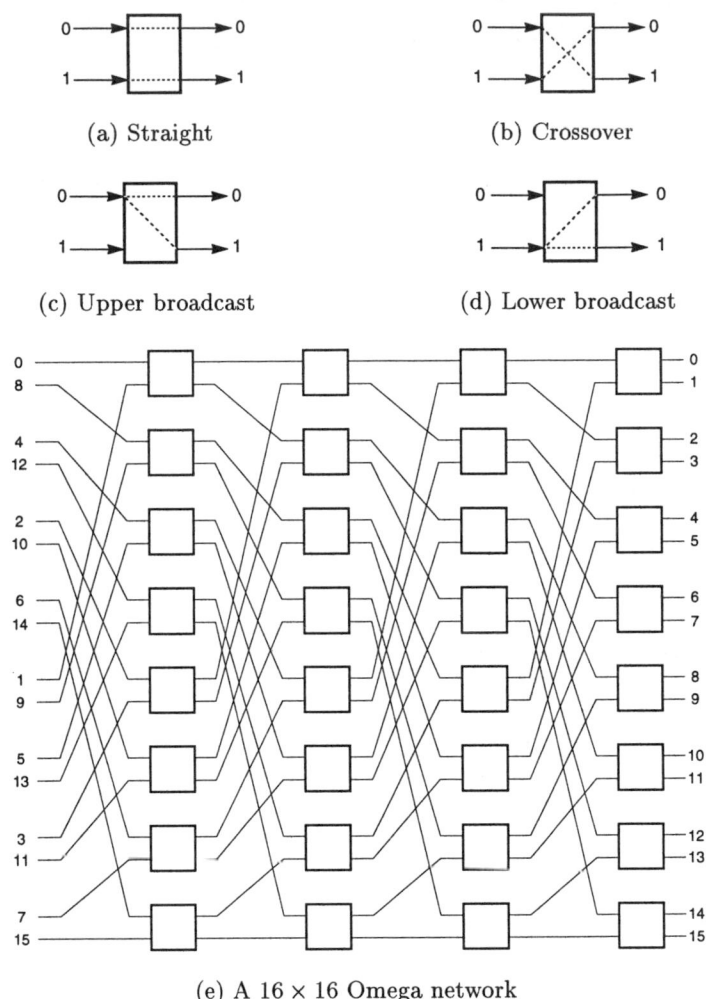

(a) Straight   (b) Crossover

(c) Upper broadcast   (d) Lower broadcast

(e) A 16 × 16 Omega network

Figure 2.24 The use of 2×2 **switches and perfect shuffle as an interstage connection pattern to construct a** 16×16 **Omega network.** (Courtesy of Duncan Lawrie; reprinted with permission from *IEEE Trans. Computers*, Dec. 1975)

In general, an n-input Omega network requires $\log_2 n$ stages of 2×2 switches. Each stage requires $n/2$ switch modules. In total, the network uses $n \log_2 n/2$ switches. Each switch module is individually controlled.

Various combinations of the switch states implement different permutations, broadcast, or other connections from the inputs to the outputs. The interconnection capabilities of the Omega and other networks will be further studied in Chapter 7.

Baseline Network   Wu and Feng (1980) have studied the relationship among a class of multistage interconnection networks. A *Baseline network* can be generated recursively as shown in Fig. 2.25a.

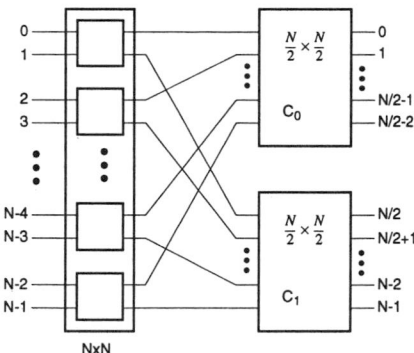

(a) Recursive construction

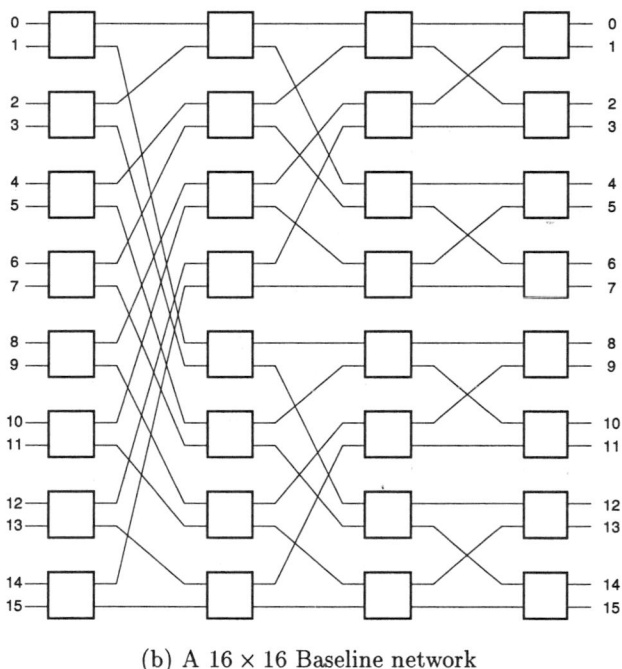

(b) A 16×16 Baseline network

Figure 2.25 Recursive construction of a Baseline network. (Courtesy of Wu and Feng; reprinted with permission from *IEEE Trans. Computers, August 1980*)

The first stage contains one $N \times N$ block, and the second stage contains two $(N/2) \times (N/2)$ subblocks, labeled C_0 and C_1. The construction process can be recursively applied to the subblocks until the $N/2$ subblocks of size 2×2 are reached.

The small boxes and the ultimate building blocks of the subblocks are the 2×2 switches, each with two legitimate connection states: *straight* and *crossover* between the two inputs and two outputs. A 16×16 Baseline network is shown in Fig. 2.25b. In Problem 2.15, readers are asked to prove the topological equivalence between the Baseline and other networks.

Crossbar Network The highest bandwidth and interconnection capability are provided by crossbar networks. A crossbar network can be visualized as a single-stage switch network. Like a telephone switchboard, the crosspoint switches provide dynamic connections between (source, destination) pairs. Each crosspoint switch can provide a dedicated connection path between a pair. The switch can be set on or off dynamically upon program demand. Two types of crossbar networks are illustrated in Fig. 2.26.

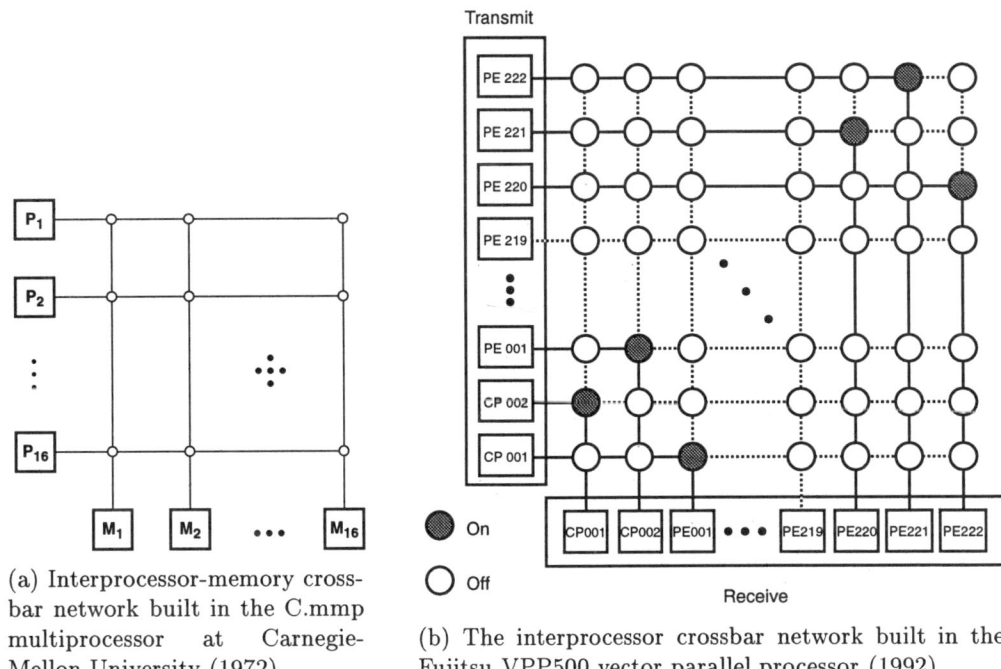

(a) Interprocessor-memory cross-bar network built in the C.mmp multiprocessor at Carnegie-Mellon University (1972)

(b) The interprocessor crossbar network built in the Fujitsu VPP500 vector parallel processor (1992)

Figure 2.26 Two crossbar switch network configurations.

To build a shared-memory multiprocessor, one can use a crossbar network between the processors and memory modules (Fig. 2.26a). This is essentially a memory-access network. The C.mmp multiprocessor (Wulf and Bell, 1972) has implemented a 16×16 crossbar network which connects 16 PDP 11 processors to 16 memory modules, each of which has a capability of 1 million words of memory cells. The 16 memory modules can be accessed by at most 16 processors simultaneously.

Note that each memory module can satisfy only one processor request at a time. When multiple requests arrive at the same memory module simultaneously, the crossbar must resolve the conflicts. The behavior of each crossbar switch is very similar to that of a bus. However, each processor can generate a sequence of addresses to access multiple memory modules simultaneously. Thus, in Fig. 2.26a, only one crosspoint switch can be set on in each column. However, several crosspoint switches can be set on simultaneously in order to support parallel (or interleaved) memory accesses.

Another type of crossbar network is for interprocessor communication and is depicted in Fig. 2.26b. This large crossbar (224 × 224) was actually built in a vector parallel processor (VPP500) by Fujitsu Inc. (1992). The PEs are processors with attached memory. The CPs stand for control processors which are used to supervise the entire system operation, including the crossbar networks. In this crossbar, only one crosspoint switch can be set on in each row and each column.

The interprocessor crossbar provides permutation connections among the processors. Only one-to-one connections are provided. Therefore, the $n \times n$ crossbar connects at most n (source,destination) pairs at a time. This is different from the interprocessor memory crossbar. The operations and purposes of the two crossbar network types are very different. We will further study these crossbar networks in Chapters 7 and 8.

Table 2.4 Summary of Dynamic Network Characteristics

Network Characteristics	Bus System	Multistage Network	Crossbar Switch
Minimum latency for unit data transfer	Constant	$O(\log_k n)$	Constant
Bandwidth per processor	$O(w/n)$ to $O(w)$	$O(w)$ to $O(nw)$	$O(w)$ to $O(nw)$
Wiring Complexity	$O(w)$	$O(nw \log_k n)$	$O(n^2 w)$
Switching Complexity	$O(n)$	$O(n \log_k n)$	$O(n^2)$
Connectivity and routing capability	Only one to one at a time.	Some permutations and broadcast, if network unblocked	All permutations, one at a time.
Representative computers	Symmetry S-1, Encore Multimax	BBN TC-2000, IBM RP3	Cray Y-MP/816, Fujitsu VPP500
Remarks	Assume n processors on the bus; bus width is w bits.	$n \times n$ MIN using $k \times k$ switches with line width of w bits.	Assume $n \times n$ crossbar with line width of w bits.

Summary In Table 2.4, we summarize the important features of buses, multistage networks, and crossbar switches in building dynamic networks. Obviously, the bus is the cheapest to build, but its drawback lies in the low bandwidth available to each processor.

Another problem with the bus is that it is prone to failure. Some fault-tolerant systems, like the Tandem multiprocessor for transaction processing, use dual buses to protect the system from single failures.

The crossbar switch is the most expensive one to build, due to the fact that its hardware complexity increases with n^2. However, the crossbar has the highest bandwidth and routing capability. For a small network size, it is the desired choice.

Multistage networks provide a compromise between the two extremes. The major advantage of MINs lies in their scalability with modular construction. However, the latency increases with $\log n$, the number of stages in the network. Also, costs due to increased wiring and switching complexity are another constraint.

For building future MPP systems, some of the static topologies are more scalable in specific applications. If optical and microelectronics technologies move more rapidly, large-scale MINs or crossbar networks may become more economical or feasible for establishing dynamic connections in general-purpose computing.

2.5 Bibliographic Notes and Exercises

The grain packing technique is based on the work of [Kruatrachue88] and Lewis. The five levels of parallelism were characterized in [Hwang87a]. [Bernstein66] revealed the conditions for program partitioning. Program flow mechanisms were discussed in [Gajski82]. The matching problem between hardware and software parallelism was discussed in tutorial notes [Hwu91]. [Sevcik89] has characterized parallelism in the context of deterministic scheduling.

The book edited by [Gaudiot91] and Bic provides a survey of dataflow computing research up to 1990. The MIT tagged-token dataflow architecture was originally reported in [Arvind83] and Iannucci. [Feng81] surveyed various interconnection architectures. [Siegel89] covers multistage networks. [Johnsson90] studied communication properties in network architectures. [Preparata79] and Vuillemin developed the CCC architecture. The concept of systolic arrays was first developed by [Kung78] and Leiserson. Applications of the perfect shuffle can be found in [Stone71].

The chordal ring was treated in [Arden81] and Lee. The fat tree was originally proposed in [Leiserson85]. [Hwang87b] and Ghosh have proposed a hierarchical hypernet architecture for constructing communication-efficient massively parallel systems. The k-ary n-cube networks are studied in [Dally90b]. The Intel/CMU iWarp is described in [Anaratone86] et al. The book by [Kung88] provides details of systolic and wavefront arrays and their applications. The Omega network was proposed in [Lawrie75]. The C.mmp was described in [Wulf72] and Bell. The Fujitsu VPP500 was described in [Fujitsu92]. The Flip network was introduced in the STARAN computer by [Batcher76]. The Baseline network was formulated by [Wu80] and Feng.

Exercises

Problem 2.1 Define the following terms related to parallelism and dependence relations:

(a) Computational granularity.

(b) Communication latency.

(c) Flow dependence.

(d) Antidependence.

(e) Output dependence.

(f) I/O dependence.

(g) Control dependence.

(h) Resource dependence.

(i) Bernstein conditions.

(j) Degree of parallelism.

Problem 2.2 Define the following terms for various system interconnect architectures:

(a) Node degree.

(b) Network diameter.

(c) Bisection bandwidth.

(d) Static connection networks.

(e) Dynamic connection networks.

(f) Nonblocking networks.

(g) Multicast and broadcast.

(h) Mesh versus torus.

(i) Symmetry in networks.

(j) Multistage networks.

(k) Crossbar networks.

(l) Digital buses.

Problem 2.3 Answer the following questions on program flow mechanisms and computer models:

(a) Compare control-flow, dataflow, and reduction computers in terms of the program flow mechanism used.

(b) Comment on the advantages and disadvantages in control complexity, potential for parallelism, and cost-effectiveness of the above computer models.

(c) What are the differences between string reduction and graph reduction machines?

Problem 2.4 Perform a data dependence analysis on each of the following Fortran program fragments. Show the dependence graphs among the statements with justification.

(a)

S1:	A = B + D
S2:	C = A × 3
S3:	A = A + C
S4:	E = A / 2

(b)

S1:	X = SIN(Y)
S2:	Z = X + W
S3:	Y = −2.5 × W
S4:	X = COS(Z)

(c) Determine the data dependences in the same and adjacent iterations of the following Do-loop.

$$\textbf{Do } 10 \text{ I} = 1, \text{ N}$$

S1:	A(I+1) = B(I−1) + C(I)
S2:	B(I) = A(I) × K
S3:	C(I) = B(I) − 1

10   **Continue**

Problem 2.5 Analyze the data dependences among the following statements in a given program:

S1:	Load R1, 1024	/R1 ← 1024/
S2:	Load R2, M(10)	/R2 ← Memory(10) /
S3:	Add R1, R2	/R1 ← (R1) + (R2)/
S4:	Store M(1024), R1	/Memory(1024) ← (R1)/
S5:	Store M((R2)), 1024	/Memory(64) ← 1024/

where (Ri) means the content of register Ri and Memory(10) contains 64 initially.

(a) Draw a dependence graph to show all the dependences.
(b) Are there any resource dependences if only one copy of each functional unit is available in the CPU?
(c) Repeat the above for the following program statements:

S1:	Load R1, M(100)	/R1 ← Memory(100)/
S2:	Move R2, R1	/R2 ← (R1) /
S3:	Inc R1	/R1 ← (R1) + 1/
S4:	Add R2, R1	/R2 ← (R2) + (R1)/
S5:	Store M(100), R1	/Memory(100) ← (R1)/

Problem 2.6 A sequential program consists of the following five statements, S1 through S5. Considering each statement as a separate process, clearly identify *input set* I_i and *output set* O_i of each process. Restructure the program using Bernstein's conditions in order to achieve maximum parallelism between processes. If any pair of processes cannot be executed concurrently, specify which of the three conditions is not satisfied.

S1:	A = B + C
S2:	C = B × D
S3:	S = 0
S4:	**Do** I = A, 100
	S = S + X(I)
	End Do
S5:	IF (S .GT. 1000) C = C × 2

Problem 2.7 Consider the execution of the following code segment consisting of seven statements. Use Bernstein's conditions to detect the maximum parallelism embedded in this code. Justify the portions that can be executed in parallel and the remaining portions that must be executed sequentially. Rewrite the code using parallel constructs such as *Cobegin* and *Coend*. No variable substitution is allowed. All statements can be executed in parallel if they are declared within the same block of a (*Cobegin, Coend*) pair.

$$
\begin{array}{ll}
\text{S1:} & A = B + C \\
\text{S2:} & C = D + E \\
\text{S3:} & F = G + E \\
\text{S4:} & C = A + F \\
\text{S5:} & M = G + C \\
\text{S6:} & A = L + C \\
\text{S7:} & A = E + A \\
\end{array}
$$

Problem 2.8 According to program order, the following six arithmetic expressions need to be executed in minimum time. Assume that all are integer operands already loaded into working registers. No memory reference is needed for the operand fetch. Also, all intermediate or final results are written back to working registers without conflicts.

$$
\begin{array}{ll}
\text{P1:} & X \leftarrow (A + B) \times (A - B) \\
\text{P2:} & Y \leftarrow (C + D)/(C - D) \\
\text{P3:} & Z \leftarrow X + Y \\
\text{P4:} & A \leftarrow E \times F \\
\text{P5:} & Y \leftarrow E - Z \\
\text{P6:} & B \leftarrow (X - F) \times A \\
\end{array}
$$

(a) Use the minimum number of working registers to rewrite the above HLL program into a minimum-length assembly language code using arithmetic opcodes *add*, *subtract*, *multiply*, and *divide* exclusively. Assume a fixed instruction format with three register fields: two for sources and one for destinations.

(b) Perform a flow analysis of the assembly code obtained in part (a) to reveal all data dependences with a dependence graph.

(c) The CPU is assumed to have two *add units*, one *multiply unit*, and one *divide unit*. Work out an optimal schedule to execute the assembly code in minimum time, assuming 1 cycle for the add unit, 3 cycles for the multiply unit, and 18 cycles for the divide unit to complete the execution of one instruction. Ignore all overhead caused by instruction fetch, decode, and writeback. No pipelining is assumed here.

Problem 2.9 Given the following assembly language code. Exploit the maximum degree of parallelism among the 16 instructions, assuming no resource conflicts and multiple functional units are available simultaneously. For simplicity, no pipelining is assumed. All instructions take one machine cycle to execute. Ignore all other overhead.

1:	Load R1, A	/R1 ← Mem(A)/
2:	Load R2, B	/R2 ← Mem(B)/
3:	Mul R3, R1, R2	/R3 ← (R1) × (R2)/
4:	Load R4, D	/R4 ← Mem(D)/
5:	Mul R5, R1, R4	/R5 ← (R1) × (R4)/
6:	Add R6, R3, R5	/R6 ← (R3) + (R5)/
7:	Store X, R6	/Mem(X) ← (R6)/
8:	Load R7, C	/R7 ← Mem(C)/
9:	Mul R8, R7, R4	/R8 ← (R7) × (R4)/
10:	Load R9, E	/R9 ← Mem(E)/
11:	Add R10, R8, R9	/R10 ← (R8) + (R9)/
12:	Store Y, R10	/Mem(Y) ← (R10)/
13:	Add R11, R6, R10	/R11 ← (R6)+(R10)/
14:	Store U, R11	/Mem(U) ← (R11)/
15:	Sub R12, R6, R10	/R12 ← (R6) − (R10)/
16:	Store V, R12	/Mem(V) ← (R12)/

(a) Draw a program graph with 16 nodes to show the flow relationships among the 16 instructions.

(b) Consider the use of a three-issue superscalar processor to execute this program fragment in minimum time. The processor can issue one memory-access instruction (Load or Store but not both), one Add/Sub instruction, and one Mul (multiply) instruction per cycle. The Add unit, Load/Store unit, and Multiply unit can be used simultaneously if there is no data dependence.

Problem 2.10 Repeat part (b) of Problem 2.9 on a dual-processor system with shared memory. Assume that the same superscalar processors are used and that all instructions take one cycle to execute.

(a) Partition the given program into two balanced halves. You may want to insert some load or store instructions to pass intermediate results generated by the two processors to each other. Show the divided program flow graph with the final output U and V generated by the two processors separately.

(b) Work out an optimal schedule for parallel execution of the above divided program by the two processors in minimum time.

Problem 2.11 You are asked to design a direct network for a multicomputer with 64 nodes using a three-dimensional torus, a six-dimensional binary hypercube, and cube-connected-cycles (CCC) with a minimum diameter. The following questions are related to the relative merits of these network topologies:

(a) Let d be the node degree, D the network diameter, and l the total number of links in a network. Suppose the *quality* of a network is measured by $(d \times D \times l)^{-1}$. Rank the three architectures according to this quality measure.

(b) A *mean internode distance* is defined as the average number of hops (links) along the shortest path for a message to travel from one node to another. The average is calculated for all (source, destination) pairs. Order the three architectures based on their mean internode distances, assuming that the probability that a node will send a message to all other nodes with distance i is $(D - i + 1) / \sum_{k=1}^{D} k$, where D is the network diameter.

Problem 2.12 Consider an Illiac mesh (8×8), a binary hypercube, and a barrel shifter, all with 64 nodes labeled $N_0, N_1, \ldots, N_{63}$. All network links are bidirectional.

(a) List all the nodes reachable from node N_0 in exactly three steps for each of the three networks.

(b) Indicate in each case the tightest upper bound on the minimum number of routing steps needed to send data from any node N_i to another node N_j.

(c) Repeat part (b) for a larger network with 1024 nodes.

Problem 2.13 Compare *buses*, *crossbar switches*, and *multistage networks* for building a multiprocessor system with n processors and m shared-memory modules. Assume a word length of w bits and 2×2 switches are used in building a multiprocessor system with n processors and m shared-memory modules. Assume a word length of w bits and that 2×2 switches are used in building the multistage networks. The comparison study is carried out separately in each of the following four categories:

(a) Hardware complexities such as switching, arbitration, wires, connector, or cable requirements.

(b) Minimum latency in unit data transfer between the processor and memory module.

(c) Bandwidth range available to each processor.

(d) Communication capabilities such as permutations, data broadcast, blocking handling, etc.

Problem 2.14 Answer the following questions related to multistage networks:

(a) How many legitimate states are there in a 4×4 switch module, including both broadcast and permutations? Justify your answer with reasoning.

(b) Construct a 64-input Omega network using 4×4 switch modules in multiple stages. How many permutations can be implemented directly in a single pass through the network without blocking?

(c) What is the percentage of one-pass permutations compared with the total number of permutations achievable in one or more passes through the network?

Problem 2.15 Topologically equivalent networks are those whose graph representations are isomorphic with the same interconnection capabilities. Prove the topological equivalence among the Omega, Flip, and Baseline networks.

(a) Prove that the Omega network (Fig. 2.24) is topologically equivalent to the Baseline network (Fig. 2.25b) .

(b) The *Flip network* (Fig. 2.27) is constructed using inverse perfect shuffle (Fig. 2.14b) for interstage connections. Prove that the Flip network is topologically equivalent to the Baseline network.

(c) Based on the results obtained in (a) and (b), prove the topological equivalence between the Flip network and the Omega network.

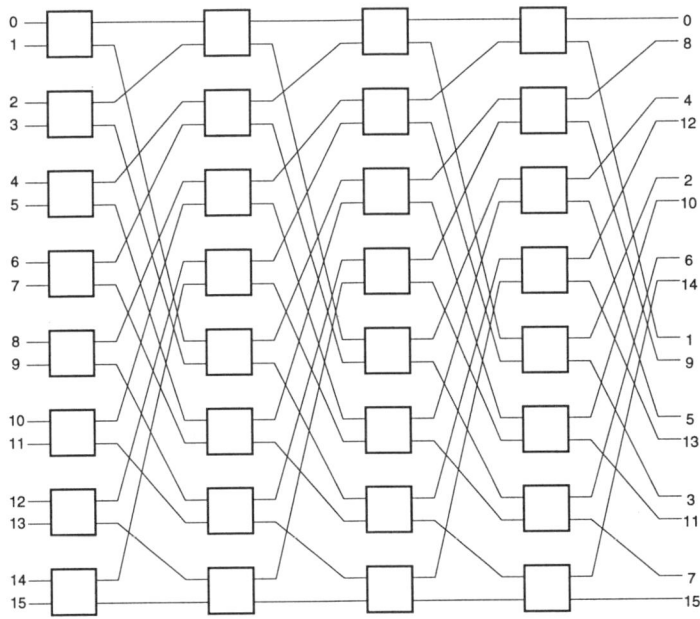

Figure 2.27 A 16 × 16 **Flip network.** (Courtesy of Ken Batcher; reprinted from *Proc. Int. Conf. Parallel Processing*, 1976)

Problem 2.16 Answer the following questions for the k-ary n-cube network:

(a) How many nodes are there?

(b) What is the network diameter?

(c) What is the bisection bandwidth?

(d) What is the node degree?

(e) Explain the graph-theoretic relationship among k-ary n-cube networks and rings, meshes, tori, binary n-cubes, and Omega networks.

(f) Explain the difference between a conventional torus and a folded torus.

(g) Under the assumption of constant wire bisection, why do low-dimensional networks (tori) have lower latency and higher hot-spot throughput than high-dimensional networks (hypercubes)?

Problem 2.17 Read the paper on fat trees by Leiserson, which appeared in *IEEE Trans. Computers*, pp. 892–901, Oct. 1985. Answer the following questions related to the organization and application of fat trees:

(a) Explain the advantages of using binary fat trees over conventional binary trees as a multiprocessor interconnection network.

(b) A *universal fat tree* is defined as a fat tree of n nodes with root capacity w, where $n^{2/3} \le w \le n$, and for each channel c_k at level k of the tree, the capacity is

$$c_k = \min \left(\lceil n/2^k \rceil, \lceil w/2^{2k/3} \rceil \right)$$

Prove that the capacities of a universal fat tree grow exponentially as we go up the tree from the leaves. The channel capacity is defined as the number of wires in a channel.

Problem 2.18 Read the paper on k-ary n-cube networks by Dally, which appeared in *IEEE Trans. Computers*, June 1990, pp. 775–785. Answer the following questions related to the network properties and applications as a VLSI communication network:

(a) Prove that the bisection width B of a k-ary n-cube with w-bit wide communication channels is

$$B(k,n) = 2w\sqrt{N}k^{n/2-1} = 2wN/k$$

where $N = k^n$ is the network size.

(b) Prove that the hot-spot throughput of a k-ary n-cube network with deterministic routing is equal to the bandwidth of a single channel $w = k - 1$, under the assumption of a constant wire cost.

Problem 2.19 Network embedding is a technique to implement a network A on a network B. Explain how to perform the following network embeddings:

(a) Embed a two-dimensional torus on an n-dimensional hypercube with $N = 2^n$ nodes where $r^2 = 2^n$.

(b) Embed the largest ring on a CCC with $N = k \times 2^k$ nodes and $k \ge 3$.

(c) Embed a complete balanced binary tree with maximum height on a mesh of $r \times r$ nodes.

Problem 2.20 Read the paper on hypernets by Hwang and Ghosh, which appeared in *IEEE Trans. Computers*, Dec. 1989. Answer the following questions related to the network properties and applications of hypernets:

(a) Explain how hypernets integrate positive features of hypercube and tree-based topologies into one combined architecture.

(b) Prove that the average node degree of a hypernet can be maintained essentially constant when the network size is increased.

(c) Discuss the application potentials of hypernets in terms of message routing complexity, cost-effective support for global as well as localized communication, I/O capabilities, and fault tolerance.

Chapter 3

Principles of Scalable Performance

We study performance measures, speedup laws, and scalability principles in this chapter. Three speedup models are presented under different computing objectives and resource constraints. These include Amdahl's law (1967), Gustafson's scaled speedup (1988), and the memory-bounded speedup by Sun and Ni (1993).

The efficiency, redundancy, utilization, and quality of a parallel computation are defined, involving the interplay between architectures and algorithms. Standard performance measures and several benchmark kernels are introduced with recently released performance data.

The performance of parallel computers relies on a design that balances hardware and software. The system architects and programmers must exploit parallelism, pipelining, and networking in a balanced approach. Toward building massively parallel systems, the scalability issues must be resolved first. Fundamental concepts of scalable systems are introduced in this chapter. Case studies can be found in subsequent chapters, especially in Chapter 9.

3.1 Performance Metrics and Measures

In this section, we first study parallelism profiles and define the asymptotic speedup factor, ignoring communication latency and resource limitations. Then we introduce the concepts of system efficiency, utilization, redundancy, and quality of parallel computations. Possible tradeoffs among these performance metrics are examined in the context of cost-effectiveness. Several commonly used performance measures, MIPS, Mflops, and TPS, are formally defined.

3.1.1 Parallelism Profile in Programs

The degree of parallelism reflects the extent to which software parallelism matches hardware parallelism. We characterize below parallelism profiles, introduce the concept of average parallelism, and define an ideal speedup with infinite machine resources. Variations on the ideal speedup factor will be presented in subsequent sections from various application viewpoints and under different system limitations.

Degree of Parallelism The execution of a program on a parallel computer may use different numbers of processors at different time periods during the execution cycle. For each time period, the number of processors used to execute a program is defined as the *degree of parallelism* (DOP). This is a discrete time function, assuming only nonnegative integer values.

The plot of the DOP as a function of time is called the *parallelism profile* of a given program. For simplicity, we concentrate on the analysis of single-program profiles. Some software tools are available to trace the parallelism profile. The profiling of multiple programs in an interleaved fashion can be easily extended from this study.

Fluctuation of the profile during an observation period depends on the algorithmic structure, program optimization, resource utilization, and run-time conditions of a computer system. The DOP was defined under the assumption of having an unbounded number of available processors and other necessary resources. The DOP may not always be achievable on a real computer with limited resources.

When the DOP exceeds the maximum number of available processors in a system, some parallel branches must be executed in chunks sequentially. However, parallelism still exists within each chunk, limited by the machine size. The DOP may be also limited by memory and by other nonprocessor resources. We consider only the limit imposed by processors in future discussions on speedup models.

Average Parallelism In what follows, we consider a parallel computer consisting of n homogeneous processors. The maximum parallelism in a profile is m. In the ideal case, $n >> m$. The *computing capacity* Δ of a single processor is approximated by the execution rate, such as MIPS or Mflops, without considering the penalties from memory access, communication latency, or system overhead. When i processors are busy during an observation period, we have DOP $= i$.

The total amount of work W (instructions or computations) performed is proportional to the area under the profile curve:

$$W = \Delta \int_{t_1}^{t_2} DOP(t) \, dt \tag{3.1}$$

This integral is often computed with the following discrete summation:

$$W = \Delta \sum_{i=1}^{m} i \cdot t_i \tag{3.2}$$

where t_i is the total amount of time that DOP $= i$ and $\sum_{i=1}^{m} t_i = t_2 - t_1$ is the total elapsed time.

The *average parallelism* A is computed by

$$A = \frac{1}{t_2 - t_1} \int_{t_1}^{t_2} DOP(t) \, dt \tag{3.3}$$

In discrete form, we have

$$A = \left(\sum_{i=1}^{m} i \cdot t_i \right) \bigg/ \left(\sum_{i=1}^{m} t_i \right) \qquad (3.4)$$

Example 3.1 Example parallelism profile and average parallelism of a divide-and-conquer algorithm (Sun and Ni, 1993)

As illustrated in Fig. 3.1, the parallelism profile of an example divide-and-conquer algorithm increases from 1 to its peak value $m = 8$ and then decreases to 0 during the observation period (t_1, t_2).

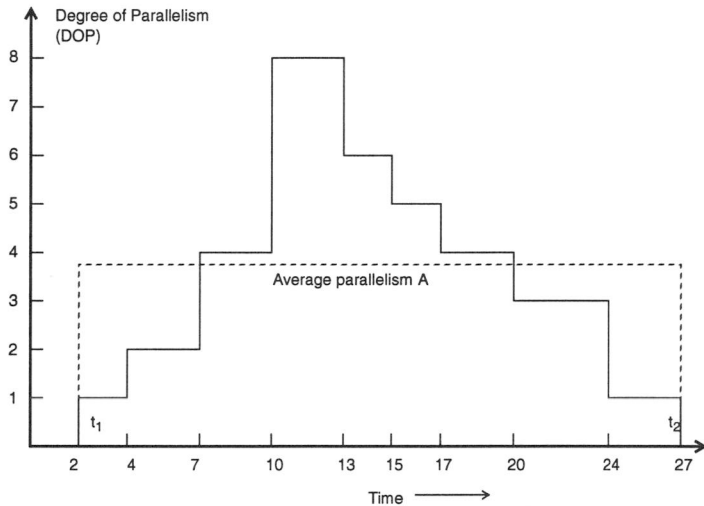

Figure 3.1 Parallelism profile of a divide-and-conquer algorithm.

In Fig. 3.1, the average parallelism $A = (1 \times 5 + 2 \times 3 + 3 \times 4 + 4 \times 6 + 5 \times 2 + 6 \times 2 + 8 \times 3)/(5 + 3 + 4 + 6 + 2 + 2 + 3) = 93/25 = 3.72$. In fact, the total workload $W = A \triangle (t_2 - t_1)$, and A is an upper bound of the asymptotic speedup to be defined below. ∎

Available Parallelism   There is a wide range of potential parallelism in application programs. Engineering and scientific codes exhibit a high DOP due to data parallelism. Manoj Kumar (1988) has reported that computation-intensive codes may execute 500 to 3500 arithmetic operations concurrently in each clock cycle in an idealized environment. Nicolau and Fisher (1984) reported that standard Fortran programs averaged about a factor of 90 parallelism available for very-long-instruction word architectures. These numbers show the optimistic side of available parallelism.

David Wall (1991) indicated that limits of instruction-level parallelism is around 5, rarely exceeding 7. Bulter et al. (1991) reported that when all constraints are removed, the DOP in programs may exceed 17 instructions per cycle. If the hardware is perfectly balanced, one can sustain from 2.0 to 5.8 instructions per cycle on a superscalar processor that is reasonably designed. These numbers show the pessimistic side of available parallelism.

The above measures of available parallelism show that computation that is less numeric than that in scientific codes has relatively little parallelism even when basic block boundaries are ignored. A *basic block* is a sequence or block of instructions in a program that has a single entry and a single exit points. While compiler optimization and algorithm redesign may increase the available parallelism in an application, limiting parallelism extraction to a basic block limits the potential instruction-level parallelism to a factor of about 2 to 5 in ordinary programs. However, the DOP may be pushed to thousands in some scientific codes when multiple processors are used simultaneously to exploit parallelism beyond the boundary of basic blocks.

Asymptotic Speedup Denote the amount of work executed with DOP $= i$ as $W_i = i \bigtriangleup t_i$ or we can write $W = \sum_{i=1}^{m} W_i$. The execution time of W_i on a single processor (sequentially) is $t_i(1) = W_i/\bigtriangleup$. The execution time of W_i on k processors is $t_i(k) = W_i/k\bigtriangleup$. With an infinite number of available processors, $t_i(\infty) = W_i/i\bigtriangleup$ for $1 \leq i \leq m$. Thus we can write the *response time* as

$$T(1) = \sum_{i=1}^{m} t_i(1) = \sum_{i=1}^{m} \frac{W_i}{\bigtriangleup} \tag{3.5}$$

$$T(\infty) = \sum_{i=1}^{m} t_i(\infty) = \sum_{i=1}^{m} \frac{W_i}{i\bigtriangleup} \tag{3.6}$$

The *asymptotic speedup* S_∞ is defined as the ratio of $T(1)$ to $T(\infty)$:

$$S_\infty = \frac{T(1)}{T(\infty)} = \frac{\sum_{i=1}^{m} W_i}{\sum_{i=1}^{m} W_i/i} \tag{3.7}$$

Comparing Eqs. 3.4 and 3.7, we realize that $S_\infty = A$ in the ideal case. In general, $S_\infty \leq A$ if communication latency and other system overhead are considered. Note that both S_∞ and A are defined under the assumption $n = \infty$ or $n >> m$.

3.1.2 Harmonic Mean Performance

Consider a parallel computer with n processors executing m programs in various modes with different performance levels. We want to define the mean performance of such multimode computers. With a weight distribution we can define a meaningful performance expression.

Different execution modes may correspond to scalar, vector, sequential, or parallel processing with different program parts. Each program may be executed with a combination of these modes. *Harmonic mean* performance provides an average performance across a large number of programs running in various modes.

Before we derive the harmonic mean performance expression, let us study the *arithmetic mean* and *geometric mean* performance expressions derived by James Smith (1988). The execution rate R_i for program i is measured in MIPS rate or Mflops rate, and so are the various performance expressions to be derived below.

Arithmetic Mean Performance   Let $\{R_i\}$ be the execution rates of programs $i = 1, 2, ..., m$. The *arithmetic mean execution rate* is defined as

$$R_a = \sum_{i=1}^{m} R_i/m \tag{3.8}$$

The expression R_a assumes equal weighting $(1/m)$ on all m programs. If the programs are weighted with a distribution $\pi = \{f_i \text{ for } i = 1, 2, ..., m\}$, we define a *weighted arithmetic mean execution rate* as follows:

$$R_a^* = \sum_{i=1}^{m} (f_i R_i) \tag{3.9}$$

Arithmetic mean execution rate is proportional to the sum of the inverses of execution times; it is not inversely proportional to the sum of execution times. Consequently, the arithmetic mean execution rate fails to represent the real times consumed by the benchmarks when they are actually executed.

Geometic Mean Performance   The *geometric mean execution rate* of m programs is defined by

$$R_g = \prod_{i=1}^{m} R_i^{1/m} \tag{3.10}$$

With a weight distribution $\pi = \{f_i | i = 1, 2, ..., m\}$, we can define a *weighted geometric mean execution rate* as follows:

$$R_g^* = \prod_{i=1}^{m} R_i^{f_i} \tag{3.11}$$

The geometric mean execution rate does not summarize the real performance either, since it does not have the inverse relation with the total time. Geometric mean has been advocated for use with performance numbers that are normalized with respect to a reference machine being compared with.

Harmonic Mean Performance   With the weakness of both arithmetic and geometric mean performance measures, we need to develop a mean performance expression based on arithmetic mean execution time. In fact, $T_i = 1/R_i$ is the mean execution time per instruction for program i. The *arithmetic mean execution time* per instruction is defined by

$$T_a = \frac{1}{m} \sum_{i=1}^{m} T_i = \frac{1}{m} \sum_{i=1}^{m} \frac{1}{R_i} \tag{3.12}$$

The *harmonic mean execution rate* across m benchmark programs is thus defined by the fact $R_h = 1/T_a$:

$$R_h = \frac{m}{\sum_{i=1}^{m} (1/R_i)} \tag{3.13}$$

Therefore, the harmonic mean performance is indeed related to the average execution time. With a weight distribution $\pi = \{f_i | i = 1, 2, ..., m\}$, we can define the *weighted harmonic mean execution rate* as:

$$R_h^* = \frac{1}{\sum_{i=1}^{m} (f_i/R_i)} \tag{3.14}$$

The above harmonic mean performance expressions correspond to the total number of operations (which have been lumped together as a constant unit time) divided by the total time. Among these means, the harmonic mean execution rate is the one closest to the real performance.

Harmonic Mean Speedup   Another way to apply the harmonic mean concept is to tie the various modes of a program to the number of processors used. Suppose a program (or a workload of multiple programs combined) is to be executed on an n-processor system. During the executing period, the program may use $i = 1, 2, ..., n$ processors in different time periods.

We say the program is executed in *mode i*, if i processors are used. The corresponding execution rate R_i is used to reflect the collective speed of i processors. Assume that $T_1 = 1/R_1 = 1$ is the sequential execution time on a uniprocessor with an execution rate $R_1 = 1$. Then $T_i = 1/R_i = 1/i$ is the execution time of using i processors with a combined execution rate of $R_i = i$ in the ideal case.

Suppose the given program is executed in n execution modes with a weight distribution $w = \{f_i \text{ for } i = 1, 2, ..., n\}$. A *weighted harmonic mean speedup* is defined as follows:

$$S = T_1/T^* = \frac{1}{(\sum_{i=1}^{n} f_i/R_i)} \tag{3.15}$$

where $T^* = 1/R_h^*$ is the *weighted arithmetic mean execution time* across the n execution modes, similar to that derived in Eq. 3.12.

Example 3.2 Harmonic mean speedup for a multiprocessor operating in n execution modes (Hwang and Briggs, 1984)

In Fig. 3.2, we plot Eq. 3.15 based on the assumption that $T_i = 1/i$ for all $i = 1, 2, \ldots, n$. This corresponds to the ideal case in which a unit-time job is done by i processors in minimum time. The assumption can also be interpreted as $R_i = i$ because the execution rate increases i times from $R_1 = 1$ when i processors are fully utilized without waste.

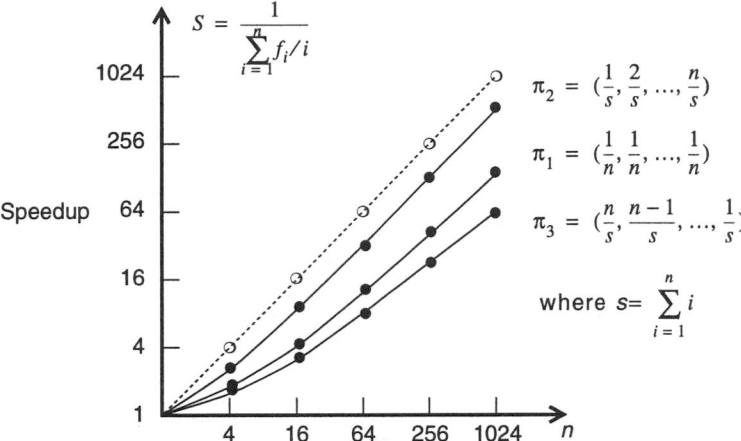

Figure 3.2 Harmonic mean speedup performance with respect to three probability distributions: π_1 for uniform distribution, π_2 in favor of using more processors, and π_3 in favor of using fewer processors

The three probability distributions π_1, π_2, and π_3 correspond to three processor utilization patterns. Let $s = \sum_{i=1}^{n} i$. $\pi_1 = (1/n, 1/n, \ldots, 1/n)$ corresponds to a uniform distribution over the n execution modes, $\pi_2 = (1/s, 2/s, \ldots, n/s)$ favors using more processors, and $\pi_3 = (n/s, (n-1)/s, \ldots, 2/s, 1/s)$ favors using fewer processors.

The ideal case corresponds to the 45° dashed line. Obviously, π_2 produces a higher speedup than π_1 does. The distribution π_1 is superior to the distribution π_3 in Fig. 3.2.

■

Amdahl's Law Using Eq. 3.15, one can derive Amdahl's law as follows: First, assume $R_i = i$, $w = (\alpha, 0, 0, \ldots, 0, 1 - \alpha)$; i.e., $w_1 = \alpha$, $w_n = 1 - \alpha$, and $w_i = 0$ for $i \neq 1$ and $i \neq n$. This implies that the system is used either in a pure sequential mode on one processor with a probability α, or in a fully parallel mode using n processors with a probability $1 - \alpha$. Substituting $R_1 = 1$ and $R_n = n$ and w into Eq. 3.15, we obtain the following speedup expression:

$$S_n = \frac{n}{1 + (n-1)\alpha} \tag{3.16}$$

This is known as Amdahl's law. The implication is that $S \to 1/\alpha$ as $n \to \infty$. In other words, under the above probability assumption, the best speedup one can expect is upper-bounded by $1/\alpha$, regardless of how many processors are employed.

In Fig. 3.3, we plot Eq. 3.16 as a function of n for four values of α. When $\alpha = 0$, the ideal speedup is achieved. As the value of α increases from 0.01 to 0.1 to 0.9, the speedup performance drops sharply.

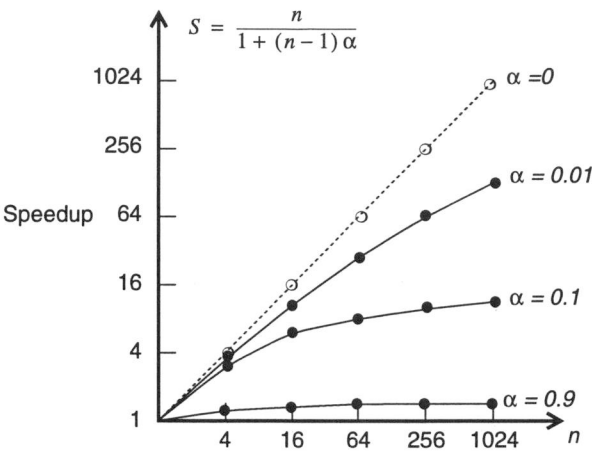

Figure 3.3 Speedup performance with respect to the probability distribution $\pi = (\alpha, 0, ..., 0, 1 - \alpha)$ where α is the fraction of sequential bottleneck

For many years, Amdahl's law has painted a very gloomy and pessimistic picture for parallel processing. That is, the system performance cannot be high as long as the serial fraction α exists. We will further examine Amdahl's law in Section 3.3.1 from the perspective of workload growth.

3.1.3 Efficiency, Utilization, and Quality

Ruby Lee (1980) has defined several parameters for evaluating parallel computations. These are fundamental concepts in parallel processing. Tradeoffs among these performance factors are often encountered in real-life applications.

System Efficiency   Let $O(n)$ be the total number of unit operations performed by an n-processor system and $T(n)$ be the execution time in unit time steps. In general, $T(n) < O(n)$ if more than one operation is performed by n processors per unit time, where $n \geq 2$. Assume $T(1) = O(1)$ in a uniprocessor system. The *speedup factor* is defined as

$$S(n) = T(1)/T(n) \tag{3.17}$$

The *system efficiency* for an n-processor system is defined by

$$E(n) = \frac{S(n)}{n} = \frac{T(1)}{nT(n)} \tag{3.18}$$

Efficiency is an indication of the actual degree of speedup performance achieved as compared with the maximum value. Since $1 \leq S(n) \leq n$, we have $1/n \leq E(n) \leq 1$.

The lowest efficiency corresponds to the case of the entire program code being executed sequentially on a single processor. The maximum efficiency is achieved when all n processors are fully utilized throughout the execution period.

Redundancy and Utilization   The *redundancy* in a parallel computation is defined as the ratio of $O(n)$ to $O(1)$:

$$R(n) = O(n)/O(1) \qquad (3.19)$$

This ratio signifies the extent of matching between software parallelism and hardware parallelism. Obviously $1 \leq R(n) \leq n$. The *system utilization* in a parallel computation is defined as

$$U(n) = R(n)E(n) = \frac{O(n)}{nT(n)} \qquad (3.20)$$

The system utilization indicates the percentage of resources (processors, memories, etc.) that was kept busy during the execution of a parallel program. It is interesting to note the following relationships: $1/n \leq E(n) \leq U(n) \leq 1$ and $1 \leq R(n) \leq 1/E(n) \leq n$.

Quality of Parallelism   The *quality* of a parallel computation is directly proportional to the speedup and efficiency and inversely related to the redundancy. Thus, we have

$$Q(n) = \frac{S(n)E(n)}{R(n)} = \frac{T^3(1)}{nT^2(n)O(n)} \qquad (3.21)$$

Since $E(n)$ is always a fraction and $R(n)$ is a number between 1 and n, the quality $Q(n)$ is always upper-bounded by the speedup $S(n)$.

Example 3.3   A hypothetical workload and performance plots

In Fig. 3.4, we compare the relative magnitudes of $S(n), E(n), R(n), U(n),$ and $Q(n)$ as a function of machine size n, with respect to a hypothetical workload characterized by $O(1) = T(1) = n^3, O(n) = n^3 + n^2 \log_2 n,$ and $T(n) = 4n^3/(n+3)$.

Substituting these measures into Eqs. 3.17 to 3.21, we obtain the following performance expressions:

$$
\begin{aligned}
S(n) &= (n+3)/4 \\
E(n) &= (n+3)/(4n) \\
R(n) &= (n + \log_2 n)/n \\
U(n) &= (n+3)(n + \log_2 n)/(4n^2) \\
Q(n) &= (n+3)^2/(16(n + \log_2 n))
\end{aligned}
$$

The relationships $1/n \leq E(n) \leq U(n) \leq 1$ and $0 \leq Q(n) \leq S(n) \leq n$ are observed where the linear speedup corresponds to the ideal case of 100% efficiency.

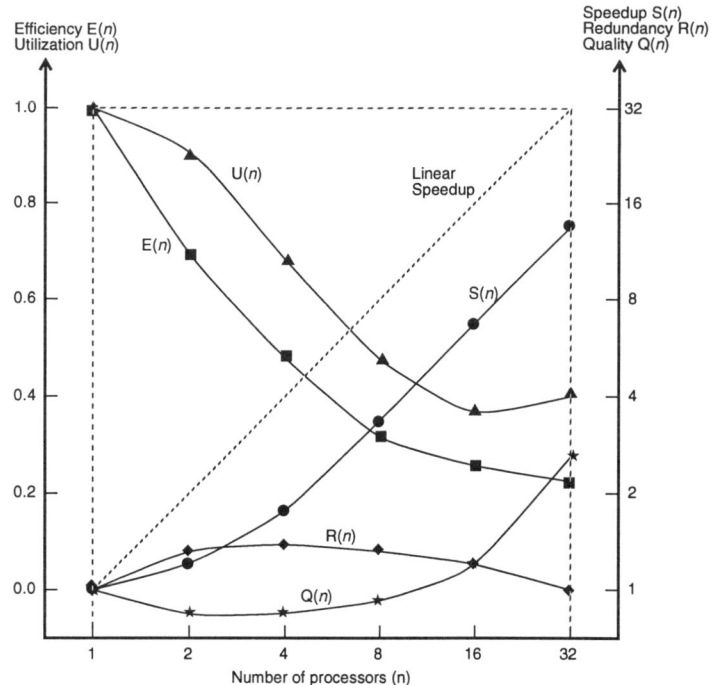

Figure 3.4 Performance measures for Example 3.3 on a parallel computer with up to 32 processors.

To summarize the above discussion on performance indices, we use the speedup $S(n)$ to indicate the degree of speed gain in a parallel computation. The efficiency $E(n)$ measures the useful portion of the total work performed by n processors. The redundancy $R(n)$ measures the extent of workload increase.

The utilization $U(n)$ indicates the extent to which resources are utilized during a parallel computation. Finally, the quality $Q(n)$ combines the effects of speedup, efficiency, and redundancy into a single expression to assess the relative merit of a parallel computation on a computer system.

The speedup and efficiency of 10 parallel computers are reported in Table 3.1 for solving a linear system of 1000 equations. The table entries are excerpts from Table 2 in Dongarra's report (1992) on LINPACK benchmark performance over a large number of computers.

Either the standard LINPACK algorithm or an algorithm based on matrix-matrix multiplication was used in these experiments. A high degree of parallelism is embedded in these experiments. Thus high efficiency (such as 0.94 for the IBM 3090/600S VF and 0.95 for the Convex C3240) was achieved. The low efficiency on the Intel Delta was based on some initial data.

Table 3.1 Speedup and Efficiency of Parallel Computers for Solving a Linear System with 1000 Unknowns

Computer Model	No. of Processors n	Uni-processor Timing T_1 (s)	Multi-processor Timing T_n (s)	Speedup $S = T_1/T_n$	Efficiency $E = S/n$
Cray Y-MP C90	16	0.77	0.069	11.12	0.69
NEC SX-3	2	0.15	0.082	1.82	0.91
Cray Y-MP/8	8	2.17	0.312	6.96	0.87
Fujitsu AP 1000	512	160.0	1.10	147.0	0.29
IBM 3090/600S VF	6	7.27	1.29	5.64	0.94
Intel Delta	512	22.0	1.50	14.7	0.03
Alliant FX/2800-200	14	22.9	2.06	11.1	0.79
nCUBE/2	1024	331.0	2.59	128.0	0.12
Convex C3240	4	14.9	3.92	3.81	0.95
Parsytec FT-400	400	1075.0	4.90	219.0	0.55

Source: Jack Dongarra, "Performance of Various Computers Using Standard Linear Equations Software," Computer Science Dept., Univ. of Tennessee, Knoxville, TN 37996-1301, March 11, 1992.

3.1.4 Standard Performance Measures

We have used MIPS and Mflops to describe the *instruction execution rate* and *floating-point capability* of a parallel computer. The MIPS rate defined in Eq. 1.3 is calculated from clock frequency and average CPI. In practice, the MIPS and Mflops ratings and other performance indicators to be introduced below should be measured from running benchmarks or real programs on real machines.

In this section, we introduce standard measures adopted by the industry to compare various computer performance, including *Mflops, MIPS, KLIPS, Dhrystone*, and *Whestone*, as often encountered in reported computer ratings.

Most computer manufacturers state peak or sustained performance in terms of MIPS or Mflops. These ratings are by no means conclusive. The real performance is always program-dependent or application-driven. In general, the MIPS rating depends on the instruction set, varies between programs, and even varies inversely with respect to performance, as observed by Hennessy and Patterson (1990).

To compare processors with different clock cycles and different instruction sets is not totally fair. Besides the native MIPS, one can define a *relative MIPS* with respect to a reference machine. We will discuss relative MIPS rating against the VAX/780 when Dhrystone performance is introduced below. For numerical computing, the LINPACK results on a large number of computers are reported in Chapter 8.

Similarly, the Mflops rating depends on the machine hardware design and on the program behavior. MIPS and Mflops ratings are not convertible because they measure different ranges of operations. The conventional rating is called the *native Mflops*, which does not distinguish unnormalized from normalized floating-point operations.

For example, a *real floating-point divide* operation may correspond to four *normalized floating-point divide* operations. One needs to use a conversion table between real and normalized floating-point operations to convert a native Mflops rating to a normalized Mflops rating.

The Dhrystone Results   This is a CPU-intensive benchmark consisting of a mix of about 100 high-level language instructions and data types found in system programming applications where floating-point operations are not used. The Dhrystone statements are balanced with respect to statement type, data type, and locality of reference, with no operating system calls and making no use of library functions or subroutines. Thus the Dhrystone rating should be a measure of the integer performance of modern processors or computers. The unit *KDhrystones/s* is often used in reporting Dhrystone results.

The Dhrystone benchmark version 1.1 was applied to a number of processors. The results were reported by National Semiconductor and are summarized in Table 3.2. The VAX 11/780 scored a 1.7 KDhrystones/s performance. This machine has been used as a reference computer with a 1-MIPS performance. Digital Equipment uses *VAX units of performance* (VUP), equivalent to one relative MIPS, which corresponds to 1.7 KDhrystones/s in the version 1.1 result.

The right column of Table 3.2 shows the *VUP* or *relative MIPS rates* of various computers with respect to the 1.0-MIPS rating for the VAX11/780. The Dhrystone benchmark varies with different versions. In a version 2.1 tested by Intel, the VAX 11/780 scored 1.6 KDhrystones/s. The relative VAX/MIPS rating is commonly accepted by the computer industry.

The Whestone Results   This is a Fortran-based synthetic benchmark assessing the floating-point performance, measured in the number of *KWhestones/s* that a system can perform. The benchmark includes both integer and floating-point operations involving array indexing, subroutine calls, parameter passing, conditional branching, and trigonometric/transcendental functions.

The Whestone benchmark does not contain any vectorizable code and shows dependence of the system's mathematics library and efficiency of the code generated by a compiler.

Table 3.3 shows the relative Whestone results with single-precision floating-point operations reported by DEC in 1986. The Whestone performance is not equivalent to the Mflops performance, although the Whestone contains a large number of scalar floating-point operations.

Both the Dhrystone and Whestone are synthetic benchmarks whose performance results depend heavily on the compilers used. As a matter of fact, the Dhrystone was originally written to test the CPU and compiler performance for a typical program. Compiler techniques, especially procedure in-lining, can significantly affect the Dhrystone performance.

Both benchmarks were criticized for being unable to predict the performance of user programs. The sensitivity to compilers is a major drawback of these benchmarks. In real-life problems, only application-oriented benchmarks will do the trick. We will examine the SPEC and the Perfect Club benchmark suites in Chapter 9.

Table 3.2 Dhrystone Version 1.1 Results and Relative MIPS Performance of Some Systems

System	KDhrystone/s	Relative MIPS
VAX 11/780	1.7	1.0
VAX 8600	6.4	3.7
Cray 1S	14.8	8.4
IBM 3081	15.0	8.5
Sun4/260(SPARC)	16.8	9.8
Cray X/MP	18.5	10.5
SPARC, 16.7 MHz	19.0	10.8
IBM 3090/200	31.2	17.8
M88100, 20 MHz	35.7	20.3
MIPS R3000, 25 MHz	43.1	24.5
Intel i860, 33.3 MHz	72.5	41.2
MC68040,25MHz	40.0	23.5
i80486, 25MHz	24.0	14.1
AMD 29000,25MHz	21.1	12.4
NS 32532,25MHz	18.3	10.7
IBM RS/6000,25MHz	60.7	35.7

Source: Weicker (1984), Cocke and Markstein (1990), and National Semiconductor (1988).

The TPS and KLIPS Ratings   On-line transaction processing applications demand rapid, interactive processing for a large number of relatively simple transactions. They are typically supported by very large databases. Automated teller machines and airline reservation systems are familiar examples.

The throughput of computers for on-line transaction processing is often measured in *transactions per second* (TPS). Each transaction may involve a database search, query answering, and database update operations. Business computers should be designed to deliver a high TPS rate. The TP1 benchmark was originally proposed in 1985 for measuring the transaction processing of business application computers. This benchmark

Table 3.3 Some Whestone Results, Single Precision, Reported by DEC in January 1986

System	KWhestones/s
HP3000/930	2.84
IBM 4381/11	2.00
DEC 11/780	1.15
Sun 3/50	0.86
IBM RT/PC	0.20

Source: DEC, *Digital Review*, January 1986; all others: company claims.

has also become a standard for gauging relational database performances. The TP1 benchmark is closely related to the Debit Credit benchmark.

Based on today's standards, any computer exceeding 100 TPS is considered rather fast for transaction processing. For example, the VAX 9000 executed 70 TPS per processor. The Sequent Symmetry multiprocessor achieved 140 TPS using a shared-memory system with 24 processors.

In artificial intelligence applications, the measure KLIPS (*kilo logic inferences per second*) is often used to indicate the reasoning power of an AI machine. For example, the high-speed inference machine developed in Japan's Fifth-Generation Computer System Project claimed a performance of 400 KLIPS.

Each logic inference operation involves about 100 assembly instructions. Thus 400 KLIPS implies approximately 40 MIPS in this sense. The conversion ratio is by no means fixed. Logic inference demands symbolic manipulations rather than numeric computations. Logic inference machines are beyond the scope of this book. Interested readers are referred to the book edited by Wah and Ramamoorthy (1990).

3.2   Parallel Processing Applications

Massively parallel processing becomes one of the frontier challenges in supercomputer applications. We introduce grand challenges in high-performance computing and communications and then assess the speed, memory, and I/O requirements to meet these challenges. Characteristics of parallel algorithms are also discussed in this context.

3.2.1   Massive Parallelism for Grand Challenges

The definition of massive parallelism varies with respect to time. Based on today's standards, any machine having hundreds or thousands of processors is a *massively parallel processing* (MPP) system. As computer technology advances rapidly, the demand for a higher degree of parallelism becomes more obvious.

The performance of most commercial computers is marked by their peak MIPS rate or peak Mflops rate. In reality, only a fraction of the peak performance is achievable in real benchmark or evaluation runs. Observing the sustained performance makes more sense in evaluating computer performance.

Grand Challenges   Based on grand challenges in scientific computing, we need to provide computers with 1 Teraflops of computing power, 1 Terabyte of main memory, and 1 Terabyte/s of I/O bandwidth. This has been called the goal of *3T performance*. Today's most powerful computers are about 100 to 10,000 times slower in speed, shorter in memory capacity, and narrower in I/O bandwidth than what can satisfy these grand challenges.

We review below the grand challenges identified in the U.S. High-Performance Computing and Communication (HPCC) program, reveal opportunities for massive parallelism, assess past developments, and comment on future trends in MPP. These grand challenge application areas are summarized in Table 3.4.

(1) The magnetic recording industry relies on the use of computers to study megne-

Table 3.4 Grand Challenge Applications posted by the US High-Performance Computing and Communication Program

Application Area	Computational Tasks and Expected Results
Magnetic recording technology	To study magnetostatic and exchange interactions to reduce noise in high-density disks.
Rational drug design	To develop drug to cure cancer or AIDS by blocking the action of HIV protease.
High-speed civil transport	To develop supersonic jets through computational fluid dynamics running on supercomputers.
Catalysis	Computer modeling of biomimetic catalysts to analyze enzymatic reactions in manufacturing process.
Fuel combustion	Designing better engine models via chemical kinetics calculations to reveal fluid mechanical effects.
Ocean modeling	Large-scale simulation of ocean activities and heat exchange with atmospherical flows.
Ozone depletion	To study chemical and dynamical mechanisms controlling the ozone depletion process.
Digital anatomy	Real-time, clinical imaging, computed tomography, magnetic resonance imaging with computers.
Air pollution	Simulated air quality models running on supercomputers to provide more understanding of atmospheric systems.
Protein structure design	3-D structural study of protein formation by using MPP system to perform computational simulations.
Image understanding	Use large supercomputers for producing the rendered images or animations in real time.
Technology linking research to education	Scientific or engineering education aided by computer simulation in heterogeneous network systems.

tostatic and exchange interactions in order to reduce noise in metallic thin films used to coat high-density disks.

(2) Rational drug design is being aided by computers in the search for a cure for cancer and acquired immunodeficiency syndrome. Using a high-performance computer, new potential agents have been identified that block the action of human immunodeficiency virus protease.

(3) High-speed civil transport aircrafts are being aided by computational fluid dynamics running on supercomputers. Fuel combustion can be made more efficient by designing better engine models through chemical kinetics calculations.

(4) Catalysts for chemical reactions are being designed with computers for many biological processes which are catalytically controlled by enzymes. Massively parallel quantum models demand large simulations to reduce the time required to design catalysts and to optimize their properties.

(5) Ocean modeling cannot be accurate without supercomputing MPP systems. Ozone depletion research demands the use of computers in analyzing the complex chemical and dynamical mechanisms involved. Both ocean and ozone modeling affect the global climate forecast.

(6) Other important areas demanding computational support include digital anato-
my in real-time medical diagnosis, air pollution reduction through computational
modeling, the design of protein structures by computational biologists, image
processing and understanding, and technology linking research to education.

Besides computer science and computer engineering, the above challenges also encourage the merging discipline of computational science and engineering. This demands
systematic application of computer systems and computational solution techniques to
mathematical models formulated to describe and to simulate phenomena of scientific
and engineering interest.

The HPCC Program has identified some grand challenge computing requirements,
as shown in Fig. 3.5. This diagram shows the levels of processing speed and memory
size required to support scientific simulation modeling, advanced *computer-aided design* (CAD), and real-time processing of large-scale database and information retrieval
operations.

Exploiting Massive Parallelism   The parallelism embedded in the instruction level
or procedural level is rather limited. Very few parallel computers can successfully execute more than two instructions per machine cycle from the same program.

Instruction parallelism is often constrained by program behavior, compiler/OS in-
capabilities, and program flow and execution mechanisms built into modern computers.
This is reflected by the fact that only a few multiprocessors have been built for commercial use in the past.

On the other hand, *data parallelism* is much higher than instruction parallelism.
Data parallelism refers to the situation where the same operation (instruction or program) executes over a large array of data (operands). Data parallelism has been implemented on pipelined vector processors, SIMD array processors, and SPMD or MPMD
multicomputer systems.

In Table 1.6, we find SIMD data parallelism over 65,536 PEs in the CM-2. One
may argue that the CM-2 is a bit-slice machine. Even if we divide the number by 64
(the word length of a typical supercomputer), we still end up with a DOP on the order
of thousands in CM-2.

The vector length can be used to determine the parallelism implementable on a
vector supercomputer. In the case of the Cray Y/MP C-90, 32 pipelines in 16 processors
can potentially achieve a DOP of $32 \times 5 = 160$ if the average pipeline has five stages.
Thus a pipelined processor can support a lower degree of data parallelism than an SIMD
computer.

On a message-passing multicomputer, the parallelism is more scalable than a shared-
memory multiprocessor. As revealed in Table 1.4, the nCUBE/2 can achieve a maximum parallelism on the order of thousands if all the node processors are kept busy
simultaneously.

The Past and the Future   MPP systems started in 1968 with the introduction of
the Illiac IV computer with 64 PEs under one controller. Subsequently, a massively

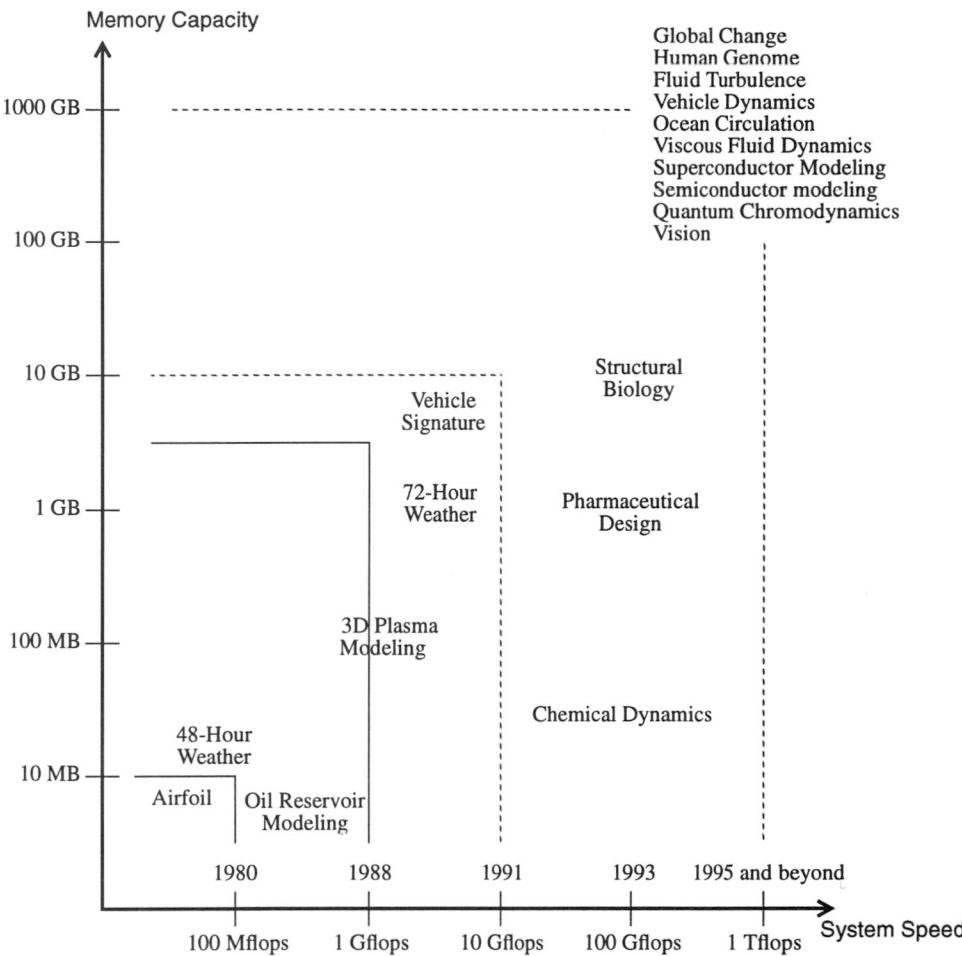

Figure 3.5 Grand challenge requirements in computing and communications. (Courtesy of U.S. High-Performance Computing and Communication Program, 1992)

parallel processor, called MPP, was built by Goodyear with 16,384 PEs. IBM built a GF11 machine with 576 PEs. The MasPar MP-1, AMT DAP610, and CM-2 are all SIMD computers.

MPP systems operating in MIMD mode include the BBN TC-2000 with a maximum configuration of 512 processors. The IBM RP-3 was designed to have 512 processors (only a 64-processor version was built). The Intel Touchstone Delta has a system with 570 processors. There is no doubt that nCUBE, MasPar, and BBN Advanced Computer are striving to extend their current models to even larger machines.

Several recently announced MPP projects include the Paragon by Intel Supercomputer Systems, the CM-5 by Thinking Machine Corporation, the KSR-1 by Kendall Square Research, the Fujitsu VPP500 System, the Tera computer, and the MIT *T system.

Recently, IBM announced an MPP project using thousands of IBM RS/6000 processors. Cray Research plans to develop an MPP system using Digital's Alpha processors as building blocks. These MPP projects are summarized in Table 3.5. We will study some of these systems in later chapters.

Table 3.5 Representative Massively Parallel Processing Systems

MPP System	Architecture, Technology, and Operational Features
Intel Paragon	A 2-D mesh-connected multicomputer, built with i860 XP processors and wormhole routers, targeted for 300 Gflops in peak performance.
IBM MPP Model	Use IBM RISC/6000 processors as building blocks, 50 Gflops peak expected for a 1024-processor configuration.
TMC CM-5	A universal architecture for SIMD/MIMD computing using SPARC PEs and custom-designed FPUs, control and data networks, 2 Tflops peak for 16K nodes.
Cray Research MPP Model	A 3D torus heterogeneous architecture using DEC Alpha chips with special communication support, global address space over physically distributed memory; first system offers 150 Gflops in a 1024-processor configuration in 1993; will eventually grow to Tflops with larger configurations.
Kendall Square Research KSR-1	An ALLCACHE ring-connected multiprocessor with custom-designed processors, 43 Gflops peak performance for a 1088-processor configuration.
Fujitsu VPP500	A crossbar-connected 222-PE MIMD vector system, with shared distributed memories using VP2000 as a host; peak performance = 355 Gflops; first delivery in September 1993.

3.2.2 Application Models of Parallel Computers

In general, if the workload or problem size s is kept unchanged as shown by curve α in Fig. 3.6a, then the efficiency E decreases rapidly as the machine size n increases. The reason is that the overhead h increases faster than the machine size. To maintain the efficiency at a desired level, one has to increase the machine size and problem size proportionally. Such a system is known as a *scalable computer* for solving *scaled problems*.

In the ideal case, we like to see a workload curve which is a linear function of n (curve γ in Fig. 3.6a). This implies linear scalability in problem size. If the linear workload curve is not achievable, the second choice is to achieve a sublinear scalability as close to linearity as possible, as illustrated by curve β in Fig. 3.6a.

Suppose that the workload follows an exponential growth pattern and becomes

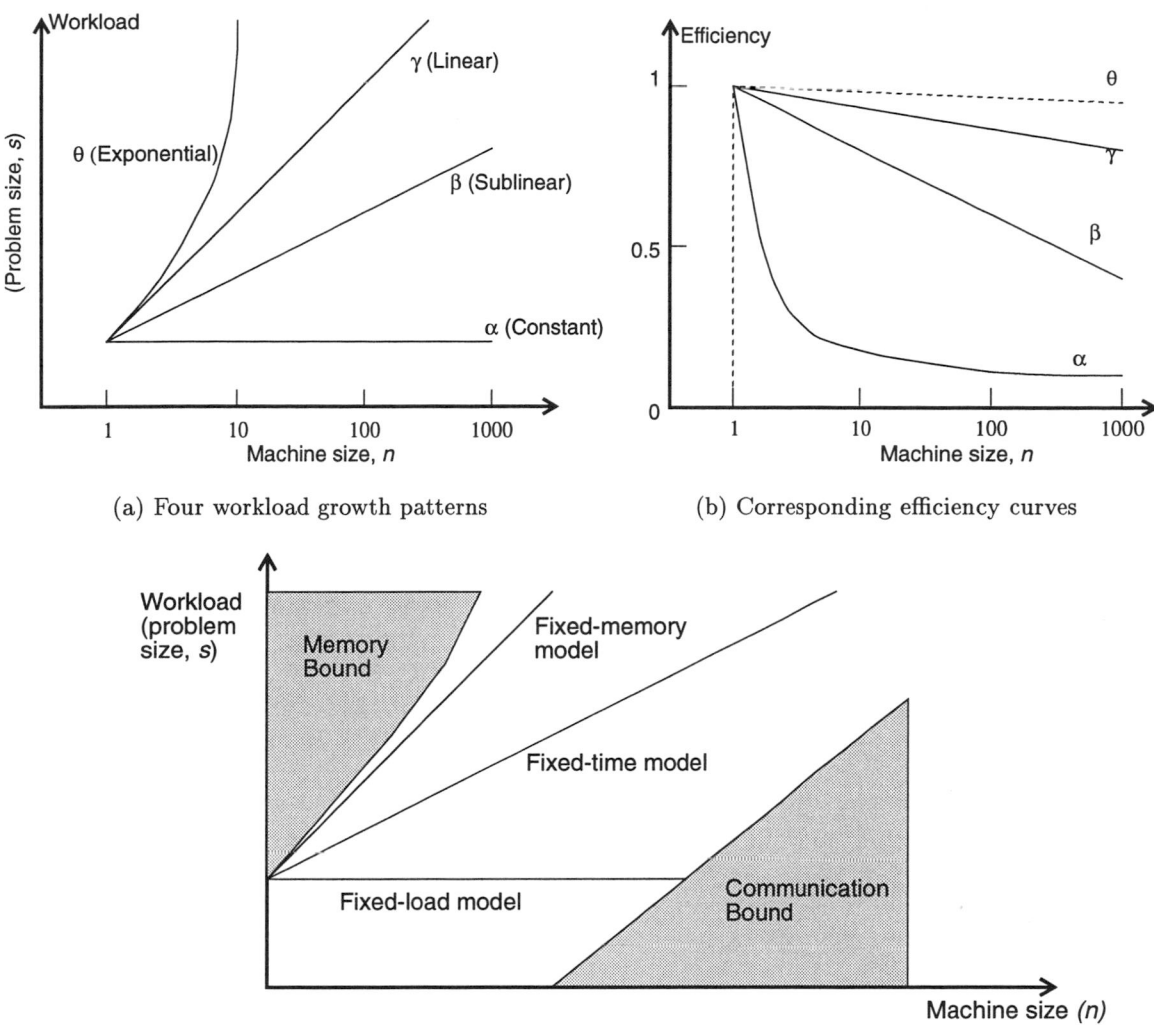

(a) Four workload growth patterns

(b) Corresponding efficiency curves

(c) Application models for parallel computers

**Figure 3.6 Workload growth, efficiency curves, and application models of par-
allel computers under resources constraints.**

enormously large, as shown by curve θ in Fig. 3.6a. The system is considered poorly
scalable in this case. The reason is that to keep a constant efficiency or a good speedup,
the increase in the problem size must be explosive to exceed the memory or I/O limits.
The problem size is allowed to grow only within the limits of the available memory on
the computer.

The Efficiency Curves   Corresponding to the four workload patterns specified in
Fig. 3.6a, four efficiency curves are shown in Fig. 3.6b, respectively. With a constant

workload, the efficiency curve (α) drops rapidly. In fact, curve α corresponds to the famous Amdahl's law. For a linear workload, the efficiency curve (γ) is almost flat, as observed by Gustafson in 1988.

The exponential workload (θ) may not be implementable due to memory shortage or I/O bounds (if real-time application is considered). Thus the θ efficiency (dashed lines) is achievable only with exponentially increased memory (or I/O) capacity. The sublinear efficiency curve (β) lies somewhere between curves α and γ.

Scalability analysis determines whether parallel processing of a given problem can offer the desired improvement in performance. The analysis should help guide the design of a massively parallel processor. It is clear that no single scalability metric suffices to cover all possible cases. Different measures will be useful in different contexts, and further research is needed along multiple dimensions.

A parallel system can be used to solve arbitrarily large problems in a fixed time if and only if its workload pattern is allowed to grow linearly. Sometimes, even if a minimum time is achieved with more processors, the system utilization (or efficiency) can be very poor.

Application Models   The workload patterns shown in Fig. 3.6a are not allowed to grow unbounded. In Fig. 3.6c, we show three models for the application of parallel computers. These models are bounded by limited memory, limited tolerance of IPC latency, or limited I/O bandwidth. These models are briefly introduced below. They lead to three speedup performance models to be formulated in Section 3.3.

The *fixed-load model* corresponds to a constant workload (curve α in Fig. 3.6a) The use of this model is eventually limited by the communication bound shown by the shaded area in Fig. 3.6c.

The *fixed-time model* demands a constant program execution time, regardless of how the workload scales up with machine size. The linear workload growth (curve γ in Fig. 3.6a) corresponds to this model. The *fixed-memory model* is limited by the memory bound, corresponding to a workload curve between γ and θ in Fig. 3.6a.

From the application point of view, the shaded areas are forbidden. The communication bound includes not only the increasing IPC overhead but also the increasing I/O demands. The memory bound is determined by main memory and disk capacities.

In practice, an algorithm designer or a parallel computer programmer may choose an application model within the above resource constraints, as shown in the unshaded application region in Fig. 3.6c.

Tradeoffs in Scalability Analysis   Computer cost c and programming overhead p (in addition to speedup and efficiency) are equally important in scalability analysis. After all, cost-effectiveness may impose the ultimate constraint on computing with a limited budget. What we have studied above was concentrated on system efficiency and fast execution of a single algorithm/program on a given parallel computer.

It would be interesting to extend the scalability analysis to multiuser environments in which multiple programs are executed concurrently by sharing the available resources. Sometimes one problem is poorly scalable, while another has good scalability characteristics. Tradeoffs exist in increasing resource utilization but not necessarily to minimize the overall execution time in an optimization process.

Exploiting parallelism for higher performance demands both scalable architectures and scalable algorithms. The architectural scalability can be limited by long communication latency, bounded memory capacity, bounded I/O bandwidth, and limited CPU speed. How to achieve a balanced design among these practical constraints is the major challenge of today's MPP designers. On the other hand, parallel algorithms and efficient data structures also need to be scalable.

3.2.3 Scalability of Parallel Algorithms

In this subsection, we analyze the scalability of parallel algorithms with respect to key machine classes. An isoefficiency concept is introduced for scalability analysis of parallel algorithms. Two examples are used to illustrate the idea. Further studies of scalability are given in Section 3.4 after we study the speedup performance laws in Section 3.3.

Algorithmic Characteristics Computational algorithms are traditionally executed sequentially on uniprocessors. *Parallel algorithms* are those specially devised for parallel computers. The idealized parallel algorithms are those written for the PRAM models if no physical constraints or communication overheads are imposed. In the real world, an algorithm is considered efficient only if it can be cost effectively implemented on physical machines. In this sense, all machine-implementable algorithms must be architecture-dependent. This means the effects of communicaiton overhead and architectural constraints cannot be ignored.

We summarize below important characteristics of parallel algorithms which are machine implementable:

(1) *Deterministic versus nondeterministic*: As defined in Section 1.4.1, only deterministic algorithms are implementable on real machines. Our study is confined to deterministic algorithms with polynomial time complexity

(2) *Computational granularity*: As introduced in Section 2.2.1, granularity decides the size of data items and program modules used in computation. In this sense, we also classify algorithms as fine-grain, medium-grain, or coarse-grain.

(3) *Parallelism profile*: The distribution of the degree of parallelism in an algorithm reveals the opportunity for parallel processing. This often affects the effectiveness of the parallel algorithms.

(4) *Communication patterns and synchronization requirements*: Communication patterns address both memory access and interprocessor communications. The patterns can be *static* or *dynamic*, depending on the algorithms. Static algorithms are more suitable for SIMD or pipelined machines, while dynamic algorithms are for MIMD machines. The *synchronization frequency* often affects the efficiency of an algorithm.

(5) *Uniformity of the operations*: This refers to the types of fundamental operations to be performed. Obviously, if the operations are uniform across the data set, the SIMD processing or pipelining may be more desirable. In other words, randomly structured algorithms are more suitable for MIMD processing. Other related issues include data types and precision desired.

(6) *Memory requirement and data structures*: In solving large-scale problems, the data sets may require huge memory space. *Memory efficiency* is affected by data structures chosen and data movement patterns in the algorithms. Both time and space complexities are key measures of the granularity of a parallel algorithm.

The Isoefficiency Concept   The workload w of an algorithm often grows in the order $O(s)$, where s is the problem size. Thus, we denote the workload $w = w(s)$ as a function of s. Kumar and Rao (1987) have introduced an *isoefficiency* concept relating workload to machine size n needed to maintain a fixed efficiency E when implementing a parallel algorithm on a parallel computer. Let h be the total communication overhead involved in the algorithm implementation. This overhead is usually a function of both machine size and problem size, thus denoted $h = h(s, n)$.

The efficiency of a parallel algorithm implemented on a given parallel computer is thus defined as

$$E = \frac{w(s)}{w(s) + h(s,n)} \tag{3.22}$$

The workload $w(s)$ corresponds to useful computations while the overhead $h(s,n)$ are useless computations attributed to synchronization and data communications delays. In general, the overhead increases with respect to both increasing values of s and n. Thus, the efficiency is always less than 1. The question is hinged on relative growth rates between $w(s)$ and $h(s,n)$.

With a fixed problem size (or fixed workload), the efficiency decreases as n increase. The reason is that the overhead $h(s,n)$ increases with n. With a fixed machine size, the overhead h grows slower than the workload w. Thus the efficiency increases with increasing problem size for a fixed-size machine. Therefore, one can expect to maintain a constant efficiency if the workload w is allowed to grow properly with increasing machine size.

For a given algorithm, the workload w might need to grow polynomially or exponentially with respect to n in order to maintain a fixed efficiency. Different algorithms may require different workload growth rates to keep the efficiency from dropping, as n is increased. The isoefficiency functions of common parallel algorithms are polynomial functions of n; i.e., they are $O(n^k)$ for some $k \geq 1$. The smaller the power of n in the isoefficiency function, the more scalable the parallel system. Here, the system includes the algorithm and architecture combination.

Isoefficiency Function   We can rewrite Eq. 3.22 as $E = 1/(1 + h(s,n)/w(s))$. In order to maintain a constant E, the workload $w(s)$ should grow in proportion to the overhead $h(s,n)$. This leads to the following condition:

$$w(s) = \frac{E}{1 - E} \times h(s,n) \tag{3.23}$$

The factor $C = E/(1 - E)$ is a constant for a fixed efficiency E. Thus we can define the *isoefficiency function* as follows:

$$f_E(n) = C \times h(s, n) \tag{3.24}$$

If the workload $w(s)$ grows as fast as $f_E(n)$ in Eq. 3.23, then a constant efficiency can be maintained for a given algorithm-architecture combination. Two examples are given below to illustrate the use of isoefficiency functions for scalability analysis.

Example 3.4 Scalability of matrix multiplication algorithms (Gupta and Kumar, 1992)

Four algorithms for matrix multiplication are compared below. The problem size s is represented by the matrix order. In other words, we consider the multiplication of two $s \times s$ matrices A and B to produce an output matrix C = A $\times$ B. The total workload involved is $w = O(s^3)$. The number of processors used is confined within $1 \leq n \leq s^3$. Some of the algorithms may use less than s^3 processors.

Table 3.6 Asymptotic Isoefficiency Functions of Four Matrix Multiplication Algorithms (Gupta and Kumar, 1992)

Matrix Multiplication Algorithm	Isoefficiency Function $f_E(n)$	Range of Applicability	Target Machine Architecture
Fox, Otto, and Hey (1987)	$O(n^{3/2})$	$1 \leq n \leq s^2$	A $\sqrt{n} \times \sqrt{n}$ torus
Berntsen (1989)	$O(n^2)$	$1 \leq n \leq s^{3/2}$	A hypercube with $n = 2^{3k}$ nodes
Gupta and Kumar (1992)	$O(n(\log n)^3)$	$1 \leq n \leq s^3$	A hypercube with $n = 2^{3k}$ nodes and $k < \frac{1}{3} \log s$
Dekel, Nassimi, and Sahni (1981)	$O(n \log n)$	$s^2 \leq n \leq s^3$	A hypercube with $n = s^3 = 2^{3k}$ nodes

Note: Two $s \times s$ matrices are multiplied.

The isoefficiency functions of the four algorithms are derived below based on equating the workload with the communication overhead (Eq. 3.23) in each algorithm. Details of these algorithms and corresponding architectures can be found in the original papers identified in Table 3.6 as well as in the paper by Gupta and Kumar (1992). The derivation of the communication overheads is left as an exercise in Problem 3.14.

The Fox-Otto-Hey algorithm has a total overhead $h(s, n) = O(n \log n + s^2 \sqrt{n})$. The workload $w = O(s^3) = O(n \log n + s^2 \sqrt{n})$. Thus we must have $O(s^3) = O(n \log n)$ and $O(s) = O(\sqrt{n})$. Combining the two, we obtain the isoefficiency function $O(s^3) = O(n^{3/2})$, where $1 \leq n \leq s^2$ as shown in the first row of Table 3.6.

Although this algorithm is written for the torus architecture, the torus can be easily embedded in a hypercube architecture. Thus we can conduct a fair comparison of the four algorithms against the hypercube architecture.

Berntsen's algorithm restricts the use of $n \leq s^{3/2}$ processors. The total overhead is $O(n^{4/3} + n \log n + s^2 n^{1/3})$. To match this with $O(s^3)$, we must have $O(s^3) = O(n^{4/3})$ and $O(s^3) = O(n)$. Thus, $O(s^3)$ must be chosen to yield the isoefficiency function $O(n^2)$.

The Gupta-Kumar algorithm has an overhead $O(n \log n + s^2 n^{1/3} \log n)$. Thus we must have $O(s^3) = O(n \log n)$ and $O(s^3) = O(s^2 n^{1/3} \log n)$. This leads to the isoefficiency function $O(n(\log n)^3)$ in the third row of Table 3.6.

The Dekel-Nassimi-Sahni algorithm has a total overhead $O(n \log n + s^3)$ besides a useful computation time of $O(s^3/n)$ for $s^2 \leq n \leq s^3$. Thus the workload growth $O(s^3) = O(n \log n)$ will yield the isoefficiency listed in the last row of Table 3.6.

∎

The above isoefficiency functions indicate the asymptotic scalabilities of the four algorithms. In practice, none of the algorithms is strictly better than the others for all possible problem sizes and machine sizes. For example, when these algorithms are implemented on a multicomputer with a long communication latency (as in Intel iPSC1), Berntsen's algorithm is superior to the others.

To map the algorithms on an SIMD computer (like the CM-2) with an extremely low synchronization overhead, the algorithm by Gupta and Kumar is inferior to the others. Hence, it is best to use the Dekel-Nassimi-Sahni algorithm for $s^2 \leq n \leq s^3$, the Fox-Otto-Hey algorithm for $s^{3/2} \leq n \leq s^2$, and Berntsen's algorithm for $n \leq s^{3/2}$ for SIMD hypercube machines.

Example 3.5 Fast Fourier transform on mesh and hypercube computers (Gupta and Kumar, 1993)

This example demonstrates the sensitivity of machine architecture on the scalability of the FFT on two different parallel computers: *mesh* and *hypercube*. We consider the Cooley-Tukey algorithm for one-dimensional s-point fast Fourier transform.

Gupta and Kumar have established the overheads: $h_1(s, n) = O(n \log n + s \log n)$ for FFT on a hypercube machine with n processors, and $h_2(s, n) = O(n \log n + s \sqrt{n})$ on a $\sqrt{n} \times \sqrt{n}$ mesh with n processors.

For an s-point FFT, the total workload involved is $w(s) = O(s \log s)$. Equating the workload with overheads, we must satisfy $O(s \log s) = O(n \log n)$ and $O(s \log s) = O(s \log n)$, leading to the isoefficiency function $f_1 = O(n \log n)$ for the hypercube machine.

Similarly, we must satisfy $O(s \log s) = O(n \log n)$ and $O(s \log s) = O(s \sqrt{n})$ by equating $w(s) = h_2(s, n)$. This leads to the isoefficiency function $f_2 = O(\sqrt{n} k^{\sqrt{n}})$ for some constant $k \geq 2$.

The above analysis leads to the conclusion that FFT is indeed very scalable on a hypercube computer. The result was plotted in Fig. 3.7a for three efficiency values.

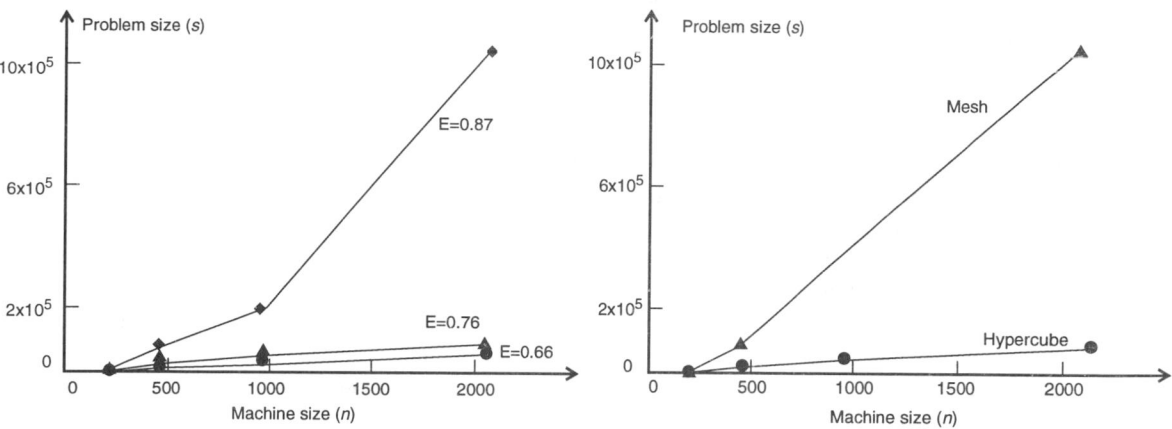

(a) Hypercube under three operating efficiencies (b) Comparison between mesh and hypercube

Figure 3.7 Isoefficiency curves for FFT on two parallel computers. (Courtesy of Gupta and Kumar, 1993)

To maintain the same efficiency, the mesh is rather poorly scalable as demonstrated in Fig. 3.7b.

This is predictable by the fact that the workload must grow exponentially in $O(\sqrt{n}k^{\sqrt{n}})$ for the mesh architecture, while the hypercube demands only $O(n \log n)$ workload increase as the machine size increases. Thus, we conclude that the FFT is scalable on a hypercube but not so on a mesh architecture.

If the bandwidth of the communication channels in a mesh architecture increases proportional to the increase of machine size, the above conclusion will change. For the design and analysis of FFT on parallel machines, readers are referred to the books by Aho, Hopcroft and Ullman (1974) and by Quinn (1987). We will further address scalability issues from the architecture standpoint in Section 3.4.

3.3 Speedup Performance Laws

Three speedup performance models are defined below. Amdahl's law (1967) is based on a fixed workload or a fixed problem size. Gustafson's law (1987) is applied to scaled problems, where the problem size increases with the increase in machine size. The speedup model by Sun and Ni (1993) is for scaled problems bounded by memory capacity.

3.3.1 Amdahl's Law for a Fixed Workload

In many practical applications that demand a real-time response, the computational workload is often fixed with a fixed problem size. As the number of processors increases in a parallel computer, the fixed load is distributed to more processors for parallel execution. Therefore, the main objective is to produce the results as soon as possible. In other words, minimal turnaround time is the primary goal. Speedup obtained for time-critical applications is called fixed-load speedup.

Fixed-Load Speedup   The ideal speedup formula given in Eq. 3.7 is based on a fixed workload, regardless of the machine size. Traditional formulations for speedup, including Amdahl's law, are all based on a fixed problem size and thus on a fixed load. The speedup factor is upper-bounded by a sequential bottleneck in this case.

We consider below both the cases of DOP $< n$ and of DOP $\geq n$. We use the ceiling function $\lceil x \rceil$ to represent the smallest integer that is greater than or equal to the positive real number x. When x is a fraction, $\lceil x \rceil$ equals 1. Consider the case where DOP $= i > n$. Assume all n processors are used to execute W_i exclusively. The execution time of W_i is

$$t_i(n) = \frac{W_i}{i\Delta} \left\lceil \frac{i}{n} \right\rceil \tag{3.25}$$

Thus the response time is

$$T(n) = \sum_{i=1}^{m} \frac{W_i}{i\Delta} \left\lceil \frac{i}{n} \right\rceil \tag{3.26}$$

Note that if $i < n$, then $t_i(n) = t_i(\infty) = W_i/i\Delta$. Now, we define the *fixed-load speedup factor* as the ratio of $T(1)$ to $T(n)$:

$$S_n = \frac{T(1)}{T(n)} = \frac{\displaystyle\sum_{i=1}^{m} W_i}{\displaystyle\sum_{i=1}^{m} \frac{W_i}{i} \left\lceil \frac{i}{n} \right\rceil} \tag{3.27}$$

Note that $S_n \leq S_\infty \leq A$, by comparing Eqs. 3.4, 3.7, and 3.27.

A number of factors we have ignored may lower the speedup performance. These include communication latencies caused by delayed memory access, interprocessor communications over a bus or a network, or operating system overhead and delay caused by interrupts. Let $Q(n)$ be the lumped sum of all system overheads on an n-processor system. We can rewrite Eq. 3.27 as follows:

$$S_n = \frac{T(1)}{T(n) + Q(n)} = \frac{\displaystyle\sum_{i=1}^{m} W_i}{\displaystyle\sum_{i=1}^{m} \frac{W_i}{i} \left\lceil \frac{i}{n} \right\rceil + Q(n)} \tag{3.28}$$

The overhead delay $Q(n)$ is certainly application-dependent as well as machine-dependent. It is very difficult to obtain a closed form for $Q(n)$. Unless otherwise specified, we assume $Q(n) = 0$ to simplify the discussion.

Amdahl's Law Revisited   In 1967, Gene Amdahl derived a fixed-load speedup for the special case where the computer operates either in sequential mode (with DOP = 1) or in perfectly parallel mode (with DOP = n). That is, $W_i = 0$ if $i \neq 1$ or $i \neq n$ in the profile. Equation 3.27 is then simplified to

$$S_n = \frac{W_1 + W_n}{W_1 + W_n/n} \qquad (3.29)$$

Amdahl's law implies that the sequential portion of the program W_i does not change with respect to the machine size n. However, the parallel portion is evenly executed by n processors, resulting in a reduced time.

Consider a normalized situation in which $W_1 + W_n = \alpha + (1 - \alpha) = 1$, with $\alpha = W_1$ and $1 - \alpha = W_n$. Equation 3.29 is reduced to Eq. 3.16, where α represents the percentage of a program that must be executed sequentially and $1 - \alpha$ corresponds to the portion of the code that can be executed in parallel.

Amdahl's law is illustrated in Fig. 3.8. When the number of processors increases, the load on each processor decreases. However, the total amount of work (workload) $W_1 + W_n$ is kept constant as shown in Fig. 3.8a. In Fig. 3.8b, the total execution time decreases because $T_n = W_n/n$. Eventually, the sequential part will dominate the performance because $T_n \to 0$ as n becomes very large and T_1 is kept unchanged.

Sequential Bottleneck   Figure 3.8c plots Amdahl's law using Eq. 3.16 over the range $0 \leq \alpha \leq 1$. The maximum speedup $S_n = n$ if $\alpha = 0$. The minimum speedup $S_n = 1$ if $\alpha = 1$. As $n \to \infty$, the limiting value of $S_n \to 1/\alpha$. This implies that the speedup is upper-bounded by $1/\alpha$, as the machine size becomes very large.

The speedup curve in Fig. 3.8c drops very rapidly as α increases. This means that with a small percentage of the sequential code, the entire performance cannot go higher than $1/\alpha$. This α has been called the *sequential bottleneck* in a program.

The problem of a sequential bottleneck cannot be solved just by increasing the number of processors in a system. The real problem lies in the existence of a sequential fraction (s) of the code. This property has imposed a very pessimistic view on parallel processing over the past two decades.

In fact, two major impacts on the parallel computer industry were observed. First, manufacturers were discouraged from making large-scale parallel computers. Second, more research attention was shifted toward developing parallelizing compilers which will reduce the value of α and in turn boost the performance.

3.3.2   Gustafson's Law for Scaled Problems

One of the major shortcomings in applying Amdahl's law is that the problem (workload) cannot scale to match the available computing power as the machine size increases. In other words, the fixed load prevents scalability in performance. Although the sequential bottleneck is a serious problem, the problem can be greatly alleviated by removing the fixed-load (or fixed-problem-size) restriction. John Gustafson (1988) has proposed a fixed-time concept which leads to a scaled speedup model.

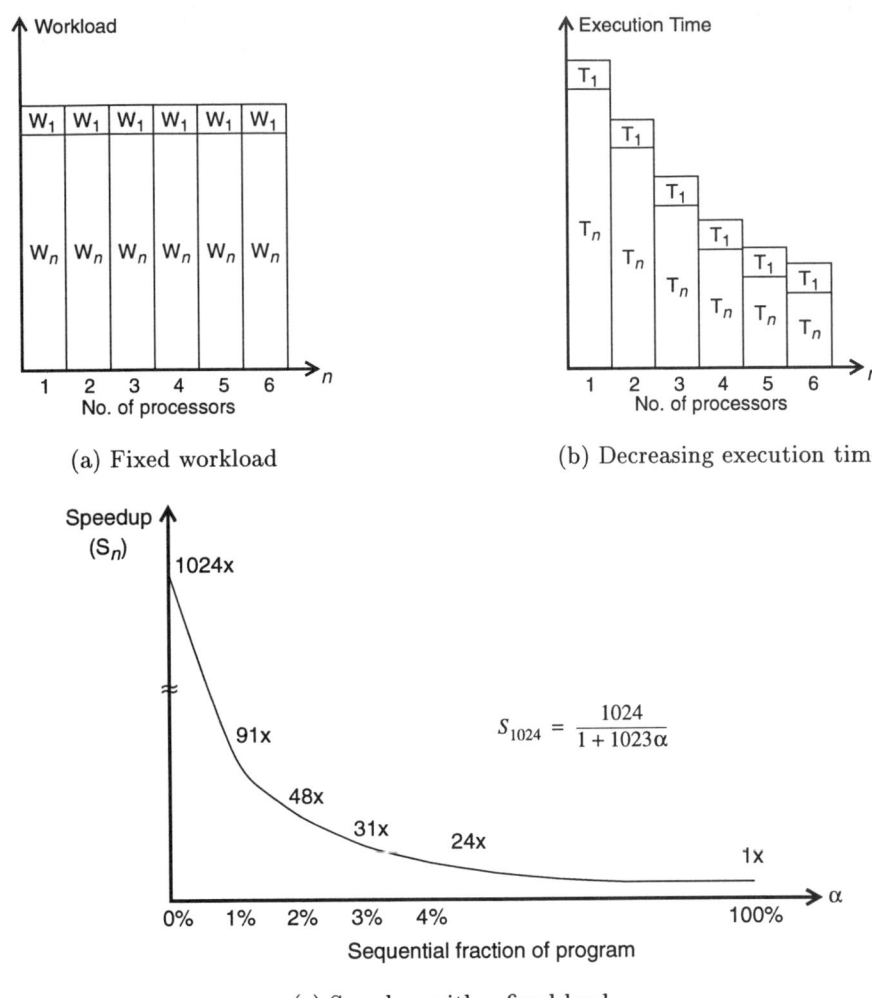

(a) Fixed workload

(b) Decreasing execution time

$$S_{1024} = \frac{1024}{1 + 1023\alpha}$$

(c) Speedup with a fixed load

Figure 3.8 Fixed-load speedup model and Amdahl's law.

Scaling for Higher Accuracy   Time-critical applications provided the major motivation leading to the development of the fixed-load speedup model and Amdahl's law. There are many other applications that emphasize accuracy more than minimum turnaround time. As the machine size is upgraded to obtain more computing power, we may want to increase the problem size in order to create a greater workload, producing more accurate solution and yet keeping the execution time unchanged.

Many scientific modeling and engineering simulation applications demand the solution of very large-scale matrix problems based on some partial differential equation (PDE) formulations discretized with a huge number of grid points. Representative examples include the use of finite-element method to perform structural analysis or the use of finite-difference method to solve computational fluid dynamics problems in weather forecasting.

Coarse grids require less computation, but finer grids require many more computations, yielding greater accuracy. The weather forecasting simulation often demands the solution of four-dimensional PDEs. If one reduces the grid spacing in each physical dimension (X, Y, and Z) by a factor of 10 and increases the time steps by the same magnitude, then we are talking about an increase of 10^4 times more grid points. The workload thus increases to at least 10,000 times greater.

With such a problem scaling, of course, we demand more computing power to yield the same execution time. The main advantage is not in saving time but in producing much more accurate weather forecasting. This problem scaling for accuracy has motivated Gustafson to develop a fixed-time speedup model. The scaled problem keeps all the increased resources busy, resulting in a better system utilization ratio.

Fixed-Time Speedup In accuracy-critical applications, we wish to solve the largest problem size possible on a larger machine with about the same execution time as for solving a smaller problem on a smaller machine. As the machine size increases, we have to deal with an increased workload and thus a new parallelism profile. Let m' be the maximum DOP with respect to the scaled problem and W_i' be the scaled workload with $\text{DOP} = i$.

Note that in general $W_i' > W_i$ for $2 \leq i \leq m'$ and $W_1' = W_1$. The fixed-time speedup is defined under the assumption that $T(1) = T'(n)$, where $T'(n)$ is the execution time of the scaled problem and $T(1)$ corresponds to the original problem without scaling. We thus obtain

$$\sum_{i=1}^{m} W_i = \sum_{i=1}^{m'} \frac{W_i'}{i} \left\lceil \frac{i}{n} \right\rceil + Q(n) \tag{3.30}$$

A general formula for fixed-time speedup is defined by $S_n' = T(1)/T'(n)$, modified from Eq. 3.28:

$$S_n' = \frac{\displaystyle\sum_{i=1}^{m'} W_i'}{\displaystyle\sum_{i=1}^{m'} \frac{W_i'}{i} \left\lceil \frac{i}{n} \right\rceil + Q(n)} = \frac{\displaystyle\sum_{i=1}^{m'} W_i'}{\displaystyle\sum_{i=1}^{m} W_i} \tag{3.31}$$

Gustafson's Law Fixed-time speedup was originally developed by Gustafson for a special parallelism profile with $W_i = 0$ if $i \neq 1$ and $i \neq n$. Similar to Amdahl's law, we can rewrite Eq. 3.31 as follows, assuming $Q(n) = 0$,

$$S'_n = \frac{\sum_{i=1}^{m'} W'_i}{\sum_{i=1}^{m} W_i} = \frac{W'_1 + W'_n}{W_1 + W_n} = \frac{W_1 + nW_n}{W_1 + W_n} \tag{3.32}$$

where $W'_n = nW_n$ and $W_1 + W_n = W'_1 + W'_n/n$, corresponding to the fixed-time condition. From Eq. 3.32, the parallel workload W'_n has been scaled to n times W_n in a linear fashion.

The relationship of a scaled workload to Gustafson's scaled speedup is depicted in Fig. 3.9. In fact, Gustafson's law can be restated as follows in terms of $\alpha = W_1$ and $1 - \alpha = W_n$ under the same assumption $W_1 + W_n = 1$ that we have made for Amdahl's law:

$$S'_n = \frac{\alpha + n(1 - \alpha)}{\alpha + (1 - \alpha)} = n - \alpha(n - 1) \tag{3.33}$$

In Fig. 3.9a, we demonstrate the workload scaling situation. Figure 3.9b shows the fixed-time execution style. Figure 3.9c plots S'_n as a function of the sequential portion α of a program running on a system with $n = 1024$ processors.

Note that the slope of the S_n curve in Fig. 3.9c is much flatter than that in Fig. 3.8c. This implies that Gustafson's law does support scalable performance as the machine size increases. The idea is to keep all processors busy by increasing the problem size. When the problem can scale to match available computing power, the sequential fraction is no longer a bottleneck.

3.3.3 Memory-Bounded Speedup Model

Xian-He Sun and Lionel Ni (1993) have developed a memory-bounded speedup model which generalizes Amdahl's law and Gustafson's law to maximize the use of both CPU and memory capacities. The idea is to solve the largest possible problem, limited by memory space. This also demands a scaled workload, providing higher speedup, higher accuracy, and better resource utilization.

Memory-Bound Problems Large-scale scientific or engineering computations often require larger memory space. In fact, many applications of parallel computers are memory-bound rather than CPU-bound or I/O-bound. This is especially true in a multicomputer system using distributed memory. The local memory attached to each node is relatively small. Therefore, each node can handle only a small subproblem.

When a large number of nodes are used collectively to solve a single large problem, the total memory capacity increases proportionally. This enables the system to solve a scaled problem through program partitioning or replication and domain decomposition of the data set.

Instead of keeping the execution time fixed, one may want to use up all the increased memory by scaling the problem size further. In other words, if you have adequate

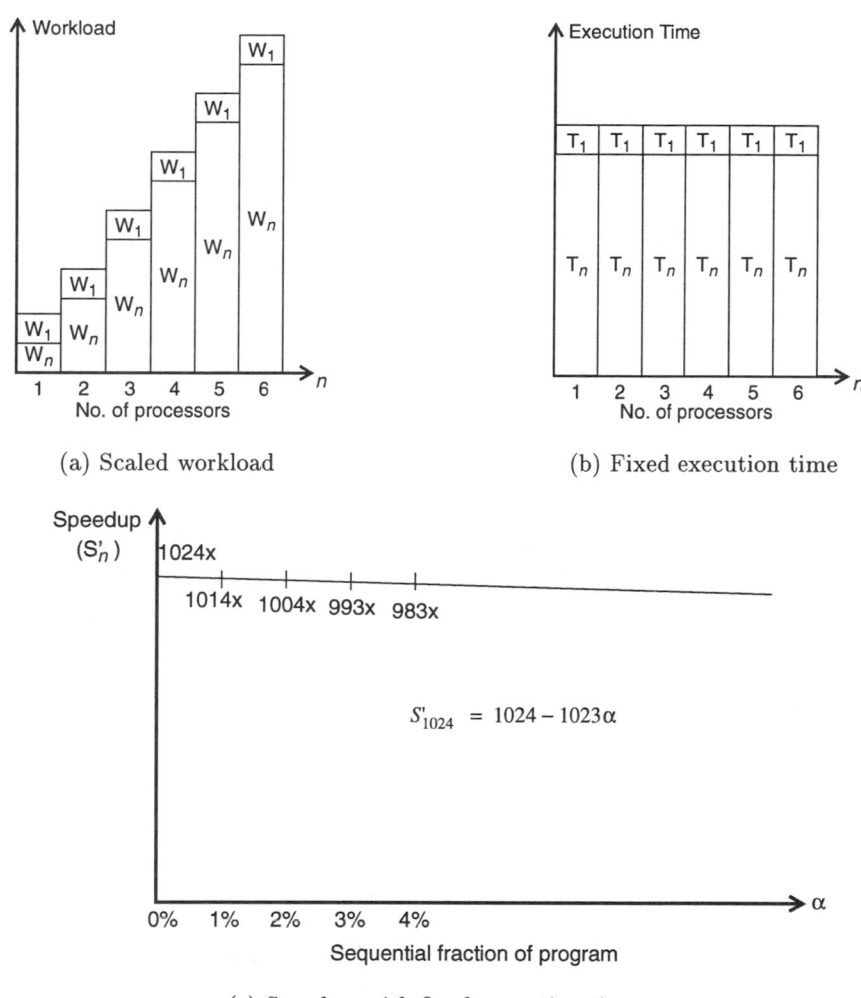

(a) Scaled workload

(b) Fixed execution time

(c) Speedup with fixed execution time

Figure 3.9 Fixed-time speedup model and Gustafson's law.

memory space and the scaled problem meets the time limit imposed by Gustafson's law, you can further increase the problem size, yielding an even better or more accurate solution.

A memory-bounded model was developed under this philosophy. The idea is to solve the largest possible problem, limited only by the available memory capacity. This model may result in a minor increase in execution time to achieve scalable performance.

Fixed-Memory Speedup   Let M be the memory requirement of a given problem and W be the computational workload. These two factors are related to each other in various ways, depending on the address space and architectural constraints. Let us write $W = g(M)$ or $M = g^{-1}(W)$, where g^{-1} is the inverse of g.

In a multicomputer, the total memory capacity increases linearly with the number of nodes available. We write $W = \sum_{i=1}^{m} W_i$ as the workload for sequential execution of the program on a single node, and $W^* = \sum_{i=1}^{m^*} W_i^*$ as the scaled workload for execution on n nodes, where m^* is the maximum DOP of the scaled problem. The memory requirement for an active node is thus bounded by $g^{-1}(\sum_{i=1}^{m} W_i)$.

A *fixed-memory speedup* is defined below similarly to that in Eq. 3.31.

$$S_n^* = \frac{\sum\limits_{i=1}^{m^*} W_i^*}{\sum\limits_{i=1}^{m^*} \frac{W_i^*}{i} \left\lceil \frac{i}{n} \right\rceil + Q(n)} \tag{3.34}$$

The workload for sequential execution on a single processor is independent of the problem size or system size. Therefore, we can write $W_1 = W_1' = W_1^*$ in all three speedup models. Let us consider the special case of two operational modes: *sequential* versus *perfectly parallel* execution. The enhanced memory is related to the scaled workload by $W_n^* = g^*(nM)$, where nM is the increased memory capacity for an n-node multicomputer.

Furthermore, we assume $g^*(nM) = G(n)g(M) = G(n)W_n$, where $W_n = g(M)$ and g^* is a homogeneous function. The factor $G(n)$ reflects the increase in workload as memory increases n times. Now we are ready to rewrite Eq. 3.34 under the assumption that $W_i = 0$ if $i \neq 1$ or n and $Q(n) = 0$:

$$S_n^* = \frac{W_1^* + W_n^*}{W_1^* + W_n^*/n} = \frac{W_1 + G(n)W_n}{W_1 + G(n)W_n/n} \tag{3.35}$$

Rigorously speaking, the above speedup model is valid under two assumptions: (1) The collection of all memory forms a global address space (in other words, we assume a shared distributed memory space); (2) All available memory areas are used up for the scaled problem. There are three special cases where Eq. 3.35 can apply:

Case 1: $G(n) = 1$. This corresponds to the case where the problem size is fixed. Thus, the fixed-memory speedup becomes equivalent to Amdahl's law; i.e., Eqs. 3.29 and 3.35 are equivalent when a fixed workload is given.

Case 2: $G(n) = n$. This applies to the case where the workload increases n times when the memory is increased n times. Thus, Eq. 3.35 is identical to Gustafson's law (Eq. 3.32) with a fixed execution time.

Case 3: $G(n) > n$. This corresponds to the situation where the computational workload increases faster than the memory requirement. Thus, the fixed-memory model (Eq. 3.35) will likely give a higher speedup than the fixed-time speedup (Eq. 3.32).

The above analysis leads to the following conclusions: Amdahl's law and Gustafson's law are special cases of the fixed-memory model. When computation grows faster than the memory requirement, as is often true in some scientific simulation and engineering applications, the fixed-memory model (Fig. 3.10) may yield an even higher speedup (i.e., $S_n^* \geq S_n' \geq S_n$) and better resource utilization.

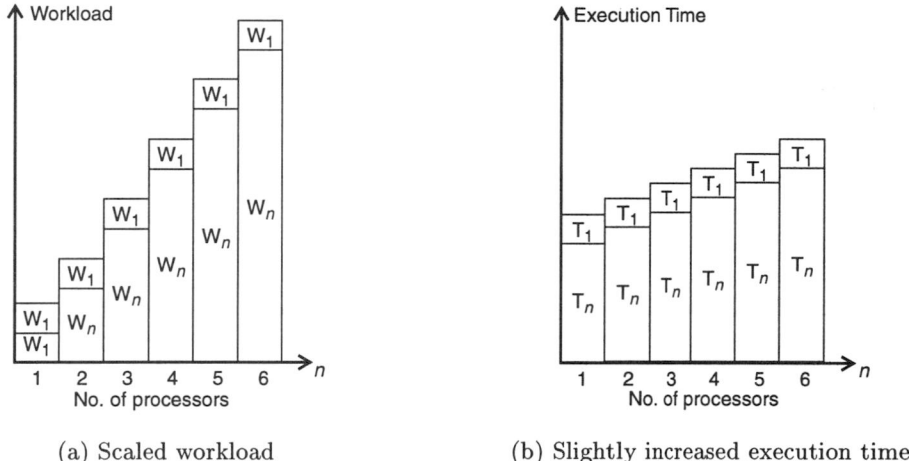

(a) Scaled workload   (b) Slightly increased execution time

Figure 3.10 Scaled speedup model using fixed memory (Courtesy of Sun and Ni; reprinted with permission from *ACM Supercomputing*, 1990)

The fixed-memory model also assumes a scaled workload and allows a slight increase in execution time. The increase in workload (problem size) is memory-bound. The growth in machine size is limited by increasing communication demands as the number of processors becomes very large. The fixed-time model can be moved very close to the fixed-memory model if available memory is fully utilized.

Example 3.6 Scaled matrix multiplication using global versus local computation models (Sun and Ni, 1993)

In scientific computations, matrix often represents some discretized data continuum. Enlarging the matrix size generally leads to a more accurate solution for the continuum. For matrices with dimension n, the number of computations involved in matrix multiplication is $2n^3$ and the memory requirement is roughly $M = 3n^2$.

As the memory increases n times in an n-processor multicomputer system, $nM = n \times 3n^2 = 3n^3$. If the enlarged matrix has a dimension of N, then $3n^3 = 3N^2$. Therefore, $N = n^{1.5}$. Thus $G(n) = n^{1.5}$, and the scaled workload $W_n^* = G(n)W_n = n^{1.5}W_n$. Using Eq. 3.35, we have

$$S^* = \frac{W_1 + n^{1.5}W_n}{W_1 + \dfrac{n^{1.5}W_n}{n}} = \frac{W_1 + n^{1.5}W_n}{W_1 + n^{0.5}W_n} \qquad (3.36)$$

under the *global computation model* illustrated in Fig. 3.11a, where all the distributed memories are used as a common memory shared by all processor nodes.

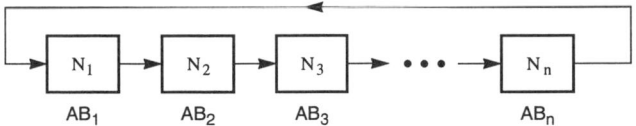

(a) Global computation with distributed shared memories

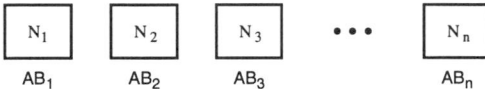

(b) Local computation with distributed private memories

Figure 3.11 Two models for the distributed matrix multiplication.

As illustrated in Fig. 3.11b, the node memories are used locally without sharing. In such a *local computation model*, $G(n) = n$, and we obtain the following speedup:

$$S_n^* = \frac{W_1 + nW_n}{W_1 + W_n} \tag{3.37}$$

∎

The above example illustrates Gustafson's scaled speedup for local computation. Comparing the above two speedup expressions, we realize that the fixed-memory speedup (Eq. 3.36) may be higher than the fixed-time speedup (Eq. 3.37). In general, many applications demand the use of a combination of local and global addressing spaces. Data may be distributed in some nodes and duplicated in other nodes.

Data duplication is added deliberately to reduce communication demand. Speedup factors for these applications depend on the ratio between the global and local computations. This reinforces the belief that $G(n)$ increases linearly in general applications. The higher the value of $G(n)$, the better the performance.

3.4   Scalability Analysis and Approaches

The performance of a computer system depends on a large number of factors, all affecting the scalability of the computer architecture and the application program involved. The simplest definition of *scalability* is that the performance of a computer system increases linearly with respect to the number of processors used for a given application.

Scalability analysis of a given computer must be conducted for a given application program/algorithm. The analysis can be performed under different constraints on the growth of the problem size (workload) and on the machine size (number of processors). A good understanding of scalability will help evaluate the performance of parallel computer architectures for large-scale applications.

3.4.1   Scalability Metrics and Goals

Scalability studies determine the degree of matching between a computer architecture and an application algorithm as illustrated in Fig. 3.12. For different (architecture, algorithm) pairs, the analysis may end up with different conclusions. A machine can be very efficient for one algorithm but bad for another, and vice versa.

Thus, a good computer architecture should be efficient in implementing a large class of application algorithms. In the ideal case, the computer performance should be linearly scalable with an increasing number of processors employed in implementing the algorithms.

Scalability metrics   Identified below are the basic metrics (Fig. 3.12a) affecting the scalability of a computer system for a given application:

- *Machine size* (n) — the number of processors employed in a parallel computer system. A large machine size implies more resources and more computing power.
- *Clock rate* (f) — the clock rate determines the basic machine cycle. We hope to build a machine with components (processors, memory, bus or network, etc.) driven by a clock which can scale up with better technology.
- *Problem size* (s) — the amount of computational workload or the number of data points used to solve a given problem. The problem size is directly proportional to the *sequential execution time* $T(s, 1)$ for a uniprocessor system because each data point may demand one or more operations.
- *CPU time* (T) — the actual CPU time (in seconds) elapsed in executing a given program on a parallel machine with n processors collectively. This is the *parallel execution time*, denoted as $T(s, n)$ and is a function of both s and n.
- *I/O demand* (d) – the input/output demand in moving the program, data, and results associated with a given application run. The I/O operations may overlap with the CPU operations in a multiprogrammed environment.
- *Memory capacity* (m) – the amount of main memory (in bytes or words) used in a program execution. Note that the memory demand is affected by the problem size, the program size, the algorithms, and the data structures used.

  The memory demand varies dynamically during program execution. Here, we refer to the maximum number of memory words demanded. Virtual memory is almost unlimited with a 64-bit address space. It is the physical memory which is limited in capacity.
- *Communication overhead* (h) – the amount of time spent for interprocessor communication, synchronization, remote memory access, etc. This overhead includes all noncompute operations which do not involve the CPUs or I/O devices. This overhead $h(s, n)$ is a function of s and n and is not part of $T(s, n)$. For a uniprocessor system, the overhead $h(s, 1) = 0$.

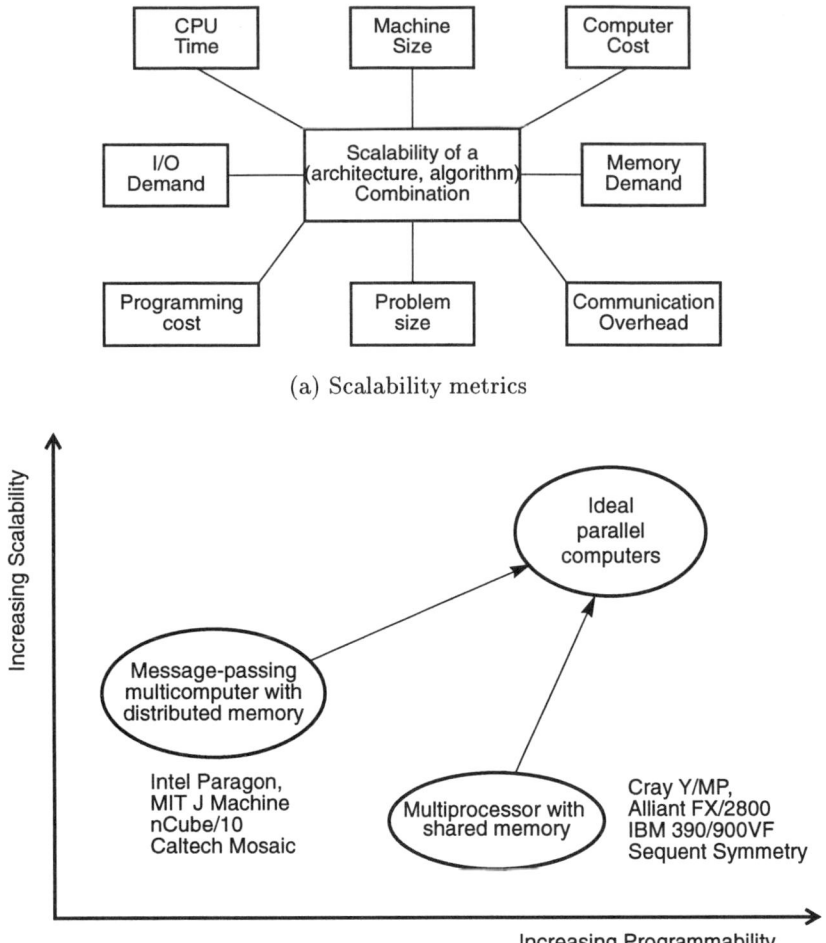

(a) Scalability metrics

(b) Programmability versus scalability

Figure 3.12 Scalability and programmability of various computer architecture and algorithm combinations.

- *Computer cost* (c) – the total cost of hardware and software resources required to carry out the execution of a program.
- *Programming overhead* (p) – the development overhead associated with an application program. Programming overhead may slow down software productivity and thus requires a high cost. Unless otherwise stated, both computer cost and programming cost are ignored in our scalability analysis.

Depending on the computational objectives and resource constraints imposed, one can fix some of the above parameters and optimize the remaining ones to achieve the highest performance with the lowest cost.

The notion of scalability is tied to the notions of speedup and efficiency. A sound definition of scalability must be able to express the effects of the architecture's interconnection network, of the communication patterns inherent to algorithms, of the physical constraints imposed by technology, and of the cost effectiveness or system efficiency. We introduce first the notion of speedup and efficiency. Then we define scalability based on the relative performance of a real machine compared with that of an idealized theoretical machine.

Speedup and Efficiency Revisited   For a given architecture, algorithm, and problem size s, the *asymptotic speedup* $S(s, n)$ is the best speedup that is attainable, varying only the number (n) of processors. Let $T(s, 1)$ be the sequential execution time on a uniprocessor, $T(s, n)$ be the minimum parallel execution time on an n-processor system, and $h(s, n)$ be the lump sum of all communication and I/O overheads. The asymptotic speedup is formally defined as follows:

$$S(s, n) = \frac{T(s, 1)}{T(s, n) + h(s, n)} \tag{3.38}$$

The problem size is the independent parameter, upon which all other metrics are based. A meaningful measurement of asymptotic speedup mandates the use of a good sequential algorithm, even it is different from the structure of the corresponding parallel algorithm. The $T(s, n)$ is minimal in the sense that the problem is solved using as many processors as necessary to achieve the minimum runtime for the given problem size.

In scalability analysis, we are mainly interested in results obtained from solving large problems. Therefore, the run times $T(s, n)$ and $(s, 1)$ should be expressed using order-of-magnitude notations, reflecting the asymptotic behavior.

The *system efficiency* of using the machine to solve a given problem is defined by the following ratio:

$$E(s, n) = \frac{S(s, n)}{n} \tag{3.39}$$

In general, the best possible efficiency is one, implying that the best speedup is linear, or $S(s, n) = n$. Therefore, an intuitive definition of scalability is: *A system is scalable if the system efficiency $E(s, n) = 1$ for all algorithms with any number of n processors and any problem size s.*

Mark Hill (1990) has indicated that this definition is too restrictive to be useful because it precludes any system from being called scalable. For this reason, a more practical efficiency or scalability definition is needed, comparing the performance of the real machine with respect to the theoretical PRAM model.

Scalability Definition   Nussbaum and Agarwal (1991) have given the following scalability definition based on a PRAM model. The scalability $\Phi(s, n)$ of a machine for a given algorithm is defined as the ratio of the asymptotic speedup $S(s, n)$ on the real machine to the asymptotic speedup $S_I(s, n)$ on the ideal realization of an EREW PRAM.

$$S_I(s, n) = \frac{T(s, 1)}{T_I(s, n)}$$

where $T_I(s, n)$ is the parallel execution time on the PRAM, ignoring all communication overhead. The scalability is defined as follows:

$$\Phi(s, n) = \frac{S(s, n)}{S_I(s, n)} = \frac{T_I(s, n)}{T(s, n)} \tag{3.40}$$

Intuitively, the larger the scalability, the better the performance that the given architecture can yield running the given algorithm. In the ideal case, $S_I(s, n) = n$, the scalability definition in Eq. 3.40 becomes identical to the efficiency definition given in Eq. 3.39.

Example 3.7 Scalability of various machine architectures for parity calculation (Nussbaum and Agarwal, 1991)

Table 3.7 shows the execution times, asymptotic speedups, and scalabilities (with respect to the EREW-PRAM model) of five representative interconnection architectures: linear array, meshes, hypercube, and Omega network, for running a parallel parity calculation.

Table 3.7 Scalability of Various Network-Based Architectures for the Parity Calculation

Metrics	Machine Architecture				
	Linear array	2-D mesh	3-D mesh	Hypercube	Omega Network
$T(s, n)$	$s^{1/2}$	$s^{1/3}$	$s^{1/4}$	$\log s$	$\log^2 s$
$S(s, n)$	$s^{1/2}$	$s^{2/3}$	$s^{3/4}$	$s / \log s$	$s / \log^2 s$
$\Phi(s, n)$	$\log s / s^{1/2}$	$\log s / s^{1/3}$	$\log s / s^{1/4}$	1	$1 / \log s$

This calculation examines s bits, determining whether the number of bits set is even or odd using a balanced binary tree. For this algorithm, $T(s, 1) = s$, $T_I(s, n) = \log s$, and $S_I(s, n) = s / \log s$ for the ideal PRAM machine.

On real architectures, the parity algorithm's performance is limited by network diameter. For example, the linear array has a network diameter equal to $n - 1$, yielding a total parallel running time of $s/n + n$. The optimal partition of the problem is to use $n = \sqrt{s}$ processors so that each processor performs the parity check on $\sqrt{s}$ bits locally. This partition gives the best match between computation costs and communication costs with $T(s, n) = s^{1/2}, S(s, n) = s^{1/2}$ and thus scalability $\Phi(s, n) = \log s / s^{1/2}$.

The 2D and 3D mesh architectures use a similar partition to match their own communication structure with the computational loads, yielding even better scalability results. It is interesting to note that the scalability increases as the communication latency decreases in a network with a smaller diameter.

The hypercube and the Omega network provide richer communication structures (and lower diameters) than meshes of lower dimensionality. The hypercube does as well as a PRAM for this algorithm, yielding $\Phi(s, n) = 1$.

The Omega network (Fig. 2.24) does not exploit locality: communication with all processors takes the same amount of time. This loss of locality hurts its performance when compared to the hypercube, but its lower diameter gives it better scalability than any of the meshes.

∎

Although performance is limited by network diameter for the above parity algorithm, for many other algorithms, the network bandwidth is the performance-limiting factor. The above analysis assumed unit communication time between directly connected communication nodes. An architecture may be scalable for one algorithm but unscalable for another. One must examine a large class of useful algorithms before drawing a scalability conclusion on a given architecture.

3.4.2 Evolution of Scalable Computers

The idea of massive parallelism is rather old, the technology is new, and the software is relatively unexplored, as observed by Cybenko and Kuck (1992). The evolutional trend is to build scalable supercomputers with distributed shared memory and standardized UNIX for parallel processing. In this section, we present the evolutional path and some scalable computer design concepts assessed by Gordon Bell (1992).

The Evolutional Path   Figure 3.13 shows the evolution of supercomputers with four-to-five-year gestation and of micro-based scalable computers with three-year gestation. This plot shows the peak performance of Cray and NEC supercomputers and of Cray, Intel, and Thinking Machines scalable computers versus the introduction year. The marked nodes correspond to machine models with increasing size and cost.

In 1988, the Cray Y-MP 8 delivered a peak of 2.8 Gflops. By 1991, the Intel Touchstone Delta, a 672-node multicomputer, and the Thinking Machine CM-2, a 2K-PE SIMD machine, both began to supply an order-of-magnitude peak power (20 Gflops) than conventional supercomputers. By mid-1992, a completely new generation of computers were introduced, including the CM-5 and Paragon.

It was predicted that by 1995, some of the announced MPP systems may exceed the Teraflops performance. The HPCC program in US and the counter program in Japan will fuel the Teraflops quest. From Fig. 3.13, the MPP approach will achieve the goal sooner. The traditional multiple vector-multiprocessor supercomputers such as Crays will not evolve to a Teraflops machine until 2000, according to Gordon Bell.

In the past, the IBM System/360 provided a 100:1 range of growth for its various models. The VAX machines existed at a range of 1000:1 over its lifetime. Based on past experiences, Gordon Bell has identified three objectives for designing scalable computers. Implications and case studies of these challenges will be further discussed in subsequent chapters.

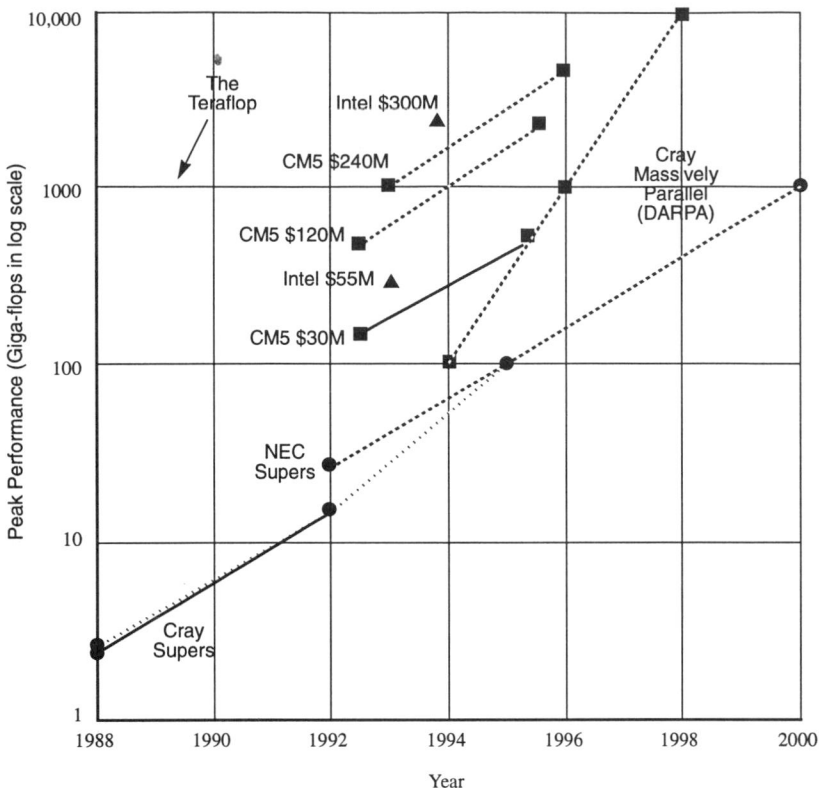

Figure 3.13 The performance (in Gflops) of various computers manufactured by Cray Research, Inc., NEC, Intel, and Thinking Machines Corporation. (Courtesy of Gordon Bell; reprinted with permission from the *Communications of ACM*, August 1992)

Size Scalability The study of system scalability started with the desire to increase the machine size. A size-scalable computer is designed to have a scaling range from a small to a large number of resource components. The expectation is to achieve linearly increased performance with incremental expansion for a well-defined set of applications. The components include computers, processors or processing elements, memories, switches, cabinets, etc.

Size scalability depends on spatial and temporal locality as well as component bottleneck. Since very large systems have inherently longer latencies than small and centralized systems, the locality behavior of program execution will help tolerate the increased latency. Locality will be characterized in Chapter 4. The bottleneck-free condition demands a balanced design among processing, storage, and I/O bandwidth.

For example, since MPPs are mostly interconnected by large networks or switches, the bandwidth of the switch should increase linearly with processor power. The I/O demand may far exceed the processing bandwidth in some real-time and large-scale applications. It was estimated that roughly 0.1 Terabytes/s or 20,000 5Mbytes/s disks are needed to operate in parallel in order to balance the 1 Teraflops computing speed.

The Cray Y-MP series scaled over a range of 16 processors (the C-90 model). The CM-2 was designed to scale between 8K and 64K processing elements. The CM-5 scaling range is 1024 to 16K computers. The KSR-1 has a range of 8 to 1088 processor-memory pairs. Size-scalability cannot be achieved alone without considering cost, efficiency, and affordability on reasonable time scale. It It is fair to conclude that no computers are truly scalable in a linear fashion.

Generation (Time) Scalability Since the basic microprocessor nodes become obsolete every three years, the time scalability is equally important as the size scalability. Not only should the hardware technology be scalable, such as the CMOS circuits and packaging technologies in building processors and memory chips, but also the software/algorithm which demands software compatibility and portability with new hardware systems.

Digital Equipment has claimed that the Alpha microprocessor is generation-scalable for 25 years. The claim is yet to be verified as time goes by. In general, all computer characteristics must scale proportionally: processing speed, memory speed and size, interconnect bandwidth and latency, I/O, and software overhead in order to be useful for a given application.

Problem Scalability The problem size corresponds to the data set size. This is the key to achieving scalable performance as the program granularity changes. A problem-scalable computer should be able to perform well as the problem size increases. The problem size can be scaled to be sufficiently large in order to operate efficiently on a computer with a given granularity.

Problems such as Monte Carlo simulation and ray tracing are "perfectly parallel", since their threads of computation almost never come together. Such an independence among threads is very much desired in using a scalable MPP system. In general, the *problem granularity* (operations on a grid point/data required from adjacent grid points) must be greater than a *machine's granularity* (node operation rate/node-to-node communication data rate) in order for a multicomputer to be effective.

Example 3.8 Problem scaling for solving Laplace equation on a distributed-memory multicomputer (Gordon Bell, 1992)

Laplace equations are often used to model physical structures. A 3-D Laplace equation is specified by

$$\bigtriangledown^2 u = \frac{\partial^2 u}{\partial x^2} + \frac{\partial^2 u}{\partial y^2} + \frac{\partial^2 u}{\partial z^2} = 0 \qquad (3.41)$$

We want to determine the problem scalability of the Laplace equation solver on a distributed -memory multicomputer with a sufficiently large number of processing nodes. Based on finite-difference method, solving Eq. 3.41 requires performing the following averaging operation iteratively across a very large grid Fig. 3.14:

$$u_{i,j,k}^{(m)} = \frac{1}{6}\left[u_{i-1,j,k}^{(m-1)} + u_{i+1,j,k}^{(m-1)} + u_{i,j-1,k}^{(m-1)} + u_{i,j+1,k}^{(m-1)} + u_{i,j,k-1}^{(m-1)} + u_{i,j,k+1}^{(m-1)}\right] \quad (3.42)$$

where $1 \leq i,j,k \leq N$ and N is the number of grid points along each dimension. In total, there are N^3 grid points in the problem domain to be evaluated during each iteration m for $1 \leq m \leq M$.

The three-dimensional domain can be partitioned into p subdomains, each having n^3 grid points such that $pn^3 = N^3$, where p is the machine size. The computations involved in each subdomain are assigned to one node of a multicomputer. Therefore, in each iteration, each node is required to perform $7n^3$ computations as specified in Eq. 3.42.

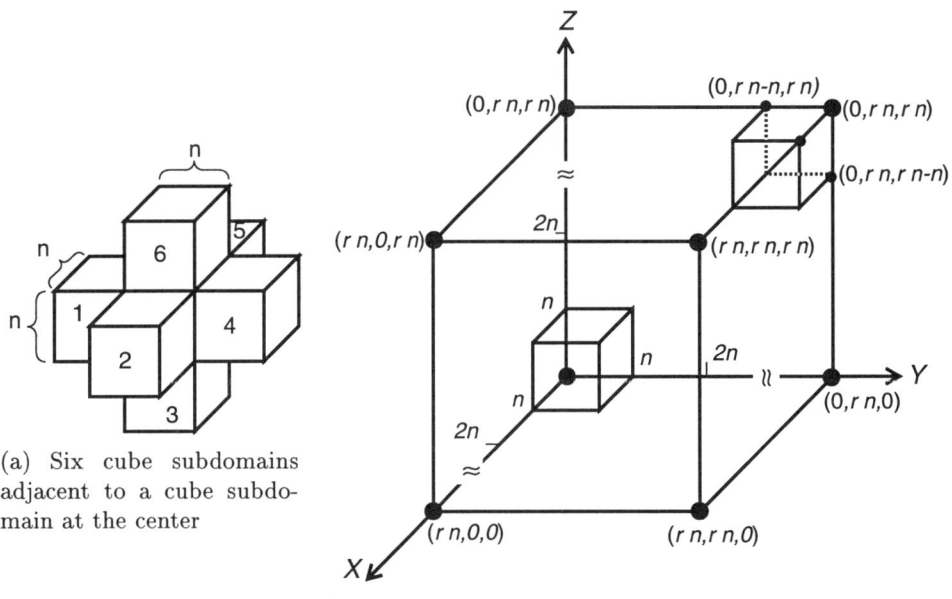

(a) Six cube subdomains adjacent to a cube subdomain at the center

(b) An $N \times N \times N$ grid partitioned into p subdomains, each being an n^3 cube, where $p = r^3 = N^3/n^3$

Figure 3.14 Partitioning of a 3D domain for solving the Laplace equation.

Each subdomain is adjacent to six other subdomains (Fig. 3.14a). Therefore, in each iteration, each node needs to exchange (send or receive) a total of $6n^2$ words of floating-point numbers with its neighbors. Assume each floating-point number is double-precision (64 bits, or 8 bytes). Each processing node has the capability of performing 100 Mflops (or 0.01 μs per floating-point operation). The internode communication latency is assumed to be 1 μs (or 1 megaword/s) for transferring a floating-point number.

For a balanced multicomputer, the computation time within each node and inter-node communication latency should be equal. Thus $0.07n^3 \mu s$ equals $6n^2 \mu s$ communication latency, implying that n has to be at least as large as 86. A node memory of capacity $86^3 \times 8 = 640K \times 8 = 5120$ Kwords = 5 megabytes is needed to hold each subdomain of data.

On the other hand, suppose each message exchange takes $2\mu s$ (one receive and one send) per word. The communication latency is doubled. We desire to scale up the problem size with an enlarged local memory of 32 megabytes. The subdomain dimension size n can be extended to at most 160, because $160^3 \times 8 = 32$ megabytes. This size problem requires 0.3 s of computation time and 2×0.15 s of send and receive time. Thus each iteration takes 0.6 (0.3 + 0.3) s, resulting in a computation rate of 50 Mflops, which is only 50% of the peak speed of each node.

If the problem size n is further increased, the effective Mflops rate and efficiency will be improved. But this cannot be achieved unless the memory capacity is further enlarged. For a fixed memory capacity, the situation corresponds to the memory-bound region shown in Fig. 1.10c. Another risk of problem scaling is to exacerbate the limited I/O capability which is not demonstrated in this example. ∎

To summarize the above studies on scalability, we realize that the machine size, problem size, and technology scalabilities are not necessarily orthogonal to each other. They must be considered jointly. In the next section, we will identify additional issues relating scalability studies to software compatibility, latency tolerance, machine programmability, and cost-effectiveness.

3.4.3   Research Issues and Solutions

Toward the development of truly scalable computers, much research needs to be done. In this section, we briefly identify several frontier research problems. Partial solutions to these problems will be studied in subsequent chapters.

The Problems   When a computer is scaled up to become an MPP system, the following difficulties will arise:

- Memory-access latency becomes too long and too nonuniformly distributed to be considered tolerable.
- The IPC complexity or synchronization overhead becomes too high to be useful.
- The multicache inconsistency problem becomes out of control.
- The processor utilization rate deteriorates as the system size becomes large.
- Message passing (or page migration) becomes too time-consuming to benefit resource sharing in a large distributed system.
- Overall system performance becomes saturated with diminishing return as system size increases further.

Some Approaches   In order to overcome the above difficulties, listed below are some approaches being pursued by most researchers:

- Searching for latency reducing and fast synchronization techniques.
- Using weaker memory consistency models.
- Developing scalable cache coherence protocols.
- Realizing shared virtual memory system.
- Integrating multithreaded architectures for improved processor utilization and system throughput.
- Expanding software portability and standardizing parallel and distributed UNIX system.

Scalability analysis can be carried out either by analytical methods or through trace-driven simulation experiments. In Chapter 9, we will study both approaches toward the development of scalable computer architectures that match program/algorithmic behaviors. Analytical tools include the use of Markov chains, Petri nets, or queueing models. A number of simulation packages have already been developed at Stanford University and at MIT.

Supporting Issues   Besides the emphases of scalability on machine size, problem size, and technology, we identify below several extended areas for continued research and development:

(1) *Software scalability*: As problem size scales in proportion to the increase in machine size, the algorithms can be optimized to match the architectural constraints. Software tools are being developed to help programmers in mapping algorithms onto a target architecture.

A perfect match between architecture and algorithm requires matching both computational and communication patterns through performance-tuning experiments in addition to simple numerical analysis. Optimizing compilers and visualization tools should be designed to reveal opportunities for algorithm/program restructuring to match with the architectural growth.

(2) *Reducing communication overhead*: Scalability analysis should concern both useful computations and available parallelism in programs. The most difficult part of the analysis is to estimate the communication overhead accurately. Excessive communication overhead, such as the time required to synchronize a large number of processors, wastes system resources. This overhead grows rapidly as machine size and problem size increase.

Furthermore, the run time conditions are often difficult to capture. How to reduce the growth of communication overhead and how to tolerate the growth of memory-access latency in very large systems are still wide-open research problems to be solved.

(3) *Enhancing programmability*: The computing community generally agrees that multicomputers are more scalable; multiprocessors are more easily programmed but are less scalable than multicomputers, as demonstrated in Fig. 3.12b. It is the centralized-memory versus distributed private-memory organization that makes the difference.

In the ideal case, we want to build machines which will retain the advantages of both architectures. This implies a system with shared distributed memory and simplified message communication among processor nodes. Heterogeneous programming paradigms are needed for future systems.

(4) *Providing longevity and generality:* Other scalability issues include *longevity*, providing an architecture with sufficiently large address space, and *generality*, supporting a wide variety of languages and binary migration of software.

Performance, scalability, programmability, and generality will be studied throughout the book for general-purpose parallel processing applications, unless otherwise noted.

3.5   Bibliographic Notes and Exercises

Previous Dhrystone results were reported in [Weicker84], [Cocke90] and Markstein and National Semiconductor [NS88]. The Whestone data were based on [DEC86]. The TP1 benchmark was first reported in *Datamation*. The KLIPS measure was introduced with Japan's FGCS Project.

[Amdahl67] developed the first speedup model using fixed workload. [Gustafson88] proposed the scaled speedup model after reevaluating Amdahl's Law. The memory-bounded speedup model was developed by [Sun93] and Ni. [Lee80] developed the formal definitions for speedup, efficiency, utilization, redundancy, and quality of parallel computation. [Eager89] et al. analyzed speedup versus efficiency in parallel systems. The performance laws are based on the work [Sun93] and Ni.

Harmonic mean performance of supercomputers was reported in [Smith88]. Example 3.2 is taken from [Hwang84] and Briggs. The Linpack benchmark results were due to [Dongarra92]. Information on the US/HPCC program can be found in [NSF92]. The isoefficiency function was first introduced by [Kumar87] and Rao. The scalability definition and Example 3.7 are from [Nussbaum91] and Agarwal. [Bell92] introduced the three scalability objectives and Example 3.8.

Parallel matrix multiplication algorithms are reported in [Gupta92] and Kumar, [Berntsen90], [Fox87] et al., and   [Dekel81] et al.   A multicoloring technique was developed by [Wang93] and Hwang for parallel solution of PDE problems. Parallel FFT algorithms are reported in [Gupta93] and Kumar , [Quinn87], and [Aho74], Hopcroft and Ullman. Available parallelism in programs was studied in [Kumar88], [Nicolau84] and Fisher, [Wall91], and [Butler91] et al.

Exercises

Problem 3.1 Consider the parallel execution of the same program in Problem 1.4 on a four-processor system with shared memory. The program can be partitioned into four equal parts (50,000 each) for balanced execution by the four processors. Due to the need for synchronization among the four program parts, 5000 extra instructions are added to each divided program part.

Assume the same instruction mix as in Problem 1.4 for each divided program part. The CPI for the memory reference (with cache miss) instructions has been increased from 8 to 12 cycles due to contentions. The CPIs for the remaining instruction types do not change.

(a) Repeat part (a) in Problem 1.4 when the program is executed on the four-processor system.

(b) Repeat part (b) in Problem 1.4 when the program is executed on the four-processor system.

(c) Calculate the speedup factor of the four-processor system over the uniprocessor system in Problem 1.4 under the respective trace statistics.

(d) Calculate the efficiency of the four-processor system by comparing the speedup factor in part (c) with the ideal case.

Problem 3.2 A uniprocessor computer can operate in either scalar or vector mode. In vector mode, computations can be performed nine times faster than in scalar mode. A certain benchmark program took time T to run on this computer. Further, it was found that 25% of T was attributed to the vector mode. In the remaining time, the machine operated in the scalar mode.

(a) Calculate the effective speedup under the above condition as compared with the condition when the vector mode is not used at all. Also calculate α, the percentage of code that has been vectorized in the above program.

(b) Suppose we double the speed ratio between the vector mode and the scalar mode by hardware improvements. Calculate the effective speedup that can be achieved.

(c) Suppose the same speedup obtained in part (b) must be obtained by compiler improvements instead of hardware improvements. What would be the new vectorization ratio α that should be supported by the vectorizing compiler for the same benchmark program?

Problem 3.3 Let α be the percentage of a program code which can be executed simultaneously by n processors in a computer system. Assume that the remaining code must be executed sequentially by a single processor. Each processor has an execution rate of x MIPS, and all the processors are assumed equally capable.

(a) Derive an expression for the effective MIPS rate when using the system for exclusive execution of this program, in terms of the parameters n, α, and x.

(b) If $n = 16$ and $x = 4$ MIPS, determine the value of α which will yield a system performance of 40 MIPS.

Problem 3.4 Consider a computer which can execute a program in two operational modes: *regular mode* versus *enhanced mode*, with a probability distribution of $\{\alpha, 1-\alpha\}$, respectively.

(a) If α varies between a and b and $0 \le a < b \le 1$, derive an expression for the *average speedup factor* using the harmonic mean concept.

(b) Calculate the speedup factor when $a \to 0$ and $b \to 1$.

Problem 3.5 Consider the use of a four-processor, shared-memory computer for the execution of a program mix. The multiprocessor can be used in four execution modes corresponding to the active use of one, two, three, and four processors. Assume that each processor has a peak execution rate of 5 MIPS.

Let f_i be the percentage of time that i processors will be used in the above program execution and $f_1 + f_2 + f_3 + f_4 = 1$. You can assume the execution rates R_1, R_2, R_3, and R_4, corresponding to the distribution (f_1, f_2, f_3, f_4), respectively.

(a) Derive an expression to show the harmonic mean execution rate R of the multiprocessor in terms of f_i and R_i for $i = 1, 2, 3, 4$. Also show an expression for the arithmetic mean execution time T in terms of R.

(b) What would be the value of the arithmetic mean execution time T of the above program mix given $f_1 = 0.4, f_2 = 0.3, f_3 = 0.2, f_4 = 0.1$ and $R_1 = 4$ MIPS, $R_2 = 8$ MIPS, $R_3 = 11$ MIPS, $R_4 = 15$ MIPS? Explain the possible causes of $R_i < 5i$ for $i = 1, 2, 3, 4$ in the above program execution.

(c) Suppose an intelligent compiler is used to enhance the degree of parallelization in the above program mix with a new distribution $f_1 = 0.1, f_2 = 0.2, f_3 = 0.3, f_4 = 0.4$. What would be the arithmetic mean execution time of the same program under the same assumption on $\{R_i\}$ as in part (b)?

Problem 3.6 Explain the applicability and the restrictions involved in using Amdahl's law, Gustafson's law, and Sun and Ni's law to estimate the speedup performance of an n-processor system compared with that of a single-processor system. Ignore all communication overheads.

Problem 3.7 The following Fortran program is to be executed on a uniprocessor, and a parallel version is to be executed on a shared-memory multiprocessor.

```
L1:              Do 10 I = 1, 1024
L2:                 SUM(I) = 0
L3:                 Do 20 J = 1, I
L4:       20            SUM(I) = SUM(I) + J
L5:       10      Continue
```

Suppose statements 2 and 4 each take two machine cycle times, including all CPU and memory-access activities. Ignore the overhead caused by the software loop control (statements L1, L3, and L5) and all other system overhead and resource conflicts.

(a) What is the total execution time of the program on a uniprocessor?

(b) Divide the *I*-loop iterations among 32 processors with prescheduling as follows: Processor 1 executes the first 32 iterations ($I = 1$ to 32), processor 2 executes the next 32 iterations ($I = 33$ to 64), and so on. What are the execution time and speedup factors compared with part (a)? (Note that the computational workload, dictated by the J-loop, is unbalanced among the processors.)

(c) Modify the given program to facilitate a balanced parallel execution of all the computational workload over 32 processors. By a balanced load, we mean an equal number of additions assigned to each processor with respect to both loops.

(d) What is the minimum execution time resulting from the balanced parallel execution on 32 processors? What is the new speedup over the uniprocessor?

Problem 3.8   Consider the multiplications of two $n \times n$ matrices $A = (a_{ij})$ and $B = (b_{ij})$ on a scalar uniprocessor and on a multiprocessor, respectively. The matrix elements are floating-point numbers, initially stored in the main memory in row-major. The resulting product matrix $C = (c_{ij})$ where $C = A \times B$, should be stored back to memory in contiguous locations.

Assume a 2-address instruction format and an instruction set of your choice. Each load/store instruction takes, on the average, 4 cycles to complete. All ALU operations must be done sequentially on the processor with 2 cycles if no memory reference is required in the instruction. Otherwise, 4 cycles are added for each memory reference to fetch an operand. Branch-type instructions require, on the average, 2 cycles.

(a) Write a minimal-length assembly-language program to perform the matrix multiplication on a scalar processor with a load-store architecture and floating-point hardware.

(b) Calculate the total instruction count, the total number of cycles needed for the program execution, and the average cycles per instruction (CPI).

(c) What is the MIPS rate of this scalar machine, if the processor is driven by a 40-MHz clock?

(d) Suggest a partition of the above program to execute the divided program parts on an N-processor shared-memory system with minimum time. Assume $n = 1000N$. Estimate the potential speedup of the multiprocessor over the uniprocessor, assuming the same type of processors are used in both systems. Ignore the memory-access conflicts, synchronization and other overheads.

(e) Sketch a scheme to perform distributed matrix computations with distributed data sets on an N-node multicomputer with distributed memory. Each node has a computer equivalent to the scalar processor used in part (a).

(f) Specify the message-passing operations required in part (e). Suppose that, on the average, each message passing requires 100 processor cycles to complete. Estimate the total execution time on the multicomputer for the distributed matrix multiplication. Make appropriate assumptions if needed in your timing analysis.

Problem 3.9   Consider the interleaved execution of the four programs in Problem 1.6 on each of the three machines. Each program is executed in a particular mode with the measured MIPS rating.

(a) Determine the arithmetic mean execution time per instruction for each machine executing the combined workload, assuming equal weights for the four programs.

(b) Determine the harmonic mean MIPS rate of each machine.

(c) Rank the machines based on the harmonic mean performance. Compare this ranking with that obtained in Problem 1.6.

Problem 3.10 Answer or prove the following statements related to speedup performance law:

(a) Derive the fixed-memory speedup expression S_n^* in Eq. 3.35 under reasonable assumptions.

(b) Derive Amdahl's law (S_n in Eq. 3.16) as a special case of the S_n^* expression.

(c) Derive Gustafson's law (S_n' in Eq. 3.33) as a special case of the S_n^* expression.

(d) Prove the relation $S_n^* \geq S_n' \geq S_n$ for solving the same problem on the same machine under different assumptions.

Problem 3.11 Prove the following relations among the speedup $S(n)$, efficiency $E(n)$, utilization $U(n)$, redundancy $R(n)$, and quality $Q(n)$ of a parallel computation, based on the definitions given by Lee (1980):

(a) Prove $1/n \leq E(n) \leq U(n) \leq 1$, where n is the number of processors used in the parallel computation.

(b) Prove $1 \leq R(n) \leq 1/E(n) \leq n$.

(c) Prove the expression for $Q(n)$ in Eq. 3.21.

(d) Verify the above relations using the hypothetical workload in Example 3.3.

Problem 3.12 Repeat Example 3.7 for sorting s numbers on five different n-processor machines using the linear array, 2D-mesh, 3D-mesh, hypercube, and Omega network as interprocessor communication architectures, respectively.

(a) Show the scalability of the five architectures as compared with the EREW-PRAM model.

(b) Compare the results obtained in part (a) with those in Example 3.7. Based on theses two benchmark results, rank the relative scalability of the five architectures. Can the results be generalized to the performance of other algorithms?

Problem 3.13 Consider the execution of two benchmark programs. The performance of three computers running these two benchmarks are given below:

Benchmark	Millions of floating-point operations	Computer 1 T_1 (sec.)	Computer 2 T_2 (sec.)	Computer 3 T_3 (sec.)
Problem 1	100	1	10	20
Problem 2	100	1000	100	20
Total time		1001	110	40

(a) Calculate R_a, R_g, and R_h for each computer under the equal-weight assumption $f_1 = f_2 = 0.5$.

(b) When benchmark 1 has a constant $R_1 = 10$ Mflops performance across the three computers, plot R_a, R_g, and R_h as a function of R_2, which varies from 1 to 100 Mflops under the assumption $f_1 = 0.8$ and $f_2 = 0.2$.

(c) Repeat part (b) for the case $f_1 = 0.2$ and $f_2 = 0.8$.

(d) From the above performance results under different conditions, can you draw a conclusion regarding the relative performance of the three machines?

Problem 3.14 In Example 3.5, four parallel algorithms are mentioned for multiplication of $s \times s$ matrices. After reading the original papers describing these algorithms, prove the following communication overheads on the target machine architectures:

(a) Prove that $h(s, n) = O(n \log n + s^2 \sqrt{n})$ when mapping the Fox-Otto-Hey algorithm on a $\sqrt{n} \times \sqrt{n}$ torus.

(b) Prove that $h(s, n) = O(n^{4/3} + n \log n + s^2 n^{1/3})$ when mapping the Berntsen's algorithm on a hypercube with $n = 2^{3k}$ nodes, where $k \leq \frac{1}{2} \log s$.

(c) Prove that $h(s, n) = O(n \log n + s^3)$ when mapping the Dekel-Nassimi-Sahni algorithm on a hypercube with $n = s^3 = 2^{3k}$ nodes.

Problem 3.15 Xian-He Sun (1992) has introduced an *isospeed* concept for scalability analysis. The concept is to maintain a fixed speed for each processor while increasing the problem size. Let W and W' be two workloads corresponding to two problem sizes. Let N and N' be two machine sizes (in terms of the number of processors). Let T_N and $T_{N'}$ be the parallel execution times using N and N' processors, respectively.

The isospeed is achieved when $W/(NT_N) = W'/(N'T_{N'})$. The *isoefficiency* concept defined by Kumar and Rao (1987) is achieved by maintaining a fixed efficiency through $S_N(W)/N = S_{N'}(W')/N'$, where $S_N(W)$ and $S_{N'}(W')$ are the corresponding speedup factors.

Prove that the two concepts are indeed equivalent if (i) the speedup factors are defined as the ratio of *parallel speed* R_N to *sequential speed* R_1 (rather than as the ratio of sequential execution time to parallel execution time, and (ii) $R_1(W) = R_1(W')$. In other words, isoefficiency is identical to isospeed when the sequential speed is fixed as the problem size is increased.

Part II

Hardware Technologies

Chapter 4
Processors and Memory Hierarchy

Chapter 5
Bus, Cache, and Shared Memory

Chapter 6
Pipelining and Superscalar Techniques

Summary

Part II contains three chapters dealing with hardware technologies underlying the development of parallel processing computers. The discussions cover advanced processors, memory hierarchy, and pipelining technologies. These hardware units must work with software, and matching hardware design with program behavior is the main theme of these chapters.

We will study RISC, CISC, scalar, superscalar, VLIW, superpipelined, vector, and symbolic processors. Digital bus, cache design, shared memory, and virtual memory technologies will be considered. Advanced pipelining principles and their applications are described for memory access, instruction execution, arithmetic computation, and vector processing. These chapters are hardware-oriented. Readers whose interest is mainly in software can skip Chapters 5 and 6 after reading Chapter 4.

The material in Chapter 4 presents the functional architectures of processors and memory hierarchy and will be of interest to both computer designers and programmers. After reading Chapter 4, one should have a clear picture of the logical structure of computers. Chapters 5 and 6 describe physical design of buses, cache operations, processor architectures, memory organizations, and their management issues.

Chapter 4

Processors and Memory Hierarchy

This chapter presents modern processor technology and the supporting memory hierarchy. We begin with a study of instruction-set architectures including CISC and RISC, and we consider typical superscalar, VLIW, superpipelined, and vector processors. The third section covers memory hierarchy and capacity planning, and the final section introduces virtual memory, address translation mechanisms, and page replacement methods.

Instruction-set processor architectures and logical addressing aspects of the memory hierarchy are emphasized at the functional level. This treatment is directed toward the programmer or computer science major. Detailed hardware designs for bus, cache, and main memory are studied in Chapter 5. Instruction and arithmetic pipelines and superscalar and superpipelined processors are further treated in Chapter 6.

4.1 Advanced Processor Technology

Architectural families of modern processors are introduced below with the underlying microelectronics/packaging technologies. The coverage spans from VLSI microprocessors used in workstations or multiprocessors to heavy-duty processors used in mainframes and supercomputers.

Major processor families to be studied include the *CISC*, *RISC*, *superscalar*, *VLIW*, *superpipelined*, *vector*, and *symbolic* processors. Scalar and vector processors are for numerical computations. Symbolic processors are developed for AI applications.

4.1.1 Design Space of Processors

Various processor families can be mapped onto a coordinated space of *clock rate* versus *cycles per instruction* (CPI), as illustrated in Fig. 4.1. As implementation technology evolves rapidly, the clock rates of various processors are gradually moving from low to higher speeds toward the right of the design space. Another trend is that processor manufacturers are trying to lower the CPI rate using hardware and software approaches.

Based on these trends, the mapping of processors in Fig. 4.1 reflects their implementation during the past decade. As time passes, some of the mapped ranges may move toward the lower right corner of the design space.

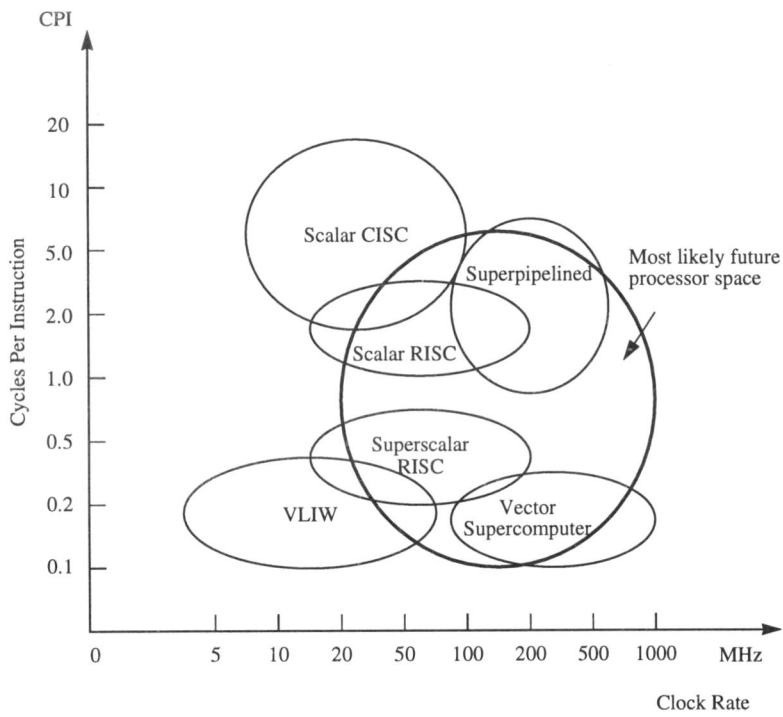

Figure 4.1 Design space of modern processor families.

The Design Space   Conventional processors like the Intel i486, M68040, VAX/8600, IBM 390, etc., fall into the family known as *complex-instruction-set computing* (CISC) architecture. The typical clock rate of today's CISC processors ranges from 33 to 50 MHz. With microprogrammed control, the CPI of different CISC instructions varies from 1 to 20. Therefore, CISC processors are at the upper left of the design space.

Today's *reduced-instruction-set computing* (RISC) processors, such as the Intel i860, SPARC, MIPS R3000, IBM RS/6000, etc., have faster clock rates ranging from 20 to 120 MHz determined by the implementation technology employed. With the use of hardwired control, the CPI of most RISC instructions has been reduced to one to two cycles.

A special subclass of RISC processors are the *superscalar processors*, which allow multiple instructions to be issued simultaneously during each cycle. Thus the effective CPI of a superscalar processor should be lower than that of a generic scalar RISC processor. The clock rate of superscalar processors matches that of scalar RISC processors.

The *very long instruction word* (VLIW) architecture uses even more functional units than that of a superscalar processor. Thus the CPI of a VLIW processor can be further lowered. Due to the use of very long instructions (256 to 1024 bits per instruction), VLIW processors have been mostly implemented with microprogrammed control. Thus the clock rate is slow with the use of *read-only memory* (ROM). A large number of microcode access cycles may be needed for some instructions.

Superpipelined processors are those that use multiphase clocks with a much increased clock rate ranging from 100 to 500 MHz. However, the CPI rate is rather high unless superpipelining is practiced jointly with multiinstruction issue. The processors in *vector supercomputers* are mostly superpipelined and use multiple functional units for concurrent scalar and vector operations.

The effective CPI of a processor used in a supercomputer should be very low, positioned at the lower right corner of the design space. However, the cost increases appreciably if a processor design is restricted to the lower right corner. The subspaces in which the various processor families will be designed in the future are by no means fixed.

Instruction Pipelines The execution cycle of a typical instruction includes four phases: *fetch, decode, execute,* and *write-back.* These instruction phases are often executed by an *instruction pipeline* as demonstrated in Fig. 4.2a. In other words, we can simply model an *instruction processor* by such a pipeline structure.

For the time being, we will use an abstract pipeline model for an intuitive explanation of various processor classes. The *pipeline,* like an industrial assembly line, receives successive instructions from its input end and executes them in a streamlined, overlapped fashion as they flow through.

A *pipeline cycle* is intuitively defined as the time required for each phase to complete its operation, assuming equal delay in all phases (pipeline stages). Introduced below are the basic definitions associated with instruction pipeline operations:

(1) *Instruction pipeline cycle* — the clock period of the instruction pipeline.

(2) *Instruction issue latency* — the time (in cycles) required between the issuing of two adjacent instructions.

(3) *Instruction issue rate* — the number of instructions issued per cycle, also called the *degree* of a superscalar processor.

(4) *Simple operation latency* — Simple operations make up the vast majority of instructions executed by the machine, such as *integer adds, loads, stores, branches, moves,* etc. On the contrary, complex operations are those requiring an order-of-magnitude longer latency, such as *divides, cache misses,* etc. These latencies are measured in number of cycles.

(5) *Resource conflicts* — This refers to the situation where two or more instructions demand use of the same functional unit at the same time.

A *base scalar processor* is defined as a machine with *one* instruction issued per cycle, a *one*-cycle latency for a simple operation, and a *one*-cycle latency between instruction issues. The instruction pipeline can be fully utilized if successive instructions can enter

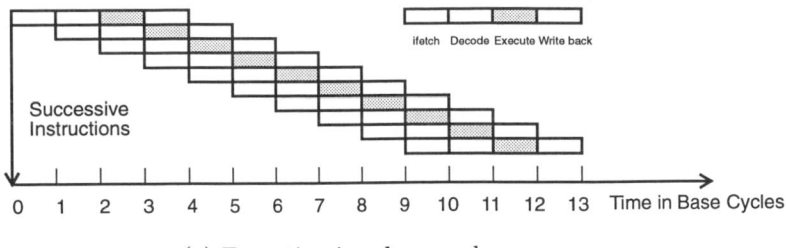

(a) Execution in a base scalar processor

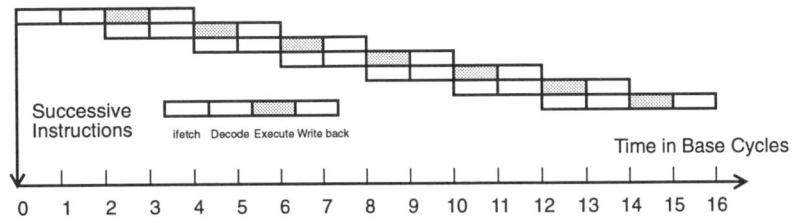

(b) Underpipelined with two cycles per instruction issue

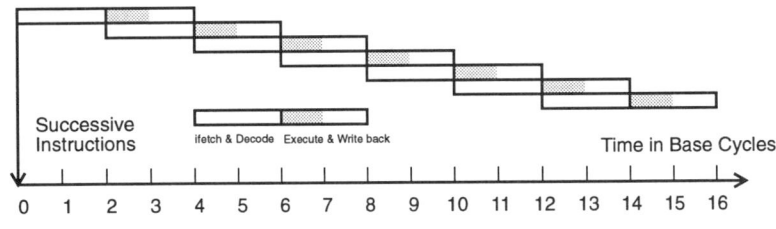

(c) Underpipelined with twice the base cycle

Figure 4.2 Pipelined execution of successive instructions in a base scalar processor and in two underpipelined cases. Courtesy of Jouppi and Wall; reprinted from *Proc. ASPLOS*, ACM Press, 1989)

it continuously at the rate of one per cycle, as shown in Fig. 4.2a.

However, the instruction issue latency can be more than one cycle for various reasons (to be discussed in Chapter 6). For example, if the instruction issue latency is two cycles per instruction, the pipeline can be underutilized, as demonstrated in Fig. 4.2b.

Another underpipelined situation is shown in Fig. 4.2c, in which the pipeline cycle time is doubled by combining pipeline stages. In this case, the *fetch* and *decode* phases are combined into one pipeline stage, and *execute* and *write-back* are combined into another stage. This will also result in poor pipeline utilization.

The effective CPI rating is 1 for the ideal pipeline in Fig. 4.2a, and 2 for the case in Fig. 4.2b. In Fig. 4.2c, the clock rate of the pipeline has been lowered by one-half. According to Eq. 1.3, either the case in Fig. 4.2b or that in Fig. 4.2c will reduce the performance by one-half, compared with the ideal case (Fig. 4.2a) for the base machine.

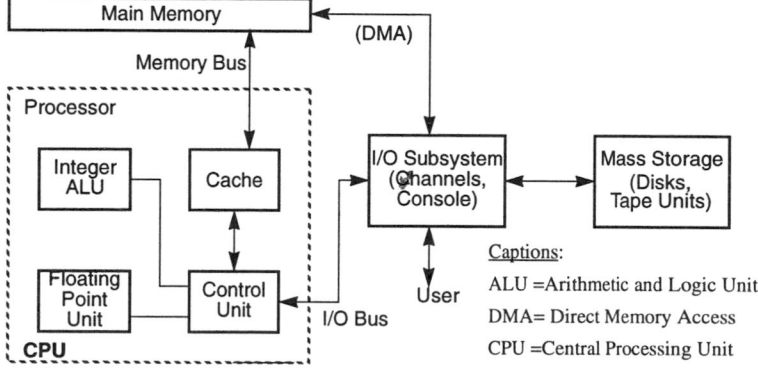

(a) CPU with built-in floating-point unit

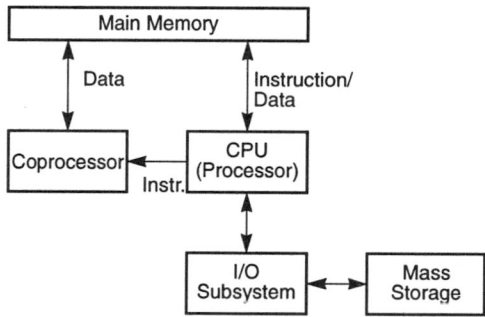

(b) CPU with an attached coprocessor

Figure 4.3 Architectural models of a basic scalar computer system.

Processors and Coprocessors   The central processor of a computer is called the *central processing unit* (CPU) as illustrated in Fig. 4.3. This CPU is essentially a scalar processor, which may consist of multiple functional units such as an integer *arithmetic and logic unit* (ALU), a floating-point accelerator, etc.

The floating-point unit can be built on a *coprocessor* (Fig. 4.3b) which is attached to the CPU. The coprocessor executes instructions dispatched from the CPU. A coprocessor may be a *floating-point accelerator* executing scalar data, *vector processor* executing vector operands, a *digital signal processor* (DSP), or a *Lisp processor* executing AI programs. Coprocessors cannot handle I/O operations.

Table 4.1 lists some processor-coprocessor pairs developed in recent years to speed up numerical computations. Coprocessors cannot be used alone. The processor and coprocessor operate with a host–back-end relationship and compatibility between them is a necessity. Some authors call coprocessors *attached processors* or *slave processors*. An attached processor may be more powerful than its host. For example, the Cray Y-MP is a back-end processor driven by a small minicomputer.

Table 4.1 Coprocessor-Processor Pairs and Characteristics

Coprocessor	Compatible Processor	Coprocessor Characteristics
Intel 8087	Intel 8086/8088	5 MHz, 70 cycles for Add and 700 cycles for Log
Intel 80287	Intel 80286	12.5 MHz, 30 cycles for Add and 264 cycles for Log
Intel 387DX	Intel 386DX	33 MHz, 12 cycles for Add and 210 cycles for Log
Intel i486	Intel i486 (the same chip)	33 MHz, 8 cycles for Add and 171 cycles for Log
Motorola MC68882	Motorola MC68020/68030	40 MHz, 56 cycles for Add and 574 cycles for Log
Weitek 3167	Intel 386DX	33 MHz, 6 cycles for Add and 365 cycles for Log by software emulation
Weitek 4167	Intel i486	33 MHz, 2 cycles for Add and not available for Log

4.1.2 Instruction-Set Architectures

In this section, we characterize computer instruction sets and examine hardware features built into generic RISC and CISC scalar processors. Distinctions between them are revealed. The boundary between RISC and CISC architectures has become blurred in recent years. Quite a few processors are now built with hybrid RISC and CISC features based on the same technology. However, the distinction is still rather sharp in instruction-set architectures.

The instruction set of a computer specifies the primitive commands or machine instructions that a programmer can use in programming the machine. The complexity of an instruction set is attributed to the instruction formats, data formats, addressing modes, general-purpose registers, opcode specifications, and flow control mechanisms used. Based on past experience in processor design, two schools of thought on instruction-set architectures have evolved, namely, CISC and RISC.

Complex Instruction Sets   In the early days of computer history, most computer families started with an instruction set which was rather simple. The main reason for being simple then was the high cost of hardware. The hardware cost has dropped and the software cost has gone up steadily in the past three decades. Furthermore, the semantic gap between HLL features and computer architecture has widened.

The net result is that more and more functions have been built into the hardware, making the instruction set very large and very complex. The growth of instruction sets was also encouraged by the popularity of microprogrammed control in the 1960s and 1970s. Even user-defined instruction sets were implemented using microcodes in some processors for special-purpose applications.

A typical CISC instruction set contains approximately 120 to 350 instructions using variable instruction/data formats, uses a small set of 8 to 24 *general-purpose registers* (GPRs), and executes a large number of memory reference operations based on more than a dozen addressing modes. Many HLL statements are directly implemented in hardware/firmware in a CISC architecture. This may simplify the compiler development, improve execution efficiency, and allow an extension from scalar instructions to vector and symbolic instructions.

Reduced Instruction Sets We started with RISC instruction sets and gradually moved to CISC instruction sets during the 1980s. After two decades of using CISC processors, computer users began to reevaluate the performance relationship between instruction-set architecture and available hardware/software technology.

Through many years of program tracing, computer scientists realized that only 25% of the instructions of a complex instruction set are frequently used about 95% of the time. This implies that about 75% of hardware-supported instructions often are not used at all. A natural question then popped up: Why should we waste valuable chip area for rarely used instructions?

With low-frequency elaborate instructions demanding long microcodes to execute them, it may be more advantageous to remove them completely from the hardware and rely on software to implement them. Even if the software implementation is slow, the net result will be still a plus due to their low frequency of appearance. Pushing rarely used instructions into software will vacate chip areas for building more powerful RISC or superscalar processors, even with on-chip caches or floating-point units.

A RISC instruction set typically contains less than 100 instructions with a fixed instruction format (32 bits). Only three to five simple addressing modes are used. Most instructions are register-based. Memory access is done by load/store instructions only. A large register file (at least 32) is used to improve fast context switching among multiple users, and most instructions execute in one cycle with hardwired control.

Because of the reduction in instruction-set complexity, the entire processor is implementable on a single VLSI chip. The resulting benefits include a higher clock rate and a lower CPI, which lead to higher MIPS ratings as reported on commercially available RISC/superscalar processors.

Architectural Distinctions Hardware features built into CISC and RISC processors are compared below. Figure 4.4 shows the architectural distinctions between modern CISC and traditional RISC. Some of the distinctions may disappear, because future processors may be designed with features from both types.

Conventional CISC architecture uses a unified cache for holding both instructions and data. Therefore, they must share the same data/instruction path. In a RISC processor, separate *instruction* and *data caches* are used with different access paths. However, exceptions do exist. In other words, CISC processors may also use split codes.

The use of microprogrammed control can be found in traditional CISC, and hard-wired control in most RISC. Thus control memory (ROM) is needed in earlier CISC processors, which may significantly slow down the instruction execution. However, modern CISC may also use hardwired control. Therefore, split caches and hardwired control are not exclusive in RISC machines.

Using hardwired control will reduce the CPI effectively to one instruction per cycle if pipelining is carried out perfectly. Some CISC processors also use split caches and hardwired control, such as the MC68040 and i586.

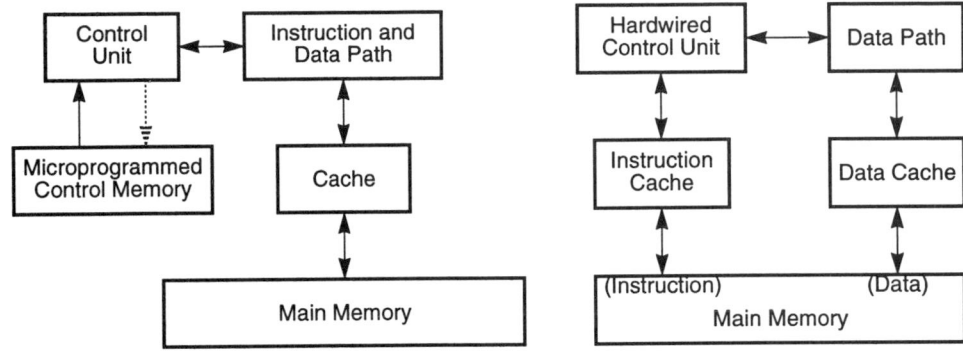

(a) The CISC architecture with micropro-grammed control and unified cache

(b) The RISC architecture with hardwired control and split instruc-tion cache and data cache.

Figure 4.4 Distinctions between typical RISC and typical CISC processor archi-tectures. (Courtesy of Gordon Bell, 1989)

In Table 4.2, we compare the main features of RISC and CISC processors. The comparison involves five areas: *instruction sets, addressing modes, register file and cache design, clock rate and expected CPI*, and *control mechanisms*.

The large number of instructions used in a CISC processor is the result of using variable-format instructions — integer, floating-point, and vector data — and of using over a dozen different addressing modes. Furthermore, with few GPRs many more instructions access the memory for operands. The CPI is thus high as a result of the long microcodes used to control the execution of some complex instructions.

On the other hand, most RISC processors use 32-bit instructions which are pre-dominantly register-based. With few simple addressing modes, the memory-access cycle is broken into pipelined access operations involving the use of caches and working reg-isters. Using a large register file and separate I- and D-caches benefits internal data forwarding and eliminates unnecessary storage of intermediate results. With hardwired control, the CPI is reduced to 1 for most RISC instructions.

Table 4.2 Characteristics of CISC and RISC Architectures

Architectural Characteristic	Complex Instruction Set Computer (CISC)	Reduced Instruction Set Computer (RISC)
Instruction-set size and instruction formats	Large set of instructions with variable formats (16–64 bits per instruction).	Small set of instructions with fixed (32-bit) format and most register-based instructions.
Addressing modes	12–24.	Limited to 3–5.
General-purpose registers and cache design	8–24 GPRs, mostly with a unified cache for instructions and data, recent designs also use split caches.	Large numbers (32–192) of GPRs with mostly split data cache and instruction cache.
Clock rate and CPI	33–50 MHz in 1992 with a CPI between 2 and 15.	50–150 MHz in 1993 with one cycle for almost all instructions and an average CPI < 1.5.
CPU Control	Most microcoded using control memory (ROM), but modern CISC also uses hardwired control.	Most hardwired without control memory.

4.1.3 CISC Scalar Processors

A scalar processor executes with scalar data. The simplest scalar processor executes integer instructions using fixed-point operands. More capable scalar processors execute both integer and floating-point operations. A modern scalar processor may possess both an integer unit and a floating-point unit in the same CPU as shown in Fig. 4.3a. Based on a complex instruction set, a *CISC scalar processor* can be built either with a single chip or with multiple chips mounted on a processor board.

In the ideal case, a CISC scalar processor should have a performance equal to that of the base scalar processor demonstrated in Fig. 4.2a. However, the processor is often underpipelined as in either of the two cases shown in Figs. 4.2b and 4.2c. Major causes of the underpipelined situations (Figs. 4.2b) include data dependence among instructions, resource conflicts, branch penalties, and logic hazards which will be studied in Chapter 6.

The case in Fig. 4.2c is caused by using a clock cycle which is greater than the simple operation latency. In subsequent sections, we will show how RISC, superscalar, and superpipelining techniques can be applied to improve pipeline performance.

Representative CISC Processors In Table 4.3, three representative CISC scalar processors are listed. The VAX 8600 processor is built on a PC board. The i486 and M68040 are single-chip microprocessors. They are still widely used at present. We use these popular architectures to explain some interesting features built into modern CISC machines. In any processor design, the designer attempts to achieve higher throughput in the processor pipelines.

Both hardware and software mechanisms have been developed to achieve these goals. Due to the complexity involved in a CISC scalar processor, the most difficult task for a designer is to shorten the clock cycle to match the simple operation latency. This problem is easier to overcome with a RISC architecture.

Example 4.1 The Digital Equipment VAX 8600 processor architecture

The VAX 8600 was introduced by Digital Equipment Corporation in 1985. This machine implements a typical CISC architecture with microprogrammed control. The instruction set contains about 300 instructions with 20 different addressing modes. As shown in Fig. 4.5, the VAX 8600 executes the same instruction set, runs the same VMS operating system, and interfaces with the same I/O buses (such as SBI and Unibus) as the VAX 11/780.

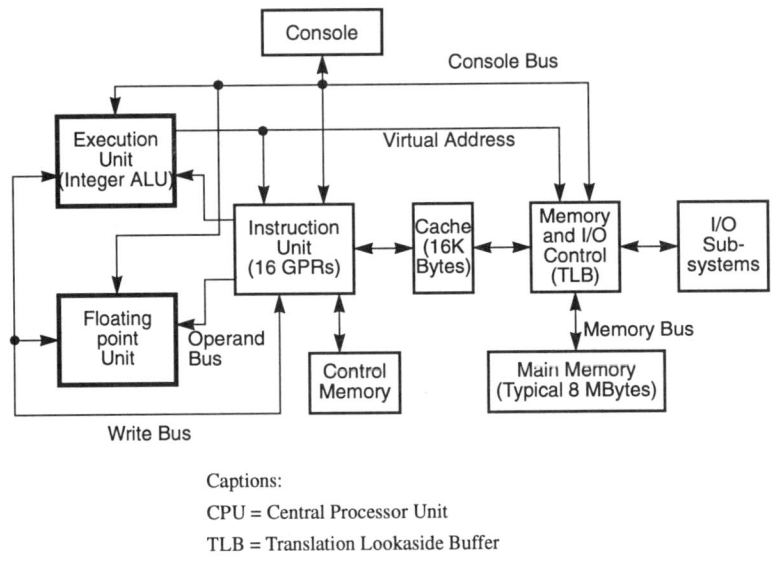

Figure 4.5 The VAX 8600 CPU, a typical CISC processor architecture. (Courtesy of Digital Equipment Corporation, 1985)

The CPU in the VAX 8600 consists of two functional units for concurrent execution of integer and floating-point instructions. The unified cache is used for holding both instructions and data. There are 16 GPRs in the instruction unit. Instruction pipelining has been built with six stages in the VAX 8600, as in most CISC machines. The instruction unit prefetches and decodes instructions, handles branching operations, and supplies operands to the two functional units in a pipelined fashion.

A *translation lookaside buffer* (TLB) is used in the memory control unit for fast generation of a physical address from a virtual address. Both integer and floating-point units are pipelined. The performance of the processor pipelines relies heavily on the cache hit ratio and on minimal branching damage to the pipeline flow.

The CPI of a VAX 8600 instruction varies within a wide range from 2 cycles to as high as 20 cycles. For example, both *multiply* and *divide* may tie up the execution unit for a large number of cycles. This is caused by the use of long sequences of microinstructions to control hardware operations.

The general philosophy of designing a CISC processor is to implement useful instructions in hardware/firmware which may result in a shorter program length with a lower software overhead. However, this advantage has been obtained at the expense of a lower clock rate and a higher CPI, which may not pay off at all.

The VAX 8600 was improved from the earlier VAX/11 Series. The system was later further upgraded to the VAX 9000 Series offering both vector hardware and multiprocessor options. All the VAX Series have used a paging technique to allocate the physical memory to user programs.

CISC Microprocessor Families   In 1971, the Intel 4004 appeared as the first microprocessor based on a 4-bit ALU. Since then, Intel has produced the 8-bit 8008, 8080, and 8085. Intel's 16-bit processors appeared in 1978 as the 8086, 8088, 80186, and 80286. In 1985, the 80386 appeared as a 32-bit machine. The 80486 and Pentium are the latest 32-bit micros in the Intel 80x86 family.

Motorola produced its first 8-bit micro, the MC6800, in 1974, then moved to the 16-bit 68000 in 1979, and then to the 32-bit 68020 in 1984. The latest are the MC68030 and MC68040 in the Motorola MC680x0 family. National Semiconductor's latest 32-bit micro is the NS 32532 introduced in 1988. These microprocessor families are widely used in the *personal computer* (PC) industry.

In recent years, the parallel computer industry has begun to build experimental systems with a large number of open-architecture microprocessors. Both CISC and RISC microprocessors have been employed in these efforts. One thing worthy of mention is the compatibility of new models with the old ones in each of the families. This makes it easier to port software along the series of models.

Example 4.2 The Motorola MC68040 microprocessor architecture

The MC68040 is a 0.8-μm HCMOS microprocessor containing more than 1.2 million transistors, comparable to the 80486. Figure 4.6 shows the MC68040 architecture. The processor implements over 100 instructions using 16 general-purpose registers, a 4-Kbyte data cache, and a 4-Kbyte instruction cache, with separate *memory management units* (MMUs) supported by an *address translation cache* (ATC), which is equivalent to the TLB used in other systems. The data formats range from 8 to 80 bits, based on the IEEE floating-point standard.

Eighteen addressing modes are supported, including register direct and indirect, indexing, memory indirect, program counter indirect, absolute, and immediate

Table 4.3 Representative CISC Scalar Processors

Feature	Intel i486	Motorola MC68040	NS 32532
Instruction-set size and word length	157 instructions, 32 bits.	113 instructions, 32 bits.	63 instructions, 32 bits.
Addressing modes	12	18	9
Integer unit and GPRs	32-bit ALU with 8 registers.	32-bit ALU with 16 registers.	32-bit ALU with 8 registers.
On-chip cache(s) and MMUs	8-KB unified cache for both code and data.	4-KB code cache 4-KB data cache with separate MMUs.	512-B code cache 1-KB data cache.
Floating-point unit, registers, and function units	On-chip with 8 FP registers adder, multiplier, shifter.	On-chip with 3 pipeline stages, 8 80-bit FP registers.	Off-chip FPU NS 32381, or WTL 3164.
Pipeline stages	5	6	4
Protection level	4	2	2
Memory organization and TLB/ATC entries	Segmented paging with 4 KB/page and 32 entries in TLB.	Paging with 4 or 8 KB/page, 64 entries in each ATC.	Paging with 4 KB/page, 64 entries.
Technology, clock rate, packaging, and year introduced	CHMOS IV, 25 MHz, 33 MHz, 1.2M transistors, 168 pins, 1989.	0.8-μm HCMOS, 1.2M transistors, 20 MHz, 40 MHz, 179 pins, 1990.	1.25-μm CMOS 370K transistors, 30 MHz, 175 pins, 1987.
Claimed performance	24 MIPS at 25 MHz,	20 MIPS at 25 MHz, 30 MIPS at 60 MHz.	15 MIPS at 30 MHz.
Successors to watch	Pentium processor.	MC68060, MC68302.	unknown.

modes. The instruction set includes data movement, integer, BCD, and floating-point arithmetic, logical, shifting, bit-field manipulation, cache maintenance, and multiprocessor communications, in addition to program and system control and memory management instructions.

The integer unit is organized in a six-stage instruction pipeline. The floating-point unit consists of three pipeline stages (details to be studied in Section 6.4.1). All instructions are decoded by the integer unit. Floating-point instructions are forwarded to the floating-point unit for execution.

Separate instruction and data buses are used to and from the instruction and data memory units, respectively. Dual MMUs allow interleaved fetch of instructions and data from the main memory. Both the address bus and the data bus are 32 bits wide.

Three simultaneous memory requests can be generated by the dual MMUs,

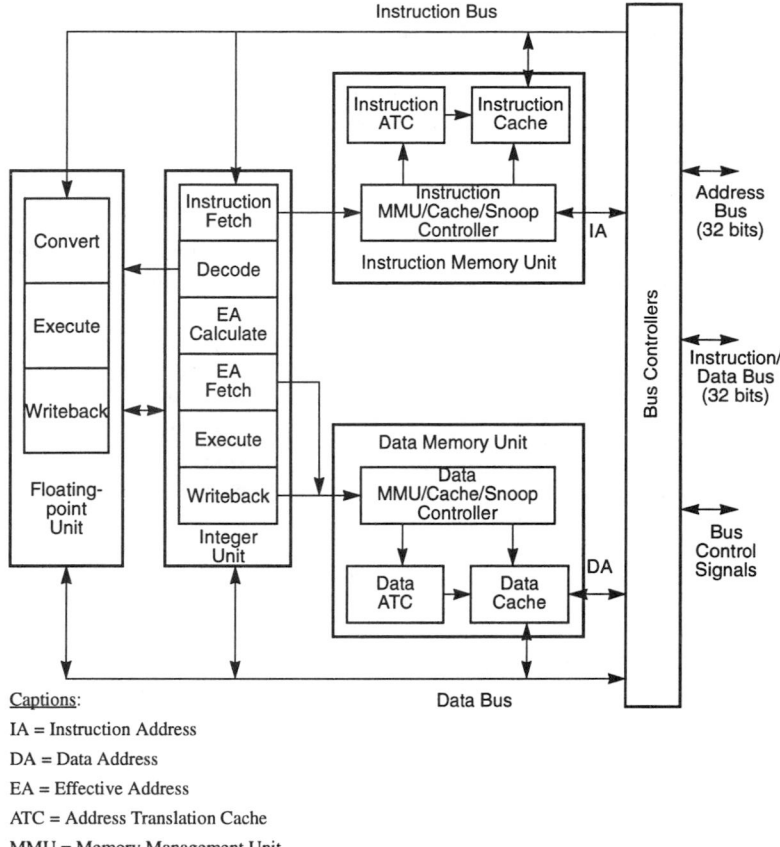

Captions:

IA = Instruction Address

DA = Data Address

EA = Effective Address

ATC = Address Translation Cache

MMU = Memory Management Unit

Figure 4.6 Architecture of the MC68040 processor. (Courtesy of Motorola Inc., 1991)

including data operand read and write and instruction pipeline refill. Snooping logic is built into the memory units for monitoring bus events for cache invalidation.

The complete memory management is provided with a virtual demand paged operating system. Each of the two ATCs has 64 entries providing fast translation from virtual address to physical address. With the CISC complexity involved, the M68040 does not provide delayed branch hardware support, which is often found in RISC processors like Motorola's M88100 microprocessor. ■

4.1.4 RISC Scalar Processors

Generic RISC processors are called *scalar RISC* because they are designed to issue one instruction per cycle, similar to the base scalar processor shown in Fig. 4.2a. In theory, both RISC and CISC scalar processors should perform about the same if they run with the same clock rate and with equal program length. However, these two assumptions are not always valid, as the architecture affects the quality and density of code generated by compilers.

The RISC design gains its power by pushing some of the less frequently used operations into software. The reliance on a good compiler is much more demanding in a RISC processor than in a CISC processor. Instruction-level parallelism is exploited by pipelining in both processor architectures.

Without a high clock rate, a low CPI, and good compilation support, neither CISC nor RISC can perform well as designed. The simplicity introduced with a RISC processor may lead to the ideal performance of the base scalar machine modeled in Fig. 4.2a.

Table 4.4 Representative RISC Scalar Processors

Feature	Sun SPARC CY7C601	Intel i860	Motorola M 88100	AMD 29000
Instruction set, formats, addressing modes.	69 instructions, 32-bit format, 7 data types, 4-stage instr. pipeline.	82 instructions, 32-bit format, 4 addressing modes.	51 instructions, 7 data types, 3 instr. formats, 4 addressing modes.	112 instructions, 32-bit format, all registers indirect addressing.
Integer unit, GPRs.	32-bit RISC/IU, 136 registers divided into 8 windows.	32-bit RISC core, 32 registers.	32-bit IU with 32 GPRs and scoreboarding.	32-bit IU with 192 registers without windows.
Caches(s), MMU, and memory organization.	Off-chip cache/MMU on CY7C604 with 64-entry TLB.	4-KB code, 8-KB data, on-chip MMU, paging with 4 KB/page.	Off-chip M88200 caches/MMUs, segmented paging, 16-KB cache.	On-chip MMU with 32-entry TLB, with 4-word prefetch buffer and 512-B branch target cache.
Floating-point unit registers and functions	Off-chip FPU on CY7C602, 32 registers, 64-bit pipeline (equiv. to TI 8848).	On-chip 64-bit FP multiplier and FP adder with 32 FP registers, 3-D graphics unit.	On-chip FPU adder, multiplier with 32 FP registers and 64-bit arithmetic.	Off-chip FPU on AMD 29027, on-chip FPU with AMD 29050.
Operation modes	Concurrent IU and FPU operations.	Allow dual instructions and dual FP operations.	Concurrent IU, FPU and memory access with delayed branch.	4-stage pipeline processor.
Technology, clock rate, packaging, and year	0.8-μm CMOS IV, 33 MHz, 207 pins, 1989.	1-μm CHMOS IV, over 1M transistors, 40 MHz, 168 pins, 1989	1-μm HCMOS, 1.2M transistors, 20 MHz, 180 pins, 1988.	1.2-μm CMOS, 30 MHz, 40 MHz, 169 pins, 1988.
Claimed performance and successor to watch	24 MIPS for 33 MHz version, 50 MIPS for 80 MHz ECL version. Up to 32 register windows can be built.	40 MIPS and 60 Mflops for 40 MHz, i860/XP announced in 1992 with 2.5M transistors.	17 MIPS and 6 Mflops at 20 MHz, up to 7 special function units can be configured.	27 MIPS at 40 MHz, new version AMD 29050 at 55 MHz in 1990.

Representative RISC Processors Four RISC-based processors, the Sun SPARC, Intel i860, Motorola M88100, and AMD 29000, are summarized in Table 4.4. All of these processors use 32-bit instructions. The instruction sets consist of 51 to 124 basic instructions. On-chip floating-point units are built into the i860 and M88100, while the SPARC and AMD use off-chip floating-point units. We consider these four processors generic scalar RISC issuing essentially only one instruction per pipeline cycle.

Among the four scalar RISC processors, we choose to examine the Sun SPARC and i860 architectures below. SPARC stands for *scalable processor architecture*. The scalability of the SPARC architecture refers to the use of a different number of *register windows* in different SPARC implementations.

This is different from the M88100, where scalability refers to the number of *special-function units* (SFUs) implementable on different versions of the M88000 processor. The Sun SPARC is derived from the original Berkeley RISC design.

Example 4.3 The Sun Microsystems SPARC architecture

The SPARC has been implemented by a number of licensed manufacturers as summarized in Table 4.5. Different technologies and window numbers are used by different SPARC manufacturers.

Table 4.5 SPARC Implementations by Licensed Manufacturers

SPARC Chip	Technology	Clock Rate (MHz)	Claimed VAX MIPS	Remarks
Cypress CY7C601 IU	$0.8-\mu m$ CMOS IV, 207 pins.	33	24	CY7C602 FPU delivers 4.5 Mflops DP Linpack, CY7C604 Cachc/MMC, CY7C157 Cache.
Fujitsu MB 86901 IU	$1.2-\mu m$ CMOS, 179 pins.	25	15	MB 86911 FPC FPC and TI 8847 FPP, MB86920 MMU, 2.7 Mflops DP Linpack by FPU.
LSI Logic L64811	$1.0-\mu m$ HCMOS, 179 pins.	33	20	L64814 FPU, L64815 MMU.
TI 8846	$0.8-\mu m$ CMOS	33	24	42 Mflops DP Linpack on TI-8847 FPP.
BIT IU B-3100	ECL family.	80	50	15 Mflops DP Linpack on FPUs: B-3120 ALU, B-3611 FP Multiply/Divide.

All of these manufacturers implement the *floating-point unit* (FPU) on a separate coprocessor chip. The SPARC processor architecture contains essentially a

RISC *integer unit* (IU) implemented with 2 to 32 register windows.

We choose to study the SPARC family chips produced by Cypress Semiconductors, Inc. Figure 4.7 shows the architecture of the Cypress CY7C601 SPARC processor and of the CY7C602 FPU. The Sun SPARC instruction set contains 69 basic instructions, a significant increase from the 39 instructions in the original Berkeley RISCII instruction set.

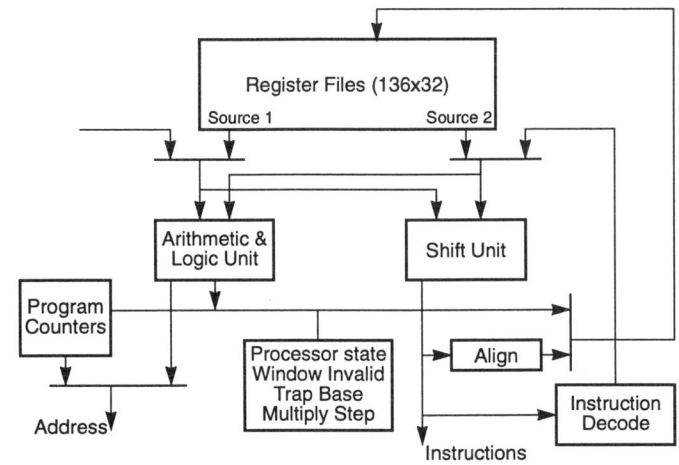

(a) The Cypress CY7C601 SPARC processor

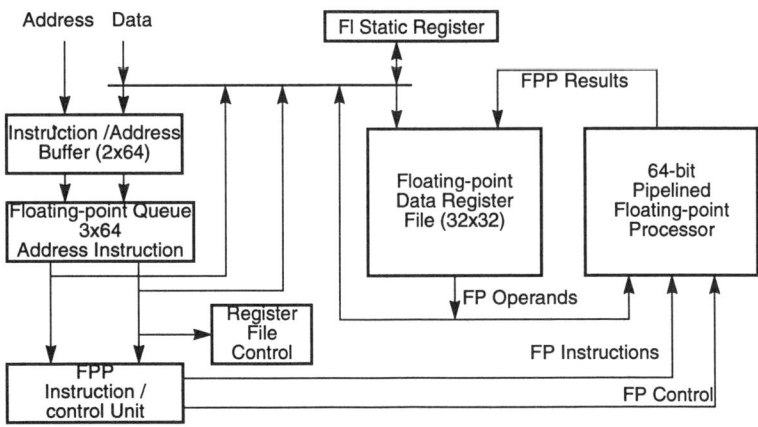

(b) The Cypress CY7C602 floating-point unit

Figure 4.7 The SPARC architecture with the processor and the floating-point unit on two separate chips. (Courtesy of Cypress Semiconductor Co., 1991)

The SPARC runs each procedure with a set of thirty-two 32-bit IU registers.

Eight of these registers are *global registers* shared by all procedures, and the remaining 24 are *window registers* associated with only each procedure. The concept of using overlapped register windows is the most important feature introduced by the Berkeley RISC architecture.

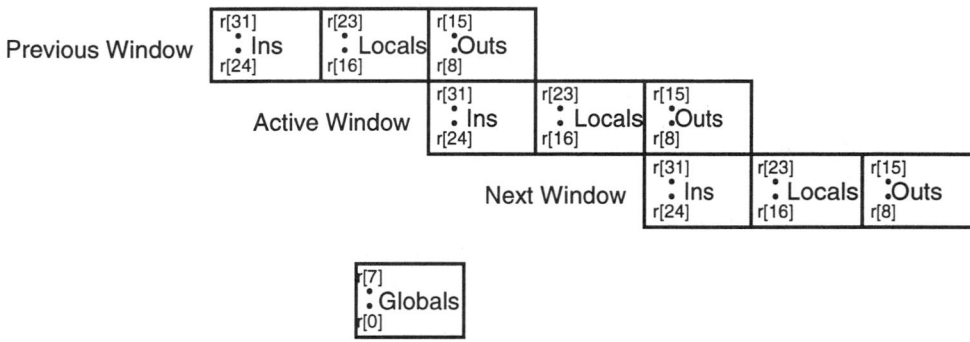

(a) Three overlapping register windows and the global registers

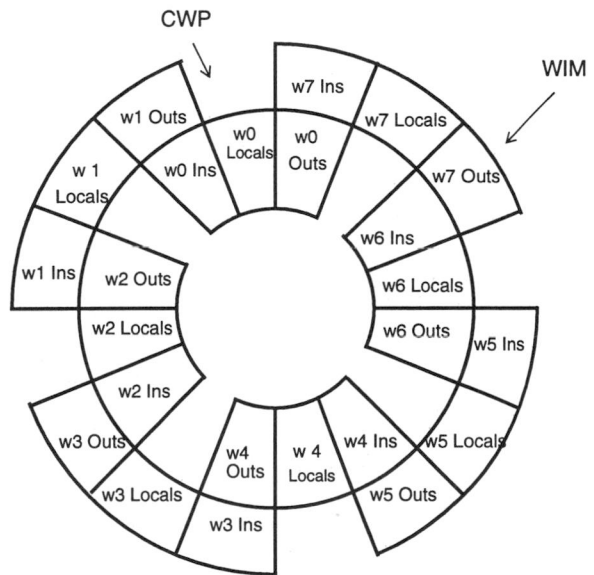

(b) Eight register windows forming a circular stack

Figure 4.8 The concept of overlapping register windows in the SPARC architecture. (Courtesy of Sun Microsystems, Inc., 1987)

The concept is illustrated in Fig. 4.8 for eight overlapping windows (formed with 64 local registers and 64 overlapped registers) and eight globals with a total of 136 registers, as implemented in the Cypress 601.

Each register window is divided into three eight-register sections, labeled *Ins*, *Locals*, and *Outs*. The local registers are only locally addressable by each procedure. The Ins and Outs are shared among procedures.

The calling procedure passes parameters to the called procedure via its Outs (r8 to r15) registers, which are the Ins registers of the called procedure. The window of the currently running procedure is called the active window pointed to by a current window pointer. A window invalid mask is used to indicate which window is invalid. The trap base register serves as a pointer to a trap handler.

A special register is used to create a 64-bit product in multiple step instructions. Procedures can also be called without changing the window. The overlapping windows can significantly save the time required for interprocedure communications, resulting in much faster context switching among cooperative procedures.

The FPU features 32 single-precision (32-bit) or 16 double-precision (64-bit) floating-point registers (Fig. 4.7b). Fourteen of the 69 SPARC instructions are for floating-point operations. The SPARC architecture implements three basic instruction formats, all using a single word length of 32 bits.

Table 4.5 shows the MIPS rate relative to that of the VAX 11/780, which has been known as a reference machine with 1 MIPS. The 50-MIPS rate is the result of ECL implementation with a 80-MHz clock. A GaAs SPARC was reported in a Hot Chips Symposium to yield a 200-MIPS peak at 200-MHz clock rate.
◼

Example 4.4 The Intel i860 processor architecture

In 1989, Intel Corporation introduced the i860 microprocessor. It is a 64-bit RISC processor fabricated on a single chip containing more than 1 million transistors. The peak performance of the i860 was designed to achieve 80 Mflops single precision or 60 Mflops double precision, or 40 MIPS in 32-bit integer operations at a 40-MHz clock rate.

A schematic block diagram of major components in the i860 is shown in Fig. 4.9. There are nine functional units (shown in nine boxes) interconnected by multiple data paths with widths ranging from 32 to 128 bits.

All external or internal address buses are 32-bit wide, and the external data path or internal data bus is 64 bits wide. However, the internal RISC integer ALU is only 32 bits wide. The instruction cache has 4 Kbytes organized as a two-way set-associative memory with 32 bytes per cache block. It transfers 64 bits per clock cycle, equivalent to 320 Mbytes/s at 40 MHz.

The data cache is a two-way set-associative memory of 8 Kbytes. It transfers 128 bits per clock cycle (640 Mbytes/s) at 40 MHz. A write-back policy is used. Cacheing can be inhibited by software, if needed. The bus control unit coordinates the 64-bit data transfer between the chip and the outside world.

The MMU implements protected 4 Kbyte paged virtual memory of 2^{32} bytes via a TLB. The paging and MMU structure of the i860 is identical to that implemented in the i486. An i860 and an i486 can be used jointly in a heterogeneous

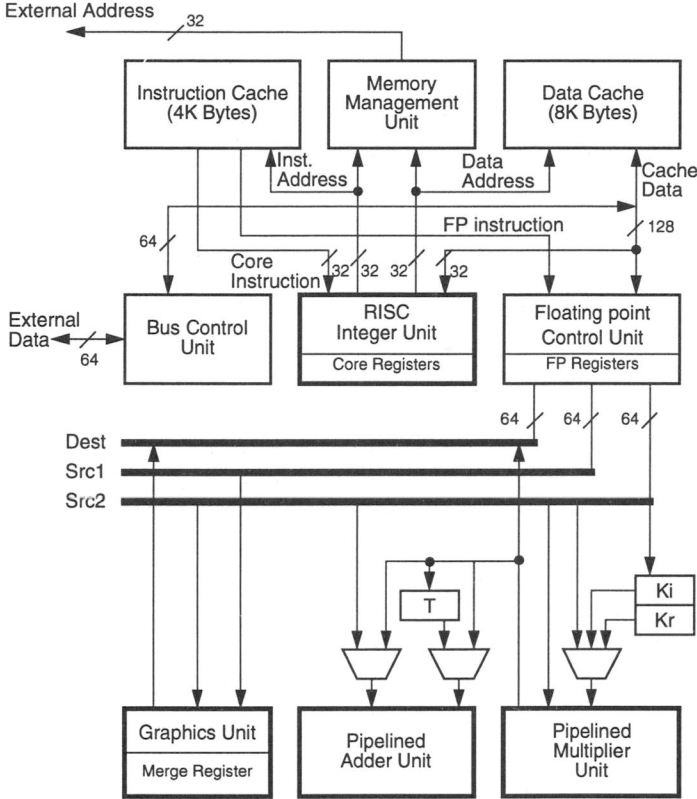

Figure 4.9 Functional units and data paths of the Intel i860 RISC microprocessor.
(Courtesy of Intel Corporation, 1990)

multiprocessor system, permitting the development of compatible OS kernels. The RISC integer unit executes *load, store, integer, bit,* and *control* instructions and fetches instructions for the floating-point control unit as well.

There are two floating-point units, namely, the *multiplier unit* and the *adder unit*, which can be used separately or simultaneously under the coordination of the floating-point control unit. Special dual-operation floating-point instructions such as *add-and-multiply* and *subtract-and-multiply* use both the multiplier and adder units in parallel (Fig. 4.10).

Furthermore, both the integer unit and the floating-point control unit can execute concurrently. In this sense, the i860 is also a superscalar RISC processor capable of executing two instructions, one integer and one floating-point, at the same time. The floating-point unit conforms to the IEEE 754 floating-point standard, operating with single-precision (32-bit) and double-precision (64-bit) operands.

The graphics unit executes integer operations corresponding to 8-, 16-, or 32-bit pixel data types. This unit supports three-dimensional drawing in a graphics frame

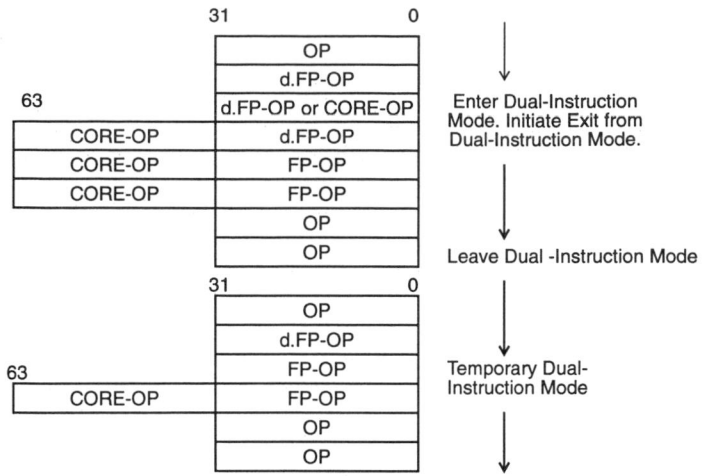

(a) Dual-instruction mode transitions

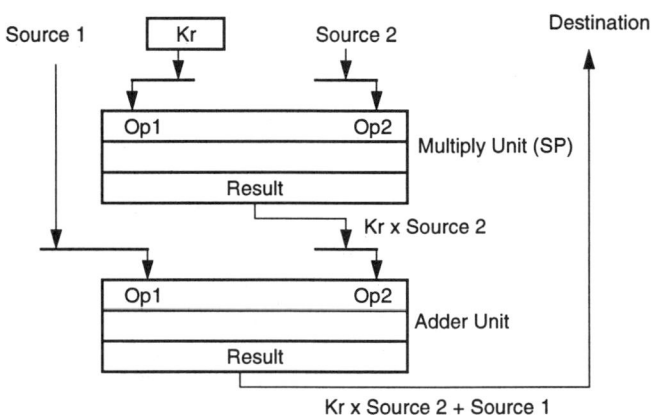

(b) Dual operations in floating-point units

Figure 4.10 Dual instructions and dual operations in the i860 processor.

buffer, with color intensity, shading, and hidden surface elimination. The merge register is used only by vector integer instructions. This register accumulates the results of multiple addition operations.

The i860 executes 82 instructions, including 42 RISC integer, 24 floating-point, 10 graphics, and 6 assembler pseudo operations. All the instructions execute in one cycle, which equals 25 ns for a 40-MHz clock rate. The i860 and its successor, the i860XP, are used in floating-point accelerators, graphics subsystems, workstations, multiprocessors, and multicomputers.

The RISC Impacts   The debate between RISC and CISC designers has lasted for more than a decade. Based on Eq. 1.3, it seems that RISC will outperform CISC if the program length does not increase dramatically. Based on one reported experiment, converting from a CISC program to an equivalent RISC program increases the code length (instruction count) by only 40%.

Of course, the increase depends on program behavior, and the 40% increase may not be typical of all programs. Nevertheless, the increase in code length is much smaller than the increase in clock rate and the reduction in CPI. Thus the intuitive reasoning in Eq. 1.3 should prevail in both cases.

A RISC processor lacks some sophisticated instructions found in CISC processors. The increase in RISC program length implies more instruction traffic and greater memory demand. Another RISC problem is caused by the use of a large register file. Although a larger register set can hold more intermediate results and reduce the data traffic between the CPU and memory, the register decoding system will be more complicated. Longer register-access time places a greater demand on the compiler to manage the register window functions. Another shortcoming of RISC lies in its hardwired control, which is less flexible and more error-prone.

RISC shortcomings are directly related to some of its claimed advantages. More benchmark and application experiments are needed to determine the optimal sizes of the register set, I-cache, and D-cache. Further processor improvements may include a 64-bit integer ALU, multiprocessor support such as snoopy logic for cache coherence control, faster interprocessor synchronization or hardware support for message passing, and special-function units for I/O interfaces and graphics support.

The boundary between RISC and CISC architectures has become blurred because both are now implemented with the same hardware technology. For example, the VAX 9000, Motorola 88100, and i586 are built with mixed features taken from both the RISC and CISC camps.

It is the applications that eventually determine the best choice of a processor architecture. In this sense, the most favorable design space for future processors is the dashed circle in the lower center of Fig. 4.1. This may require merging some of the positive features from different processor families.

4.2   Superscalar and Vector Processors

A CISC or a RISC scalar processor can be improved with a *superscalar* or *vector* architecture. Scalar processors are those executing one instruction per cycle. Only one instruction is issued per cycle, and only one completion of instruction is expected from the pipeline per cycle.

In a superscalar processor, multiple instruction pipelines are used. This implies that multiple instructions are issued per cycle and multiple results are generated per cycle. A vector processor executes vector instructions on arrays of data. Thus each instruction involves a string of repeated operations, which are ideal for pipelining with one result per cycle.

4.2.1   Superscalar Processors

Superscalar processors are designed to exploit more instruction-level parallelism in user programs. Only independent instructions can be executed in parallel without causing a wait state. The amount of instruction-level parallelism varies widely depending on the type of code being executed.

It has been observed that the average value is around 2 for code without loop unrolling. Therefore, for these codes there is not much benefit gained from building a machine that can issue more then three instructions per cycle. The *instruction-issue degree* in a superscalar processor has thus been limited to 2 to 5 in practice.

Pipelining in Superscalar Processors   The fundamental structure of a superscalar pipeline is illustrated in Fig. 4.11. This diagram shows the use of three instruction pipelines in parallel for a triple-issue processor. Superscalar processors were originally developed as an alternative to vector processors.

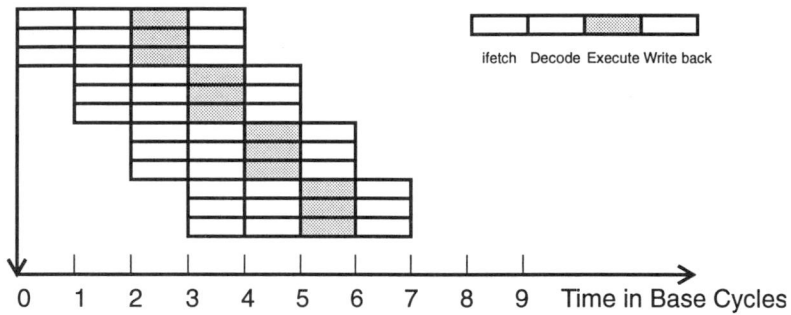

Figure 4.11 A superscalar processor of degree m = 3.

A superscalar processor of degree m can issue m instructions per cycle. In this sense, the base scalar processor, implemented either in RISC or CISC, has $m = 1$. In order to fully utilize a superscalar processor of degree m, m instructions must be executable in parallel. This situation may not be true in all clock cycles. In that case, some of the pipelines may be stalling in a wait state.

In a superscalar processor, the simple operation latency should require only one cycle, as in the base scalar processor. Due to the desire for a higher degree of instruction-level parallelism in programs, the superscalar processor depends more on an optimizing compiler to exploit parallelism.

In theory, a superscalar processor can attain the same performance as a machine with vector hardware. A superscalar machine that can issue a fixed-point, floating-point, load, and branch all in one cycle achieves the same effective parallelism as a *vector machine* which executes a vector load, chained into a vector add, with one element loaded and added per cycle. This will become more evident as we discuss *pipeline chaining* for vector processors in Chapters 6 and 8. A typical superscalar architecture

Table 4.6 Representative Superscalar Processors

Feature	Intel i960CA	IBM RS/6000	DEC 21064
Technology, clock rate, year	25 MHz 1986.	1-μm CMOS technology, 30 MHz, 1990.	0.75-μm CMOS, 150 MHz, 431 pins, 1992.
Functional units and multiple instruction issues	Issue up to 3 instructions (register, memory, and control) per cycle, seven functional units available for concurrent use.	POWER architecture, issue 4 instructions (1 FXU, 1 FPU, and 2 ICU operations) per cycle.	Alpha architecture, issue 2 instructions per cycle, 64-bit IU and FPU, 128-bit data bus, and 34-bit address bus implemented in initial version.
Registers, caches, MMU, address space	1-KB I-cache, 1.5-KB RAM, 4-channel I/O with DMA, parallel decode, multiported registers.	32 32-bit GPRs, 8-KB I-cache, 64-KB D-cache with separate TLBs.	32 64-bit GPRs, 8-KB I-cache, 8-KB D-cache, 64-bit virtual space designed, 43-bit address space implemented in initial version.
Floating-point unit and functions	On-chip FPU, fast multimode interrupt, multitask control.	On-chip FPU 64-bit multiply, add, divide, subtract, IEEE 754 standard.	On-chip FPU, 32 64-bit FP registers, 10-stage pipeline, IEEE and VAX FP standards.
Claimed performance and remarks	30 VAX/MIPS peak at 25 MHz, real-time embedded system control, and multiprocessor applications.	34 MIPS and 11 Mflops at 25 MHz on POWERstation 530.	300 MIPS peak and 150 Mflops peak at 150 MHz, multiprocessor and cache coherence support.

Note: KB = Kbytes, FP = floating point.

for a RISC processor is shown in Fig. 4.12.

Multiple instruction pipelines are used. The instruction cache supplies multiple instructions per fetch. However, the actual number of instructions issued to various functional units may vary in each cycle. The number is constrained by data dependences and resource conflicts among instructions that are simultaneously decoded. Multiple functional units are built into the integer unit and into the floating-point unit.

Multiple data buses exist among the functional units. In theory, all functional units can be simultaneously used if conflicts and dependences do not exist among them during a given cycle.

Representative Superscalar Processors A number of commercially available processors have been implemented with the superscalar architecture. Notable ones include the IBM RS/6000, DEC 21064, and Intel i960CA processors as summarized in Table 4.6.

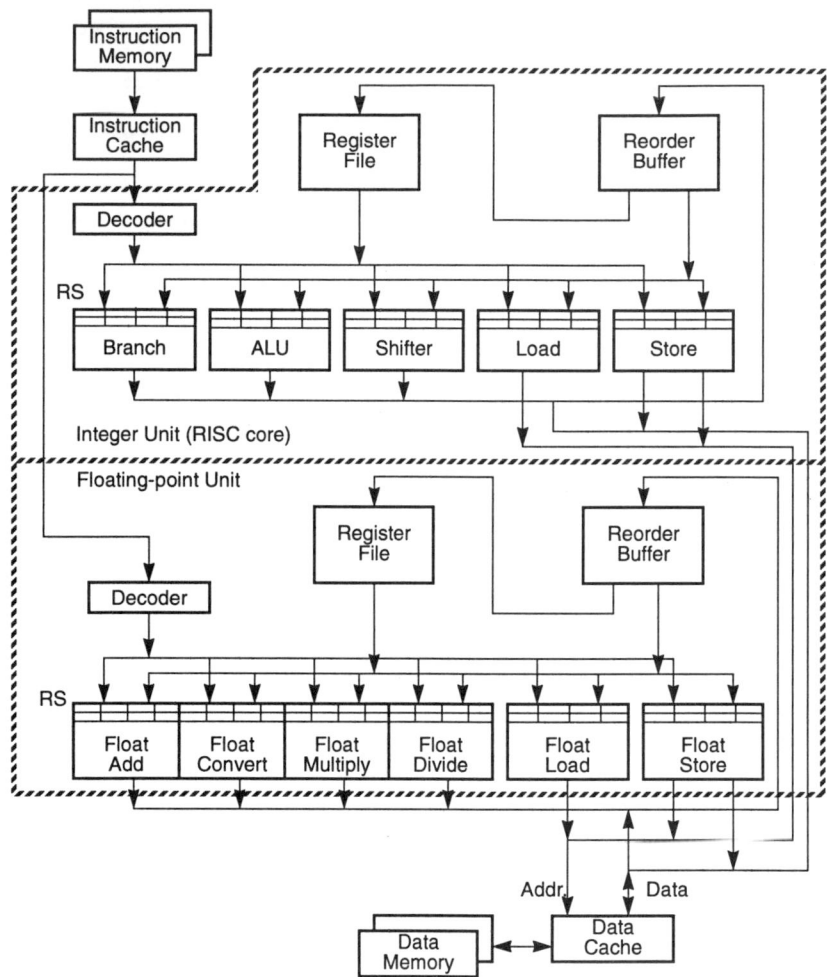

Figure 4.12 A typical superscalar RISC processor architecture consisting of an integer unit and a floating-point unit. (Courtesy of M. Johnson, 1991; reprinted with permission from Prentice-Hall, Inc.)

Due to the reduced CPI and higher clock rates used, most superscalar processors outperform scalar processors.

The maximum number of instructions issued per cycle ranges from two to five in these four superscalar processors. Typically, the register files in the IU and FPU each have 32 registers. Most superscalar processors implement both the IU and the FPU on the same chip. The superscalar degree is low due to limited instruction parallelism that can be exploited in ordinary programs.

Besides the register files, *reservation stations* and *reorder buffers* can be used to establish *instruction windows*. The purpose is to support instruction lookahead and internal data forwarding, which are needed to schedule multiple instructions through the multiple pipelines simultaneously. We will discuss the use of these mechanisms in Chapter 6, where advanced pipelining techniques are studied.

Example 4.5 The IBM RS/6000 architecture

In early 1990, IBM announced the RISC System 6000. It is a superscalar processor as illustrated in Fig. 4.13. There are three functional units called the *branch processor, fixed-point unit*, and *floating-point units*, which can operate in parallel.

The branch processor can arrange the execution of up to five instructions per cycle. These include one *branch* instruction in the branch processor, one *fixed-point* instruction in the FXU, one *condition-register* instruction in the branch processor, and one *floating-point multiply-add* instruction in the FPU, which can be counted as two floating-point operations.

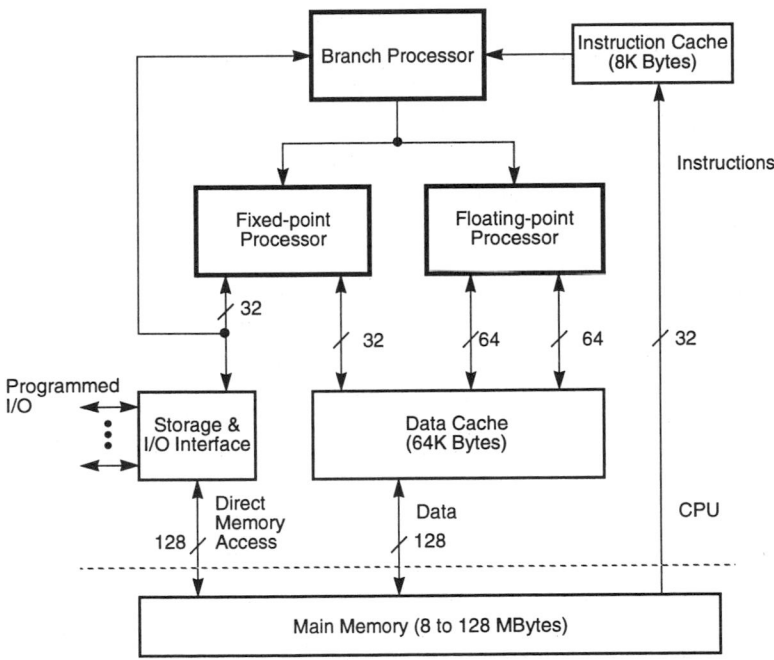

Figure 4.13 The POWER architecture of the IBM RISC System/6000 superscalar processor. (Courtesy of International Business Machines Corporation, 1990)

The RS/6000 is hardwired rather than microcoded. The system uses a number of wide buses ranging from one word (32 bits) for the FXU to two words (64 bits) for the FPU, and four words for the I-cache and D-cache, respectively. These wide buses provide the high instruction and data bandwidths required for superscalar implementation.

The RS/6000 design is optimized to perform well in numerically intensive scientific and engineering applications, as well as in multiuser commercial environments. A number of RS/6000-based workstations and servers are produced by IBM. For example, the POWERstation 530 has a clock rate of 25 MHz with performance benchmarks reported as 34.5 MIPS and 10.9 Mflops.

◼

4.2.2 The VLIW Architecture

The VLIW architecture is generalized from two well-established concepts: horizontal microcoding and superscalar processing. A typical VLIW (*very long instruction word*) machine has instruction words hundreds of bits in length. As illustrated in Fig. 4.14a, multiple functional units are used concurrently in a VLIW processor. All functional units share the use of a common large register file. The operations to be simultaneously executed by the functional units are synchronized in a VLIW instruction, say, 256 or 1024 bits per instruction word, as implemented in the Multiflow computer models.

The VLIW concept is borrowed from horizontal microcoding. Different fields of the long instruction word carry the opcodes to be dispatched to different functional units. Programs written in conventional short instruction words (say 32 bits) must be compacted together to form the VLIW instructions. This code compaction must be done by a compiler which can predict branch outcomes using elaborate heuristics or run-time statistics.

Pipelining in VLIW Processors The execution of instructions by an ideal VLIW processor was shown in Fig. 4.14b. Each instruction specifies multiple operations. The effective CPI becomes 0.33 in this particular example. VLIW machines behave much like superscalar machines with three differences: First, the decoding of VLIW instructions is easier than that of superscalar instructions.

Second, the code density of the superscalar machine is better when the available instruction-level parallelism is less than that exploitable by the VLIW machine. This is because the fixed VLIW format includes bits for nonexecutable operations, while the superscalar processor issues only executable instructions.

Third, a superscalar machine can be object-code-compatible with a large family of nonparallel machines. On the contrary, a VLIW machine exploiting different amounts of parallelism would require different instruction sets.

Instruction parallelism and data movement in a VLIW architecture are completely specified at compile time. Run-time resource scheduling and synchronization are thus completely eliminated. One can view a VLIW processor as an extreme of a superscalar processor in which all independent or unrelated operations are already synchronously compacted together in advance. The CPI of a VLIW processor can be even lower than that of a superscalar processor. For example, the Multiflow trace computer allows up

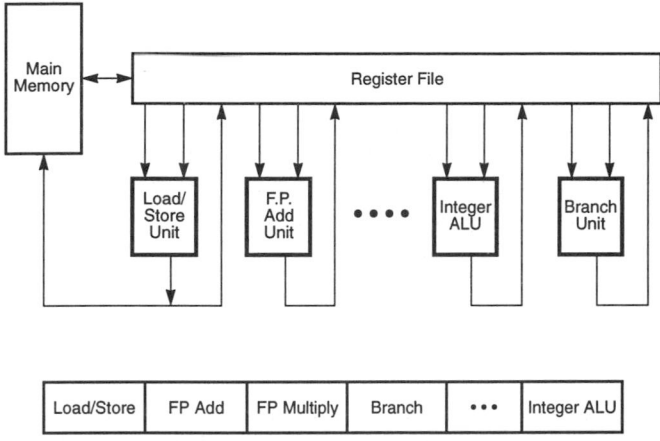

(a) A typical VLIW processor and instruction format

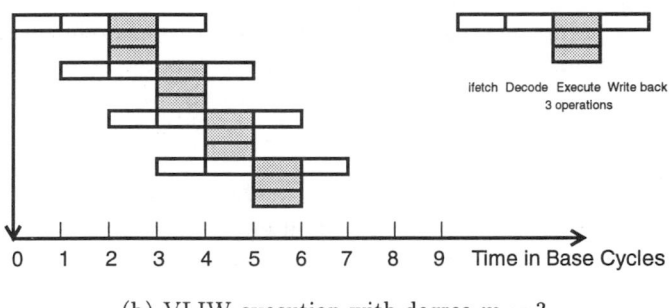

(b) VLIW execution with degree $m = 3$

Figure 4.14 The architecture of a very long instruction word (VLIW) processor and its pipeline operations. (Courtesy of Multiflow Computer, Inc., 1987)

to seven operations to be executed concurrently with 256 bits per VLIW instruction.

VLIW Opportunities   In a VLIW architecture, random parallelism among scalar operations is exploited instead of regular or synchronous parallelism as in a vectorized supercomputer or in an SIMD computer. The success of a VLIW processor depends heavily on the efficiency in code compaction. The architecture is totally incompatible with that of any conventional general-purpose processor.

Furthermore, the instruction parallelism embedded in the compacted code may require a different latency to be executed by different functional units even though the instructions are issued at the same time. Therefore, different implementations of the same VLIW architecture may not be binary-compatible with each other.

By explicitly encoding parallelism in the long instruction, a VLIW processor can eliminate the hardware or software needed to detect parallelism. The main advantage of VLIW architecture is its simplicity in hardware structure and instruction set. The VLIW processor can potentially perform well in scientific applications where the program behavior (branch predictions) is more predictable.

In general-purpose applications, the architecture may not be able to perform well. Due to its lack of compatibility with conventional hardware and software, the VLIW architecture has not entered the mainstream of computers. Although the idea is academically sound, the dependence on trace-scheduling compiling and code compaction has prevented it from gaining acceptance in the commercial world.

4.2.3 Vector and Symbolic Processors

By definition, a *vector processor* is a coprocessor specially designed to perform vector computations. A *vector* instruction involves a large array of operands. In other words, the same operation will be performed over a string of data. Vector processors are often used in a multipipelined supercomputer.

A vector processor can assume either a *register-to-register* architecture or a *memory-to-memory* architecture. The former uses shorter instructions and vector register files. The latter uses memory-based instructions which are longer in length, including memory addresses.

Vector Instructions   Register-based vector instructions appear in most register-to-register vector processors like Cray supercomputers. Denote a *vector register* of length n as V_i, a *scalar register* as s_i, and a *memory array* of length n as $M(1:n)$. Typical register-based vector operations are listed below, where a vector operator is denoted by a small circle "$\circ$":

$$
\begin{array}{llll}
V_1 \circ V_2 & \longrightarrow & V_3 & \text{(binary vector)} \\
s_1 \circ V_1 & \longrightarrow & V_2 & \text{(scaling)} \\
V_1 \circ V_2 & \longrightarrow & s_1 & \text{(binary reduction)} \\
M(1:n) & \longrightarrow & V_1 & \text{(vector load)} \\
V_1 & \longrightarrow & M(1:n) & \text{(vector store)} \\
\circ\, V_1 & \longrightarrow & V_2 & \text{(unary vector)} \\
\circ\, V_1 & \longrightarrow & s_1 & \text{(unary reduction)}
\end{array} \tag{4.1}
$$

It should be noted that the vector length should be equal in all operands used in a vector instruction. The reduction is an operation on one or two vector operands, and the result is a scalar — such as the *dot product* between two vectors and the *maximum* of all components in a vector.

In all cases, these vector operations are performed by dedicated pipeline units, including *functional pipelines* and *memory-access pipelines*. Long vectors exceeding the register length n must be segmented to fit the vector registers n elements at a time.

Memory-based vector operations are found in memory-to-memory vector processors such as those in the Cyber 205. Listed below are a few examples:

$$
\begin{aligned}
M_1(1:n) &\circ M_2(1:n) &\longrightarrow& M(1:n) \\
s_1 &\circ M_1(1:n) &\longrightarrow& M_2(1:n) \\
&\circ M_1(1:n) &\longrightarrow& M_2(1:n) \\
M_1(1:n) &\circ M_2(1:n) &\longrightarrow& M(k)
\end{aligned}
\tag{4.2}
$$

where $M_1(1:n)$ and $M_2(1:n)$ are two vectors of length n and $M(k)$ denotes a scalar quantity stored in memory location k. Note that the vector length is not restricted by register length. Long vectors are handled in a streaming fashion using *superwords* cascaded from many shorter memory words.

Vector Pipelines   Vector processors take advantage of unrolled-loop-level parallelism. The vector pipelines can be attached to any scalar processor, whether it is superscalar, superpipelined, or both.

Dedicated vector pipelines will eliminate some software overhead in looping control. Of course, the effectiveness of a vector processor relies on the capability of an optimizing compiler that vectorizes sequential code for vector pipelining.

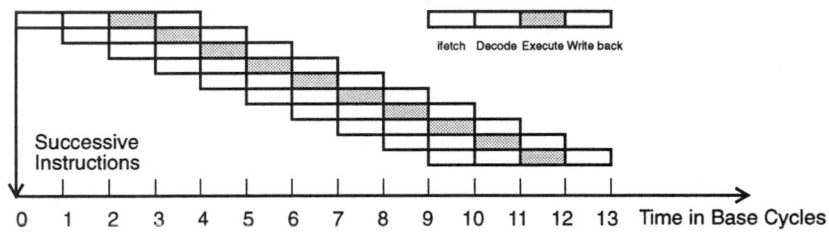

(a) Scalar pipeline execution (Fig. 4.2a redrawn)

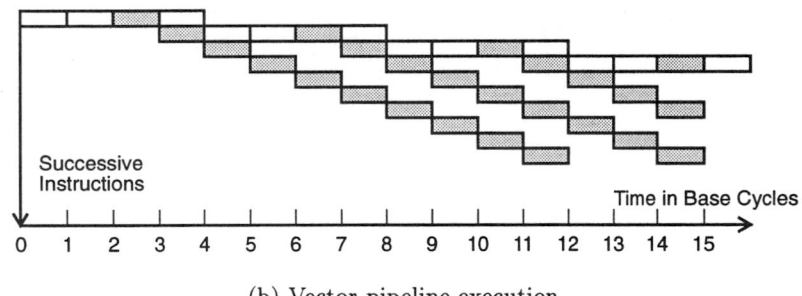

(b) Vector pipeline execution

Figure 4.15 Pipelined execution in a base scalar processor and in a vector processor, respectively. (Courtesy of Jouppi and Wall; reprinted from *Proc. ASP-LOS*, ACM Press, 1989)

The pipelined execution in a vector processor is compared with that in a scalar processor in Fig. 4.15. Figure 4.15a is a redrawing of Fig. 4.2a in which each scalar instruction executes only one operation over one data element. For clarity, only serial issue and parallel execution of vector instructions are illustrated in Fig. 4.2b. Each vector instruction executes a string of operations, one for each element in the vector.

We will study vector processors and SIMD architectures in Chapter 8. Various functional pipelines and their chaining or networking schemes will be introduced for the execution of compound vector functions. Many of the above vector instructions also have equivalent counterparts in an SIMD computer. Vector processing is achieved through efficient pipelining in vector supercomputers and through spatial or data parallelism in an SIMD computer.

Symbolic Processors Symbolic processing has been applied in many areas, including theorem proving, pattern recognition, expert systems, knowledge engineering, text retrieval, cognitive science, and machine intelligence. In these applications, data and knowledge representations, primitive operations, algorithmic behavior, memory, I/O and communications, and special architectural features are different than in numerical computing. Symbolic processors have also been called *prolog processors*, *Lisp processors*, or *symbolic manipulators*.

Table 4.7 Characteristics of Symbolic Processing

Attributes	Characteristics
Knowledge Representations	Lists, relational databases, scripts, semantic nets, frames, blackboards, objects, production systems.
Common Operations	Search, sort, pattern matching, filtering, contexts, partitions, transitive closures, unification, text retrieval, set operations, reasoning.
Memory Requirements	Large memory with intensive access pattern. Addressing is often content-based. Locality of reference may not hold.
Communication Patterns	Message traffic varies in size and destination; granularity and format of message units change with applications.
Properties of Algorithms	Nondeterministic, possibly parallel and distributed computations. Data dependences may be global and irregular in pattern and granularity.
Input/Output requirements	User-guided programs; intelligent person-machine interfaces; inputs can be graphical and audio as well as from keyboard; access to very large on-line databases.
Architecture Features	Parallel update of large knowledge bases, dynamic load balancing; dynamic memory allocation; hardware-supported garbage collection; stack processor architecture; symbolic processors.

For example, a Lisp program can be viewed as a set of functions in which data are passed from function to function. The concurrent execution of these functions forms the basis for parallelism. The applicative and recursive nature of Lisp requires an environment that efficiently supports stack computations and function calling. The use of linked lists as the basic data structure makes it possible to implement an automatic garbage collection mechanism.

Table 4.7 summarizes the major characteristics of symbolic processing. Instead of dealing with numerical data, symbolic processing deals with logic programs, symbolic lists, objects, scripts, blackboards, production systems, semantic networks, frames, and artificial neural networks.

Primitive operations for artificial intelligence include *search, compare, logic inference, pattern matching, unification, filtering, context, retrieval, set operations, transitive closure,* and *reasoning operations.* These operations demand a special instruction set containing *compare, matching, logic,* and *symbolic manipulation* operations. Floating-point operations are not often used in these machines.

Example 4.6   The Symbolics 3600 Lisp processor

The processor architecture of the Symbolics 3600 is shown in Fig. 4.16. This is a stack-oriented machine. The division of the overall machine architecture into layers allows the use of a pure stack model to simplify instruction-set design, while implementation is carried out with a stack-oriented machine. Nevertheless most operands are fetched from the stack, so the stack buffer and scratch-pad memories are implemented as fast caches to main memory.

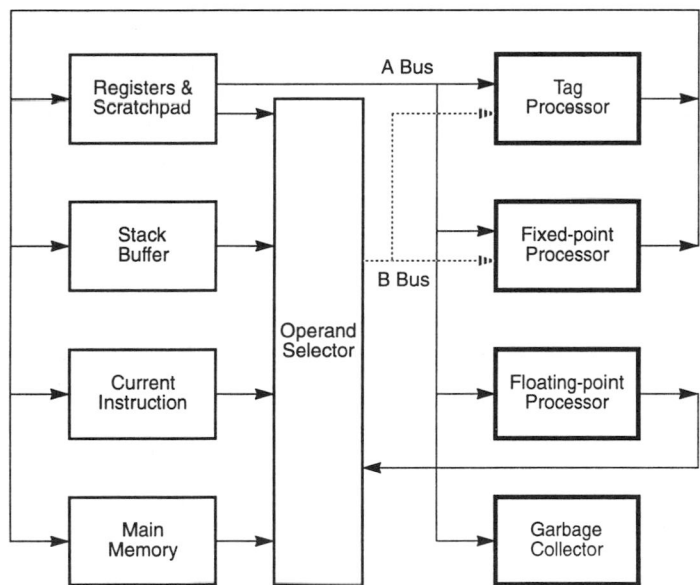

Figure 4.16 The architecture of the Symbolics 3600 Lisp processor. (Courtesy of Symbolics, Inc., 1985)

The Symbolics 3600 executes most Lisp instructions in one machine cycle. Integer instruction fetch operands form the stack buffer and the duplicate top of the stack in the scratch-pad memory. Floating-point addition, garbage collection, data type checking by the tag processor, and fixed-point addition can be carried out in parallel.

∎

4.3 Memory Hierarchy Technology

In a typical computer configuration, the cost of memory, disks, printers, and other peripherals has far exceeded that of the central processor. We briefly introduce below the memory hierarchy and peripheral technology.

4.3.1 Hierarchical Memory Technology

Storage devices such as *registers*, *caches*, *main memory*, *disk devices*, and *tape units* are often organized as a hierarchy as depicted in Fig. 4.17. The memory technology and storage organization at each level are characterized by five parameters: the *access time* (t_i), *memory size* (s_i), *cost per byte* (c_i), *transfer bandwidth* (b_i), and *unit of transfer* (x_i).

The access time t_i refers to the round-trip time from the CPU to the ith-level memory. The memory size s_i is the number of bytes or words in level i. The cost of the ith-level memory is estimated by the product $c_i s_i$. The bandwidth b_i refers to the rate at which information is transferred between adjacent levels. The unit of transfer x_i refers to the grain size for data transfer between levels i and $i + 1$.

Memory devices at a lower level are faster to access, smaller in size, and more expensive per byte, having a higher bandwidth and using a smaller unit of transfer as compared with those at a higher level. In other words, we have $t_{i-1} < t_i$, $s_{i-1} < s_i$, $c_{i-1} > c_i$, $b_{i-1} > b_i$, and $x_{i-1} < x_i$, for $i = 1, 2, 3$, and 4, in the hierarchy where $i = 0$ corresponds to the CPU register level. The cache is at level 1, main memory at level 2, the disks at level 3, and the tape unit at level 4. The physical memory design and operations of these levels are studied in subsequent sections and in Chapter 5.

Registers and Caches The register and the cache are parts of the processor complex, built either on the processor chip or on the processor board. Register assignment is often made by the compiler. Register transfer operations are directly controlled by the processor after instructions are decoded. Register transfer is conducted at processor speed, usually in one clock cycle.

Therefore, many designers would not consider registers a level of memory. We list them here for comparison purposes. The cache is controlled by the MMU and is programmer-transparent. The cache can also be implemented at one or multiple levels, depending on the speed and application requirements.

Main Memory The main memory is sometimes called the primary memory of a computer system. It is usually much larger than the cache and often implemented by

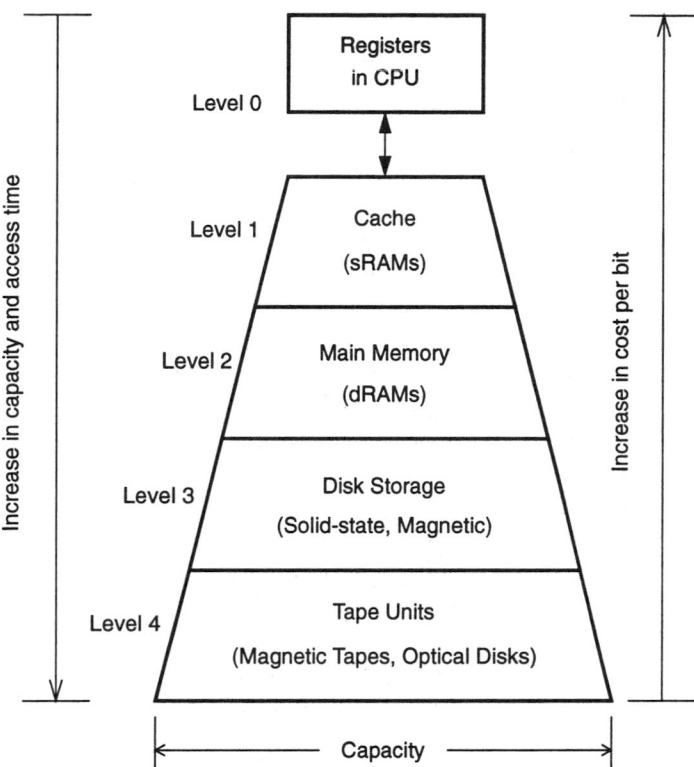

Figure 4.17 A four-level memory hierarchy with increasing capacity and decreasing speed and cost from low to high levels.

the most cost-effective RAM chips, such as the 4-Mbit DRAMs used in 1991 and the 64-Mbit DRAMs projected for 1995.

The main memory is managed by a MMU in cooperation with the operating system. Options are often provided to extend the main memory by adding more memory boards to a system. Sometimes, the main memory is itself divided into two sublevels using different memory technologies.

Disk Drives and Tape Units   Disk drives and tape units are handled by the OS with limited user intervention. The disk storage is considered the highest level of on-line memory. It holds the system programs such as the OS and compilers and some user programs and their data sets. The magnetic tape units are off-line memory for use as backup storage. They hold copies of present and past user programs and processed results and files.

A typical workstation computer has the cache and main memory on a processor board and hard disks in an attached disk drive. In order to access the magnetic tape units, user intervention is needed. Table 4.8 presents representative values of memory parameters for a typical 32-bit mainframe computer built in 1993.

Table 4.8 Memory Characteristics of a Typical Mainframe Computer in 1993

Memory level Characteristics	Level 0 CPU Registers	Level 1 Cache	Level 2 Main Memory	Level 3 Disk Storage	Level 4 Tape Storage
Device technology	ECL	256K-bit SRAM	4M-bit DRAM	1-Gbyte magnetic disk unit	5-Gbyte magnetic tape unit
Access time, t_i	10 ns	25–40 ns	60–100 ns	12–20 ms	2–20 min (search time)
Capacity, s_i (in bytes)	512 bytes	128 Kbytes	512 Mbytes	60–228 Gbytes	512 Gbytes– 2 Tbytes
Cost, c_i (in cents/KB)	18,000	72	5.6	0.23	0.01
Bandwidth, b_i (in MB/s)	400–800	250–400	80–133	3–5	0.18–0.23
Unit of transfer, x_i	4–8 bytes per word	32 bytes per block	0.5–1 Kbytes per page	5–512 Kbytes per file	Backup storage
Allocation management	Compiler assignment	Hardware control	Operating system	Operating system/user	Operating system/user

Peripheral Technology   Besides disk drives and tape units, peripheral devices include printers, plotters, terminals, monitors, graphics displays, optical scanners, image digitizers, output microfilm devices, etc. Some I/O devices are tied to special-purpose or multimedia applications.

The technology of peripheral devices has improved rapidly in recent years. For example, we used dot-matrix printers in the past. Now, as laser printers become so popular, in-house publishing becomes a reality. The high demand for multimedia I/O such as image, speech, video, and sonar will further upgrade I/O technology in the future.

4.3.2   Inclusion, Coherence, and Locality

Information stored in a memory hierarchy $(M_1, M_2, \cdots, M_n)$ satisfies three important properties: *inclusion, coherence,* and *locality* as illustrated in Fig. 4.18. We consider cache memory the innermost level M_1, which directly communicates with the CPU registers. The outermost level M_n contains all the information words stored. In fact, the collection of all addressable words in M_n forms the virtual address space of a computer. Program locality is characterized below as the foundation for using a memory hierarchy effectively.

Inclusion Property   The *inclusion property* is stated as $M_1 \subset M_2 \subset M_3 \subset \cdots \subset M_n$.

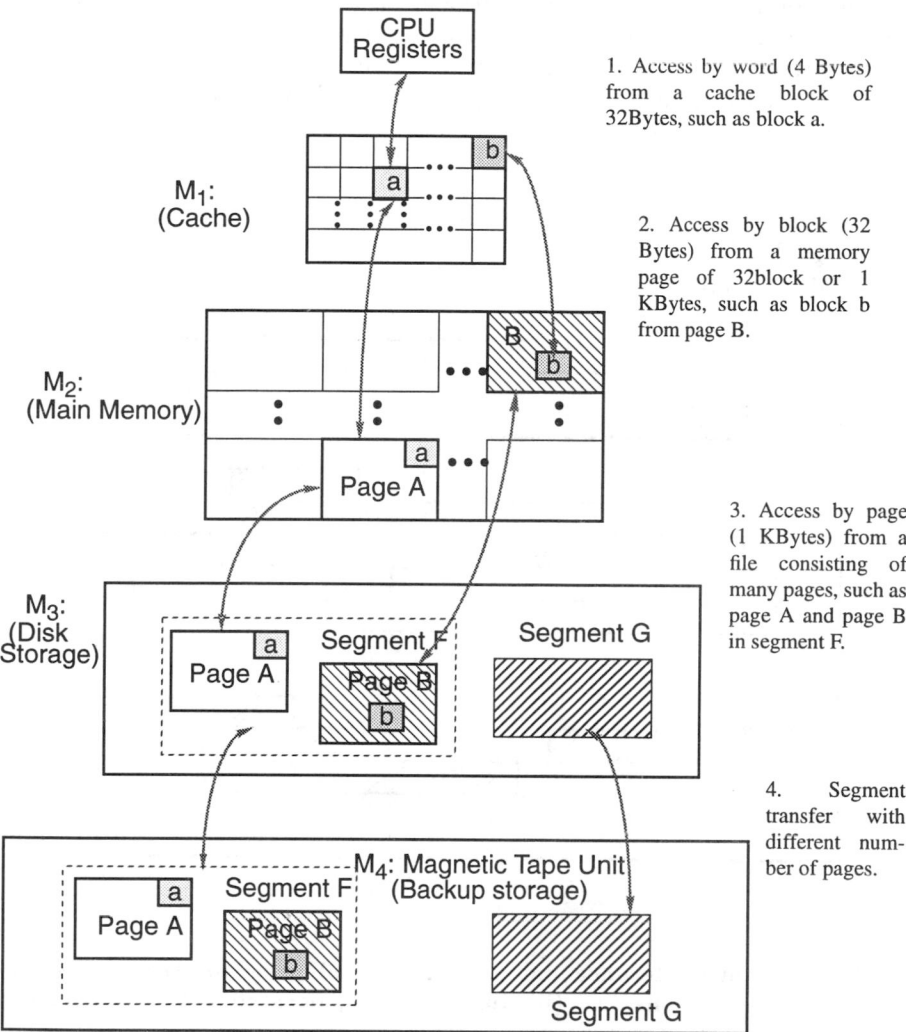

Figure 4.18 The inclusion property and data transfers between adjacent levels of a memory hierarchy.

The set inclusion relationship implies that all information items are originally stored in the outermost level M_n. During the processing, subsets of M_n are copied into M_{n-1}. Similarly, subsets of M_{n-1} are copied into M_{n-2}, and so on.

In other words, if an information word is found in M_i, then copies of the same word can be also found in all upper levels $M_{i+1}, M_{i+2}, ..., M_n$. However, a word stored in M_{i+1} may not be found in M_i. A *word miss* in M_i implies that it is also missing from all lower levels $M_{i-1}, M_{i-2}, ..., M_1$. The highest level is the backup storage, where everything can be found.

Information transfer between the CPU and cache is in terms of *words* (4 or 8 bytes each depending on the word length of a machine). The cache (M_1) is divided into *cache blocks*, also called *cache lines* by some authors. Each block is typically 32 bytes (8 words). Blocks (such as "a" and "b" in Fig. 4.18) are the units of data transfer between the cache and main memory.

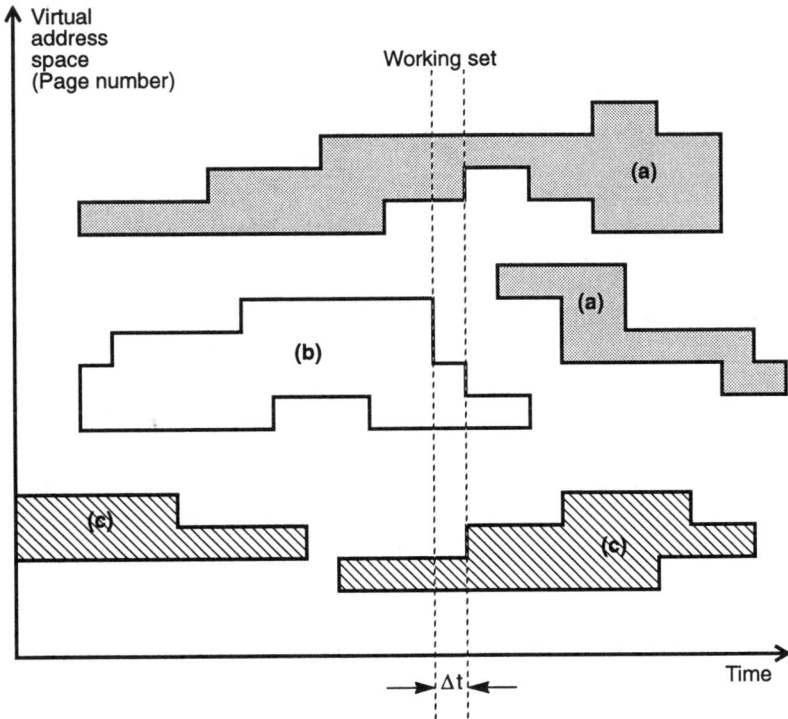

Figure 4.19 Memory reference patterns in typical program trace experiments, where regions (a), (b), and (c) are generated with the execution of three software processes.

The main memory (M_2) is divided into *pages*, say, 4 Kbytes each. Each page contains 128 blocks for the example in Fig. 4.18. Pages are the units of information transferred between disk and main memory.

Scattered pages are organized as a segment in the disk memory, for example, segment F contains page A, page B, and other pages. The size of a segment varies depending on the user's needs. Data transfer between the disk and the tape unit is handled at the file level, such as segments F and G illustrated in Fig. 4.18.

Coherence Property The *coherence property* requires that copies of the same information item at successive memory levels be consistent. If a word is modified in the cache, copies of that word must be updated immediately or eventually at all higher levels. The hierarchy should be maintained as such. Frequently used information is often found in the lower levels in order to minimize the effective access time of the memory hierarchy. In general, there are two strategies for maintaining the coherence in a memory hierarchy.

The first method is called *write-through* (WT), which demands immediate update in M_{i+1} if a word is modified in M_i, for $i = 1, 2, ..., n-1$.

The second method is *write-back* (WB), which delays the update in M_{i+1} until the word being modified in M_i is replaced or removed from M_i. Memory replacement policies are studied in Section 4.4.3.

Locality of References　　The memory hierarchy was developed based on a program behavior known as *locality of references*. Memory references are generated by the CPU for either instruction or data access. These accesses tend to be clustered in certain regions in time, space, and ordering.

In other words, most programs act in favor of a certain portion of their address space at any time window. Hennessy and Patterson (1990) have pointed out a 90-10 rule which states that a typical program may spend 90% of its execution time on only 10% of the code such as the innermost loop of a nested looping operation.

There are three dimensions of the locality property: *temporal, spatial*, and *sequential*. During the lifetime of a software process, a number of pages are used dynamically. The references to these pages vary from time to time, however, they follow certain access patterns as illustrated in Fig. 4.19. These memory reference patterns are caused by the following locality properties:

(1)　*Temporal locality* — Recently referenced items (instructions or data) are likely to be referenced again in the near future. This is often caused by special program constructs such as iterative loops, process stacks, temporary variables, or subroutines. Once a loop is entered or a subroutine is called, a small code segment will be referenced repeatedly many times. Thus temporal locality tends to cluster the access in the recently used areas.

(2)　*Spatial locality* — This refers to the tendency for a process to access items whose addresses are near one another. For example, operations on tables or arrays involve accesses of a certain clustered area in the address space. Program segments, such as routines and macros, tend to be stored in the same neighborhood of the memory space.

(3)　*Sequential locality* — In typical programs, the execution of instructions follows a sequential order (or the program order) unless branch instructions create out-of-order executions. The ratio of in-order execution to out-of-order execution is roughly 5 to 1 in ordinary programs. Besides, the access of a large data array also follows a sequential order.

Memory Design Implications　　The sequentiality in program behavior also contributes to the spatial locality because sequentially coded instructions and array elements are often stored in adjacent locations. Each type of locality affects the design of the memory hierarchy.

The temporal locality leads to the popularity of the *least recently used* (LRU) replacement algorithm, to be defined in Section 4.4.3. The spatial locality assists us in determining the size of unit data transfers between adjacent memory levels. The temporal locality also helps determine the size of memory at successive levels.

The sequential locality affects the determination of grain size for optimal scheduling (grain packing). Prefetch techniques are heavily affected by the locality properties. The principle of localities will guide us in the design of cache, main memory, and even virtual memory organization.

The Working Sets   Figure 4.19 shows the memory reference patterns of three programs or three software processes. As a function of time, the virtual address space (identified by page numbers) is clustered into regions due to the locality of references. The subset of addresses (or pages) referenced within a given time window $(t, t + \Delta t)$ is called the *working set* by Denning (1968).

During the execution of a program, the working set changes slowly and maintains a certain degree of continuity as demonstrated in Fig. 4.19. This implies that the working set is often accumulated at the innermost (lowest) level such as the cache in the memory hierarchy. This will reduce the effective memory-access time with a higher hit ratio at the lowest memory level. The time window Δt is a critical parameter set by the OS kernel which affects the size of the working set and thus the desired cache size.

4.3.3   Memory Capacity Planning

The performance of a memory hierarchy is determined by the *effective access time* T_{eff} to any level in the hierarchy. It depends on the *hit ratios* and *access frequencies* at successive levels. We formally define these terms below. Then we discuss the issue of how to optimize the capacity of a memory hierarchy subject to a cost constraint.

Hit Ratios   Hit ratio is a concept defined for any two adjacent levels of a memory hierarchy. When an information item is found in M_i, we call it a *hit*, otherwise, a *miss*. Consider memory levels M_i and M_{i-1} in a hierarchy, $i = 1, 2, ..., n$. The *hit ratio* h_i at M_i is the probability that an information item will be found in M_i. It is a function of the characteristics of the two adjacent levels M_{i-1} and M_i. The *miss ratio* at M_i is defined as $1 - h_i$.

The hit ratios at successive levels are a function of memory capacities, management policies, and program behavior. Successive hit ratios are independent random variables with values between 0 and 1. To simplify the future derivation, we assume $h_0 = 0$ and $h_n = 1$, which means the CPU always accesses M_1 first and the access to the outermost memory M_n is always a hit.

The *access frequency* to M_i is defined as $f_i = (1 - h_1)(1 - h_2) \cdots (1 - h_{i-1})h_i$. This is indeed the probability of successfully accessing M_i when there are $i - 1$ misses at the lower levels and a hit at M_i. Note that $\sum_{i=1}^{n} f_i = 1$ and $f_1 = h_1$.

Due to the locality property, the access frequencies decrease very rapidly from low to high levels; that is, $f_1 \gg f_2 \gg f_3 \gg \cdots \gg f_n$. This implies that the inner levels of memory are accessed more often than the outer levels.

Effective Access Time   In practice, we wish to achieve as high a hit ratio as possible at M_1. Every time a miss occurs, a penalty must be paid to access the next higher level of memory. The misses have been called *block misses* in the cache and *page faults* in the main memory because blocks and pages are the units of transfer between these levels.

The time penalty for a page fault is much longer than that for a block miss due to the fact that $t_1 < t_2 < t_3$. Stone (1990) has pointed out that a cache miss is 2 to 4 times as costly as a cache hit, but a page fault is 1000 to 10,000 times as costly as a page hit.

Using the access frequencies f_i for $i = 1, 2, ..., n$, we can formally define the *effective access time* of a memory hierarchy as follows:

$$
\begin{aligned}
T_{\textit{eff}} &= \sum_{i=1}^{n} f_i \cdot t_i \\
&= h_1 t_1 + (1 - h_1)h_2 t_2 + (1 - h_1)(1 - h_2)h_3 t_3 + \cdots + \\
&\quad (1 - h_1)(1 - h_2)\cdots(1 - h_{n-1})t_n
\end{aligned}
\tag{4.3}
$$

The first several terms in Eq. 4.3 dominate. Still, the effective access time depends on the program behavior and memory design choices. Only after extensive program trace studies can one estimate the hit ratios and the value of $T_{\textit{eff}}$ more accurately.

Hierarchy Optimization   The total cost of a memory hierarchy is estimated as follows:

$$
C_{total} = \sum_{i=1}^{n} c_i \cdot s_i
\tag{4.4}
$$

This implies that the cost is distributed over n levels. Since $c_1 > c_2 > c_3 > \cdots c_n$, we have to choose $s_1 < s_2 < s_3 < \cdots s_n$. The optimal design of a memory hierarchy should result in a $T_{\textit{eff}}$ close to the t_1 of M_1 and a total cost close to the c_n of M_n. In reality, this is difficult to achieve due to the tradeoffs among n levels.

The optimization process can be formulated as a linear programming problem, given a ceiling C_0 on the total cost — that is, a problem to minimize

$$
T_{\textit{eff}} = \sum_{i=1}^{n} f_i \cdot t_i
\tag{4.5}
$$

subject to the following constraints:

$$
s_i > 0, \; t_i > 0 \qquad \text{for } i = 1, 2, ..., n
$$

$$
C_{total} = \sum_{i=1}^{n} c_i \cdot s_i < C_0
\tag{4.6}
$$

As shown in Table 4.7, the unit cost c_i and capacity s_i at each level M_i depend on the speed t_i required. Therefore, the above optimization involves tradeoffs among t_i, c_i,

s_i, and f_i or h_i at all levels $i = 1, 2, ..., n$. The following example shows such a tradeoff design.

Example 4.7 The design of a memory hierarchy

Consider the design of a three-level memory hierarchy with the following specifications for memory characteristics:

Memory level	Access time	Capacity	Cost/Kbyte
Cache	$t_1 = 25$ ns	$s_1 = 512$ Kbytes	$c_1 = \$1.25$
Main memory	$t_2 = $ unknown	$s_2 = 32$ Mbytes	$c_2 = \$0.2$
Disk array	$t_3 = 4$ ms	$s_3 = $ unknown	$c_3 = \$0.0002$

The design goal is to achieve an effective memory-access time $t = 10.04$ μs with a cache hit ratio $h_1 = 0.98$ and a hit ratio $h_2 = 0.9$ in the main memory. Also, the total cost of the memory hierarchy is upper-bounded by \$15,000. The memory hierarchy cost is calculated as

$$C = c_1 s_1 + c_2 s_2 + c_3 s_3 \leq 15,000 \qquad (4.7)$$

The maximum capacity of the disk is thus obtained as $s_3 = 39.8$ Gbytes without exceeding the budget.

Next, we want to choose the access time (t_2) of the RAM to build the main memory. The effective memory-access time is calculated as

$$t = h_1 t_1 + (1 - h_1)h_2 t_2 + (1 - h_1)(1 - h_2)h_3 t_3 \leq 10.04 \qquad (4.8)$$

Substituting all known parameters, we have $10.04 \times 10^{-6} = 0.98 \times 25 \times 10^{-9} + 0.02 \times 0.9 \times t_2 + 0.02 \times 0.1 \times 1 \times 4 \times 10^{-3}$. Thus $t_2 = 903$ ns.

Suppose one wants to double the main memory to 64 Mbytes at the expense of reducing the disk capacity under the same budget limit. This change will not affect the cache hit ratio. But it may increase the hit ratio in the main memory if a proper page replacement algorithm is used. Also, the effective memory-access time will be enhanced. ∎

4.4 Virtual Memory Technology

In this section, we introduce two models of virtual memory. We study address translation mechanisms and page replacement policies for memory management. Physical memory such as caches and main memory will be studied in Chapter 5.

4.4.1 Virtual Memory Models

The main memory is considered the *physical memory* in which many programs want to reside. However, the limited-size physical memory cannot load in all programs simultaneously. The *virtual memory* concept was introduced to alleviate this problem. The idea is to expand the use of the physical memory among many programs with the help of an auxiliary (backup) memory such as disk arrays.

Only active programs or portions of them become residents of the physical memory at one time. The vast majority of programs or inactive programs are stored on disk. All programs can be loaded in and out of the physical memory dynamically under the coordination of the operating system. To the users, virtual memory provides them with almost unbounded memory space to work with. Without virtual memory, it would have been impossible to develop the multiprogrammed or time-sharing computer systems that are in use today.

Address Spaces Each word in the physical memory is identified by a unique *physical address*. All memory words in the main memory form a *physical address space*. *Virtual addresses* are generated by the processor during compile time. In UNIX systems, each process created is given a *virtual address space* which contains all virtual addresses generated by the compiler.

The virtual addresses must be translated into physical addresses at run time. A system of translation tables and mapping functions are used in this process. The address translation and memory management policies are affected by the virtual memory model used and by the organization of the disk arrays and of the main memory.

The use of virtual memory facilitates sharing of the main memory by many users on a dynamic basis. It also facilitates software portability and allows users to execute programs requiring much more memory than the physical memory.

Only currently executable programs are brought into the main memory. This permits the relocation of code and data, makes it possible to implement protection in the OS kernel, and allows high-level optimization of memory allocation and management.

Address Mapping Let V be the set of virtual addresses generated by a program (or by a software process) running on a processor. Let M be the set of physical addresses allocated to run this program. A virtual memory system demands an automatic mechanism to implement the following mapping:

$$f_t : V \rightarrow M \cup \{\emptyset\} \tag{4.9}$$

This mapping is a time function which varies from time to time because the physical memory is dynamically allocated and deallocated. Consider any virtual address $v \in V$. The mapping f_t is formally defined as follows:

$$f_t(v) = \begin{cases} m, & \text{if } m \in M \text{ has been allocated to store the} \\ & \quad \text{data identified by virtual address } v \\ \emptyset, & \text{if data } v \text{ is missing in } M \end{cases} \tag{4.10}$$

In other words, the mapping $f_t(v)$ uniquely translates the virtual address v into a physical address m if there is a *memory hit* in M. When there is a *memory miss*, the value returned, $f_t(v) = \emptyset$, signals that the referenced item (instruction or data) has not been brought into the main memory yet.

The efficiency of the address translation process affects the performance of the virtual memory. Processor support is needed for precise interrupts and translator updates. Virtual memory is more difficult to implement in a multiprocessor, where additional problems such as coherence, protection, and consistency become much more involved. Two virtual memory models are studied below.

Private Virtual Memory   The first model uses a *private virtual memory space* associated with each processor, as seen in the VAX/11 and in most UNIX systems (Fig. 4.20a). Each private virtual space is divided into pages. Virtual pages from different virtual spaces are mapped into the same physical memory shared by all processors.

The advantages of using private virtual memory include the use of a small processor address space (32 bits), protection on each page or on a per-process basis, and the use of private memory maps, which require no locking.

The shortcoming lies in the *synonym problem*, in which different virtual addresses in different/same virtual spaces point to the same physical page. Another problem is that the same virtual address in different virtual spaces may point to different pages in the main memory.

Shared Virtual Memory   This model combines all the virtual address spaces into a single globally *shared virtual space* (Fig. 4.20b). Each processor is given a portion of the shared virtual memory to declare their addresses. Different processors may use disjoint spaces. Some areas of virtual space can be also shared by multiple processors.

Examples of machines using shared virtual memory include the IBM801, RT, RP3, System 38, the HP Spectrum, the Stanford Dash, MIT Alewife, Tera, etc. We will further study shared virtual memory in Chapter 9. Until then, all virtual memory systems discussed are assumed private unless otherwise specified.

The advantages in using shared virtual memory include the fact that all addresses are unique. However, each processor must be allowed to generate addresses larger than 32 bits, such as 46 bits for a 64-Tbyte (2^{46}-byte) address space. Synonyms are not allowed in a globally shared virtual memory.

The page table must allow shared accesses. Therefore, *mutual exclusion* (locking) is needed to enforce protected access. Segmentation is built on top of the paging system to confine each process to its own address space (segments). Global virtual memory makes the address translation process even longer.

4.4.2   TLB, Paging, and Segmentation

Both the virtual memory and physical memory are partitioned into fixed-length pages as illustrated in Fig. 4.18. The purpose of memory allocation is to allocate *pages* of virtual memory to the *page frames* of the physical memory.

Address Translation Mechanisms   The process demands the translation of virtual addresses into physical addresses. Various schemes for virtual address translation are summarized in Fig. 4.21a. The translation demands the use of *translation maps* which can be implemented in various ways.

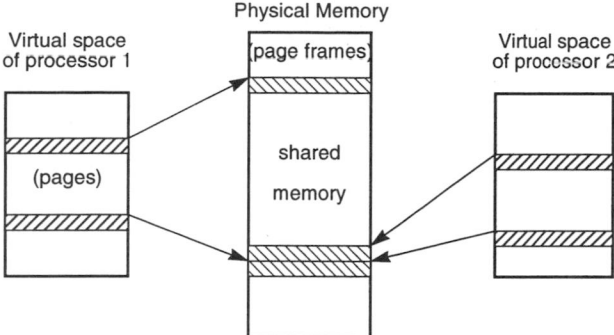

(a) Private virtual memory spaces in different processors

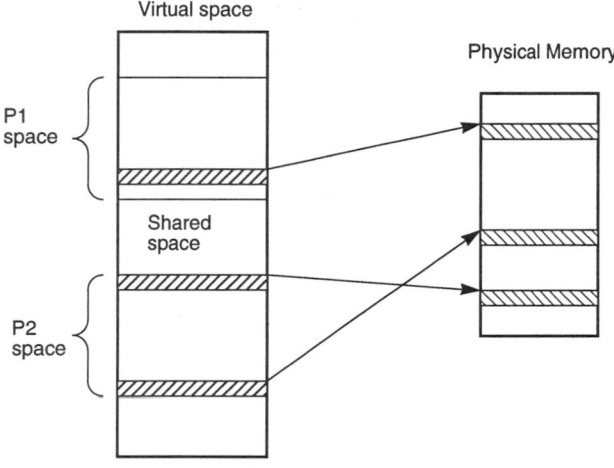

(b) Globally shared virtual memory space

Figure 4.20 Two virtual memory models for multiprocessor systems. (Courtesy of Dubois and Briggs, tutorial, *Annual Symposium on Computer Architecture*, 1990)

Translation maps are stored in the cache, in associative memory, or in the main memory. To access these maps, a mapping function is applied to the virtual address. This function generates a pointer to the desired translation map. This mapping can be implemented with a *hashing* or *congruence* function.

Hashing is a simple computer technique for converting a long page number into a short one with fewer bits. The hashing function should randomize the virtual page number and produce a unique hashed number to be used as the pointer. The congruence function provides hashing into a linked list.

Translation Lookaside Buffer   Translation maps appear in the form of a *translation*

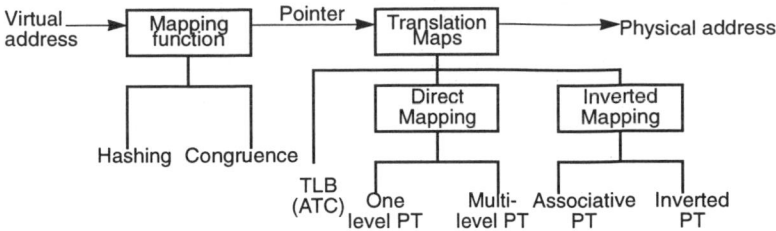

(a) Virtual address translation schemes (PT = page table)

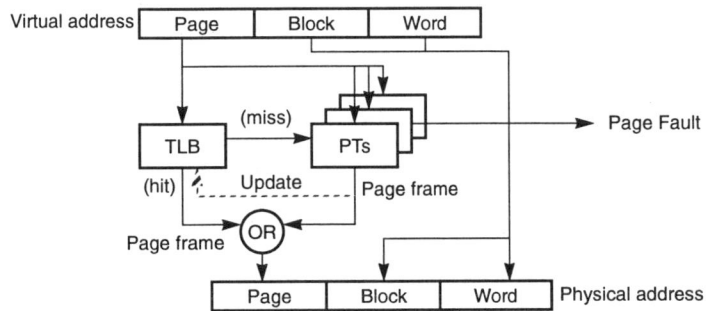

(b) Use of a TLB and PTs for address translation

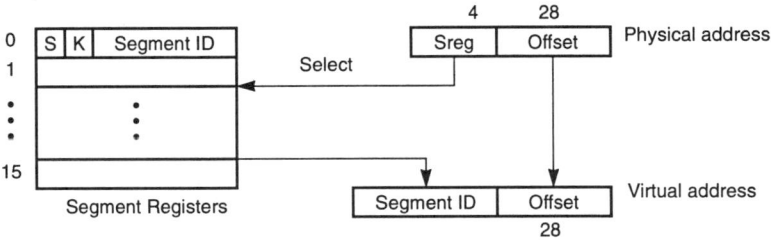

(c) Inverted address mapping

Figure 4.21 Address translation mechanisms using a TLB and various forms of page tables.

lookaside buffer (TLB) and *page tables* (PTs). Based on the principle of locality in memory references, a particular *working set* of pages is referenced within a given context or time window.

The TLB is a high-speed lookup table which stores the most recently or likely referenced page entries. A *page entry* consists of essentially a (virtual page number, page frame number) pair. It is hoped that pages belonging to the same working set will be directly translated using the TLB entries.

The use of a TLB and PTs for address translation is shown in Fig 4.21b. Each virtual address is divided into three fields: The leftmost field holds the *virtual page number*, the middle field identifies the *cache block number*, and the rightmost field is the *word address* within the block.

Our purpose is to produce the physical address consisting of the page frame number, the block number, and the word address. The first step of the translation is to use the virtual page number as a key to search through the TLB for a match. The TLB can be implemented with a special associative memory (content-addressable memory) or use part of the cache memory.

In case of a match (a *hit*) in the TLB, the page frame number is retrieved from the matched page entry. The cache block and word address are copied directly. In case the match cannot be found (a *miss*) in the TLB, a hashed pointer is used to identify one of the page tables where the desired page frame number can be retrieved.

Paged Memory   Paging is a technique for partitioning both the physical memory and virtual memory into fixed-size pages. Exchange of information between them is conducted at the page level as described before. Page tables are used to map between pages and page frames. These tables are implemented in the main memory upon creation of user processes in application programs. Since many user processes may be created dynamically, the number of PTs maintained in the main memory can be very large. The *page table entries* (PTEs) are similar to the TLB entries, containing essentially (virtual page, page frame) address pairs.

Note that both TLB entries and PTEs need to be dynamically updated to reflect the latest memory reference history. Only "snapshots" of the history are maintained in these translation maps.

If the demanded page cannot be found in the PT, a *page fault* is declared. A page fault implies that the referenced page is not resident in the main memory. When a page fault occurs, the running process is suspended. A *context switch* is made to another ready-to-run process while the missing page is transferred from the disk or tape unit to the physical memory.

This direct page mapping can be extended with multiple levels of page tables. Multilevel paging takes a longer time to generate the desired physical address because multiple memory references are needed to access a sequence of page tables. The reason for multilevel paging is to expand the memory space and to provide more sophisticated protection of page access at different levels of the memory hierarchy.

Segmented Memory   A large number of pages can be shared by segmenting the virtual address space among multiple user programs simultaneously. A *segment* of scattered pages is formed logically in the virtual memory space. Segments are defined by users in order to declare a portion of the virtual address space.

In a *segmented memory system*, user programs can be logically structured as *segments*. Segments can invoke each other. Unlike pages, segments can have variable lengths. The management of a segmented memory system is much more complex due to the nonuniform segment size.

Segments are a user-oriented concept, providing logical structures of programs and data in the virtual address space. On the other hand, paging facilitates the management of physical memory. In a paged system, all page addresses form a linear address space within the virtual space.

The segmented memory is arranged as a two-dimensional address space. Each virtual address in this space has a prefix field called the *segment number* and a postfix field called the *offset* within the segment. The offset addresses within each segment form one dimension of the contiguous addresses. The segment numbers, not necessarily contiguous to each other, form the second dimension of the address space.

Paged Segments   The above two concepts of paging and segmentation can be combined to implement a type of virtual memory with *paged segments*. Within each segment, the addresses are divided into fixed-size pages. Each virtual address is thus divided into three fields. The upper field is the *segment number*, the middle one is the *page number*, and the lower one is the *offset* within each page.

Paged segments offer the advantages of both paged memory and segmented memory. For users, program files can be better logically structured. For the OS, the virtual memory can be systematically managed with fixed-size pages within each segment. Tradeoffs do exist among the sizes of the segment field, the page field, and the offset field. This sets limits on the number of segments that can be declared by users, the segment size (the number of pages within each segment), and the page size.

Inverted Paging   The direct paging described above works well with a small virtual address space such as 32 bits. In modern computers, the virtual address space is very large, such as 52 bits in the IBM RS/6000. A large virtual address space demands either large PTs or multilevel direct paging which will slow down the address translation process and thus lower the performance.

Besides direct mapping, address translation maps can also be implemented with inverted mapping (Fig. 4.21c). An *inverted page table* is created for each page frame that has been allocated to users. Any virtual page number can be paired with a given physical page number.

Inverted page tables are accessed either by an associative search or by the use of a hashing function. The IBM 801 prototype and subsequently the IBM RT/PC have implemented inverted mapping for page address translation. In using an inverted PTE, only virtual pages that are currently resident in physical memory are included. This provides a significant reduction in the size of the page tables.

The generation of a long virtual address from a short physical address is done with the help of segment registers, as demonstrated in Fig. 4.21c. The leading 4 bits (denoted *sreg*) of a 32-bit address name a segment register. The register provides a *segment id* that replaces the 4-bit sreg to form a long virtual address.

This effectively creates a single long virtual address space with segment boundaries at multiples of 256 Mbytes (2^{28} bytes). The IBM RT/PC has a 12-bit segment id (4096 segments) and a 40-bit virtual address space.

Either associative page tables or inverted page tables can be used to implement inverted mapping. The inverted page table can also be assisted with the use of a TLB. An inverted PT avoids the use of a large page table or a sequence of page tables.

Given a virtual address to be translated, the hardware searches the inverted PT for that address and, if it is found, uses the table index of the matching entry as the address of the desired page frame. A hashing table is used to search through the inverted PT. The size of an inverted PT is governed by the size of the physical space, while that of traditional PTs is determined by the size of the virtual space. Because of limited physical space, no multiple levels are needed for the inverted page table.

Example 4.8 Paging and segmentation in the Intel i486 processor

As with its predecessor in the 80x86 family, the i486 features both segmentation and paging compatibilities. Protected mode increases the linear address from 4 Gbytes (2^{32} bytes) to 64 Tbytes (2^{46} bytes) with four levels of protection. The maximal memory size in real mode is 1 Mbyte (2^{20} bytes). Protected mode allows the i486 to run all software from existing 8086, 80286, and 80386 processors. A segment can have any length from 1 byte to 4 Gbytes, the maximal physical memory size.

A segment can start at any base address, and storage overlapping between segments is allowed. The virtual address (Fig. 4.22a) has a 16-bit segment selector to determine the base address of the *linear address space* to be used with the i486 paging system.

The 32-bit offset specifies the internal address within a segment. The segment descriptor is used to specify access rights and segment size besides selection of the address of the first byte of the segment.

The paging feature is optional on the i486. It can be enabled or disabled by software control. When paging is enabled, the virtual address is first translated into a linear address and then into the physical address. When paging is disabled, the linear address and physical address are identical. When a 4-Gbyte segment is selected, the entire physical memory becomes one large segment, which means the segmentation mechanism is essentially disabled.

In this sense, the i486 can be used with four different memory organizations, *pure paging*, *pure segmentation*, *segmented paging*, or *pure physical addressing* without paging and segmentation.

A 32-entry TLB (Fig 4.22b) is used to convert the linear address directly into the physical address without resorting to the two-level paging scheme (Fig 4.22c). The standard page size on the i486 is 4 Kbytes = 2^{12} bytes. Four control registers are used to select between regular paging and page fault handling.

The page table directory (4 Kbytes) allows 1024 page directory entries. Each page table at the second level is 4 Kbytes and holds up to 1024 PTEs. The upper 20 linear address bits are compared to determine if there is a hit. The hit ratios of the TLB and of the page tables depend on program behavior and the efficiency of the update (page replacement) policies. A 98% hit ratio has been observed in TLB operations in the past.

∎

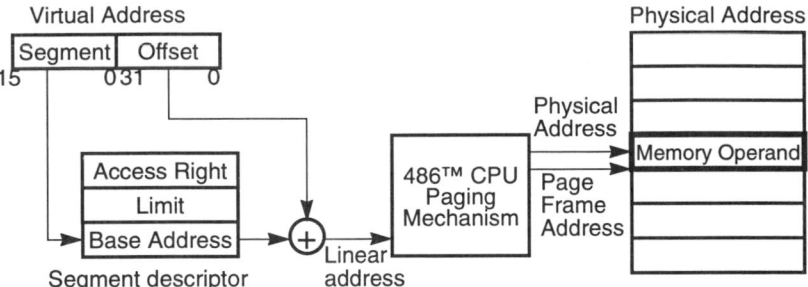

(a) Segmentation to produce the linear address

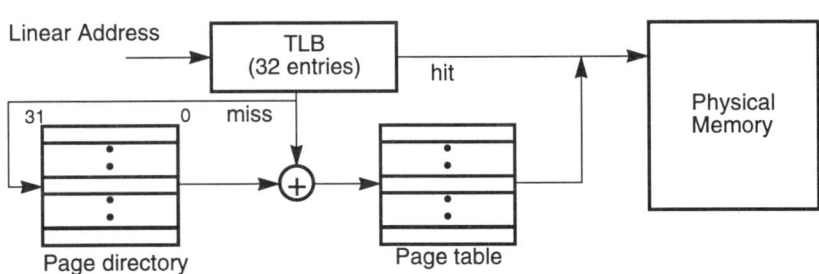

(b) The TLB operations

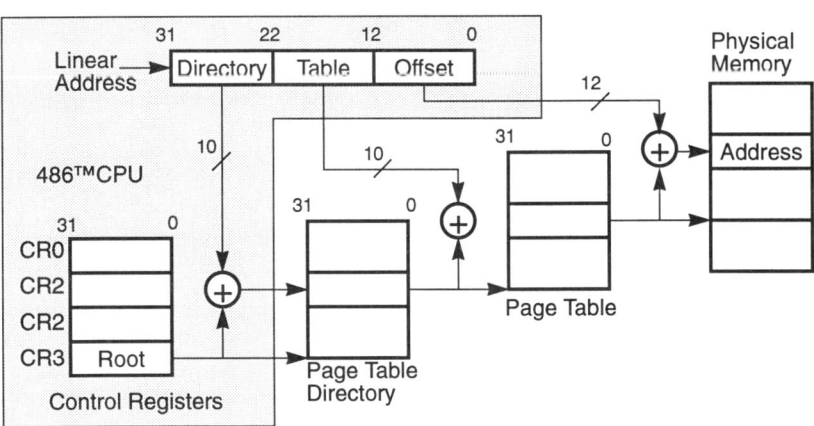

(c) A two-level paging scheme

Figure 4.22 Paging and segmentation mechanisms built into the Intel i486 CPU
(Courtesy of Intel Corporation, 1990)

4.4.3 Memory Replacement Policies

Memory management policies include the allocation and deallocation of memory pages to active processes and the replacement of memory pages. We will study allocation and deallocation problems in Section 5.3.3 after we discuss main memory organization in Section 5.3.1.

In this section, we study page replacement schemes which are implemented with demand paging memory systems. *Page replacement* refers to the process in which a resident page in main memory is replaced by a new page transferred from the disk.

Since the number of available page frames is much smaller (10^3 to 10^5 fewer) than the number of pages, the frames will eventually be fully occupied. In order to accommodate a new page, one of the resident pages must be replaced. Different policies have been suggested for page replacement. These policies are specified and compared below.

The goal of a page replacement policy is to minimize the number of possible page faults so that the effective memory-access time can be reduced. The effectiveness of a replacement algorithm depends on the program behavior and memory traffic patterns encountered. A good policy should match the program locality property. The policy is also affected by page size and by the number of available frames.

Page Traces To analyze the performance of a paging memory system, page trace experiments are often performed. A *page trace* is a sequence of *page frame numbers* (PFNs) generated during the execution of a given program. To simplify the analysis, we ignore the cache effect.

Each PFN corresponds to the prefix portion of a physical memory address. By tracing the successive PFNs in a page trace against the resident page numbers in the page frames, one can determine the occurrence of page hits or of page faults. Of course, when all the page frames are taken, a certain replacement policy must be applied to swap the pages. A page trace experiment can be performed to determine the *hit ratio* of the paging memory system. A similar idea can also be applied to perform *block traces* on cache behavior.

Consider a page trace $P(n) = r(1)r(2) \cdots r(n)$ consisting of n PFNs requested in discrete time from 1 to n, where $r(t)$ is the PFN requested at time t. We define two reference distances between the repeated occurrences of the same page in $P(n)$.

The *forward distance* $f_t(x)$ for page x is the number of time slots from time t to the first repeated reference of page x in the future:

$$f_t(x) = \begin{cases} k, & \text{if } k \text{ is the smallest integer such that} \\ & \quad r(t+k) = r(t) = x \text{ in } P(n) \\ \infty, & \text{if } x \text{ does not reappear in } P(n) \text{ beyond time } t \end{cases} \quad (4.11)$$

Similarly, we define a *backward distance* $b_t(x)$ as the number of time slots from time t to the most recent reference of page x in the past:

$$b_t(x) = \begin{cases} k, & \text{if } k \text{ is the smallest integer such that} \\ & \quad r(t-k) = r(t) = x \text{ in } P(n) \\ \infty, & \text{if } x \text{ never appeared in } P(n) \text{ in the past} \end{cases} \quad (4.12)$$

Let $R(t)$ be the *resident set* of all pages residing in main memory at time t. Let $q(t)$ be the page to be replaced from $R(t)$ when a page fault occurs at time t.

Page Replacement Policies   The following page replacement policies are specified in a demand paging memory system for a page fault at time t.

(1) *Least recently used* (LRU) — This policy replaces the page in $R(t)$ which has the longest backward distance:

$$q(t) = y, \quad \text{iff} \quad b_t(y) = \max_{x \in R(t)} \{b_t(x)\} \tag{4.13}$$

(2) *Optimal (OPT) algorithm* — This policy replaces the page in $R(t)$ with the longest forward distance:

$$q(t) = y, \quad \text{iff} \quad f_t(y) = \max_{x \in R(t)} \{f_t(x)\} \tag{4.14}$$

(3) *First-in-first-out* (FIFO) — This policy replaces the page in $R(t)$ which has been in memory for the longest time.

(4) *Least frequently used* (LFU) — This policy replaces the page in $R(t)$ which has been least referenced in the past.

(5) *Circular FIFO* — This policy joins all the page frame entries into a circular FIFO queue using a pointer to indicate the front of the queue. An *allocation bit* is associated with each page frame. This bit is set upon initial allocation of a page to the frame.

When a page fault occurs, the queue is circularly scanned from the pointer position. The pointer skips and resets the allocated page frames and replaces the very first unallocated page frame. When all frames are allocated, the front of the queue is replaced, as in the FIFO policy.

(6) *Random replacement* — This is a trivial algorithm which chooses any page for replacement randomly.

Example 4.9 Page tracing experiments and interpretation of results

Consider a paged virtual memory system with a two-level hierarchy: main memory M_1 and disk memory M_2. For clarity of illustration, assume a page size of four words. The number of page frames in M_1 is 3, labeled a, b and c; and the number of pages in M_2 is 10, identified by $0, 1, 2, ..., 9$. The ith page in M_2 consists of word addresses $4i$ to $4i + 3$ for all $i = 0, 1, 2, ..., 9$.

A certain program generates the following sequence of word addresses which are grouped (underlined) together if they belong to the same page. The sequence of page numbers so formed is the *page trace*:

Word trace:	0,1,2,3,	4,5,6,7,	8,	16,17,	9,10,11,	12,	28,29,30,	8,9,10,	4,5,	12,	4,5
	↓	↓	↓	↓	↓	↓	↓	↓	↓	↓	↓
Page trace:	0	1	2	4	2	3	7	2	1	3	1

Page tracing experiments are described below for three page replacement policies: LRU, OPT, and FIFO, respectively. The successive pages loaded in the page frames (PFs) form the trace entries. Initially, all PFs are empty.

	PF	0	1	2	4	2	3	7	2	1	3	1	Hit Ratio
LRU	a	0	0	0	4	4	4	7	7	7	3	3	$\frac{3}{11}$
	b		1	1	1	1	3	3	3	1	1	1	
	c			2	2	2	2	2	2	2	2	2	
	Faults	*	*	*	*		*	*		*	*		
OPT	a	0	0	0	4	4	3	7	7	7	3	3	$\frac{4}{11}$
	b		1	1	1	1	1	1	1	1	1	1	
	c			2	2	2	2	2	2	2	2	2	
	Faults	*	*	*	*		*	*			*		
FIFO	a	0	0	0	4	4	4	4	2	2	2	2	$\frac{2}{11}$
	b		1	1	1	1	3	3	3	1	1	1	
	c			2	2	2	2	7	7	7	3	3	
	Faults	*	*	*	*		*	*	*	*	*		

The above results indicate the superiority of the OPT policy over the others. However, the OPT cannot be implemented in practice. The LRU policy performs better than the FIFO due to the locality of references. From these results, we realize that the LRU is generally better than the FIFO. However, exceptions still exist due to the dependence on program behavior. ∎

Relative Performance The performance of a page replacement algorithm depends on the page trace (program behavior) encountered. The best policy is the OPT algorithm. However, the OPT replacement is not realizable because no one can predict the future page demand in a program.

The LRU algorithm is a popular policy and often results in a high hit ratio. The FIFO and random policies may perform badly because of violation of the program locality.

The circular FIFO policy attempts to approximate the LRU with a simple circular queue implementation. The LFU policy may perform between the LRU and the FIFO policies. However, there is no fixed superiority of any policy over the others because of the dependence on program behavior and run-time status of the page frames.

In general, the page fault rate is a monotonic decreasing function of the size of the resident set $R(t)$ at time t because more resident pages result in a higher hit ratio in the main memory.

Block Replacement Policies The relationship between the cache block frames and cache blocks is similar to that between page frames and pages on a disk. Therefore, those page replacement policies can be modified for *block replacement* when a cache miss occurs.

Different cache organizations (Section 5.1) may offer different flexibilities in implementing some of the replacement algorithms. The cache memory is often associatively searched, while the main memory is randomly addressed.

Due to the difference between page allocation in main memory and block allocation in the cache, the cache hit ratio and memory page hit ratio are affected by the replacement policies differently. *Cache traces* are often needed to evaluate the cache performance. These considerations will be further discussed in Chapter 5.

4.5   Bibliographic Notes and Exercises

Advanced microprocessors were surveyed in the book by [Tabak91]. A tutorial on RISC computers was given by [Stallings90]. Superscalar and superpipelined machines were characterized by Jouppi and Wall [Jouppi89]. [Johnson91] provided an excellent book on superscalar processor design. The VLIW architecture was first developed by [Fisher83].

The Berkeley RISC was reported by Patterson and Sequin in [Patterson82]. A MIPS R2000 overview can be found in [Kane88]. The HP precision architecture has been assessed in [Lee89]. Optimizing compilers for SPARC appears in [Muchnick88]. A further description of the M68040 can be found in [Edenfield90].

A good source of information on the i860 can be found in [Margulis90]. The DEC Alpha architecture is described in [DEC92]. The latest MIPS R4000 was reported by Mirapuri, Woodacre, and Vasseghi [Mirapuri92]. The IBM RS/6000 was discussed in [IBM90]. The Hot-Chips Symposium Series [Hotchips91] usually present the latest developments in high-performance processor chips.

The virtual memory models were based on the tutorial by Dubois and Briggs [Dubois90c]. The book by [Hennessy90] and Patterson has treated the memory hierarchy design based on extensive program trace data. Distributed shared virtual memory was surveyed in [Nitzberg91] and Lo.

The concept of a working set was introduced by [Denning68]. A linear programming optimization of the memory hierarchy was reported in [Chow74]. [Crawford90] explained the paging and segmentation schemes in the i486. Inverted paging was described by [Chang88] and Mergen. [Cragon92b] has discussed memory systems for pipeline processor design.

Exercises

Problem 4.1   Define the following basic terms related to modern processor technology:

(a) Processor design space.   (f) Processor versus coprocessor.
(b) Instruction issue latency.   (g) General-purpose registers.
(c) Instruction issue rate.   (h) Addressing modes.
(d) Simple operation latency.   (i) Unified versus split caches.
(e) Resource conflicts.   (j) Hardwired versus microcoded control.

Problem 4.2 Define the following basic terms associated with memory hierarchy design:

(a) Virtual address space.

(b) Physical address space.

(c) Address mapping.

(d) Cache blocks.

(e) Multilevel page tables.

(f) Hit ratio.

(g) Page fault.

(h) Hashing function.

(i) Inverted page table.

(j) Memory replacement policies.

Problem 4.3 Answer the following questions on designing scalar RISC or superscalar RISC processors:

(a) Why do most RISC integer units use 32 general-purpose registers? Explain the concept of register windows implemented in the SPARC architecture.

(b) What are the design tradeoffs between a large register file and a large D-cache? Why are reservation stations or reorder buffers needed in a superscalar processor?

(c) Explain the relationship between the integer unit and the floating-point unit in most RISC processors with scalar or superscalar organization.

Problem 4.4 Based on the discussion of advanced processors in Section 4.1, answer the following questions on RISC, CISC, superscalar, and VLIW architectures.

(a) Compare the instruction-set architecture in RISC and CISC processors in terms of instruction formats, addressing modes, and cycles per instruction (CPI).

(b) Discuss the advantages and disadvantages in using a common cache or separate caches for instructions and data. Explain the support from data paths, MMU and TLB, and memory bandwidth in the two cache architectures.

(c) Distinguish between scalar RISC and superscalar RISC in terms of instruction issue, pipeline architecture, and processor performance.

(d) Explain the difference between superscalar and VLIW architectures in terms of hardware and software requirements.

Problem 4.5 Explain the structures and operational requirements of the instruction pipelines used in CISC, scalar RISC, superscalar RISC, and VLIW processors. Comment on the cycles per instruction expected from these processor architectures.

Problem 4.6 Study the Intel i486 instruction set and the CPU architecture reported by J. H. Crawford in *IEEE Micro*, February 1990, and answer the following questions:

(a) What are the instruction formats and data formats?

(b) What are the addressing modes?

(c) What are the instruction categories? Describe one example instruction in each category.

(d) What are the HLL support instructions and assembly directives?

(e) What are the interrupt, testing, and debug features?

(f) Explain the difference between real and virtual mode execution.

(g) Explain how to disable paging in the i486 and what kind of application may benefit from this option.

(h) Explain how to disable segmentation in the i486 and what kind of application may use this option.

(i) What kind of protection mechanisms are built into the i486?

(j) Compare the i486 architecture with the Intel Pentium processor. Identify the improvements made in Pentium over i486.

Problem 4.7   Answer the following questions after studying Example 4.4, the i860 instruction set, and the architecture of the i860 and its successor, the i860XP:

(a) Repeat parts (a), (b), and (c) in Problem 4.6 for the i860/i860XP.

(b) What multiprocessor support instructions are added in the i860XP?

(c) Explain the dual-instruction mode and the dual-operation instructions in i860 processors.

(d) Explain the address translation and paged memory organization of the i860.

Problem 4.8   The SPARC architecture can be implemented with two to eight register windows, for a total of 40 to 132 GPRs in the integer unit. Explain how the GPRs are organized into overlapping windows in each of the following designs:

(a) Use 40 GPRs to construct two windows.

(b) Use 72 GPRs to construct four windows.

(c) In what sense is the SPARC considered a scalable architecture?

(d) Explain how to use the overlapped windows for parameter passing between the calling procedure and the called procedure.

Problem 4.9   Study Section 4.2 and also the paper by Jouppi and Wall (1989) and answer the following questions:

(a) What causes a processor pipeline to be underpipelined?

(b) What are the factors limiting the degree of superscalar design?

Problem 4.10   Answer the following questions related to vector processing:

(a) What are the differences between scalar instructions and vector instructions?

(b) Compare the pipelined execution style in a vector processor with that in a base scalar processor (Fig. 4.15). Analyze the speedup gain of the vector pipeline over the scalar pipeline for long vectors.

(c) Suppose parallel issue is added to vector pipeline execution. What would be the further improvement in throughput, compared with parallel issue in a superscalar pipeline of the same degree?

Problem 4.11 Consider a two-level memory hierarchy, M_1 and M_2. Denote the hit ratio of M_1 as h. Let c_1 and c_2 be the costs per kilobyte, s_1 and s_2 the memory capacities, and t_1 and t_2 the access times, respectively.

(a) Under what conditions will the average cost of the entire memory system approach c_2?

(b) What is the effective memory-access time t_a of this hierarchy?

(c) Let $r = t_2/t_1$ be the speed ratio of the two memories. Let $E = t_1/t_a$ be the *access efficiency* of the memory system. Express E in terms of r and h.

(d) Plot E against h for $r = 5, 20$, and 100, respectively, on grid paper.

(e) What is the required hit ratio h to make $E > 0.95$ if $r = 100$?

Problem 4.12 You are asked to perform capacity planning for a two-level memory system. The first level, M_1, is a cache with three capacity choices of 64 Kbytes, 128 Kbytes, and 256 Kbytes. The second level, M_2, is a main memory with a 4-Mbyte capacity. Let c_1 and c_2 be the costs per byte and t_1 and t_2 the access times for M_1 and M_2, respectively. Assume $c_1 = 20c_2$ and $t_2 = 10t_1$. The cache hit ratios for the three capacities are assumed to be 0.7, 0.9, and 0.98, respectively.

(a) What is the average access time t_a in terms of $t_1 = 20$ ns in the three cache designs? (Note that t_1 is the time from CPU to M_1 and t_2 is that from CPU to M_2, not from M_1 to M_2).

(b) Express the average byte cost of the entire memory hierarchy if $c_2 = \$0.2/\text{Kbyte}$.

(c) Compare the three memory designs and indicate the order of merit in terms of average costs and average access times, respectively. Choose the optimal design based on the product of average cost and average access time.

Problem 4.13 Compare the advantages and shortcomings in implementing private virtual memories and a globally shared virtual memory in a multicomputer system. This comparative study should consider the latency, coherence, page migration, protection, implementation, and application problems in the context of building a scalable multicomputer system with distributed shared memories.

Problem 4.14 Explain the *inclusion property* and *memory coherence* requirements in a multilevel memory hierarchy. Distinguish between *write-through* and *write-back* policies in maintaining the coherence in adjacent levels. Also explain the basic concepts of *paging* and *segmentation* in managing the physical and virtual memories in a hierarchy.

Problem 4.15 A two-level memory system has eight virtual pages on a disk to be mapped into four page frames (PFs) in the main memory. A certain program generated the following page trace:

$$1, 0, 2, 2, 1, 7, 6, 7, 0, 1, 2, 0, 3, 0, 4, 5, 1, 5, 2, 4, 5, 6, 7, 6, 7, 2, 4, 2, 7, 3, 3, 2, 3$$

(a) Show the successive virtual pages residing in the four page frames with respect to the above page trace using the LRU replacement policy. Compute the hit ratio in the main memory. Assume the PFs are initially empty.

(b) Repeat part (a) for the circular FIFO page replacement policy. Compute the hit ratio in the main memory.

(c) Compare the hit ratio in parts (a) and (b) and comment on the effectiveness of using the circular FIFO policy to approximate the LRU policy with respect to this particular page trace.

Problem 4.16

(a) Explain the temporal locality, spatial locality, and sequential locality associated with program/data access in a memory hierarchy.

(b) What is the working set? Comment on the sensitivity of the observation window size to the size of the working set. How will this affect the main memory hit ratio?

(c) What is the 90-10 rule and its relationship to the locality of references?

Problem 4.17

Consider a two-level memory hierarchy, M_1 and M_2, with access times t_1 and t_2, costs per byte c_1 and c_2, and capacities s_1 and s_2, respectively. The cache hit ratio $h_1 = 0.95$ at the first level. (Note that t_2 is the access time between the CPU and M_2, not between M_1 and M_2).

(a) Derive a formula showing the effective access time t_{eff} of this memory system.

(b) Derive a formula showing the total cost of this memory system.

(c) Suppose $t_1 = 20$ ns, t_2 is unknown, $s_1 = 512$ Kbytes, s_2 is unknown, $c_1 = \$0.01$/byte, and $c_2 = \$0.0005$/byte. The total cost of the cache and main memory is upper-bounded by $\$15,000$.

  (i) How large a capacity of M_2 ($s_2 =$?) can you acquire without exceeding the budget limit?

  (ii) How fast a main memory ($t_2 =$?) do you need to achieve an effective access time of $t_{eff} = 40$ ns in the entire memory system under the above hit ratio assumptions?

Problem 4.18

Distinguish between numeric processing and symbolic processing computers in terms of data objects, common operations, memory requirements, communication patterns, algorithmic properties, I/O requirements, and processor architectures.

Chapter 5

Bus, Cache, and Shared Memory

This chapter describes the design and operational principles of bus, cache, and shared-memory organization. Backplane bus systems are studied first, including the latest Futurebus+ specifications. Cache addressing models and implementation schemes are described. We study memory interleaving, allocation schemes, and the sequential and weak consistency models for shared-memory systems. Other relaxed memory consistency models are given in Chapter 9.

5.1 Backplane Bus Systems

The system bus of a computer system operates on a contention basis. Several active devices such as processors may request use of the bus at the same time. However, only one of them can be granted access at a time. The *effective bandwidth* available to each processor is inversely proportional to the number of processors contending for the bus.

For this reason, most bus-based commercial multiprocessors are small in size. The simplicity and low cost of a bus system have made it attractive in building small multiprocessors ranging from 4 to 16 processors based on today's technology.

In this section, we specify system buses which are confined to a single backplane chassis. We concentrate on logical specification instead of physical implementation. Standard bus specifications should be both technology-independent and architecture-independent.

5.1.1 Backplane Bus Specification

A backplane bus interconnects processors, data storage, and peripheral devices in a tightly coupled hardware configuration. The system bus must be designed to allow communication between devices on the bus without disturbing the internal activities of all the devices attached to the bus. Timing protocols must be established to arbitrate among multiple requests. Operational rules must be set to ensure orderly data transfers on the bus.

Signal lines on the backplane are often functionally grouped into several buses as depicted in Fig. 5.1. The four groups shown here are very similar to those proposed in the 64-bit VME bus specification (VITA, 1990).

Various functional boards are plugged into slots on the backplane. Each slot is provided with one or more connectors for inserting the boards as demonstrated by the vertical arrows in Fig. 5.1. For example, one or two 96-pin connectors are used per slot on the VME backplane.

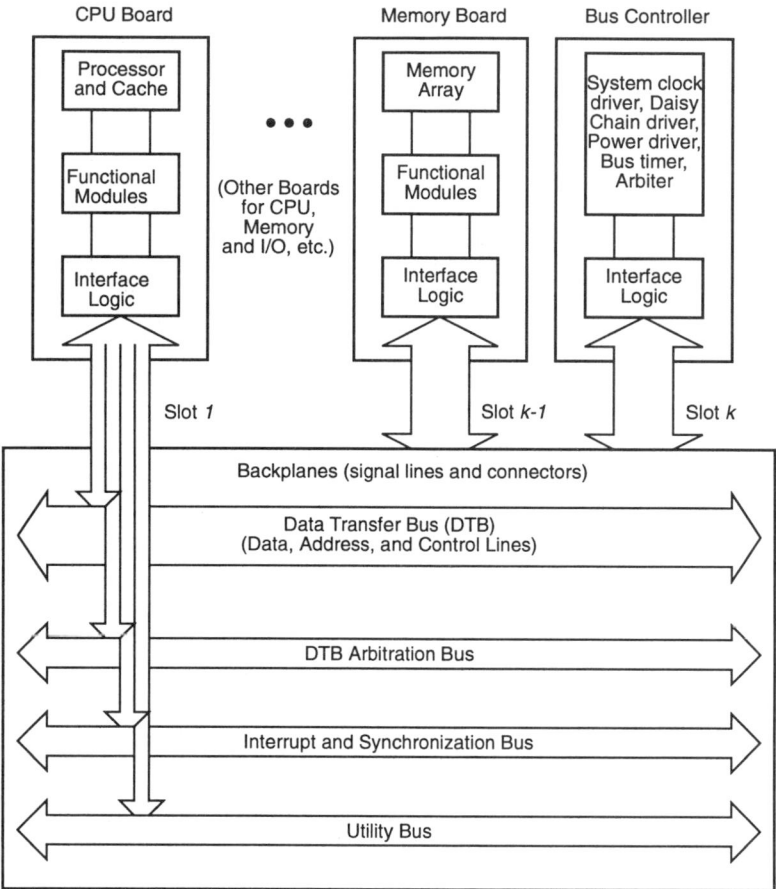

Figure 5.1 Backplane buses, system interfaces, and slot connections to various functional boards in a multiprocessor system.

Data Transfer Bus   Data, address, and control lines form the *data transfer bus* (DTB) in a VME bus. The addressing lines are used to broadcast the data and device address. The number of address lines is proportional to the logarithm of the size of the address space. Address modifier lines can be used to define special addressing modes.

The data lines are often proportional to the memory word length.

For example, the revised VME bus specification has 32 address lines and 32 (or 64) data lines. Besides being used in addressing, the 32 address lines can be multiplexed to serve as the lower half of the 64-bit data during data transfer cycles. The DTB control lines are used to indicate read/write, timing control, and bus error conditions.

Bus Arbitration and Control The process of assigning control of the DTB to a requester is called *arbitration*. Dedicated lines are reserved to coordinate the arbitration process among several requesters. The requester is called a *master*, and the receiving end is called a *slave*.

Interrupt lines are used to handle interrupts, which are often prioritized. Dedicated lines may be used to synchronize parallel activities among the processor modules. Utility lines include signals that provide periodic timing (clocking) and coordinate the power-up and power-down sequences of the system.

The backplane is made of signal lines and connectors. A special bus controller board is used to house the backplane control logic, such as the system clock driver, arbiter, bus timer, and power driver.

Functional Modules A *functional module* is a collection of electronic circuitry that resides on one functional board (Fig. 5.1) and works to achieve special bus control functions. Special functional modules are introduced below:

An *arbiter* is a functional module that accepts bus requests from the requester module and grants control of the DTB to one requester at a time.

A *bus timer* measures the time each data transfer takes on the DTB and terminates the DTB cycle if a transfer takes too long.

An *interrupter* module generates an interrupt request and provides status/ID information when an *interrupt handler* module requests it.

A *location monitor* is a functional module that monitors data transfers over the DTB. A *power monitor* watches the status of the power source and signals when power becomes unstable.

A *system clock driver* is a module that provides a clock timing signal on the utility bus. In addition, *board interface logic* is needed to match the signal line impedance, the propagation time, and termination values between the backplane and the plug-in boards.

Physical Limitations Due to electrical, mechanical, and packaging limitations, only a limited number of boards can be plugged into a single backplane. Multiple backplane buses can be mounted on the same backplane chassis.

For example, the VME chassis can house one to three backplane buses. Two can be used as a shared bus among all processors and memory boards, and the third as a local bus connecting a host processor to additional memory and I/O boards. Means of extending a single-bus system to build larger multiprocessors will be studied in Section 7.1.1. The bus system is difficult to scale, mainly limited by packaging constraints.

5.1.2 Addressing and Timing Protocols

There are two types of printed-circuit boards connected to a bus: *active* and *passive*. Active boards like processors can act as bus masters or as slaves at different times. Passive boards like memory boards can act only as slaves.

The master can initiate a bus cycle, and the slaves respond to requests by a master. Only one master can control the bus at a time. However, one or more slaves can respond to the master's request at the same time.

Bus Addressing   The backplane bus is driven by a digital clock with a fixed cycle time called the *bus cycle*. The bus cycle is determined by the electrical, mechanical, and packaging characteristics of the backplane. Signals traveling on bus lines may experience unequal delays from the source to the destination.

The backplane is designed to have a limited physical size which will not skew information with respect to the associated strobe signals. To speed up the operations, cycles on parallel lines in different buses may overlap in time. Factors affecting the bus delay include the source's line drivers, the destination's receivers, slot capacitance, line length, and the bus loading effects (the number of boards attached).

Not all the bus cycles are used for data transfers. To optimize performance, the bus should be designed to minimize the time required for request handling, arbitration, addressing, and interrupts so that most bus cycles are used for useful data transfer operations.

Each board can be identified with a *slot number*. When the slot number matches the contents of high-order address lines, the board is selected as a slave. This geographical (slot) addressing allows the allocation of a logical board address under software control, which will increase the application flexibility.

Broadcall and Broadcast   Most bus transactions involve only one master and one slave. However, a *broadcall* is a read operation involving multiple slaves placing their data on the bus lines. Special AND or OR operations over these data are performed on the bus from the selected slaves.

Broadcall operations are used to detect multiple interrupt sources. A *broadcast* is a write operation involving multiple slaves. This operation is essential in implementing multicache coherence on the bus.

Timing protocols are needed to synchronize master (source) and slave (destination) operations. Figure 5.2 shows a typical timing sequence when information is transferred over a bus from a source to a destination. Most bus timing protocols implement this sequence.

Synchronous Timing   All bus transaction steps take place at fixed clock edges as shown in Fig. 5.3a. The clock signals are broadcast to all potential masters and slaves. The clock cycle time is determined by the slowest device connected to the bus.

Once the data becomes stabilized on the data lines, the master uses a *data-ready* pulse to initiate the transfer. The slave uses a *data-accept* pulse to signal completion of the bit information being transferred.

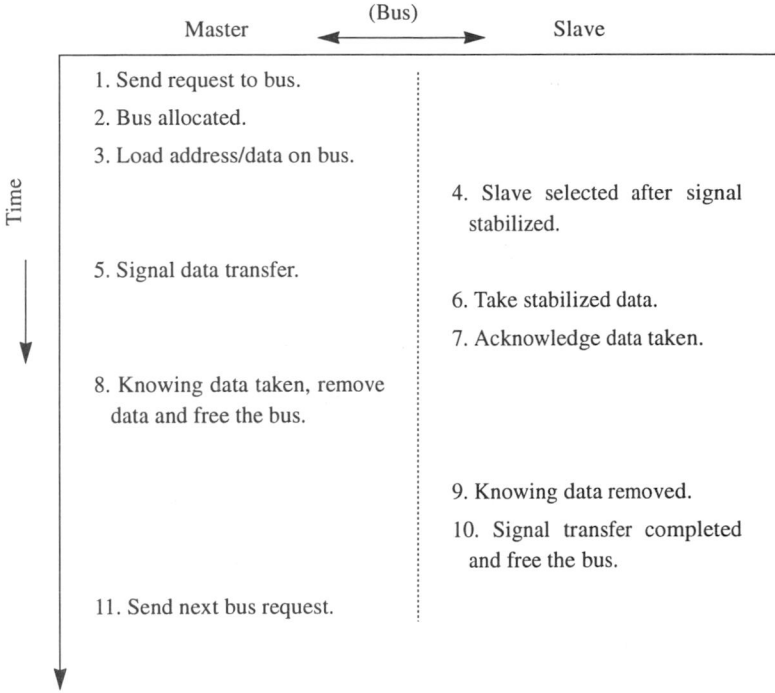

Figure 5.2 Typical time sequence for information transfer between a master and a slave over a system bus.

A synchronous bus is simple to control, requires less control circuitry, and thus costs less. It is suitable for connecting devices having relatively the same speed. Otherwise, the slowest device will slow down the entire bus operation.

Asynchronous Timing Asynchronous timing is based on a handshaking or interlocking mechanism as illustrated in Fig. 5.3b. No fixed clock cycle is needed. The rising edge (1) of the *data-ready* signal from the master triggers the rising (2) of the *data-accept* signal from the slave. The second signal triggers the falling (3) of the *data-ready* clock and the removal of data from the bus. The third signal triggers the trailing edge (4) of the *data-accept* clock. This four-edge handshaking (interlocking) process is repeated until all the data are transferred.

The advantage of using an asynchronous bus lies in the freedom of using variable-length clock signals for different-speed devices. This does not impose any response-time restrictions on the source and destination. It allows fast and slow devices to be connected on the same bus, and it is less prone to noise. Overall, an asynchronous bus offers better application flexibility at the expense of increased complexity and costs.

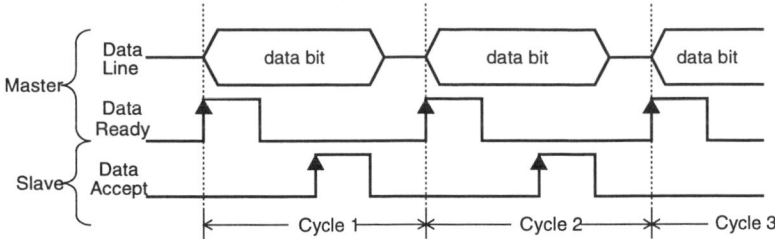

(a) Synchronous bus timing with fixed-length clock signals for all devices

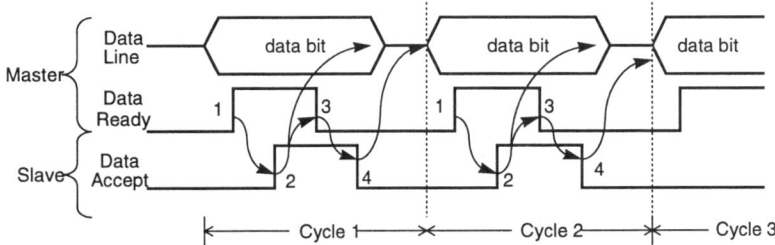

(b) Asynchronous bus timing using a four-edge handshaking (interlocking) with variable-length signals for different-speed devices

Figure 5.3 Synchronous versus asynchronous bus timing protocols.

5.1.3 Arbitration, Transaction, and Interrupt

The process of selecting the next bus master is called *arbitration.* The duration of a master's control of the bus is called *bus tenure.* This arbitration process is designed to restrict tenure of the bus to one master at a time. Competing requests must be arbitrated on a fairness or priority basis.

Arbitration competition and bus transactions may take place concurrently on a parallel bus with separate lines for both purposes.

Central Arbitration As illustrated in Fig. 5.4a, a central arbitration scheme uses a central arbiter. Potential masters are daisy-chained in a cascade. A special signal line is used to propagate a *bus-grant* signal level from the first master (at slot 1) to the last master (at slot n).

Each potential master can send a bus request. However, all requests share the same *bus-request* line. As shown in Fig. 5.4b, the *bus-request* signals the rise of the *bus-grant* level, which in turn raises the *bus-busy* level.

A fixed priority is set in a daisy chain from left to right. Only when the devices on the left do not request bus control can a device be granted bus tenure. When the bus transaction is complete, the *bus-busy* level is lowered, which triggers the falling of the *bus grant* signal and the subsequent rising of the *bus-request* signal.

The advantage of this arbitration scheme is its simplicity. Additional devices can

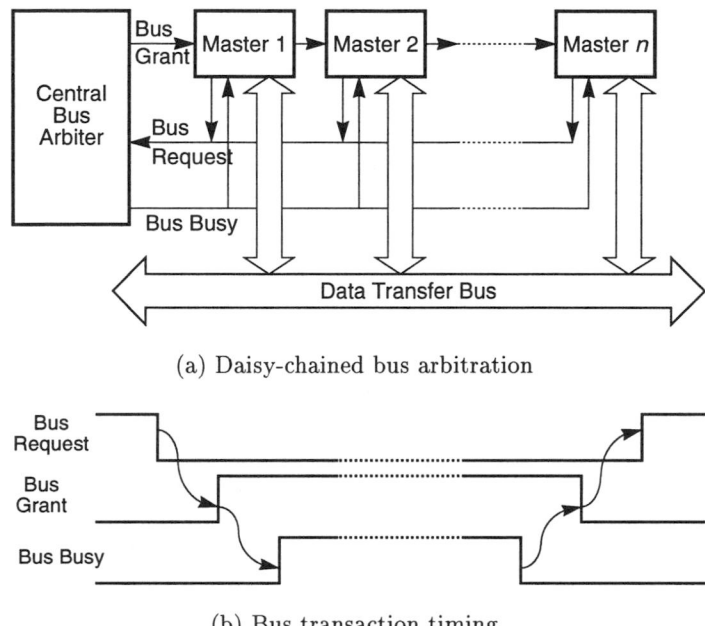

(a) Daisy-chained bus arbitration

(b) Bus transaction timing

Figure 5.4 Central bus arbitration using shared requests and daisy-chained bus grants with a fixed priority.

be added anywhere in the daisy chain by sharing the same set of arbitration lines. The disadvantage is a fixed-priority sequence violating the fairness practice. Another drawback is its slowness in propagating the *bus-grant* signal along the daisy chain.

Whenever a higher-priority device fails, all the lower-priority devices on the right of the daisy chain cannot use the bus. Bypassing a failing device or a removed device on the daisy chain is desirable. Some new bus standards are specified with such a capability.

Independent Requests and Grants Instead of using shared request and grant lines as in Fig. 5.4, multiple *bus-request* and *bus-grant* signal lines can be independently provided for each potential master as in Fig. 5.5a. No daisy-chaining is used in this scheme.

The arbitration among potential masters is still carried out by a central arbiter. However, any priority-based or fairness-based bus allocation policy can be implemented. A multiprocessor system usually uses a priority-based policy for I/O transactions and a fairness-based policy among the processors.

In some asymmetric multiprocessor architectures, the processors may be assigned different functions, such as serving as a front-end host, an executive processor, or a back-end slave processor. In such cases, a priority policy can also be used among the processors.

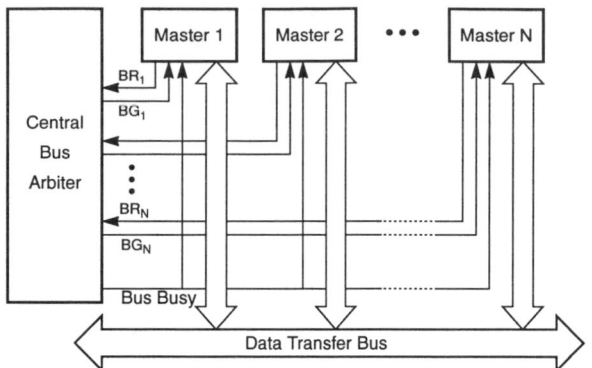

Legends: BR$_i$ (Bus request from master i) BG$_i$ (Bus grant to master i)

(a) Independent requests with a central arbiter

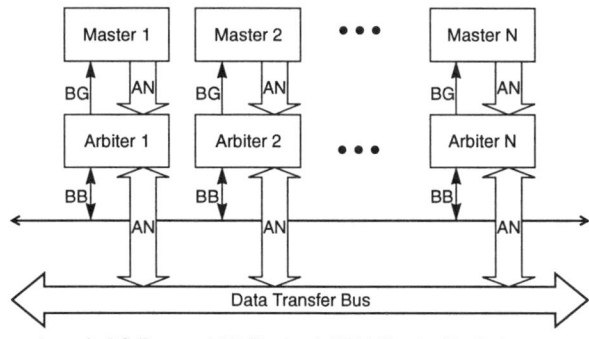

Legends: BG (Bus grant) BB (Bus busy) AN (Arbitration Number)

(b) Using distributed arbiters

Figure 5.5 Two bus arbitration schemes using independent requests and distributed arbiters, respectively.

The advantage of using independent requests and grants in bus arbitration is their flexibility and faster arbitration time compared with the daisy-chained policy. The drawback is the large number of arbitration lines used.

Distributed Arbitration   The idea of using distributed arbiters is depicted in Fig. 5.5b. Each potential master is equipped with its own arbiter and a unique *arbitration number*. The arbitration number is used to resolve the arbitration competition. When two or more devices compete for the bus, the winner is the one whose arbitration number is the largest.

Parallel contention arbitration is used to determine which device has the highest arbitration number. All potential masters can send their arbitration numbers to the shared-bus request/grant (SBRG) lines on the arbitration bus via their respective arbiters.

Each arbiter compares the resulting number on the SBRG lines with its own arbitration number. If the SBRG number is greater, the requester is dismissed. At the end, the winner's arbitration number remains on the arbitration bus. After the current bus transaction is completed, the winner seizes control of the bus.

Clearly, the distributed arbitration policy is priority-based. The Multibus II and the proposed Futurebus+ have adopted such a distributed arbitration scheme. Besides distributed arbiters, the Futurebus+ standard also provides options for a separate central arbiter.

Transaction Modes An *address-only transfer* consists of an address transfer followed by no data. A *compelled data transfer* consists of an address transfer followed by a block of one or more data transfers to one or more contiguous addresses. A *packet data transfer* consists of an address transfer followed by a fixed-length block of data transfers (*packet*) from a set of continuous addresses.

Data transfers and priority interrupts handling are two classes of operations regularly performed on a bus. A bus transaction consists of a request followed by a response. A *connected transaction* is used to carry out a master's request and a slave's response in a single bus transaction. A *split transaction* splits the request and response into separate bus transactions. Three data transfer modes are specified below.

Split transactions allow devices with a long data latency or access time to use the bus resources in a more efficient way. A complete split transaction may require two or more connected bus transactions. Split transactions across multiple bus sequences are performed to achieve cache coherence in a large multiprocessor system.

Interrupt Mechanisms An *interrupt* is a request from I/O or other devices to a processor for service or attention. A priority interrupt bus is used to pass the interrupt signals. The interrupter must provide status and identification information. A functional module can be used to serve as an interrupt handler.

Priority interrupts are handled at many levels. For example, the VME bus uses seven *interrupt-request* lines. Up to seven interrupt handlers can be used to handle multiple interrupts.

Interrupts can also be handled by message passing using the data bus lines on a time-sharing basis. The saving of dedicated interrupt lines is obtained at the expense of requiring some bus cycles for handling message-based interrupts.

The use of time-shared data bus lines to implement interrupts is called *virtual interrupt*. The Futurebus+ was proposed not to include dedicated interrupt lines because virtual interrupts can be effectively implemented with the data transfer bus.

5.1.4 The IEEE Futurebus+ Standards

A large number of backplane bus standards have been developed by various computer manufacturers in cooperation with the relevant IEEE standard committees. Among the well-known ones are those for the VME bus (VITA and IEEE Standard 1014-1987), the Multibus II (Intel and IEEE Standard 1296-1987), the Nubus (Texas Instruments, 1983), the Fastbus (Gustavason, 1986), and the Nanobus by Encore Computer Systems.

Some of these buses have been used in building multiprocessors, however, each has its own limitations. Most of them support only a data path of 32 bits, and none of them support an efficient cache coherence protocol or fast interprocessor synchronization.

The evolving Futurebus+ standard is being developed under the cooperative effort of the VME International Trade Association, Multibus Manufacturers Group, U.S. Navy Next Generation Computer Resources Program, IEEE Microcomputer Standards Committee, and experts from companies and universities.

The objective is to develop a truly open bus standard that can support a 64-bit address space and the throughput required by multi-RISC or future generations of multiprocessor architectures.

The standards must be expandable upward or scalable and be independent of particular architecture and processor technologies. The key features of the IEEE Futurebus+ Standard 896.1-1991 are presented below. We expect continued evolution and refinement of this bus standard in the future.

Standard Requirements The major objectives of the Futurebus+ standards committee were to create a bus standard that would provide a significant step forward in improving the facilities and performance available to the designers of future multiprocessor systems. This will provide a stable platform on which several generations of computer systems can be based. Summarized below are design requirements set by the IEEE 896.1-1991 Standards Committee:

(1) Architecture-, processor-, and technology-independent toward an open standard available to all designers.

(2) A fully asynchronous (compelled) timing protocol for data transfer with handshaking flow control.

(3) An optional source-synchronized (packet) protocol for high-speed block data transfers.

(4) Fully distributed parallel arbitration protocols to support a rich variety of bus transactions including broadcast, broadcall, and three-party transactions.

(5) Support of high reliability and fault-tolerant applications with provisions for live card insertion/removal, parity checks on all lines and feedback checking, and no daisy-chained signals to facilitate dynamic system reconfiguration in the event of module failure.

(6) Use of multilevel mechanisms for the locking of modules and avoidance of deadlock or livelock.

(7) Circuit-switched and split transaction protocols plus support for memory commands for implementing remote lock and SIMD-like operations.

(8) Support of real-time mission-critical computations with multiple priority levels and consistent priority treatment, plus support of a distributed clock synchronization protocol.

(9) Support of 32- or 64-bit addressing with dynamically sized data buses from 32 to 64, 128, and 256 bits to satisfy different bandwidth demands.

(10) Direct support of snoopy cache-based multiprocessors with recursive protocols to support large systems interconnected by multiple buses.

(11) Compatible message-passing protocols with multicomputer connections and special application profiles and interface design guides provided.

Example 5.1 Signal lines in the proposed Futurebus+ standard

As illustrated in Fig. 5.6, the Futurebus+ consists of information, synchronization, bus arbitration, and handshake lines that can match different system designs. The 64-bit *address lines* are multiplexed with the lower-order 64-bit *data lines*. Additional data lines can be added to form a data path up to 256 bits wide. The *tag lines* are optional for extending the addressing/data modes.

The *command lines* carry command information from the master to one or more slaves. The *status lines* are used by slaves to respond to the master's commands. The *capability lines* are used to declare special bus transactions. Every byte of lines is protected by at least one parity-check line.

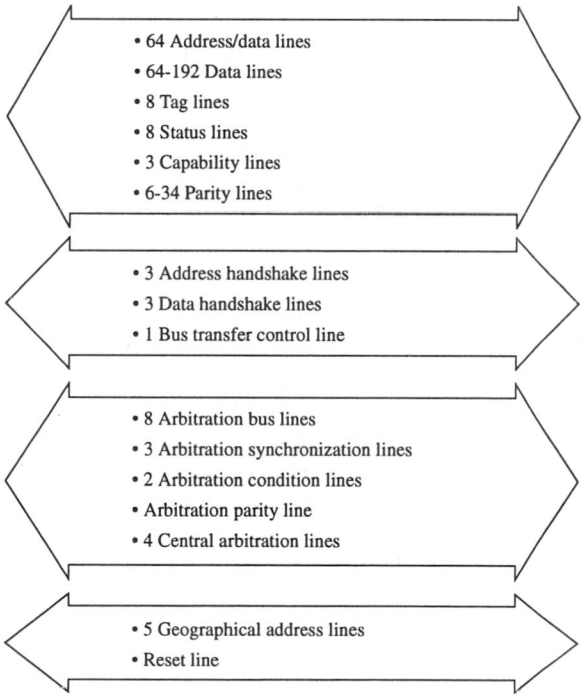

Figure 5.6 The Futurebus+ organization. (Reprinted with permission from IEEE Standard 896.1-1991, copyright © 1991 by IEEE, Inc.)

Synchronization signals are used to coordinate exchange of the address, command, capability status, and data during a bus transaction. The *address handshake lines* and the *data handshake lines* are used by both master and slaves. The *bus tenure line* is used to coordinate transfer of bus control.

The *arbitration bus lines* carry a number which signifies the precedence of competitors during the bus arbitration process. *Synchronization* and *condition lines* are used to coordinate handshaking and special conditions.

The *central arbitration lines* are used by a central arbiter in case central bus control is desired. The *miscellaneous lines* are needed to indicate geographical addresses (slot addresses) and to initialize the buses during system reset or after live insertion of a card. The total number of bus lines ranges from 91, 127, 199, and 343 for 32-, 64-, 128-, and 256-bit Futurebus+ widths, respectively. Additional lines may be needed to provide utility, clock, and power connections.

■

Technology/Architecture Independence   The Futurebus+ standard achieves *technology independence* through basing the protocols on fundamental protocol and physics principles and optimizing them for maximum communication efficiency rather than a particular generation or type of processor. Timing and handshake protocols are governed by operational constraints rather than limitations of technology such as device delays and capture windows.

The notion of *broadcast* facilitates the ability of slow, fast, old, or new modules to be attached to the same Futurebus+ segment. The standard specification may be implemented with any logic family (TTL, BTL, CMOS, ECL, GaAs), provided that physical implementation meets the Futurebus+ signaling requirements.

Architecture independence provides a flexible general-purpose solution to cache consistency within which other cache protocols will operate compatibly while at the same time providing an elegant unification with the message-passing protocols used in a multicomputer environment. Large systems can be built from smaller subsystems in a hierarchical manner using the Futurebus+.

Examples of this approach include the choice of split or interlocked transactions and a fully recursive, hierarchical cacheing paradigm. The Futurebus+ has been designed with possible VLSI implementation of the bus interface in mind. The architecture independence increases the application flexibility of a multiprocessor system built around the new bus standard.

5.2   Cache Memory Organizations

This section deals with physical address caches, virtual address caches, cache implementation using direct, fully associative, set-associative, and sector mapping. Finally, cache performance issues are analyzed based on some reported trace results. Multicache coherence protocols will be studied in Chapter 7.

5.2.1 Cache Addressing Models

Most multiprocessor systems use private caches associated with different processors as depicted in Fig. 5.7. Caches can be addressed using either a physical address or a virtual address. This leads to the two different cache design models presented below.

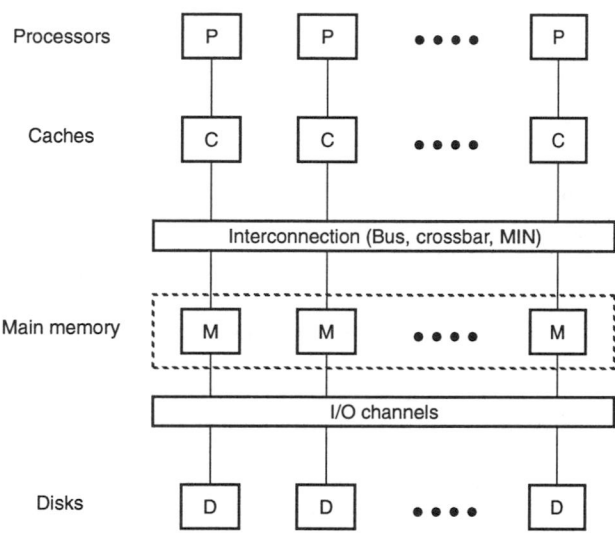

Figure 5.7 A memory hierarchy for a shared-memory multiprocessor.

Physical Address Caches When a cache is accessed with a physical memory address, it is called a *physical address cache*. Physical address cache models are illustrated in Fig. 5.8. In Fig. 5.8a, the model is based on the experience of using a unified cache as in the VAX 8600 and the Intel i486.

In this case, the cache is indexed and tagged with the physical address. Cache lookup must occur after address translation in the TLB or MMU. No aliasing is allowed so that the address is always uniquely translated without confusion.

A *cache hit* occurs when the addressed data/instruction is found in the cache. Otherwise, a *cache miss* occurs. After a miss, the cache is loaded with the data from the memory. Since a whole cache block is loaded at one time, unwanted data may also be loaded. Locality of references will find most of the loaded data useful in subsequent instruction cycles.

Data is written through the main memory immediately via a *write-through* (WT) cache, or delayed until block replacement by using a *write-back* (WB) cache. A WT cache requires more bus or network cycles to access the main memory, while a WB cache allows the CPU to continue without waiting for the memory to cycle.

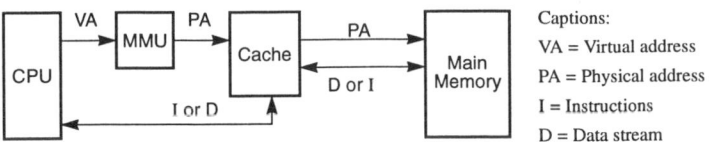

(a) A unified cache accessed by physical address

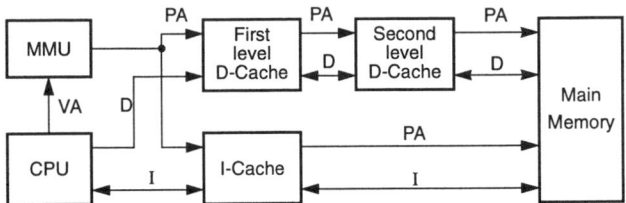

(b) Split caches accessed by physical address in the Silicon Graphics workstation

Figure 5.8 Physical address models for unified and split caches.

Example 5.2 Cache design in a Silicon Graphics workstation

Figure 5.8b demonstrates the split cache design using the MIPS R3000 CPU in the Silicon Graphics 4-D Series workstation. Both data cache and instruction cache are accessed with a physical address issued from the on-chip MMU. A two-level data cache is implemented in this design.

The first level uses 64 Kbytes of WT D-cache. The second level uses 256 Kbytes of WB D-cache. The single-level I-cache is 64 Kbytes. By the inclusion property, the first-level cache is always a subset of the second-level cache. Most manufacturers put the first-level caches on the processor chip and leave the second-level caches as options on the processor board. Hardware consistency needs to be enforced between the two cache levels.

The major advantages of physical address caches include no need to perform cache flushing, no aliasing problems, and thus fewer cache bugs in the OS kernels. The shortcoming is the slowdown in accessing the cache until the MMU/TLB finishes translating the address. This motivates the use of a virtual address cache. Integration of the MMU and caches on the same VLSI chip can alleviate some of these problems.

Most conventional system designs use a physical address cache because of its simplicity and because it requires little intervention from the OS kernel. When physical address caches are used in a UNIX environment, no flushing of data caches is needed if bus watching is provided to monitor the system bus for DMA requests from I/O devices or from other CPUs. Otherwise, the cache must be flushed for every I/O without proper bus watching.

Virtual Address Caches When a cache is indexed or tagged with a virtual address as shown in Fig. 5.9, it is called a *virtual address cache*. In this model, both cache and MMU translation or validation are done in parallel. The physical address generated by the MMU can be saved in tags for later write-back but is not used during the cache lookup operations. The virtual address cache is motivated with its enhanced efficiency to access the cache faster, overlapping with the MMU translation as exemplified below.

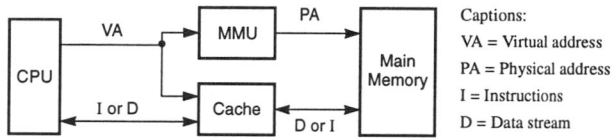

(a) A unified cache accessed by virtual address

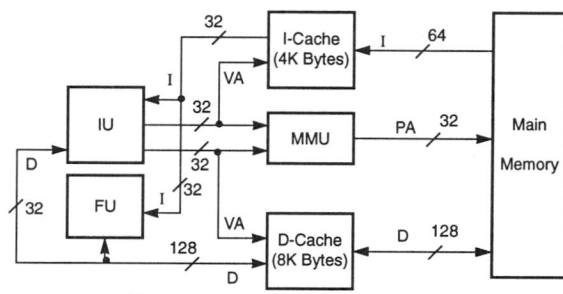

(b) A split cache accessed by virtual address as in the Intel i860 processor

Figure 5.9 Virtual address models for unified and split caches. (Courtesy of Intel Corporation, 1989)

Example 5.3 The virtual addressed split cache design in Intel i860

Figure 5.9b shows the virtual address design in the Intel i860 using split caches for data and instructions. Instructions are 32 bits wide. Virtual addresses generated by the integer unit (IU) are 32 bits wide, and so are the physical addresses generated by the MMU. The data cache is 8 Kbytes with a block size of 32 bytes. A two-way, set-associative cache organization (Section 5.2.3) is implemented with 128 sets in the D-cache and 64 sets in the I-cache.

The Aliasing Problem The major problem associated with a virtual address cache is *aliasing*, when different logically addressed data have the same index/tag in the cache. Multiple processes may use the same range of virtual addresses. This aliasing problem may create confusion if two or more processes access the same physical cache location.

One way to solve the aliasing problem is to flush the entire cache whenever aliasing occurs.

Large amounts of *flushing* may result in a poor cache performance, with a low hit ratio and too much time wasted in flushing. When a virtual address cache is used with UNIX, flushing is needed after each context switching. Before I/O writes or after I/O reads, the cache must be flushed. Furthermore, aliasing between the UNIX kernel and user data is a serious problem. All of these problems will introduce additional system overhead.

Flushing the cache will not overcome the aliasing problem completely when using a shared memory with mapped files and copy-on-write as in UNIX systems. These UNIX operations may not benefit from virtual caches. In each entry/exit to or from the UNIX kernel, the cache must be flushed upon every system call and interrupt.

The virtual space must be divided between kernel and user modes by tagging kernel and user data separately. This implies that a virtual address cache may lead to a lower performance unless most processes are compute-bound.

With a frequently flushed cache, the debugging of programs is almost impossible to perform. Two commonly used methods to get around the virtual address cache problems are to apply special tagging with a *process key* or with a *physical address*. For example, the SUN 3/200 Series has used a virtual address, write-back cache with the capability of being noncacheable. Three-bit keys are used in the cache to distinguish among eight simultaneous contexts.

The flushing can be done at the page, segment, or context level. In this case, context switching does not need to flush the cache but needs to change the current process key. Thus cached shared memory must be shared on fixed-size boundaries. Other memory can be shared with noncacheability. Flushing and special tagging may be traded for performance/cost reasons.

5.2.2 Direct Mapping and Associative Caches

The transfer of information from main memory to cache memory is conducted in units of cache blocks or cache lines. Four block placement schemes are presented below. Each placement scheme has its own merits and shortcomings. The ultimate performance depends on the cache-access patterns, cache organization, and management policy used.

Blocks in caches are called *block frames* in order to distinguish them from the corresponding *blocks* in main memory. Block frames are denoted as $\overline{B_i}$ for $i = 0, 1, 2, \ldots, m$. Blocks are denoted as B_j for $j = 0, 1, 2, \ldots, n$. Various mappings can be defined from set $\{B_j\}$ to set $\{\overline{B_i}\}$. It is also assumed that $n >> m$, $n = 2^s$, and $m = 2^r$.

Each block (or block frame) is assumed to have b words, where $b = 2^w$. Thus the cache consists of $m \cdot b = 2^{r+w}$ words. The main memory has $n \cdot b = 2^{s+w}$ words addressed by $(s + w)$ bits. When the block frames are divided into $v = 2^t$ *sets*, $k = m/v = 2^{r-t}$ blocks are in each set.

Direct-Mapping Cache This cache organization is based on a direct mapping of $n/m = 2^{s-r}$ memory blocks, separated by equal distances, to one block frame in the cache. The placement is defined below using a modulo-m function. Block B_j is mapped to block frame $\overline{B_i}$:

$$B_j \to \overline{B_i}, \qquad \text{if } i = j \text{ (modulo } m) \tag{5.1}$$

There is a *unique* block frame $\overline{B_i}$ that each B_j can load into. There is no way to implement a block replacement policy. This direct mapping is very rigid but is the simplest cache organization to implement. Direct mapping is illustrated in Fig. 5.10a for the case where each block contains four words ($w = 2$ bits).

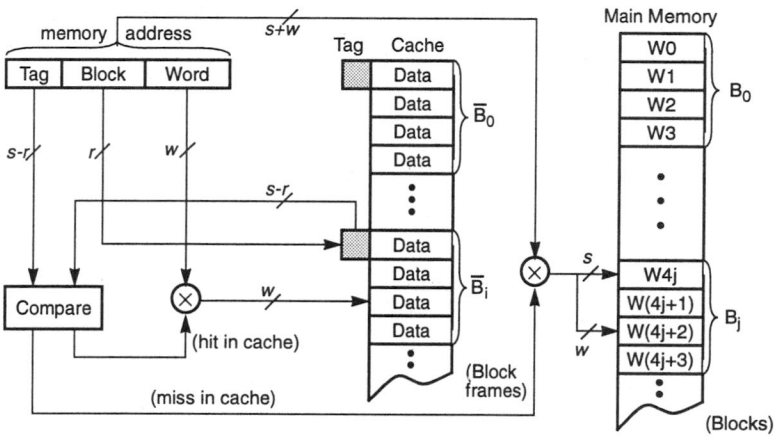

(a) The cache/memory addressing

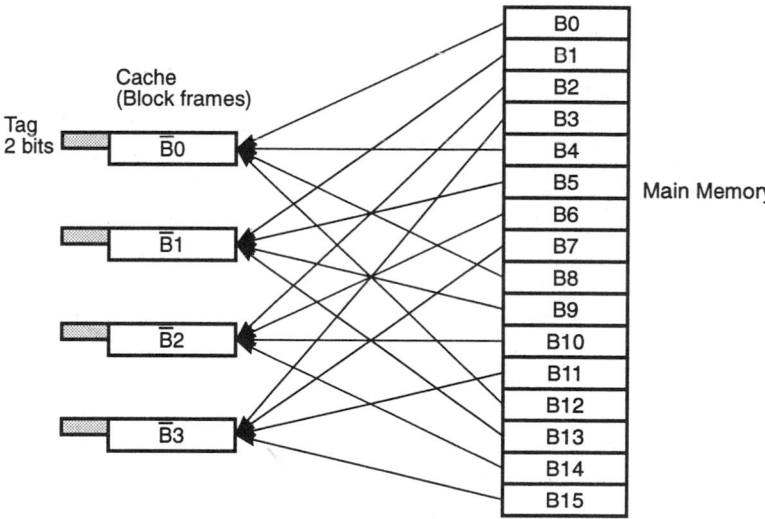

(b) Block B_j can be mapped to block frame $\bar{B_i}$ if $i = j$ (modulo 4)

Figure 5.10 Direct-mapping cache organization and a mapping example.

The memory address is divided into three fields: The lower w bits specify the *word offset* within each block. The upper s bits specify the *block address* in main memory, while the leftmost $(s - r)$ bits specify the *tag* to be matched. The *block* field (r bits) is used to implement the (modulo-m) placement, where $m = 2^r$. Once the block $\overline{B_i}$ is uniquely identified by this field, the tag associated with the addressed block is compared with the tag in the memory address.

A cache hit occurs when the two tags match. Otherwise a cache miss occurs. In case of a cache hit, the word offset is used to identify the desired data word within the addressed block. When a miss occurs, the entire memory address ($s + w$ bits) is used to access the main memory. The first s bits locate the addressed block, and the lower w bits locate the word within the block.

Example 5.4 Direct-mapping cache design and block mapping

An example mapping is given in Fig. 5.10b, where $n = 16$ blocks are mapped to $m = 4$ block frames, with four possible sources mapping into one destination using modulo-4 mapping. The direct-mapping cache has been implemented in the IBM System/370 Model 158 and in the VAX/8800.

∎

Cache Design Parameters   In practice, the two parameters n and m differ by at least two to three orders of magnitude. A typical cache block has 32 bytes corresponding to eight 32-bit words. Thus $w = 3$ bits if the machine is word-addressable. If the machine is byte-addressable, one may even consider $w = 5$ bits.

Consider a cache with 64 Kbytes. This implies $m = 2^{11} = 2048$ block frames with $r = 11$ bits. Consider a main memory with 32 Mbytes. Thus $n = 2^{20}$ blocks with $s = 20$ bits, and the memory address needs $s + w = 20 + 3 = 23$ bits for word addressing and 25 bits for byte addressing. In this case, $2^{s-r} = 2^9 = 512$ blocks are possible candidates to be mapped into a single block frame in a direct-mapping cache.

Advantages of a direct-mapping cache include simplicity in hardware, no associative search needed, no page replacement algorithm needed, and thus lower cost and higher speed.

However, the rigid mapping may result in a poorer hit ratio than with the associative mappings to be introduced next. The scheme also prohibits parallel virtual address translation. The hit ratio may drop sharply if many addressed blocks have to map into the same block frame. For this reason, direct-mapped caches tend to use a larger cache size with more block frames to avoid the contention.

Fully Associative Cache   Unlike direct mapping, this cache organization offers the most flexibility in mapping cache blocks. As illustrated in Fig. 5.11a, each block in main memory can be placed in any of the available block frames. Because of this flexibility, an s-bit tag is needed in each cache block. As $s > r$, this represents a significant increase in tag length.

The name *fully associative cache* is derived from the fact that an m-way associative search requires the tag to be compared with all block tags in the cache. This scheme offers the greatest flexibility in implementing block replacement policies for a higher hit ratio.

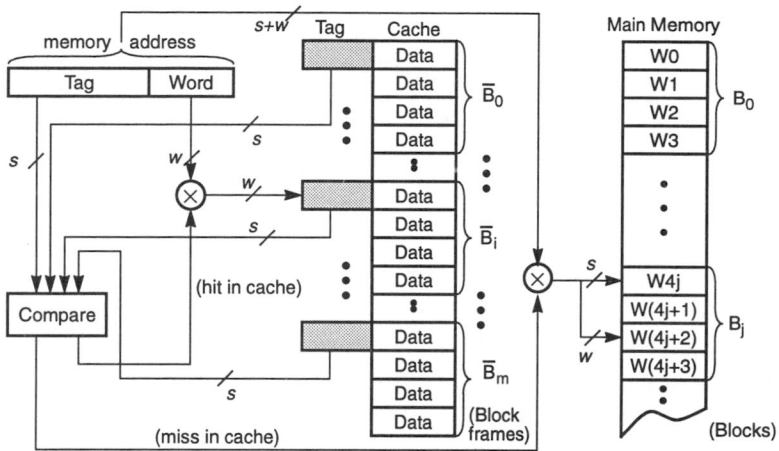

(a) Associative search with all block tags

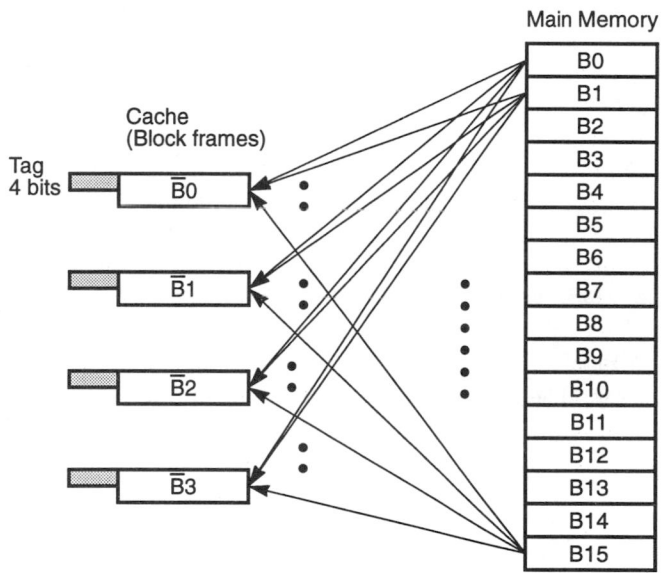

(b) Every block is mapped to any of the four block frames identified by the tag

Figure 5.11 Fully associative cache organization and a mapping example.

The m-way comparison of all tags is very time-consuming if the tags are compared sequentially using RAMs. Thus an *associative memory* (content-addressable memory, CAM) is needed to achieve a parallel comparison with all tags simultaneously. This demands a higher implementation cost for the cache. Therefore, a fully associative cache has been implemented only in moderate size, such as those used in microprocessor-based computer systems.

Figure 5.11b shows a four-way mapping example using a fully associative search. The tag is 4 bits long because 16 possible cache blocks can be destined for the same block frame. The major advantage of using full associativity is to allow the implementation of a better block replacement policy with reduced block contention. The major drawback lies in the expensive search process requiring a higher hardware cost.

5.2.3   Set-Associative and Sector Caches

Set-associative caches are the most popular cache designs built into commercial computers. Sector mapping caches offer a design alternative to set-associative caches. These two types of cache design are described below.

Set-Associative Cache   This design offers a compromise between the two extreme cache designs based on direct mapping and full associativity. If properly designed, this cache may offer the best performance-cost ratio. Most high-performance computer systems are based on this approach. The idea is illustrated in Fig. 5.12.

In a *k-way associative* cache, the m cache block frames are divided into $v = m/k$ sets, with k blocks per set. Each set is identified by a d-bit *set number*, where $2^d = v$. The cache block tags are now reduced to $s - d$ bits. In practice, the *set size* k, or *associativity*, is chosen as 2, 4, 8, 16, or 64, depending on a tradeoff among block size w, cache size m, and other performance/cost factors.

Fully associative mapping can be visualized as having a single set (i.e., $v = 1$) or an m-way associativity. In a k-way associative search, the *tag* needs to be compared only with the k tags within the identified set, as shown in Fig. 5.12a. Since k is rather small in practice, the k-way associative search is much more economical than the full associativity.

In general, a block B_j can be mapped into any one of the available frames $\overline{B_f}$ in a set S_i defined below. The matched tag identifies the current block which resides in the frame.

$$B_j \rightarrow \overline{B_f} \in S_i \qquad \text{if} \quad j(\text{modulo } v) = i \tag{5.2}$$

Design Tradeoffs   The set size (associativity) k and the number of sets v are inversely related by

$$m = v \times k \tag{5.3}$$

For a fixed cache size there exists a tradeoff between the set size and the number of sets.

The advantages of the set-associative cache include the following:

First, the block replacement algorithm needs to consider only a few blocks in the same set. Thus the replacement policy can be more economically implemented with limited choices, as compared with the fully associative cache.

Second, the k-way associative search is easier to implement, as mentioned earlier.

Third, many design tradeoffs can be considered (Eq. 5.3) to yield a higher hit ratio in the cache. The cache operation is often used together with TLB.

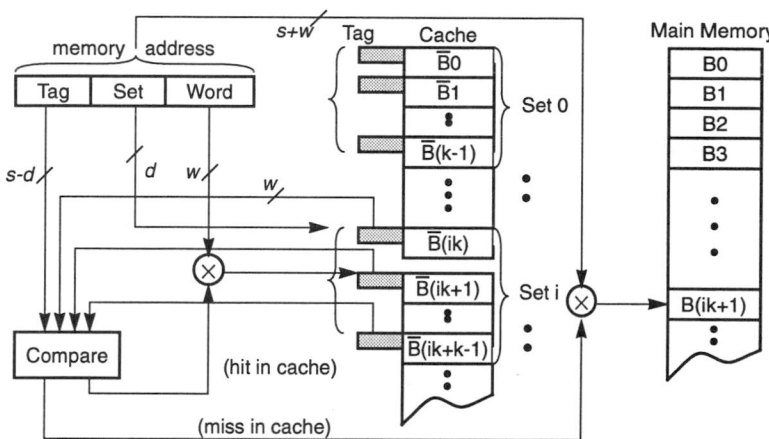

(a) A k-way associative search within each set of k cache blocks

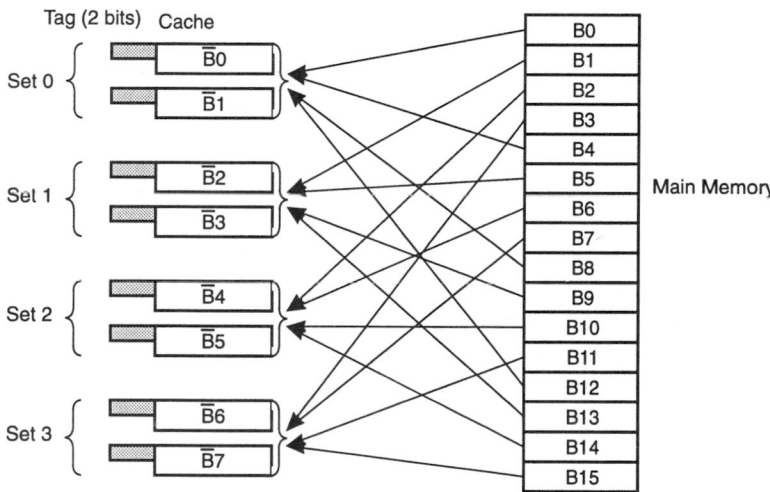

(b) Mapping cache blocks in a two-way associative cache with four sets

Figure 5.12 Set-associative cache organization and a two-way associative mapping example.

Example 5.5 Set-associative cache design and block mapping

An example is shown in Fig. 5.12b for the mapping of $n = 16$ blocks from main memory into a two-way associative cache ($k = 2$) with $v = 4$ sets over $m = 8$ block frames. For the i860 example in Fig. 5.9b, both the D-cache and I-cache are two-way associative ($k = 2$). There are 128 sets in the D-cache and 64 sets in the I-cache, with 256 and 128 block frames, respectively. Table 5.1 presents other example systems using set-associative caches. Note that the associativity ranges from 2 to 16 in these systems.

Table 5.1 Representative Set-Associative Cache Organizations

System	Cache Size, m	Set Size, k	No. of Sets, v
DEC VAX 11/780	1024	2	512
Amdahl 470/V6	512	2	256
Intel i860 D-cache	256	2	128
Honeywell 66/60	512	4	128
Amdahl 470/V7	2048	4	512
IBM 370/168	1024	8	128
IBM 3033	1024	16	64
Motorola 88110 I-cache	256	2	128

Sector Mapping Cache   This block placement scheme is generalized from the above schemes. The idea is to partition both the cache and main memory into fixed-size *sectors*. Then a fully associative search is applied. That is, each sector can be placed in any of the available sector frames.

The memory requests are destined for blocks, not for sectors. This can be filtered out by comparing the sector tag in the memory address with all sector tags using a fully associative search. If a matched sector frame is found (a cache hit), the block field is used to locate the desired block within the sector frame.

If a cache miss occurs, only the missing block is fetched from the main memory and brought into a congruent block frame in an available sector. That is, the ith block in a sector must be placed into the ith block frame in a destined sector frame. A *valid bit* is attached to each block frame to indicate whether the block is *valid* or *invalid*.

When the contents of a block frame are replaced, the remaining block frames in the same sector are marked invalid. Only the most recently replaced block frame in a sector is marked valid for reference. However, multiple valid bits can be used to record other block states. The sector mapping just described can be modified to yield other designs, depending on the block replacement policy being implemented.

Compared with fully associative or set-associative caches, the sector mapping cache offers the advantages of being flexible to implement various block replacement algorithms and being economical to perform a fully associative search across a limited number of sector tags.

The sector partitioning offers more freedom in grouping cache lines at both ends of the mapping. Making design choice between set-associative and sector mapping caches requires more trace and simulation evidence.

Example 5.6 Sector mapping cache design

Figure 5.13 shows an example of sector mapping with a sector size of four blocks. Note that each sector can be mapped to any of the sector frames with full associativity at the sector level.

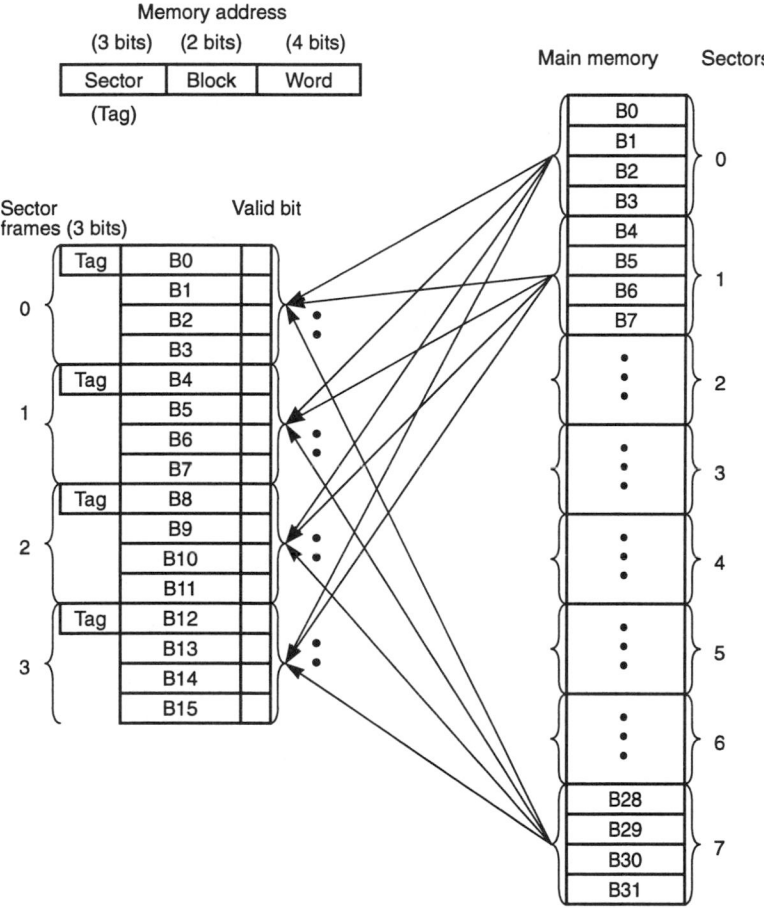

Figure 5.13 A four-way sector mapping cache organization.

This scheme was first implemented in the IBM System/360 Model 85. In the

Model 85, there are 16 sectors, each having 16 blocks. Each block has 64 bytes, giving a total of 1024 bytes in each sector and a total cache capacity of 16 Kbytes using a LRU block replacement policy.

∎

5.2.4 Cache Performance Issues

The performance of a cache design concerns two related aspects: the *cycle count* and the *hit ratio*. The cycle count refers to the number of basic machine cycles needed for cache access, update, and coherence control. The hit ratio determines how effectively the cache can reduce the overall memory-access time. Tradeoffs do exist between these two aspects. Key factors affecting cache speed and hit ratio are discussed below.

Program trace-driven simulation and *analytical modeling* are two complementary approaches to studying cache performance. Both have to be applied together in order to provide a credible performance assessment.

Simulation studies present snapshots of program behavior and cache responses but they suffer from having a microscopic perspective.

Analytical models may deviate from reality under simplification. However, they provide some macroscopic and intuitive insight into the underlying processes.

Agreement between results generated from the two approaches allows one to draw a more credible conclusion. However, the generalization of any conclusion is limited by the finite-sized address traces and by the assumptions about address trace patterns. Simulation results can be used to verify the theoretical results, and analytical formulation can guide simulation experiments on a wider range of parameters.

Cycle Counts   The cache speed is affected by the underlying static or dynamic RAM technology, the cache organization, and the cache hit ratios. The total cycle count should be predicated with appropriate cache hit ratios. This will affect various cache design decisions, as already seen in previous sections.

The cycle counts are not credible unless detailed simulation of all aspects of a memory hierarchy is performed. The write-through or write-back policies also affect the cycle count. Cache size, block size, set number, and associativity all affect the cycle count as illustrated in Fig. 5.14.

The cycle count is directly related to the hit ratio, which decreases almost linearly with increasing values of the above cache parameters. But the decreasing trend becomes flat and after a certain point turns into an increasing trend (the dashed line in Fig. 5.14a). This is caused primarily by the effect of the block size on the hit ratio, which will be discussed below.

Hit Ratios   The cache hit ratio is affected by the cache size and by the block size in different ways. These effects are illustrated in Figs. 5.14b and 5.14c, respectively. Generally, the hit ratio increases with respect to increasing cache size (Fig. 5.14b).

When the cache size approaches infinity, a 100% hit ratio should be expected. However, this will never happen because the cache size is always bounded by a limited budget. The initial cache loading and changes in locality also prevent such an ideal

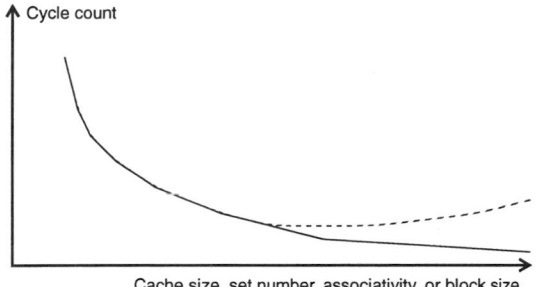

(a) The total cycle count for cache access (Courtesy of S. A. Przybylski; reprinted with permission from *Cache and Memory Hierarchy Design*, Morgan Kaufmann Publishers, 1990)

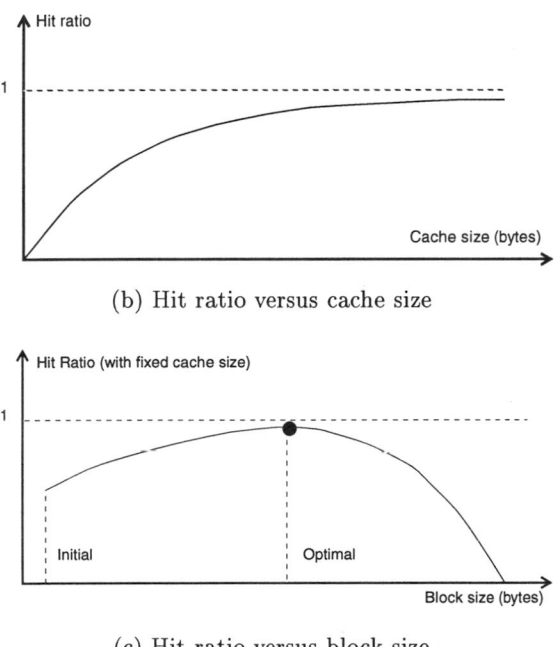

(b) Hit ratio versus cache size

(c) Hit ratio versus block size

Figure 5.14 Cache performance versus design parameters used.

performance. The curves in Fig. 5.14b can be approximated by $1 - C^{-0.5}$, where C is the total cache size.

Effect of Block Size   With a fixed cache size, cache performance is rather sensitive to block size. Figure 5.14c illustrates the rise and fall of the hit ratio as the cache block varies from small to large. Initially, we assume a block size (such as 32 bytes per block). This block size is determined mainly by the temporal locality in typical programs.

As the block size increases, the hit ratio improves because of spatial locality in referencing larger instruction/data blocks. The increase reaches its peak at a certain *optimum block size*. After this point, the hit ratio decreases with increasing block size. This is caused by the mismatch between program behavior and block size.

As a matter of fact, as the block size becomes very large, many words fetched into the cache may never be used. Also, the temporal locality effects are gradually lost with larger block size. Finally, the hit ratio approaches zero when the block size equals the entire cache size.

For a bus-based system, Smith (1987) has determined that the optimum block size should be chosen to minimize the effective memory-access time. This optimum size depends on the ratio of the access latency and the bus cycle time (data transfer rate). He has identified design targets for the hit ratio, bus traffic, and average delay per reference based on an empirical model derived from a wide variety of benchmark simulations.

Effects of Set Number   In a set-associative cache, the effects of set number are obvious. For a fixed cache capacity, the hit ratio may decrease as the number of sets increases. As the set number increases from 32 to 64, 128, and 256, the decrease in the hit ratio is rather small based on Smith's 1982 report. When the set number increases to 512 and beyond, the hit ratio decreases faster. Also, the tradeoffs between block size and set number should not be ignored (Eq. 5.3).

Other Performance Factors   In a performance-directed design, tradeoffs exist among the cache size, set number, block size, and memory speed. Independent blocks, fetch sizes, and fetch strategies also affect the performance in various ways.

Multilevel cache hierarchies offer options for expanding the cache effects. Very often, a write-through policy is used in the first-level cache, and a write-back policy in the second-level cache. As in the memory hierarchy, an optimal cache hierarchy design must match special program behavior in the target application domain.

The distribution of cache references for instruction, loads, and writes of data will affect the hierarchy design. Based on some previous program traces, 63% instruction fetches, 25% loads, and 12% writes were reported. This affects the decision to split the instruction cache from the data cache.

Of course, the optimal hierarchy design must be based on the access time and memory technology used. Pipelined access to the cache is also very much desired to enhance the performance.

5.3   Shared-Memory Organizations

Memory interleaving provides a higher bandwidth for pipelined access of contiguous memory locations. Methods for allocating and deallocating main memory to multiple user programs are considered for optimizing memory utilization. Memory bandwidth analysis and fault tolerance issues are discussed below.

5.3.1 Interleaved Memory Organization

Various organizations of the physical memory are studied in this section. In order to close up the speed gap between the CPU/cache and main memory built with RAM modules, an *interleaving* technique is presented below which allows pipelined access of the parallel memory modules.

The memory design goal is to broaden the *effective memory bandwidth* so that more memory words can be accessed per unit time. The ultimate purpose is to match the memory bandwidth with the bus bandwidth and with the processor bandwidth.

Memory Interleaving   The main memory is built with multiple modules. These memory modules are connected to a system bus or a switching network to which other resources such as processors or I/O devices are also connected.

Once presented with a memory address, each memory module returns with one word per cycle. It is possible to present different addresses to different memory modules so that parallel access of multiple words can be done simultaneously or in a pipelined fashion. Both parallel access and pipelined access are forms of parallelism practiced in a parallel memory organization.

Consider a main memory formed with $m = 2^a$ memory modules, each containing $w = 2^b$ words of memory cells. The total memory capacity is $m \cdot w = 2^{a+b}$ words. These memory words are assigned linear addresses. Different ways of assigning linear addresses result in different memory organizations.

Besides random access, the main memory is often block-accessed at consecutive addresses. Block access is needed for fetching a sequence of instructions or for accessing a linearly ordered data structure. Each block access may correspond to the size of a cache block (cache line) or to that of several cache blocks. Therefore, it is desirable to design the memory to facilitate block access of contiguous words.

Figure 5.15a shows two address formats for memory interleaving. *Low-order interleaving* spreads contiguous memory locations across the m modules horizontally (Fig. 5.15a). This implies that the low-order a bits of the memory address are used to identify the memory module. The high-order b bits are the word addresses (displacement) within each module. Note that the same word address is applied to all memory modules simultaneously. A module address decoder is used to distribute module addresses.

High-order interleaving (Fig. 5.15b) uses the high-order a bits as the module address and the low-order b bits as the word address within each module. Contiguous memory locations are thus assigned to the same memory module. In each memory cycle, only one word is accessed from each module. Thus the high-order interleaving cannot support block access of contiguous locations.

On the other hand, the low-order m-way interleaving does support block access in a pipelined fashion. Unless otherwise specified, we consider only low-order memory interleaving in subsequent discussions.

Pipelined Memory Access   Access of the m memory modules can be overlapped in a pipelined fashion. For this purpose, the memory cycle (called the *major cycle*) is

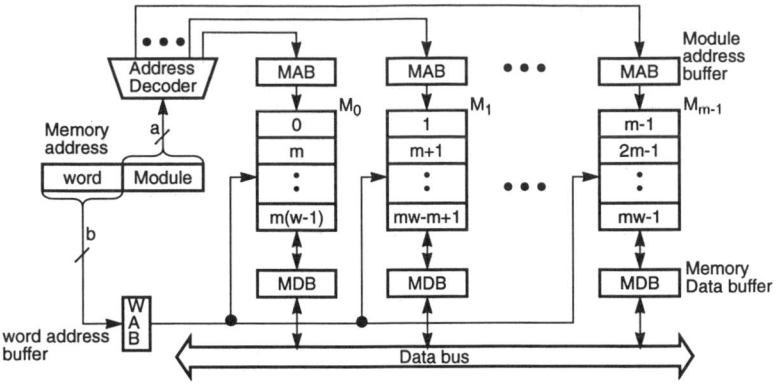

(a) Low-order m-way interleaving (the C-access memory scheme)

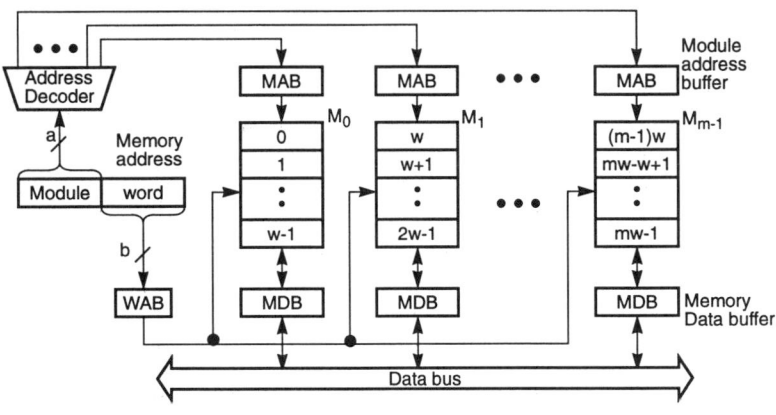

(b) High-order m-way interleaving

Figure 5.15 Two interleaved memory organizations with $m = 2^a$ modules and $w = 2^b$ words per module (word addresses shown in boxes).

subdivided into m *minor cycles*.

An eight-way interleaved memory (with $m = 8$ and $w = 8$ and thus $a = b = 3$) is shown in Fig. 5.16a. Let θ be the major cycle and τ the minor cycle. These two cycle times are related as follows:

$$\tau = \frac{\theta}{m} \tag{5.4}$$

where m is the *degree of interleaving*. The timing of the pipelined access of the eight contiguous memory words is shown in Fig. 5.16b. This type of *concurrent access* of contiguous words has been called a *C-access* memory scheme. The major cycle θ is the total time required to complete the access of a single word from a module. The minor cycle τ is the actual time needed to produce one word, assuming overlapped access of successive memory modules separated in every minor cycle τ.

Note that the pipelined access of the block of eight contiguous words is sandwiched between other pipelined block accesses before and after the present block. Even though the total block access time is 2θ, the *effective access time* of each word is reduced to τ as the memory is contiguously accessed in a pipelined fashion.

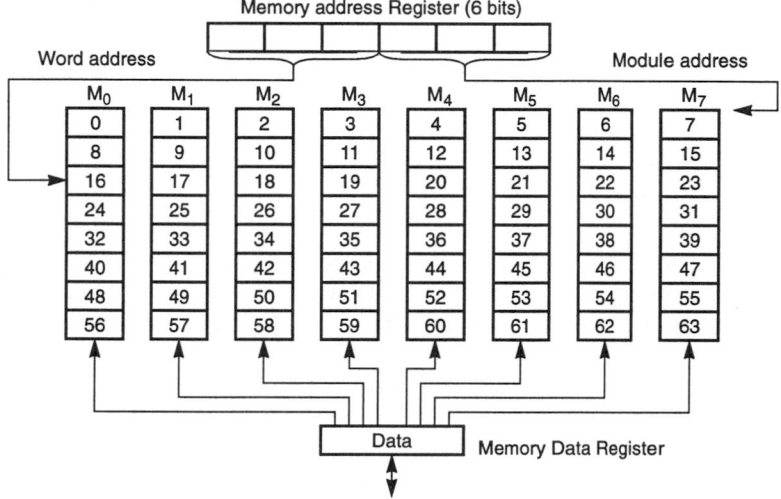

(a) Eight-way low-order interleaving (absolute address shown in each memory word)

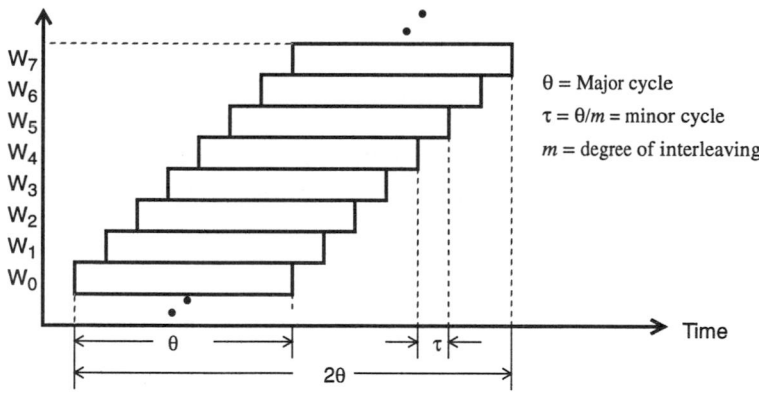

θ = Major cycle

$\tau = \theta/m$ = minor cycle

m = degree of interleaving

(b) Pipelined access of eight consecutive words in a C-access memory

Figure 5.16 Multiway interleaved memory organization and the C-access timing chart.

5.3.2 Bandwidth and Fault Tolerance

Hellerman (1967) has derived an equation to estimate the effective increase in memory bandwidth through multiway interleaving. A single memory module is assumed to deliver one word per memory cycle and thus has a bandwidth of 1.

Memory Bandwidth   The *memory bandwidth B* of an m-way interleaved memory is upper-bounded by m and lower-bounded by 1. The Hellerman estimate of B is

$$B = m^{0.56} \simeq \sqrt{m} \qquad (5.5)$$

where m is the number of interleaved memory modules. This equation implies that if 16 memory modules are used, then the effective memory bandwidth is approximately four times that of a single module.

This pessimistic estimate is due to the fact that block access of various lengths and access of single words are randomly mixed in user programs. Hellerman's estimate was based on a single-processor system. If memory-access conflicts from multiple processors (such as the hot spot problem) are considered, the effective memory bandwidth will be further reduced.

In a vector processing computer, the access time of a long vector with n elements and stride distance 1 has been estimated by Cragon (1992) as follows: It is assumed that the n elements are stored in contiguous memory locations in an m-way interleaved memory system. The average time t_1 required to access one element in a vector is estimated by

$$t_1 = \frac{\theta}{m}\left(1 + \frac{m-1}{n}\right) \qquad (5.6)$$

When $n \to \infty$ (very long vector), $t_1 \to \theta/m = \tau$ as derived in Eq. 5.4. As $n \to 1$ (scalar access), $t_1 \to \theta$. Equation 5.6 conveys the message that interleaved memory appeals to pipelined access of long vectors; the longer the better.

Fault Tolerance   High- and low-order interleaving can be combined to yield many different interleaved memory organizations. Sequential addresses are assigned in the high-order interleaved memory in each memory module.

This makes it easier to isolate faulty memory modules in a *memory bank* of m memory modules. When one module failure is detected, the remaining modules can still be used by opening a window in the address space. This fault isolation cannot be carried out in a low-order interleaved memory, in which a module failure may paralyze the entire memory bank. Thus low-order interleaving memory is not fault-tolerant.

Example 5.7 Memory banks, fault tolerance, and bandwidth tradeoffs

In Fig. 5.17, two alternative memory addressing schemes are shown which combines the high- and low-order interleaving concepts. These alternatives offer a better bandwidth in case of module failure. A four-way low-order interleaving is organized in each of two memory banks in Fig. 5.17a.

On the other hand, two-way low-order interleaving is depicted in Fig. 5.17b with the memory system divided into four memory banks. The high-order bits are used to identify the memory banks. The low-order bits are used to address the modules for memory interleaving.

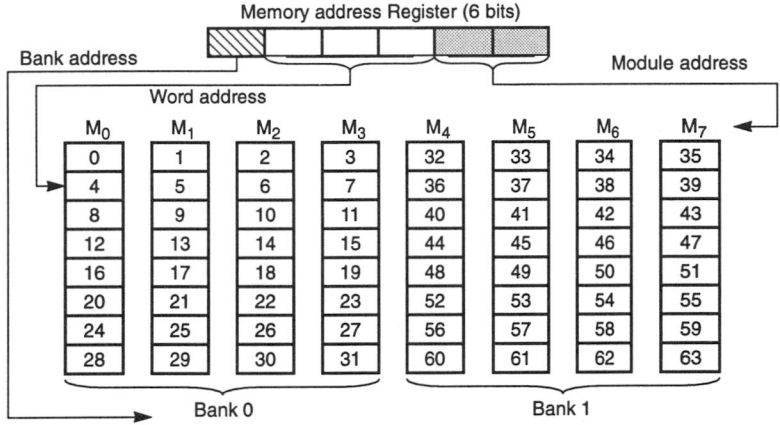

(a) Four-way interleaving within each memory bank

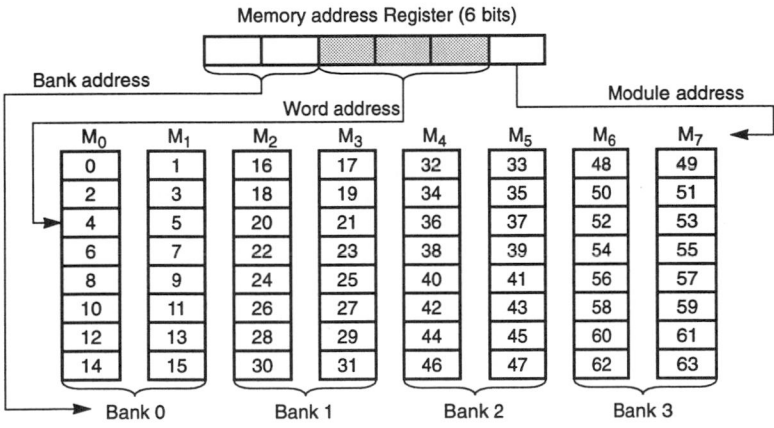

(b) Two-way interleaving within each memory bank

Figure 5.17 Bandwidth analysis of two alternative interleaved memory organizations over eight memory modules. (Absolute address shown in each memory bank.)

In case of single module failure, the maximum memory bandwidth of the eight-way interleaved memory (Fig. 5.16a) is reduced to zero because the entire memory bank must be abandoned. For the four-way two-bank design (Fig. 5.17a), the maximum bandwidth is reduced to four words per memory cycle because only one of the two faulty banks is abandoned.

In the two-way design in Fig. 5.17b, the gracefully degraded memory system may still have three working memory banks; thus a maximum bandwidth of six words is expected. The higher the degree of interleaving, the higher the potential memory bandwidth if the system is fault-free.

■

If fault tolerance is an issue which cannot be ignored, then tradeoffs do exist between the degree of interleaving and the number of memory banks used. Each memory bank is essentially self-enclosed, is independent of the conditions of other banks, and thus offers better fault isolation in case of failure.

5.3.3 Memory Allocation Schemes

The idea of virtual memory is to allow many software processes time-shared use of the main memory, which is a precious resource with limited capacity. The portion of the OS kernel which handles the allocation and deallocation of main memory to executing processes is called the *memory manager*. The memory manager monitors the amount of available main memory and decides which processes should reside in main memory and which should be put back to disk if the main memory reaches its limit.

In this section, we study the basic concepts of memory swapping, either at the process level or at the individual page level. Both swapping systems and demand paging systems are introduced, based on the development of the memory management subsystem in UNIX. Possible extensions of these memory allocation schemes are discussed along with some performance issues.

Allocation Policies   *Memory swapping* is the process of moving blocks of information between the levels of a memory hierarchy. For simplicity, we concentrate on swapping between the main memory and the disk memory. Several key concepts or design alternatives in implementing memory swapping are introduced below.

First, the swapping policy can be made either nonpreemptive or preemptive. In *nonpreemptive allocation*, the incoming block can be placed only in a free region of the main memory. A *preemptive allocation* scheme allows the placement of an incoming block in a region presently occupied by another process. In either case, the memory manager should try to allocate the free space first.

When the main memory space is fully allocated, the nonpreemptive scheme swaps out some of the allocated processes (or pages) to vacate space for the incoming block. On the other hand, a preemptive scheme has the freedom to preempt an executing process. The nonpreemptive scheme is easier to implement, but it may not yield the best memory utilization.

The preemptive scheme offers more flexibility, but it requires mechanisms be established to determine which pages or processes are to be swapped out and to avoid *thrashing* caused by an excessive amount of swapping between memory levels. This implies that preemptive allocation schemes are more complex and more expensive to implement.

In addition, an allocation policy can be made either local or global. A *local allocation* policy involves only the resident working set of the faulty process. A *global allocation* policy considers the history of the working sets of all resident processes in making a swapping decision. Most computers use the local policy.

Swapping Systems   This refers to memory systems which allow swapping to occur only at the entire process level. A *swap device* is a configurable section of a disk which is set aside for temporary storage of information being swapped out of the main memory. The portion of the disk memory space set aside for a swap device is called the *swap space*, as depicted in Fig. 5.18.

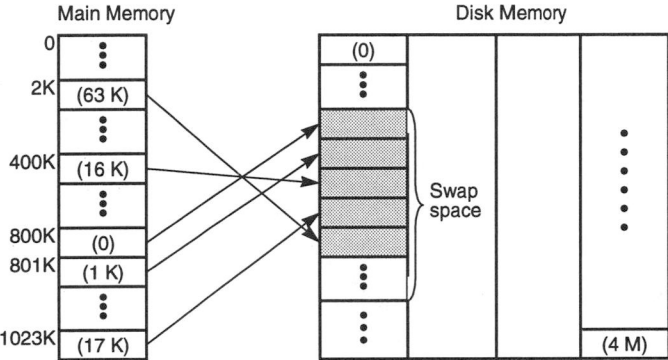

(a) Moving a process (or pages) onto the swap space on a disk

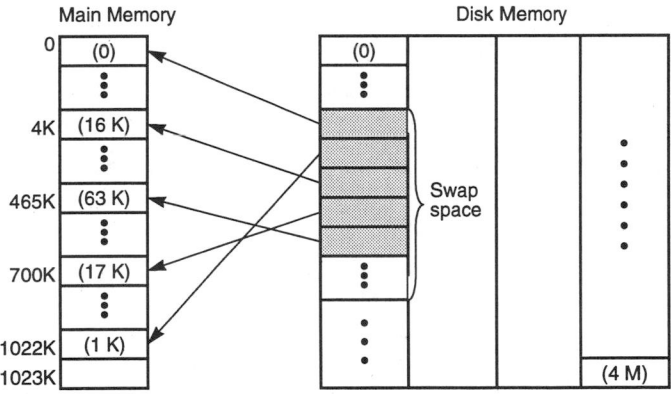

(b) Swapping in a process (or pages) to the memory

Figure 5.18 The concept of memory swapping in a virtual memory hierarchy (virtual page addresses are identified by numbers within parentheses, assuming a page size of 1K words).

The memory manager allocates disk space for program files one block at a time, but it allocates space on the swap device in groups of contiguous blocks. For simplicity, we consider blocks as fixed-size pages. The virtual address space of a process may occupy a number of pages. The size of a process address space is limited by the amount of physical memory available on a swapping system.

The swapping system was used in the PDP-11 and in early UNIX systems. It transfers the entire process between main memory and the swap device. It does not transfer parts (pages) of a process separately. For example, the PDP-11 allows a maximum process size of 64 Kbytes. The entire process must reside in main memory to execute.

A simple example is shown in Fig. 5.18 to illustrate the concepts of *swapping out* and *swapping in* a process consisting of five resident pages identified by virtual page addresses 0, 1K, 16K, 17K, and 63K with an assumed page size of 1K words (or 4 Kbytes for a 32-bit word length).

Figure 5.18a shows the allocation of physical memory before the swapping. The main memory is assumed to have 1024 page frames, and the disk can accommodate 4M pages. The five resident pages, scattered around the main memory, are swapped out to the swap device in contiguous pages as shown by the shaded boxes.

Later on, the entire process may be required to swap back into the main memory, as depicted in Fig. 5.18b. Different page frames may be allocated to accommodate the returning pages. The reason why contiguous blocks are mapped in the swap device is to enable faster I/O in one multiblock data transfer rather than in several single-block transfer operations.

It should be noted that only the assigned pages are swapped out and in, not the entire process address space. In the example process, the entire process address space is assumed to be 64K. The unassigned pages include the two gaps of virtual addresses between 2K and 16K and between 18K and 63K, respectively.

These *empty* spaces are not involved in the swapping process. When the memory manager swaps the process back into memory, the virtual address map should be able to identify the virtual addresses required for the returning process.

Swapping in UNIX   In the early UNIX/OS, the kernel swaps out a process to create free memory space under the following system calls:

(1) The allocation of space for a child process being created.
(2) The increase in the size of a process address space.
(3) The increased space demand by the stack for a process.
(4) The demand for space by a returning process swapped out previously.

In UNIX, a special process 0 is reserved as a *swapper*. The swapper must swap the process into main memory before the kernel can schedule it for execution. In fact, process 0 is the only process which can do so. Only when there are processes to swap in and eligible processes to swap out can the swapper do its work. Otherwise, the swapper goes to sleep.

However, the kernel periodically wakes the swapper up when the situation demands. The swapper should be designed to avoid thrashing, especially in swapping out a process which has not been executed yet.

Demand Paging Systems A paged memory system often uses a *demand paging* memory allocation policy. This policy allows only pages (instead of processes) to be transferred between the main memory and the swap device. In Fig. 5.18, individual pages of a process are allowed to be independently swapped out and in, and we have a demand paging system.

The UNIX BSD 4.0 release was the first implementation of the demand paging policy. The recent UNIX System V also supports demand paging. In a demand paging system, the entire process does not have to move into main memory to execute. The pages are brought into main memory only upon demand.

This allows the process address space to be larger than the physical address space. The major advantage of demand paging is that it offers the flexibility to dynamically accommodate a large number of processes in the physical memory on a time-sharing or multiprogrammed basis with significantly enlarged address spaces.

The idea of demand paging matches nicely with the working-set concept. Only the working sets of active processes are resident in memory. Back (1986) has defined the *working set* of a process as the set of pages referenced by the process during the last n memory references, where n is the *window size* of the working set.

Example 5.8 Working sets generated with a page trace

In the following page trace, the successive contents of the working set of a process are shown for a window of size $n = 3$:

Page trace	7	24	7	15	24	24	8	1	1	8	9	24	8	1
Working set	7	7	7	7	7	7	8	8	8	8	8	8	8	8
		24	24	24	24	24	24	24	24	24	9	9	9	1
				15	15	15	15	1	1	1	1	24	24	24

∎

If the kernel keeps only the working sets with a sufficiently large window in the main memory, many more active processes can concurrently reside in the memory than the swapping system can provide. This potentially increases the system throughput and reduces the swapping traffic. In other words, undemanded pages are not involved in the swapping process.

Hybrid Memory Systems The VAX/VMS and UNIX System V have implemented *hybrid memory systems* combining the advantages of both swapping and demand paging. When several processes simultaneously are in the ready-to-run-but-swapped situation, the swapper may choose to swap out several processes entirely to vacate the needed space. This scheme may lower the page fault rate and reduce thrashing.

Other virtual memory systems may use *anticipatory paging*, which prefetches pages based on anticipation. This scheme is rather difficult to implement. Unless memory reference patterns can be predicted at the time when the compiler generates the addresses, this scheme cannot demonstrate its power. A short-range memory reference pattern is a lot easier to predict due to the locality properties.

5.4   Sequential and Weak Consistency Models

This section studies shared-memory behavior in relation to program execution order and memory-access order. The sequential consistency and weak consistency memory models are characterized and their potential for improving performance is assessed. In Chapter 9, we will introduce the processor consistency and release consistency models for building scalable multiprocessor systems.

5.4.1   Atomicity and Event Ordering

The problem of memory inconsistency arises when the memory-access order differs from the program execution order. As illustrated in Fig. 5.19a, a uniprocessor system maps an SISD sequence into a similar execution sequence. Thus memory accesses (for instructions and data) are consistent with the program execution order. This property has been called *sequential consistency* (Lamport, 1979).

In a shared-memory multiprocessor, there are multiple instruction sequences in different processors as shown in Fig. 5.19b. Different ways of interleaving the MIMD instruction sequences into a global memory-access sequence lead to different shared memory behaviors.

How these two sequences are made consistent distinguishes the memory behavior in strong and weak models. The quality of a memory model is indicated by hardware/software efficiency, simplicity, usefulness, and bandwidth performance.

Memory Consistency Issues   The behavior of a shared-memory system as observed by processors is called a *memory model*. Specification of the memory model answers three fundamental questions: (1) What behavior should a programmer/compiler expect from a shared-memory multiprocessor? (2) How can a definition of the expected behavior guarantee coverage of all contingencies? (3) How must processors and the memory system behave to ensure consistent adherence to the expected behavior of the multiprocessor?

In general, choosing a memory model involves making a compromise between a strong model minimally restricting software and a weak model offering efficient implementation. The use of *partial order* in specifying memory events gives a formal description of special memory behavior.

Primitive memory operations for multiprocessors include *load* (*read*), *store* (*write*), and one or more synchronization operations such as *swap* (atomic *load-store*) or *conditional store*. For simplicity, we consider one representative synchronization operation *swap*, besides the *load* and *store* operations.

Event Orderings   On a multiprocessor, concurrent instruction streams (or threads) executing on different processors are *processes*. Each process executes a code segment. The order in which shared memory operations are performed by one process may be used by other processes. *Memory events* correspond to shared-memory accesses. Consistency models specify the order by which the events from one process should be observed by other processes in the machine.

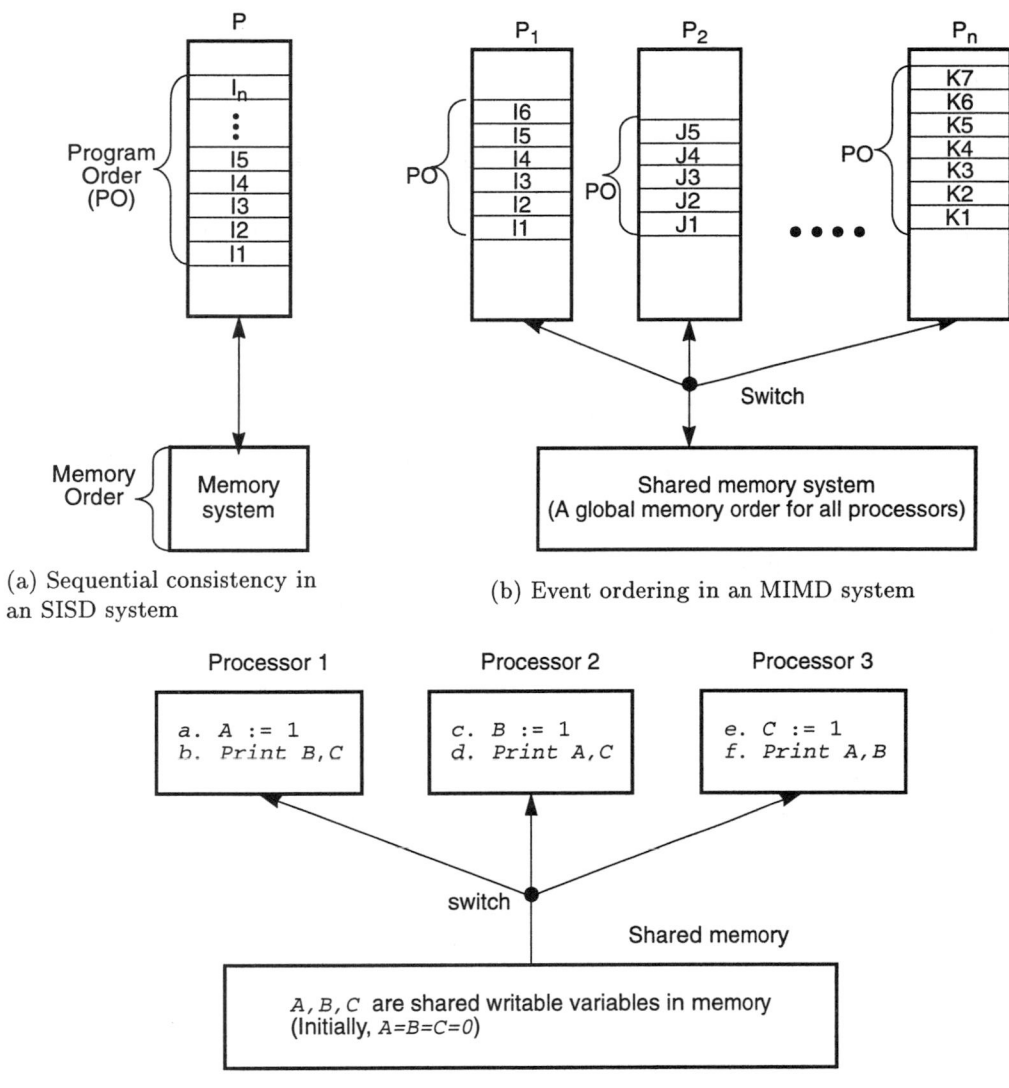

(a) Sequential consistency in an SISD system

(b) Event ordering in an MIMD system

(c) A parallel program for Example 5.9

Figure 5.19 The access ordering of memory events in a uniprocessor and in a multiprocessor, respectively. (Courtesy of Dubois and Briggs, *Tutorial Notes on Shared-Memory Multiprocessors*, Int. Symp. Computer Arch., May 1990)

The *event ordering* can be used to declare whether a memory event is legal or illegal, when several processes are accessing a common set of memory locations. A *program order* is the order by which memory accesses occur for the execution of a single process, provided that no program reordering has taken place. Dubois et al. (1986) have defined three primitive memory operations for the purpose of specifying memory consistency models:

(1) A *load* by processor P_i is considered *performed* with respect to processor P_k at a point of time when the issuing of a *store* to the same location by P_k cannot affect the value returned by the *load*.

(2) A *store* by P_i is considered *performed* with respect to P_k at one time when an issued *load* to the same address by P_k returns the value by this *store*.

(3) A *load* is *globally performed* if it is performed with respect to all processors and if the *store* that is the source of the returned value has been performed with respect to all processors.

As illustrated in Fig. 5.19a, a processor can execute instructions out of program order using a compiler to resequence instructions in order to boost performance. A uniprocessor system allows these out-of-sequence executions provided that hardware interlock mechanisms exist to check data and control dependences between instructions.

When a processor in a multiprocessor system executes a concurrent program as illustrated in Fig. 5.19b, local dependence checking is necessary but may not be sufficient to preserve the intended outcome of a concurrent execution.

Maintaining the correctness and predictability of the execution results is rather complex on an MIMD system for the following reasons:

(a) The order in which instructions belonging to different streams are executed is not fixed in a parallel program. If no synchronization among the instruction streams exists, then a large number of different instruction interleavings is possible.

(b) If for performance reasons the order of execution of instructions belonging to the same stream is different from the program order, then an even larger number of instruction interleavings is possible.

(c) If accesses are not atomic with multiple copies of the same data coexisting as in a cache-based system, then different processors can individually observe different interleavings during the same execution. In this case, the total number of possible execution instantiations of a program becomes even larger.

Example 5.9 Event ordering in a three-processor system (Dubois, Scheurich, and Briggs, 1988)

To illustrate the possible ways of interleaving concurrent program executions among multiple processors updating the same memory, we examine the simultaneous and asynchronous executions of three program segments on the three processors in Fig. 5.19c.

The shared variables are initially set as zeros, and we assume a *Print* statement reads both variables indivisibly during the same cycle to avoid confusion. If the outputs of all three processors are concatenated in the order P_1, P_2, and P_3, then the output forms a 6-tuple of binary vectors.

There are $2^6 = 64$ possible output combinations. If all processors execute instructions in their own program orders, then the execution interleaving a, b, c, d, e, f is possible, yielding the output 001011. Another interleaving, a, c, e, b, d, f, also preserves the program orders and yields the output 111111.

If processors are allowed to execute instructions out of program order, assuming that no data dependences exist among reordered instructions, then the interleaving b, d, f, e, a, c is possible, yielding the output 000000.

Out of $6! = 720$ possible execution interleavings, 90 preserve the individual program order. From these 90 interleavings not all 6-tuple combinations can result. For example, the outcome 000000 is not possible if processors execute instructions in program order only. As another example, the outcome 011001 is possible if different processors can observe events in different orders, as can be the case with replicated memories.

■

Atomicity From the above example, multiprocessor memory behavior can be described in three categories:

(1) Program order preserved and uniform observation sequence by all processors.
(2) Out-of-program-order allowed and uniform observation sequence by all processors.
(3) Out-of-program-order allowed and nonuniform sequences observed by different processors.

This behavioral categorization leads to two classes of shared-memory systems for multiprocessors: The first allows *atomic memory accesses*, and the second allows *nonatomic memory accesses*. A shared-memory access is atomic if the memory updates are known to all processors at the same time. Thus a *store* is atomic if the value stored becomes readable to all processors at the same time. Thus a necessary and sufficient condition for an atomic memory to be sequentially consistent is that all memory accesses must be performed to preserve all individual program orders.

In a multiprocessor with nonatomic memory accesses, having individual program orders that conform is not a sufficient condition for sequential consistency. In a cache/network-based multiprocessor, the system can be nonatomic if an invalidation signal does not reach all processors at the same time. Thus a *store* is inherently nonatomic in such an architecture unless special hardware mechanisms are provided to assure atomicity. Only in atomic systems can the ordering of memory events be strongly ordered to make the program order consistent with the memory-access order.

With a nonatomic memory system, the multiprocessor cannot be strongly ordered. Thus weak ordering is very much desired in a multiprocessor with nonatomic memory accesses. The above discussions lead to the division between strong and weak consistency models to be described in the next two subsections.

5.4.2   Sequential Consistency Model

The *sequential consistency* (SC) memory model is widely understood among multiprocessor designers. In this model, the *loads, stores,* and *swaps* of all processors appear to execute serially in a single global memory order that conforms to the individual program orders of the processors, as illustrated in Fig. 5.20. Two definitions of SC model are given below.

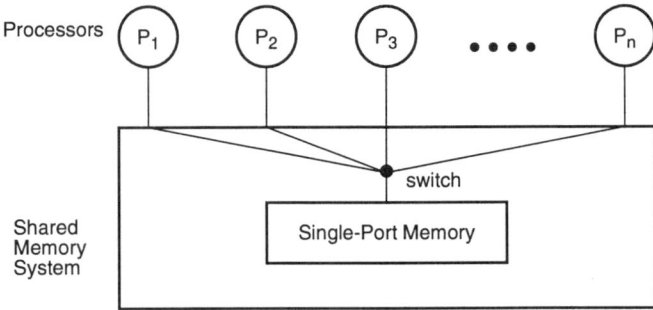

Figure 5.20 Sequential consistency memory model. (Courtesy of Sindhu, Frailong, and Cekleov; reprinted with permission from *Scalable Shared-Memory Multiprocessors*, p. 31, Kluwer Academic Publishers, 1992)

Lamport's Definition   Lamport (1979) defined *sequential consistency* as follows: A multiprocessor system is *sequentially consistent* if the result of any execution is the same as if the operations of all the processors were executed in some sequential order, and the operations of each individual processor appear in this sequence in the order specified by its program.

Dubois, Scheurich, and Briggs (1986) have provided the following two sufficient conditions to achieve sequential consistency in shared-memory access:

(a) Before a *load* is allowed to perform with respect to any other processor, all previous *load* accesses must be globally performed and all previous *store* accesses must be performed with respect to all processors.

(b) Before a *store* is allowed to perform with respect to any other processor, all previous *load* accesses must be globally performed and all previous *store* accesses must be performed with respect to all processors.

Sindhu, Frailong, and Cekleov (1992) have specified the sequential consistency memory model with the following five axioms:

(1) A *load* by a processor always returns the value written by the latest *store* to the same location by other processors.

(2) The memory order conforms to a total binary order in which shared memory is

accessed in real time over all *loads* and *stores* with respect to all processor pairs and location pairs.

(3) If two operations appear in a particular program order, then they appear in the same memory order.

(4) The *swap* operation is atomic with respect to other *stores*. No other *store* can intervene between the *load* and *store* parts of a *swap*.

(5) All *stores* and *swaps* must eventually terminate.

Lamport's definition sets the basic spirit of sequential consistency. The memory-access constraints imposed by Dubois et al. are refined from Lamport's definition with respect to atomicity. The conditions on sequential consistency specified by Sindhu et al. are further refined with respect to partial ordering relations. Implementation requirements of these constraints are discussed below.

Implementation Considerations   Figure 5.20 shows that the shared memory consists of a single port that is able to service exactly one operation at a time, and a switch that connects this memory to one of the processors for the duration of each memory operation. The order in which the switch is thrown from one processor to another determines the global order of memory-access operations.

The sequential consistency model implies total ordering of *stores/loads* at the instruction level. This should be transparent to all processors. In other words, sequential consistency must hold for any processor in the system.

A conservative multiprocessor designer may prefer the sequential consistency model, in which consistency is enforced by hardware on-the-fly. Memory accesses are atomic and strongly ordered, and confusion can be avoided by having all processors/caches wait sufficiently long for unexpected events.

Strong ordering of all shared-memory accesses in the sequential consistency model preserves the program order in all processors. A sequentially consistent multiprocessor cannot determine whether the system is a multitasking uniprocessor or a multiprocessor. Interprocessor communication can be implemented with simple *loads/stores*, such as Dekker's algorithm for synchronized entry into a critical section by multiple processors. All memory accesses must be globally performed in program order.

A processor cannot issue another access until the most recently shared writable memory access by a processor has been globally performed. This may require the propagation of all shared-memory accesses to all processors, which is rather time-consuming and costly.

Most multiprocessors have implemented the sequential consistency model because of its simplicity. However, the model may lead to rather poor memory performance due to the imposed strong ordering of memory events. This is especially true when the system becomes very large. Thus sequential consistency reduces the scalability of a multiprocessor system.

5.4.3   Weak Consistency Models

The multiprocessor memory model may range anywhere from strong (or sequential) consistency to various degrees of weak consistency. In this section, we describe the *weak consistency* model introduced by Dubois et al. (1986) and a TSO model introduced with the SPARC architecture.

The DSB Model   Dubois, Scheurich, and Briggs (1986) have derived a weak consistency memory model by relating memory request ordering to synchronization points in the program. We call this the DSB model specified by the following three conditions:

(1) All previous *synchronization* accesses must be performed, before a *load* or a *store* access is allowed to perform with respect to any other processor.

(2) All previous *load* and *store* accesses must be performed, before a *synchronization* access is allowed to perform with respect to any other processor

(3) *Synchronization* accesses are sequentially consistent with respect to one another.

These conditions provide a weak ordering of memory-access events in a multiprocessor. The dependence conditions on shared variables are weaker in such a system because they are only limited to hardware-recognized synchronizing variables. Buffering is allowed in *write buffers* except for operations on hardware-recognized synchronizing variables. Buffering memory accesses in multiprocessors can enhance the shared memory performance.

With different restrictions on the memory-access ordering, many different weak memory models can be similarly defined. The following is another weak consistency model, called the TSO (total store order), developed by the SPARC architecture group in Sun Microsystems.

Example 5.10   The TSO weak consistency model used in SPARC architecture (Sun Microsystems, Inc., 1990 and Sindhu et al., 1992)

Figure 5.21 shows the weak consistency TSO model developed by Sun Microsystems' SPARC architecture group (1990). Sindhu et al. described that the *stores* and *swaps* issued by a processor are placed in a dedicated store buffer for the processor, which is operated as first-in-first-out. Thus the order in which memory executes these operations for a given processor is the same as the order in which the processor issued them (in program order).

The memory order corresponds to the order in which the switch is thrown from one processor to another. This was described by Sindhu et al. as follows: A *load* by a processor first checks its store buffer to see if it contains a *store* to the same location. If it does, then the *load* returns the value of the most recent such *store*. Otherwise, the *load* goes directly to memory. Since not all *loads* go to memory immediately, *loads* in general do not appear in memory order. A processor is logically blocked from issuing further operations until the *load* returns a value. A *swap* behaves like a *load* and a *store*. It is placed in the store buffer like a *store*, and it blocks the processor like a *load*. In other words, the *swap* blocks until the store buffer is empty and then proceeds to the memory.

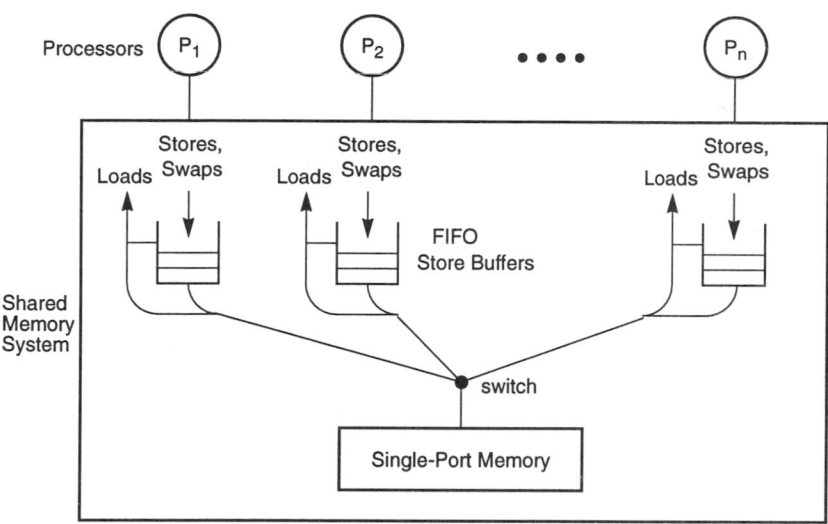

Figure 5.21 The TSO Weak consistency memory model. (Courtesy of Sindhu, Frailong, and Cekleov; reprinted with permission from *Scalable Shared-Memory Multiprocessors*, p. 31, Kluwer Academic Publishers, 1992)

A TSO Formal Specification   Sindhu, Frailong, and Cekleov (1992) have specified the TSO weak consistency model with six behavioral axioms. Only an intuitive description of their axioms is abstracted below:

(1) A *load* access is always returned with the latest *store* to the same memory location issued by any processor in the system.

(2) The memory order is a total binary relation over all pairs of *store* operations.

(3) If two *stores* appear in a particular program order, then they must also appear in the same memory order.

(4) If a memory operation follows a *load* in program order, then it must also follow the *load* in memory order.

(5) A *swap* operation is atomic with respect to other *stores*. No other *store* can interleave between the *load* and *store* parts of a *swap*.

(6) All *stores* and *swaps* must eventually terminate.

Note that the above axioms (5) and (6) are identical to axioms (4) and (5) for the sequential consistency model. Axiom (1) covers the effects of both local and remote processors. Both axioms (2) and (3) are weakened from the corresponding axioms (2) and (3) for the sequential consistency model. Axiom (4) states that the *load* operations do not have to be weakened, as far as ordering is concerned. For a formal axiomatic specification, the reader is referred to the original paper by Sindhu et al. (1992).

Comparison of Memory Models   In summary, the weak consistency model may offer better performance than the sequential consistency model at the expense of more complex hardware/software support and more programmer awareness of the imposed restrictions. The relative merits of the strong and weak memory models are still being debated in the multiprocessor research community.

The DSB and the TSO are two different weak-consistency memory models. The DSB model is weakened by enforcing sequential consistency at synchronization points. The TSO model is weakened by treating *reads*, *stores*, and *swaps* differently using FIFO store buffers. The TSO model has been implemented in some SPARC architectures, while the DSB model has not been implemented in real systems yet.

Sindhu et al. (1992) have identified four system-level issues which also affect the choice of memory model. First, they suggest that one should consider extending the memory model from the processor level to the process level. To do this, one must maintain a process switch sequence. The second issue is the incorporation of I/O locations into the memory model. I/O operations may introduce even more side effects in addition to the normal semantics of *loads* and *stores*.

The third issue is code modification. In the SPARC architecture, synchronization of code modification is accomplished through the use of a *flush* instruction defined in the TSO model. Finally, they feel that the memory model should address the issue of how to incorporate pipelined processors and processors with noncoherent internal caches. The interested reader is referred to their original paper and the SPARC Architecture Manual for details of these issues.

Strong memory ordering introduces unnecessary processor/cache waiting time and reduces the amount of concurrency. Weak consistency has the potential to eliminate these shortcomings. Besides the sequential consistency, the DSB and the TSO weak memory models, other memory consistency models, such as *processor consistency* and *release consistency*, will be treated in Chapter 9.

5.5   Bibliographic Notes and Exercises

There are several books dealing with cache and memory architecture. Readers are referred to [Przybylski90], [Dubois92b] and Thakkar, and the thesis of [Scheurich89]. The paper by [Dubois88], Scheurich, and Briggs has characterized synchronization, coherence, and event-ordering problems in multiprocessors. [Cragon92b] treated memory systems for pipelined processors.

[Smith82] has provided a survey of cache memories developed up to the early 1980s. Private and shared virtual memory models have been described in tutorial notes by [Dubois90a] and Briggs. Recent developments in cache architecture can be found in [Dubois90b] and Thakkar and in [Stone90].

The DSB weak consistency memory models was originally developed by [Dubois86], Scheurich, and Briggs. The formal specifications of the TSO and PSO (partial store order) weak models were given in [Sindhu92], Frailong, and Cekleov.   [Adve90] and Hill have also studied the weak ordering issue. Besides, memory consistency models are discussed in [Goodman90], [Gharachorloo90] et al., [Dubois92b], [Gharachorloo92a] et al., and [Hagersten92]. The weak consistency model TSO was developed for the SPARC architecture by Sun Microsystems. [SPARC90] The pioneering work on event ordering was attributed to [Lamport78] and [Lamport79].

Exercises

Problem 5.1 This is an example of a design specification of a backplane bus for a shared-memory multiprocessor with 4 processor boards and 16 memory boards under the following assumptions:

- Bus clock rate = 20 MHz.
- Memory word length = 64 bits; processors always request data in blocks of four words.
- Memory access time = 100 ns.
- Shared address space = 2^{40} words.
- Maximum number of signal lines available on the backplane is 96.
- Synchronous timing protocol.
- Neglect buffer and propagation delays.

Specify the following in your design of the bus system:

(a) Maximum bus bandwidth.
(b) Effective bus bandwidth (worst case).
(c) Arbitration scheme.
(d) Name and functionality of each of the signal lines.
(e) Number of slots required on the backplane.

Problem 5.2 Describe the daisy-chaining (Fig. 5.4) and the *distributed arbiter* (Fig. 5.5b) for bus arbitration in a multiprocessor system. State the advantages and shortcomings of each case from both the implementational and operational points of view.

Problem 5.3 Read the paper by Mudge et al. (1987) on multiple-bus systems and solve the following problems:

(a) Find the maximum bandwidth for a multiprocessor system using b buses, where $b > m$ and m is the number of memory modules and the system has n processors.
(b) Prove that $BW_b < np$, where $p > 0$ is the probability that an arbitrary processor will generate a request to access the shared memory at the start of a memory cycle.

Problem 5.4 Estimate the effective MIPS rate of a bus-connected multiprocessor system under the following assumptions. The system has 16 processors, each connected to an on-board private cache which is connected to a common bus. Globally shared memory is also connected to the bus. The private cache and the shared memory form a two-level access hierarchy.

Each processor is rated with 10 MIPS if a 100% cache hit ratio is assumed. On the average, each instruction needs 0.2 memory access. The read access and write access are assumed equally probable.

For a crude approximation, consider only the penalty caused by shared-memory access and ignore all other overheads. The cache is targeted to maintain a hit ratio of 0.95. A cache access on a read-hit takes 20 ns; that on a write-hit takes 60 ns with a write-back scheme, and with a write-through scheme it needs 400 ns.

When a cache block is to be replaced, the probability that it is dirty is estimated as 0.1. An average block transfer time between the cache and shared memory via the bus is 400 ns.

(a) Derive the effective memory-access times per instruction for the write-through and write-back caches separately.

(b) Calculate the effective MIPS rate for each processor. Determine an upper bound on the effective MIPS rate of the 16-processor system. Discuss why the upper bound cannot be achieved by considering the memory penalty alone.

Problem 5.5 Explain the following terms associated with cache and memory architectures.

(a) Low-order memory interleaving.
(b) Physical address cache versus virtual address cache.
(c) Atomic versus nonatomic memory accesses.
(d) Memory bandwidth and fault tolerance.

Problem 5.6 Explain the following terms associated with cache design:

(a) Write-through versus write-back caches.
(b) Cacheable versus noncacheable data.
(c) Private caches versus shared caches.
(d) Cache flushing policies.
(e) Factors affecting cache hit ratios.

Problem 5.7 Consider the simultaneous execution of the three programs on the three processors shown in Fig. 5.19c. Answer the following questions with reasoning or supported by computer simulation results:

(a) List the 90 execution interleaving orders of the six instructions $\{a, b, c, d, e, f\}$ which will preserve the individual program orders. The corresponding output patterns (6-tuples) should be listed accordingly.

(b) Can all 6-tuple combinations be generated out of the 720 non-program-order interleavings? Justify the answer with reasoning and examples.

(c) We have assumed atomic memory access in this example. Explain why the output 011001 is not possible in an atomic memory multiprocessor system if individual program orders are preserved.

(d) Suppose nonatomic memory access is allowed in the above multiprocessor. For example, an invalidation does not reach all private caches at the same time. Prove that 011001 is possible even if all instructions were executed in program order but other processors did not observe them in program order.

Problem 5.8   The main memory of a computer is organized as 64 blocks, with a block size of eight words. The cache has eight block frames. In parts (a) through (d), show the mappings from the numbered blocks in main memory to the block frames in the cache. Draw all lines showing the mappings as clearly as possible.

(a) Show the direct mapping and the address bits that identify the tag field, the block number, and the word number.

(b) Show the fully associative mapping and the address bits that identify the tag field and the word number.

(c) Show the two-way set-associative mapping and the address bits that identify the tag field, the set number, and the word number.

(d) Show the sector mapping with four blocks per sector and the address bits that identify the sector number, the block number, and the word number.

Problem 5.9   Consider a cache (M_1) and memory (M_2) hierarchy with the following characteristics:

> M_1:   16K words, 50 ns access time
> M_2:   1M words, 400 ns access time

Assume eight-word cache blocks and a set size of 256 words with set-associative mapping.

(a) Show the mapping between M_2 and M_1.

(b) Calculate the effective memory-access time with a cache hit ratio of $h = 0.95$.

Problem 5.10   Consider a main memory consisting of four memory modules with 256 words per module. Assume 16 words in each cache block. The cache has a total capacity of 256 words. Set-associative mapping is used to allocate cache blocks to block frames. The cache is divided into four sets.

(a) Show the address assignment for all 1024 words in a four-way low-order interleaved organization of the main memory.

(b) How many blocks are there in the main memory? How many block frames are there in the cache?

(c) Explain the bit fields needed for addressing each word in the two-level memory system.

(d) Show the mapping from the blocks in the main memory to the sets in the cache and explain how to use the tag field to locate a block frame within each set.

Problem 5.11

(a) A uniprocessor system uses separate instruction and data caches with hit ratios h_i and h_d, respectively. The access time from the processor to either cache is c clock cycles, and the block transfer time between the caches and main memory is b clock cycles.

Among all memory references made by the CPU, f_i is the percentage of references to instructions. Among blocks replaced in the data cache, f_{dir} is the percentage of *dirty* blocks. (*Dirty* means that the cache copy is different from the memory copy.)

Assuming a write-back policy, determine the effective memory-access time in terms of h_i, h_d, c, b, f_i, and f_{dir} for this memory system.

(b) The processor memory system described in part (a) is used to construct a bus-based shared-memory multiprocessor. Assume that the hit ratio and access times remain the same as in part (a). However, the effective memory-access time will be different because every processor must now handle cache invalidation in addition to reads and writes.

Let f_{inv} be the fraction of data references that cause invalidation signals to be sent to other caches. The processor sending the invalidation signal requires i clock cycles to complete the invalidation operation. Other processors are not involved in the invalidation process. Assuming a write-back policy again, determine the effective memory-access time for this multiprocessor.

Problem 5.12 A computer system has a 128-byte cache. It uses four-way set-associative mapping with 8 bytes in each block. The physical address size is 32 bits, and the smallest addressable unit is 1 byte.

(a) Draw a diagram showing the organization of the cache and indicating how physical addresses are related to cache addresses.

(b) To what block frames of the cache can the address $000010AF_{16}$ be assigned?

(c) If the addresses $000010AF_{16}$ and $FFFF7Axy_{16}$ can be simultaneously assigned to the same cache set, what values can the address digits x and y have?

Problem 5.13 Consider a shared-memory multiprocessor system with p processors. Let m be the average number of global memory references per instruction execution on a typical processor.

Let t be the average access time to the shared memory and x be the MIPS rate of a uniprocessor using local memory. Consider the execution of n instructions on each processor of the multiprocessor.

(a) Determine the effective MIPS rate of the multiprocessor in terms of the parameters m, t, x, n, and p.

(b) Suppose a multiprocessor has $p = 32$ RISC processors, $m = 0.4$, and $t = 1$ μs. What is the MIPS rate of each processor (i.e., $x = ?$) needed to achieve a multiprocessor performance of 56 MIPS effectively?

(c) Suppose $p = 32$ CISC processors with $x = 2$ MIPS each are used in the above multiprocessor system with $m = 1.6$ and $t = 1$ μs. What will be the effective MIPS rate?

Problem 5.14 Consider a RISC-based shared-memory multiprocessor with p processors, each having an off-chip instruction cache and data cache. The peak performance rating of each processor (assuming a 100% hit ratio in both caches) is x MIPS. You are required to derive a performance formula, taking into account cache misses, shared-memory accesses, and synchronization overhead.

Assume that on the average α percent of the instructions executed are for synchronization purpose, and the penalty for each synchronization operation is an additional t_s μs. The number of memory accesses per instruction is m. Among all memory references made by the CPU, f_i is the percentage of references to instructions. Assume that the instruction cache and data cache have hit ratios h_i and h_d, respectively, after a long period of program tracing on the machine. On cache misses, instructions and data are accessed from the shared memory with an average access time t_m μs.

(a) Derive an expression for approximating the effective MIPS rate of this multiprocessor in terms of $p, x, m, f_i, h_i, h_d, t_m, \alpha$, and t_s. Note that f_i, h_i, h_d, and α are all fractions and t_m and t_s are measured in μs. Ignore the cache-access time and other system overheads in your derivation.

(b) Suppose $m = 0.4, f_i = 0.5, h_i = 0.95, h_d = 0.7, \alpha = 0.05, x = 5, t_m = 0.5$ μs, and $t_s = 5$ μs. Determine the minimum number of processors needed in the above multiprocessor system in order to achieve an effective MIPS rate of 25.

(c) Suppose the total cost of all the caches and shared-memory is upper-bounded by $25,000. The cache memory costs $4.70/Kbyte, and the shared memory costs $0.4/Kbyte. With $p = 16$ processors, each having an instruction cache of capacity $S_i = 32$ Kbytes and a data cache of capacity $S_d = 64$ Kbytes, what is the maximum shared-memory capacity C_m (in Mbytes) that can be acquired within the budget limit?

Problem 5.15 Consider the following three interleaved memory designs for a main memory system with 16 memory modules. Each module is assumed to have a capacity of 1 Mbyte. The machine is byte-addressable.

> *Design 1*: 16-way interleaving with one memory bank.
> *Design 2*: 8-way interleaving with two memory banks.
> *Design 3*: 4-way interleaving with four memory banks.

(a) Specify the address formats for each of the above memory organizations.
(b) Determine the maximum memory bandwidth obtained if only one memory module fails in each of the above memory organizations.
(c) Comment on the relative merits of the three interleaved memory organizations.

Problem 5.16 Consider a memory system for the Cray 1 computer. There are $m = 16$

interleaved modules. The access time of a module is $t_a = 50$ ns and the memory cycle time is $t_c = 12.5$ ns. We know that for this memory system the maximum memory bandwidth of 80M words per second is achieved for vector loads/stores except when the stride is a multiple of 16 (bandwidth: 20M words per second) or a multiple of 8 (but not 16) (bandwidth: 40M words per second).

(a) Find the bandwidth for all strides for similar systems but with the following parameters: $t_c = 12.5$ ns, $t_a = 50$ ns, $m = 17$.

(b) Repeat part (a) for the following parameters: $t_c = 12.5$ ns, $t_a = 50$ ns, $m = 8$.

Problem 5.17 Consider the concurrent execution of two programs by two processors with a shared memory. Assume that A, B, C, D are initialized to 0 and that a *Print* statement prints both arguments indivisibly at the same cycle. The output forms a 4-tuple as either $ADBC$ or $BCAD$.

P_0:
a. $A = 1$
b. $B = 1$
c. Print A, D

P_1:
d. $C = 1$
e. $D = 1$
f. Print B, C

(a) List all execution interleaving orders of six statements which will preserve the individual program order.

(b) Assume program orders are preserved and all memory accesses are atomic; i.e., a store by one processor is immediately seen by all the remaining processors. List all the possible 4-tuple output combinations.

(c) Assume program orders are preserved but memory accesses are nonatomic; i.e., a store by one processor may be buffered so that some other processors may not immediately observe the update. List all possible 4-tuple output combinations.

Problem 5.18 Compare the relative merits of the four cache memory organizations:

(1) Direct-mapping cache
(2) Fully associative cache
(3) Set-associative cache
(4) Sector mapping cache

Answer the following questions with reasoning:

(a) In terms of hardware complexity and implementation cost, rank the four cache organizations with justification.

(b) With respect to flexibility in implementing block replacement algorithms, rank the four cache organizations and justify the ranking order.

(c) With each cache organization, explain the effects of block mapping policies on the hit ratio issues.

(d) Explain the effects of block size, set number, associativity, and cache size on the performance of a set-associative cache organization.

Problem 5.19 Explain the following terms associated with memory management:

(a) The role of a memory manager in an OS kernel.
(b) Preemptive versus nonpreemptive memory allocation policies.
(c) Swapping memory system and examples.
(d) Demand paging memory system and examples.
(e) Hybrid memory system and examples.

Problem 5.20 Compare the memory-access constraints in the following memory consistency models:

(a) Determine the similarities and subtle differences among the conditions on sequential consistency imposed by Lamport (1979), by Dubois et al. (1986), and by Sindhu et al. (1992), respectively.
(b) Repeat question (a) between the DSB model and the TSO model for weak consistency memory systems.
(c) A PSO (partial store order) model for weak consistency has been refined from the TSO model. Study the PSO specification in the paper by Sindhu et al. (1992) and compare the relative merits between the TSO and the PSO memory models.

Chapter 6

Pipelining and Superscalar Techniques

This chapter deals with advanced pipelining and superscalar design in processor development. We begin with a discussion of conventional linear pipelines and analyze their performance. A generalized pipeline model is introduced to include nonlinear interstage connections. Collision-free scheduling techniques are described for performing dynamic functions.

Specific techniques for building instruction pipelines, arithmetic pipelines, and memory-access pipelines are presented. The discussion includes instruction prefetching, internal data forwarding, software interlocking, hardware scoreboarding, hazard avoidance, branch handling, and instruction-issuing techniques. Both static and multifunctional arithmetic pipelines are designed. Superpipelining and superscalar design techniques are studied along with a performance analysis.

6.1 Linear Pipeline Processors

A *linear pipeline processor* is a cascade of processing stages which are linearly connected to perform a fixed function over a stream of data flowing from one end to the other. In modern computers, linear pipelines are applied for instruction execution, arithmetic computation, and memory-access operations.

6.1.1 Asynchronous and Synchronous Models

A linear pipeline processor is constructed with k processing stages. External inputs (operands) are fed into the pipeline at the first stage S_1. The processed results are passed from stage S_i to stage S_{i+1}, for all $i = 1, 2, \ldots, k-1$. The final result emerges from the pipeline at the last stage S_k.

Depending on the control of data flow along the pipeline, we model linear pipelines in two categories: *asynchronous* and *synchronous*.

Asynchronous Model   As shown in Fig. 6.1a, data flow between adjacent stages in an asynchronous pipeline is controlled by a handshaking protocol. When stage S_i is ready to transmit, it sends a *ready* signal to stage S_{i+1}. After stage S_{i+1} receives the incoming data, it returns an *acknowledge* signal to S_i.

Asynchronous pipelines are useful in designing communication channels in message-passing multicomputers where pipelined wormhole routing is practiced (see Chapter 9). Asynchronous pipelines may have a variable throughput rate. Different amounts of delay may be experienced in different stages.

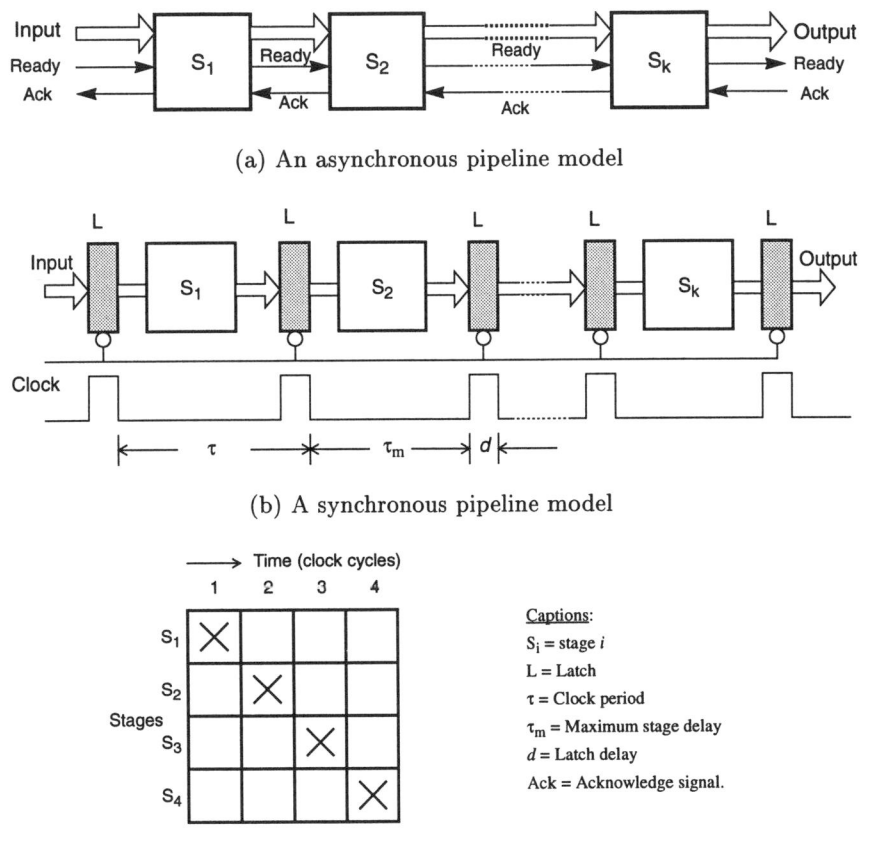

(a) An asynchronous pipeline model

(b) A synchronous pipeline model

Captions:

S_i = stage i

L = Latch

τ = Clock period

τ_m = Maximum stage delay

d = Latch delay

Ack = Acknowledge signal.

(c) Reservation table of a four-stage linear pipeline

Figure 6.1 Two models of linear pipeline units and the corresponding reservation table.

Synchronous Model   Synchronous pipelines are illustrated in Fig. 6.1b. Clocked latches are used to interface between stages. The latches are made with master-slave flip-flops, which can isolate inputs from outputs. Upon the arrival of a clock pulse, all latches transfer data to the next stage simultaneously.

The pipeline stages are combinational logic circuits. It is desired to have approximately equal delays in all stages. These delays determine the clock period and thus the speed of the pipeline. Unless otherwise specified, only synchronous pipelines are studied in this book.

The utilization pattern of successive stages in a synchronous pipeline is specified by a *reservation table*. For a linear pipeline, the utilization follows the diagonal streamline pattern shown in Fig. 6.1c. This table is essentially a space-time diagram depicting the precedence relationship in using the pipeline stages. For a k-stage linear pipeline, k clock cycles are needed to flow through the pipeline.

Successive tasks or operations are initiated one per cycle to enter the pipeline. Once the pipeline is filled up, one result emerges from the pipeline for each additional cycle. This throughput is sustained only if the successive tasks are independent of each other.

6.1.2 Clocking and Timing Control

The *clock cycle* τ of a pipeline is determined below. Let τ_i be the time delay of the circuitry in stage S_i and d the time delay of a latch, as shown in Fig. 6.1b.

Clock Cycle and Throughput   Denote the *maximum stage delay* as τ_m, and we can write τ as

$$\tau = \max_i \{\tau_i\}_1^k + d = \tau_m + d \tag{6.1}$$

At the rising edge of the clock pulse, the data is latched to the master flip-flops of each latch register. The clock pulse has a width equal to d. In general, $\tau_m >> d$ for one to two orders of magnitude. This implies that the maximum stage delay τ_m dominates the clock period.

The *pipeline frequency* is defined as the inverse of the clock period:

$$f = \frac{1}{\tau} \tag{6.2}$$

If one result is expected to come out of the pipeline per cycle, f represents the *maximum throughput* of the pipeline. Depending on the initiation rate of successive tasks entering the pipeline, the *actual throughput* of the pipeline may be lower than f. This is because more than one clock cycle has elapsed between successive task initiations.

Clock Skewing   Ideally, we expect the clock pulses to arrive at all stages (latches) at the same time. However, due to a problem known as *clock skewing*, the same clock pulse may arrive at different stages with a time offset of s. Let t_{max} be the time delay of the longest logic path within a stage and t_{min} that of the shortest logic path within a stage.

To avoid a race in two successive stages, we must choose $\tau_m \geq t_{max} + s$ and $d \leq t_{min} - s$. These constraints translate into the following bounds on the clock period when clock skew takes effect:

$$d + t_{max} + s \leq \tau \leq \tau_m + t_{min} - s \tag{6.3}$$

In the ideal case $s = 0$, $t_{max} = \tau_m$, and $t_{min} = d$. Thus, we have $\tau = \tau_m + d$, consistent with the definition in Eq. 6.1 without the effect of clock skewing.

6.1.3   Speedup, Efficiency, and Throughput

Ideally, a linear pipeline of k stages can process n tasks in $k + (n - 1)$ clock cycles, where k cycles are needed to complete the execution of the very first task and the remaining $n - 1$ tasks require $n - 1$ cycles. Thus the total time required is

$$T_k = [k + (n - 1)]\tau \tag{6.4}$$

where τ is the clock period. Consider an equivalent-function nonpipelined processor which has a *flow-through delay* of $k\tau$. The amount of time it takes to execute n tasks on this nonpipelined processor is $T_1 = nk\tau$.

Speedup Factor   The speedup factor of a k-stage pipeline over an equivalent non-pipelined processor is defined as

$$S_k = \frac{T_1}{T_k} = \frac{nk\tau}{k\tau + (n - 1)\tau} = \frac{nk}{k + (n - 1)} \tag{6.5}$$

Example 6.1   Pipeline speedup versus stream length

The maximum speedup is $S_k \to k$ as $n \to \infty$. This maximum speedup is very difficult to achieve because of data dependences between successive tasks (instructions), program branches, interrupts, and other factors to be studied in subsequent sections.

Figure 6.2a plots the speedup factor as a function of n, the number of tasks (operations or instructions) performed by the pipeline. For small values of n, the speedup can be very poor. The smallest value of S_k is 1 when $n = 1$.

The larger the number k of subdivided pipeline stages, the higher the potential speedup performance. When $n = 64$, an eight-stage pipeline has a speedup value of 7.1 and a four-stage pipeline has a speedup of 3.7. However, the number of pipeline stages cannot increase indefinitely due to practical constraints on costs, control complexity, circuit implementation, and packaging limitations. Furthermore, the stream length n also affects the speedup; the longer the better in using a pipeline.

■

Optimal Number of Stages   The finest level of pipelining is called *micropipelining*, with a subdivision of pipeline stages at the logic gate level. In practice, most pipelining is staged at the functional level with $2 \le k \le 15$. Very few pipelines are designed to exceed 10 stages in real computers.

On the other hand, the coarse level for pipeline stages can be conducted at the processor level, called *macropipelining*. The optimal choice of the number of pipeline stages should be able to maximize a performance/cost ratio.

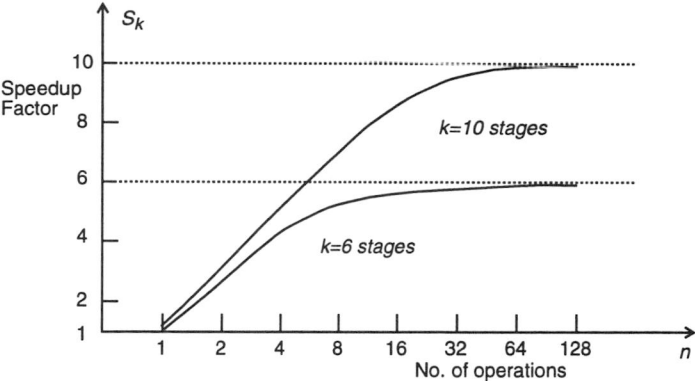

(a) Speedup factor as a function of the number of operations (Eq. 6.5)

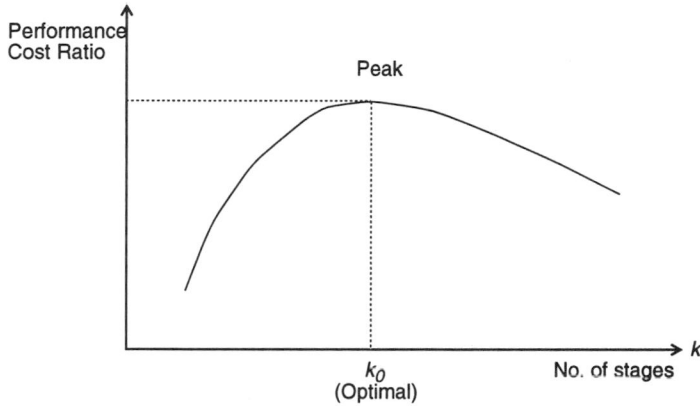

(b) Optimal number of pipeline stages (Eqs. 6.6 and 6.7)

Figure 6.2 Speedup factors and the optimal number of pipeline stages for a linear pipeline unit.

Let t be the total time required for a nonpipelined sequential program of a given function. To execute the same program on a k-stage pipeline with an equal flow-through delay t, one needs a clock period of $p = t/k + d$, where d is the latch delay. Thus, the pipeline has a maximum throughput of $f = 1/p = 1/(t/k+d)$. The total pipeline cost is roughly estimated by $c + kh$, where c covers the cost of all logic stages and h represents the cost of each latch. A pipeline *performance/cost ratio* (PCR) has been defined by Larson (1973):

$$PCR = \frac{f}{c + kh} = \frac{1}{(t/k + d)(c + kh)} \tag{6.6}$$

Figure 6.2b plots the PCR as a function of k. The peak of the PCR curve corresponds to an optimal choice for the number of desired pipeline stages:

$$k_0 = \sqrt{\frac{t \cdot c}{d \cdot h}} \tag{6.7}$$

where t is the total flow-through delay of the pipeline. The total stage cost c, the latch delay d, and the latch cost h can be adjusted to achieve the optimal value k_0.

Efficiency and Throughput   The *efficiency E_k* of a linear k-stage pipeline is defined as

$$E_k = \frac{S_k}{k} = \frac{n}{k + (n - 1)} \tag{6.8}$$

Obviously, the efficiency approaches 1 when $n \to \infty$, and a lower bound on E_k is $1/k$ when $n = 1$. The *pipeline throughput H_k* is defined as the number of tasks (operations) performed per unit time:

$$H_k = \frac{n}{[k + (n - 1)]\tau} = \frac{nf}{k + (n - 1)} \tag{6.9}$$

The *maximum throughput f* occurs when $E_k \to 1$ as $n \to \infty$. This coincides with the speedup definition given in Chapter 3. Note that $H_k = E_k \cdot f = E_k/\tau = S_k/k\tau$.

6.2   Nonlinear Pipeline Processors

A *dynamic pipeline* can be reconfigured to perform variable functions at different times. The traditional linear pipelines are static pipelines because they are used to perform fixed functions.

A dynamic pipeline allows feedforward and feedback connections in addition to the streamline connections. For this reason, some authors call such a structure a *nonlinear pipeline*.

6.2.1   Reservation and Latency Analysis

In a static pipeline, it is easy to partition a given function into a sequence of linearly ordered subfunctions. However, function partitioning in a dynamic pipeline becomes quite involved because the pipeline stages are interconnected with loops in addition to streamline connections.

A multifunction dynamic pipeline is shown in Fig. 6.3a. This pipeline has three stages. Besides the *streamline connections* from S_1 to S_2 and from S_2 to S_3, there is a *feedforward connection* from S_1 to S_3 and two *feedback connections* from S_3 to S_2 and from S_3 to S_1.

These feedforward and feedback connections make the scheduling of successive events into the pipeline a nontrivial task. With these connections, the output of the pipeline is not necessarily from the last stage. In fact, following different dataflow patterns, one can use the same pipeline to evaluate different functions.

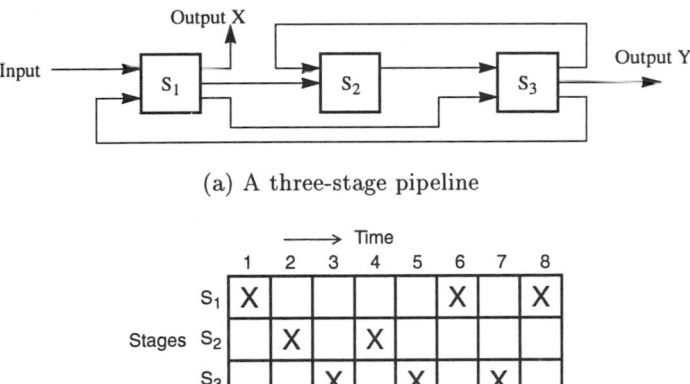

(a) A three-stage pipeline

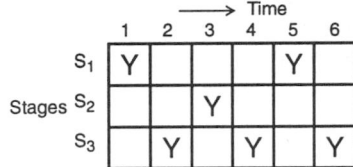

(b) Reservation table for function X

(c) Reservation table for function Y

Figure 6.3 A dynamic pipeline with feedforward and feedback connections for two different functions.

Reservation Tables   The reservation table for a static linear pipeline is trivial in the sense that dataflow follows a linear streamline. The *reservation table* for a dynamic pipeline becomes more interesting because a nonlinear pattern is followed. Given a pipeline configuration, multiple reservation tables can be generated for the evaluation of different functions.

Two reservation tables are given in Figs. 6.3b and 6.3c, corresponding to a function X and a function Y, respectively. Each function evaluation is specified by one reservation table. A static pipeline is specified by a single reservation table. A dynamic pipeline may be specified by more than one reservation table.

Each reservation table displays the time-space flow of data through the pipeline for one function evaluation. Different functions may follow different paths on the reservation table. A number of pipeline configurations may be represented by the same reservation table. There is a many-to-many mapping between various pipeline configurations and different reservation tables.

The number of columns in a reservation table is called the *evaluation time* of a given function. For example, the function X requires eight clock cycles to evaluate, and function Y requires six cycles, as shown in Figs. 6.3b and 6.3c, respectively.

A pipeline *initiation* table corresponds to each function evaluation. All initiations to a static pipeline use the same reservation table. On the other hand, a dynamic pipeline may allow different initiations to follow a mix of reservation tables. The checkmarks in each row of the reservation table correspond to the time instants (cycles) that a particular stage will be used.

There may be multiple checkmarks in a row, which means repeated usage of the same stage in different cycles. Contiguous checkmarks in a row simply imply the extended usage of a stage over more than one cycle. Multiple checkmarks in a column mean that multiple stages are used in parallel during a particular clock cycle.

Latency Analysis   The number of time units (clock cycles) between two initiations of a pipeline is *the latency* between them. Latency values must be nonnegative integers. A latency of k means that two initiations are separated by k clock cycles. Any attempt by two or more initiations to use the same pipeline stage at the same time will cause a *collision*.

A collision implies resource conflicts between two initiations in the pipeline. Therefore, all collisions must be avoided in scheduling a sequence of pipeline initiations. Some latencies will cause collisions, and some will not. Latencies that cause collisions are called *forbidden latencies*. In using the pipeline in Fig. 6.3 to evaluate the function X, latencies 2 and 5 are forbidden, as illustrated in Fig. 6.4.

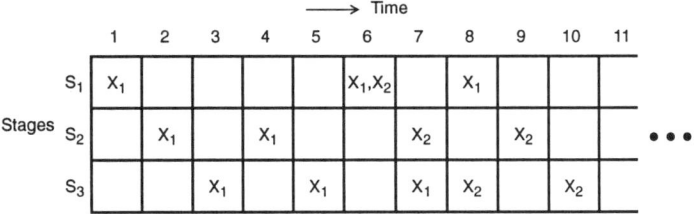

(a) Collision with scheduling latency 2

(b) Collision with scheduling latency 5

Figure 6.4 Collisions with forbidden latencies 2 and 5 in using the pipeline in Fig. 6.3 to evaluate the function X.

The ith initiation is denoted as X_i in Fig. 6.4. With latency 2, initiations X_1 and X_2 collide in stage 2 at time 4. At time 7, these initiations collide in stage 3. Similarly, other collisions are shown in times 5, 6, 8, ..., etc. The collision patterns for latency 5 are shown in Fig. 6.4b, where X_1 and X_2 are scheduled 5 clock cycles apart. Their first collision occurs at time 6.

To detect a forbidden latency, one needs simply to check the distance between any two checkmarks in the same row of the reservation table. For example, the distance between the first mark and the second mark in row S_1 in Fig. 6.3b is 5, implying that 5 is a forbidden latency.

Similarly, latencies 2, 4, 5, and 7 are all seen to be forbidden from inspecting the same reservation table. From the reservation table in Fig. 6.3c, we discover the forbidden latencies 2 and 4 for function Y. A *latency sequence* is a sequence of permissible nonforbidden latencies between successive task initiations.

A *latency cycle* is a latency sequence which repeats the same subsequence (cycle) indefinitely. Figure 6.5 illustrates latency cycles in using the pipeline in Fig. 6.3 to evaluate the function X without causing a collision. For example, the latency cycle (1, 8) represents the infinite latency sequence $1, 8, 1, 8, 1, 8, \ldots$. This implies that successive initiations of new tasks are separated by one cycle and eight cycles alternately.

	1	2	3	4	5	6	7	8	9	10	11	12	13	14	15	16	17	18	19	20	21
S_1	X1	X2				X_1	X_2	X_1	X_2	X_3	X_4				X_3	X_4	X_3	X_4	X_5	X_6	
S_2		X1	X2	X1	X_2					X_3	X_4	X_3	X_4							X_5	$\bullet\bullet\bullet$
S_3			X1	X2	X_1		X_1	X_2		X_3	X_4	X_3				X_3					

(a) Latency cycle $(1,8) = 1, 8, 1, 8, 1, 8, \ldots$, with an average latency of 4.5

	1	2	3	4	5	6	7	8	9	10	11	12	13	14	15	16	17	18	19	20	21
S_1	X1			X2		X_1	X_3	X_1	X_2	X_4	X_2	X_3	X_5	X_3	X_4	X_6	X_4	X_5	X_7	X_5	
S_2		X1		X1	X_2		X_2	X_3		X_3	X_4		X_4	X_5		X_5	X_6		X_6	X_7	$\bullet\bullet\bullet$
S_3			X1			X_1	X2	X_1	X_2	X_3	X_2	X_3	X_4	X_3	X_4	X_5	X_4	X_5	X_6	X_5	X_6

(b) Latency cycle $(3) = 3, 3, 3, 3, \ldots$, with an average latency of 3

	1	2	3	4	5	6	7	8	9	10	11	12	13	14	15	16	17	18	19	20	21
S_1	X1					X_1	X2	X_1				X_2	X_3	X_2				X_3	X_4	X_3	
S_2		X1		X1				X_2		X_2				X_3		X_3				X_4	$\bullet\bullet\bullet$
S_3			X1		X_1		X_1		X2		X_2		X_2		X_3	X_3		X_3		X_3	

(c) Latency cycle $(6) = 6, 6, 6, 6, \ldots$, with an average latency of 6

Figure 6.5 A dynamic pipeline with feedforward and feedback connections for two different functions.

The *average latency* of a latency cycle is obtained by dividing the sum of all latencies by the number of latencies along the cycle. The latency cycle (1,8) thus has an average latency of $(1 + 8)/2 = 4.5$. A *constant cycle* is a latency cycle which contains only one latency value. Cycles (3) and (6) in Figs. 6.5b and 6.5c are both constant cycles. The average latency of a constant cycle is simply the latency itself. In the next section, we describe how to obtain these latency cycles systematically.

6.2.2 Collision-Free Scheduling

When scheduling events in a pipeline, the main objective is to obtain the shortest average latency between initiations without causing collisions. In what follows, we present a systematic method for achieving such collision-free scheduling.

We study below *collision vectors, state diagrams, single cycles, greedy cycles,* and *minimal average latency* (MAL). These pipeline design theory were originally developed by Davidson (1971) and his students.

Collision Vectors   By examining the reservation table, one can distinguish the set of permissible latencies from the set of forbidden latencies. For a reservation table with n columns, the *maximum forbidden latency* $m \leq n - 1$. The permissible latency p should be as small as possible. The choice is made in the range $1 \leq p \leq m - 1$.

A permissible latency of $p = 1$ corresponds to the ideal case. In theory, a latency of 1 can always be achieved in a static pipeline which follows a linear (diagonal or streamlined) reservation table as shown in Fig. 6.1c.

The combined set of permissible and forbidden latencies can be easily displayed by a collision vector, which is an m-bit binary vector $C = (C_m C_{m-1} \cdots C_2 C_1)$. The value of $C_i = 1$ if latency i causes a collision and $C_i = 0$ if latency i is permissible. Note that it is always true that $C_m = 1$, corresponding to the maximum forbidden latency.

For the two reservation tables in Fig. 6.3, the collision vector $C_X = (1011010)$ is obtained for function X, and $C_Y = (1010)$ for function Y. From C_X, we can immediately tell that latencies $7, 5, 4$, and 2 are forbidden and latencies $6, 3$, and 1 are permissible. Similarly, 4 and 2 are forbidden latencies and 3 and 1 are permissible latencies for function Y.

State Diagrams   From the above collision vector, one can construct a *state diagram* specifying the permissible state transitions among successive initiations. The collision vector, like C_X above, corresponds to the *initial state* of the pipeline at time 1 and thus is called an *initial collision vector*. Let p be a permissible latency within the range $1 \leq p \leq m - 1$.

The *next state* of the pipeline at time $t + p$ is obtained with the assistance of an m-bit right shift register as in Fig. 6.6a. The initial collision vector C is initially loaded into the register. The register is then shifted to the right. Each 1-bit shift corresponds to an increase in the latency by 1. When a 0 bit emerges from the right end after p shifts, it means p is a permissible latency. Likewise, a 1 bit being shifted out means a collision, and thus the corresponding latency should be forbidden.

Logical 0 enters from the left end of the shift register. The next state after p shifts is thus obtained by bitwise-ORing the initial collision vector with the shifted register contents. For example, from the initial state $C_X = (1011010)$, the next state (1111111) is reached after one right shift of the register, and the next state (1011011) is reached after three shifts or six shifts.

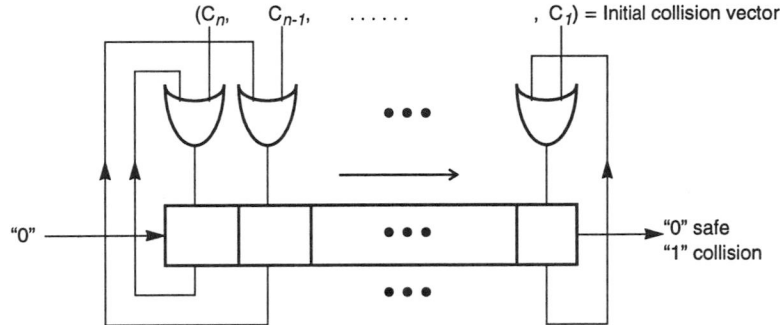

(a) State transition using an n-bit right shift register, where n is the maximum forbidden latency

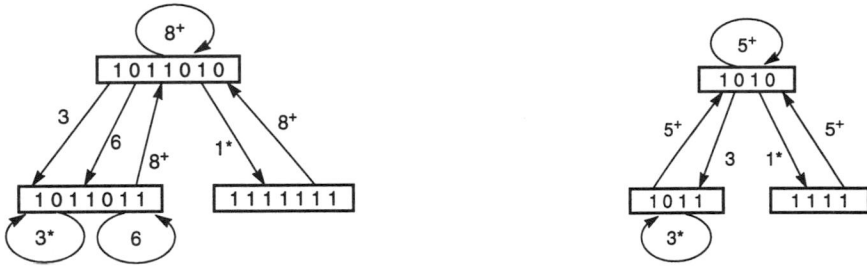

(b) State diagram for function X   (c) State diagram for function Y

Figure 6.6 Two state diagrams obtained from the two reservation tables in Fig. 6.3, respectively.

Example 6.2 The state transition diagram for a pipeline unit

A *state diagram* is obtained in Fig. 6.6b for function X. From the initial state (1011010), only three outgoing transitions are possible, corresponding to the three permissible latencies 6, 3, and 1 in the initial collision vector. Similarly, from state (1011011), one reaches the same state after either three shifts or six shifts.

When the number of shifts is $m+1$ or greater, all transitions are redirected back to the initial state. For example, after eight or more (denoted as 8^+) shifts, the next state must be the initial state, regardless of which state the transition starts from. In Fig. 6.6c, a state diagram is obtained for the reservation table in Fig. 6.3c using a 4-bit shift register. Once the initial collision vector is determined, the corresponding state diagram is uniquely determined. Different reservation tables may result in the same or different initial collision vectors(s).

This implies that even different reservation tables may produce the same state diagram. However, different reservation tables may produce different collision vectors and thus different state diagrams.

■

The 0's and 1's in the present state, say at time t, of a state diagram indicate the permissible and forbidden latencies, respectively, at time t. The bitwise ORing of the shifted version of the present state with the initial collision vector is meant to prevent collisions from future initiations starting at time $t + 1$ and onward.

Thus the state diagram covers all permissible state transitions that avoid collisions. All latencies equal to or greater than m are permissible. This implies that collisions can always be avoided if events are scheduled far apart (with latencies of m^+). However, such long latencies are not tolerable from the viewpoint of pipeline throughput.

Greedy Cycles From the state diagram, we can determine optimal latency cycles which result in the MAL. There are infinitely many latency cycles one can trace from the state diagram. For example, (1, 8), (1, 8, 6, 8), (3), (6), (3, 8), (3, 6, 3) ..., are legitimate cycles traced from the state diagram in Fig. 6.6b. Among these cycles, only *simple cycles* are of interest.

A simple cycle is a latency cycle in which each state appears only once. In the state diagram in Fig. 6.6b, only (3), (6), (8), (1, 8), (3, 8), and (6, 8) are simple cycles. The cycle (1, 8, 6, 8) is not simple because it travels through the state (1011010) twice. Similarly, the cycle (3, 6, 3, 8, 6) is not simple because it repeats the state (1011011) three times.

Some of the simple cycles are *greedy cycles*. A greedy cycle is one whose edges are all made with minimum latencies from their respective starting states. For example, in Fig. 6.6b the cycles (1, 8) and (3) are greedy cycles. Greedy cycles in Fig. 6.6c are (1, 5) and (3). Such cycles must first be simple, and their average latencies must be lower than those of other simple cycles. The greedy cycle (1, 8) in Fig. 6.6b has an average latency of $(1 + 8)/2 = 4.5$, which is lower than that of the simple cycle (6, 8) $= (6 + 8)/2 = 7$. The greedy cycle (3) has a constant latency which equals the MAL for evaluating function X without causing a collision.

The MAL in Fig. 6.6c is 3, corresponding to either of the two greedy cycles. The minimum-latency edges in the state diagrams are marked with asterisks.

At least one of the greedy cycles will lead to the MAL. The collision-free scheduling of pipeline events is thus reduced to finding greedy cycles from the set of simple cycles. The greedy cycle yielding the MAL is the final choice.

6.2.3 Pipeline Schedule Optimization

An optimization technique based on the MAL is given below. The idea is to insert noncompute delay stages into the original pipeline. This will modify the reservation table, resulting in a new collision vector and an improved state diagram. The purpose is to yield an optimal latency cycle, which is absolutely the shortest.

Bounds on the MAL In 1972, Shar determined the following bounds on the *minimal average latency* (MAL) achievable by any control strategy on a statically reconfigured pipeline executing a given reservation table:

(1) The MAL is lower-bounded by the maximum number of checkmarks in any row of the reservation table.

(2) The MAL is lower than or equal to the average latency of any greedy cycle in the state diagram.

(3) The average latency of any greedy cycle is upper-bounded by the number of 1's in the initial collision vector plus 1. This is also an upper bound on the MAL.

Interested readers may refer to Shar (1972) or find proofs of these bounds in Kogge (1981). These results suggest that the optimal latency cycle must be selected from one of the lowest greedy cycles. However, a greedy cycle is not sufficient to guarantee the optimality of the MAL. The lower bound guarantees the optimality. For example, the MAL = 3 for both function X and function Y and has met the lower bound of 3 from their respective reservation tables.

From Fig. 6.6b, the upper bound on the MAL for function X is equal to $4 + 1 = 5$, a rather loose bound. On the other hand, Fig. 6.6c shows a rather tight upper bound of $2 + 1 = 3$ on the MAL. Therefore, all greedy cycles for function Y lead to the optimal latency value of 3, which cannot be lowered further.

To optimize the MAL, one needs to find the lower bound by modifying the reservation table. The approach is to reduce the maximum number of checkmarks in any row. The modified reservation table must preserve the original function being evaluated. Patel and Davidson (1976) have suggested the use of noncompute delay stages to increase pipeline performance with a shorter MAL. Their technique is described below.

Delay Insertion The purpose of delay insertion is to modify the reservation table, yielding a new collision vector. This leads to a modified state diagram, which may produce greedy cycles meeting the lower bound on the MAL.

Before delay insertion, the three-stage pipeline in Fig. 6.7a is specified by the reservation table in Fig. 6.7b. This table leads to a collision vector $C = (1011)$, corresponding to forbidden latencies 1, 2, and 4. The corresponding state diagram (Fig. 6.7c) contains only one self-reflecting state with a greedy cycle of latency 3 equal to the MAL.

Based on the given reservation table, the maximum number of checkmarks in any row is 2. Therefore, the MAL = 3 so obtained in Fig. 6.7c is not optimal.

Example 6.3 Inserting noncompute delays to reduce the MAL

To insert a noncompute stage D_1 after stage S_3 will delay both X_1 and X_2 operations one cycle beyond time 4. To insert yet another noncompute stage D_2 after the second usage of S_1 will delay the operation X_2 by another cycle.

These delayed operations, as grouped in Fig. 6.7b, result in a new pipeline configuration in Fig. 6.8a. Both delay elements D_1 and D_2 are inserted as extra stages, as shown in Fig. 6.8b with an enlarged reservation table having $3 + 2 = 5$ rows and $5 + 2 = 7$ columns.

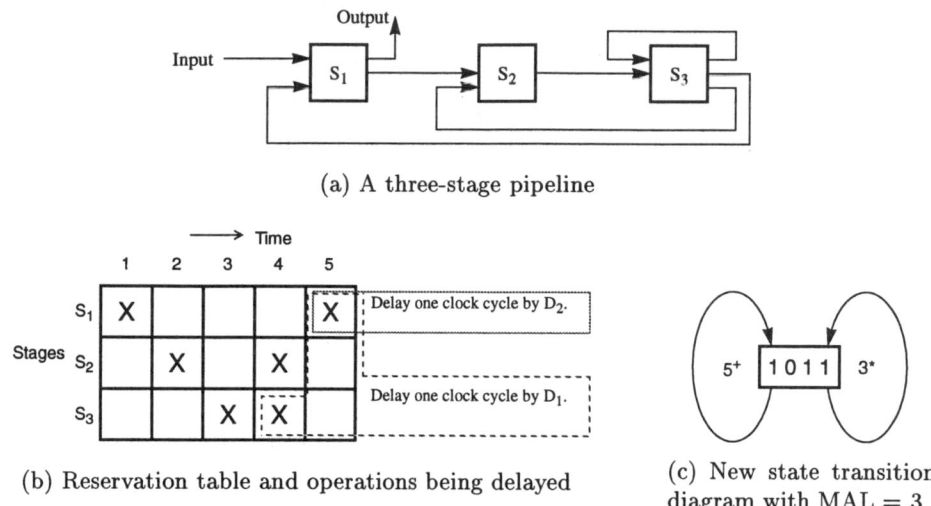

(a) A three-stage pipeline

(b) Reservation table and operations being delayed

(c) New state transition diagram with MAL = 3

Figure 6.7 A pipeline with a minimum average latency of 3.

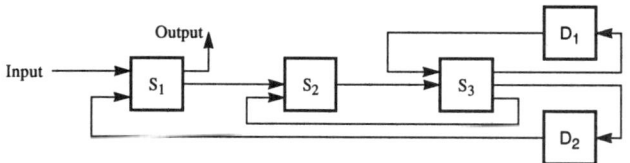

(a) Insertion of two noncompute delay stages

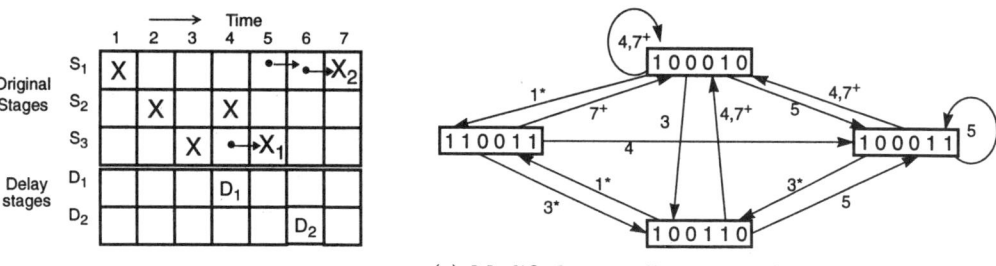

(b) Modified reservation table

(c) Modified state diagram with a reduced MAL = $(1+3)/2 = 2$

Figure 6.8 Insertion of two delay stages to obtain an optimal MAL for the pipeline in Fig. 6.7.

In total, the operation X_1 has been delayed one cycle from time 4 to time 5 and the operation X_2 has been delayed two cycles from time 5 to time 7. All remaining operations (marked as X in Fig. 6.8b) are unchanged. This new table leads to a new collision vector (100010) and a modified state diagram in Fig. 6.8c.

This diagram displays a greedy cycle $(1, 3)$, resulting in a reduced MAL = $(1 + 3)/2 = 2$. The delay insertion thus improves the pipeline performance, yielding a lower bound for the MAL.

∎

Pipeline Throughput　This is essentially the initiation rate or the average number of task initiations per clock cycle. If N tasks are initiated within n pipeline cycles, then the *initiation rate* or *pipeline throughput* is measured as N/n. This rate is determined primarily by the inverse of the MAL adapted. Therefore, the scheduling strategy does affect the pipeline performance.

In general, the shorter the adapted MAL, the higher the throughput that can be expected. The highest achievable throughput is one task initiation per cycle, when the MAL equals 1 since $1 \leq$ MAL $\leq$ the shortest latency of any greedy cycle. Unless the MAL is reduced to 1, the pipeline throughput becomes a fraction.

Pipeline Efficiency　Another important measure is *pipeline efficiency*. The percentage of time that each pipeline stage is used over a sufficiently long series of task initiations is the *stage utilization*. The accumulated rate of all stage utilizations determines the pipeline efficiency.

Let us reexamine latency cycle (3) in Fig. 6.5b. Within each latency cycle of three clock cycles, there are two pipeline stages, S_1 and S_3, which are completely and continuously utilized after time 6. The pipeline stage S_2 is used for two cycles and is idle for one cycle.

Therefore, the entire pipeline can be considered $8/9 = 88.8\%$ efficient for latency cycle (3). On the other hand, the pipeline is only $14/27 = 51.8\%$ efficient for latency cycle $(1, 8)$ and $8/16 = 50\%$ efficient for latency cycle (6), as illustrated in Figs. 6.5a and 6.5c, respectively. Note that none of the three stages is fully utilized with respect to two initiation cycles.

The pipeline throughput and pipeline efficiency are related to each other. Higher throughput results from a shorter latency cycle. Higher efficiency implies less idle time for pipeline stages. The above example demonstrates that higher throughput also accompanies higher efficiency. Other examples may show a contrary conclusion. The relationship between the two measures is a function of the reservation table and of the initiation cycle adopted.

At least one stage of the pipeline should be fully (100%) utilized at the steady state in any acceptable initiation cycle; otherwise, the pipeline capability has not been fully explored. In such cases, the initiation cycle may not be optimal and another initiation cycle should be examined for improvement.

6.3 Instruction Pipeline Design

A stream of instructions can be executed by a pipeline in an overlapped manner. We describe below instruction pipelines for CISC and RISC scalar processors. Topics to be studied include instruction prefetching, data forwarding, hazard avoidance, interlocking for resolving data dependences, dynamic instruction scheduling, and branch handling techniques for improving pipelined processor performance.

6.3.1 Instruction Execution Phases

A typical instruction execution consists of a sequence of operations, including instruction fetch, decode, operand fetch, execute, and write-back phases. These phases are ideal for overlapped execution on a linear pipeline. Each phase may require one or more clock cycles to execute, depending on the instruction type and processor/memory architecture used.

Pipelined Instruction Processing A typical instruction pipeline is depicted in Fig. 6.9. The *fetch stage* (F) fetches instructions from a cache memory, presumably one per cycle.. The *decode stage* (D) reveals the instruction function to be performed and identifies the resources needed. Resources include general-purpose registers, buses, and functional units. The *issue stage* (I) reserves resources. Pipeline control interlocks are maintained at this stage. The operands are also read from registers during the issue stage.

The instructions are executed in one or several *execute stages* (E). Three execute stages are shown in Fig. 6.9a. The last *writeback stage* (W) is used to write results into the registers. Memory load or store operations are treated as part of execution. Figure 6.9 shows the flow of machine instructions through a typical pipeline. These eight instructions are for pipelined execution of the high-level language statements X = Y + Z and A = B × C. Assume *load* and *store* instructions take four execution clock cycles, while floating-point *add* and *multiply* operations take three cycles.

The above timing assumptions represent typical values used in a CISC processor. In many RISC processors, fewer clock cycles are needed. On the other hand, Cray 1 requires 11 cycles for a load and a floating-point addition takes six. With in-order instruction issuing, if an instruction is blocked from issuing due to a data or resource dependence, all instructions following it are blocked.

Figure 6.9b illustrates the issue of instructions following the original program order. The shaded boxes correspond to idle cycles when instruction issues are blocked due to resources latency or conflicts or due to data dependences. The first two *load* instructions issue on consecutive cycles. The *add* is dependent on both *loads* and must wait three cycles before the data (Y and Z) are loaded in.

Similarly, the *store* of the sum to memory location X must wait three cycles for the *add* to finish due to a flow dependence. There are similar blockages during the calculation of A. The total time required is 17 clock cycles. This time is measured beginning at cycle 4 when the first instruction starts execution until cycle 20 the last instruction starts execution. This timing measure eliminates the unduly effects of the

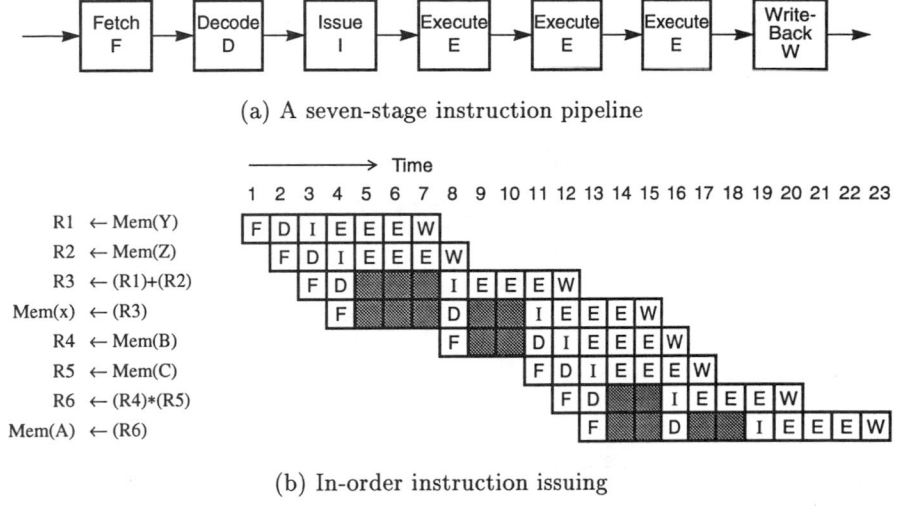

(a) A seven-stage instruction pipeline

(b) In-order instruction issuing

(c) Reordered instruction issuing

Figure 6.9 Pipelined execution of X = Y + Z and A = B × C. (Courtesy of James Smith; reprinted with permission from *IEEE Computer*, July 1989)

pipeline "startup" or "draining" delays.

Figure 6.9c shows an improved timing after the instruction issuing order is changed to eliminate unnecessary delays due to dependence. The idea is to issue all four *load* operations in the beginning. Both the *add* and *multiply* instructions are blocked fewer cycles due to this data prefetching. The reordering should not change the end results. The time required is being reduced to 11 cycles, measured from cycle 4 to cycle 14.

Example 6.4 The MIPS R4000 instruction pipeline (MIPS Computer Systems, 1992)

The R4000 is a superpipelined 64-bit processor using separate instruction and data caches and a higher-speed pipeline for executing register-based instructions. As illustrated in Fig. 6.10, the CPU's eight-stage pipeline design is targeted to achieve an execution rate approaching one instruction per cycle.

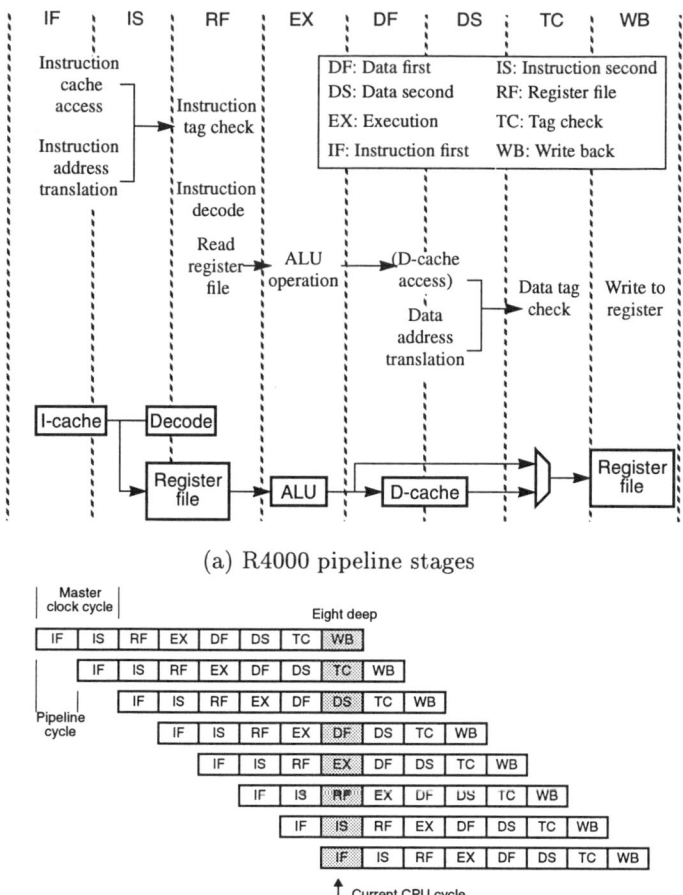

(a) R4000 pipeline stages

(b) R4000 instruction overlapping in pipeline

Figure 6.10 The architecture of the MIPS R4000 instruction pipeline. (Courtesy of MIPS Computer Systems, 1992)

The execution of each R4000 instruction consists of eight major steps as summarized in Fig. 6.10a. Each of these steps requires approximately one clock cycle. Superpipelining has split the instruction and data memory references across two stages. The single-cycle ALU stage takes slightly more time than each of the cache-access stages.

The overlapped execution of successive instructions is shown in Fig. 6.10b. This pipeline operates efficiently because different CPU resources, such as address and bus access, ALU operations, register accesses, and so on, are utilized simultaneously on a noninterfering basis.

The internal pipeline clock rate (100 MHz) of the R4000 is twice the external input or master clock frequency. Figure 6.10b shows the optimal pipeline movement, completing one instruction every internal clock cycle. Load and branch instructions introduce extra delays.

■

6.3.2 Mechanisms for Instruction Pipelining

We introduce instruction buffers and describe the use of cacheing, collision avoidance, multiple functional units, register tagging, and internal forwarding to smooth pipeline flow and to remove bottlenecks and unnecessary memory-access operations.

Prefetch Buffers   Three types of buffers can be used to match the instruction fetch rate to the pipeline consumption rate. In one memory-access time, a block of consecutive instructions are fetched into a prefetch buffer as illustrated in Fig. 6.11. The block access can be achieved using interleaved memory modules or using a cache to shorten the effective memory-access time as demonstrated in the MIPS R4000.

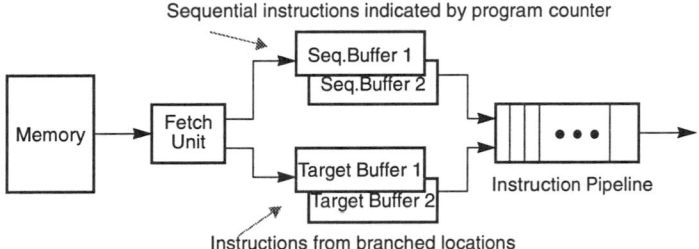

Figure 6.11 The use of sequential and target buffers.

Sequential instructions are loaded into a pair of *sequential buffers* for in-sequence pipelining. Instructions from a branch target are loaded into a pair of *target buffers* for out-of-sequence pipelining. Both buffers operate in a first-in-first-out fashion. These buffers become part of the pipeline as additional stages.

A conditional branch instruction causes both sequential buffers and target buffers to fill with instructions. After the branch condition is checked, appropriate instructions are taken from one of the two buffers, and instructions in the other buffer are discarded. One can use one buffer to load instructions from memory and use another buffer to feed instructions into the pipeline. The two buffers alternate to prevent a collision between instructions flowing into and out of the pipeline.

A third type of prefetch buffer is known as a *loop buffer*. This buffer holds sequential instruction contained in a small loop. The loop buffers are maintained by the fetch stage of the pipeline. Prefetched instructions in the loop body will be executed repeatedly until all iterations complete execution. The loop buffer operates in two steps. First, it contains instructions sequentially ahead of the current instruction. This saves the instruction fetch time from memory. Second, it recognizes when the target of a branch falls within the loop boundary. In this case, unnecessary memory accesses can be avoided if the target instruction is already in the loop buffer. The CDC 6600 and Cray 1 have used loop buffers.

Multiple Functional Units   Sometimes a certain pipeline stage becomes the bottleneck. This stage corresponds to the row with the maximum number of checkmarks in the reservation table. This bottleneck problem can be alleviated by using multiple copies of the same stage simultaneously. This leads to the use of multiple execution units in a pipelined processor design (Fig. 6.12).

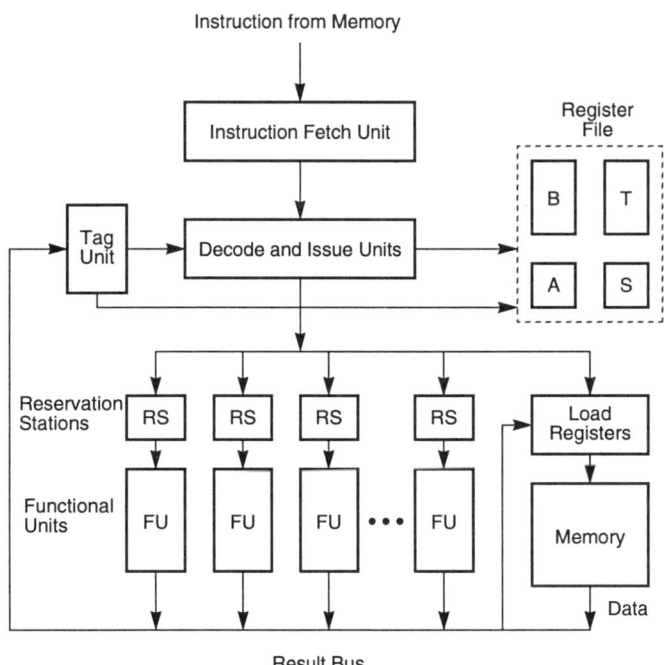

Figure 6.12 A pipelined processor with multiple functional units and distributed reservation stations supported by tagging. (Courtesy of G. Sohi; reprinted with permission from *IEEE Transactions on Computers*, March 1990)

Sohi (1990) used a model architecture for a pipelined scalar processor containing multiple functional units (Fig. 6.12). In order to resolve data or resources dependences among the successive instructions entering the pipeline, the *reservation stations* (RS) are used with each functional unit. Operands can wait in the RS until its data dependences have been resolved. Each RS is uniquely identified by a *tag*, which is monitored by a *tag unit*.

The tag unit keeps checking the tags from all currently used registers or RSs. This register tagging technique allows the hardware to resolve conflicts between source and destination registers assigned for multiple instructions. Besides resolving conflicts, the RSs also serve as buffers to interface the pipelined functional units with the decode and issue units. The multiple functional units are supposed to operate in parallel, once the dependences are resolved. This will alleviate the bottleneck in the execution stages of the instruction pipeline.

Internal Data Forwarding   The throughput of a pipelined processor can be further improved with internal data forwarding among multiple functional units. First, some memory-access operations can be replaced by register transfer operations. The idea is described in Fig. 6.13.

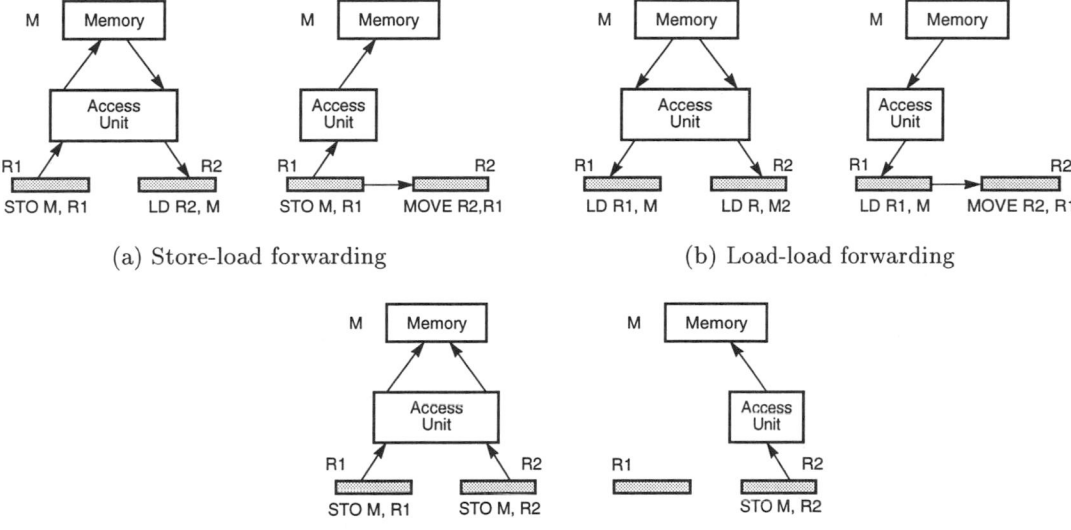

(a) Store-load forwarding   (b) Load-load forwarding

(c) Store-store forwarding due to overlapping

Figure 6.13 Internal data forwarding by replacing memory-access operations with register transfer operations.

A *store-load forwarding* is shown in Fig. 6.13a in which the *load operation* (LD R2, M) from memory to register R2 can be replaced by the *move* operation (MOVE R2, R1) from register R1 to register R2. Since register transfer is faster than memory access, this data forwarding will reduce memory traffic and thus results in a shorter execution time.

Similarly, *load-load forwarding* (Fig. 6.13b) eliminates the second *load* operation (LD R2, M) and replaces it with the *move* operation (MOVE R2, R1). In Fig. 6.13c a *store-store forwarding* is shown. The two *stores* are executed immediately, one after another. Therefore, the second *store* overwrites the first *store*. The first *store* becomes redundant and thus can be eliminated without affecting the outcome. The general guideline in these forwarding techniques is to eliminate unnecessary memory-access (for either *load* or *store*) operations.

Example 6.5 Implementing the dot-product operation with internal data forwarding between a multiply unit and an add unit

One can feed the output of a multiplier directly to the input of an adder (Fig. 6.14) for implementing the following dot-product operation:

$$s = \sum_{i=1}^{n} a_i \times b_i \qquad (6.10)$$

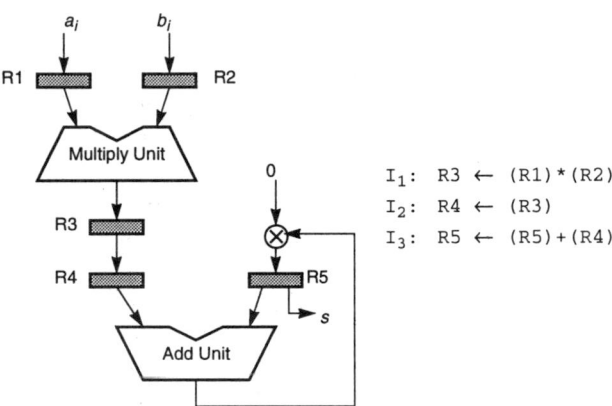

(a) Without data forwarding

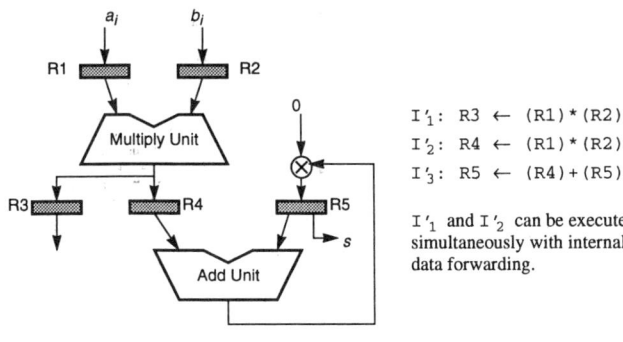

(b) With internal data forwarding

Figure 6.14 Internal data forwarding for implementing the dot-product operation.

Without internal data forwarding between the two functional units, the three instructions must be sequentially executed in a looping structure (Fig. 6.14a). With data forwarding, the output of the multiplier is fed directly into the input register R4 of the adder (Fig. 6.14b). At the same time, the output of the multiplier is also routed to register R3. Internal data forwarding between the two functional units thus reduces the total execution time through the pipelined processor.

∎

Hazard Avoidance The *read* and *write* of shared variables by different instructions in a pipeline may lead to different results if these instructions are executed out of order. As illustrated in Fig. 6.15, three types of logic *hazards* are possible.

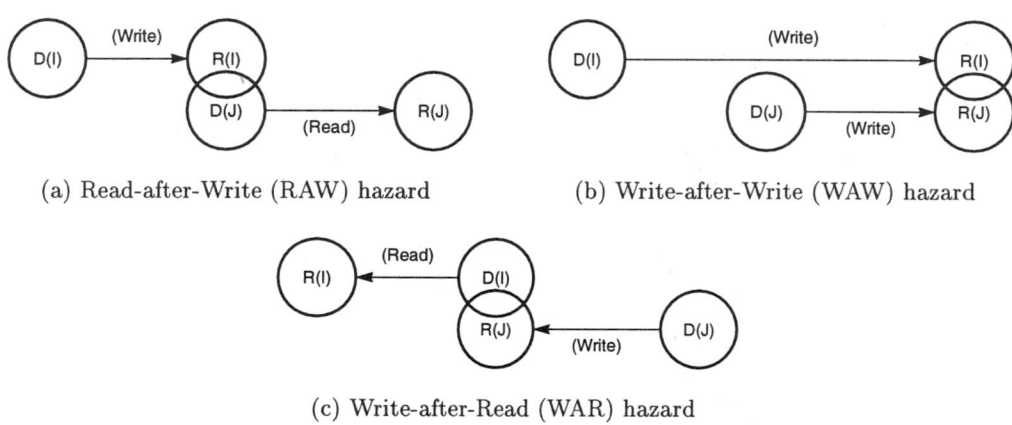

(a) Read-after-Write (RAW) hazard   (b) Write-after-Write (WAW) hazard

(c) Write-after-Read (WAR) hazard

Figure 6.15 Possible hazards between read and write operations in an instruction pipeline (instruction I is ahead of instruction J in program order).

Consider two instructions I and J. Instruction J is assumed to logically follow instruction I according to program order. If the actual execution order of these two instructions violates the program order, incorrect results are read or written, thereby producing hazards.

Hazards should be prevented before these instructions enter the pipeline, such as by holding instruction J until the dependence on instruction I is resolved. We use the notation D(I) and R(I) for the *domain* and *range* of an instruction I.

The domain contains the *input set* (such as operands in registers or in memory) to be used by instruction I. The range corresponds to the *output set* of instruction I. Listed below are conditions under which possible hazards can occur:

$$
\begin{aligned}
R(I) \cap D(J) &\neq \emptyset \text{ for RAW hazard} \\
R(I) \cap R(J) &\neq \emptyset \text{ for WAW hazard} \\
D(I) \cap R(J) &\neq \emptyset \text{ for WAR hazard}
\end{aligned}
\tag{6.11}
$$

These conditions are necessary but not sufficient. This means the hazard may not appear even if one or more of the conditions exist. The RAW hazard corresponds to the flow dependence, WAR to the antidependence, and WAW to the output dependence introduced in Section 2.1. The occurrence of a logic hazard depends on the order in which the two instructions are executed. As long as the order is right, the hazard will not occur.

The resolution of hazard conditions can be checked by special hardware while instructions are being loaded into the prefetch buffer. A special *tag bit* can be used with each operand register to indicate safe or hazard-prone. Successive *read* or *write* operations are allowed to set or reset the tag bit to avoid hazards.

6.3.3   Dynamic Instruction Scheduling

In this section, we described three methods for scheduling instructions through an instruction pipeline. The *static scheduling* scheme is supported by an optimizing compiler. *Dynamic scheduling* is achieved with Tomasulo's *register-tagging* scheme built in the IBM 360/91, or using the *scoreboarding* scheme built in the CDC 6600 processor.

Static Scheduling   Data dependences in a sequence of instructions create interlocked relationships among them. Interlocking can be resolved through a compiler-based static scheduling approach. A compiler or a postprocessor can be used to increase the separation between interlocked instructions.

Consider the execution of the following code fragment. The *multiply* instruction cannot be initiated until the preceding *load* is complete. This data dependence will stall the pipeline for three clock cycles since the two *loads* overlap by one cycle.

Stage delay:	Instruction:		
2 cycles	Add	R0, R1	/ R0 ← (R0)+(R1) /
1 cycle	Move	R1, R5	/ R1 ← (R5) /
2 cycles	Load	R2, M(α)	/ R2 ← (Memory (α)) /
2 cycles	Load	R3, M(β)	/ R3 ← (Memory (β)) /
3 cycles	Multiply	R2, R3	/ R2 ← (R2) × (R3) /

The two *loads*, since they are independent of the *add* and *move*, can be moved ahead to increase the spacing between them and the *multiply* instruction. The following program is obtained after this modification:

Load R2, M(α)	2 to 3 cycles
Load R3, M(β)	2 cycles due to overlapping
Add R0, R1	2 cycles
Move R1, R5	1 cycle
Multiply R2, R3	3 cycles

Through this code rearrangement, the data dependences and program semantics are preserved, and the *multiply* can be initiated without delay. While the operands are being loaded from memory cells α and β into registers R2 and R3, the two instructions *add* and *move* consume three cycles and thus pipeline stalling is avoided.

Most of today's pipelined computers use some form of static scheduling by the compiler. The compiler-based software interlocking is cheaper to implement and flexible to apply. However, in high-performance computers, we need special hardware support to achieve dynamic instruction scheduling as explained below.

Tomasulo's Algorithm   This hardware dependence-resolution scheme was first implemented with multiple floating-point units of the IBM 360/91 processor. The hardware platform was abstracted in Fig. 6.12. For the Model 91 processor, three RSs are used in a floating-point adder and two pairs in a floating-point multiplier. The scheme resolves resource conflicts as well as data dependences using register tagging to allocate or deallocate the source and destination registers.

An issued instruction whose operands are not available is forwarded to an RS associated with the functional unit it will use. It waits until its data dependences have been resolved and its operands become available. The dependence is resolved by monitoring the result bus (called *common data bus* in Model 91). When all operands for an instruction are available, it is dispatched to the functional unit for execution. All working registers are tagged. If a source register is busy when an instruction reaches the issue stage, the tag for the source register is forwarded to an RS. When the register becomes available, the tag can signal the availability.

Example 6.6   Tomasulo's algorithm for dynamic instruction scheduling

Tomasulo's algorithm was applied to work with processors having a few floating-point registers. In the case of Model 91, only four registers were available. Figure 6.16a shows a minimum-register machine code for computing X = Y + Z and A = B × C. The pipeline timing with Tomasulo's algorithm appears in Fig. 6.16b. Here, the total execution time is 13 cycles, counting from cycle 4 to cycle 15 by ignoring the pipeline startup and draining times.

Memory is treated as a special functional unit. When an instruction has completed execution, the result (along with its tag) appears on the result bus. The registers as well as the RSs monitor the result bus and update their contents (and ready/busy bits) when a matching tag is found. Details of the algorithm can be found in the original paper by Tomasulo (1967).

∎

CDC Scoreboarding   The CDC 6600 was an early high-performance computer that used dynamic instruction scheduling hardware. Figure 6.17a shows a CDC 6600-like processor, in which multiple functional units appeared as multiple execution pipelines. Parallel units allow instructions to complete out of the original program order. The processor had instruction buffers for each execution unit. Instructions are issued to available functional units regardless of whether register input data were available.

The instruction's control information would then wait in a buffer for its data to be produced by other instructions. To control the correct routing of data between execution units and registers, the CDC 6600 used a centralized control units known as

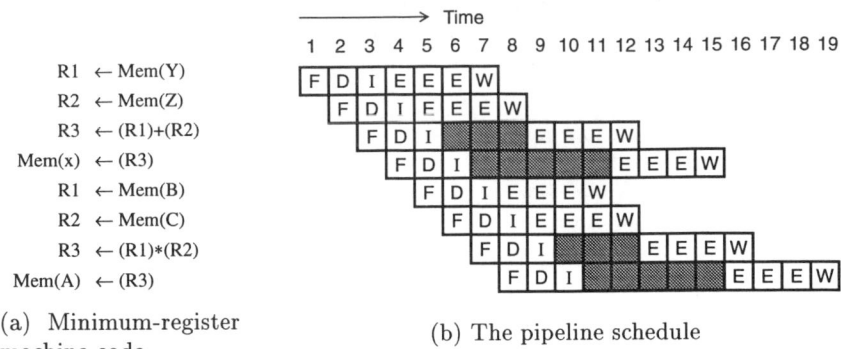

(a) Minimum-register machine code

(b) The pipeline schedule

Figure 6.16 Dynamic instruction scheduling using Tomasulo's algorithm on the processor in Fig. 6.12. (Courtesy of James Smith; reprinted with permission from *IEEE Computer*, July 1989)

the *scoreboard*. This unit kept track of the registers needed by instructions waiting for the various functional units. When all registers had valid data, the scoreboard enabled the instruction execution. Similarly, when a functional unit finished, it signaled the scoreboard to release the resources.

Example 6.7 Pipelined operations using hardware scoreboarding on the CDC 6600-like processors (James Smith, 1989)

Figure 6.17b shows the pipeline schedule based on scoreboard issue logic. The schedule corresponds to the execution of the same machine code for X = Y + Z and A = B × C. The pipeline latencies are the same as those resulting from Tomasulo's algorithm. The *add* instruction is issued to its functional unit before its registers are ready. It then waits for its input register operands.

The scoreboard routes the register values to the adder unit when they become available. In the meantime, the issue stage is not blocked, so other instructions can bypass the blocked add. Performance is thus improved in a way similar to the static software interlocking. It takes 13 clock cycles to perform the operations. Details of the CDC scoreboarding can be found in the book by Thornton (1970). ■

The scoreboard is a centralized control logic which keeps track of the status of registers and multiple functional units. When functional units generate new results, some data dependences can be resolved and thus a higher degree of parallelism can be explored with scoreboarding. Scoreboarding in modern microprocessors like MC88000 uses forwarding logic and register tagging. In a way, scoreboarding implements a kind of *data-driven* mechanism to achieve dataflow computations, as described in Section 2.3.

Dynamic instruction scheduling was implemented only in high-end mainframes or supercomputers in the past. Most microprocessors used static scheduling. But the

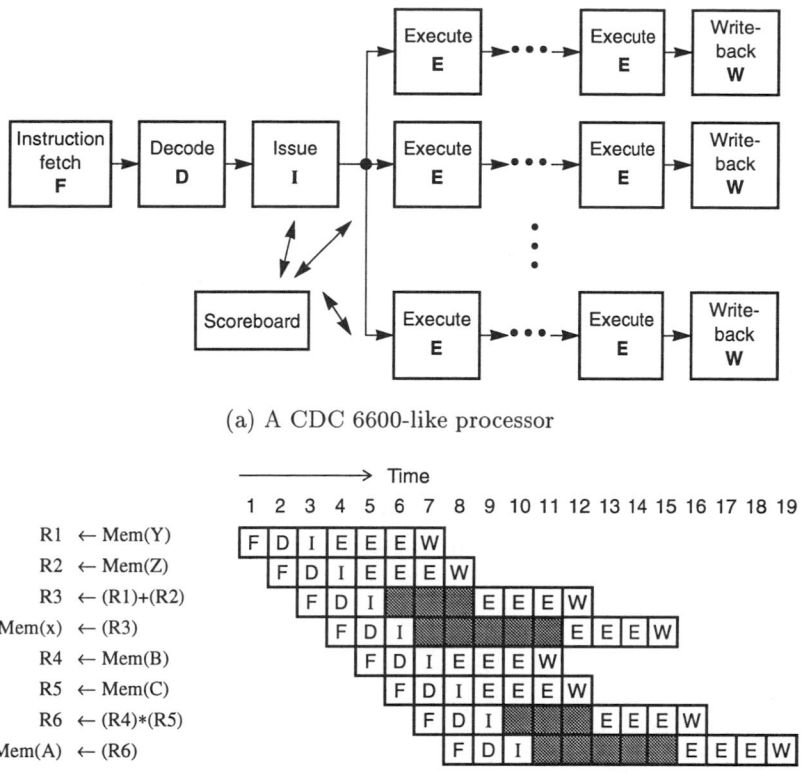

(a) A CDC 6600-like processor

(b) The improved schedule from Fig. 6.9b

Figure 6.17 Hardware scoreboarding for dynamic instruction scheduling. (Courtesy of James Smith; reprinted with permission from *IEEE Computer*, July 1989)

trend is gradually changing. More and more RISC and superscalar processors are built with hardware support of dynamic scheduling at runtime. However, dynamic and static instruction scheduling can be used jointly with two stages of optimization. Toward this goal, processor and compiler designers have to work together to achieve an efficient design. Significant trace-driven data are needed to optimize the pipelined processor design. Multiple-issue instruction pipelines, which are much more complicated than single-issue instruction pipelines, will be studied in Section 6.5.

6.3.4 Branch Handling Techniques

The performance of pipelined processors is limited by data dependences and branch instructions. In previous sections, we have studied the effects of data dependence. Various instruction issuing logic and resource monitoring schemes were described. In this subsection, we study the effects of branching. Various branching strategies are reviewed. The evaluation of branching strategies can be performed either on specific pipeline architecture using trace data, or by applying analytic models. We provide below a simple performance analysis. For an in-depth treatment of the subject, readers are referred to the book *Branch Strategy Taxonomy and Performance Models* by Harvey Cragon (1992).

Effect of Branching Three basic terms are introduced below for the analysis of branching effects: The action of fetching a nonsequential or remote instruction after a branch instruction is called *branch taken*. The instruction to be executed after a branch taken is called a *branch target*. The number of pipeline cycles wasted between a branch taken and its branch target is called the *delay slot*, denoted by b. In general, $0 \leq b \leq k - 1$, where k is the number of pipeline stages.

When a branch taken occurs, all the instructions following the branch in the pipeline become useless and will be drained from the pipeline. This implies that a branch taken causes the pipeline to be flushed, losing a number of useful cycles.

These terms are illustrated in Fig. 6.18, where a branch taken causes I_{b+1} through I_{b+k-1} to be drained from the pipeline. Let p be the probability of a conditional branch instruction in a typical instruction stream and q the probability of a successfully executed conditional branch instruction (a branch taken). Typical values of $p = 20\%$ and $q = 60\%$ have been observed in some programs.

The penalty paid by branching is equal to $pqnb\tau$ because each branch taken costs $b\tau$ extra pipeline cycles. Based on Eq. 6.4, we thus obtain the total execution time of n instructions, including the effect of branching, as follows:

$$T_{eff} = k\tau + (n - 1)\tau + pqnb\tau$$

Modifying Eq. 6.9, we define the following *effective pipeline throughput* with the influence of branching:

$$H_{eff} = \frac{n}{T_{eff}} = \frac{nf}{k + n - 1 + pqnb} \tag{6.12}$$

When $n \to \infty$, the tightest upper bound on the effective pipeline throughput is obtained when $b = k - 1$:

$$H_{eff}^* \doteq \frac{f}{pq(k - 1) + 1} \tag{6.13}$$

When $p = q = 0$ (no branching), the above bound approaches the maximum throughput $f = 1/\tau$, same as in Eq. 6.2. Suppose $p = 0.2, q = 0.6$, and $b = k - 1 = 7$. We define the following *performance degradation factor*:

$$D = \frac{f - H_{eff}^*}{f} = 1 - \frac{1}{pq(k - 1) + 1} = \frac{pq(k - 1)}{pq(k - 1) + 1} = \frac{0.84}{1.84} = 0.46 \tag{6.14}$$

The above analysis implies that the pipeline performance can be degraded by 46% with branching when the instruction stream is sufficiently long. This analysis demonstrates the degree of performance degradation caused by branching in an instruction pipeline.

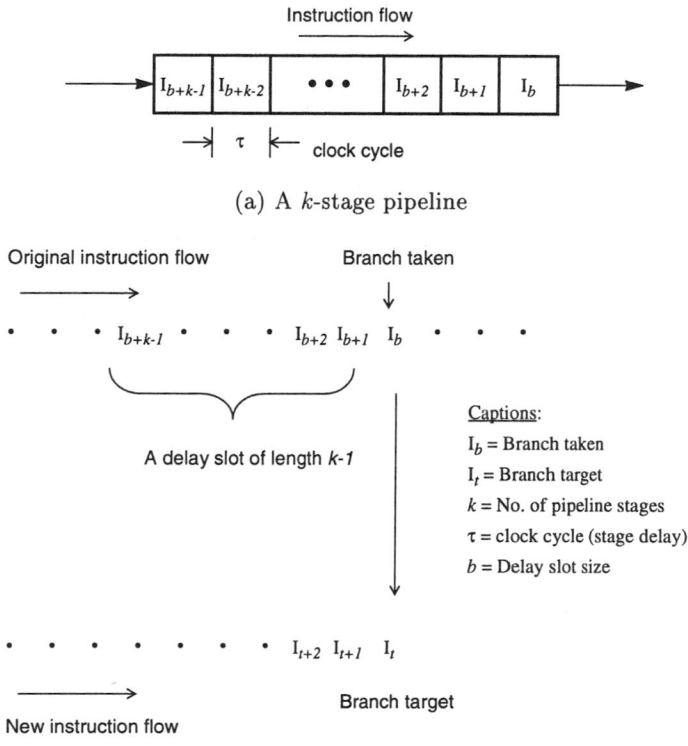

(a) A *k*-stage pipeline

(b) An instruction stream containing a branch taken

Figure 6.18 **The decision of a branch taken at the last stage of an instruction pipeline causes $b \leq k - 1$ previously loaded instructions to be drained from the pipeline.**

Branch Prediction Branch can be predicted either based on branch code types statically or based on branch history during program execution. The probability of branch with respect to a particular branch instruction type can be used to predict branch. This requires collecting the frequency and probabilities of branch taken and branch types across a large number of program traces. Such a *static branch strategy* may not be always accurate.

The static prediction direction (*taken* or *not taken*) is usually wired into the processor. According to past experience, the best performance is given by predicting *taken*. This results from the fact that most conditional branch instructions are taken in program execution. The wired-in static prediction cannot be changed once committed to the hardware. However, the scheme can be modified to allow the programmer or compiler to select the direction of each branch on a *semi-static* prediction basis.

A *dynamic branch strategy* uses recent branch history to predict whether or not the branch will be taken next time when it occurs. To be accurate, one may need to use the entire history of the branch to predict the future choice. This is infeasible to implement. Therefore, most dynamic prediction is determined with limited recent history, as illustrated in Fig. 6.19.

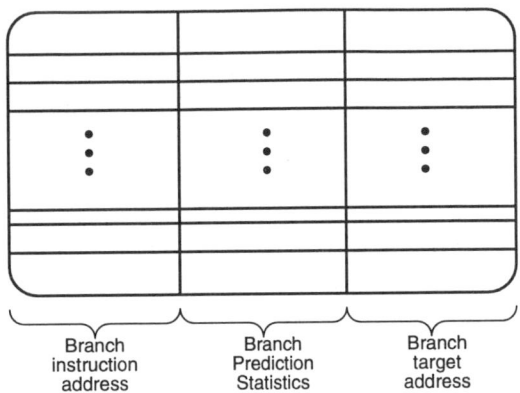

(a) Branch target buffer organization

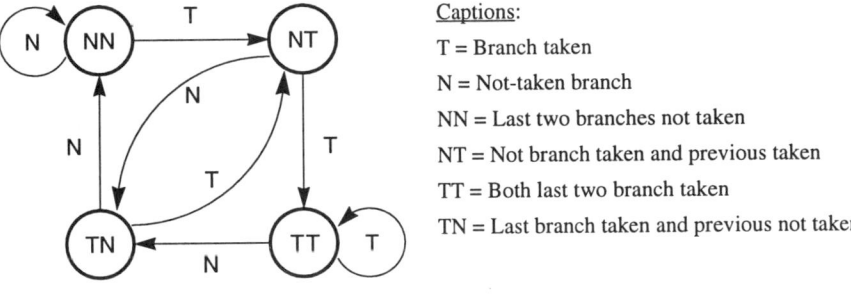

(b) A typical state diagram

Figure 6.19 Branch history buffer and a state transition diagram used in dynamic branch prediction. (Courtesy of Lee and Smith, *IEEE Computer*, 1984)

Cragon (1992) has classified dynamic branch strategies into three major classes: One class predicts the branch direction based upon information found at the decode stage. The second class uses a cache to store target addresses at the stage the effective address of the branch target is computed. The third scheme uses a cache to store target instructions at the fetch stage. All dynamic predictions are adjusted dynamically as a program is executed.

Dynamic prediction demands additional hardware to keep track of the past behavior of the branch instructions at run time. The amount of history recorded should be small. Otherwise, the prediction logic becomes too costly to implement.

Lee and Smith (1984) have shown the use of a *branch target buffer* (BTB) to implement branch prediction (Fig. 6.19a). The BTB is used to hold recent branch information including the address of the branch target used. The address of the branch instruction locates its entry in the BTB.

For example, a state transition diagram (Fig. 6.19b) has been given by Lee and Smith for backtracking the last two branches in a given program. The BTB entry contains the backtracking information which will guide the prediction. Prediction information is updated upon completion of the current branch.

The BTB can be extended to store not only the branch target address but also the target instruction itself and a few of its successor instructions, in order to allow zero delay in converting conditional branches to unconditional branches. The *taken* (T) and *not-taken* (N) labels in the state diagram correspond to actual program behavior. Different programs may use different state diagrams which are modified dynamically according to historical program events.

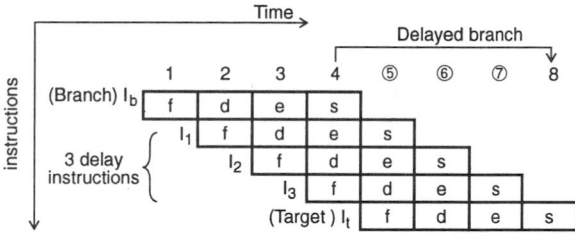

(a) A delayed branch for 2 cycles when the branch condition is resolved at the decode stage

(b) A delayed branch for 3 cycles when the branch condition is resolved at the execute stage

(c) A delayed branch for 4 cycles when the branch condition is resolved at the store stage

Figure 6.20 The concept of delayed branch by moving independent instructions or NOP fillers into the delay slot of a four-stage pipeline.

Delayed Branches   Examining the branch penalty, we realize that the branch penalty would be reduced significantly if the delay slot could be shortened or minimized to a zero penalty. The purpose of delayed branches is to make this possible, as illustrated in Fig. 6.20.

The idea was originally used to reduce the branching penalty in coding microinstructions. A *delayed branch* of d cycles allows at most $d - 1$ useful instructions to be executed following the branch taken. The execution of these instructions should be independent of the outcome of the branch instruction. Otherwise, a zero branching penalty cannot be achieved.

The technique is similar to that used for software interlocking. NOPs can be used as fillers if needed. The probability of moving one instruction ($d = 2$ in Fig. 6.20a) into the delay slot is greater than 0.6, that of moving two instructions ($d = 3$ in Fig. 6.20b) is about 0.2, and that of moving three instructions ($d = 4$ in Fig. 6.20c) is less than 0.1, according to some program trace results.

Example 6.8 A delayed branch with code motion into a delay slot

Code motion across branches can be used to achieve a delayed branch, as illustrated in Fig. 6.21. Consider the execution of a code fragment in Fig. 6.21a. The original program is modified by moving the useful instruction I1 into the delay slot after the branch instruction I3. By so doing, instructions I1, I4, and I5 are executed regardless of the branch outcome.

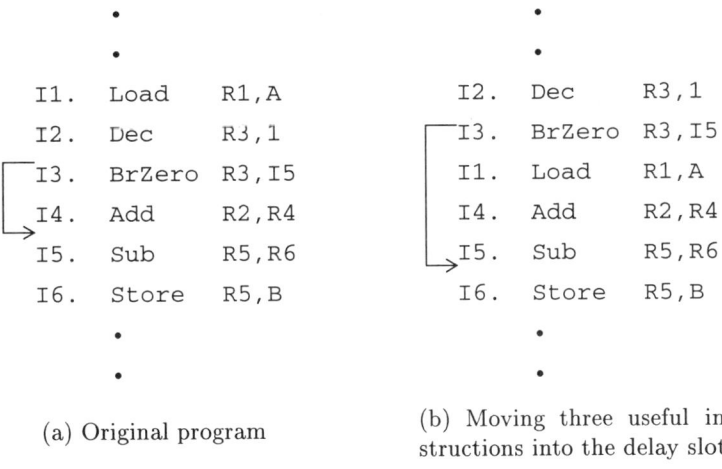

(a) Original program

(b) Moving three useful instructions into the delay slot

Figure 6.21 Code motion across a branch to achieve a delayed branch with a reduced penalty to pipeline performance.

In case the branch is not taken, the execution of the modified program produces the same results as the original program. In case the branch is taken in the modified program, execution of the delayed instructions I1 and I5 is needed anyway.

Only one cycle was wasted in executing instruction I4, which is not needed. Therefore, the delay slot has been reduced to zero for a branch not taken and reduced to one for a branch taken in this example.

In general, data dependence between instructions moving across the branch and the remaining instructions being scheduled must be analyzed. Since instructions I1 and I4 are independent of the remaining instructions (I2, I3, I5, and I6), leaving them in the delay slot will not create logic hazards or data dependences.

∎

Sometimes NOP fillers can be inserted in the delay slot if no useful instructions can be found. However, inserting NOP fillers does not save any cycles in the delayed branch operation. From the above analysis, one sees that delayed branching is effective in short instruction pipelines with about four stages. Delayed branching has been built into most RISC processors, including the MIPS R4000 and Motorola MC88110.

6.4   Arithmetic Pipeline Design

Pipelining techniques can be applied to speed up numerical arithmetic computations. We start with a review of arithmetic principles and standards. Then we consider arithmetic pipelines with fixed functions.

A fixed-point multiply pipeline design and the MC68040 floating-point unit are used as examples to illustrate the design techniques involved. A multifunction arithmetic pipeline is studied with the TI-ASC arithmetic processor as an example.

6.4.1   Computer Arithmetic Principles

In a digital computer, arithmetic is performed with *finite precision* due to the use of fixed-size memory words or registers. Fixed-point or integer arithmetic offers a fixed range of numbers that can be operated upon. Floating-point arithmetic operates over a much increased dynamic range of numbers.

In modern computers, fixed-point and floating-point arithmetic operations are performed by separate hardware. Advanced RISC microprocessors usually have the hardware for both operations on the same processor chip.

Finite precision implies that numbers exceeding the limit must be truncated or rounded to provide a precision within the number of significant bits allowed. In the case of floating-point numbers, exceeding the exponent range means error conditions, called *overflow* or *underflow*. The Institute of Electrical and Electronics Engineers (IEEE) has developed standard formats for 32- and 64-bit floating numbers known as the *IEEE 754 Standard*. This standard has been adopted for most of today's computers.

Fixed-Point Operations   Fixed-point numbers are represented internally in machines in *sign-magnitude, one's complement,* or *two's complement* notation. Most computers use the two's complement notation because of its unique representation of all numbers (including zero). One's complement notation introduces a second zero representation called *dirty zero*.

Add, subtract, multiply, and *divide* are the four primitive arithmetic operations. For fixed-point numbers, the add or subtract of two n-bit integers (or fractions) produces an n-bit result with at most one carry-out.

The multiplication of two n-bit numbers produces a $2n$-bit result which may require use of two memory words or two registers to hold the full-precision result.

The division of an n-bit number by another may create an arbitrarily long quotient and a remainder. Only an approximate result is expected in fixed-point division with rounding or truncation. However, one can always expand the precision by using a $2n$-bit dividend and a $2n$-bit divisor to yield a $2n$-bit quotient.

Floating-Point Numbers   A floating-point number X is represented by a pair (m, e), where m is the *mantissa* (or *fraction*) and e is the *exponent* with an implied *base* (or *radix*). The algebraic value is represented as $X = m \times r^e$. The sign of X can be embedded in the mantissa.

Example 6.9   The IEEE 754 floating-point standard

A 32-bit floating-point number is specified in the IEEE 754 Standard as follows:

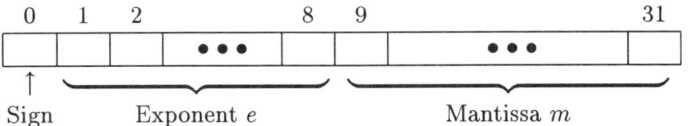

A binary base is assumed with $r = 2$. The 8-bit exponent e field uses an *excess-127* code. The dynamic range of e is $(-127, 128)$, internally represented as $(0, 255)$. The sign s and the 23-bit mantissa field m form a 25-bit sign-magnitude fraction, including an implicit or "hidden" 1 bit to the left of the binary point. Thus the complete mantissa actually represents the value $1.m$.

This hidden bit is not stored with the number. If $0 < e < 255$, then a nonzero normalized number represents the following algebraic value:

$$X = (-1)^s \times 2^{e-127} \times (1.m) \qquad (6.15)$$

When $e = 255$ and $m \neq 0$, a *not-a-number* (NaN) is represented. NaNs can be caused by dividing a zero by a zero or taking the square root of a negative number, among many other nondeterminate cases. When $e = 255$ and $m = 0$, an infinite number $X = (-1)^s \infty$ is represented. Note that $+\infty$ and $-\infty$ are represented differently.

When $e = 0$ and $m \neq 0$, the number represented is $X = (-1)^s 2^{-126}(0.m)$. When $e = 0$ and $m = 0$, a zero is represented as $X = (-1)^s 0$. Again, $+0$ and -0 are possible.

The 64-bit (double-precision) floating point can be defined similarly using an excess-1023 code in the exponent field and a 52-bit mantissa field. A number which is nonzero, finite, non-NaN, and normalized, has the following value:

$$X = (-1)^s \times 2^{e-1023} \times (1.m) \tag{6.16}$$

Special rules are given in the standard to handle overflow or underflow conditions. Interested readers may check the published IEEE standards for details. ∎

Floating-Point Operations The four primitive arithmetic operations are defined below for a pair of floating-point numbers represented by $X = (m_x, e_x)$ and $Y = (m_y, e_y)$. For clarity, we assume $e_x \leq e_y$ and base $r = 2$.

$$X + Y = (m_x \times 2^{e_x - e_y} + m_y) \times x^{e_y} \tag{6.17}$$

$$X - Y = (m_x \times 2^{e_x - e_y} - m_y) \times x^{e_y} \tag{6.18}$$

$$X \times Y = (m_x \times m_y) \times 2^{e_x + e_y} \tag{6.19}$$

$$X \div Y = (m_x \div m_y) \times 2^{e_x - e_y} \tag{6.20}$$

The above equations clearly identify the number of arithmetic operations involved in each floating-point function. These operations can be divided into two halves: One half is for exponent operations such as comparing their relative magnitudes or adding/subtracting them; the other half is for mantissa operations, including four types of fixed-point operations.

Floating-point units are ideal for pipelined implementation. The two halves of the operations demand almost twice as much hardware as that required in a fixed-point unit. Arithmetic shifting operations are needed for equalizing the two exponents before their mantissas can be added or subtracted.

Shifting a binary fraction m to the right k places corresponds to the weighting $m \times 2^{-k}$, and shifting k places to the left corresponds to $m \times 2^k$. In addition, normalization of a floating-point number also requires left shifts to be performed.

Elementary Functions Elementary functions include trigonometric, exponential, logarithmic, and other transcendental functions. Truncated polynomials or power series can be used to evaluate the elementary functions, such as $\sin x$, $\ln x$, e^x, $\cosh x$, $\tan^{-1} y$, $\sqrt{x}$, x^3, etc.

Some CORDIC (coordinate rotation digital computer) algorithms have been developed to calculate trigonometric functions and to convert between binary and mixed radix number systems. Interested readers may refer to the book by Hwang (1979) for details of computer arithmetic functions and their hardware implementation.

It should be noted that computer arithmetic can be implemented by hardwired random logic circuitry as well as by table lookup using ROMs or RAMs in memory. Frequently used constants and special function values can be easily generated by table lookup. Hashing can provide fast access to these tables.

6.4.2 Static Arithmetic Pipelines

Most of today's arithmetic pipelines are designed to perform fixed functions. These *arithmetic/logic units* (ALUs) perform fixed-point and floating-point operations separately. The fixed-point unit is also called the integer unit. The floating-point unit can be built either as part of the central processor or on a separate coprocessor.

These arithmetic units perform scalar operations involving one pair of operands at a time. The pipelining in scalar arithmetic pipelines is controlled by software loops. Vector arithmetic units can be designed with pipeline hardware directly under firmware or hardwired control.

Scalar and vector arithmetic pipelines differ mainly in the areas of register files and control mechanisms involved. Vector hardware pipelines are often built as add-on options to a scalar processor or as an attached processor driven by a control processor. Both scalar and vector processors are used in modern supercomputers.

Arithmetic Pipeline Stages Depending on the function to be implemented, different pipeline stages in an arithmetic unit require different hardware logic. Since all arithmetic operations (such as *add, subtract, multiply, divide, squaring, square rooting, logarithm*, etc.) can be implemented with the basic add and shifting operations, the core arithmetic stages require some form of hardware to add or to shift.

For example, a typical three-stage floating-point adder includes a first stage for exponent comparison and equalization which is implemented with an integer adder and some shifting logic; a second stage for fraction addition using a high-speed carry lookahead adder; and a third stage for fraction normalization and exponent readjustment using a shifter and another addition logic.

Arithmetic or logical shifts can be easily implemented with *shift registers*. High-speed addition requires either the use of a *carry-propagation adder* (CPA) which adds two numbers and produces an arithmetic sum as shown in Fig. 6.22a, or the use of a *carry-save adder* (CSA) to "add" three input numbers and produce one sum output and a carry output as exemplified in Fig. 6.22b.

In a CPA, the carries generated in successive digits are allowed to propagate from the low end to the high end, using either ripple carry propagation or some carry lookahead technique.

In a CSA, the carries are not allowed to propagate but instead are saved in a carry vector. In general, an n-bit CSA is specified as follows: Let X, Y, and Z be three n-bit input numbers, expressed as $X = (x_{n-1}, x_{n-2}, \ldots, x_1, x_0)$. The CSA performs bitwise operations simultaneously on all columns of digits to produce two n-bit output numbers, denoted as $S^b = (0, S_{n-1}, S_{n-2}, \ldots, S_1, S_0)$ and $C = (C_n, C_{n-1}, \ldots, C_1, 0)$.

Note that the leading bit of the *bitwise sum* S^b is always a 0, and the tail bit of the *carry vector* C is always a 0. The input-output relationships are expressed below:

$$\begin{aligned} S_i &= x_i \oplus y_i \oplus z_i \\ C_{i+1} &= x_i y_i \vee y_i z_i \vee z_i x_i \end{aligned} \qquad (6.21)$$

for $i = 0, 1, 2, \ldots, n-1$, where $\oplus$ is the exclusive OR and $\vee$ is the logical OR operation. Note that the arithmetic sum of three input numbers, i.e., $S = X + Y + Z$, is obtained by adding the two output numbers, i.e., $S = S^b + C$, using a CPA. We use the CPA and CSAs to implement the pipeline stages of a fixed-point multiply unit as follows.

Multiply Pipeline Design Consider the multiplication of two 8-bit integers $A \times B =$

e.g. n=4

$$
\begin{array}{rl}
A = & 1\ 0\ 1\ 1 \\
+)\quad B = & 0\ 1\ 1\ 1 \\
\hline
S = 1\ 0\ 0\ 1\ 0 & = A + B
\end{array}
$$

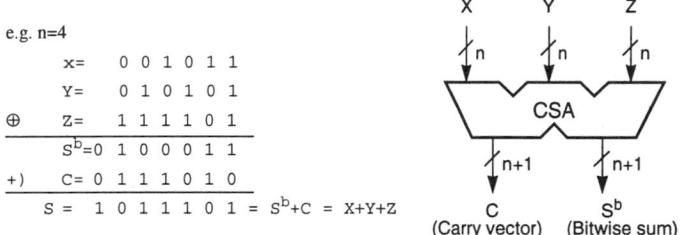

(a) An n-bit carry-propagate adder (CPA) which allows either carry propagation or applies the carry-lookahead technique

e.g. n=4

$$
\begin{array}{rl}
X = & 0\ 0\ 1\ 0\ 1\ 1 \\
Y = & 0\ 1\ 0\ 1\ 0\ 1 \\
\oplus\quad Z = & 1\ 1\ 1\ 1\ 0\ 1 \\
\hline
S^b = 0 & 1\ 0\ 0\ 0\ 1\ 1 \\
+)\quad C = 0 & 1\ 1\ 1\ 0\ 1\ 0 \\
\hline
S = & 1\ 0\ 1\ 1\ 1\ 0\ 1 = S^b + C = X + Y + Z
\end{array}
$$

(b) An n-bit carry-save adder (CSA), where S^b is the bitwise sum of X, Y, and Z, and C is a carry vector generated without carry propagation between digits

Figure 6.22 Distinction between a carry-propagate adder (CPA) and a carry-save adder (CSA).

P, where P is the 16-bit product in double precision. This fixed-point multiplication can be written as the summation of eight partial products as shown below: $P = A \times B = P_0 + P_1 + P_2 + \cdots + P_7$, where $\times$ and $+$ are arithmetic multiply and add operations, respectively.

					1	0	1	1	0	1	0	1	=	A
				$\times)$	1	0	0	1	0	0	1	1	=	B
					1	0	1	1	0	1	0	1	=	P_0
				1	0	1	1	0	1	0	1	0	=	P_1
			0	0	0	0	0	0	0	0	0	0	=	P_2
		0	0	0	0	0	0	0	0	0	0	0	=	P_3
	1	0	1	1	0	1	0	1	0	0	0	0	=	P_4
0	0	0	0	0	0	0	0	0	0	0	0	0	=	P_5
0	0	0	0	0	0	0	0	0	0	0	0	0	=	P_6
$+)$ 1	0	1	1	0	1	0	1	0	0	0	0	0	=	P_7

$$
0\ 1\ 1\ 0\ 0\ 1\ 1\ 1\ 1\ 1\ 1\ 0\ 1\ 1\ 1\ 1 = P
$$

Note that the partial product P_j is obtained by multiplying the multiplicand A by the jth bit of B and then shifting the result j bits to the left for $j = 0, 1, 2, \ldots, 7$. Thus P_j is $(8 + j)$ bits long with j trailing zeros. The summation of the eight partial products is done with a *Wallace tree* of CSAs plus a CPA at the final stage, as shown in Fig. 6.23.

The first stage (S_1) generates all eight partial products, ranging from 8 bits to 15 bits, simultaneously. The second stage (S_2) is made up of two levels of four CSAs, which essentially merges eight numbers into four numbers ranging from 13 to 15 bits. The third stage (S_3) consists of two CSAs, which merges four numbers into two 16-bit numbers. The final stage (S_4) is a CPA, which adds up the last two numbers to produce the final product P.

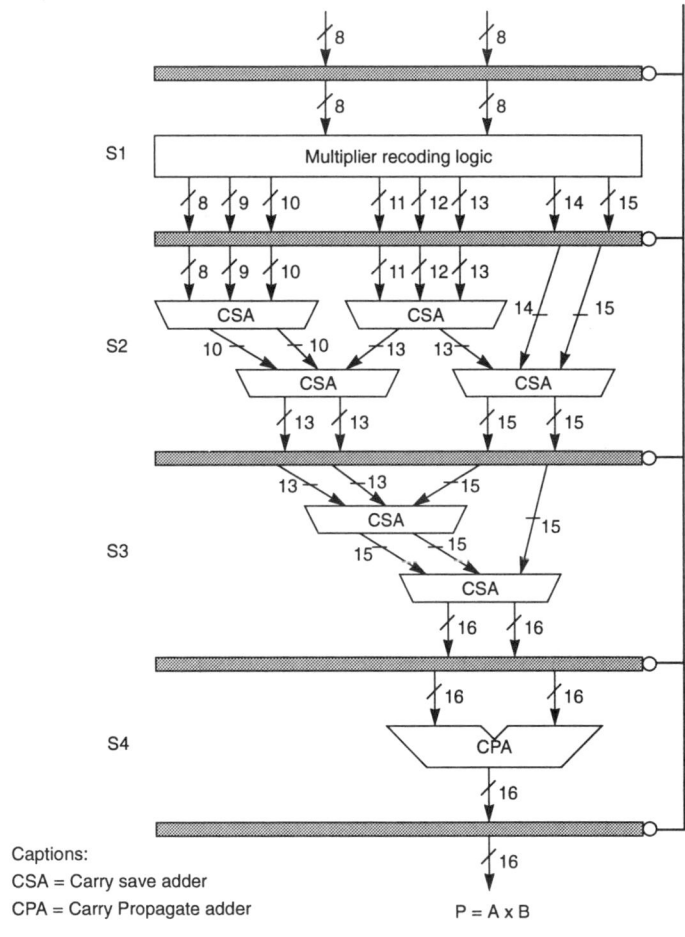

Figure 6.23 A pipeline unit for fixed-point multiplication of 8-bit integers. (The number along each line indicates the line width.)

For a maximum width of 16 bits, the CPA is estimated to need four gate levels of delay. Each level of the CSA can be implemented with a two-gate-level logic. The delay of the first stage (S_1) also involves two gate levels. Thus all the pipeline stages have an approximately equal amount of delay.

The matching of stage delays is crucial to the determination of the number of pipeline stages, as well as the clock period (Eq. 6.1). If the delay of the CPA stage can be further reduced to match that of a single CSA level, then the pipeline can be divided into six stages with a clock rate twice as fast.

Example 6.10 The floating-point unit in the Motorola MC68040

Figure 6.24 shows the design of a pipelined floating-point unit built as an on-chip feature in the Motorola M68040 processor.

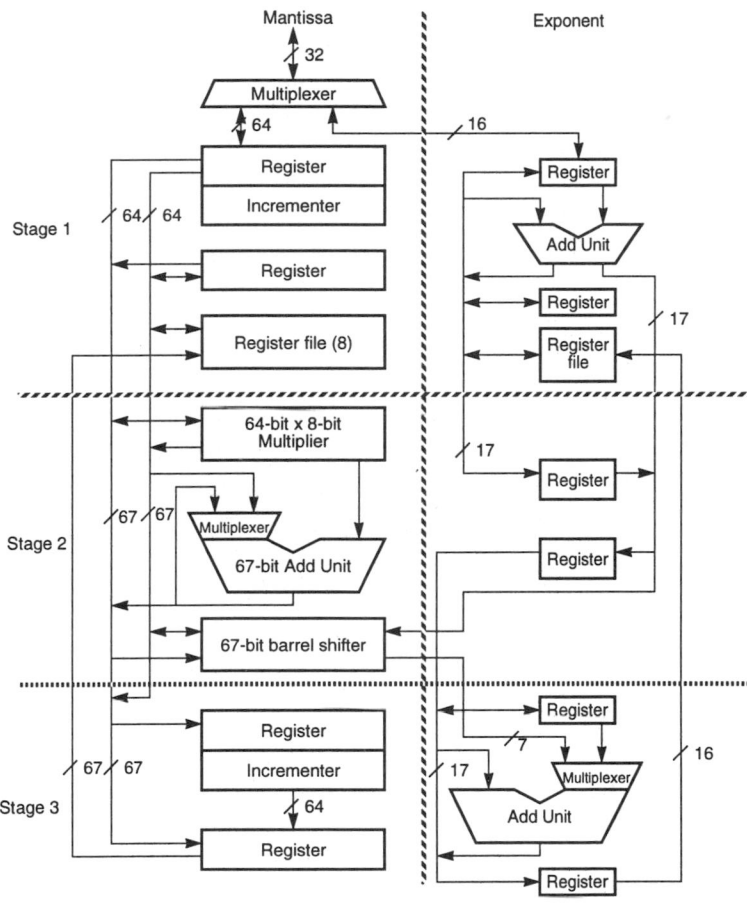

Figure 6.24 Pipelined floating-point unit of the Motorola MC68040 processor. (Courtesy of Motorola, Inc., 1992)

This arithmetic pipeline has three stages. The mantissa section and exponent section are essentially two separate pipelines. The mantissa section can perform floating-point add or multiply operations, either single-precision (32 bits) or double-precision (64 bits).

In the mantissa section, stage 1 receives input operands and returns with computation results; 64-bit registers are used in this stage. Note that all three stages are connected to two 64-bit data buses. Stage 2 contains the array multiplier (64 × 8) which must be repeatedly used to carry out a long multiplication of the two mantissas.

The 67-bit adder performs the addition/subtraction of two mantissas, the barrel shifter is used for normalization. Stage 3 contains registers for holding results before they are loaded into the register file in stage 1 for subsequent use by other instructions.

On the exponent side, a 16-bit bus is used between stages. Stage 1 has an exponent adder for comparing the relative magnitude of two exponents. The result of stage 1 is used to equalize the exponents before mantissa addition can be performed. Therefore, a shift count (from the output of the exponent adder) is sent to the barrel shifter for mantissa alignment.

After normalization of the final result (getting rid of leading zeros), the exponent needs to be readjusted in stage 3 using another adder. The final value of the resulting exponent is fed from the register in stage 3 to the register file in stage 1, ready for subsequent usage.

■

Convergence Division Division can be carried out by repeated multiplications. Mantissa division is carried out by a *convergence method*. This convergence division obtains the quotient $Q = M/D$ of two normalized fractions $0.5 \leq M < D < 1$ in two's complement notation by performing two sequences of chain multiplications as follows:

$$Q = \frac{M \times R_1 \times R_2 \times \cdots \times R_k}{D \times R_1 \times R_2 \times \cdots \times R_k} \tag{6.22}$$

where the successive multipliers

$$R_i = 1 + \delta^{2^{i-1}} = 2 - D^{(i)} \qquad \text{for } i = 1, 2, \ldots, k \quad \text{and} \quad D = 1 - \delta$$

The purpose is to choose R_i such that the denominator $D^{(k)} = D \times R_1 \times R_2 \times \cdots \times R_k \to 1$ for a sufficient number of k iterations, and then the resulting numerator $M \times R_1 \times R_2 \times \cdots \times R_k \to Q$.

Note that the multiplier R_i can be obtained by finding the two's complement of the previous chain product $D^{(i)} = D \times R_1 \times \cdots \times R_{i-1} = 1 - \delta^{2^{i-1}}$ because $2 - D^{(i)} = R_i$. The reason why $D^{(k)} \to 1$ for large k is that

$$
\begin{aligned}
D^{(i)} &= (1-\delta)(1+\delta)(1+\delta^2)(1+\delta^4)\cdots(1+\delta^{2^{i-1}}) \\
&= (1-\delta^2)(1+\delta^2)(1+\delta^4)\cdots(1+\delta^{2^{i-1}}) \\
&= (1-\delta^{2^i}) \qquad \text{for } i = 1, 2, \cdots, k
\end{aligned}
\tag{6.23}
$$

Since $0 < \delta = 1 - D \leq 0.5$, $\delta^{2^i} \to 0$ as i becomes sufficiently large, say, $i = k$ for some k; thus $D^{(k)} = 1 - \delta^{2^k} = 1$ for large k. The end result is

$$Q = M \times (1 + \delta) \times (1 + \delta^2) \times \cdots \times (1 + \delta^{2^{k-1}}) \tag{6.24}$$

The above two sequences of chain multiplications are carried out alternately between the numerator and denominator through the pipeline stages. To summarize, division is carried out by repeated multiplications. Thus divide and multiply can share the same hardware pipeline.

Example 6.11 The IBM 360/Model 91 floating-point unit design

In the history of building scientific computers, the Model 91 was certainly a milestone. Many of the pipeline design features introduced in previous sections were implemented in this machine. Therefore, it is worth the effort to examine the architecture of Model 91. In particular, we describe how floating-point add and multiply/divide operations were implemented in this machine.

As shown in Fig. 6.25, the floating-point execution unit in Model 91 consists of two separate functional pipelines: the *add unit* and the *multiply/divide unit*, which can be used concurrently. The former is a two-stage pipeline, and the latter is a six-stage pipeline. A single-precision floating-point operand has 32 bits, and a double-precision, 64 bits.

The floating-point operation stack is a kind of prefetch buffer holding eight floating-point instructions for subsequent execution through the two functional pipelines. Input operands may come from the storage unit (main memory), and the floating-point buffers are used to hold these operands.

Other operands may come from the floating-point registers which are connected via the common data bus to the output bus. Results from the two functional units can be sent back to the memory via the store data buffers, or they can be routed back to the FLR or to the reservation stations at the input ends.

The add unit allows three pairs of operands to be loaded into three pairs of reservation stations. Only one pair will be used at a time. The other two pairs hold operands for subsequent use. The use of these reservation stations makes the add unit behave like three virtual functional units.

Similarly, the two pairs at the input end of the multiply/divide unit make it behave like two virtual units. Internal data forwarding in Model 91 is accomplished using source tags on all registers and reservation stations. Divide is implemented in Model 91 based on the convergence method.

Every source of an input operand is uniquely identified with a 4-bit tag. Every destination of an input operand has an associated tag register that holds the tag naming the source of data if the destination is busy. Through this *register tagging* technique, operands/results can be directly passed among the virtual functional units. This forwarding can significantly cut down the data flow time between them.

Dynamic scheduling logic has been built into Model 91 using Tomasulo's algorithm to resolve the data dependence problem. Either the add unit or the multi-

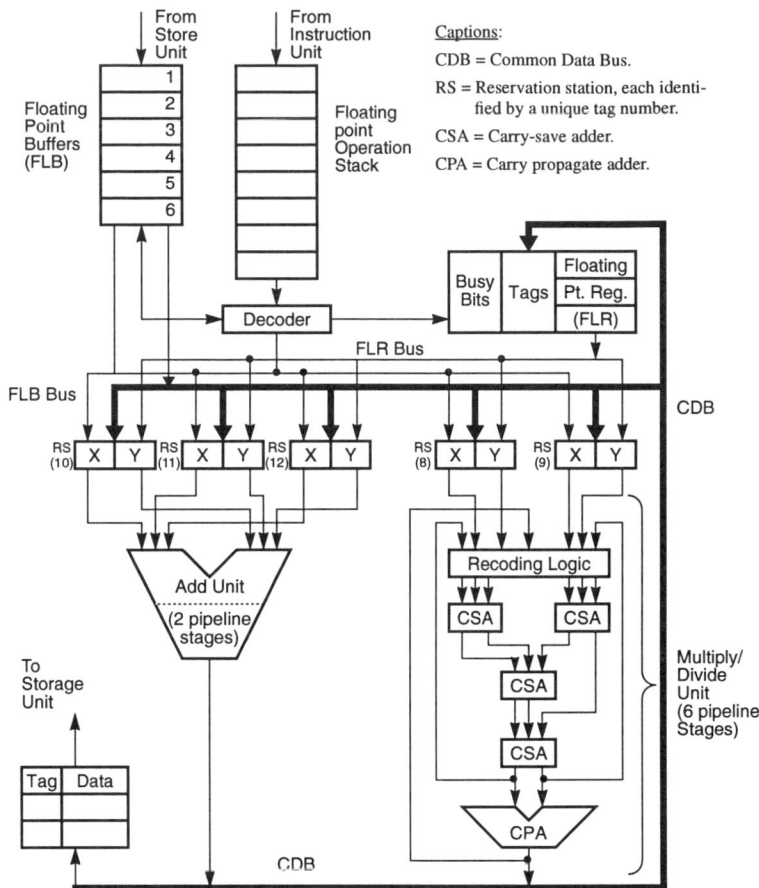

Figure 6.25 The IBM 360 Model 91 floating-point unit. (Courtesy of IBM Corporation, 1967)

ply/divide unit can execute an operation using operands from one pair of reservation stations.

The data dependences are preserved by copying source tags when the sources are busy. When data is generated by a source, it passes its identification and the data onto the common data bus. Awaiting destinations continuously monitor the bus in a tag watch.

When the source tag matches, the destination takes in the data from the bus. Other variations of Tomasulo's algorithm can be made to store the source tags within the destinations, to use a special tag (such as 0000) to indicate nonbusy register/buffers, or to use direct-mapped tags to avoid associative hardware. ∎

Besides the IBM 360/370, the CDC 6600/7600 also implemented this convergence division. It takes two pipeline cycles to perform the floating-point add, six cycles to multiply, and 18 cycles to divide in the IBM System/360 Model 91 due to five iterations involved in the convergence division process.

6.4.3 Multifunctional Arithmetic Pipelines

Static arithmetic pipelines are designed to perform a fixed function and are thus called *unifunctional*. When a pipeline can perform than one function, it is called *multifunctional*. A multifunctional pipeline can be either *static* or *dynamic*. Static pipelines perform one function at a time, but different functions can be performed at different times. A dynamic pipeline allows several functions to be performed simultaneously through the pipeline, as long as there are no conflicts in the shared usage of pipeline stages. In this section, we study a static multifunctional pipeline designed into the TI Advanced Scientific Computer (ASC).

Example 6.12 The TI/ASC arithmetic processor design

There are four pipeline arithmetic units built into the TI-ASC system, as shown in Fig. 6.26. The instruction-processing unit handles the fetching and decoding of instructions. There are a large number of working registers in the processor which also controls the operations of the memory buffer unit and of the arithmetic units.

There are two sets of operand buffers, $\{X, Y, Z\}$ and $\{X', Y', Z'\}$, in each arithmetic unit. X', X, Y', and Y are used for input operands, and Z' and Z are used to output results. Note that intermediate results can be also routed from Z-registers to either X- or Y-registers. Both processors and memory buffers access the main memory for instructions and operands/results, respectively.

Each pipeline arithmetic unit has eight stages as shown in Fig. 6.27a. The PAU is a static multifunction pipeline which can perform only one function at a time. Figure 6.26a shows all the possible interstage connections for performing arithmetic, logical, shifting, and data conversion functions.

Both fixed-point and floating-point arithmetic functions can be performed by this pipeline. The PAU also supports vector in addition to scalar arithmetic operations. It should be noted that different functions require different pipeline stages and different interstage connection patterns.

For example, fixed-point multiplication requires the use of only segments S_1, S_6, S_7, and S_8 as shown in Fig. 6.27b. On the other hand, the floating-point dot product function, which performs the dot product operation between two vectors, requires the use of all segments with the complex connections shown in Fig. 6.27c. This dot product is implemented by essentially the following accumulated summation of a sequence of multiplications through the pipeline:

$$Z \leftarrow A_i \times B_i + Z \tag{6.25}$$

where the successive operands (A_i, B_i) are fed through the X- and Y-buffers, and the accumulated sums through the Z-buffer recursively.

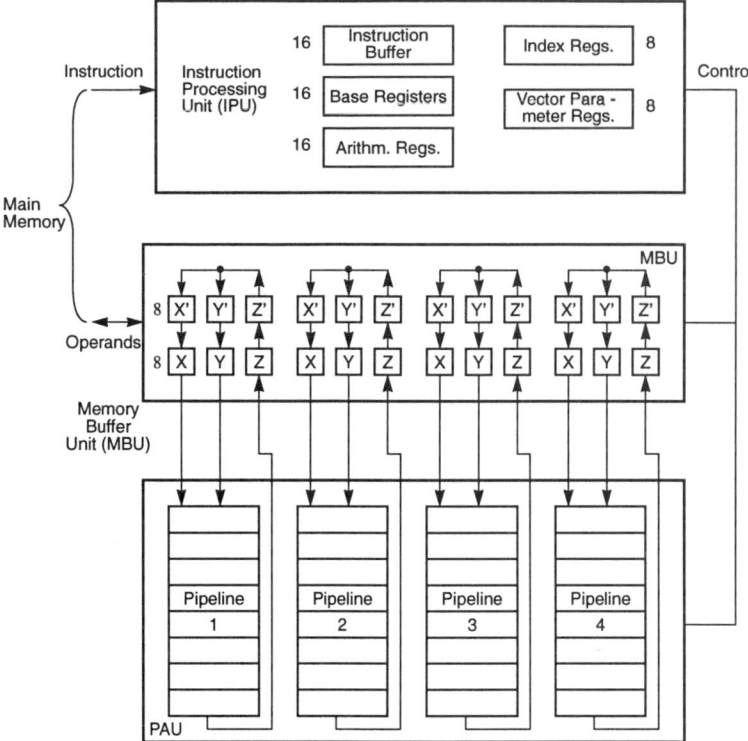

Figure 6.26 The architecture of the TI Advanced Scientific Computer (ASC) (Courtesy of Texas Instruments, Inc.)

The entire pipeline can perform the *multiply* ($\times$) and the *add* ($+$) in a single flow through the pipeline. The two levels of buffer registers will isolate the loading and fetching of operands to or from the PAU, respectively, as in the concept of using a pair in the prefetch buffers described in Fig. 6.11.

Even though the TI-ASC is no longer in production, the system provided a unique design for multifunction arithmetic pipelines. Today, most supercomputers implement arithmetic pipelines with dedicated functions for much simplified control circuitry.

■

6.5   Superscalar and Superpipeline Design

We present two architectural approaches to improving pipeline performance using the base scalar pipeline as a reference machine. There is a duality of symmetry between the two architectures. Practical design constraints are discussed in regard to

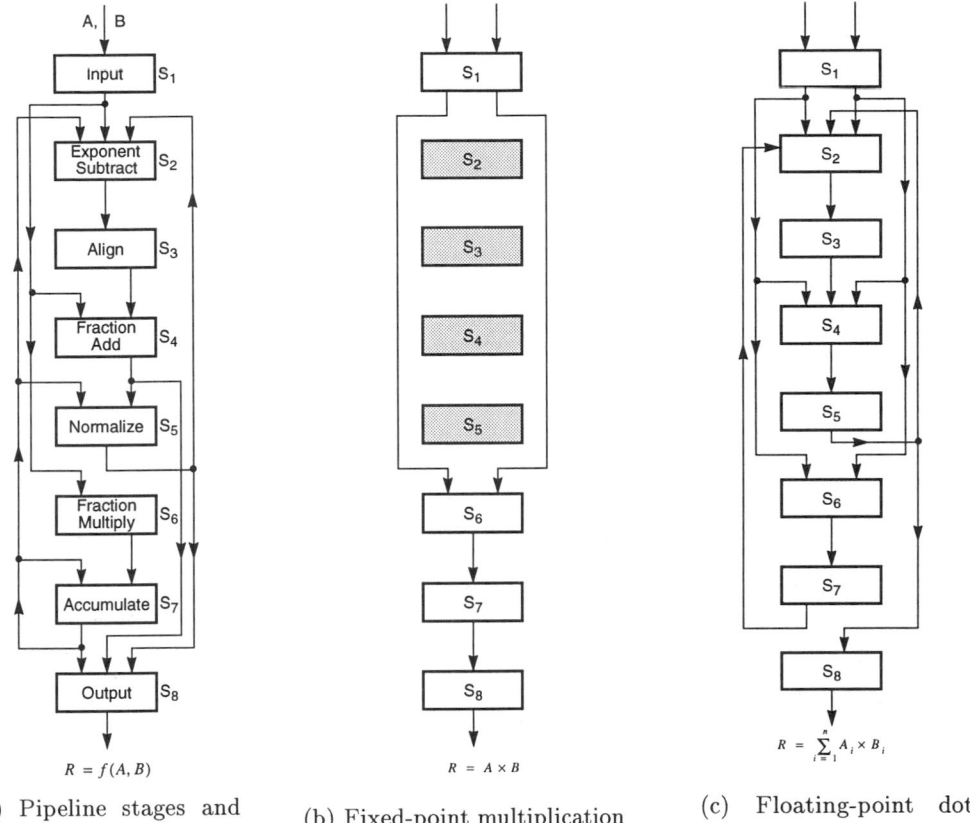

(a) Pipeline stages and interconnections

(b) Fixed-point multiplication

(c) Floating-point dot product

Figure 6.27 The multiplication arithmetic pipeline of the TI Advanced Scientific Computer and the interstage connections of two representative functions. (Shaded stages are unutilized).

implementation technology.

Pipeline Design Parameters   Some parameters used in designing the scalar base machine and superscalar machines are summarized in Table 6.1 for four types of pipeline processors to be studied below. All pipelines discussed are assumed to have k stages.

The machine *pipeline cycle* for the scalar base machine is assumed to be 1 time unit, called the *base cycle*. We defined the instruction *issue rate*, *issue latency*, and *simple operation latency* in Section 4.1.1. The *instruction-level parallelism* (ILP) is the maximum number of instructions that can be simultaneously executed in the pipeline.

For the base machine, all of these parameters have a value of 1. All machine types are designed relative to the base machine. The ILP is needed to fully utilize a given pipeline machine.

Table 6.1 Design Parameters for Pipeline Processors

Machine type	Scalar base machine of k pipeline stages	Superscalar machine of degree m	Superpipelined machine of degree n	Superpipelined superscalar machine of degree (m, n)
Machine pipeline cycle	1 (base cycle)	1	$1/n$	$1/n$
Instruction issue rate	1	m	1	m
Instruction issue latency	1	1	$1/n$	$1/n$
Simple operation latency	1	1	$1(= n \cdot \frac{1}{n})$	$1(= n \cdot \frac{1}{n})$
ILP to fully utilize the pipeline	1	m	n	mn

Note: All timing is relative to the base cycle for the scalar base machine. ILP: Instruction level parallelism.

6.5.1   Superscalar Pipeline Design

We study below the structure of superscalar pipelines, the data dependence problem, the factors causing pipeline stalling, and multiinstruction-issuing mechanisms for achieving parallel pipelining operations. For a superscalar machine of degree m, m instructions are issued per cycle and the ILP should be m in order to fully utilize the pipeline. As a matter of fact, the scalar base machine can be considered a degenerate case of a superscalar machine of degree 1.

Superscalar Pipeline Structure   In an m-issue superscalar processor, the instruction decoding and execution resources are increased to form essentially m pipelines operating concurrently. At some pipeline stages, the functional units may be shared by multiple pipelines.

This resource-shared multiple-pipeline structure is illustrated by a design example in Fig. 6.28a. In this design, the processor can issue two instructions per cycle if there is no resource conflict and no data dependence problem. There are essentially two pipelines in the design. Both pipelines have four processing stages labeled fetch, decode, execute, and store, respectively.

Each pipeline essentially has its own fetch unit, decode unit, and store unit. The two instruction streams flowing through the two pipelines are retrieved from a single source stream (the I-cache). The fan-out from a single instruction stream is subject to resource constraints and a data dependence relationship among the successive instructions.

For simplicity, we assume that each pipeline stage requires one cycle, except the

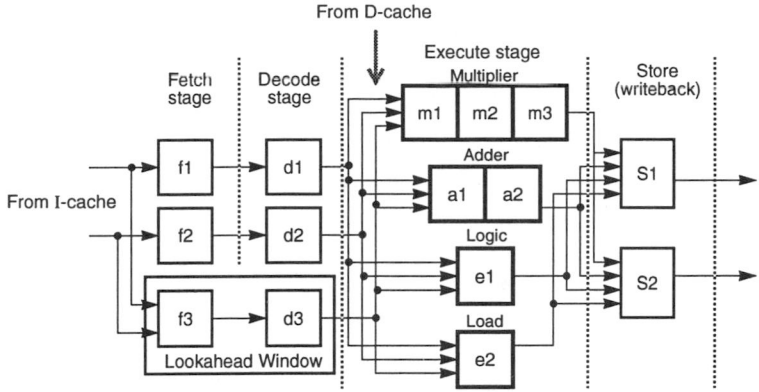

(a) A dual-pipeline, superscalar processor with four functional units in the execution stage and a lookahead window producing out-of-order issues

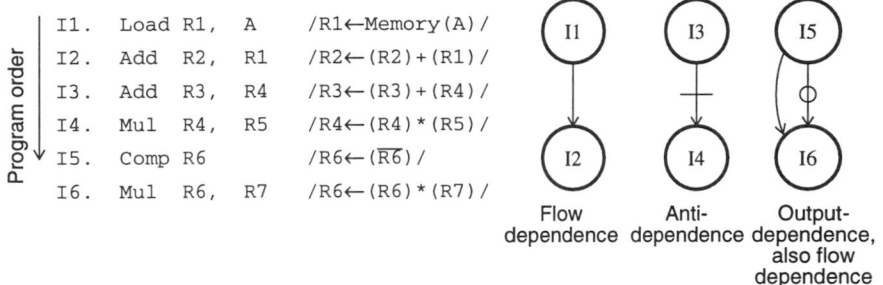

(b) A sample program and its dependence graph, where I2 and I3 share the adder and I4 and I6 share the same multiplier

Figure 6.28 A two-issue superscalar processor and a sample program for parallel execution.

execute stage which may require a variable number of cycles. Four functional units, multiplier, adder, logic unit, and load unit, are available for use in the execute stage. These functional units are shared by the two pipelines on a dynamic basis. The multiplier itself has three pipeline stages, the adder has two stages, and the others each have only one stage.

The two store units (S1 and S2) can be dynamically used by the two pipelines, depending on availability at a particular cycle. There is a *lookahead window* with its own fetch and decoding logic. This window is used for instruction lookahead in case out-of-order instruction issue is desired to achieve better pipeline throughput.

It is difficult to schedule multiple pipelines simultaneously, especially when the instructions are retrieved from the same source. We have to avoid pipeline stalling and minimize pipeline idle time.

Data Dependences Consider the example program in Fig. 6.28b. A dependence
graph is drawn to indicate the relationship among the instructions. Because the register
content in R1 is loaded by I1 and then used by I2, we have flow dependence: I1 → I2.

Because the result in register R4 after executing I4 may affect the operand register
R4 used by I3, we have antidependence: I3 ↦ I4. Since both I5 and I6 modify the
register R6 and R6 supplies an operand for I6, we have both flow and output dependence:
I5 → I6 and I5 ↤↦ I6 as shown in the dependence graph.

To schedule instructions through one or more pipelines, these data dependences
must not be violated. Otherwise, erroneous results may be produced. These data
dependences are detected by a compiler and made available at pipeline scheduling time.

Pipeline Stalling This is a problem which may seriously lower pipeline utilization.
Proper scheduling avoids pipeline stalling. The problem exists in both scalar and superscalar processors. However, it is more serious in a superscalar pipeline. Stalling can be
caused by data dependences or by resource conflicts among instructions already in the
pipeline or about to enter the pipeline. We use an example to illustrate the conditions
causing pipeline stalling.

Consider the scheduling of two instruction pipelines in a two-issue superscalar processor. Figure 6.29a shows the case of no data dependence on the left and flow dependence (I1 → I2) on the right. Without data dependence, all pipeline stages are utilized
without idling.

With dependence, instruction I2 entering the second pipeline must wait for two
cycles (shaded time slots) before entering the execution stages. This delay may also
pass to the next instruction I4 entering the pipeline.

In Fig. 6.29b, we show the effect of branching (instruction I2). A delay slot of four
cycles results from a branch taken by I2 at cycle 5. Therefore, both pipelines must be
flushed before the target instructions I3 and I4 can enter the pipelines from cycle 6.
Here, delayed branch or other amending operations are not taken.

In Fig. 6.29c, we show a combined problem involving both resource conflicts and
data dependence. Instructions I1 and I2 need to use the same functional unit, and
I2 → I4 exists.

The net effect is that I2 must be scheduled one cycle behind because the two
pipeline stages (e_1 and e_2) of the same functional unit must be used by I1 and I2 in an
overlapped fashion. For the same reason, I3 is also delayed by one cycle. Instruction I4
is delayed for two cycles due to the flow dependence on I2. The shaded boxes in all the
timing charts correspond to idle stages.

Multipipeline Scheduling Instruction issue and completion policies are critical to
superscalar processor performance. Three scheduling policies are introduced below.
When instructions are issued in program order, we call it *in-order issue*. When program
order is violated, *out-of-order issue* is being practiced.

Similarly, if the instructions must be completed in program order, it is called *in-
order completion*. Otherwise, *out-of-order completion* may result. In-order issue is easier
to implement but may not yield the optimal performance. In-order issue may result in
either in-order or out-of-order completion.

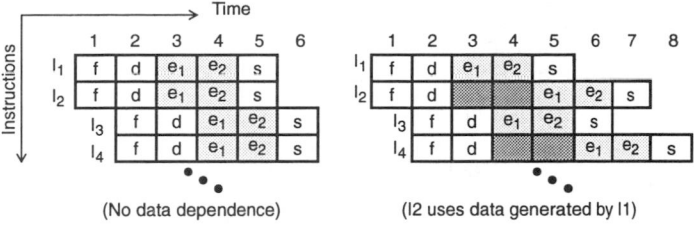

(a) Data dependence stalls the second pipeline in shaded cycles

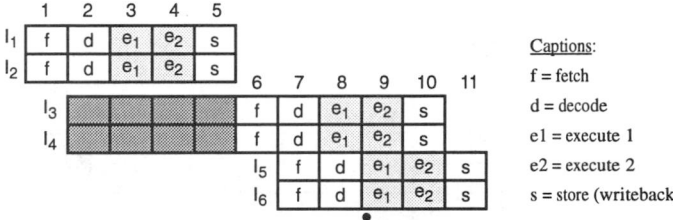

(b) Branch instruction I2 causes a delay slot of length 4 in both pipelines

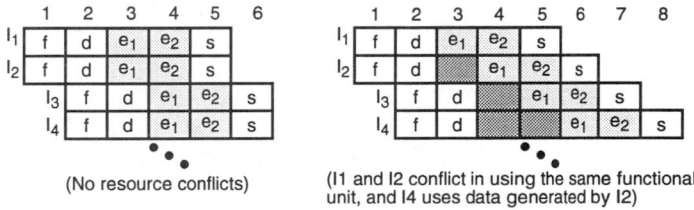

(c) Resource conflicts and data dependences cause the stalling of pipeline operations for some cycles

Figure 6.29 Dependences and resource conflicts may stall one or two pipelines in a two-issue superscalar processor.

Out-of-order issue usually ends up with out-of-order completion. The purpose of out-of-order issue and completion is to improve performance. These three scheduling policies are illustrated in Fig. 6.30 by execution of the example program in Fig. 6.28b on the dual-pipeline hardware in Fig. 6.28a.

It is demonstrated that performance can be improved from an in-order to an out-of-order schedule. The performance is often indicated by the total execution time and the utilization rate of pipeline stages. Not all programs can be scheduled out of order. Data dependence and resource conflicts do impose constraints.

In-Order Issue Figure 6.30a shows a schedule for the six instructions being issued in program order I1, I2, ..., I6. Pipeline 1 receives I1, I3, and I5, and pipeline 2 receives

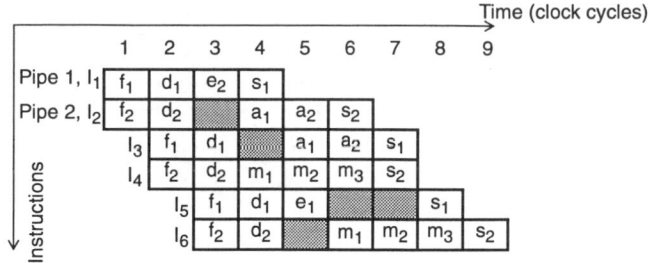

(a) In-order issue with in-order completion in nine cycles

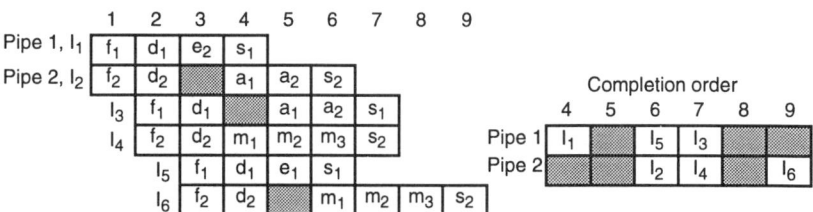

(b) In-order issue and out-of-order completion in nine cycles

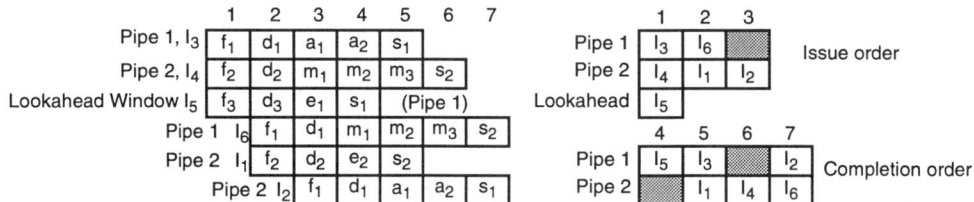

(c) Out-of-order issue and out-of-order completion in seven cycles
using an instruction lookahead window in the recoding process

Figure 6.30 Instruction issue and completion policies for a superscalar processor with and without instruction lookahead support. (Timing charts correspond to parallel execution of the program in Fig. 6.28)

instructions I2, I4, and I6 in three consecutive cycles. Due to I1 → I2, I2 has to wait one cycle to use the data loaded in by I1.

I3 is delayed one cycle for the same adder used by I2. I6 has to wait for the result of I5 before it can enter the multiplier stages. In order to maintain in-order completion, I5 is forced to wait for two cycles to come out of pipeline 1. In total, nine cycles are needed and five idle cycles (shaded boxes) are observed.

In Fig. 6.30b, out-of-order completion is allowed even if in-order issue is practiced. The only difference between this out-of-order schedule and the in-order schedule is that I5 is allowed to complete ahead of I3 and I4, which are totally independent of I5. The total execution time does not improve. However, the pipeline utilization rate does.

Only three idle cycles were observed. Note that in Figs. 6.29a and 6.29b, we did not use the lookahead window. In order to shorten the total execution time, the window can be used to reorder the instruction issues.

Out-of-Order Issue By using the lookahead window, instruction I5 can be decoded in advance because it is independent of all the other instructions. The six instructions are issued in three cycles as shown: I5 is fetched and decoded by the window, while I3 and I4 are decoded concurrently.

It is followed by issuing I6 and I1 at cycle 2, and I2 at cycle 3. Because the issue is out of order, the completion is also out of order as shown in Fig. 6.30c. Now, the total execution time has been reduced to seven cycles with no idle stages during the execution of these six instructions.

The in-order issue and completion is the simplest one to implement. It is rarely used today even in a conventional scalar processor due to some unnecessary delays in maintaining program order. However, in a multiprocessor environment, this policy is still attractive. Allowing out-of-order completion can be found in both scalar and superscalar processors.

Some long-latency operations, such as loads (AMD 29000) and floating-point operations (MC88100), can be hidden in out-of-order completion to achieve a better performance. Output dependence and antidependence are the two relations preventing out-of-order completion. Out-of-order issue gives the processor more freedom to exploit parallelism, and thus pipeline efficiency is enhanced.

The above example clearly demonstrates the advantages of instruction lookahead and of out-of-order issue and completion as far as pipeline optimization is concerned. It should be noted that multiple-pipeline scheduling is an NP-complete problem. Optimal scheduling is very expensive to obtain.

Simple data dependence checking, a small lookahead window, and scoreboarding mechanisms are needed, along with an optimizing compiler, to exploit instruction parallelism in a superscalar processor.

Motorola 88110 Architecture The Motorola 88110 is a superscalar RISC microprocessor. It combines the three-chip set, one CPU (88100) chip and two cache (88200) chips, in a single-chip implementation, with additional improvements. The 88110 employs advanced techniques for exploiting instruction-level parallelism, including instruction issue, out-of-order instruction completion, speculative execution, dynamic instruction rescheduling, and two on-chip caches. The unit also supports demanding graphics and digital signal processing applications.

The 88110 employs a symmetrical superscalar instruction dispatch unit which dispatches two instructions each clock cycle into an array of 10 concurrent units. The design fully implements interlocked pipelines and a precise execution model. It allows out-of-order instruction completion and some out-of-order instruction issue, and branch prediction with speculative execution past branches.

The instruction set of the 88110 extends that of the 88100 in integer and floating-point operations. It added a new set of capabilities to support 3-D, color graphics image rendering. The 88110 has separate, independent instruction and data paths, along with split caches for instructions and data. The instruction cache is 8K-byte, 2-way, set-associative with 128 sets, two blocks for each set, and 32 bytes (8 instructions) per block. The data cache resembles that of the instruction set.

The 88110 employs the MESI cache coherence protocol. A write-invalidate procedure guarantees that one processor on the bus has a modified copy of any cache block at the same time. The 88110 is implemented with 1.3 million transistors in a 299-pin package and driven by a 50-MHz clock. Interested readers may refer to Diefendorff and Allen (1992) for details.

Superscalar Performance   To compare the relative performance of a superscalar processor with that of a scalar base machine, we estimate the ideal execution time of N independent instructions through the pipeline.

The time required by the scalar base machine is

$$T(1,1) = k + N - 1 \quad \text{(base cycles)} \tag{6.26}$$

The ideal execution time required by an m-issue superscalar machine is

$$T(m,1) = k + \frac{N-m}{m} \quad \text{(base cycles)} \tag{6.27}$$

where k is the time required to execute the first m instructions through the m pipelines simultaneously, and the second term corresponds to the time required to execute the remaining $N - m$ instructions, m per cycle, through m pipelines.

The ideal speedup of the superscalar machine over the base machine is

$$S(m,1) = \frac{T(1,1)}{T(m,1)} = \frac{N+k-1}{N/m+k-1} = \frac{m(N+k-1)}{N+m(k-1)} \tag{6.28}$$

As $N \to \infty$, the speedup limit $S(m,1) \to m$, as expected.

6.5.2   Superpipelined Design

In a superpipelined processor of degree n, the pipeline cycle time is $1/n$ of the base cycle. As a comparison, while a fixed-point add takes one cycle in the base scalar processor, the same operation takes n short cycles in a superpipelined processor implemented with the same technology.

Figure 6.31a shows the execution of instructions with a superpipelined machine of degree $n = 3$. In this case, only one instruction is issued per cycle, but the cycle time is one-third of the base cycle. Single-operation latency is n pipeline cycles, equivalent to one base cycle. The ILP required to fully utilize the machine is n instructions, as shown in the third column of Table 6.1.

Superpipelined machines have been around for a long time. Cray has built both the CDC 7600 and the Cray 1 as superpipelined machines with a latency of $n = 3$ cycles for a fixed-point add. Superpipelining is not possible without a high-speed clocking mechanism.

Superpipeline Performance   The minimum time required to execute N instructions

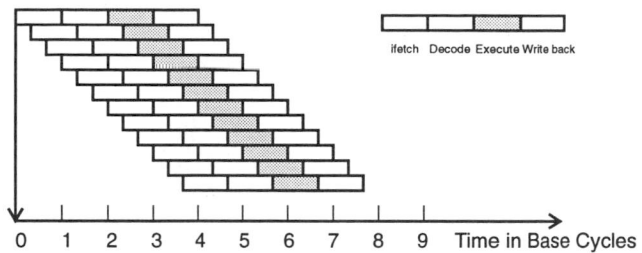

(a) Superpipelined execution with degree $n = 3$

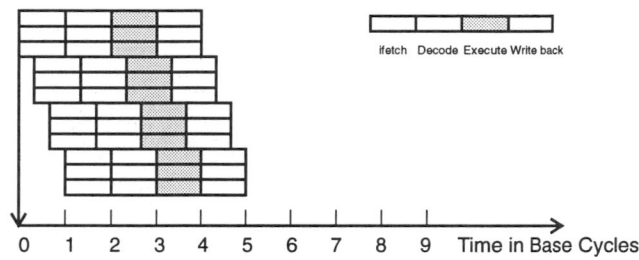

(b) Superpipelined superscalar execution with degree $m = n = 3$

Figure 6.31 Superpipelined processor architectures without and with multiple instruction issues, respectively.

for a superpipelined machine of degree n with k stages in the pipeline is

$$T(1, n) = k + \frac{1}{n}(N - 1) \quad \text{(base cycles)} \tag{6.29}$$

Thus, the potential speedup of a superpipelined machine over the base machine is

$$S(1, n) = \frac{T(1, 1)}{T(1, n)} = \frac{k + N - 1}{k + (N - 1)/n} = \frac{n(k + N - 1)}{nk + N - 1} \tag{6.30}$$

The speedup $S(1, n) \to n$, as $N \to \infty$.

The superpipeline and superscalar approaches can be combined in building so-called *superpipelined superscalar processors*. Good examples include Digital Equipment's Alpha processor and the MIPS R4000 processor.

Superpipelined Superscalar Design   A superpipelined superscalar processor of *degree* (m, n) is shown in Fig. 6.31b with $(m, n) = (3, 3)$. This machine executes m instructions every cycle with a pipeline cycle $1/n$ of the base cycle. Simple operation latency is n pipeline cycles. The level of parallelism required to fully utilize this machine is mn instructions.

As an example, the DEC Alpha processor offers an $m = 2$ instruction issue rate. Simple operations are directly supported by four functional units for integer, floating-point, load/store, and branch operations. The initial Alpha runs with a 150-MHz clock. Compared with a scalar processor with a clock rate of 25 MHz, this implies a super-pipelining degree of $n = 150$ MHz$/25$ MHz $= 6$. Thus we can approximately characterize the Alpha as having a degree of $(m, n) = (2, 6)$.

Similarly, the R4000 applies superpipelining to split the instruction and data memory references across two pipeline stages. This effectively stretches from R3000's five-stage pipeline to the eight-stage pipeline in the R4000. The ALU stage requires only one cycle. The R3000 runs with a 25-MHz clock, and the R4000 is driven by a 75-MHz clock. Therefore one can assume $n = 3$ for the R4000, compared with the R3000 which has $n = 1$. Both the R3000 and the R4000 are single-issue processors with $m = 1$ per cycle.

The superscalar approach appears to better match CMOS technology than the superpipelined approach. Superscalar designs rely primarily on spatial parallelism — multiple operations running concurrently on separate hardware — achieved by duplicating hardware resources, such as execution units and register file ports. Superpipelined designs, on the other hand, emphasize temporal parallelism — overlapping multiple operations on a common piece of hardware — achieved through more deeply pipelined execution units with faster clock cycles.

In general, superscalar machines require more transistors, whereas superpipelined designs require faster transistors and more careful circuit design to minimize the effects of clock skews. The CMOS technology generally favors replicating circuits over increasing clock cycle rates. Superpipelining has been implemented in Cray, NEC, and Fujitsu supercomputers. Recently some RISC processors have attempted to apply the superpipelining approach to increase the clock rates.

Superpipelined Superscalar Performance   Similar to the above derivations, we estimate the minimum time needed to execute N independent instructions on a superpipelined superscalar machine of degree (m, n) as

$$T(m, n) = k + \frac{N - m}{mn} \quad \text{(base cycles)} \tag{6.31}$$

Thus the speedup over the base machine is

$$S(m, n) = \frac{T(1, 1)}{T(m, n)} = \frac{k + N - 1}{k + (N - m)/(mn)} = \frac{mn(k + N - 1)}{mnk + N - m} \tag{6.32}$$

This speedup limit $S(m, n) \rightarrow mn$ as $N \rightarrow \infty$, as expected.

Example 6.13 The DEC 21064 superpipelined superscalar architecture

As illustrated in Fig. 6.32, this is a 64-bit superpipelined superscalar processor. The design emphasizes speed, multiple-instruction issue, multiprocessor applications, software migration from the VAX/VMS and MIPS Ultrix, and a long list of usable features. The reported clock rate is 150 MHz with the first chip implementation.

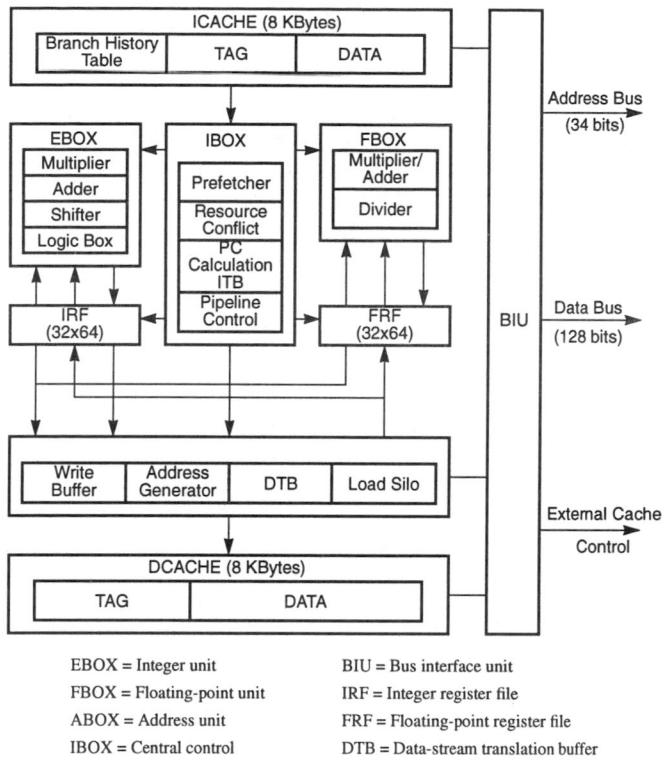

EBOX = Integer unit   BIU = Bus interface unit
FBOX = Floating-point unit   IRF = Integer register file
ABOX = Address unit   FRF = Floating-point register file
IBOX = Central control   DTB = Data-stream translation buffer

Figure 6.32 The architecture of the DEC 21064-A microprocessor. (Courtesy of Digital Equipment Corporation, 1992)

Unlike others, the Alpha architecture has thirty-two 64-bit integer registers and thirty-two 64-bit floating-point registers. The integer pipeline has 7 stages, and the floating-point pipeline has 10 stages in the initial Alpha design. All Alpha instructions are 32 bits. The instructions interact with each other only by one instruction writing into a register or memory and another one reading from the same place.

The first Alpha implementation issues two instructions per cycle. It is expected to increase the number of issues in later implementations. Pipeline timing hazards, load delay slots, and branch delay slots are all minimized by hardware support. The Alpha is designed to support fast multiprocessor interlocking and interrupts.

A privileged library of software has been developed to run full VMS and to run OSF/1 using different versions of the software library that mirror many of the VAX/OS and MIPS/OS features, respectively. This library makes Alpha an attractive architecture for multiple operating systems. The processor is designed to have a 300-MIPS peak and a 150-Mflops peak at 150 MHz.

6.5.3   Supersymmetry and Design Tradeoffs

Figure 6.33 compares a superscalar processor with a superpipelined processor, both of degree 3 (i.e., $m = n = 3$), issuing a basic block of six independent instructions. The issue of three instructions per base cycle is the rate for the superscalar machine, while the superpipelined machine takes only one-third of the base cycle to issue each instruction. In the steady state, both machines will execute the same number of instructions during the same time interval.

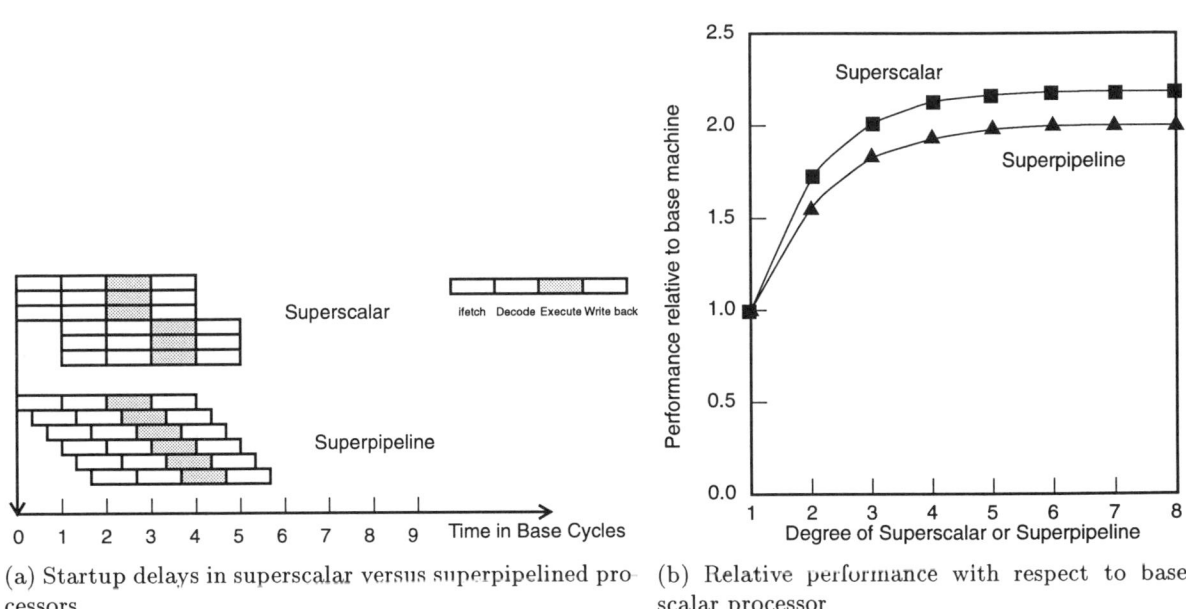

(a) Startup delays in superscalar versus superpipelined processors

(b) Relative performance with respect to base scalar processor

Figure 6.33 Relative performance of superscalar and superpipelined processors.
(Courtesy of Jouppi and Wall; reprinted from *Proc. ASPLOS*, ACM press, 1989)

Supersymmetry The superpipelined machine has a longer startup delay and lags behind the superscalar machine at the start of the program (Fig. 6.33a). Furthermore, a branch may create more damage on the superpipelined machine than on the superscalar machine. This effect diminishes as the degree of superpipelining increases, and all issuable instructions are issued closer together.

Figure 6.33b shows the performance of superscalar and of superpipelined processors relative to the base scalar processor. There is a duality of latency and a parallel issue of instructions.

These two curves were obtained by Jouppi and Wall (1989) after simulation runs of eight benchmarks on an ideal base machine, on a superscalar machine of degree m, and on a superpipelined machine of degree n, where $2 \leq m, n \leq 8$.

These simulation results suggest that the superpipelined machine has a lower performance than the superscalar machine. However, the performance gap decreases with increasing degree. This confirms the extra startup overhead and higher branching damages reported.

The shorter latency in issuing instructions is traded for more superpipelined cycles, but both types of machines have the same simple operation latency. Thus, they should perform equally in the steady state.

Design Tradeoffs We have plotted in Fig 6.34 the speedup curve $S(m, n)$ in Eq. 6.32 for $k = 8$ pipeline stages. Four cases are plotted with respect to single-issue, dual-issue, triple-issue, and quad-issue machines ($m = 1, 2, 3$ and 4, respectively). All speedup curves are functions of N, the number of instructions being executed.

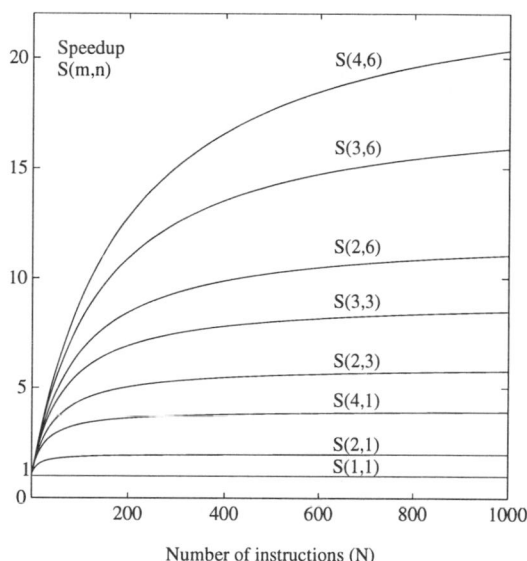

Figure 6.34 Ideal speedup of superpipelined and/or superscalar machines of degrees (m,n), compared with the scalar base machine with a unity speedup S(1,1) = 1.

As the degree of superpipelining increases, the speedup curves show steady increases. The improvement of pipeline performance is similar when the instruction issue rate increases in these plots. Design tradeoffs do exist in regard to the choices of (m, n).

Two important limitations were observed. First, the superscalar degree m is limited by the small ILP encountered in programs. Second, the superpipeline clock cycle is limited by the multiphase clocking technology available for distributing clock phases and by the long setup time of registers.

Until compiler technology is improved to increase the ILP, or multiphase clocking technology is further advanced, we cannot expect a significant increase in (m, n). Reasonable ranges are given in Fig. 6.34 for $1 \leq m \leq 4$ and $1 \leq n \leq 6$. When data dependence and branch penalty are considered, the ideal speedup curves will drop in all cases.

6.6 Bibliographic Notes and Exercises

A classic reference on pipeline architecture is [Kogge81]. [Smith89] surveyed instruction scheduling techniques. Branch strategies and performance models for pipeline processors are treated in [Lee84] and Smith and summarized in a research monograph by [Cragon92a]. Also, a book on pipeline processors is in preparation by [Cragon92b]. Pipeline scheduling theory started with the pioneer work reported in [Davidson71], [Davidson75] et al., [Shar72], and [Patel78]. [Sohi90] studied multiple-functional-unit pipelined processors. [Larson73] studied the optimal number of pipeline stages.

Superscalar and superpipelining techniques were treated in [Jouppi89] and Wall. Michael Johnson's book [Johnson91] also provides valuable insights on superscalar processor design.

Pipeline chaining and networking techniques were investigated by [Hwang88] and Xu. The paper by [Emma87] and Davidson described superpipeline clocking requirements. Compiler techniques for pipeline architecture were reported in [Gross83].

A survey of techniques for reducing branch penalties can be found in [Lilja88]. Instruction issue logic was studied by [Weiss84] and Smith. Dynamic instruction scheduling was pioneered by [Tomasulo67]. Recently, [Lam88] studied software pipelining techniques for VLIW architecture. [Acosta86], Kjelstrup, and Torng have examined the multiple-instruction issuing problem.

Computer arithmetic was treated in [Hwang78]. The CSA tree was reported in [Wallace64]. The details of the IBM 360/91 were reported in [Anderson67]. The TI-ASC was discussed by [Cragon89] and Watson. The CDC 6600 was reported in [Thornton70].

The IEEE floating-point standard can be ordered from [IEEE85]. MIPS R4000 was described in [Mirapuri92]. The DEC Alpha was described in [DEC92]. Information on the Motorola MC88110 can be found in [Diefendorff92] and Allen.

Exercises

Problem 6.1 Consider the execution of a program of 15,000 instructions by a linear pipeline processor with a clock rate of 25 MHz. Assume that the instruction pipeline has five stages and that one instruction is issued per clock cycle. The penalties due to branch instructions and out-of-sequence executions are ignored.

(a) Calculate the speedup factor in using this pipeline to execute the program as compared with the use of an equivalent nonpipelined processor with an equal amount of flow-through delay.

(b) What are the efficiency and throughput of this pipelined processor?

Problem 6.2   Study the DEC Alpha architecture in Example 6.13, find more information in the DEC Alpha handbook, and then answer the following questions with reasoning:

(a) Analyze the scalability of the Alpha processor implementation in terms of superscalar degree and superpipeline degree.
(b) Analyze the scalability of an Alpha-based multiprocessor system in terms of address space and multiprocessor support.

Problem 6.3   Study Section 6.5 and the paper by Jouppi and Wall (1989) and answer the following questions:

(a) Explain why a superpipelined processor performs less effectively than a superscalar processor of small degree.
(b) What is the duality of latency and parallel instruction issue under ideal circumstances?

Problem 6.4   Find the optimal number of pipeline stages k_0 given in Eq. 6.7 using the performance/cost ratio (PCR) given in Eq. 6.6.

Problem 6.5   Prove the lower bound and upper bound on the minimal average latency (MAL) specified in page 277.

Problem 6.6   Consider the following reservation table for a four-stage pipeline with a clock cycle $\tau = 20$ ns.

	1	2	3	4	5	6
S1	X					X
S2		X		X		
S3			X			
S4				X	X	

(a) What are the forbidden latencies and the initial collision vector?
(b) Draw the state transition diagram for scheduling the pipeline.
(c) Determine the MAL associated with the shortest greedy cycle.
(d) Determine the pipeline throughput corresponding to the MAL and given τ.
(e) Determine the lower bound on the MAL for this pipeline. Have you obtained the optimal latency from the above state diagram?

Problem 6.7   You are allowed to insert one noncompute delay stage into the pipeline in Problem 6.6 to make a latency of 1 permissible in the shortest greedy cycle. The purpose is to yield a new reservation table leading to an optimal latency equal to the lower bound.

(a) Show the modified reservation table with five rows and seven columns.
(b) Draw the new state transition diagram for obtaining the optimal cycle.
(c) List all the simple cycles and greedy cycles from the state diagram.
(d) Prove that the new MAL equals the lower bound.
(e) What is the optimal throughput of this pipeline? Indicate the percentage of throughput improvement compared with that obtained in part (d) of Problem 6.6.

Problem 6.8 Consider an adder pipeline with four stages as shown below. The pipeline consists of input lines X and Y and output line Z. The pipeline has a register R at its output where the temporary result can be stored and fed back to S1 at a later point in time. The inputs X and Y are multiplexed with the outputs R and Z.

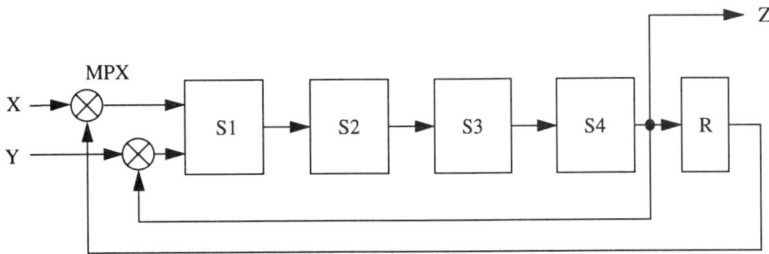

(a) Assume the elements of the vector A are fed into the pipeline through input X, one element per cycle. What is the minimum number of clock cycles required to compute the sum of an N-element vector A: $s = \sum_{I=1}^{N} A(I)$? In the absence of an operand, a value of 0 is input into the pipeline by default. Neglect the setup time for the pipeline.
(b) Let τ be the clock period of the pipelined adder. Consider an equivalent non-pipelined adder with a flow-through delay of 4τ. Find the actual speedup $S_4(64)$ and the efficiency $\eta_4(64)$ of using the above pipeline adder for $N = 64$.
(c) Find the maximum speedup $S_4(\infty)$ and the efficiency $\eta_4(\infty)$ when N tends to infinity.
(d) Find $N_{1/2}$, the minimum vector length required to achieve half of the maximum speedup.

Problem 6.9 Consider the following pipeline reservation table.

	1	2	3	4
S1	X			X
S2		X		
S3			X	

(a) What are the forbidden latencies?

(b) Draw the state transition diagram.

(c) List all the simple cycles and greedy cycles.

(d) Determine the optimal constant latency cycle and the minimal average latency.

(e) Let the pipeline clock period be $\tau = 20$ ns. Determine the throughput of this pipeline.

Problem 6.10 Consider the five-stage pipelined processor specified by the following reservation table:

	1	2	3	4	5	6
S1	X					X
S2		X			X	
S3			X			
S4				X		
S5		X				X

(a) List the set of forbidden latencies and the collision vector.

(b) Draw a state transition diagram showing all possible initial sequences (cycles) without causing a collision in the pipeline.

(c) List all the simple cycles from the state diagram.

(d) Identify the greedy cycles among the simple cycles.

(e) What is the minimum average latency (MAL) of this pipeline?

(f) What is the minimum allowed constant cycle in using this pipeline?

(g) What will be the maximum throughput of this pipeline?

(h) What will be the throughput if the minimum constant cycle is used?

Problem 6.11 The following assembly code is to be executed in a three-stage pipelined processor with hazard detection and resolution in each stage. The stages are instruction fetch, operand fetch (one or more as required), and execution (including a write-back operation). Explain all possible hazards in the execution of the code.

Inc R0	/ R0 ← (R0) + 1 /
Mul ACC, R0	/ ACC ← (ACC) × (R0) /
Store R1, ACC	/ R1 ← (ACC) /
Add ACC, R0	/ ACC ← (ACC) + (R0) /
Store M, ACC	/ M ← (ACC) /

Problem 6.12 This problem compares the performance of a superpipelined superscalar processor of degree (m, n) with that of a base scalar processor of degree $(1, 1)$. Analyze the speedup expression $S(m, n)$ in Eq. 6.32 for the following limiting cases:

(a) Within the range $1 \leq m \leq 4$ and $1 \leq n \leq 6$, what is the optimal number of pipeline stages that would maximize the speedup $S(m, n)$?

(b) What are the practical limitations preventing the growth of the superscalar degree m?

(c) What are the practical limitations preventing the growth of the superpipeline degree n?

Problem 6.13 Consider the following pipelined processor with four stages. This pipeline has a total evaluation time of six clock cycles. All successor stages must be used after each clock cycle.

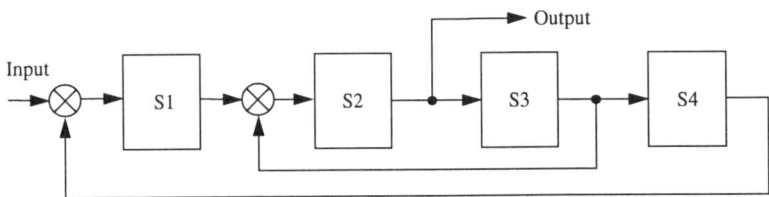

(a) Specify the reservation table for this pipeline with six columns and four rows.

(b) List the set of forbidden latencies between task initiations.

(c) Draw the state diagram which shows all possible latency cycles.

(d) List all greedy cycles from the state diagram.

(e) What is the value of the minimal average latency?

(f) What is the maximal throughput of this pipeline?

Problem 6.14 Three functional pipelines f_1, f_2, and f_3 are characterized by the following reservation tables. Using these three pipelines, a composite pipeline network is formed below:

f_1:

	1	2	3	4
S1	X			
S2		X		
S3			X	X

f_2:

	1	2	3	4
T1	X			X
T2		X		
T3			X	

f_3:

	1	2	3	4
U1	X		X	
U2				X
U3		X		

Each task going through this composite pipeline uses the pipeline in the following order: f_1 first, f_2 and f_3 next, f_1 again, and then the output is obtained. The dual multiplexer selects a pair of inputs, (A,B) or (X,Y), and feeds them into the input of f_1. The use of the composite pipeline is described by the combined reservation table.

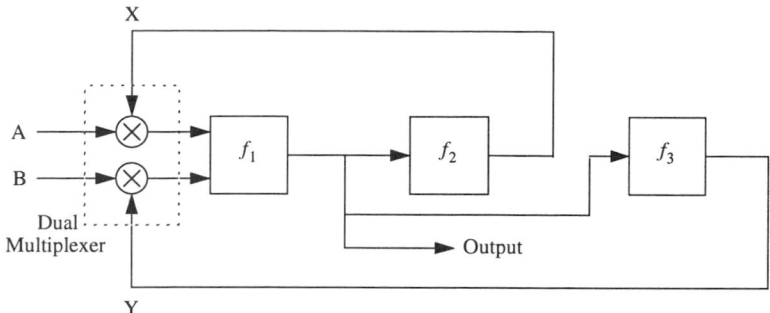

(a) Complete the following reservation table for this composite pipeline.

	1	2	3	4	5	6	7	8	9	10	11	12
S1	X											
S2										X		
S3				X								
T1												
T2												
T3							X					
U1					X							
U2												
U3												

(b) Write the forbidden list and the initial collision vector.

(c) Draw a state diagram clearly showing all latency cycles.

(d) List all simple cycles and greedy cycles.

(e) Calculate the MAL and the maximal throughput of this composite pipeline.

Problem 6.15   A nonpipelined processor X has a clock rate of 25 MHz and an average CPI (cycles per instruction) of 4. Processor Y, an improved successor of X, is designed with a five-stage linear instruction pipeline. However, due to latch delay and clock skew effects, the clock rate of Y is only 20 MHz.

(a) If a program containing 100 instructions is executed on both processors, what is the speedup of processor Y compared with that of processor X?

(b) Calculate the MIPS rate of each processor during the execution of this particular program.

Problem 6.16   Design a binary integer multiply pipeline with five stages. The first stage is for partial product generation. The last stage is a 36-bit carry-lookahead adder. The middle three stages are made of 16 carry-save adders (CSAs) of appropriate lengths.

(a) Prepare a schematic design of the five-stage multiply pipeline. All line widths and interstage connections must be shown.

(b) Determine the maximal clock rate of the pipeline if the stage delays are $\tau_1 = \tau_2 = \tau_3 = \tau_4 = 90$ ns, $\tau_5 = 45$ ns, and the latch delay is 20 ns.

(c) What is the maximal throughput of this pipeline in terms of the number of 36-bit results generated per second?

Problem 6.17 Consider a four-stage floating-point adder with a 10-ns delay per stage which equals the pipeline clock period.

(a) Name the appropriate functions to be performed by the four stages.

(b) Find the minimum number of periods required to add 100 floating-point numbers $A_1 + A_2 + \cdots + A_{100}$ using this pipeline adder, assuming that the output Z of stage S_4 can be routed back to either of the two inputs X or Y of the pipeline with delays equal to a multiple of the clock period.

Problem 6.18 Consider two four-stage pipeline adders and a number of noncompute delay elements. Each delay element has a one-unit time delay.

(a) Use the available adders and delays to construct a composite pipeline unit for evaluating the following expression: $b(i) = a(i) + a(i-1) + a(i-2) + a(i-3)$ for all $i = 4, 5, ..., n$. The composite pipeline receives $a(i)$ for $i = 1, 2, ..., n$, as the successive inputs.

(b) Consider a third four-stage pipeline adder. Augment the design in part (a) with this third adder to compute the following recursive expression: $x(i) = a(i) + x(i-1)$, for all $i = 4, 5, ..., n$. Note that $x(i) = a(i) + x(i-1) = a(i) + [a(i-1) + x(i-2)] = \cdots = b(i) + x(i-4)$, where $b(i)$ is generated by the composite pipeline in part (a).

Part III

Parallel and Scalable Architectures

Chapter 7
Multiprocessors and Multicomputers

Chapter 8
Multivector and SIMD Computers

Chapter 9
Scalable, Multithreaded, and Dataflow Architectures

Summary

Part III consists of three chapters dealing with parallel, vector, and scalable architectures for building high-performance computers. The multiprocessor system interconnects studied include crossbar switches, multistage networks, hierarchical buses, and multidimensional ring, mesh, and torus architectures. Three generations of multicomputer developments are reviewed. Then we consider message-passing mechanisms.

Vector supercomputers appear either as pipelined multiprocessors or as SIMD data-parallel computers. We study the architectures of the Cray Y-MP, C-90, Cray/MPP, NEC SX, Fujitsu VP-2000, VPP500, VAX 9000, Hitachi S-820, Stardent 3000, CM-2, MasPar MP-1, and CM5 for concurrent scalar/vector processing.

Chapter 9 introduces scalable architectures for massively parallel processing applications. These include both von Neumann, fine-grain, multithreaded, and dataflow architectures. Various latency-hiding techniques are described, including the principles of multithreading. Case studies include the Intel Paragon, Stanford Dash, MIT Alewife, J-Machine and *T, Tera computer, KSR-1, Wisconsin Multicube, USC/OMP, ETL EM4, etc.

Chapter 7

Multiprocessors and Multicomputers

In this chapter, we study system architectures of multiprocessors and multicomputers. Various cache coherence protocols, synchronization methods, crossbar switches, multiport memory, and multistage networks are described for building multiprocessor systems. Then we discuss multicomputers with distributed memories which are not globally shared. The Intel Paragon is used as a case study. Message-passing mechanisms built into current multicomputers are also reviewed. Single-address-space multicomputers will be studied in Chapter 9.

7.1 Multiprocessor System Interconnects

Parallel processing demands the use of efficient system interconnects for fast communication among multiple processors and shared memory, I/O, and peripheral devices. Hierarchical buses, crossbar switches, and multistage networks are often used for this purpose.

A generalized multiprocessor system is depicted in Fig. 7.1. This architecture combines features from the UMA, NUMA, and COMA models introduced in Section 1.4.1. Each processor P_i is attached to its own local memory and private cache. Multiple processors are connected to shared-memory modules through an interprocessor-memory network (IPMN).

The processors share the access of I/O and peripheral devices through a processor-I/O network (PION). Both IPMN and PION are necessary in a shared-resource multiprocessor. Direct interprocessor communications are supported by an optional interprocessor communication network (IPCN) instead of through the shared memory.

Network Characteristics Each of the above types of networks can be designed with many choices. The choices are based on the topology, timing protocol, switching method, and control strategy. Dynamic networks are used in multiprocessors in which the interconnections are under program control. Timing, switching, and control are

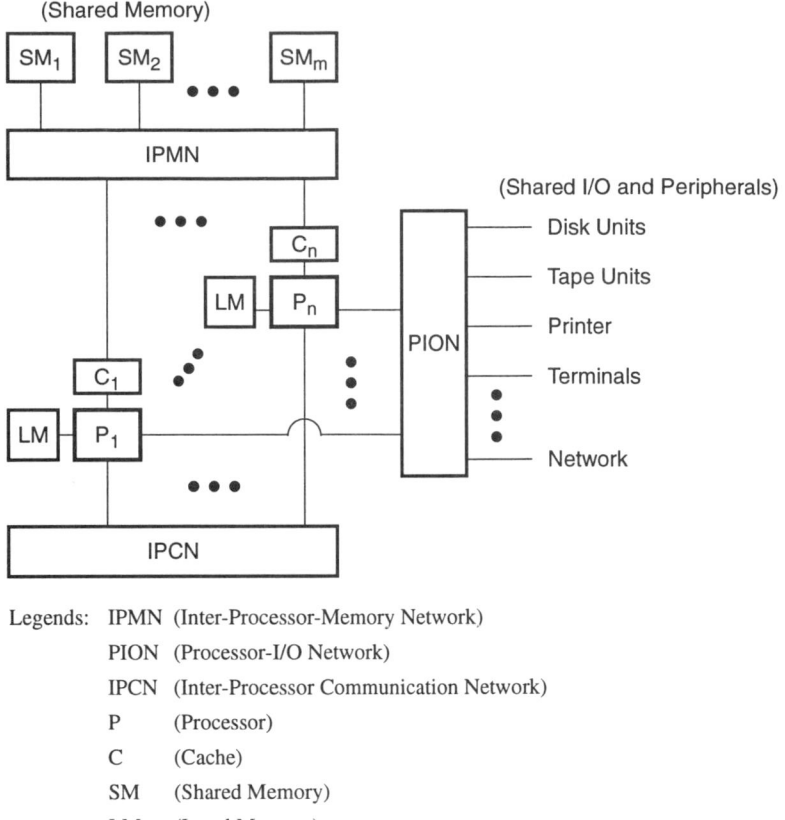

Figure 7.1 Interconnection structures in a generalized multiprocessor system with local memory, private caches, shared memory, and shared peripherals.

three major operational characteristics of an interconnection network. The timing control can be either *synchronous* or *asynchronous*. Synchronous networks are controlled by a global clock that synchronizes all network activities. Asynchronous networks use handshaking or interlocking mechanisms to coordinate fast and slow devices requesting use of the same network.

A network can transfer data using either *circuit switching* or *packet switching*. In circuit switching, once a device is granted a path in the network, it occupies the path for the entire duration of the data transfer. In packet switching, the information is broken into small packets individually competing for a path in the network.

Network control strategy is classified as *centralized* or *distributed*. With centralized control, a global controller receives requests from all devices attached to the network and grants the network access to one or more requesters. In a distributed system, requests are handled by local devices independently.

7.1.1 Hierarchical Bus Systems

A *bus system* consists of a hierarchy of buses connecting various system and sub system components in a computer. Each bus is formed with a number of signal, control, and power lines. Different buses are used to perform different interconnection functions.

In general, the hierarchy of bus systems are packaged at different levels as depicted in Fig. 7.2, including local buses on boards, backplane buses, and I/O buses.

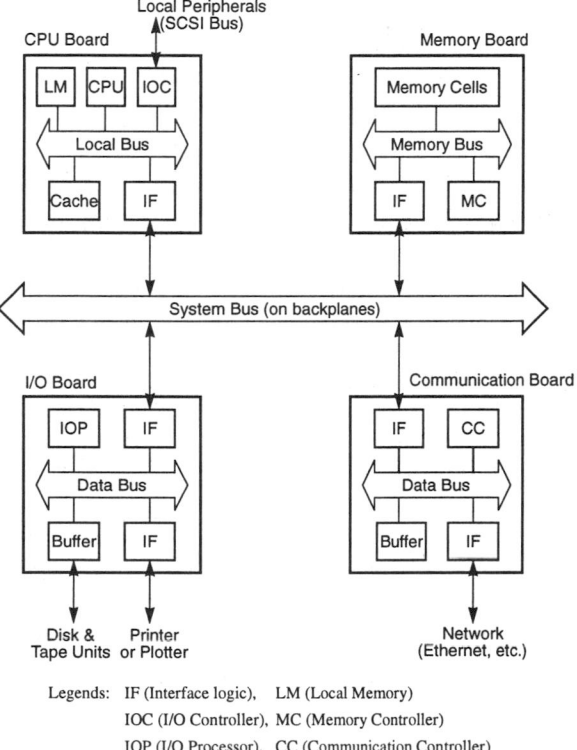

Figure 7.2 Bus systems at board level, backplane level, and I/O level.

Local Bus   Buses implemented on *printed-circuit* boards are called *local buses*. On a processor board one often finds a local bus which provides a common communication path among major components (chips) mounted on the board. A memory board uses a *memory bus* to connect the memory with the interface logic.

An I/O board or network interface board uses a *data bus*. Each of these board buses consists of signal and utility lines. With the sharing of the lines by many I/O devices, the layout of these lines may be at different layers of the PC board.

Backplane Bus   A *backplane* is a printed circuit on which many connectors are used to plug in functional boards. A *system bus*, consisting of shared signal paths and utility lines, is built on the backplane. This system bus provides a common communication path among all plug-in boards.

Several backplane bus standards have been developed such as the VME bus (IEEE Standard 1014-1987), Multibus II (IEEE Standard 1296-1987), and Futurebus+ (IEEE Standard 896.1-1991) as introduced in Chapter 5.

I/O Bus   Input/output devices are connected to a computer system through an *I/O bus* such as the SCSI (small computer systems interface) bus. This bus is made of coaxial cables with taps connecting disks, printer, and tape unit to a processor through an I/O controller (Fig. 7.2). Special interface logic is used to connect various board types to the backplane bus.

Complete specifications for a bus system include logical, electrical, and mechanical properties, various application profiles, and interface requirements. Our study will be confined to the logical and application aspects of state-of-the-art system buses. Emphasis will be placed on the scalability and bus support for cache coherence and fast synchronization.

To give the reader some real design data, we introduce below several digital buses built into commercial machines. The core of the Encore Multimax multiprocessor is the Nanobus, consisting of 20 slots, a 32-bit address, a 64-bit data path, and a 14-bit vector bus, and operating at a clock rate of 12.5 MHz with a total memory bandwidth of 100 Mbytes/s.

The Sequent multiprocessor bus has a 64-bit data path, a 10-MHz clock rate, and a 32-bit address and yields a channel bandwidth of 80 Mbytes/s. A write-back private cache is used to reduce the bus traffic by 50% .

Digital buses are the fundamental interconnects adopted in commercial systems ranging from workstations to minicomputers, mainframes, and multiprocessors. Hierarchical bus systems can be used to build medium-sized multiprocessors with less than 100 processors. The bus approach is limited mainly by the packaging technology employed. The development of the Futurebus is expected to expand the limits imposed by conventional buses.

Hierarchical Buses and Caches   Wilson (1987) has proposed a hierarchical cache/-bus architecture as shown in Fig. 7.3. This is a multilevel tree structure in which the leaf nodes are processors and their private caches (denoted P_j and C_{1j} in Fig. 7.3). These are divided into several clusters, each of which is connected through a cluster bus.

An intercluster bus is used to provide communications among the clusters. Second-level caches (denoted as C_{2i}) are used between each cluster bus and the intercluster bus. Each second-level cache must have a capacity that is at least an order of magnitude larger than the sum of the capacities of all first-level caches connected beneath it.

Each single cluster operates as a single-bus system. Snoopy bus coherence protocols can be used to establish consistency among first-level caches belonging to the same cluster. Second-level caches are used to extend consistency from each local cluster to the upper level.

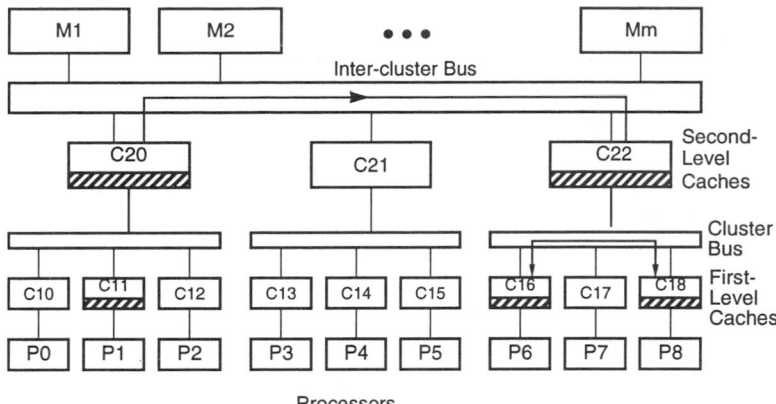

Figure 7.3 A hierarchical cache/bus architecture for designing a scalable multi-processor. (Courtesy of Wilson; reprinted from *Proc. of Annual Int. Symp. on Computer Architecture*, 1987)

The upper-level caches form another level of shared memory between each cluster and the main memory modules connected to the intercluster bus. Most memory requests should be satisfied at the lower-level caches. Intercluster cache coherence is controlled among the second-level caches and the resulting effects are passed to the lower level.

Example 7.1 Encore's Ultramax multiprocessor architecture

The Ultramax has a two-level hierarchical-bus architecture as depicted in Fig. 7.4. The Ultramax architecture is very similar to that characterized by Wilson, except that the global Nanobus is used only for intercluster communications.

The shared memories are distributed to all clusters instead of being connected to the intercluster bus. The cluster caches form the second-level caches and perform the same filtering and cache coherence control for remote accesses as in Wilson's scheme. After an access request reaches the top bus, it is routed down to the cluster memory that matches it with the reference address.

∎

The idea of using *bridges* between multiprocessor clusters is to allow transactions initiated on a local bus to be completed on a remote bus. As exemplified in Fig. 7.5, multiple Futurebuses+ are used to build a very large system consisting of three multiprocessor clusters.

Bridges are used to interface these clusters. The main functions of a bridge include communication protocol conversion, interrupt handling in split transactions, and serving as cache and memory agents. Besides bridging between multiprocessor clusters built around Futurebus+, the standard also provides bridge specifications for the VME bus (IEEE P1014.2) and Multibus II (IEEE P1296.2). These will allow Futurebus+ to be compatible with other bus architectures.

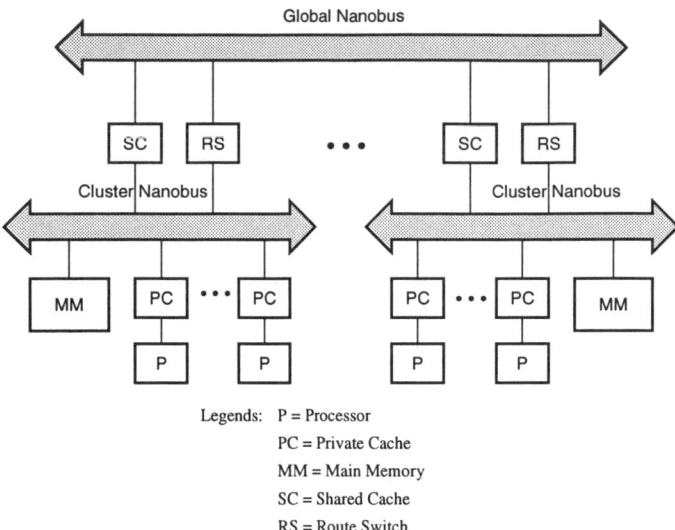

Figure 7.4 The Ultramax multiprocessor architecture using hierarchical buses with multiple clusters. (Courtesy of Encore Computer Corporation, 1987)

7.1.2   Crossbar Switch and Multiport Memory

Switched networks provide dynamic interconnections between the inputs and outputs. Major classes of switched networks are specified below, based on the number of stages and blocking or nonblocking. We describe the crossbar networks and multiport memory structures first and then the multistage networks. Crossbar networks are mostly used in small or medium-size systems. The multistage networks can be extended to larger systems if the increased latency problem can be solved.

Network Stages   Depending on the interstage connections used, a *single-stage network* is also called a *recirculating network* because data items may have to recirculate through the single stage many times before reaching their destination. A single-stage network is cheaper to build, but multiple passes may be needed to establish certain connections. The crossbar switch and multiport memory organization are both single-stage networks.

A multistage network consists of more than one stage of switch boxes. Such a network should be able to connect from any input to any output. We will study unidirectional multistage networks in Section 7.1.3. The choice of interstage connection patterns determines the network connectivity. These patterns may be the same or different at different stages, depending the class of networks to be designed. The Omega network, Flip network, and Baseline networks are all multistage networks.

Blocking versus Nonblocking Networks   A multistage network is called *blocking* if

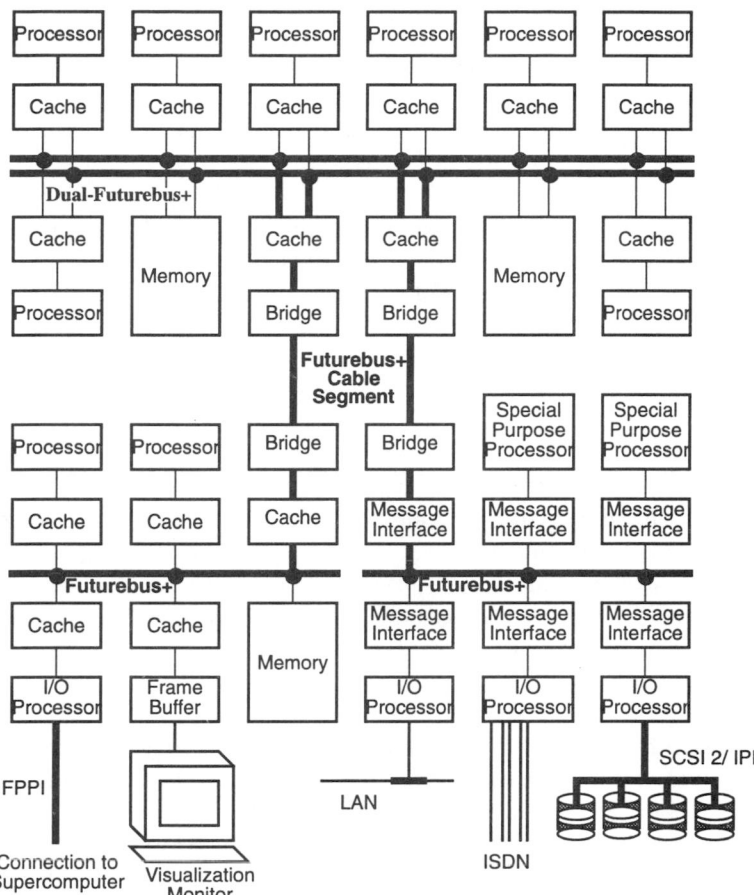

Figure 7.5 A large multiprocessor system using multiple Futurebus+ segments.
(Reprinted with permission from IEEE Standard 896.1-1991, copyright © 1991 by IEEE, Inc.)

the simultaneous connections of some multiple input-output pairs may result in conflicts in the use of switches or communication links.

Examples of blocking networks include the Omega (Lawrie, 1975), Baseline (Wu and Feng, 1980), Banyan (Goke and Lipovski, 1973), and Delta networks (Patel, 1979). Some blocking networks are equivalent after graph transformations. In fact, most multistage networks are blocking in nature. In a blocking network, multiple passes through the network may be needed to achieve certain input-output connections.

A multistage network is called *nonblocking* if it can perform all possible connections between inputs and outputs by rearranging its connections. In such a network, a connection path can always be established between any input-output pair. The Benes networks (Benes, 1965) have such a capability. However, Benes networks require almost twice the number of stages to achieve the nonblocking connections. The Clos networks (Clos, 1953) can also perform all permutations in a single pass without blocking. Certain subclasses of blocking networks can also be made nonblocking if extra stages are added or connections are restricted. The blocking problem can be avoided by using combining networks to be described in the next section.

Crossbar Networks   In a *crossbar network*, every input port is connected to a free output port through a crosspoint switch (circles in Fig. 2.26a) without blocking. A crossbar network is a single-stage network built with unary switches at the crosspoints.

Once the data is read from the memory, its value is returned to the requesting processor along the same crosspoint switch. In general, such a crossbar network requires the use of $n \cdot m$ crosspoint switches. A square crossbar ($n = m$) can implement any of the $n!$ permutations without blocking.

As introduced earlier, a crossbar switch network is a single-stage, nonblocking, permutation network. Each crosspoint in a crossbar network is a unary switch which can be set open or closed, providing a point-to-point connection path between the source (processor) and destination (memory).

All processors can send memory requests independently and asynchronously. This poses the problem of multiple requests destined for the same memory module at the same time. In such cases, only one of the requests is serviced at a time. Let us characterize below the crosspoint switching operations.

Crosspoint Switch Design   Out of n crosspoint switches in each column of an $n \times m$ crossbar mesh, only one can be connected at a time. To resolve the contention for each memory module, each crosspoint switch must be designed with extra hardware.

Furthermore, each crosspoint switch requires the use of a large number of connecting lines accommodating address, data path, and control signals. This means that each crosspoint has a complexity almost matching that of a single bus.

For an $n \times n$ crossbar network, this implies that n^2 sets of crosspoint switches and a large number of lines are needed. What this amounts to is a crossbar network requiring extensive hardware when n is very large. So far only small crossbar networks with $n \leq 16$ have been built into commercial machines.

On each row of the crossbar mesh, multiple crosspoint switches can be connected simultaneously. Simultaneous data transfers can take place in a crossbar between n pairs of processors and memories.

Figure 7.6 shows the schematic design of a crosspoint switch in a single crossbar network. Multiplexer modules are used to select one of n *read* or *write* requests for service. Each processor sends in an independent request, and the arbitration logic makes the selection based on certain fairness or priority rules.

For example, a 4-bit control signal will be generated for $n = 16$ processors. Note that n sets of data, address, and read/write lines are connected to the input of the multiplexer tree. Based on the control signal received, only one out of n sets of information lines is selected as the output of the multiplexer tree.

The memory address is entered for both *read* and *write* access. In the case of *read*,

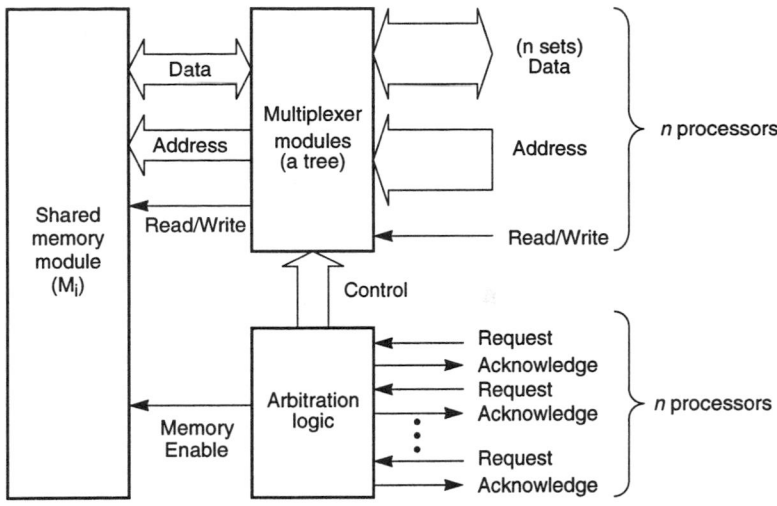

Figure 7.6 Schematic design of a crosspoint switch in a crossbar network.

the data fetched from memory are returned to the selected processor in the reverse direction using the data path established at the corresponding crosspoint. In the case of *write*, the data on the data path are stored in memory.

Acknowledge signals are used to indicate the arbitration result to all requesting processors. These signals initiate data transfer and are used to avoid conflicts. Note that the data path established is bidirectional, in order to serve both *read* and *write* requests at different memory cycles.

Crossbar Limitations A single processor can send many requests to multiple memory modules. For an $n \times n$ crossbar network, at most n memory words can be delivered to at most n processors in each memory cycle. The memory modules can be n-way interleaved to allow overlapped access using n minor cycles within each major cycle of the memory.

The crossbar network offers the highest bandwidth of n data transfers per memory cycle, as compared with only one data transfer per bus cycle. Since all necessary switching and conflict resolution logic are built into the crosspoint switch, the processor interface and memory port logic are much simplified and cheaper. A crossbar network is cost-effective only for small multiprocessors with a few processors accessing a few memory modules. A single-stage crossbar network is not expandable once it is built.

Redundancy or parity-check lines can be built into each crosspoint switch to enhance the fault tolerance and reliability of the crossbar network. Besides building a crossbar network using electronic means, the research community is presently exploring the possibility of using electrooptical means to build much larger and faster crossbar networks. An optical crossbar network has not yet appeared as a commercial product.

Multiport Memory   Because building a crossbar network into a large system is cost-prohibitive, many mainframe multiprocessors use a multiport memory organization. The idea is to move all crosspoint arbitration and switching functions associated with each memory module into the memory controller.

Thus the memory module becomes more expensive due to the added access ports and associated logic as demonstrated in Fig. 7.7a. The circles in the diagram represent n switches tied to n input ports of a memory module. Only one of n processor requests will be honored at a time.

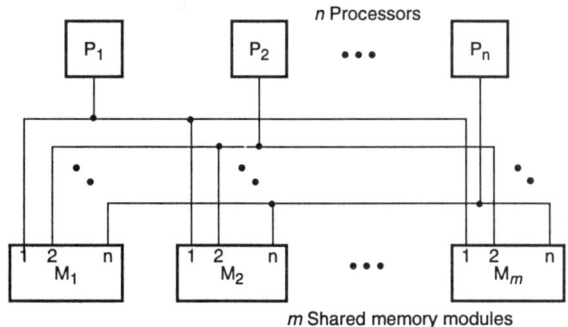

(a) n-port memory modules used

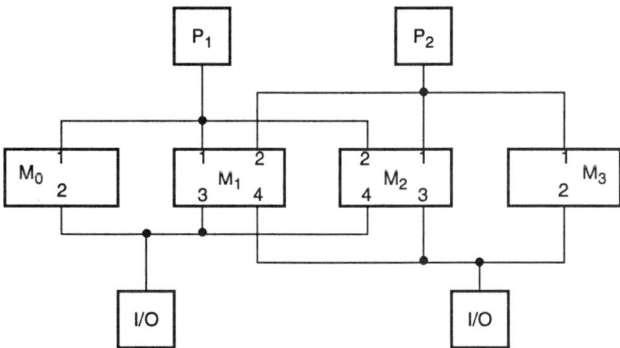

(b) Memory ports prioritized or privileged in each module by numbers

Figure 7.7 Multiport memory organizations for multiprocessor systems. (Courtesy of P. H. Enslow, *ACM Computing Surveys*, March 1977)

The multiport memory organization is a compromise solution between a low-cost, low-performance bus system and a high-cost, high-bandwidth crossbar system. The contention bus is time-shared by all processors and device modules attached. The multiport memory must resolve conflicts among processors.

This memory structure becomes expensive when m and n become very large. A typical mainframe multiprocessor configuration uses $n = 4$ processors and $m = 16$ memory modules. A multiport memory multiprocessor is not scalable because once the ports are fixed, no more processors can be added without redesigning the memory controller.

Another drawback is the need for a large number of interconnection cables and connectors when the configuration becomes large. The ports of each memory module in Fig. 7.7b are prioritized. Some of processors are CPUs, some are I/O processors, and some are connected to dedicated processors.

For example, the Univac 1100/94 multiprocessor consists of four CPUs, four I/O processors, and two scientific vector processors which are connected to four shared-memory modules, each of which is 10-way ported. The access to these ports is prioritized under operating system control. In other multiprocessors, part of the memory module can be made private with ports accessible only to the owner processors.

7.1.3 Multistage and Combining Networks

Multistage networks are used to build larger multiprocessor systems. We describe two multistage networks, the Omega network and the Butterfly network, that have been built into commercial machines. We will study a special class of multistage networks, called combining networks, for resolving access conflicts automatically through the network. The combining network has been built into the NYU's Ultracomputer.

Routing in Omega Network   We have defined the Omega network in Chapter 2. In what follows, we describe the message-routing algorithm and broadcast capability of Omega network. This class of network has been built into the Illinois Cedar multiprocessor (Kuck et al., 1987), into the IBM RP3 (Pfister et al., 1985), and into the NYU Ultracomputer (Gottlieb et al., 1983). An 8-input Omega network is shown in Fig. 7.8.

In general, an n-input Omega network has $\log_2 n$ stages. The stages are labeled from 0 to $\log_2 n - 1$ from the input end to the output end. Data routing is controlled by inspecting the destination code in binary. When the ith high-order bit of the destination code is a 0, a 2×2 switch at stage i connects the input to the upper output. Otherwise, the input is directed to the lower output.

Two switch settings are shown in Figs. 7.8a and b with respect to permutations $\pi_1 = (0, 7, 6, 4, 2)\ (1, 3)(5)$ and $\pi_2 = (0, 6, 4, 7, 3)\ (1, 5)(2)$, respectively.

The switch settings in Fig. 7.8a are for the implementation of π_1, which maps $0 \rightarrow 7,\ 7 \rightarrow 6,\ 6 \rightarrow 4,\ 4 \rightarrow 2,\ 2 \rightarrow 0,\ 1 \rightarrow 3,\ 3 \rightarrow 1,\ 5 \rightarrow 5$. Consider the routing of a message from input 001 to output 011. This involves the use of switches A, B, and C. Since the most significant bit of the destination 011 is a "zero," switch A must be set straight so that the input 001 is connected to the upper output (labeled 2). The middle bit in 011 is a "one," thus input 4 to switch B is connected to the lower output with a "crossover" connection. The least significant bit in 011 is a "one," implying a flat connection in switch C. Similarly, the switches A, E, and D are set for routing a message from input 101 to output 101. There exists no conflict in all the switch settings needed to implement the permutation π_1 in Fig. 7.8a.

Now consider implementing the permutation π_2 in the 8-input Omega network

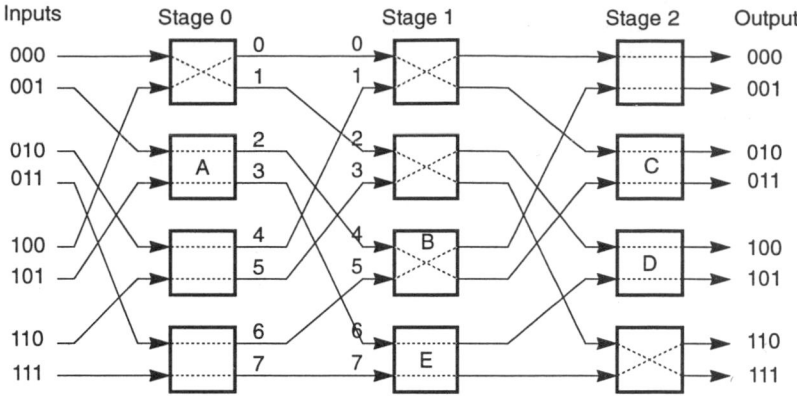

(a) Permutation $\pi_1 = (0,7,6,4,2)(1,3)(5)$ implemented on an Omega network without blocking

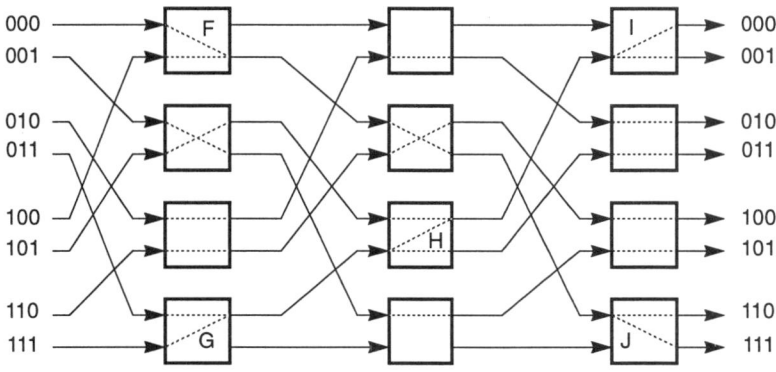

(b) Permutation $\pi_2 = (0,6,4,7,3)(1,5)(2)$ blocked at switches marked F, G, and H

Figure 7.8 Two switch settings of an 8×8 Omega network built with 2×2 switches.

(Fig. 7.8b). Conflicts in switch settings do exist in three switches identified as F, G, and H. The conflicts occurring at F are caused by the desired routings $000 \to 110$ and $100 \to 111$. Since both destination addresses have a leading bit 1, both inputs to switch F must be connected to the lower output. To resolve the conflicts, one request must be rejected.

Similarly, we see conflicts at switch G between $011 \to 000$ and $111 \to 011$, and at switch H between $101 \to 001$ and $011 \to 000$. At switches I and J, broadcast is used from one input to two outputs, which is allowed if the hardware is built to have four legitimate states as shown in Fig. 2.24a. The above example indicates the fact that not all permutations can be implemented in one pass through the Omega network.

The Omega network is a kind of blocking network. In case of blocking, one can establish the conflicting connections in several passes. For the example π_2, we can connect $000 \rightarrow 110$, $001 \rightarrow 101$, $010 \rightarrow 010$, $101 \rightarrow 001$, $110 \rightarrow 100$ in the first pass and $011 \rightarrow 000$, $100 \rightarrow 111$, $111 \rightarrow 011$ in the second pass. In general, if 2×2 switch boxes are used, an n-input Omega network can implement $n^{n/2}$ permutations in a single pass. There are $n!$ permutations in total.

For $n = 8$, this implies that only $8^4/8! = 4096/40320 = 0.1016 = 10.16\%$ of all permutations are implementable in a single pass through an 8-input Omega network. All others will cause blocking and demand up to three passes to be realized. In general, a maximum of $\log_2 n$ passes are needed for an n-input Omega. Blocking is not a desired feature in any multistage network. It may lower the effective bandwidth.

The Omega network can also be used to broadcast data from one source to many destinations, as exemplified in Fig. 7.9a, using the upper broadcast or lower broadcast switch settings. In Fig. 7.9a, the message at input 001 is being broadcast to all eight outputs through a binary tree connection.

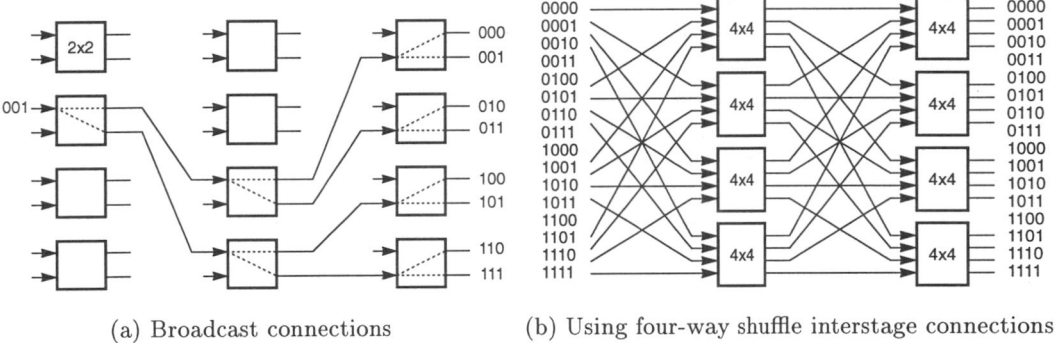

(a) Broadcast connections   (b) Using four-way shuffle interstage connections

Figure 7.9 Broadcast capability of an Omega network built with 4×4 switches.

The two-way shuffle interstage connections can be replaced by four-way shuffle interstage connections when 4×4 switch boxes are used as building blocks, as exemplified in Fig. 7.9b for a 16-input Omega network with $\log_4 16 = 2$ stages.

Note that a four-way shuffle corresponds to dividing the 16 inputs into four equal subsets and then shuffling them evenly among the four subsets. When $k \times k$ switch boxes are used, one can define a k-way shuffle function to build an even larger Omega network with $\log_k n$ stages.

Routing in Butterfly Networks   This class of networks is constructed with crossbar switches as building blocks. Figure 7.10 shows two Butterfly networks of different sizes. Figure 7.10a shows a 64-input Butterfly network built with two stages ($2 = \log_8 64$) of 8×8 crossbar switches. The eight-way shuffle function is used to establish the interstage connections between stage 0 and stage 1. In Fig. 7.10b, a three-stage Butterfly network is constructed for 512 inputs, again with 8×8 crossbar switches. Each of the 64×64 boxes in Fig. 7.10b is identical to the two-stage Butterfly network in Fig. 7.10a.

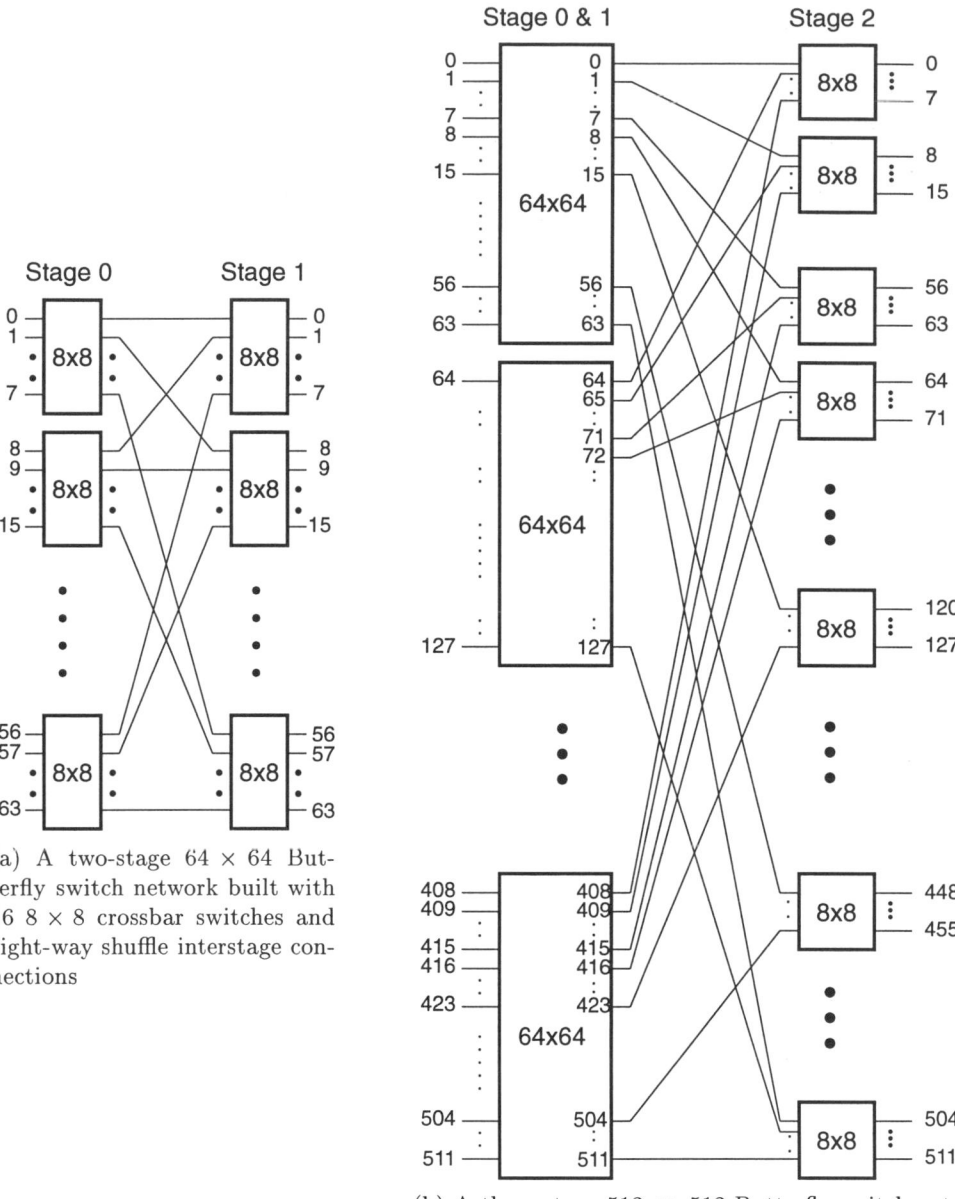

(a) A two-stage 64×64 Butterfly switch network built with 16 8×8 crossbar switches and eight-way shuffle interstage connections

(b) A three-stage 512×512 Butterfly switch network built with 192 8×8 crossbar switches

Figure 7.10 Modular construction of Butterfly switch networks with 8×8 crossbar switches. (Courtesy of BBN Advanced Computers, Inc., 1990)

In total, sixteen 8×8 crossbar switches are used in Fig. 7.10a and $16 \times 8 + 8 \times 8 = 192$ are used in Fig. 7.10b. Larger Butterfly networks can be modularly constructed using more stages. Note that no broadcast connections are allowed in a Butterfly network, making these networks a restricted subclass of Omega networks.

The Hot-Spot Problem   When the network traffic is nonuniform, a *hot spot* may appear corresponding to a certain memory module being excessively accessed by many processors at the same time. For example, a semaphore variable being used as a synchronization barrier may become a hot spot since it is shared by many processors.

Hot spots may degrade the network performance significantly. In the NYU Ultra-computer and the IBM RP3 multiprocessor, a combining mechanism has been added to the Omega network. The purpose was to combine multiple requests heading for the same destination at switch points where conflicts are taking place.

An atomic read-modify-write primitive Fetch&Add(x, e), has been developed to perform parallel memory updates using the combining network.

Fectch&Add   This atomic memory operation is effective in implementing an N-way synchronization with a complexity independent of N. In a Fetch&Add(x, e) operation, x is an integer variable in shared memory and e is an integer increment. When a single processor executes this operation, the semantics is

$$
\begin{aligned}
&\text{Fetch\&Add}(x, e) \\
&\quad \{ \ temp \quad \leftarrow \quad x; \\
&\qquad\quad x \quad \leftarrow \quad temp + e; \\
&\quad \text{return} \quad temp \}
\end{aligned}
\qquad (7.1)
$$

When N processes attempt to Fetch&Add(x, e) the same memory word simultaneously, the memory is updated only once following a *serialization principle*. The sum of the N increments, $e_1 + e_2 + \cdots + e_N$, is produced in any arbitrary serialization of the N requests.

This sum is added to the memory word x, resulting in a new value $x + e_1 + e_2 + \cdots + e_N$. The values returned to the N requests are all unique, depending on the serialization order followed. The net result is similar to a sequential execution of N Fetch&Adds but is performed in one indivisible operation. Two simultaneous requests are combined in a switch as illustrated in Fig. 7.11.

One of the following operations will be performed if processor P_1 executes Ans$_1 \leftarrow$ Fetch&Add(x, e_1) and P_2 executes Ans$_2 \leftarrow$ Fetch&Add(x, e_2) simultaneously on the shared variable x. If the request from P_1 is executed ahead of that from P_2, the following values are returned:

$$
\begin{aligned}
\text{Ans}_1 \quad &\leftarrow \quad x \\
\text{Ans}_2 \quad &\leftarrow \quad x + e_1
\end{aligned}
\qquad (7.2)
$$

If the execution order is reversed, the following values are returned:

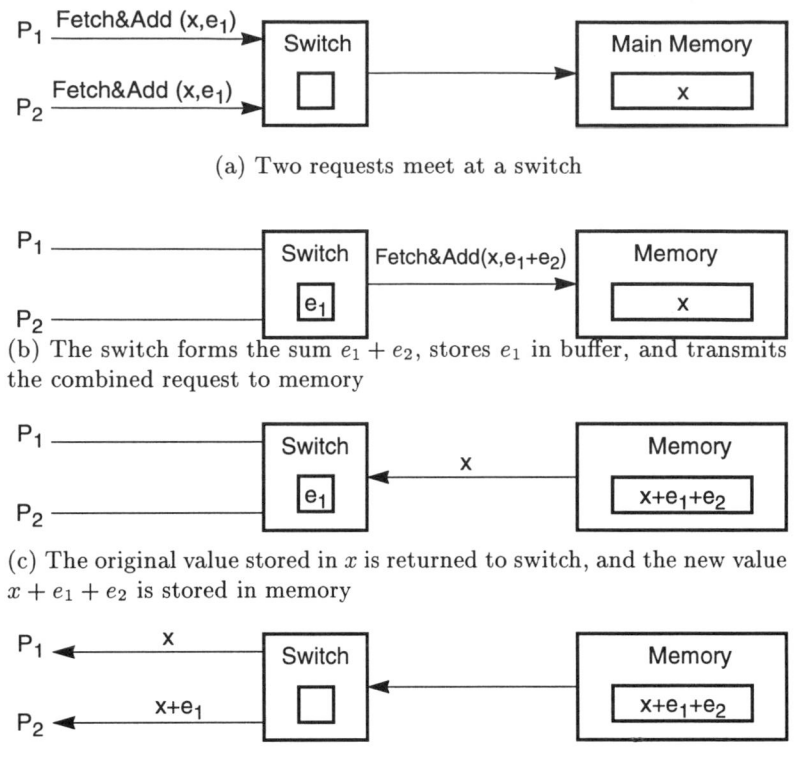

(a) Two requests meet at a switch

(b) The switch forms the sum $e_1 + e_2$, stores e_1 in buffer, and transmits the combined request to memory

(c) The original value stored in x is returned to switch, and the new value $x + e_1 + e_2$ is stored in memory

(d) The values x and $x + e_1$ are returned to P_1 and P_2, respectively

Figure 7.11 Two Fetch&Add operations are combined to access a shared variable simultaneously via a combining network.

$$\text{Ans}_1 \quad \leftarrow \quad x + e_2$$
$$\text{Ans}_2 \quad \leftarrow \quad x \tag{7.3}$$

Regardless of the executing order, the value $x + e_1 + e_2$ is stored in memory. It is the responsibility of the switch box to form the sum $e_1 + e_2$, transmit the combined request Fetch&Add$(x, e_1 + e_2)$, store the value e_1 (or e_2) in a wait buffer of the switch, and return the values x and $x + e_1$ to satisfy the original requests Fetch&Add(x, e_1) and Fetch&Add(x, e_2), respectively, as illustrated in Fig. 7.11 in four steps.

Applications and Drawbacks   The Fetch&Add primitive is very effective in accessing sequentially allocated queue structures in parallel, or in forking out parallel processes with identical code that operate on different data sets.

Consider the parallel execution of N independent iterations of the following Do loop by p processors:

> **Doall** $N = 1$ to 100
> $\langle$ Code using N $\rangle$
> **Endall**

Each processor executes a Fetch&Add on N before working on a specific iteration of the loop. In this case, a unique value of N is returned to each processor, which is used in the code segment. The code for each processor is written as follows, with N being initialized as 1:

> $n \leftarrow$ Fetch&Add$(N, 1)$
> While $(n \le 100)$ **Doall**
> $\quad \langle$ Code using n $\rangle$
> $\quad n \leftarrow$ Fetch&Add$(N, 1)$
> **Endall**

The advantage of using a combining network to implement the Fetch&Add operation is achieved at a significant increase in network cost. According to NYU Ultracomputer experience, message queueing and combining in each bidirectional 2×2 switch box increased the network cost by a factor of at least 6 or more.

Additional switch cycles are also needed to make the entire operation an atomic memory operation. This may increase the network latency significantly. Multistage combining networks have the potential of supporting large-scale multiprocessors with thousands of processors. The problem of increased cost and latency may be alleviated with the use of faster and cheaper switching technology in the future.

Multistage Networks in Real Systems The IBM RP3 was designed to include 512 processors using a high-speed Omega network for reads or writes and a combining network for synchronization using Fetch&Adds. A 128-port Omega network in the RP3 yields a bandwidth of 13 Gbytes/s using a 50-MHz clock.

Multistage Omega networks are also built into the Cedar multiprocessor (Kuck et al., 1986) at the University of Illinois and in the Ultracomputer (Gottlieb et al., 1983) at New York University.

The BBN Butterfly processor (TC2000) uses 8×8 crossbar switch modules to build a two-stage 64×64 Butterfly network for a 64-processor system, and a three-stage 512×512 Butterfly switch (see Fig. 7.10) for a 512-processor system in the TC2000 Series. The switch hardware is clocked at 38 MHz with a 1-byte data path. The maximum interprocessor bandwidth for a 64-processor TC2000 is designed at 2.4 Gbytes/s.

The Cray Y-MP multiprocessor uses 64-, 128-, or 256-way interleaved memory banks, each of which can be accessed via four ports. Crossbar networks are used between the processors and memory banks in all Cray multiprocessors. The Alliant FX/2800 uses crossbar interconnects between seven four-processor (i860) boards plus one I/O board and eight shared, interleaved cache boards which are connected to the physical memory via a memory bus.

7.2   Cache Coherence and Synchronization Mechanisms

Cache coherence protocols for coping with the multicache inconsistency problem are considered below. Snoopy protocols are designed for bus-connected systems. Directory-based protocols apply to network-connected systems. Finally, we study hardware support for fast synchronization. Software-implemented synchronization will be discussed in Chapter 11.

7.2.1   The Cache Coherence Problem

In a memory hierarchy for a multiprocessor system, data inconsistency may occur between adjacent levels or within the same level. For example, the cache and main memory may contain inconsistent copies of the same data object. Multiple caches may possess different copies of the same memory block because multiple processors operate asynchronously and independently.

Caches in a multiprocessing environment introduce the *cache coherence problem.* When multiple processors maintain locally cached copies of a unique shared-memory location, any local modification of the location can result in a globally inconsistent view of memory. Cache coherence schemes prevent this problem by maintaining a uniform state for each cached block of data. Cache inconsistencies caused by data sharing, process migration, or I/O are explained below.

Inconsistency in Data Sharing   The cache inconsistency problem occurs only when multiple private caches are used. In general, three sources of the problem are identified: *sharing of writable data, process migration,* and *I/O activity.* Figure 7.12 illustrates the problems caused by the first two sources. Consider a multiprocessor with two processors, each using a private cache and both sharing the main memory. Let X be a shared data element which has been referenced by both processors. Before update, the three copies of X are consistent.

If processor P_1 *writes* new data X' into the cache, the same copy will be written immediately into the shared memory using a *write-through* policy. In this case, inconsistency occurs between the two copies (X' and X) in the two caches (Fig. 7.12a).

On the other hand, inconsistency may also occur when a *write-back* policy is used, as shown on the right in Fig. 7.12a. The main memory will be eventually updated when the modified data in the cache are replaced or invalidated.

Process Migration and I/O   Figure 7.12b shows the occurrence of inconsistency after a process containing a shared variable X migrates from processor 1 to processor 2 using the write-back cache on the right. In the middle, a process migrates from processor 2 to processor 1 when using write-through caches.

In both cases, inconsistency appears between the two cache copies, labeled X and X'. Special precautions must be exercised to avoid such inconsistencies. A coherence protocol must be established before processes can safely migrate from one processor to another.

Inconsistency problems may occur during I/O operations that bypass the caches.

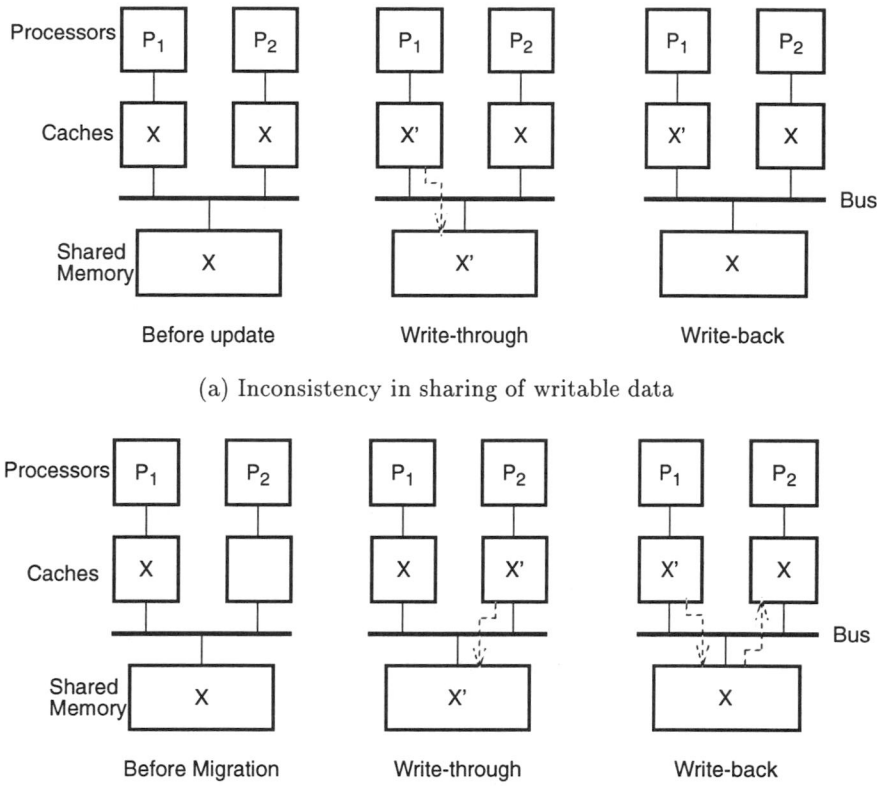

(a) Inconsistency in sharing of writable data

(b) Inconsistency after process migration

Figure 7.12 Cache coherence problems in data sharing and in process migration. (Adapted from Dubois, Scheurich, and Briggs 1988)

When the I/O processor *loads* a new data X' into the main memory, bypassing the write-through caches (middle diagram in Fig. 7.13a), inconsistency occurs between cache 1 and the shared memory. When outputting a data directly from the shared memory (bypassing the caches), the write-back caches also create inconsistency.

One possible solution to the I/O inconsistency problem is to attach the I/O processors (IOP_1 and IOP_2) to the private caches (C_1 and C_2), respectively, as shown in Fig. 7.13b. This way I/O processors share caches with the CPU. The I/O consistency can be maintained if cache-to-cache consistency is maintained via the bus. An obvious shortcoming of this scheme is the likely increase in cache perturbations and the poor locality of I/O data, which may result in higher miss ratios.

Two Protocol Approaches Many of today's commercially available multiprocessors use bus-based memory systems. A bus is a convenient device for ensuring cache coherence because it allows all processors in the system to observe ongoing memory

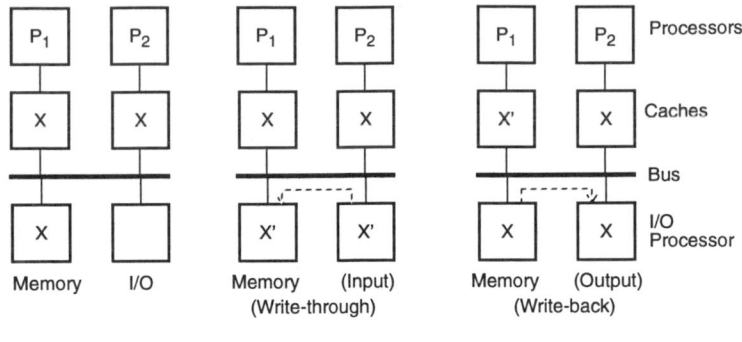

(a) I/O operations bypassing the cache

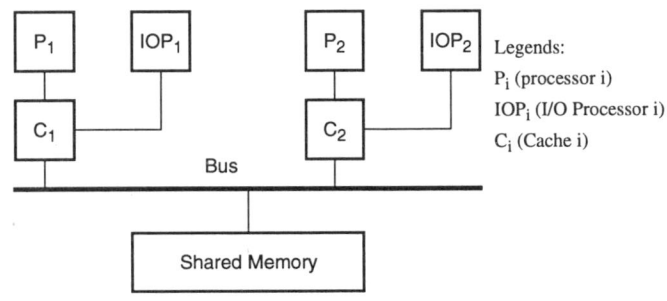

(b) A possible solution

Figure 7.13 Cache inconsistency after an I/O operation and a possible solution.
(Adapted from Dubois, Scheurich, and Briggs, 1988)

transactions. If a bus transaction threatens the consistent state of a locally cached object, the cache controller can take appropriate actions to invalidate the local copy. Protocols using this mechanism to ensure coherence are called *snoopy protocols* because each cache snoops on the transactions of other caches.

On the other hand, scalable multiprocessor systems interconnect processors using short point-to-point wires in direct or multistage networks. Unlike the situation in buses, the bandwidth of these networks increases as more processors are added to the system. However, such networks do not have a convenient snooping mechanism and do not provide an efficient broadcast capability. In such systems, the cache coherence problem can be solved using some variant of directory schemes.

In general, a cache coherence protocol consists of the set of possible states in the local caches, the state in the shared memory, and the state transitions caused by the messages transported through the interconnection network to keep memory coherent. In what follows, we first describe the snoopy protocols and then the directory-based protocols. These protocols rely on software, hardware, or a combination of both for implementation. Cache coherence can also be enforced in the TLB or assisted by the compiler. Other approaches to designing a scalable cache coherence interface will be studied in Chapter 9.

7.2.2   Snoopy Bus Protocols

In using private caches associated with processors tied to a common bus, two approaches have been practiced for maintaining cache consistency: *write-invalidate* and *write-update* policies. Essentially, the write-invalidate policy will invalidate all remote copies when a local cache block is updated. The write-update policy will broadcast the new data block to all caches containing a copy of the block.

Snoopy protocols achieve data consistency among the caches and shared memory through a bus watching mechanism. As illustrated in Fig. 7.14, two snoopy bus protocols create different results. Consider three processors (P_1, P_2, and P_n) maintaining consistent copies of block X in their local caches (Fig. 7.14a) and in the shared-memory module marked X.

Using a *write-invalidate protocol*, the processor P_1 modifies (writes) its cache from X to X', and all other copies are invalidated via the bus (denoted I in Fig. 7.14b). Invalidated blocks are sometimes called *dirty*, meaning they should not be used. The *write-update protocol* (Fig. 7.14c) demands the new block content X' be broadcast to all cache copies via the bus. The memory copy is also updated if write-through caches are used. In using write-back caches, the memory copy is updated later at block replacement time.

Write-Through Caches   The states of a cache block copy change with respect to *read, write*, and *replacement* operations in the cache. Figure 7.15 shows the state transitions for two basic write-invalidate snoopy protocols developed for write-through and write-back caches, respectively. A block copy of a write-through cache i attached to processor i can assume one of two possible cache states: *valid* or *invalid* (Fig. 7.15a).

A remote processor is denoted j, where $j \neq i$. For each of the two cache states, six possible events may take place. Note that all cache copies of the same block use the same transition graph in making state changes.

In a *valid* state (Fig. 7.15a), all processors can *read* ($R(i), R(j)$) safely. Local processor i can also *write* ($W(i)$) safely in a *valid* state. The *invalid* state corresponds to the case of the block either being invalidated or being replaced ($Z(i)$ or $Z(j)$).

Wherever a remote processor *writes* ($W(j)$) into its cache copy, all other cache copies become invalidated. The cache block in cache i becomes *valid* whenever a successful read ($R(i)$) or write ($W(i)$) is carried out by a local processor i.

The fraction of *write cycles* on the bus is higher than the fraction of *read cycles* in a write-through cache, due to the need for request invalidations. The *cache directory* (registration of cache states) can be made in dual copies or dual-ported to filter out most invalidations. In case *locks* are cached, an atomic Test&Set must be enforced.

Write-Back Caches   The *valid* state of a write-back cache can be further split into two cache states, labeled RW (*read-write*) and RO (*read-only*) in Fig. 7.15b. The INV (invalidated or not-in-cache) cache state is equivalent to the *invalid* state mentioned

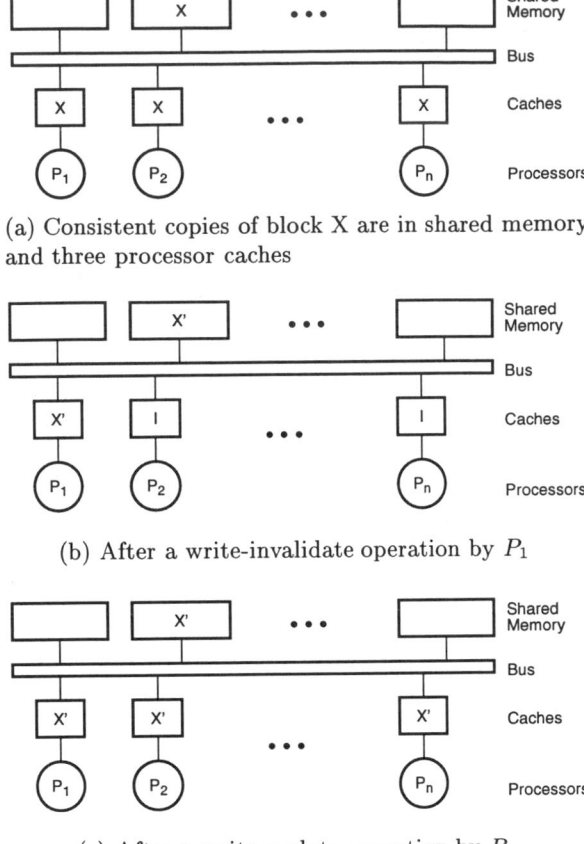

(a) Consistent copies of block X are in shared memory and three processor caches

(b) After a write-invalidate operation by P_1

(c) After a write-update operation by P_1

Figure 7.14 Write-invalidate and write-update coherence protocols for write-through caches (I: invalidate).

before. This three-state coherence scheme corresponds to an *ownership protocol*.

When the memory owns a block, caches can contain only the RO copies of the block. In other words, multiple copies may exist in the RO state and every processor having a copy (called a *keeper* of the copy) can *read* $(R(i), R(j))$ the copy safely.

The INV state is entered whenever a remote processor *writes* $(W(j))$ its local copy or the local processor *replaces* $(Z(i))$ its own block copy. The RW state corresponds to only one cache copy existing in the entire system owned by the local processor i. *Read* $(R(i))$ and *write* $(W(i))$ can be safely performed in the RW state. From either the RO state or the INV state, the cache block becomes uniquely owned when a local *write* $(W(i))$ takes place.

Other state transitions in Fig. 7.15b can be similarly figured out. Before a block is modified, ownership for exclusive access must first be obtained by a *read-only* bus

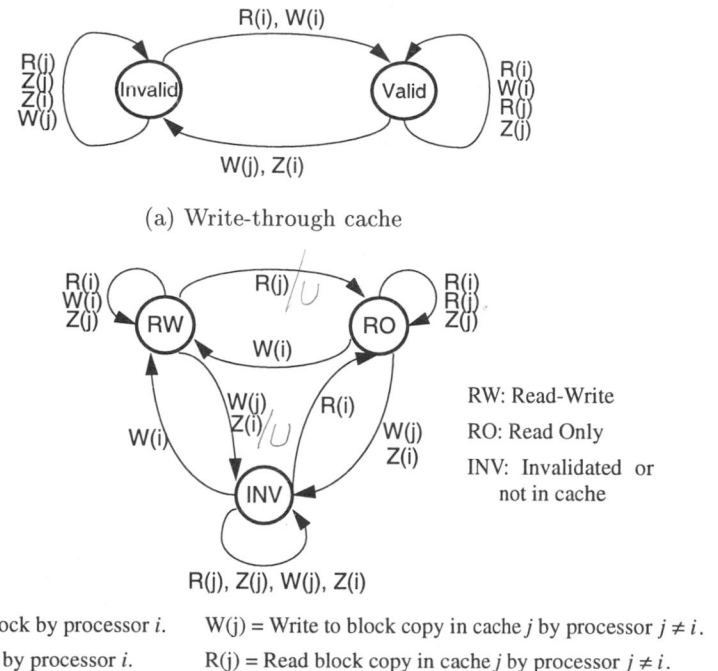

(a) Write-through cache

RW: Read-Write

RO: Read Only

INV: Invalidated or
not in cache

W(i) = Write to block by processor *i*.   W(j) = Write to block copy in cache *j* by processor *j ≠ i*.

R(i) = Read block by processor *i*.   R(j) = Read block copy in cache *j* by processor *j ≠ i*.

Z(i) = Replace block in cache *i*.   Z(j) = Replace block copy in cache *j ≠ i*.

(b) Write-back cache

**Figure 7.15 Two state-transition graphs for a cache block using write-invalidate
snoopy protocols.** (Adapted from Dubois, Scheurich, and Briggs, 1988)

transaction which is broadcast to all caches and memory. If a modified block copy
exists in a remote cache, memory must first be updated, the copy invalidated, and
ownership transferred to the requesting cache.

Write-once Protocol   James Goodman (1983) has proposed a cache coherence protocol for bus-based multiprocessors. This scheme combines the advantages of both
write-through and write-back invalidations. In order to reduce bus traffic, the very first
write of a cache block uses a write-through policy.

This will result in a consistent memory copy while all other cache copies are invalidated. After the first *write*, shared memory is updated using a write-back policy. This
scheme can be described by the four-state transition graph shown in Fig. 7.16. The four
cache states are defined below:

- *Valid*: The cache block, which is consistent with the memory copy, has been *read*
  from shared memory and has not been modified.
- *Invalid*: The block is not found in the cache or is inconsistent with the memory

copy.

- *Reserved*: Data has been *written* exactly *once* since being *read* from shared memory. The cache copy is consistent with the memory copy, which is the only other copy.
- *Dirty*: The cache block has been modified (*written*) more than once, and the cache copy is the only one in the system (thus inconsistent with all other copies).

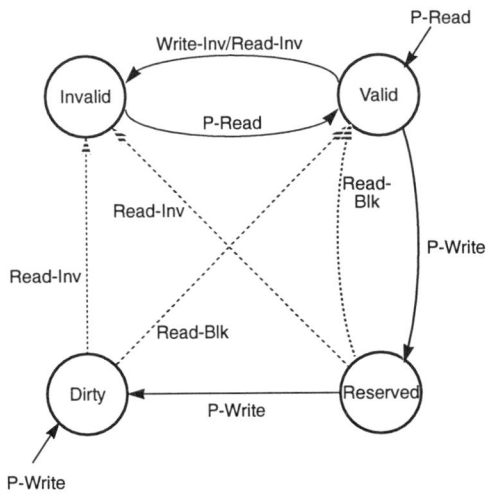

Solid lines: Command issued by local processor

Dashed lines: Commands issued by remote processors via the system bus.

Figure 7.16 Goodman's write-once cache coherence protocol using the write-invalidate policy on write-back caches. (Adapted from James Goodman 1983, reprinted from Stenström, *IEEE Computer*, June 1990)

To maintain consistency, the protocol requires two different sets of commands. The solid lines in Fig. 7.16 correspond to access commands issued by a local processor labeled *read-miss*, *write-hit*, and *write-miss*. Whenever a *read-miss* occurs, the *valid* state is entered.

The first *write-hit* leads to the *reserved* state. The second *write-hit* leads to the *dirty* state, and all future *write-hits* stay in the *dirty* state. Whenever a *write-miss* occurs, the cache block enters the *dirty* state.

The dashed lines correspond to invalidation commands issued by remote processors via the snoopy bus. The *read-invalidate* command reads a block and invalidates all other copies. The *write-invalidate* command invalidates all other copies of a block. The *bus-read* command corresponds to a normal memory *read* by a remote processor via the bus.

Cache Events and Actions   The memory-access and invalidation commands trigger the following events and actions:

- *Read-miss*: When a processor wants to read a block that is not in the cache, a *read-miss* occurs. A *bus-read* operation will be initiated. If no *dirty* copy exists, then main memory has a consistent copy and supplies a copy to the requesting cache. If a *dirty* copy does exist in a remote cache, that cache will inhibit the main memory and send a copy to the requesting cache. In all cases, the cache copy will enter the *valid* state after a read-miss.

- *Write-hit*: If the copy is in the *dirty* or *reserved* state, the *write* can be carried out locally and the new state is *dirty*. If the new state is *valid*, a *write-invalidate* command is broadcast to all caches, invalidating their copies. The shared memory is *written through*, and the resulting state is *reserved* after this first *write*.

- *Write-miss*: When a processor fails to write in a local cache, the copy must come either from the main memory or from a remote cache with a dirty block. This is accomplished by sending a *read-invalidate* command which will invalidate all cache copies. The local copy is thus updated and ends up in a *dirty* state.

- *Read-hit*: Read-hits can always be performed in a local cache without causing a state transition or using the snoopy bus for invalidation.

- *Block Replacement*: If a copy is *dirty*, it has to be written back to main memory by block replacement. If the copy is *clean* (i.e., in either the *valid*, *reserved*, or *invalid* state), no replacement will take place.

Goodman's write-once protocol demands special bus lines to inhibit the main memory when the memory copy is invalid, and a *bus-read* operation is needed after a *read-miss*. Most standard buses cannot support this inhibition operation.

The IEEE Futurebus+ proposed to include this special bus provision. Using a write-through policy after the first *write* and using a write-back policy in all additional *writes* will eliminate unnecessary invalidations.

Snoopy cache protocols are popular in bus-based multiprocessors because of their simplicity of implementation. The write-invalidate policies have been implemented on the Sequent Symmetry multiprocessor and on the Alliant FX multiprocessor.

Besides the DEC Firefly multiprocessor, the Xerox Palo Alto Research Center has implemented another write-update protocol for its Dragon multiprocessor workstation. The Dragon protocol avoids updating memory until replacement, in order to improve the efficiency of intercache transfers.

The Futurebus+ Protocol   The Futurebus+ parallel protocols support both connected and split transactions. Generally speaking, connected transactions are cheaper and easier to implement on a single bus segment. Split transactions are more complex and expensive but provide greater concurrency in building large hierarchical-bus multiprocessors. This section explains the parallel arbitration mechanism and cache coherence developed for Futurebus+-based multiprocessors. The Futurebus+ is well suited to shared-memory multiprocessors. The types of bus transactions are tuned to drive the state of various cache modules to conform with almost any cache coherence protocol. A unified cache model, called MESI, has been developed to maintain cache consistency in an environment with multiple logical Futurebuses connected in a hierarchy.

Bus repeaters are used to support split transactions, which insist that a copy of a transaction must be repeated on adjacent buses having the same cache block. This implies that the bus repeater contains a cache buffer which can record the passage of a cache block to a neighboring bus. The cache coherence protocol proposed for Futurebus+ is modified from Goodman's protocol.

The snoopy coherence protocol works for connected transactions. For split transactions over multiple bus segments, additional interface modules are needed. Figure 7.17 shows a split-transaction, shared-memory multiprocessor architecture.

This system is an extension of the shared-memory architecture with *cache agent* and *memory agent* modules. A cache agent uses split transactions to assume all the rights and responsibilities of a number of remote cache modules. Cache agents must maintain tags indicating the cache block states.

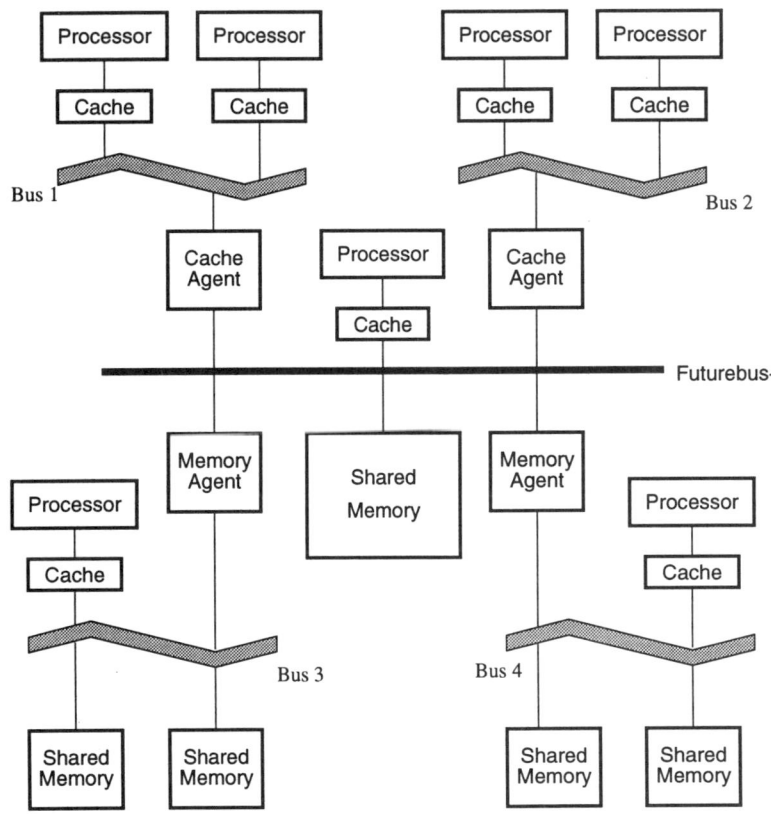

Figure 7.17 Split transactions on a multiprocessor. (Reprinted with permission from IEEE Standard 896.1-1991, copyright © 1991 by IEEE, Inc.)

Similarly, a memory agent uses split transactions to assume all the rights and responsibilities of some remote memory modules. A memory agent splits a transaction to assign *read* and *write* permissions to the location it is handling for the bus segment it is on. It should also pass external invalidations or interventions to the bus segment. Split transactions have been used to delay invalidation completions. Various cache coherence commands are needed on the buses.

Multilevel Cache Coherence To maintain consistency among cache copies at various levels, Wilson proposed an extension to the write-invalidate protocol used on a single bus. Consistency among cache copies at the same level is maintained in the same way as described above. Consistency of caches at different levels is illustrated in Fig. 7.3.

An invalidation must propagate vertically up and down in order to invalidate all copies in the shared caches at level 2. Suppose processor P_1 issues a *write* request. The *write* request propagates up to the highest level and invalidates copies in C_{20}, C_{22}, C_{16}, and C_{18}, as shown by the arrows from C_{11} to all the shaded copies.

High-level caches such as C_{20} keep track of dirty blocks beneath them. A subsequent *read* request issued by P_7 will propagate up the hierarchy because no copies exist. When it reaches the top level, cache C_{20} issues a flush request down to cache C_{11} and the dirty copy is supplied to the private cache associated with processor P_7. Note that higher-level caches act as filters for consistency control. An invalidation command or a read request will not propagate down to clusters that do not contain a copy of the corresponding block. The cache C_{21} acts in this manner.

Protocol Performance Issues The performance of any snoopy protocol depends heavily on the workload patterns and implementation efficiency. The main motivation for using the snooping mechanism is to reduce bus traffic, with a secondary goal of reducing the effective memory-access time. The block size is very sensitive to cache performance in write-invalidate protocols, but not in write-update protocols.

For a uniprocessor system, bus traffic and memory-access time are mainly contributed by cache misses. The miss ratio decreases when block size increases. However, as the block size increases to a *data pollution* point, the miss ratio starts to increase. For larger caches, the data pollution point appears at a larger block size.

For a system requiring extensive process migration or synchronization, the write-invalidate protocol will perform better. However, a cache miss can result for an invalidation initiated by another processor prior to the cache access. Such *invalidation misses* may increase bus traffic and thus should be reduced.

Extensive simulation results have suggested that bus traffic in a multiprocessor may increase when the block size increases. Write-invalidate also facilitates the implementation of synchronization primitives. Typically, the average number of invalidated cache copies is rather small (one or two) in a small multiprocessor.

The write-update protocol requires a bus broadcast capability. This protocol also can avoid the ping-pong effect on data shared between multiple caches. Reducing the sharing of data will lessen bus traffic in a write-update multiprocessor. However, write-update cannot be used with long write bursts. Only through extensive program traces (trace-driven simulation) can one reveal the cache behavior, hit ratio, bus traffic, and effective memory-access time.

7.2.3   Directory-Based Protocols

A write-invalidate protocol may lead to heavy bus traffic caused by *read-misses*, resulting from the processor updating a variable and other processors trying to read the same variable. On the other hand, the write-update protocol may update data items in remote caches which will never be used by other processors. In fact, these problems pose additional limitations in using buses to build large multiprocessors.

When a multistage network is used to build a large multiprocessor with hundreds of processors, the snoopy cache protocols must be modified to suit the network capabilities. Since broadcasting is very expensive to perform in a multistage network, consistency commands will be sent only to those caches that keep a copy of the block. This leads to *directory-based protocols* for network-connected multiprocessors.

Directory Structures   In a multistage network, cache coherence is supported by using cache directories to store information on where copies of cache blocks reside. Various directory-based protocols differ mainly in how the directory maintains information and what information it stores.

Tang (1976) proposed the first directory scheme, which used a *central directory* containing duplicates of all cache directories. This central directory, providing all the information needed to enforce consistency, is usually very large and must be associatively searched, like the individual cache directories. Contention and long search times are two drawbacks in using a central directory for a large multiprocessor.

A distributed-directory scheme was proposed by Censier and Feautrier (1978). Each memory module maintains a separate directory which records the state and presence information for each memory block. The state information is local, but the presence information indicates which caches have a copy of the block.

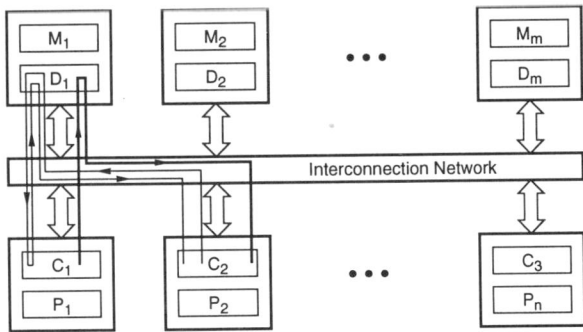

Figure 7.18 Basic concept of a directory-based cache coherence scheme. (Courtesy of Censier and Feautrier, *IEEE Trans. Computers*, Dec. 1978)

In Fig. 7.18, a *read-miss* (thin lines) in cache 2 results in a request sent to the memory module. The memory controller retransmits the request to the dirty copy in cache 1. This cache *writes back* its copy. The memory module can supply a copy to the requesting cache. In the case of a *write-hit* at cache 1 (bold lines), a command is sent to the memory controller, which sends invalidations to all caches (cache 2) marked in the presence vector residing in the directory D_1.

A cache-coherence protocol that does not use broadcasts must store the locations of all cached copies of each block of shared data. This list of cached locations, whether centralized or distributed, is called a *cache directory*. A directory entry for each block of data contains a number of *pointers* to specify the locations of copies of the block. Each directory entry also contains a dirty bit to specify whether a unique cache has permission to write the associated block of data.

Different types of directory protocols fall under three primary categories: *full-map directories, limited directories,* and *chained directories.* Full-map directories store enough data associated with each block in global memory so that every cache in the system can simultaneously store a copy of any block of data. That is, each directory entry contains N pointers, where N is the number of processors in the system.

Limited directories differ from full-map directories in that they have a fixed number of pointers per entry, regardless of the system size. Chained directories emulate the full-map schemes by distributing the directory among the caches. The following descriptions of the three classes of cache directories are based on the original classification by Chaiken, Fields, Kwihara, and Agarwal (1990):

Full-Map Directories   The full-map protocol implements directory entries with one bit per processor and a dirty bit. Each bit represents the status of the block in the corresponding processor's cache (present or absent). If the dirty bit is set, then one and only one processor's bit is set and that processor can write into the block.

A cache maintains two bits of state per block. One bit indicates whether a block is valid, and the other indicates whether a valid block may be written. The cache coherence protocol must keep the state bits in the memory directory and those in the cache consistent.

Figure 7.19a illustrates three different states of a full-map directory. In the first state, location X is missing in all of the caches in the system. The second state results from three caches (C1, C2, and C3) requesting copies of location X. Three pointers (processor bits) are set in the entry to indicate the caches that have copies of the block of data. In the first two states, the dirty bit on the left side of the directory entry is set to clean (C), indicating that no processor has permission to write to the block of data. The third state results from cache C3 requesting write permission for the block. In the final state, the dirty bit is set to dirty (D), and there is a single pointer to the block of data in cache C3.

Let us examine the transition from the second state to the third state in more detail. Once processor P3 issues the write to cache C3, the following events will take place:

(1) Cache C3 detects that the block containing location X is valid but that the processor does not have permission to write to the block, indicated by the block's write-permission bit in the cache.

(2) Cache C3 issues a write request to the memory module containing location X

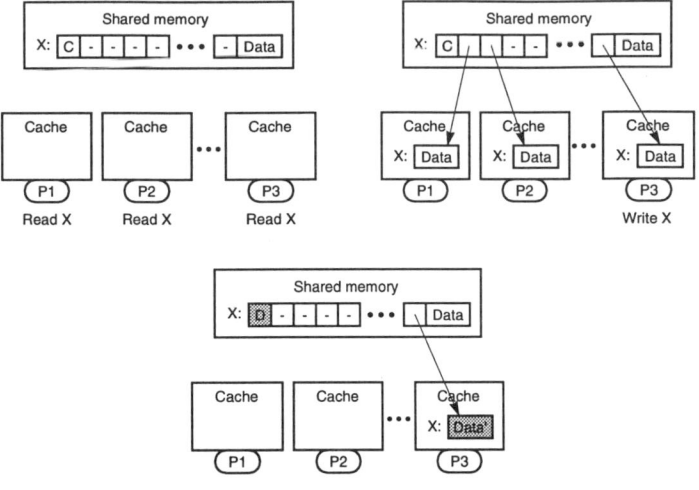

(a) Three states of a full-map directory

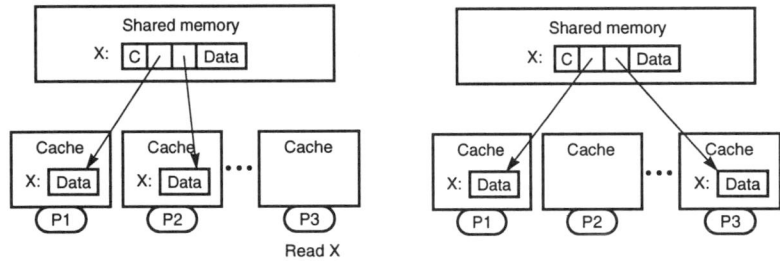

(b) Eviction in a limited directory

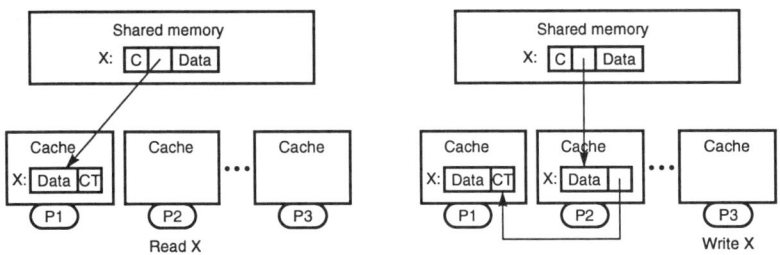

(c) The chained directory

Figure 7.19 Three types of cache directory protocols. (Courtesy of Chaiken et al., *IEEE Computer*, June 1990)

and stalls processor P3.

(3) The memory module issues invalidate requests to caches C1 and C2.

(4) Caches C1 and C2 receive the invalidate requests, set the appropriate bit to indicate that the block containing location X is invalid, and send acknowledgments back to the memory module.

(5) The memory module receives the acknowledgments, sets the dirty bit, clears the pointers to caches C1 and C2, and sends write permission to cache C3.

(6) Cache C3 receives the write permission message, updates the state in the cache, and reactivates processor P3.

The memory module waits to receive the acknowledgments before allowing processor P3 to complete its write transaction. By waiting for acknowledgments, the protocol guarantees that the memory system ensures sequential consistency. The full-map protocol provides a useful upper bound for the performance of centralized directory-based cache coherence. However, it is not scalable due to excessive memory overhead.

Because the size of the directory entry associated with each block of memory is proportional to the number of processors, the memory consumed by the directory is proportional to the size of memory $O(N)$ multiplied by the size of the directory $O(N)$. Thus, the total memory overhead scales as the square of the number of processors $O(N^2)$.

Limited Directories   Limited directory protocols are designed to solve the directory size problem. Restricting the number of simultaneously cached copies of any particular block of data limits the growth of the directory to a constant factor.

A directory protocol can be classified as $Dir_i X$ using the notation from Agarwal et al. (1988). The symbol i stands for the number of pointers, and X is NB for a scheme with no broadcast. A full-map scheme without broadcast is represented as $Dir_N NB$. A limited directory protocol that uses $i < N$ pointers is denoted $Dir_i NB$. The limited directory protocol is similar to the full-map directory, except in the case when more than i caches request read copies of a particular block of data.

Figure 7.19b shows the situation when three caches request read copies in a memory system with a $Dir_2 NB$ protocol. In this case, we can view the two-pointer directory as a two-way set-associative cache of pointers to shared copies. When cache C3 requests a copy of location X, the memory module must invalidate the copy in either cache C1 or cache C2. This process of pointer replacement is called *eviction*. Since the directory acts as a set-associative cache, it must have a pointer replacement policy.

If the multiprocessor exhibits processor locality in the sense that in any given interval of time only a small subset of all the processors access a given memory word, then a limited directory is sufficient to capture this small worker set of processors.

Directory pointers in a $Dir_i NB$ protocol encode binary processor identifiers, so each pointer requires $\log_2 N$ bits of memory, where N is the number of processors in the system. Given the same assumptions as for the full-map protocol, the memory overhead of limited directory schemes grows as $O(N \log_2 N)$.

These protocols are considered scalable with respect to memory overhead because the resource required to implement them grows approximately linearly with the number of processors in the system. $Dir_i B$ protocols allow more than i copies of each block of data to exist, but they resort to a broadcast mechanism when more than i cached copies of a block need to be invalidated. However, point-to-point interconnection networks do not provide an efficient systemwide broadcast capability. In such networks, it is difficult to determine the completion of a broadcast to ensure sequential consistency.

Chained Directories   Chained directories realize the scalability of limited directories without restricting the number of shared copies of data blocks. This type of cache coherence scheme is called a *chained* scheme because it keeps track of shared copies of data by maintaining a chain of directory pointers.

The simpler of the two schemes implements a singly linked chain, which is best described by example (Fig. 7.19c). Suppose there are no shared copies of location X. If processor P1 reads location X, the memory sends a copy to cache C1, along with a *chain termination* (CT) pointer. The memory also keeps a pointer to cache C1. Subsequently, when processor P2 reads location X, the memory sends a copy to cache C2, along with the pointer to cache C1. The memory then keeps a pointer to cache C2.

By repeating the above step, all of the caches can cache a copy of the location X. If processor P3 writes to location X, it is necessary to send a data invalidation message down the chain. To ensure sequential consistency, the memory module denies processor P3 write permission until the processor with the chain termination pointer acknowledges the invalidation of the chain. Perhaps this scheme should be called a *gossip* protocol (as opposed to a snoopy protocol) because information is passed from individual to individual rather than being spread by covert observation.

The possibility of cache block replacement complicates chained-directory protocols. Suppose that caches C1 through CN all have copies of location X and that location X and location Y map to the same (direct-mapped) cache line. If processor P_i reads location Y, it must first evict location X from its cache with the following possibilities:

(1) Send a message down the chain to cache C_{i-1} with a pointer to cache C_{i+1} and splice C_i out of the chain, or

(2) Invalidate location X in cache C_{i+1} through cache C_N.

The second scheme can be implemented by a less complex protocol than the first. In either case, sequential consistency is maintained by locking the memory location while invalidations are in progress. Another solution to the replacement problem is to use a doubly linked chain. This scheme maintains forward and backward chain pointers for each cached copy so that the protocol does not have to traverse the chain when there is a cache replacement. The doubly linked directory optimizes the replacement condition at the cost of a larger average message block size (due to the transmission of extra directory pointers), twice the pointer memory in the caches, and a more complex coherence protocol.

Although the chained protocols are more complex than the limited directory protocols, they are still scalable in terms of the amount of memory used for the directories. The pointer sizes grow as the logarithm of the number of processors, and the number of pointers per cache or memory block is independent of the number of processors.

Cache Design Alternatives   The relative merits of physical address caches and virtual address caches have to be judged based on the access time, the aliasing problem, the flushing problem, OS kernel overhead, special tagging at the process level, and cost/performance considerations. Beyond the use of private caches, three design alternatives are suggested below.

Each of the design alternatives has its own advantages and shortcomings. There exists insufficient evidence to determine whether any of the alternatives is better or worse than the use of private caches. More research and trace data are needed to apply these cache architectures in designing future high-performance multiprocessors.

Shared Caches   An alternative approach to maintaining cache coherence is to completely eliminate the problem by using *shared caches* attached to shared-memory modules. No private caches are allowed in this case. This approach will reduce the main memory access time but contributes very little to reducing the overall memory-access time and to resolving access conflicts.

Shared caches can be built as second-level caches. Sometimes, one can make the second-level caches partially shared by different clusters of processors. Various cache architectures are possible if private and shared caches are both used in a memory hierarchy. The use of shared cache alone may be against the scalability of the entire system. Tradeoffs between using private caches, caches shared by multiprocessor clusters, and shared main memory are interesting topics for further research.

Noncacheable Data   Another approach is not to cache shared writable data. Shared data are *noncacheable*, and only instructions or private data are *cacheable* in local caches. Shared data include locks, process queues, and any other data structures protected by critical sections.

The compiler must tag data as either *cacheable* or *noncacheable*. Special hardware tagging must be used to distinguish them. Caches with cacheable and noncacheable blocks demand more programmer effort, in addition to support from hardware and compilers. Again, this opens up new research issues between programmability and scalability.

Cache Flushing   A third approach is to use *cache flushing* every time a synchronization primitive is executed. This may work well with transaction processing multiprocessor systems. Cache flushes are slow unless special hardware is used. This approach does not solve I/O and process migration problems.

Flushing can be made very selective by programmers or by the compiler in order to increase efficiency. Cache flushing at synchronization, I/O, and process migration may be carried out unconditionally or selectively. Cache flushing is more often used with virtual address caches.

7.2.4   Hardware Synchronization Mechanisms

Synchronization is a special form of communication in which control information is exchanged, instead of data, between communicating processes residing in the same or different processors. Synchronization enforces correct sequencing of processors and ensures mutually exclusive access to shared writable data. Synchronization can be implemented in software, firmware, and hardware through controlled sharing of data and control information in memory.

Multiprocessor systems use hardware mechanisms to implement low-level or primitive synchronization operations, or use software (operating system)-level synchronization mechanisms such as *semaphores* or *monitors*. Only hardware synchronization mechanisms are studied below. Software approaches to synchronization will be treated in Chapter 10.

Atomic Operations   Most multiprocessors are equipped with hardware mechanisms for enforcing atomic operations such as memory *read*, *write*, or *read-modify-write* operations which can be used to implement some synchronization primitives. Besides atomic memory operations, some interprocessor interrupts can be used for synchronization purposes. For example, the synchronization primitives, Test&Set (*lock*) and Reset (*lock*), are defined below:

$$
\begin{aligned}
\text{Test\&Set} \quad & (lock) \\
& temp \leftarrow lock; \quad lock \leftarrow 1; \\
& \text{return } temp \\
\text{Reset} \quad & (lock) \\
& lock \leftarrow 0
\end{aligned}
\tag{7.4}
$$

Test&Set is implemented with atomic *read-modify-write* memory operations. To synchronize concurrent processes, the software may repeat Test&Set until the returned value (*temp*) becomes 0. This synchronization primitive may tie up some bus cycles while a processor enters busy-waiting on the *spin lock*. To avoid spinning, interprocessor interrupts can be used.

A lock tied to an interrupt is called a *suspend lock*. Using such a lock, a process does not relinquish the processor while it is waiting. Whenever the process fails to open the lock, it records its status and disables all interrupts aiming at the lock. When the lock is open, it signals all waiting processors through an interrupt. A similar primitive, Compare&Swap, has been implemented in IBM 370 mainframes.

Synchronization on Futurebus+   The asynchronous (source-synchronized) protocols used in Futurebus+ allow the synchronization domain of the *sender* to extend along the bus segment, presenting only one synchronization interface between the *bus* and the *receiver*.

A centrally synchronized bus requires three separate synchronization domains (*bus*, *sender*, *receiver*). Since Futurebus+ uses only one synchronization interface, it eliminates unnecessary delays and spatial skews caused by resynchronization in other interfaces. Thus higher performance is expected using a sender-synchronized protocol for synchronization.

Concurrent processes residing in different processors can be synchronized using *barriers*. A barrier can be implemented by a shared-memory word which keeps counting the number of processes reaching the barrier. After all processes have updated the barrier counter, the synchronization point has been reached. No processor can execute beyond the barrier until the synchronization process is complete.

Wired Barrier Synchronization A wired-NOR logic is shown in Fig. 7.20 for implementing a barrier mechanism for fast synchronization. Each processor uses a dedicated control vector $X = (X_1, X_2, ..., X_m)$ and accesses a common monitor vector $Y = (Y_1, Y_2, ..., Y_m)$ in shared memory, where m corresponds to the barrier lines used.

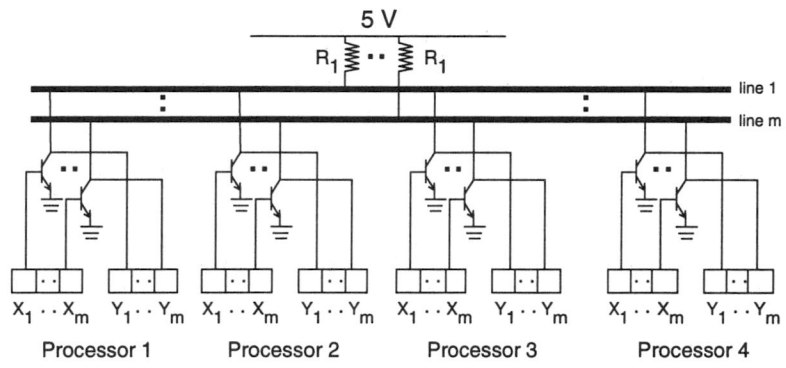

(a) Barrier lines and interface logic

Step 1: Forking (use of one barrier line)

	Processor 1	Processor 2	Processor 3	Processor 4
Line	1			
X	1	1	1	1
Y	0	0	0	0

Step 2: Process 1 and Process 3 reach the synchronization point

X	0	1	0	1
Y	0	0	0	0
	Process 1	Process 2	Process 3	Process 4

Step 3: All processes reach the synchronization point

X	0	0	0	0
Y	1	1	1	1
	Process 1	Process 2	Process 3	Process 4

(b) Synchronization steps

Figure 7.20 The synchronization of four independent processes on four processors using one wired-NOR barrier line. (Adapted from Hwang and Shang, *Proc. Int. Conf. Parallel Processing*, 1991)

The number of barrier lines needed for synchronization depends on the multiprogramming degree and the size of the multiprocessor system. Each control bit X_i is connected to the base (input) of a probing transistor. The monitor bit Y_i checks the collector voltage (output) of the transistor.

Each barrier line is wired-NOR to n transistors from n processors. Whenever bit X_i is raised to high (1), the corresponding transistor is closed, pulling down (0) the level of barrier line i. The wired-NOR connection implies that line i will be high (1) only if control bits X_i from all processors are low (0).

This demonstrates the ability to use the control bit X_i to signal the completion of a process on processor i. The bit X_i is set to 1 when a process is initiated and reset to 0 when the process finishes its execution.

When all processes finish their jobs, the X_i bits from the participating processors are all set to 0; and the barrier line is then raised to high (1), signaling the synchronization barrier has been crossed. This timing is watched by all processors through snooping on the Y_i bits. Thus only one barrier line is needed to monitor the initiation and completion of a single synchronization involving many concurrent processes.

Multiple barrier lines can be used simultaneously to monitor several synchronization points. Figure 7.20 shows the synchronization of four processes residing on four processors using one barrier line. Note that other barrier lines can be used to synchronize other processes at the same time in a multiprogrammed multiprocessor environment.

Example 7.2 Wired barrier synchronization of five partially ordered processes (Hwang and Shang, 1991)

If the synchronization pattern is predicted after compile time, then one can follow the precedence graph of a partially ordered set of processes to perform multiple synchronization as demonstrated in Fig. 7.21.

Here five processes $(P_1, P_2, ..., P_5)$ are synchronized by snooping on five barrier lines corresponding to five synchronization points labeled a, b, c, d, e. At step 0 the control vectors need to be initialized. All five processes are synchronized at point a. The crossing of barrier a is signaled by monitor bit Y_1, which is observable by all processors.

Barriers b and c can be monitored simultaneously using two lines as shown in steps 2a and 2b. Only four steps are needed to complete the entire process. Note that only one copy of the monitor vector Y is maintained in the shared memory. The bus interface logic of each processor module has a copy of Y for local monitoring purposes as shown in Fig. 7.21c.

■

Separate control vectors are used in local processors. The above dynamic barrier synchronization is possible only if the synchronization pattern is predicted at compile time and process preemption is not allowed. One can also use the barrier wires along with counting semaphores in memory to support multiprogrammed multiprocessors in which preemption is allowed.

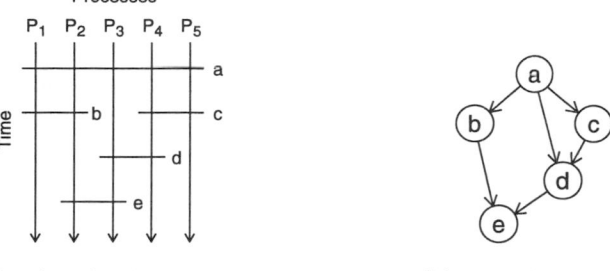

(a) Synchronization patterns   (b) Precedence graph

Step 0: Initializing the control vectors (use 5 barrier lines)

Processor 1   Processor 2   Processor 3   Processor 4   Processor 5

X 1 1 0 0 0   1 1 0 0 1   1 0 0 1 1   1 0 1 1 0   1 0 1 0 0
Y 0 0 0 0 0   0 0 0 0 0   0 0 0 0 0   0 0 0 0 0   0 0 0 0 0

Step 1: Synchronization at barrier a

0 1 0 0 0   0 1 0 0 1   0 0 0 1 1   0 0 1 1 0   0 0 1 0 0
1 0 0 0 0   1 0 0 0 0   1 0 0 0 0   1 0 0 0 0   1 0 0 0 0

Step 2a: Synchronization at barrier b

0 0 0 0 0   0 0 0 0 1   0 0 0 1 1   0 0 1 1 0   0 0 1 0 0
1 1 0 0 0   1 1 0 0 0   1 1 0 0 0   1 1 0 0 0   1 1 0 0 0

Step 2b: Synchronization at barrier c

0 0 0 0 0   0 0 0 0 1   0 0 0 1 1   0 0 0 1 0   0 0 0 0 0
1 1 1 0 0   1 1 1 0 0   1 1 1 0 0   1 1 1 0 0   1 1 1 0 0

Step 3: Synchronization at barrier d

0 0 0 0 0   0 0 0 0 1   0 0 0 0 1   0 0 0 0 0   0 0 0 0 0
1 1 1 1 0   1 1 1 1 0   1 1 1 1 0   1 1 1 1 0   1 1 1 1 0

Step 4: Synchronization at barrier e

0 0 0 0 0   0 0 0 0 0   0 0 0 0 0   0 0 0 0 0   0 0 0 0 0
1 1 1 1 1   1 1 1 1 1   1 1 1 1 1   1 1 1 1 1   1 1 1 1 1

(c) Synchronization steps

Figure 7.21 The synchronization of five partially ordered processes using wired-NOR barrier lines. (Adapted from Hwang and Shang, *Proc. Int. Conf. Parallel Processing*, 1991)

7.3   Three Generations of Multicomputers

Three generations of multicomputers are reviewed in this section. Experiences from Intel, nCUBE, MIT, and Caltech are examined. In particular, we present the Intel Paragon system in some detail. The generic multicomputer model shown in Fig. 1.9 and various network topologies presented in Section 2.3 form the background needed for reading this section.

7.3.1   Design Choices in the Past

So far, message-passing multicomputers have gone through two generations of development, corresponding to the past and present, and a new generation will be emerging in the next few years. Before we examine these developments, let us identify the major design choices made so far in building multicomputers, as compared with the development of other types of parallel computers.

As illustrated in Fig. 7.22, the choices made involve the selection of processors, memory structure, interconnection schemes, and control strategy.

Design Choices   In selecting a processor technology, a multicomputer designer typically chooses low-cost medium-grain processors as building blocks. In fact, the majority of parallel computers have been built with standard off-the-shelf microprocessors. Even the custom-designed processors used in the AMT DAP, nCUBE, TMC/CM-2, and IBM RP3 computers were low-cost medium- or fine-grain processors.

The next step was to choose distributed memory for multicomputers rather than using shared memory which would limit the scalability. Each processor has its own local memory to address. Scalability becomes more feasible without shared resources. With distributed memory, a new programming model and tools are needed for multicomputers.

Multicomputers have message-passing, point-to-point, direct networks as an interconnection scheme rather than the address-switching networks used in NUMA multiprocessors like the IBM RP3 and BBN Butterfly. A message-passing network routes messages between nodes. Any node can send a message to another. Send/receive semantics must be incorporated to guarantee consistent programming with uniform messaging speeds.

In selecting a control strategy, designers of multicomputers choose the asynchronous MIMD, MPMD, and SPMD operations, rather than the SIMD lockstep operations as in the CM-2 and DAP. Even though both support massive parallelism, the SIMD approach offers little or no opportunity to utilize existing multiprocessor code because radical changes must be made in the programming style.

On the other hand, multicomputers allow the use of existing software with minor changes from that developed for multiprocessors or for other types of parallel computers.

First Generation   Caltech's Cosmic Cube (Seitz, 1983) was the first of the first-generation multicomputers. The Intel iPSC/1, Ametek S/14, and nCUBE/10 were various evolutions of the original Cosmic Cube.

For example, the iPSC/1 uses i80286 processors with 512 Kbytes of local memory per node. Each node is implemented on a single printed-circuit board with eight I/O ports. Seven I/O ports are used to form a seven-dimensional hypercube. The eighth port is used for an Ethernet connection from each node to the host.

Table 7.1 summarizes the important parameters used in designing the three generations of multicomputers. The communication latency (for a 100-byte message) was rather long in the early 1980s. The 3-to-1 ratio between remote and local communication latencies was caused by the use of a *store-and-forward* routing scheme where the

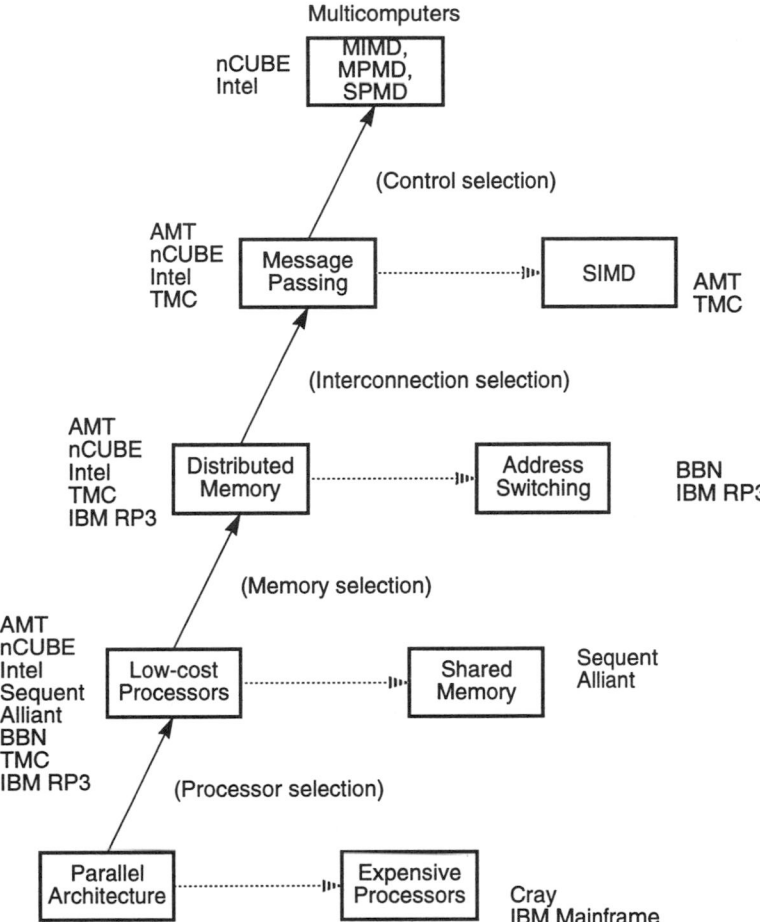

Figure 7.22 Design choices made in the past for developing message-passing multicomputers compared to those made for other parallel computers. (Courtesy of Intel Scientific Computers, 1988)

latency is proportional to the number of hops between two communicating nodes.

Vector hardware was added on a separate board attached to each processing node board. Or one can use the second board to hold extended local memory. The host used in the iPSC/1 was an Intel 310 microprocessor. All I/O must be done through the host.

7.3.2 Present and Future Development

The second and third generations of multicomputers are introduced below. The Intel Paragon is presented as a case study of a current system. Future trends are identified for developing the next generation of multicomputers. They include the evolution

Table 7.1 Three Generations of Multicomputer Development

Generation	First	Second	Third
Years	1983-87	1988-92	1993-97
Typical node			
MIPS	1	10	100
Mflops scalar	0.1	2	40
Mflops vector	10	40	200
Memory (Mbytes)	0.5	4	32
Typical system			
N (nodes)	64	256	1024
MIPS	64	2560	100K
Mflops scalar	6.4	512	40K
Mflops vector	640	10K	200K
Memory (Mbytes)	32	1K	32K
Communication latency (100-byte message)			
Neighbor (microseconds)	2000	5	0.5
Nonlocal (microseconds)	6000	5	0.5

(Modified from Athas and Seitz, "Multicomputers: Message-Passing Concurrent Computers", *IEEE Computer*, August 1988).

from medium- to fine-grain multicomputers using a globally shared virtual memory.

The Second Generation   These are the multicomputers which are still in use at present. A major improvement of the second generation includes the use of better processors, such as the i386 in the iPSC/2 and the i860 in the iPSC/860 and in the Delta. The nCUBE/2 implements 64 custom-designed VLSI processors on a single PC board. The memory per node has also increased to 10 times that of the first generation.

Most importantly, hardware-supported routing, such as *wormhole routing*, reduces the communication latency significantly from 6000 μs to less than 5 μs. In fact, the latency for remote and local communications has become the same, independent of the number of hops between any two nodes.

The architecture of a typical second-generation multicomputer is shown in Fig. 7.23. This corresponds to a 16-node mesh-connected architecture. Mesh routing chips (MRCs) are used to establish the four-neighbor mesh network. All the mesh communication channels and MRCs are built on a backplane.

Each node is implemented on a PC board plugged into the backplane at the proper MRC position. All I/O devices, graphics, and the host are connected to the periphery (boundary) of the mesh. The Intel Delta system has such a mesh architecture.

Another representative system is the nCUBE/2 which implements a hypercube with up to 8192 nodes with a total of 512 Gbytes of distributed memory. Note that some parameters in Table 7.1 have been updated from the conservative estimates made by Atlas and Seitz in 1988.

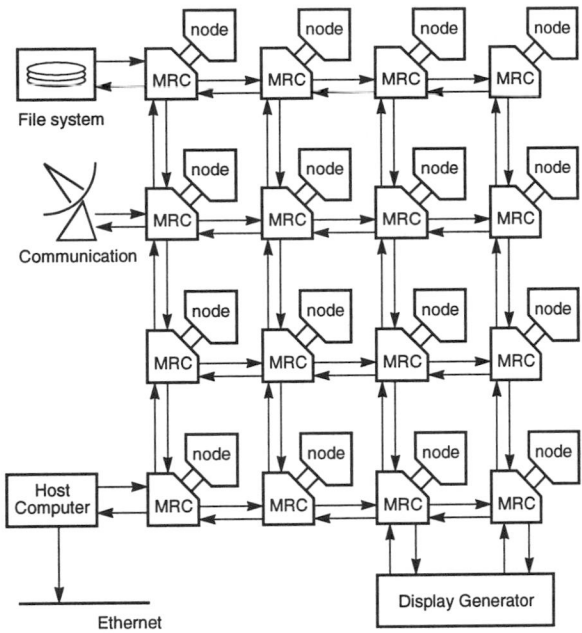

Legends: MRC = Mesh routing chip

Figure 7.23 The architecture of a second-generation multicomputer using a hardware-routed mesh interconnect. (Courtesy of Charles Seitz; reprinted with permission from "Concurrent Architectures", p. 31, *VLSI and Parallel Computation*, edited by Suaya and Birtwistle, Morgan Kaufmann Publishers, 1990)

The SuperNode 1000 is a transputer-based multicomputer produced by Parsystem Ltd., England. Another second-generation system is Ametek's Series 2010, made with 25-MHz M68020 processors using a mesh-routed architecture with 225-Mbytes/s channels.

The Third Generation   This will be the next generation of multicomputers. Caltech has an ongoing Mosaic C Project designed to use VLSI-implemented nodes, each containing a 14-MIPS processor, 20-Mbytes/s routing channels, and 16 Kbytes of RAM integrated on a single chip.

The full size of the Mosaic is targeted to have a total of 16,384 nodes organized in a three-dimensional mesh architecture. MIT has built a J-machine which it plans to extend to a 65K-node multicomputer with VLSI nodes interconnected by a three-dimensional mesh network. We will study the J-machine experience in Section 9.3.2.

The communication latency in future systems is expected to be reduced to less than 500 ns using superpipelining communication technologies. The J-machine plans to use message-driven processors which will reduce the message handling overhead to $< 1~\mu$s.

Each processor chip will contain a 512-Kbit DRAM, a 32-bit processor, a floating-point unit, and a communication controller.

The significant reduction of overhead in communication and synchronization will permit the execution of much shorter tasks with grain sizes of 5 μs per processor in the J-machine, as opposed to executing tasks of 100 μs in the iPSC/1. This implies that concurrency may increase from 10^2 in the iPSC/1 to 10^5 in the J-machine.

The first two generations of multicomputers have been called *medium-grain systems*. With a significant reduction in communication latency, the third generation systems will be called *fine-grain multicomputers*. They represent the research frontier of multicomputer architectures and programming.

Fine-grain message-driven programs require very small private memory objects, often less than 100 bytes. Exploiting fine-grain parallelism is much more difficult than the medium-grain approach. Fine-grain processor nodes demand more support from the language, compilers, and OS management of system resources.

Research is also underway to combine the private virtual address spaces distributed over the nodes into a globally shared virtual memory in the next generation of MPP multicomputers. Instead of page-oriented message passing, the fine-grain system may require block-level cache communications.

This fine-grain and shared virtual memory approach will combine the relative merits of multiprocessors and multicomputers in a *heterogeneous processing* (HP) environment.

7.3.3 The Intel Paragon System

In the 1980s, hypercube multicomputers were made with homogeneous nodes because all I/O functions were given to the host. This limited the I/O bandwidth, and thus these computers could not be used in solving large-scale problems with efficiency or high throughput. The Intel Paragon was designed to overcome this difficulty. The usage model turns the multicomputer into an applications server with multiuser access in a network environment.

Ever since the introduction of the iPSC/2 CFS, parallel I/O has been possible with dedicated disk nodes in addition to the computing nodes. The iPSC/860 further pushed the idea of using heterogeneous node types. The Paragon system has gone further by making it a host-free multicomputer. We explain below the various node types used in the Paragon and present the hardware router design.

The architecture of the Intel Paragon System is shown in Fig. 7.24. This system is driven by applications which require solving general sparse matrix problems, performing parallel data manipulation, or making scientific predictions through simulation modeling.

These difficult problems demand heterogeneous node types for numeric, service, I/O (disk and tape), and network gateways, as demonstrated in the schematic diagram of the Paragon system. The mesh architecture of the Paragon is divided into three sections.

The middle section, called the compute partition, is a mesh of numeric nodes implemented with Intel i860XP microprocessors. This array has an aggregate of 8.8 Gbytes of distributed memory.

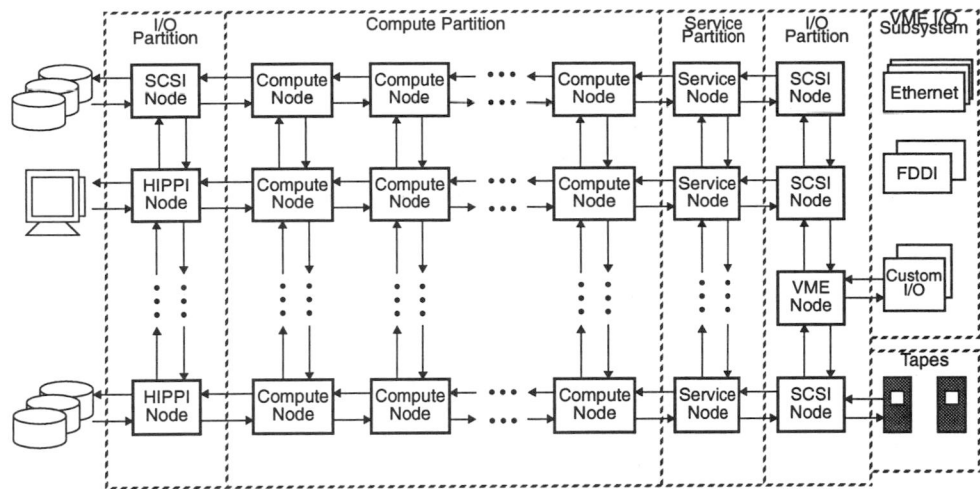

Figure 7.24 The Intel Paragon system architecture. (Courtesy of Intel Supercomputer Systems Division, 1991)

The system has a potential performance of 5 to 300 Gflops collectively. This mesh architecture eliminates the power-of-2 upgrade requirement of a hypercube architecture. All I/O are handled by the two disk I/O columns at the left and right edges of the mesh. Each column is a 16 × 1 array of 16 disk nodes. The aggregate I/O bandwidth has reached 48 Mbytes/s with a total of 27.4 Gbytes per disk I/O column.

The processors used in the I/O columns are the Intel i386's which supervise the massive data transfers between the disk arrays and the computational array during I/O operations. The system I/O column is made up of six *service nodes*, two tape nodes, two Ethernet nodes, and a HIPPI node. The service nodes are used for system diagnosis and handling of interrupts. The tape nodes are used for backup storage.

The Ethernet and HIPPI nodes are used for fast gateway connections with the outside world. Collectively, a 17,000-MIPS performance was claimed possible on the 570 numeric and disk I/O nodes involved in program execution. The system was designed to run iPSC/860-compatible software.

Node and Router Architecture   The Paragon was designed as an experimental system. Only one unit was built and delivered to Caltech in May 1991 for research use by a consortium of 13 national laboratories and universities. The Paragon is still a medium-grain multicomputer with the typical node architecture shown in Fig. 7.25.

Each node takes a separate board. For numeric nodes, the processor and floating-point units are on the same i860 chip. The local memory takes up most of the board space. The external I/O interface is implemented only on the boundary nodes with a computational array. The message I/O interface is required for message passing between local nodes and the mesh network. The *mesh-connected router* is shown in Fig. 7.26.

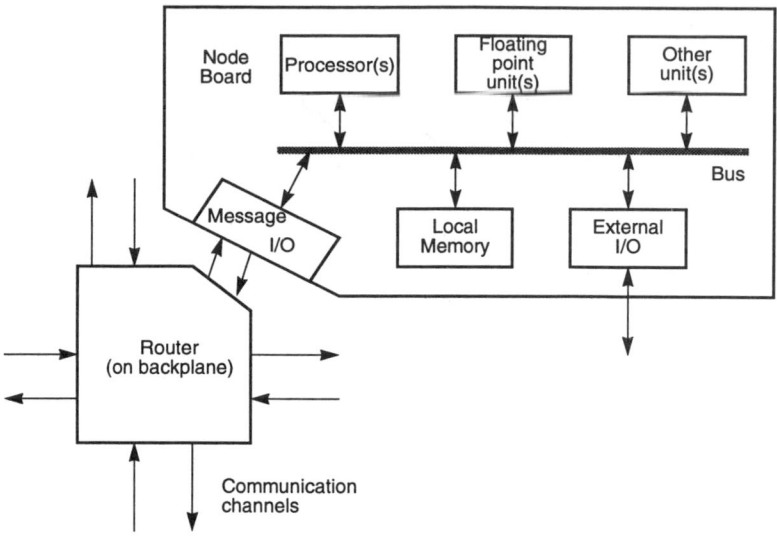

Figure 7.25 Node architecture of the Paragon multicomputer.

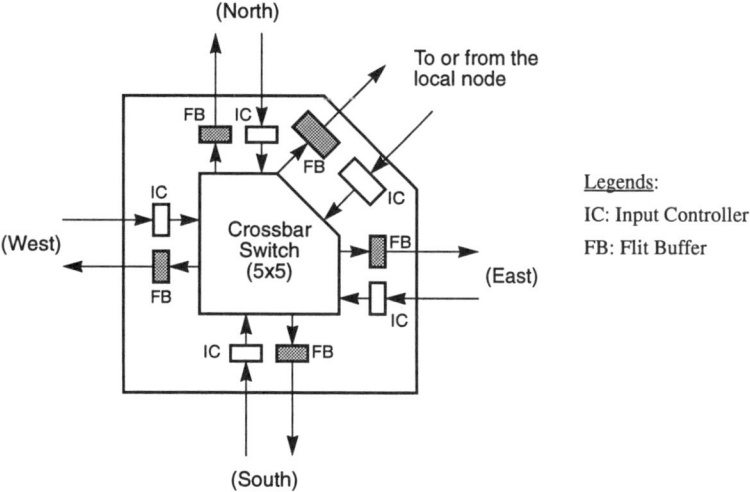

Figure 7.26 The structure of a mesh-connected router with four pairs of I/O channels connected to neighboring routers.

Each router has 10 I/O ports, 5 for input and 5 for output. Four pairs of I/O channels are used for mesh connection to the four neighbors at the north, south, east, and west nodes.

Flow control digits (flits) buffers are used at the end of input channels to hold the incoming flits. The concept of flits will be clarified in the next section. Besides four pairs of external channels, a fifth pair is used for internal connection between the router and local nodes. A 5×5 crossbar switch is used to establish a connection between any input channel and any output channel.

The functions of the hardware router include pipelined message routing at the flit level and resolving buffer or channel deadlock situations to achieve deadlock-free routing. In the next section, we will explain various routing mechanisms and deadlock avoidance schemes.

All the I/O channels shown in Figs. 7.25 and 7.26 are *physical channels* which allow only one message (flit) to pass at a time. Through time-sharing, one can also implement *virtual channels* to multiplex the use of physical channels as described in the next section.

7.4 Message-Passing Mechanisms

Message passing in a multicomputer network demands special hardware and software support. In this section, we study the store-and-forward and wormhole routing schemes and analyze their communication latencies. We introduce the concept of virtual channels. Deadlock situations in a message-passing network are examined. We show how to avoid deadlocks using virtual channels.

Both deterministic and adaptive routing algorithms are presented for achieving deadlock-free message routing. We first study deterministic dimension-ordering routing schemes such as E-cube routing for hypercubes and X-Y routing for two-dimensional meshes. Then we discuss adaptive routing using virtual channels or virtual subnets. Besides one-to-one unicast routing, we will consider one-to-many multicast and one-to-all broadcast operations using virtual subnets and greedy routing algorithms.

7.4.1 Message-Routing Schemes

Message formats are introduced below. Refined formats led to the improvement from store-and-forward to wormhole routing in two generations of multicomputers. A handshaking protocol is described for asynchronous pipelining of successive routers along a communication path. Finally, latency analysis is conducted to show the time difference between the two routing schemes presented.

Message Formats Information units used in message routing are specified in Fig. 7.27. A *message* is the logical unit for internode communication. It is often assembled by an arbitrary number of fixed-length packets, thus it may have a variable length.

A *packet* is the basic unit containing the destination address for routing purposes. Because different packets may arrive at the destination asynchronously, a sequence number is needed in each packet to allow reassembly of the message transmitted.

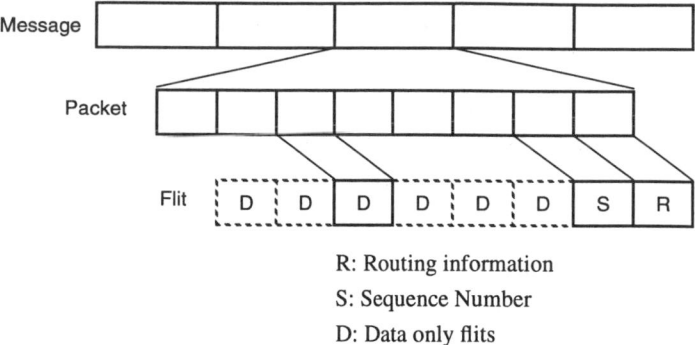

R: Routing information
S: Sequence Number
D: Data only flits

Figure 7.27 The format of message, packets, and flits (control flow digits) used as information units of communication in a message-passing network.

A packet can be further divided into a number of fixed-length *flits* (flow control digits). Routing information (destination) and sequence number occupy the header flits. The remaining flits are the data elements of a packet.

In multicomputers with store-and-forward routing, packets are the smallest unit of information transmission. In wormhole-routed networks, packets are further subdivided into flits. The flit length is often affected by the network size. A 256-node network requires 8 bits per flit.

The packet length is determined by the routing scheme and network implementation. Typical packet lengths range from 64 to 512 bits. The sequence number may occupy one to two flits depending on the message length. Other factors affecting the choice of packet and flit sizes include channel bandwidth, router design, network traffic intensity, etc.

Store-and-Forward Routing   Packets are the basic unit of information flow in a *store-and-forward* network. The concept is illustrated in Fig. 7.28a. Each node is required to use a packet buffer. A packet is transmitted from a source node to a destination node through a sequence of intermediate nodes.

When a packet reaches an intermediate node, it is first stored in the buffer. Then it is forwarded to the next node if the desired output channel and a packet buffer in the receiving node are both available.

The latency in store-and-forward networks is directly proportional to the distance (the number of hops) between the source and the destination. This routing scheme was implemented in the first-generation of multicomputers.

Wormhole Routing   By subdividing the packet into smaller flits, newer multicomputers implement the *wormhole routing* scheme, as illustrated in Fig. 7.28b. Flit buffers are used in the hardware routers attached to nodes. The transmission from the source node to the destination node is done through a sequence of routers.

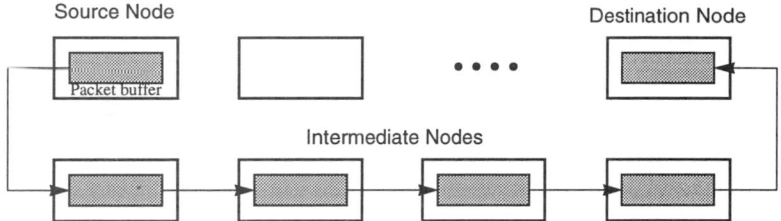

(a) Store-and-forward routing using packet buffers in successive nodes

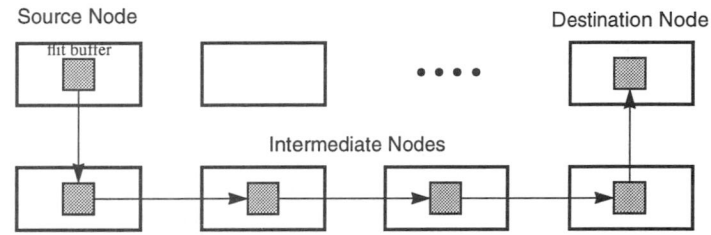

(b) Wormhole routing using flit buffers in successive routers

Figure 7.28 Store-and-forward routing and wormhole routing. (Courtesy of Lionel Ni, 1991)

All the flits in the same packet are transmitted in order as inseparable companions in a pipelined fashion. The packet can be visualized as a railroad train with an engine car (the header flit) towing a long sequence of box cars (data flits).

Only the header flit knows where the train (packet) is going. All the data flits (box cars) must follow the header flit. Different packets can be interleaved during transmission. However, the flits from different packets cannot be mixed up. Otherwise they may be towed to the wrong destinations.

We prove below that wormhole routing has a latency almost independent of the distance between the source and the destination.

Asynchronous Pipelining   The pipelining of successive flits in a packet is done asynchronously using a handshaking protocol as shown in Fig. 7.29. Along the path, a 1-bit *ready/request* (R/A) line is used between adjacent routers.

When the receiving router (D) is ready (Fig. 7.29a) to receive a flit (i.e., the flit buffer is available), it pulls the R/A line low. When the sending router (S) is ready (Fig. 7.29b), it raises the line high and transmits flit i through the channel.

While the flit is being received by D (Fig. 7.29c), the R/A line is kept high. After flit i is removed from D's buffer (i.e., is transmitted to the next node) (Fig. 7.29d), the cycle repeats itself for the transmission of the next flit $i + 1$ until the entire packet is received.

Asynchronous pipelining can be very efficient, and the clock used can be faster than

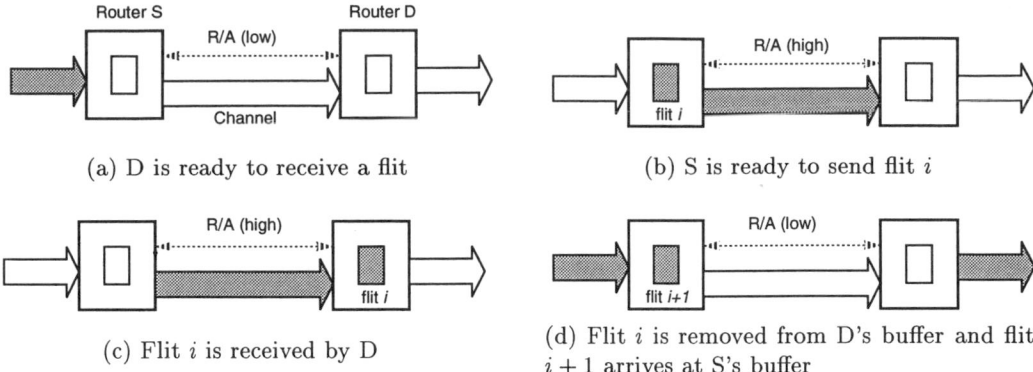

(a) D is ready to receive a flit   (b) S is ready to send flit i

(c) Flit i is received by D   (d) Flit i is removed from D's buffer and flit $i+1$ arrives at S's buffer

Figure 7.29 Handshaking protocol between two wormhole routers. (Courtesy of Lionel Ni, 1991)

that used in a synchronous pipeline. However, the pipeline can be stalled if flit buffers or successive channels along the path are not available during certain cycles. Should that happen, the packet can be buffered, blocked, dragged, or detoured. We will discuss these flow control methods in Section 7.4.3.

Latency Analysis   A time comparison between store-and-forward and wormhole-routed networks is given in Fig. 7.30. Let L be the packet length (in bits), W the channel bandwidth (in bits/s), D the distance (number of nodes traversed minus 1), and F the flit length (in bits).

The communication latency T_{SF} for a store-and-forward network is expressed by

$$T_{SF} = \frac{L}{W}(D+1) \tag{7.5}$$

The latency T_{WH} for a wormhole-routed network is expressed by

$$T_{WH} = \frac{L}{W} + \frac{F}{W} \times D \tag{7.6}$$

Equation 7.5 implies that T_{SF} is directly proportional to D. In Eq. 7.6, $T_{WH} = L/W$ if $L >> F$. Thus the distance D has a negligible effect on the routing latency.

We have ignored the network startup latency and block time due to resource shortage (such as channels being busy or buffers being full, etc.) The channel propagation delay has also been ignored because it is much smaller than those terms in T_{SF} or T_{WH}.

According to the estimate given in Table 7.1, a typical value of T_{SF} is between 2000 and 6000 μs, while a typical value of T_{WH} is 5 μs or less. The store-and-forward scheme is often controlled by software, while the wormhole scheme is entirely hardware-routed in a pipelined fashion.

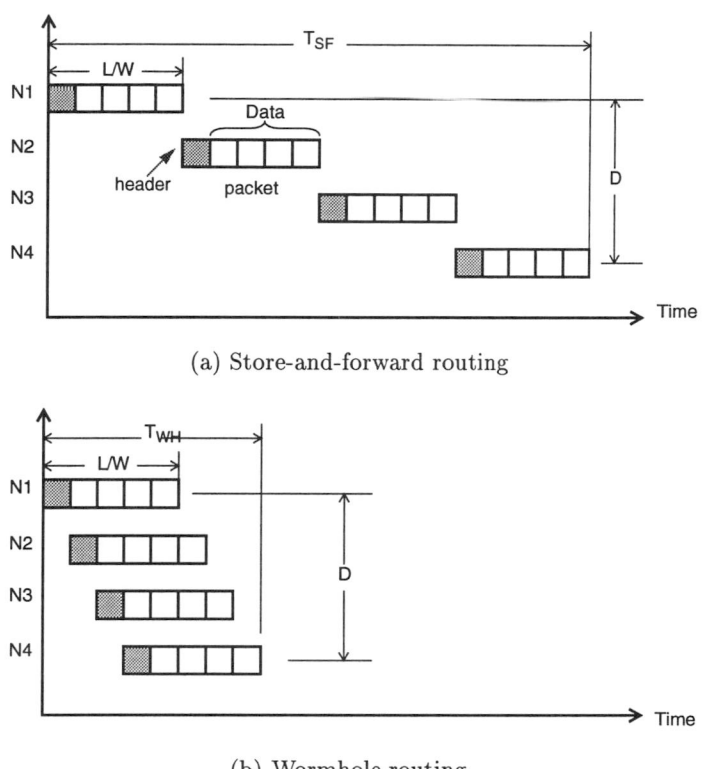

(a) Store-and-forward routing

(b) Wormhole routing

Figure 7.30 Time comparison between the two routing techniques.

7.4.2 Deadlock and Virtual Channels

The communication channels between nodes in a wormhole-routed multicomputer network are actually shared by many possible source and destination pairs. The sharing of a physical channel leads to the concept of virtual channels.

We introduce below the concept and explain its applications in avoiding deadlocks in this section and in facilitating network partitioning for multicasting in Section 7.4.4.

Virtual Channels   A virtual channel is a logical link between two nodes. It is formed by a flit buffer in the source node, a physical channel between them, and a flit buffer in the receiver node. Figure 7.31 shows the concept of four virtual channels sharing a single physical channel.

Four flit buffers are used at the source node and receiver node, respectively. One source buffer is paired with one receiver buffer to form a virtual channel when the physical channel is allocated for the pair.

In other words, the physical channel is time-shared by all the virtual channels. Besides the buffers and channel involved, some channel states (such as R/A signals)

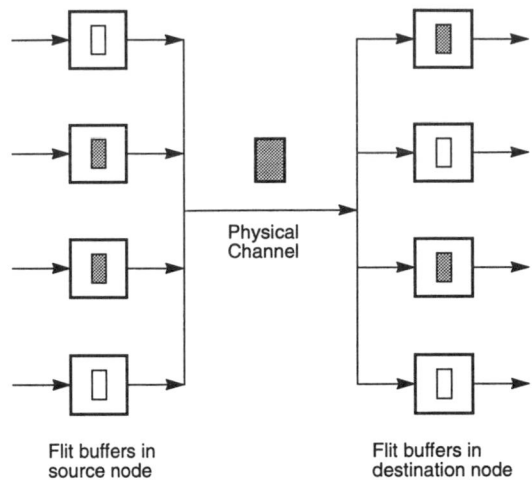

Physical
Channel

Flit buffers in
source node

Flit buffers in
destination node

Figure 7.31 Four virtual channels sharing a physical channel with time-multiplexing on a flit-by-flit basis.

must be identified with different virtual channels. The source buffers hold flits awaiting use of the channel. The receiver buffers hold flits just transmitted over the channel. The channel (wires or fibers) provides a communication medium between them.

Comparing the setup in Fig. 7.31 with that in Fig. 7.29, the difference lies in the added buffers at both ends and added R/A lines for controlling the states of virtual channels. Crossbar switch control, a multiplexer, and a demultiplexer are needed to implement the virtual channels. The sharing of a physical channel by a set of virtual channels is conducted by time-multiplexing on a flit-by-flit basis.

Example 7.3 The deadlock situations caused by circular waits at buffers or at channels

As illustrated in Fig. 7.32, two types of deadlock situations are caused by a circular wait at buffers or channels. A *buffer deadlock* is shown in Fig. 7.32a for a store-and-forward network. A circular wait situation results from four packets occupying four buffers in four nodes. Unless one packet is discarded or misrouted, the deadlock cannot be broken. In Fig. 7.32b, a *channel deadlock* results from four messages being simultaneously transmitted along four channels in a mesh-connected network using wormhole routing. ∎

Four flits from four messages occupy the four channels simultaneously. If none of the channels in the cycle is freed, the deadlock situation will continue. Circular waits are further illustrated in Fig. 7.33 using a *channel-dependence graph*.

The channels involved are represented by nodes, and directed arrows are used to

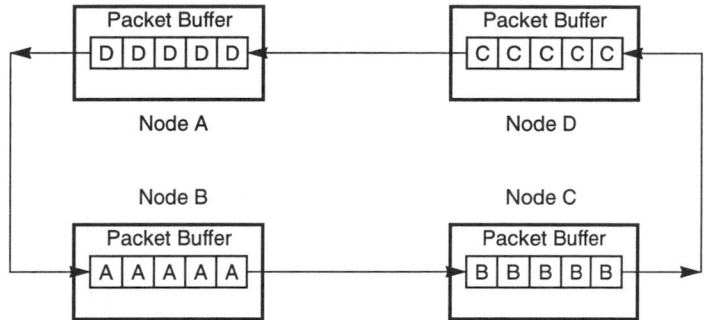

(a) Buffer deadlock among four nodes with store-and-forward routing

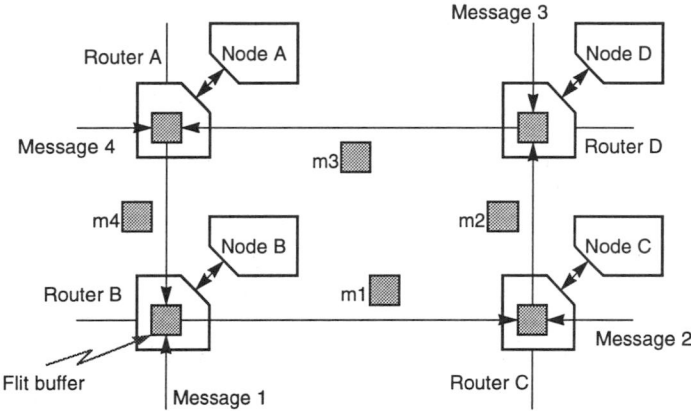

(b) Channel deadlock among four nodes with wormhole routing; shaded boxes are flit buffers

Figure 7.32 Deadlock situations caused by a circular wait at buffers or at communication channels.

show the dependence relations among them. A deadlock avoidance scheme is presented using virtual channels.

Deadlock Avoidance By adding two virtual channels, V_3 and V_4 in Fig. 7.33c, one can break the deadlock cycle. A modified channel-dependence graph is obtained by using the virtual channels V_3 and V_4, after the use of channel C_2, instead of reusing C_3 and C_4.

The cycle in Fig. 7.33b is being converted to a spiral, thus avoiding a deadlock. Channel multiplexing can be done at the flit level or at the packet level if the packet length is sufficiently short.

Virtual channels can be implemented with either *unidirectional channels* or *bidirectional channels*. Combining two unidirectional channels into a single bidirectional

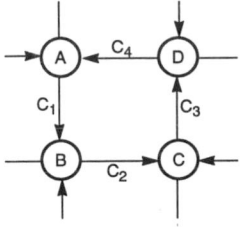

(a) Channel deadlock

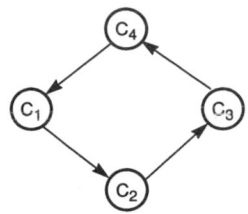

(b) Channel-dependence graph containing a cycle

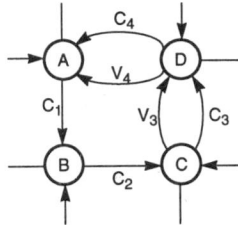

(c) Adding two virtual
channels (V_3, V_4)

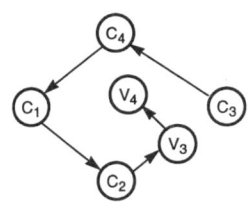

(d) A modified channel-dependence
graph using the virtual channels

**Figure 7.33 Deadlock avoidance using virtual channels to convert a cycle to a spiral
on a channel-dependence graph.**

channel will increase not only the utilization rate but may also double the channel bandwidth.

However, arbitration in a bidirectional channel is more involved. A special *arbitration line* is needed between adjacent nodes interconnected by a bidirectional channel. This line is used to control the direction of information flow.

In practice, bidirectional channels may introduce more delay due to direction arbitration and higher costs due to increased control complexity, as compared with unidirectional channels. When the network traffic is light, a bidirectional channel can be more efficient.

The use of virtual channels may reduce the effective channel bandwidth available to each request. There exists a tradeoff between network throughput and communication latency in determining the degree of using virtual channels. High-speed multiplexing is required for implementing a large number of virtual channels.

7.4.3   Flow Control Strategies

In this section, we examine various strategies developed to control smooth network traffic flow without causing congestion or deadlock situations. When two or more packets collide at a node when competing for buffer or channel resources, policies must be set regarding how to resolve the conflict.

Based on these policies, we describe below deterministic and adaptive routing algorithms developed for one-to-one communications.

Packet Collision Resolution   In order to move a flit between adjacent nodes in a pipeline of channels, three elements must be present: (1) the source buffer holding the flit, (2) the channel being allocated, and (3) the receiver buffer accepting the flit.

When two packets reach the same node, they may request the same receiver buffer or the same outgoing channel. Two arbitration decisions must be made: (i) Which packet will be allocated the channel? and (ii) What will be done with the packet being denied the channel? These decisions lead to the four methods illustrated in Fig. 7.34 for coping with the packet collision problem.

Figure 7.34 illustrates four methods for resolving the conflict between two packets competing for the use of the same outgoing channel at an intermediate node. Packet 1 is being allocated the channel, and packet 2 being denied. A *buffering* method has been proposed with the *virtual cut-through routing* scheme devised by Kermani and Kleinrock (1979).

Packet 2 is temporarily stored in a packet buffer. When the channel becomes available later, it will be transmitted then. This buffering approach has the advantage of not wasting the resources already allocated. However, it requires the use of a large buffer to hold the entire packet.

Furthermore, the packet buffers along the communication path should not form a cycle as shown in Fig. 7.32a. The packet buffer cannot be built into a router chip. It may require the use of local memory to implement the packet buffer, which may cause significant storage delay. The virtual cut-through method offers a compromise by combining the store-and-forward and wormhole routing schemes. When collisions do not occur, the scheme should perform as well as wormhole routing. In the worst case, it will behave like a store-and-forward network.

Pure wormhole routing uses a *blocking* policy in case of packet collision, as illustrated in Fig. 7.34b. The second packet is being blocked from advancing, however, it is not being abandoned. Figure 7.34c shows the *discard* policy, which simply drops the packet being blocked from passing through.

The fourth policy is called *detour* (Fig. 7.34d). The blocked packet is misrouted to a detour channel. The blocking policy is economical to implement but may result in the idling of resources allocated to the blocked packet.

The discard policy may result in a severe waste of resources, and it demands packet retransmission and acknowledgment. Otherwise, a packet may be lost after discarding. This policy is rarely used now because of its unstable packet delivery rate. The BBN Butterfly network has used this discard policy.

Detour routing offers more flexibility in packet routing. However, the detour may waste more channel resources than necessary to reach the destination. Furthermore, a misrouted packet may enter a cycle of *livelock*, which wastes network resources. Both the Connection Machine and the Denelcor HEP have used this detour policy.

In practice, some multicomputer networks use hybrid policies which may combine the advantages of some of the above flow control policies.

Dimension-Order Routing   Packet routing can be conducted deterministically or

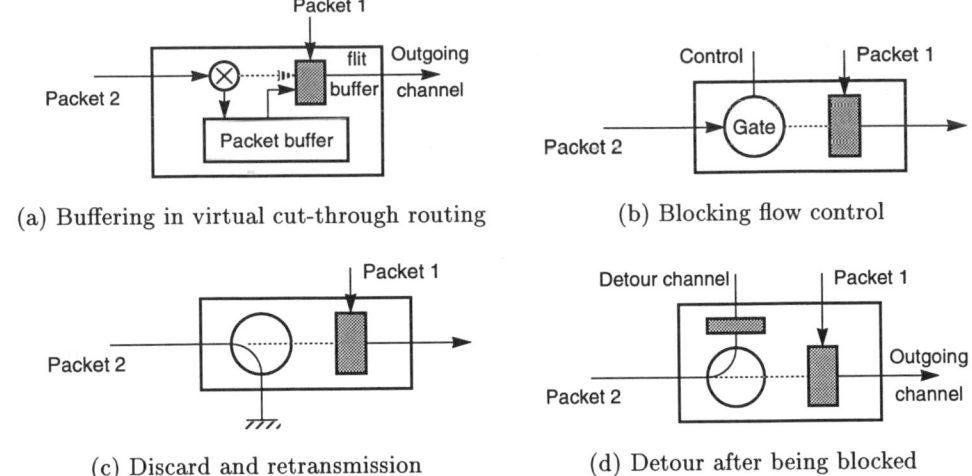

(a) Buffering in virtual cut-through routing   (b) Blocking flow control

(c) Discard and retransmission   (d) Detour after being blocked

Figure 7.34 Flow control methods for resolving a collision between two packets requesting the same outgoing channel (packet 1 being allocated the channel and packet 2 being denied).

adaptively. In *deterministic routing*, the communication path is completely determined by the source and destination addresses. In other words, the routing path is uniquely predetermined in advance, independent of network condition.

Adaptive routing may depend on network conditions, and alternate paths are possible. In both types of routing, deadlock-free algorithms are desired. Two such deterministic routing algorithms are given below, based on a concept called *dimension ordering*.

Dimension-ordering routing requires the selection of successive channels to follow a specific order based on the dimensions of a multidimensional network. In the case of a two-dimensional mesh network, the scheme is called *X-Y routing* because a routing path along the X-dimension is decided first before choosing a path along the Y-dimension. For hypercube (or n-cube) networks, the scheme is called *E-cube routing* as originally proposed by Sullivan and Bashkow (1977). These two routing algorithms are described below by presenting examples.

E-cube Routing on Hypercube   Consider an n-cube with $N = 2^n$ nodes. Each node b is binary-coded as $b = b_{n-1}b_{n-2}\cdots b_1 b_0$. Thus the source node is $s = s_{n-1}\cdots s_1 s_0$ and the destination node is $d = d_{n-1}\cdots d_1 d_0$. We want to determine a route from s to d with a minimum number of steps.

We denote the n dimensions as $i = 1, 2, \ldots, n$, where the ith dimension corresponds to the $(i-1)$st bit in the node address. Let $v = v_{n-1}\cdots v_1 v_0$ be any node along the route. The route is uniquely determined as follows:

1. Compute the direction bit $r_i = s_{i-1} \oplus d_{i-1}$ for all n dimensions $(i = 1, \ldots, n)$.

Start the following with dimension $i = 1$ and $v = s$.

2. Route from the current node v to the next node $v \oplus 2^{i-1}$ if $r_i = 1$. Skip this step if $r_i = 0$.

3. Move to dimension $i + 1$ (i.e., $i \leftarrow i + 1$). If $i \leq n$, go to step 2, else quit.

Example 7.4 E-cube routing on a four-dimensional hypercube

The above E-cube routing algorithm is illustrated with the example in Fig. 7.35. Now $n = 4$, $s = 0110$, and $d = 1101$. Thus $r = r_4 r_3 r_2 r_1 = 1011$. Route from s to $s \oplus 2^0 = 0111$ since $r_1 = 0 \oplus 1 = 1$. Route from $v = 0111$ to $v \oplus 2^1 = 0101$ since $r_2 = 1 \oplus 0 = 1$. Skip dimension $i = 3$ because $r_3 = 1 \oplus 1 = 0$. Route from $v = 0101$ to $v \oplus 2^3 = 1101 = d$ since $r_4 = 1$.

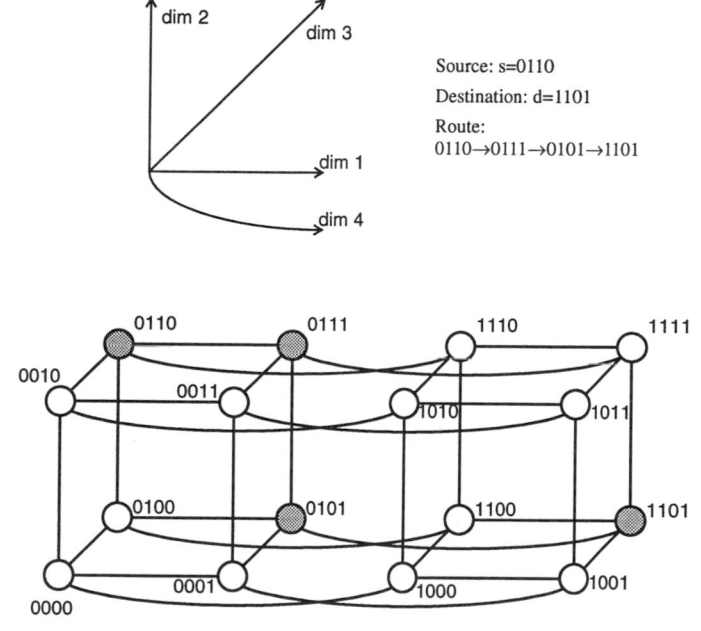

Figure 7.35 E-cube routing on a hypercube computer with 16 nodes.

The route selected is shown in Fig. 7.35 by arrows. Note that the route is determined from dimension 1 to dimension 4 in order. If the ith bit of s and d agree, no routing is needed along dimension i. Otherwise, move from the current node to the other node along the same dimension. The procedure is repeated until the destination is reached.

X-Y Routing on a 2D Mesh   The same idea is applicable to mesh-connected networks. X-Y routing is illustrated by the example in Fig. 7.36. From any source node $s = (x_1y_1)$ to any destination node $d = (x_2y_2)$, route from s along the X-axis first until it reaches the column Y_2, where d is located. Then route to d along the Y-axis.

There are four possible X-Y routing patterns corresponding to the east-north, east-south, west-north, and west-south paths chosen.

Example 7.5   X-Y routing on a 2D mesh-connected multicomputer

Four (source, destination) pairs are shown in Fig. 7.36 to illustrate the four possible routing patterns on a two-dimensional mesh.

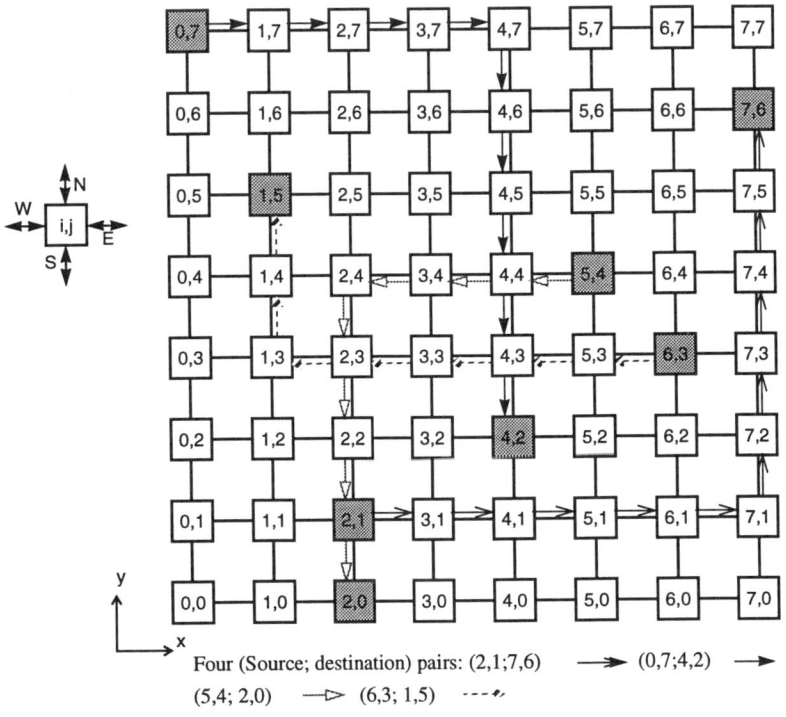

Figure 7.36 X-Y routing on a 2D mesh computer with $8 \times 8 = 64$ nodes.

An east-north route is needed from node $(2, 1)$ to node $(7, 6)$. An east-south route is set up from node $(0, 7)$ to node $(4, 2)$. A west-south route is needed from node $(5, 4)$ to $(2, 0)$. The fourth route is west-north bound from node $(6, 3)$ to node $(1, 5)$. If the X-dimension is always routed first and then the Y-dimension, a deadlock or circular wait situation will not exist.

X-Y routing can be easily extended to an *n*-dimensional mesh following dimension ordering. Using a three-dimensional mesh as an example, an X-Y-Z routing can be similarly specified. It is left as an exercise for the reader to prove that both E-cube and X-Y schemes result in deadlock-free routing. Both can be applied in either store-and-forward or wormhole-routed networks, resulting in a minimal route with the shortest distance between source and destination.

However, the same dimension ordering routing scheme cannot produce minimal routes for torus networks. Nonminimal routing algorithms, producing deadlock-free routes, allow packets to traverse through longer paths, sometimes to reduce network traffic or for other reasons.

Adaptive Routing   The main purpose of using adaptive routing is to avoid deadlock. The concept of virtual channels makes adaptive routing more economical and feasible to implement. We have shown in Fig. 7.33 how to apply virtual channels for this purpose. The idea can be further extended by having virtual channels in all connections along the same dimension of a mesh-connected network (Fig. 7.37).

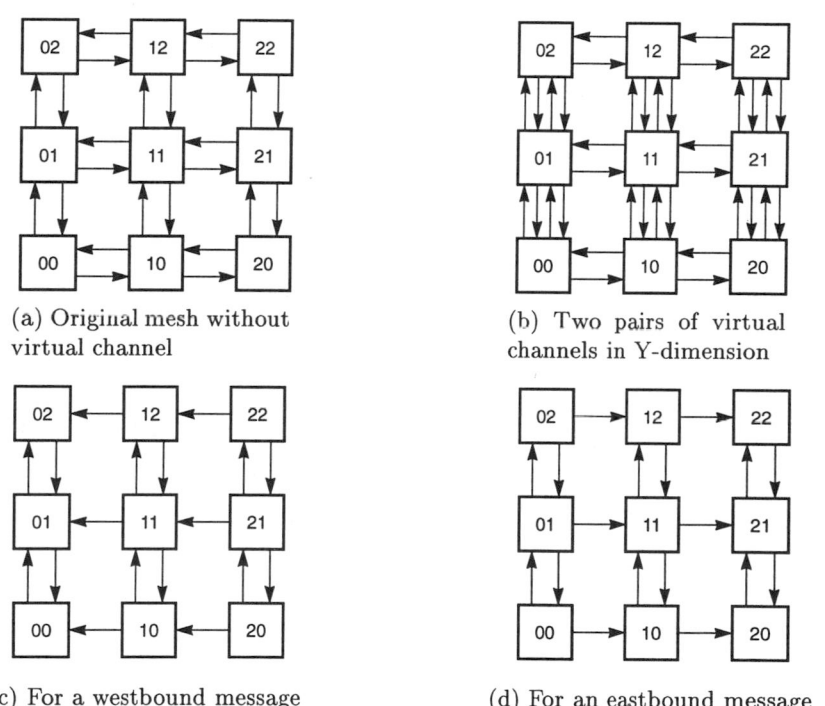

(a) Original mesh without virtual channel

(b) Two pairs of virtual channels in Y-dimension

(c) For a westbound message

(d) For an eastbound message

Figure 7.37 Adaptive X-Y routing using virtual channels to avoid deadlock; only westbound and eastbound traffic are deadlock-free. (Courtesy of Lionel Ni, 1991)

Example 7.6 Adaptive X-Y routing using virtual channels

This example uses two pairs of virtual channels in the Y-dimension of a mesh using X-Y routing.

For westbound traffic, the *virtual network* in Fig. 7.37c can be used to avoid deadlock because all eastbound X-channels are not in use. Similarly, the virtual network in Fig. 7.37d supports only eastbound traffic using a different set of virtual Y-channels.

The two virtual networks are used at different times; thus deadlock can be adaptively avoided. This concept will be further elaborated for achieving deadlock-free multicast routing in the next section.

■

7.4.4 Multicast Routing Algorithms

Various communication patterns are specified below. Routing efficiency is defined. The concept of virtual networks and network partitioning are applied to realize the complex communication patterns with efficiency.

Communication Patterns   Four types of communication patterns may appear in multicomputer networks. What we have implemented in previous sections is the one-to-one *unicast* pattern with one source and one destination.

A *multicast* pattern corresponds to one-to-many communication in which one source sends the same message to multiple destinations.

A *broadcast* pattern corresponds to the case of one-to-all communication. The most generalized pattern is the many-to-many *conference* communication.

In what follows, we consider the requirements for implementing multicast, broadcast, and conference communication patterns. Of course, all patterns can be implemented with multiple unicasts sequentially, or even simultaneously if resource conflicts can be avoided.

Special hardware or routing schemes must be used to implement these multi-destination patterns.

Routing Efficiency   Two commonly used efficiency parameters are *channel traffic* and *communication latency*. The channel traffic at any time instant (or during any time period) is indicated by the number of channels used to deliver the messages involved. The latency is indicated by the longest packet transmission time involved.

An optimally routed network should achieve both minimum traffic and minimum latency for the communication patterns involved. However, these two parameters are not totally independent. Achieving minimum traffic may not necessarily achieve minimum latency at the same time, and vice versa.

Depending on the switching technology used, latency is the more important issue in a store-and-forward network, while traffic demand affects efficiency more in a wormhole-routed network.

Example 7.7 Multicast and broadcast on a mesh-connected computer

Multicast routing is implemented on a 3×3 mesh in Fig. 7.38. The source node is identified as s, which transmits a packet to five destinations labeled D_i for $i = 1, 2, \ldots, 5$.

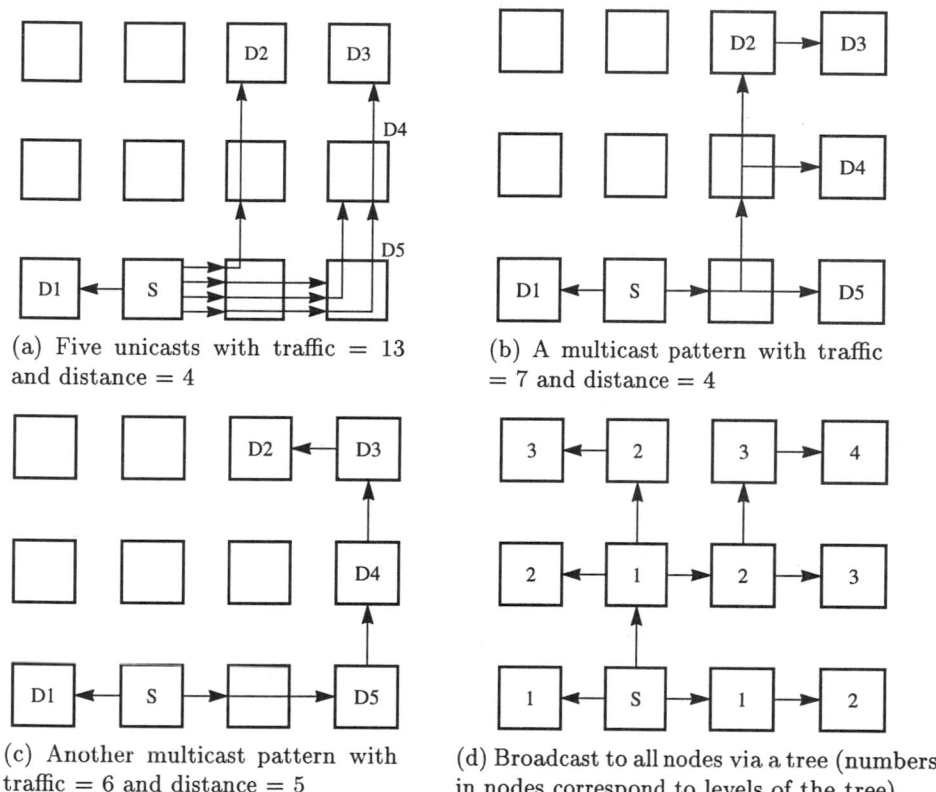

(a) Five unicasts with traffic = 13 and distance = 4

(b) A multicast pattern with traffic = 7 and distance = 4

(c) Another multicast pattern with traffic = 6 and distance = 5

(d) Broadcast to all nodes via a tree (numbers in nodes correspond to levels of the tree)

Figure 7.38 Multiple unicasts, multicast patterns, and a broadcast tree on a 3 × 4 mesh computer.

This five-destination multicast can be implemented by five unicasts, as shown in Fig. 7.38a. The X-Y routing traffic requires the use of $1 + 3 + 4 + 3 + 2 = 13$ channels, and the latency is 4 for the longest path leading to D3.

A multicast can be implemented by replicating the packet at an intermediate node, and multiple copies of the packet reach their destinations with significantly reduced channel traffic.

Two multicast routes are given in Figs. 7.38b and 7.38c, resulting in traffic of 7 and 6, respectively. On a wormhole-routed network, the multicast route in Fig. 7.38c is better. For a store-and-forward network, the route in Fig. 7.38b is better and has a shorter latency.

A four-level spanning tree is used from node S to broadcast a packet to all the mesh nodes in Fig. 7.38d. Nodes reached at level i of the tree have latency i. This broadcast tree should result in minimum latency as well as in minimum traffic.

Example 7.8 Multicast and broadcast on a hypercube computer

To broadcast on an n-cube, a similar spanning tree is used to reach all nodes within a latency of n. This is illustrated in Fig. 7.39a for a 4-cube rooted at node 0000. Again, minimum traffic should result with a broadcast tree for a hypercube.

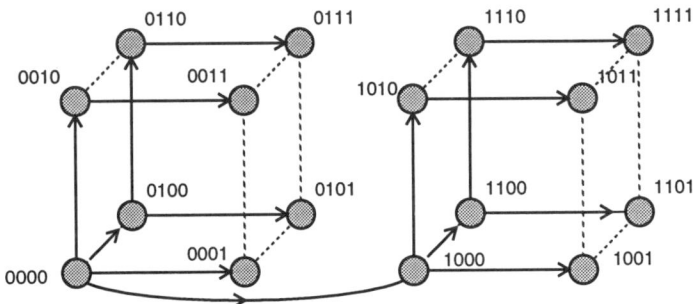

(a) Broadcast tree for a 4-cube rooted at node 0000

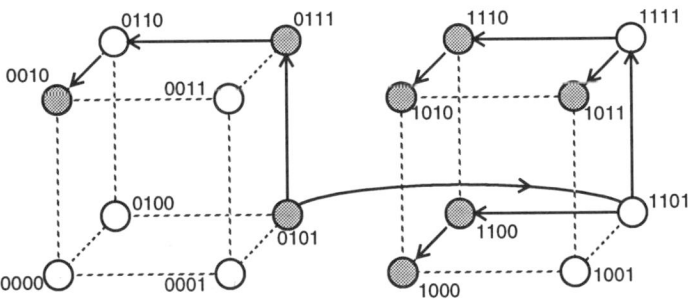

(b) A multicast tree from node 0101 to seven destination nodes
1100, 0111, 1010, 1110, 1011, 1000, and 0010

Figure 7.39 Broadcast tree and multicast tree on a 4-cube using a greedy algorithm. (Lan, Esfahnian, and Ni, 1990)

A greedy multicast tree is shown in Fig. 7.39b for sending a packet from node 0101 to seven destination nodes. The greedy multicast algorithm is based on sending the packet through the dimension(s) which can reach the most number of remaining destinations.

Starting from the source node $S = 0101$, there are two destinations via dimension 2 and five destinations via dimension 4. Therefore, the first-level channels used are $0101 \rightarrow 0111$ and $0101 \rightarrow 1101$.

From node 1101, there are three destinations reachable in dimension 2 and four destinations via dimension 1. Thus the second-level channels used include $1101 \rightarrow 1111, 1101 \rightarrow 1100$, and $0111 \rightarrow 0110$.

Similarly, the remaining destinations can be reached with third-level channels $1111 \rightarrow 1110, 1111 \rightarrow 1011, 1100 \rightarrow 1000$, and $0110 \rightarrow 0010$, and fourth-level channel $1110 \rightarrow 1010$.

∎

Extending the multicast tree, one should compare the reachability via all dimensions before selecting certain dimensions to obtain a minimum cover set for the remaining nodes. In case of a tie between two dimensions, selecting any one of them is sufficient. Therefore, the tree may not be uniquely generated.

It has been proved that this greedy multicast algorithm requires the least number of traffic channels compared with multiple unicasts or a broadcast tree. To implement multicast operations on wormhole-routed networks, the router in each node should be able to replicate the data in the flit buffer.

In order to synchronize the growth of a multicast tree or a broadcast tree, all outgoing channels at the same level of the tree must be ready before transmission can be pushed one level down. Otherwise, additional buffering is needed at intermediate nodes.

Virtual Networks   Consider a mesh with dual virtual channels along both dimensions as shown in Fig. 7.40a.

These virtual channels can be used to generate four possible virtual networks. For west-north traffic, the virtual network in Fig. 7.40b should be used.

Similarly, one can construct three other virtual nets for other traffic orientations. Note that no cycle is possible on any of the virtual networks. Thus deadlock can be completely avoided when X-Y routing is implemented on these networks.

If both pairs between adjacent nodes are physical channels, then any two of the four virtual networks can be simultaneously used without conflict. If only one pair of physical channels is shared by the dual virtual channels between adjacent nodes, then only (b) and (e) or (c) and (d) can be used simultaneously.

Other combinations, such as (b) and (c), or (b) and (d), or (c) and (e), or (d) and (e), cannot coexist at the same time due to a shortage of channels.

Obviously, adding channels to the network will increase the adaptivity in making routing decisions. However, the increased cost can be appreciable and thus prevent the use of redundancy.

Network Partitioning   The concept of virtual networks leads to the partitioning of a given physical network into logical subnetworks for multicast communications. The idea is illustrated in Fig. 7.41.

Suppose source node (4,2) wants to transmit to a subset of nodes in the 6×8 mesh.

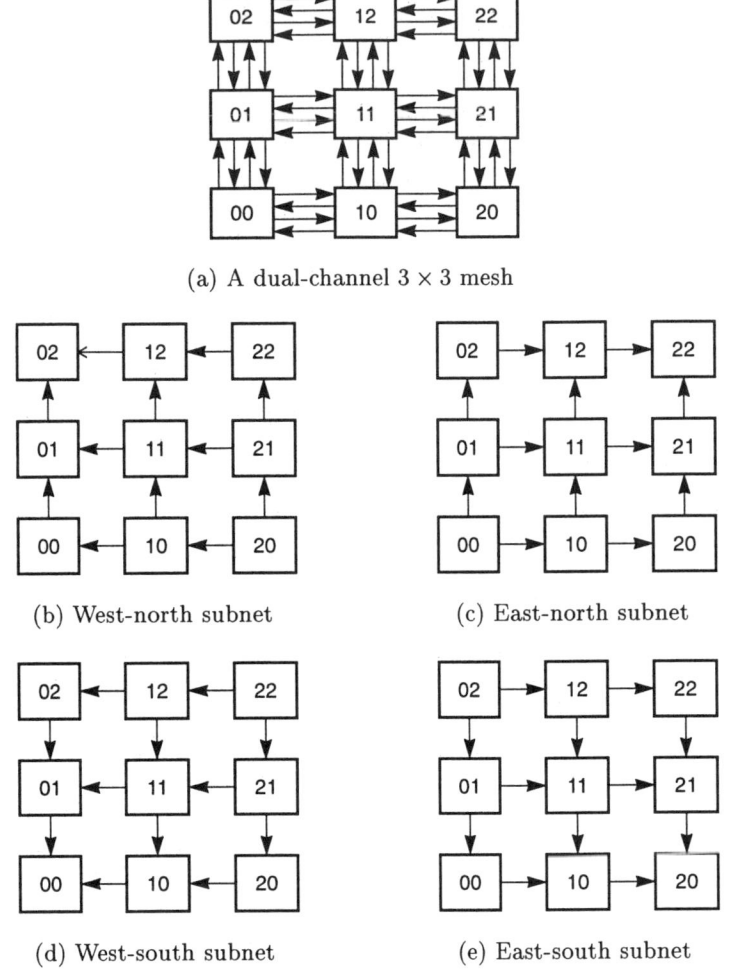

(a) A dual-channel 3 × 3 mesh

(b) West-north subnet   (c) East-north subnet

(d) West-south subnet   (e) East-south subnet

Figure 7.40 Four virtual networks implementable from a dual-channel mesh.

The mesh is partitioned into four logical subnets. All traffic heading for east and north uses the subnet at the upper right corner. Similarly, one constructs three other subnets at the remaining corners of the mesh.

Nodes in the fifth column and third row are along the boundary between subnets. Essentially, the traffic is being directed outward from the center node (4,2). There is no deadlock if an X-Y multicast is performed in this partitioned mesh.

Similarly, one can partition a binary n-cube into 2^{n-1} subcubes to provide deadlock-free adaptive routing. Each subcube has $n + 1$ levels with 2^n virtual channels per level for the bidirectional network. The number of required virtual channels increases rapidly with n. It has been shown that for low-dimensional cubes ($n = 2$ to 4), this method is

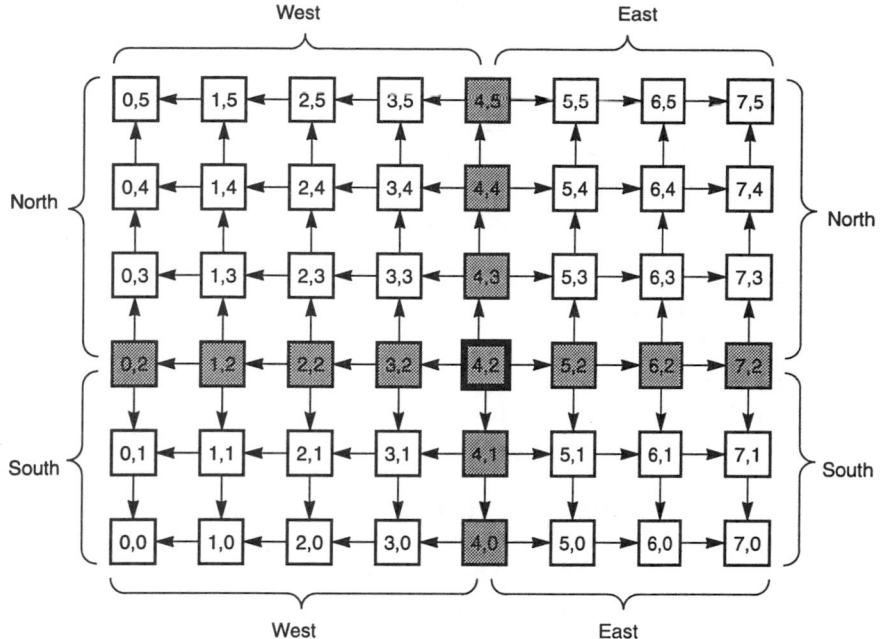

Figure 7.41 Partitioning of a 6 × 8 mesh into four subnets for a multicast from source node (4,2). Shaded nodes are along the boundary of adjacent subnets. (Courtesy of Lin, McKinly, and Ni, 1991)

best for general-purpose routing.

7.5   Bibliographic Notes and Exercises

For early surveys of multiprocessor systems, the reader is referred to [Enslow74], [Sayanarayanan80], and [Jones80] and Schwarz. More recent coverage of multiprocessors can be found in [Tabak90], [Trew91] and Wilson, and [Dubois92b] and Thakkar. The theoretical performance of multiprocessors has been treated in [Marson88] et al. using queueing, Markov, and Petri net models. Several survey articles on multiprocessors appeared in [Gajski85] and Peir, [Hwang87a], and [Dubois88] et al.

Multiple-bus systems for multiprocessors were discussed in [Mudge87] et al. Bandwidth analysis of bus and crossbar systems can be found in [Lang82] et al. and [Goyal84] and Agerwala. Other information on backplane buses can be found in NuBus specification [TI83], Multibus II [Intel84], SCSI bus [SCSI84], Fastbus [Gustavson86], Nanobus [Encore87], VME bus [VITA90], Futurebus+ specification [IEEE91], and [Parker91]. Hierarchical buses are treated in [Wilson87] and [Nassi87].

The Omega network was first proposed in [Lawrie75] and Delta networks in [Patel81]. Multistage networks were examined in [Wu80] and Feng, [Chin84] and Hwang, and

[Bhuyan83] and Agrawal. Combining networks were treated in [Gottlieb83], [Pfister85b] and Norton, and [Almasi89] and Gottlieb. The design of the Butterfly switch network was reported in [BBN89].

Recent surveys on cache coherence schemes were reported in [Archibald86] and Baer, in [Stenström90], and in [Adve91]. [Dubois90b] and Thakkar have edited a special issue on cache architectures in tightly coupled multiprocessors in *IEEE Computer*. [Patel82] has analyzed multiprocessors using private caches. The write-once coherence protocol was proposed in [Goodman83]. [Censier78] and Feautrier presented the directory-based protocol for network-based multiprocessors. The Firefly cache coherence protocol was reviewed in [Archibald86] and Baer . Directory-based cache coherence was discussed in [Chaiken90] et al.

Hierarchical cache coherence control was treated in [Wilson87]. The SCI structure was reported in [James90a]. The snoopy wires for barrier synchronization were introduced by [Hwang91] and Shang. MIMD synchronization and coherence problems have also been studied in [Bitar91] and [Bitar86] and Despain. The NYU Ultracomputer was reported in [Gottlieb83] et al.

Good assessments of multicomputer technology can be found in [Seitz90], [Dally90a], and [Athas88] and Seitz. The Caltech Cosmic Cube was reported in [Seitz85]. A dedicated treatment of the hypercube system and its applications was given in [Shih89] and Fier. Hypercube computers were also studied in [Gustafson86], [Hayes86], [Heath87], and [nCube90].

Virtual cut-through switching was proposed by [Kermani79] and Kleinrock. Wormhole routing was proposed by [Dally86] and Seitz. Wormhole routing was adopted in the Ametek 2010 [Seitz88], nCube 6400 [nCube90], Intel Paragon system [Intel91], and MIT J-Machine [Dally92]. The Intel/iWarp [Borkar90] and Transputer IMS T9000 family [Homewood87] also support wormhole routing.

E-cube routing for hypercube was proposed by [Sullivan77] and Bashkow, and The k-ary n-cubes were discussed in [Dally87b] and Seitz. [Linder91] and Harden extended the work to adaptive routing using virtual channels. [Glass92] and Ni have developed a Turn model for adaptive routing. The concept of virtual channels was introduced in [Dally90c]. Multicast wormhole routing was described in [Lin91b] and Ni, [Lan90] et al., and [Lin91a] et al. Other works on virtual channels can be found in [Borkar90] and [Dally87c] and Song.

Exercises

Problem 7.1 Consider a multiprocessor with n processors and m shared-memory modules, all connected to the same backplane bus with a central arbiter as depicted below:

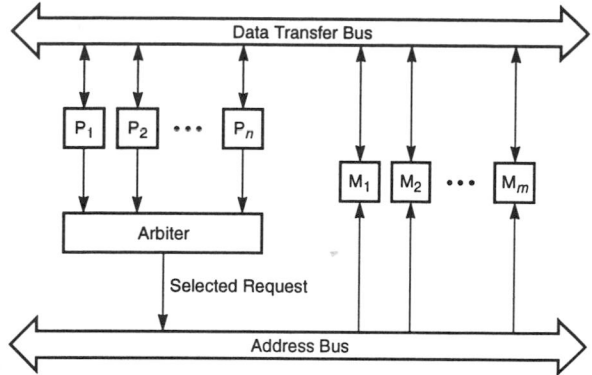

Assume $m > n$ and all memory modules are equally accessible to each processor. In other words, each processor generates a request for any module with probability $1/m$. The address bus and the DTB can be used at the same time to serve different requests. Both buses take one cycle to pass the address of a request or to transfer one word of 4 bytes between memory and processor. At each bus cycle (τ), the arbiter randomly selects one of the requests from the processors.

Once a memory module is identified at the end of the address cycle (one bus cycle), it takes a memory cycle (which equals c bus cycles) to retrieve the addressed word from the memory module, and another bus cycle to transfer the word to the requesting processor via the data transfer bus.

Until a memory cycle is completed, the arbiter will not issue another request to the same module. All rejected requests are ignored and resubmitted in subsequent bus cycles until being selected.

(a) Calculate the memory bandwidth defined as the average number of memory words transferred per second over the DTB if $n = 8$, $m = 16$, $\tau = 100$ ns, and $c \cdot \tau = 4\tau = 400$ ns.

(b) Calculate the memory utilization defined as the average number of requests accepted by all memory modules per memory cycle using the same set of parameters used in part (a).

Problem 7.2 Use two-input AND and OR gates (no wired-OR) to construct an $n \times n$ crossbar switch network between n processors and n memory modules. Let the width of each crosspoint be w bits (or a word) in each direction.

(a) Prepare a schematic design of a typical crosspoint switch using c_{ij} as the enable signal for the switch in the ith row and jth column. Estimate the total number of AND and OR gates needed as a function of n and w.

(b) Assume that processor P_i has higher priority over processor P_j if $i < j$ when they are competing for access to the same memory module. Let $k = \log_2 n$ be the address width. Design an arbiter which generates all the crosspoint enable signals c_{ij}, again using only two-input AND and OR gates and some inverters if needed. The memory address decoder is assumed available from each processor and thus is not included in the arbiter design. Indicate the complexity of the arbiter design as a function of n and k.

Problem 7.3 Consider a dual-processor (P1 and P2) system using write-back private caches and a shared memory, all connected to a common contention bus. Each cache has four block frames labeled below as 0, 1, 2, 3.

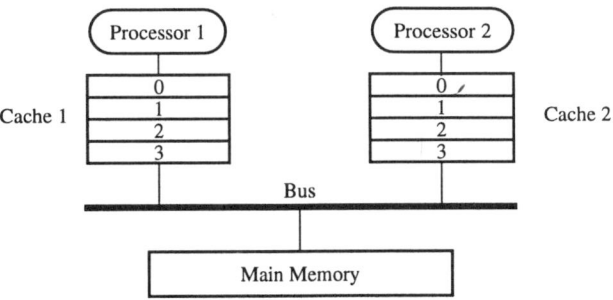

The shared memory is divided into eight cache blocks as 0, 1, ..., 7. To maintain cache coherence, the system uses a three-state (RO, RW, and invalid) snoopy protocol based on the write-invalidate policy described in Fig. 7.12b.

Assume the same clock drives the processors and the memory bus. Within each cycle, any processor can submit a request to access the bus. In case of simultaneous bus requests from both processors, the request from P1 is granted and P2 must wait one or more cycles to access the bus.

In all cases, the bus allows only one transaction per cycle. Once a bus access is granted, the transaction must be completed before the next request is granted. When there is no bus contention, memory-access events from each processor may require one to two cycles to complete, as specified below separately:

- Read-hit in cache requires one cycle and no bus request at all.
- Read-miss in cache requires two cycles without contention: one for block fetch and one for CPU read from cache.
- Write-hit requires one cycle for CPU write and bus invalidation simultaneously.
- Write-miss requires two cycles: one for block fetch and bus invalidation, and one for CPU write.
- Replacement of a dirty block requires one cycle to update memory via the bus.

(a) In the case of bus contention, one additional cycle is needed for bus arbitration in all the above cases except a read-hit.

  (i) Show how to map the eight cache blocks to four cache block frames using a direct-mapping cache organization.

  (ii) Show how to map the eight cache block frames using a two-way set-associative cache organization.

(b) Consider the following two asynchronous sequences of memory-access events, where boldface numbers are for write and the remaining are for read.

Processor #1: 0,0,0,1,1,4,3,3,5,5,5
Processor #2: 2,2,0,0,7,5,5,5,7,7,0

(i) Trace the execution of these two sequences on the two processors by executing the successive blocks. Both caches are initially flushed (empty). Assume a direct-mapping organization in both caches. Indicate the state (RO or RW) of each valid cache block and mark cache miss and bus utilization (busy or idle) in the block trace for each cycle. Assume that the very first memory-access events from both processors take place in cycle 1 simultaneously. Calculate the hit ratio of cache 1 and cache 2, respectively.

(ii) Assume a two-way set-associative cache organization and a LRU cache block replacement policy.

Problem 7.4 Consider the execution of 24 code segments, S_1 through S_{24}, following a given precedence graph on a multiprocessor with four processors and six memory modules as shown below. Assume all segments have the same gain size and execute with equal time. When two or more processors try to access the same memory module at the same time, the request of the lowest numbered processor is granted and the rest of the requests are deferred to later segment time steps.

A processor waiting from an earlier memory-access rejection has seniority priority over new requests to access the same memory module. No processor should wait for more than three steps to access any given memory module. Each code segment takes a fixed unit time to access a memory and to execute. Assume that the four processors are synchronized in each segment execution period.

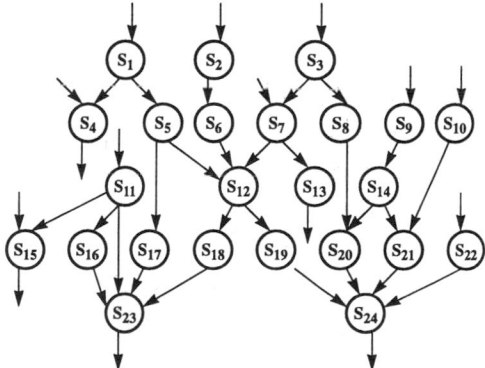

In some cases, a single segment may require to access several memory modules simultaneously. Ignore the contention problem in the interconnection network. The four processors operate in MIMD mode, and different instructions can be executed by different processors during the same cycle.

What is the average memory bandwidth in words per unit time? Try to achieve the minimum execution time by maximizing the degree of parallelism at all steps.

Note that at each step some of the memory modules may be idle. The highest possible memory bandwidth is six words per step. Some segments may require a wait of no more than three steps before granting of the memory access requested. But such a waiting period should be minimized.

Code segment	Processor					Code segment	Processor			
	P_1	P_2	P_3	P_4			P_1	P_2	P_3	P_4
S_1	M_1		M_5	M_1		S_{13}		M_1		M_6
S_2	M_1	M_2	M_2	M_2		S_{14}		M_4	M_5	M_4
S_3			M_3	M_3		S_{15}	M_3	M_3	M_3	
S_4	M_5	M_3	M_2	M_4		S_{16}	M_2	M_2	M_2	M_4
S_5	M_1	M_6	M_2			S_{17}	M_1			
S_6		M_2	M_1	M_3		S_{18}		M_2	M_5	
S_7			M_6	M_5		S_{19}	M_2	M_2	M_2	M_1
S_8		M_2	M_3			S_{20}		M_3	M_3	M_4
S_9		M_3	M_4	M_4		S_{21}	M_2			M_4
S_{10}	M_1	M_3	M_4	M_4		S_{22}	M_3	M_1		M_6
S_{11}	M_2	M_4	M_5	M_1		S_{23}	M_1	M_2	M_5	M_3
S_{12}		M_2	M_6	M_5		S_{24}			M_3	M_4

Problem 7.5   This problem is based on Fig. 7.11 which combines multiple Fetch&Add requests to the same shared variable in a common memory.

(a) Show the necessary combining network components needed to combine four Fetch& Add(x, e_i) for $i = 1, 2, 3, 4$.

(b) Show the successive snapshots and variations in switch and memory contents, as in Fig. 7.11, for combining the four requests.

Problem 7.6   You have learned about a two-way shuffle (perfect shuffle) in Fig. 2.14 and a four-way shuffle in Fig. 7.9. Generalize the mappings to an m-way shuffle over n objects, where $m \times k = n$ for some integer $k \geq 2$, for the construction of the class of Delta networks introduced by Patel (1980).

(a) Show how to perform a four-way shuffle over 12 objects.

(b) Use a minimum number of 4×3 switch modules and a four-way shuffle mapping as an interstage connection pattern to build a 64-input, 27-output Delta network in three stages.

(c) In general, an n-stage $a^n \times b^n$ Delta network is implemented with $a \times b$ switch modules as shown in Fig. 2.23. Calculate the total number of switch modules needed and specify the interstage connection pattern from b^n inputs to a^n outputs.

(d) Figure out a simple routing scheme to control the switch settings from stage to stage in an $a^n \times b^n$ Delta network with n stages.

(e) What is the relationship between Omega networks and Delta networks?

Problem 7.7   Prove the following properties associated with multistage Omega networks using different-sized building blocks:

(a) Prove that the number of legitimate states (connections) in a $k \times k$ switch module equals k^k.

(b) Determine the percentage of permutations that can be realized in one pass through a 64-input Omega network built with 2×2 switch modules.

(c) Repeat part (b) for a 64-input Omega network built with 8×8 switch modules.

(d) Repeat part (b) for a 512-input Omega network built with 8×8 switch modules.

Problem 7.8 Consider the interleaved execution of k programs in a multiprogrammed multiprocessor using m wired-NOR synchronization lines on n processors as described in Fig. 7.20a.

In general, the number m_i of barrier lines needed for a program i is estimated as $m_i = b_i \lceil l_i/P_i \rceil + 1$, where $b_i =$ the number of barriers demanded in program i, $l_i =$ the number of processes created in program i, and $P_i =$ the number of processors allocated to program i.

Thus $m = m_1 + m_2 + \cdots + m_k$. For simplicity, assume $b_i = b$ and $l_i = l$ for $i = 1, 2, \ldots, k$, and $P_i = \min(n/k, l)$ processors are allocated to each program i.

Prove that m can be approximated by $b \cdot l \cdot k^2/n + k$, or that the degree of multiprogramming is $k \leq (-n + \sqrt{n^2 + 4blmn})/(2bl)$ in such a multiprocessor system. Note that bl represents the number of required synchronization points, which depends on the parallelism profiles in user programs. For fixed values of bl and n, the maximally allowed multiprogramming degree k increases with respect to $\sqrt{m}$.

Problem 7.9 Wilson (1987) proposed a hierarchical cache/bus architecture (Fig. 7.3) and outlined how multilevel cache coherence can be enforced by extending the write-invalidate protocol. Can you figure out a write-broadcast protocol for achieving multilevel cache coherence on the same hardware platform? Comment on the relative merits of the two protocols. Feel free to modify the hardware in Fig. 7.3 if needed to implement the write-broadcast protocol on the hierarchical bus/cache architecture.

Problem 7.10 Answer the following questions on design choices of multicomputers made in the past:

(a) Why were low-cost processors chosen over expensive processors as processing nodes?

(b) Why was distributed memory chosen over shared memory?

(c) Why was message passing chosen over address switching?

(d) Why was MIMD, MPMD, or SPMD control chosen over SIMD data parallelism?

Problem 7.11 Explain the following terms associated with multicomputer networks and message-passing mechanisms:

(a) Message, packets, and flits.

(b) Store-and-forward routing at packet level.

(c) Wormhole routing at flit level.

(d) Virtual channels versus physical channels.

(e) Buffer deadlock versus channel deadlock.

(f) Buffering flow control using virtual cut-through routing.

(g) Blocking flow control in wormhole routing.

(h) Discard and retransmission flow control.

(i) Detour flow control after being blocked.

(j) Virtual networks and subnetworks.

Problem 7.12

(a) Draw a 16-input Omega network using 2×2 switches as building blocks.

(b) Show the switch settings for routing a message from node 1011 to node 0101 and from node 0111 to node 1001 simultaneously. Does blocking exist in this case?

(c) Determine how many permutations can be implemented in one pass through this Omega network. What is the percentage of one-pass permutations among all permutations?

(d) What is the maximum number of passes needed to implement any permutation through the network?

Problem 7.13 Explain the following terms as applied to communication patterns in a message-passing network:

(a) Unicast versus multicast

(b) Broadcast versus conference

(c) Channel traffic or network traffic

(d) Network communication latency

(e) Network partitioning for multicasting communications

Problem 7.14 Determine the optimal routing paths in the following mesh and hypercube multicomputers.

(a) Consider a 64-node hypercube network. Based on the E-cube routing algorithm, show how to route a message from node (101101) to node (011010). All intermediate nodes must be identified on the routing path.

(b) Determine two optimal routes for multicast on an 8×8 mesh, subject to the following constraints separately. The source node is (3, 5), and there are 10 destination nodes (1, 1), (1, 2), (1, 6), (2, 1), (4, 1), (5, 5), (5, 7), (6, 1), (7, 1), (7, 5). (i) The first multicast route should be implemented with a minimum number of channels. (ii) The second multicast route should result in minimum distances from the source to each of the 10 destinations.

(c) Based on the greedy algorithm (Fig. 7.39), determine a suboptimal multicast route, with minimum distances from the source to all destinations using as few traffic channels as possible, on a 16-node hypercube network. The source node is (1010), and there are 9 destination nodes (0000), (0001), (0011), (0100), (0101), (0111), (1111), (1101), and (1001).

Problem 7.15 Prove the following statements with reasoning or analysis or counter-

examples:

(a) Prove that E-cube routing is deadlock-free on a wormhole-routed hypercube with a pair of opposite unidirectional channels between adjacent nodes.
(b) Prove that X-Y routing is deadlock-free on a 2D mesh.
(c) Prove that E-cube routing on the 3D mesh (k-ary n-cube) used in the J-Machine is deadlock-free with wormhole routing and blocking flow control.

Problem 7.16 Study the Turn model for adaptive routing proposed by Glass and Ni (1992) in the 1992 *Annual International Symposium on Computer Architecture*. Answer the following questions:

(a) Why is the Turn model deadlock-free from having cycles?
(b) How can the Turn model be applied on an n-dimensional mesh to prevent deadlock?
(c) How can the Turn model be applied on a k-ary n-cube to prevent deadlock?

Problem 7.17 The following assignments are related to the greedy algorithm for multicast routing on a wormhole-routed hypercube network.

(a) Formulate the successive steps of the greedy algorithm (Example 7.8) as a minimum-cover problem, similar to that practiced in Karnaugh maps.
(b) Prove that the greedy algorithm always yields the minimum network traffic and minimum distance from the source to any of the destinations.

Problem 7.18 Consider the implementation of Goodman's write-once cache coherence protocol in a bus-connected multiprocessor system. Specify the use of additional bus lines to inhibit the main memory when the memory copy is invalid. Also specify all other hardware mechanisms and software support needed for an economical and fast implementation of the Goodman protocol.

Explain why this protocol will reduce bus traffic and how unnecessary invalidations can be eliminated. Consult the IEEE Futurebus+ specifications or the two related papers published by Goodman in 1983 and 1990.

Problem 7.19 Study the paper by Archibald and Baer (1986) which evaluated various cache coherence protocols using a multiprocessor simulation model. Explain the Dragon protocol implemented in the Dragon multiprocessor workstation at the Xerox Palo Alto Research Center. Compare the relative merits of the Goodman protocol, the Firefly protocol, and the Dragon protocol in the context of implementation requirements and expected performance.

Problem 7.20 The Cedar multiprocessor at Illinois is built with a clustered Omega network as shown below. Four 8×4 crossbar switches are used in the first stage and four 4×8 crossbar switches are used in the second stage. There are 32 processors and 32 memory modules, divided into four clusters with eight of each per cluster.

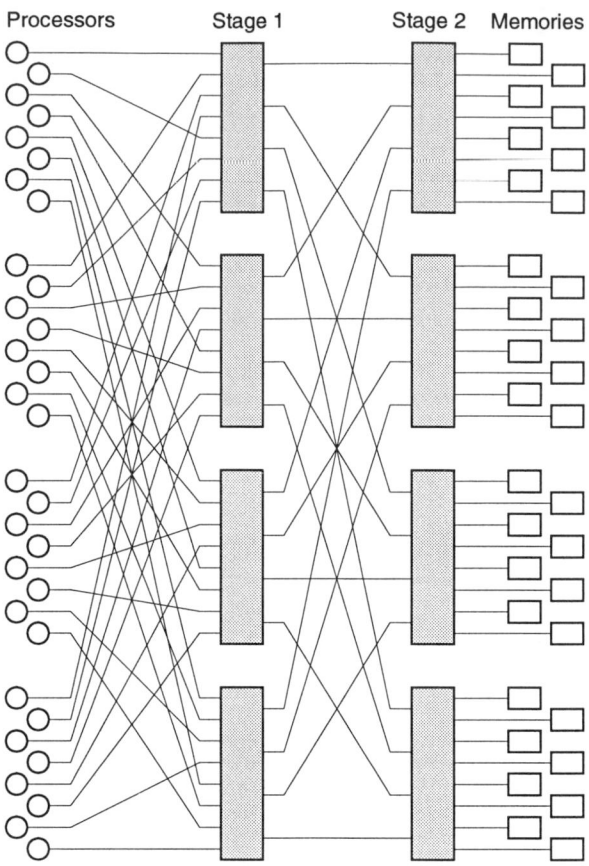

(a) Figure out a fixed priority scheme to avoid conflicts in using the crossbar switches for nonblocking connections. For simplicity, consider only the forward connections from the processors to the memory modules.

(b) Suppose both stages use 8×8 crossbar switches. Design a two-stage Cedar network to provide switched connections between 64 processors and 64 memory modules, again in a clustered manner similar to the above Cedar network design.

(c) Further expand the Cedar network to three stages using 8×8 crossbar switches as building blocks to connect 512 processors and 512 memory modules. Show the schematic interconnections in all three stages from the input end to the output end.

Chapter 8

Multivector and SIMD Computers

By definition, supercomputers are the fastest computers available at any specific time. The value of supercomputing has been identified by Buzbee (1983) in three areas: *knowledge acquisition*, *computational tractability*, and *promotion of productivity*. Computing demand, however, is always ahead of computer capability. Today's supercomputers are still one generation behind the computing requirements in most application areas.

In this chapter, we study the architectures of pipelined multivector supercomputers and of SIMD array processors. Both types of machines perform vector processing over large volumes of data. Besides discussing basic vector processors, we describe compound vector functions and multipipeline chaining and networking techniques for developing higher-performance vector multiprocessors.

The evolution from SIMD and MIMD computers to hybrid SIMD/MIMD computer systems is also considered. The Connection Machine CM-5 reflects this architectural trend. This hybrid approach to designing reconfigurable computers opens up new opportunities for exploiting coordinated parallelism in complex application problems.

8.1 Vector Processing Principles

Vector instruction types, memory-access schemes for vector operands, and an overview of supercomputer families are given in this section.

8.1.1 Vector Instruction Types

Basic concepts behind vector processing are defined below. Then we discuss major types of vector instructions encountered in a modern vector machine. The intent is to acquaint the reader with the instruction-set architectures of available vector supercomputers.

Vector Processing Definitions   A *vector* is a set of scalar data items, all of the same type, stored in memory. Usually, the vector elements are ordered to have a fixed addressing increment between successive elements called the *stride*.

A *vector processor* is an ensemble of hardware resources, including vector registers, functional pipelines, processing elements, and register counters, for performing vector operations. *Vector processing* occurs when arithmetic or logical operations are applied to vectors. It is distinguished from scalar processing which operates on one or one pair of data. The conversion from scalar code to vector code is called *vectorization*.

In general, vector processing is faster and more efficient than scalar processing. Both pipelined processors and SIMD computers can perform vector operations. Vector processing reduces software overhead incurred in the maintenance of looping control, reduces memory-access conflicts, and above all matches nicely with the pipelining and segmentation concepts to generate one result per each clock cycle continuously.

Depending on the *speed ratio* between vector and scalar operations (including startup delays and other overheads) and on the *vectorization ratio* in user programs, a vector processor executing a well-vectorized code can easily achieve a speedup of 10 to 20 times, as compared with scalar processing on conventional machines.

Of course, the enhanced performance comes with increased hardware and compiler costs, as expected. A compiler capable of vectorization is called a *vectorizing compiler* or simply a *vectorizer*. For successful vector processing, one needs to make improvements in vector hardware, vectorizing compilers, and programming skills specially targeted at vector machines.

Vector Instruction Types   We briefly introduced basic vector instructions in Chapter 4. What are characterized below are vector instructions for register-based, pipelined vector machines. Six types of vector instructions are illustrated in Figs. 8.1 and 8.2. We define these vector instruction types by mathematical mappings between their working registers or memory where vector operands are stored.

(1) *Vector-vector instructions* — As shown in Fig. 8.1a, one or two vector operands are fetched from the respective vector registers, enter through a functional pipeline unit, and produce results in another vector register. These instructions are defined by the following two mappings:

$$f_1 : V_i \rightarrow V_j \tag{8.1}$$

$$f_2 : V_j \times V_k \rightarrow V_i \tag{8.2}$$

Examples are $V_1 = \sin(V_2)$ and $V_3 = V_1 + V_2$ for the mappings f_1 and f_2, respectively, where V_i for $i = 1, 2$, and 3 are vector registers.

(2) *Vector-scalar instructions* — Figure 8.1b shows a vector-scalar instruction corresponding to the following mapping:

$$f_3 : s \times V_i \rightarrow V_j \tag{8.3}$$

An example is a scalar product $s \times V_1 = V_2$, in which the elements of V_1 are each multiplied by a scalar s to produce vector V_2 of equal length.

(3) *Vector-memory instructions* — This corresponds to vector load or vector store (Fig. 8.1c), element by element, between the vector register (V) and the memory

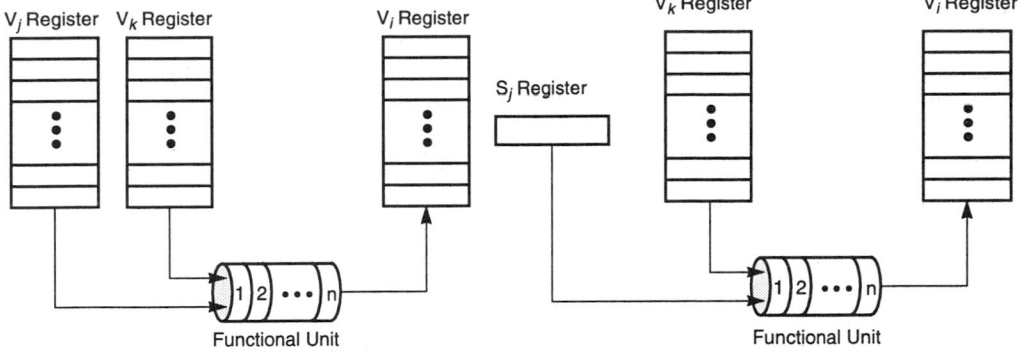

(a) Vector-vector instruction   (b) Vector-scalar instruction

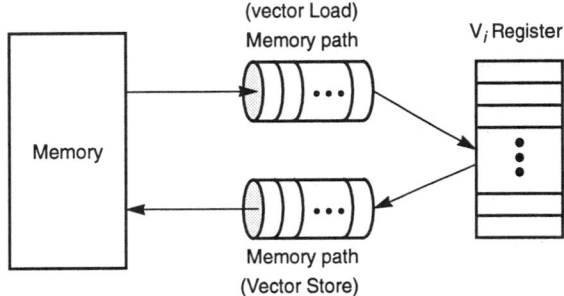

(c) Vector-memory instructions

Figure 8.1 Vector instruction types in Cray-like computers.

(M) as defined below:

$$f_4 : M \to V \qquad \qquad \textit{Vector load} \qquad \qquad (8.4)$$

$$f_5 : V \to M \qquad \qquad \textit{Vector store} \qquad \qquad (8.5)$$

(4) *Vector reduction instructions* — These correspond to the following mappings:

$$f_6 : V_i \to s_j \qquad \qquad (8.6)$$

$$f_7 : V_i \times V_j \to s_k \qquad \qquad (8.7)$$

Examples of f_6 include finding the *maximum*, *minimum*, *sum*, and *mean value* of all elements in a vector. A good example of f_7 is the *dot product*, which performs $s = \sum_{i=1}^{n} a_i \times b_i$ from two vectors $A = (a_i)$ and $B = (b_i)$.

(5) *Gather and scatter instructions* — These instructions use two vector registers to gather or to scatter vector elements randomly throughout the memory, corresponding to the following mappings:

$$f_8 : M \to V_1 \times V_0 \qquad \qquad \textit{Gather} \qquad \qquad (8.8)$$

$$f_9 : V_1 \times V_0 \to M \qquad\qquad \textit{Scatter} \qquad\qquad (8.9)$$

Gather is an operation that fetches from memory the nonzero elements of a sparse vector using indices that themselves are indexed. *Scatter* does the opposite, storing into memory a vector in a sparse vector whose nonzero entries are indexed. The vector register V_1 contains the data, and the vector register V_0 is used as an index to gather or scatter data from or to random memory locations as illustrated in Figs. 8.2a and 8.2b, respectively.

(6) *Masking instructions* — This type of instruction uses a *mask vector* to compress or to expand a vector to a shorter or longer index vector, respectively, corresponding to the following mappings:

$$f_{10} : V_0 \times V_m \to V_1 \qquad\qquad (8.10)$$

The following example will clarify the meaning of *gather*, *scatter*, and *masking* instructions.

Example 8.1 Gather, scatter, and masking instructions in the Cray Y-MP (Cray Research, Inc., 1990)

The gather instruction (Fig. 8.2a) transfers the contents (600, 400, 250, 200) of nonsequential memory locations (104, 102, 107, 100) to four elements of a vector register $V1$. The base address (100) of the memory is indicated by an address register $A0$. The number of elements being transferred is indicated by the contents (4) of a *vector length* register VL.

The offsets (indices) from the base address are retrieved from the vector register $V0$. The effective memory addresses are obtained by adding the base address to the indices.

The scatter instruction reverses the mapping operations, as illustrated in Fig. 8.2b. Both the VL and $A0$ registers are embedded in the instruction.

The masking instruction is shown in Fig. 8.2c for compressing a long vector into a short index vector. The contents of vector register $V0$ are tested for zero or nonzero elements. A *masking register* (VM) is used to store the test results. After testing and forming the *masking vector* in VM, the corresponding nonzero indices are stored in the $V1$ register. The VL register indicates the length of the vector being tested.

∎

The *gather*, *scatter*, and *masking* instructions are very useful in handling sparse vectors or sparse matrices often encountered in practical vector processing applications. Sparse matrices are those in which most of the entries are zeros. Advanced vector processors implement these instructions directly in hardware.

The above instruction types cover the most important ones. A given specific machine may implement an instruction set containing only a subset or even a superset of the above instructions.

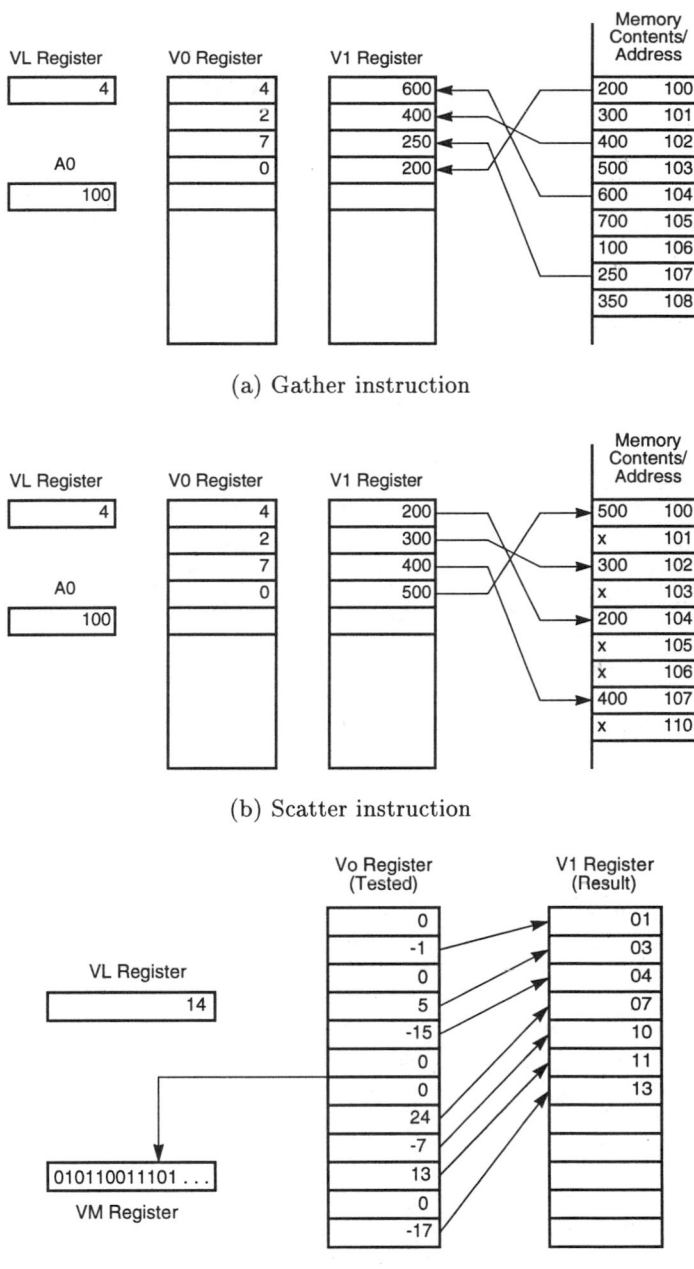

(a) Gather instruction

(b) Scatter instruction

(c) Masking instruction

Figure 8.2 Gather, scatter and masking operations on the Cray Y-MP. (Courtesy of Cray Research Inc., 1990)

8.1.2 Vector-Access Memory Schemes

The flow of vector operands between the main memory and vector registers is usually pipelined with multiple access paths. In this section, we specify vector operands and describe three vector-access schemes from interleaved memory modules allowing overlapped memory accesses.

Vector Operand Specifications   Vector operands may have arbitrary length. Vector elements are not necessarily stored in contiguous memory locations. For example, the entries in a matrix may be stored in row major or in column major. Each row, column, or diagonal of the matrix can be used as a vector.

When row elements are stored in contiguous locations with a unit stride, the column elements must be stored with a stride of n, where n is the matrix order. Similarly, the diagonal elements are also separated by a stride of $n + 1$.

To access a vector in memory, one must specify its *base address*, *stride*, and *length*. Since each vector register has a fixed number of component registers, only a segment of the vector can be loaded into the vector register in a fixed number of cycles. Long vectors must be segmented and processed one segment at a time.

Vector operands should be stored in memory to allow pipelined or parallel access. The memory system for a vector processor must be specifically designed to enable fast vector access. The access rate should match the pipeline rate. In fact, the access path is often itself pipelined and is called an *access pipe*. These vector-access memory organizations are described below.

C-Access Memory Organization   The m-way low-order interleaved memory structure shown in Figs. 5.15a and 5.16 allows m memory words to be accessed concurrently in an overlapped manner. This *concurrent access* has been called *C-access* as illustrated in Fig. 5.16b.

The access cycles in different memory modules are staggered. The low-order a bits select the modules, and the high-order b bits select the word within each module, where $m = 2^a$ and $a + b = n$ is the address length.

To access a vector with a stride of 1, successive addresses are latched in the address buffer at the rate of one per cycle. Effectively it takes m minor cycles to fetch m words, which equals one (major) memory cycle as stated in Eq. 5.4 and Fig. 5.16b.

If the stride is 2, the successive accesses must be separated by two minor cycles in order to avoid access conflicts. This reduces the memory throughput by one-half. If the stride is 3, there is no module conflict and the maximum throughput (m words) results. In general, C-access will yield the maximum throughput of m words per memory cycle if the stride is relatively prime to m, the number of interleaved memory modules.

S-Access Memory Organization   The low-order interleaved memory can be rearranged to allow *simultaneous access*, or *S-access*, as illustrated in Fig. 8.3a. In this case, all memory modules are accessed simultaneously in a synchronized manner. Again the high-order $(n - a)$ bits select the same offset word from each module.

At the end of each memory cycle (Fig. 8.3b), $m = 2^a$ consecutive words are latched in the data buffers simultaneously. The low-order a bits are then used to multiplex the m words out, one per each minor cycle. If the minor cycle is chosen to be $1/m$ of the major memory cycle (Eq. 5.4), then it takes two memory cycles to access m consecutive words.

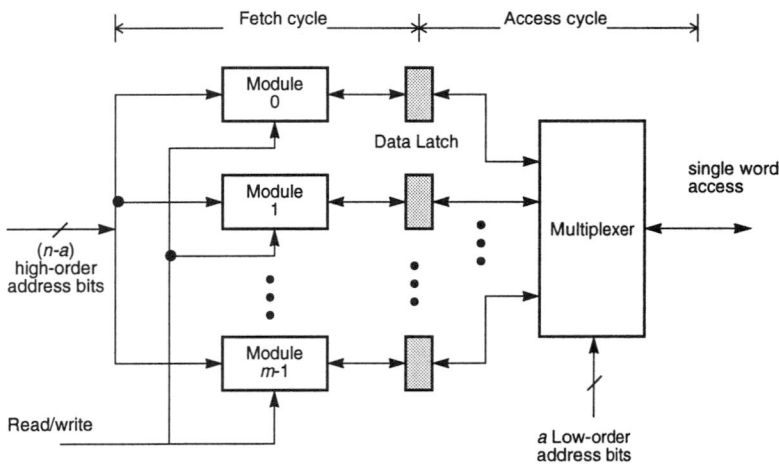

(a) S-access organization for an m-way interleaved memory

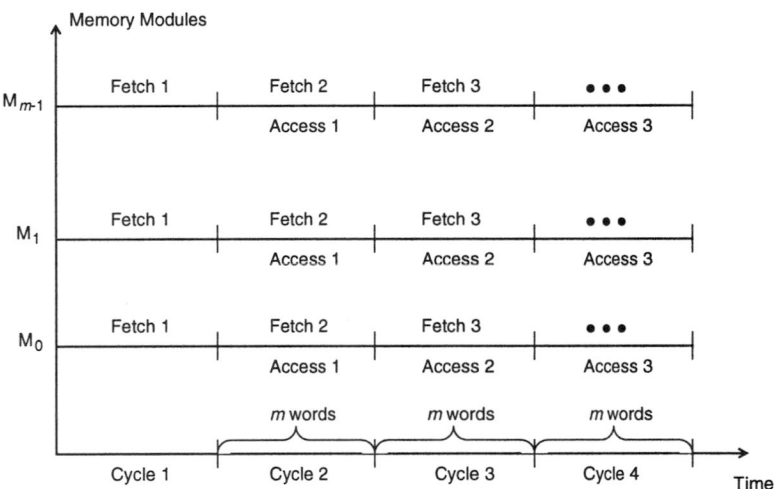

(b) Successive vector accesses using overlapped fetch and access cycles

Figure 8.3 The S-access interleaved memory for vector operands access.

However, if the access phase of the last access is overlapped with the fetch phase of the current access (Fig. 8.3b), effectively m words take only one memory cycle to access.

If the stride is greater than 1, then the throughput decreases, roughly proportionally to the stride.

C/S-Access Memory Organization   A memory organization in which the C-access and S-access are combined is called *C/S-access*. This scheme is shown in Fig. 8.4, where n access buses are used with m interleaved memory modules attached to each bus. The m modules on each bus are m-way interleaved to allow C-access. The n buses operate in parallel to allow S-access. In each memory cycle, at most $m \cdot n$ words are fetched if the n buses are fully used with pipelined memory accesses.

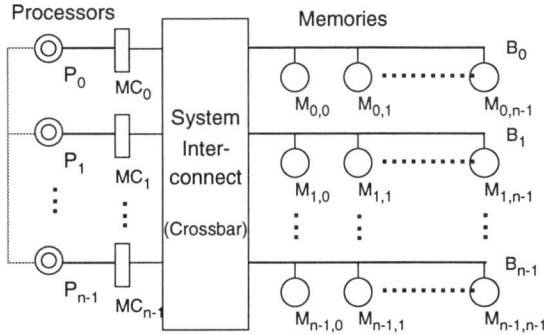

Figure 8.4 The C/S memory organization. (Courtesy of D.K. Panda, 1990)

The C/S-access memory is suitable for use in vector multiprocessor configurations. It provides parallel pipelined access of a vector data set with high bandwidth. A special *vector cache* design is needed within each processor in order to guarantee smooth data movement between the memory and multiple vector processors.

8.1.3   Past and Present Supercomputers

This section introduces full-scale supercomputers manufactured recently in the United States and in Japan. Five supercomputer families are reviewed, including the Cray Research Series, the CDC/ETA Series, the Fujitsu VP Series, the NEC SX Series, and the Hitachi 820 Series (Table 8.1). The relative performance of these machines for vector processing are compared with scalar processing at the end.

The Cray Research Series   Seymour Cray founded Cray Research, Inc. in 1972. Since then, about 400 units of Cray supercomputers have been produced and installed worldwide.

The Cray 1 was introduced in 1975. An enhanced version, the Cray 1S, was produced in 1979. It was the first ECL-based supercomputer with a 12.5-ns clock cycle. High degrees of pipelining and vector processing were the major features of these machines.

Table 8.1 Summary of Full-Scale Supercomputers Made in the United States and in Japan

System model	Maximum configuration, clock rate, OS/Compiler	Unique features and remarks
Cray 1S	Uniprocessor with 10 pipelines, 12.5 ns, COS/CF77 2.1.	First ECL-based super, introduced in 1976.
Cray 2S /4-256	4 processors with 256M-word memory, 4.1 ns, COS or UNIX/CF77 3.0.	16K-word local memory, ported UNIX V introduced in 1985.
Cray X-MP 416	4 processors with 16M-word memory, and 128M-word SSD, 8.5 ns, COS CF77 5.0.	Using shared register clusters for IPC, introduced 1983.
Cray Y-MP 832	8 processors with 128M-word memory, 6 ns, CF77 5.0.	Enhanced from X-MP, introduced 1988.
Cray Y-MP C-90	16 processors with 2 vector pipes per processor, 4.2 ns, UNICOS/CF 77 5.0.	The largest Cray machine, introduced in 1991.
CDC Cyber 205	Uniprocessor with 4 pipelines, 20 ns, virtual OS/FTN 200.	Memory-to-memory architecture, introduced in 1982.
ETA 10 E	Uniprocessor with 10.5 ns, ETAV/FTN 200	Successor to Cyber 205, introduced in 1985.
NEC SX-X/44	4 processors with 4 sets of pipelines per processor, 2.9 ns, F77SX.	Succeeded by SX-X Series, introduced in 1991.
Fujitsu VP2600/10	Uniprocessor with 5 vector pipes and dual scalar processors, 3.2 ns, MSP-EX/F77 EX/VP.	Use reconfigurable vector registers and masking, introduced in 1991.
Hitachi 820/80	18 functional pipelines in a uniprocessor with 512 Mbytes memory 4 ns, FORT 77/HAP V23-OC.	Introduced in 1987 with 64 I/O channels providing a maximum of 288 Mbytes/s transfer.

Ten functional pipelines can run simultaneously in the Cray 1S to achieve a computing power equivalent to that of 10 IBM 3033's or CDC Cyber 7600's. Only batch processing with a single user was allowed when the Cray 1 was initially introduced using the Cray Operating System (COS) with a Fortran 77 compiler (CF 77 Version 2.1).

The Cray X-MP Series introduced multiprocessor configurations in 1983. Steve Chen led the effort at Cray Research in developing this series using one to four Cray-1-equivalent CPUs with shared memory. A unique feature introduced with the X-MP models was shared register clusters for fast interprocessor communications without going through the shared memory.

Besides 128 Mbytes of shared memory, the X-MP system has 1 Gbyte of *solid-state storage* (SSD) as extended shared memory. The clock rate was also reduced to 8.5 ns. The peak performance of the X-MP-416 is 840 Mflops when eight vector pipelines for add and multiply are used simultaneously across four processors.

The successor to the Cray X-MP was the Cray Y-MP introduced in 1988 with up to eight processors in a single system using a 6-ns clock rate and 256 Mbytes of shared memory.

The Cray Y-MP C-90 was introduced in 1990 to offer an integrated system with 16 processors using a 4.2-ns clock. We will study models Y-MP 816 and C-90 in detail in the next section.

Another product line is the Cray 2S introduced in 1985. The system allows up to four processors with 2 Gbytes of shared memory and a 4.1-ns superpipelined clock. A major contribution of the Cray 2 was to switch from the batch processing COS to multiuser UNIX System V on a supercomputer. This led to the UNICOS operating system, derived from the UNIX/V and Berkeley 4.3 BSD, currently in use in most Cray supercomputers.

The Cyber/ETA Series   Control Data Corporation (CDC) introduced its first supercomputer, the STAR-100, in 1973. Cyber 205 was the successor produced in 1982. The Cyber 205 runs at a 20-ns clock rate, using up to four vector pipelines in a uniprocessor configuration.

Different from the register-to-register architecture used in Cray and other supercomputers, the Cyber 205 and its successor, the ETA 10, have memory-to-memory architecture with longer vector instructions containing memory addresses.

The largest ETA 10 consists of 8 CPUs sharing memory and 18 I/O processors. The peak performance of the ETA 10 was targeted for 10 Gflops. Both the Cyber and the ETA Series are no longer in production but are still in use at several supercomputer centers.

Japanese Supercomputers   NEC produced the SX-X Series with a claimed peak performance of 22 Gflops in 1991. Fujitsu produced the VP-2000 Series with a 5-Gflops peak performance at the same time. These two machines use 2.9- and 3.2-ns clocks, respectively.

Shared communication registers and reconfigurable vector registers are special features in these machines. Hitachi offers the 820 Series providing a 3-Gflops peak performance. Japanese supercomputers are strong in high-speed hardware and interactive vectorizing compilers.

The NEC SX-X 44   NEC claims that this machine is the fastest vector supercomputer (22 Gflops peak) ever built up to 1992. The architecture is shown in Fig. 8.5. One of the major contributions to this performance is the use of a 2.9-ns clock cycle based on VLSI and high-density packaging.

There are four arithmetic processors communicating through either the shared registers or via the shared memory of 2 Gbytes. There are four sets of vector pipelines per processor, each set consisting of two add/shift and two multiply/logical pipelines. Therefore, 64-way parallelism is observed with four processors, similar to that in the C-90.

Besides the vector unit, a high-speed scalar unit employs RISC architecture with 128 scalar registers. Instruction reordering is supported to exploit higher parallelism. The main memory is 1024-way interleaved. The extended memory of up to 16 Gbytes

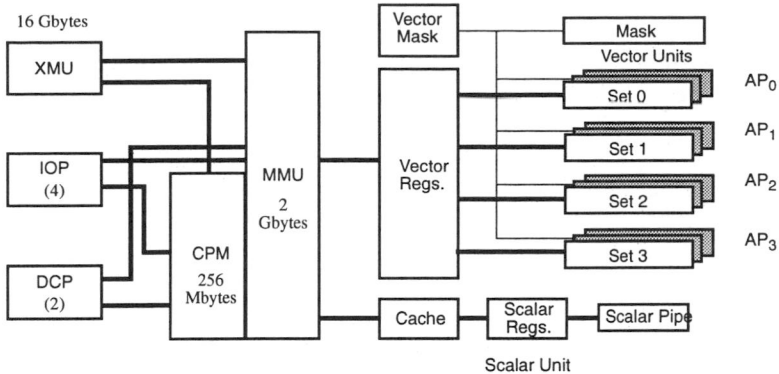

Captions :
XMU : Extended memory unit
IOP : I/O processors (4)
DCP : Data central processors (2)
AP : Arithmetic processors (4)
MMU : Main memory unit
CPM : Data central processor memory
Each set consists of 4 pipelines for add/shift
and multiply/logical vector operations

Figure 8.5 The NEC SX-X 44 vector supercomputer architecture. (Courtesy of NEC, 1991)

provides a maximum transfer rate of 2.75 Gbytes/s.

A maximum of four I/O processors can be configured to accommodate a 1-Gbyte/s data transfer rate per I/O processor. The system can provide a maximum of 256 channels for high-speed network, graphic, and peripheral operations. The support includes 100-Mbytes/s channels.

Relative Vector/Scalar Performance   Let r be the vector/scalar speed ratio and f the vectorization ratio. By Amdahl's law in Section 3.3.1, the following *relative performance* can be defined:

$$P = \frac{1}{(1-f) + f/r} = \frac{r}{(1-f)r + f} \qquad (8.11)$$

This relative performance indicates the speedup performance of vector processing over scalar processing. The hardware speed ratio r is the designer's choice. The vectorization ratio f reflects the percentage of code in a user program which is vectorized.

The relative performance is rather sensitive to the value of f. This value can be increased by using a better vectorizing compiler or through user program transformations. The following example shows the IBM experience in vector processing.

Example 8.2 The vector/scalar relative performance of the IBM 3090/VF

Figure 8.6 plots the relative performance P as a function of r with f as a running parameter. The higher the value of f, the higher the relative speedup. The IBM 3090 with vector facility (VF) was a high-end mainframe with add-on vector hardware.

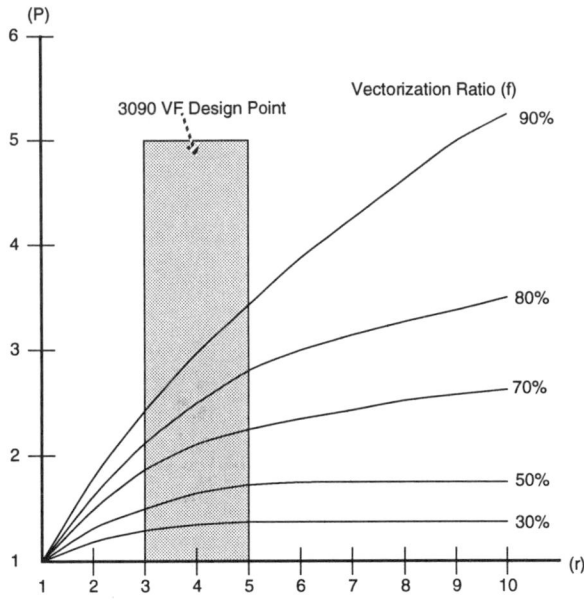

Figure 8.6 Speedup performance of vector processing over scalar processing in the IBM 3090/VF. (Courtesy of IBM Corporation, 1986)

The designer of the 3090/VF chose a speed ratio in the range $3 \le r \le 5$ because IBM wanted a balance between business and scientific applications. When the program is 70% vectorized, one expects a maximum speedup of 2.2. However, for $f \le 30\%$, the speedup is reduced to less than 1.3.

The IBM designers did not choose a high speed ratio because they did not expect user programs to be highly vectorizable. When f is low, the speedup cannot be high, even with a very high r. In fact, the limiting case is $P \to 1$ if $f \to 0$.

On the other hand, $P \to r$ when $f \to 1$. Scientific supercomputer designers like Cray and Japanese manufacturers often choose a much higher speed ratio, say, $10 \le r \le 25$, because they expect a higher vectorization ratio f in user programs, or they use better vectorizers to increase the ratio to a desired level.

8.2   Multivector Multiprocessors

The architectural design of supercomputers must be upgraded based on past experience with current technology. Design rules are provided for high performance, and we review these rules in case studies on existing full-scale supers, high-end mainframes, and minisupers. The trends toward scalable architectures in building MPP systems for supercomputing are also assessed.

8.2.1   Performance-Directed Design Rules

Supercomputers are targeted toward large-scale scientific and engineering problems. They should provide the highest performance constrained only by current technology. In addition, they must be programmable and accessible in a multiuser environment.

Supercomputer architecture design rules are presented below. These rules are driven by the desire to offer the highest available performance in a variety of respects, including processor, memory, and I/O performance, capacities, and bandwidths in all subsystems.

Architecture Design Goals   Smith, Hsu, and Hsiung (1990) have identified the following four major challenges in the development of future general-purpose supercomputers:

- Maintaining a good vector/scalar performance balance.
- Supporting scalability with an increasing number of processors.
- Increasing memory system capacity and performance.
- Providing high-performance I/O and an easy-access network.

Balanced Vector/Scalar Ratio   In a supercomputer, separate hardware resources with different speeds are dedicated to concurrent vector and scalar operations. Scalar processing is indispensable for general-purpose architectures. Vector processing is needed for regularly structured parallelism in scientific and engineering computations. These two types of computations must be balanced.

The *vector balance point* is defined as the percentage of vector code in a program required to achieve equal utilization of vector and scalar hardware. In other words, we expect equal time spent in vector and scalar hardware so that no resources will be idle.

Example 8.3 Vector/scalar balance point in supercomputer design (Smith, Hsu, and Hsiung, 1990)

If a system is capable of 9 Mflops in vector mode and 1 Mflops in scalar mode, equal time will be spent in each mode if the code is 90% vector and 10% scalar, resulting in a vector balance point of 0.9.

It may not be optimal for a system to spend equal time in vector and scalar modes. However, the vector balance point should be maintained sufficiently high, matching the level of vectorization in user programs.

■

Vector performance can be enhanced with replicated functional unit pipelines in each processor. Another approach is to apply superpipelining on vector units with a double or triple clock rate with respect to scalar pipeline operations.

Longer vectors are required to really achieve the target performance. It was projected that by the year 2000, an 8-Gflops peak will be achievable with multiple functional units running simultaneously in a processor.

Vector/Scalar Performance   In Figs. 8.7a and 8.7b, the single-processor vector performance and scalar performance are shown, based on running Livermore Fortran loops on Cray Research and Japanese supercomputers. The scalar performance of all supercomputers increases along the dashed lines in the figure.

Japanese supercomputers certainly outperform Cray Research machines in vector capability. One of the contributing factors is the high clock rate, and other factors include use of a better compiler and the optimization support provided.

Table 8.2 compares the vector and scalar performances in seven supercomputers. Most supercomputers have a 90% or higher vector balance point. The higher the vector/scalar ratio, the heavier the dependence on a high degree of vectorization in the object code.

Table 8.2 Vector and Scalar Performance of Various Supercomputers

Machine	Cray 1S	Cray 2S	Cray X-MP	Cray Y-MP	Hitachi S820	NEC SX2	Fujitsu VP400
Vector performance (Mflops)	85.0	151.5	143.3	201.6	737.3	424.2	207.1
Scalar performance (Mflops)	9.8	11.2	13.1	17.0	17.8	9.5	6.6
Vector balance point	0.90	0.93	0.92	0.92	0.98	0.98	0.97

Source: J.E. Smith et al., Future General-Purpose Supercomputing Conference, *IEEE Supercomputing Conference*, 1990.

The above approach is quite different from the design in IBM vector machines which maintain a low vector/scalar ratio between 3 and 5. The idea was to make a good compromise between the demands of scalar and vector processing for general-purpose applications.

I/O and Networking Performance   With the aggregate speed of supercomputers increasing at least three to five times each generation, problem size has been increasing accordingly, as have I/O bandwidth requirements. Figure 8.7c illustrates the aggregate I/O bandwidths supported by current and past supercomputer systems.

The I/O is defined as the transfer of data between the mainframe and peripherals

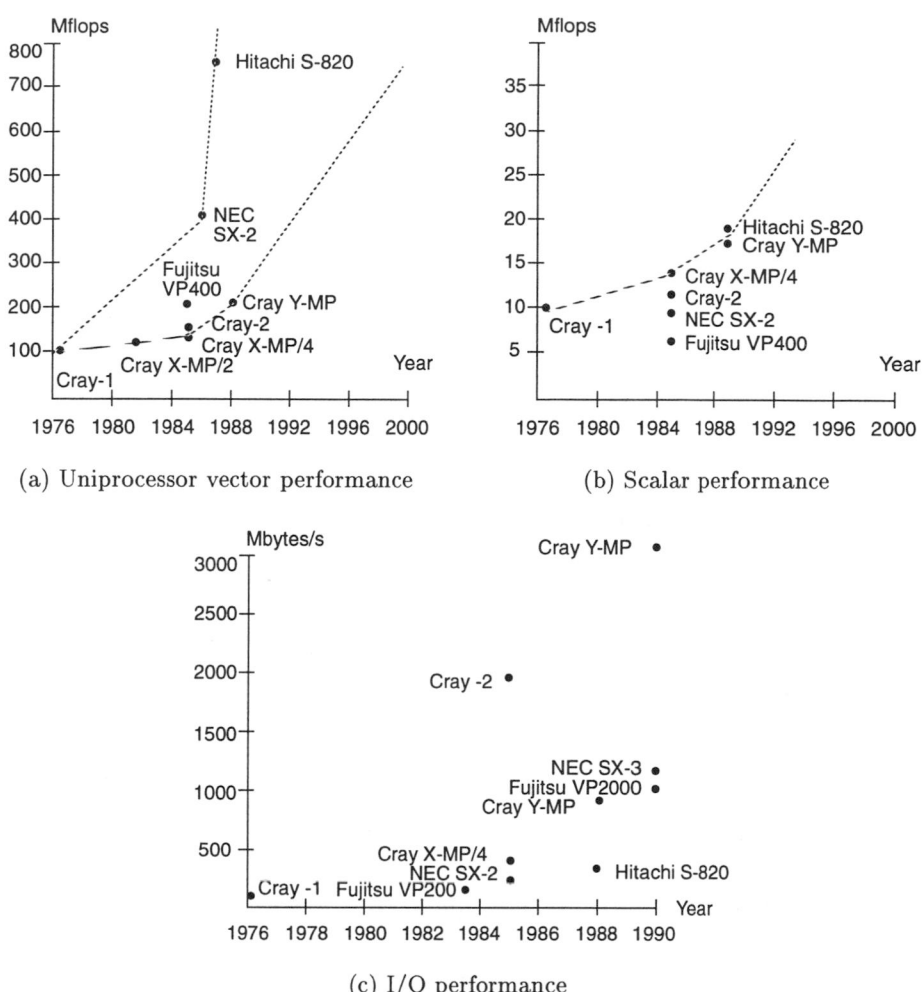

(a) Uniprocessor vector performance

(b) Scalar performance

(c) I/O performance

Figure 8.7 Some reported supercomputer performance data (Source: Smith, Hsu, and Hsiung, *IEEE Supercomputing Conference*, 1990)

or a network. Over the past generation, I/O bandwidths have not always been well correlated with computational performance. I/O processor architectures were implemented by Cray Research with two different approaches.

The first approach is exemplified by the Cray Y-MP I/O subsystem, which uses I/O processors that are very flexible and can do complex processing. The second is used in the Cray 2, where a simple front-end processor controls high-speed channels with most of the I/O management being done by the mainframe's operating system.

It is expected that in the next 5 to 10 years, more than a 50-Gbytes/s I/O transfer rate will be needed in supercomputers connected to high-speed disk arrays and networks. Support for high-speed networking will become a major component of the I/O architecture in supercomputers.

Memory Demand   The main memory sizes and extended memory sizes of past and present supercomputers are shown in Fig. 8.8. A large-scale memory system must provide a low latency for scalar processing, a high bandwidth for vector and parallel processing, and a large size for grand challenge problems and throughput.

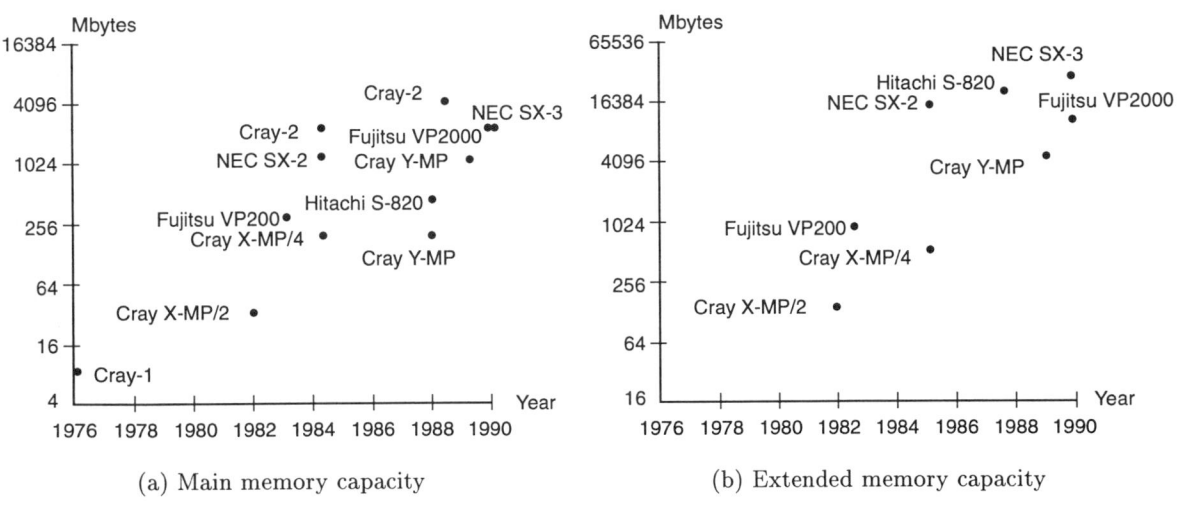

(a) Main memory capacity   (b) Extended memory capacity

Figure 8.8 Supercomputer memory capacities (Source: Smith, Hsu, and Hsiung, *IEEE Supercomputing Conference*, 1990)

To achieve the above goals, an effective memory hierarchy is necessary. A typical hierarchy may consist of data files or disks, extended memory in dynamic RAMs, a fast shared memory in static RAMs, and a cache/local memory using RAM on arrays.

It is expected that main memory will reach 32 Gbytes with an extended memory four times as large in the next decade. This will match the expected growth of super-computer architectures of 64 processors by the mid-1990s and 1024 processors by the end of the century.

Supporting Scalability   Multiprocessor supercomputers must be designed to support the triad of scalar, vector, and parallel processing. The dominant scalability problem involves support of shared memory with an increasing number of processors and memory ports. Increasing memory-access latency and interprocessor communication overhead impose additional constraints on scalability.

Scalable architectures include multistage interconnection networks in flat systems, hierarchical clustered systems, and multidimensional spanning buses, ring, mesh, or torus networks with a distributed shared memory. Table 8.3 summarizes the key features of three representative multivector supercomputers built today.

8.2.2   Cray Y-MP, C-90, and MPP

We study below the architectures of the Cray Research Y-MP, C-90, and recently announced MPP. Besides architectural features, we examine the operating systems, languages/compilers, and target performance of these machines.

Table 8.3 Architectural Characteristics of Three Current Supercomputers

Machine Characteristics	Cray Y-MP C90/16256	NEC SX-X Series	Fujitsu VP-2000 Series
Number of processors	16 CPUs	4 arithmetic processors	1 for VP2600/10, 2 for VP2400/40
Machine cycle time	4.2 ns	2.9 ns	3.2 ns
Max. memory	256M words (2 Gbytes).	2 Gbytes, 1024-way interleaving.	1 or 3 Gbytes of SRAM.
Optional SSD memory	512M, 1024M, or 2048M words (16 Gbytes).	16 Gbytes with 2.75 Gbytes/s transfer rate.	32 Gbytes of extended memory.
Processor architecture: vector pipelines, functional and scalar units	Two vector pipes and two functional units per CPU, delivering 64 vector results per clock period.	Four sets of vector pipelines per processor, each set consists of two adder/shift and two multiply/logical pipelines. A separate scalar pipeline.	Two load/store pipes and 5 functional pipes per vector unit, 1 or 2 vector units, 2 scalar units can be attached to each vector unit.
Operating system	UNICOS derived from UNIX/V and BSD.	Super-UX based on UNIX System V and 4.3 BSD.	UXP/M and MSP/EX enhanced for vector processing.
Front ends	IBM, CDC, DEC, Univac, Apollo, Honeywell.	Built-in control processor and 4 I/O processors.	IBM-compatible hosts.
Vectorizing languages / compilers	Fortran 77, C, CF77 5.0, Cray C release 3.0	Fortran 77/SX, Vectorizer/XS, Analyzer/SX.	Fortran 77 EX/VP, C/VP compiler with interactive vectorizer.
Peak performance and I/O bandwidth	16 Gflops, 13.6 Gbytes/s.	22 Gflops, 1 Gbyte/s per I/O processor.	5 Gflops, 2 Gbyte/s with 256 channels.

The Cray Y-MP 816   A schematic block diagram of the Y-MP 8 is shown in Fig. 8.9. The system can be configured to have one, two, four, and eight processors. The eight CPUs of the Y-MP share the central memory, the I/O section, the interprocessor communication section, and the real-time clock.

The central memory is divided into 256 interleaved banks. Overlapping memory access is made possible through memory interleaving via four memory-access ports per CPU. A 6-ns clock period is used in the CPU design.

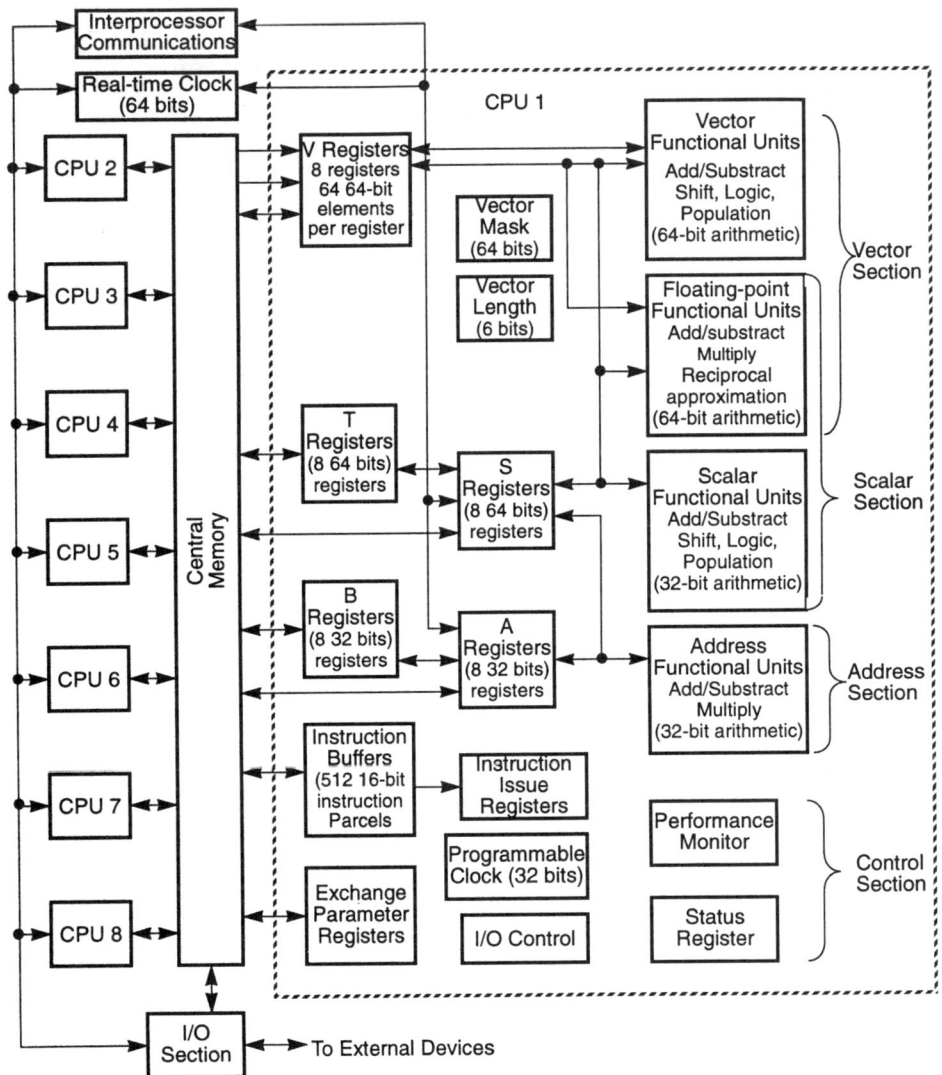

Figure 8.9 Cray Y-MP 816 system organization. (Courtesy of Cray Research, Inc., 1991)

The central memory offers 16M-, 32M-, 64M-, and 128M-word options with a maximum size of 1 Gbyte. The SSD options are from 32M to 512M words or up to 4

Gbytes.

The four memory-access ports allow each CPU to perform two scalar and vector *fetches*, one *store*, and one independent *I/O* simultaneously. These parallel memory accesses are also pipelined to make the *vector read* and *vector write* possible.

The system has built-in resolution hardware to minimize the delays caused by memory conflicts. To protect data, single-error correction/double-error detection (SECDED) logic is used in central memory and on the data channels to and from central memory.

The CPU computation section consists of 14 functional units divided into vector, scalar, address, and control sections (Fig. 8.9). Both scalar and vector instructions can be executed in parallel. All arithmetic is register-to-register. Eight out of the 14 functional units can be used by vector instructions.

Large numbers of address, scalar, vector, intermediate, and temporary registers are used. Flexible chaining of functional pipelines is made possible through the use of registers and multiple memory-access and arithmetic/logic pipelines. Both 64-bit floating-point and 64-bit integer arithmetic are performed. Large instruction caches (buffers) are used to hold 512 16-bit instruction parcels at the same time.

The interprocessor communication section of the mainframe contains clusters of shared registers for fast synchronization purposes. Each cluster consists of shared address, shared scalar, and semaphore registers. Note that vector data communication among the CPUs is done through the shared memory.

The real-time clock consists of a 64-bit counter that advances one count each clock period. Because the clock advances synchronously with program execution, it can be used to time the execution to an exact clock count.

The I/O section supports three channel types with transfer rates of 6 Mbytes/s, 100 Mbytes/s, and 1 Gbyte/s. The IOS and SSD are high-speed data transfer devices designed to support the mainframe processing by eight caches.

Example 8.4 The multistage crossbar network in the Cray Y-MP 816

The interconnections between the 8 CPUs and 256 memory banks in the Cray Y-MP 816 are implemented with a multistage crossbar network, logically depicted in Fig. 8.10. The building blocks are 4×4 and 8×8 crossbar switches and 1×8 demultiplexers.

The network is controlled by a form of circuit switching where all conflicts are worked out early in the memory-access process and all requests from a given port return to the port in order.

■

The use of a multistage network instead of a single-stage crossbar for interprocessor memory connections will certainly promote scalability in the building of even larger systems with 64 or 1024 processors in the future.

Crossbar networks work only for small systems. To enhance scalability, emphasis should be given to data routing, heavier reliance on processor-based local memory (as in the Cray 2), or the use of clustered structures (as in the Cedar multiprocessor) to offset any increased latency when system size increases.

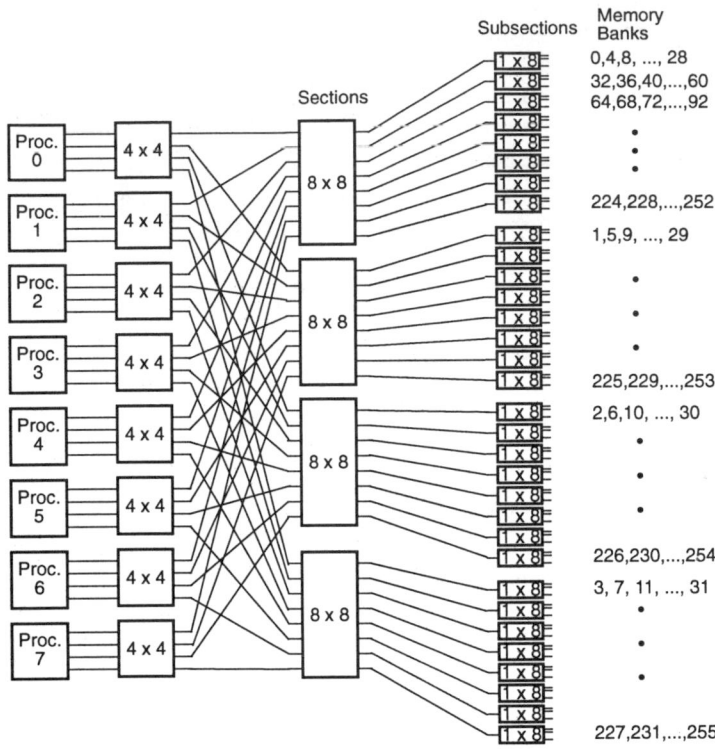

Figure 8.10 Schematic logic diagram of the crossbar network between 8 processors and 256 memory banks in the Cray Y-MP 816.

The C-90 and Clusters The C-90 is further enhanced in technology and scaled in size from the Y-MP Series. The architectural features of C-90/16256 are summarized in Table 8.3. The system is built with 16 CPUs, each of which is similar to that used in the Y-MP. The system uses up to 256 megawords (2 Gbytes) of shared main memory among the 16 processors. Up to 16 Gbytes of SSD memory is available as optional secondary main memory. In each cycle, two vector pipes and two functional units can operate in parallel, producing four vector results per clock. This implies a four-way parallelism within each processor. Thus 16 processors can deliver a maximum of 64 vector results per clock cycle.

The C-90 applies the UNICOX operating system, which was extended from the UNIX system V and Berkeley BSD 4.3. The C-90 can be driven by a number of host machines. Vectorizing compilers are available for Fortran 77 and C on the system. The 64-way parallelism, coupled with a 4.2-ns clock cycle, leads to a peak performance of 16 Gflops. The system has a maximum I/O bandwidth of 13.6 Gbytes/s.

Multiple C-90's can be used in a clustered configuration in order to solve large-scale problems. As illustrated in Fig. 8.11, four C-90 clusters are connected to a group of SSDs via 1000 Mbytes/s channels. Each C-90 cluster is allowed to access only its own main memory. However, they share the access of the SSDs. In other words, large data sets in the SSD can be shared by four clusters of C-90's. The clusters can also communicate with each other through a shared semaphore unit. Only synchronization and control information is passed via the semaphore unit. In this sense, the C-90 clusters are loosely coupled, but collectively they can provide a maximum of 256-way parallelism.

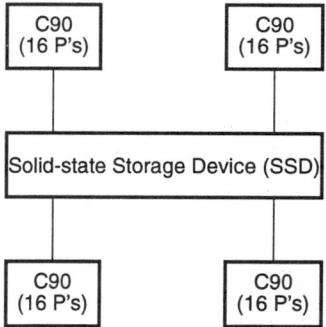

Figure 8.11 Four Cray Y-MP C-90's connected to a common SSD forming a loosely coupled 64-way parallel system.

If the computations are well partitioned and balanced among the clusters, a maximum peak performance of 64 Gflops would be possible for a four-cluster configuration. The clustered configuration was experimented with successfully at Cray Research Laboratories. Readers can obtain benchmark results directly from Cray reports.

The Cray/MPP System Massively parallel processing (MPP) systems have the potential for tackling highly parallel problems. Standard off-the-shelf microprocessors have deficiencies when used as building blocks of an MPP system. What is needed is a balanced system that matches fast processor speed with fast I/O, fast memory access, and capable software. Cray Research, Inc., announced its MPP development in October 1992. Although the first Cray/MPP system will not be available until 1993, the intended development plan may shed some light on the future of MPP from the standpoint of a major supercomputer manufacturer.

Most of today's RISC microprocessors lack the communication, memory, and synchronization features needed for efficient MPP systems. Cray Research plans to circumvent these shortcomings by surrounding the RISC chip with powerful communications hardware, besides exploiting Cray's expertise in supercomputer packaging and cooling. In this way, thousands of commodity RISC processors will be transformed into a supercomputer-class MPP system that can address terabytes of memory, minimize communication overhead, and provide flexible, lightweight synchronization in a UNIX environment.

Cray's first MPP system is code-named T3D because a three-dimensional, dense torus network will be used to interconnect the machine resources. The heart of Cray's T3D is a scalable macroarchitecture that combines the DEC Alpha microprocessors through a low-latency interconnect network that will have a bisection bandwidth an order of magnitude greater than that of existing MPP systems. The T3D system is designed to work jointly with the Cray Y-MP C-90 or the large-memory M-90 in a closely coupled fashion. Specific features of the MPP macroarchitecture are summarized below:

(1) The T3D is an MIMD machine that can be dynamically partitionable to emulate SIMD or multicomputer MIMD operations. The 3-D torus operates at a 150-MHz clock matching that of the Alpha chips. High-speed bidirectional switching nodes are built into the T3D network so that interprocessor communications can be handled without interrupting the PEs attached to the nodes. The T3D network is designed to be scalable from tens to thousands of PEs.

(2) The system uses a globally addressable, physically distributed memory. Because the memory is logically shared, any PE can access the memory of any other processing element without explicit message passing and without involving the remote PE. As a result, the system can be scaled to address terabytes of memory. Latency hiding (to be studied in Chapter 9) is supported by data prefetching, fast synchronization, and parallel I/O. These are supported by dedicated hardware. For example, special remote-access hardware is provided to hide the long latency in memory accesses. Fast synchronization support includes special primitives for data-parallel and message-passing programming paradigms.

(3) The Cray/MPP will use a Mach-based microkernel operating system. Each PE will have a microkernel that manages communications with other PEs and with the closely coupled Y-MP vector processors. Software portability is a major design goal in the Cray/MPP Series. Software-configurable redundant hardware will be included so that processing can continue should a PE fail.

(4) The Cray CFT77 compiler will be modified with extended directives for MPP applications. Program debugging and performance tools will be developed. Cray also plans to focus on software conversion on key MPP applications.

Cray/MPP Development Phases The Cray/MPP program was planned to have three phases as illustrated in Fig. 8.12. The T3D/MPP is attached to the Cray Y-MP as a back-end accelerator engine. Plans call for delivering the first-phase MPP system in 1993, with a 150-Gflops peak performance in a 1024-processor configuration, scalable to a 300-Gflops peak in a 2048-processor configuration.

The second-phase system will be developed in the mid-1990s, with a peak performance of 1 Tflops, and the third-phase system in 1997, with a sustained Teraflops performance. Besides hardware development, the biggest challenge in any MPP development is the software environment and availability. The Cray T3D programming model is based on an MIMD-oriented concept. Both the Connection Machine CM-5 (to be described in Section 8.5) and the Cray T3D emphasize this model, in order to broaden the application spectrum for their machines.

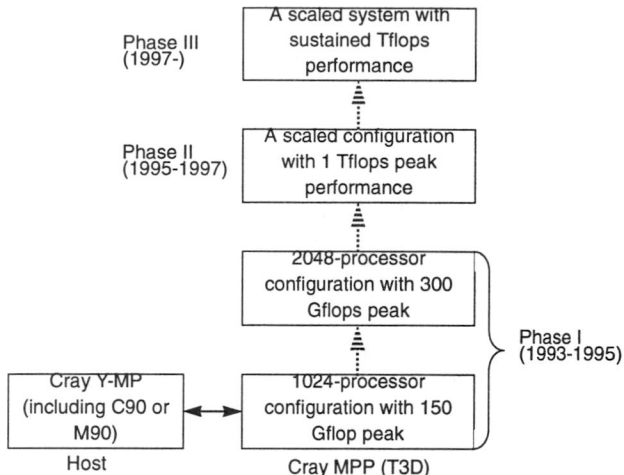

Figure 8.12 The development phases of the Cray/MPP system. (Courtesy of Cray Research, Inc. 1992)

8.2.3 Fujitsu VP2000 and VPP500

Multivector multiprocessors from Fujitsu Ltd. are reviewed in this section as supercomputer design examples. The VP2000 Series offers one- or two-processor configurations. The VPP500 Series offers from 7 to 222 processing elements (PEs) in a single MPP system. The two systems can be used jointly in solving large-scale problems. We describe below the functional specifications and technology bases of the Fujitsu supercomputers.

The Fujitsu VP2000 Figure 8.13 shows the architecture of the VP-2600/10 uniprocessor system. The system can be expanded to have dual processors (the VP-2400/40). The system clock is 3.2 ns, the main memory unit has 1 or 2 Gbytes, and the system storage unit provides up to 32 Gbytes of extended memory.

Each vector processing unit consists of two load/store pipelines, three functional pipelines, and two mask pipelines. Two scalar units can be attached to each vector unit, making a maximum of four scalar units in the dual-processor configuration. The maximum vector performance ranges from 0.5 to 5 Gflops across 10 different models of the VP2000 Series.

Example 8.5 Reconfigurable vector register file in the Fujitsu VP2000

Vector registers in Cray and Fujitsu machines are illustrated in Fig. 8.14. Cray machines use 8 vector registers, and each has a fixed length of 64 component registers. Each component register is 64 bits wide as demonstrated in Fig. 8.14a.

A component counter is built within each Cray vector register to keep track

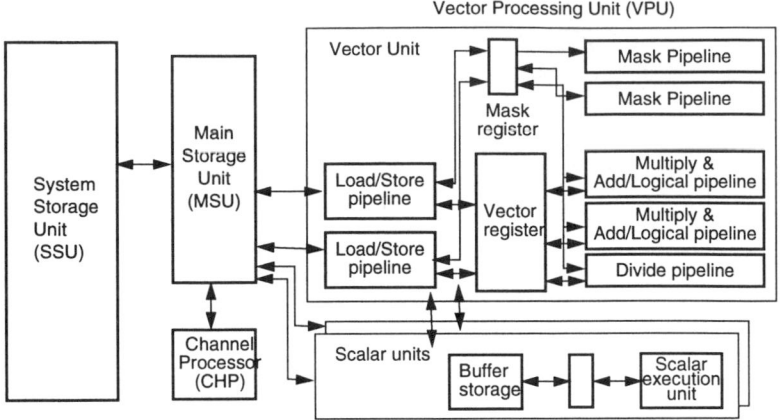

Figure 8.13 The Fujitsu VP2000 Series supercomputer architecture. (Courtesy of Fujitsu Ltd., 1991)

of the number of vector elements fetched or processed. A segment of a 64-element subvector is held as a package in each vector register. Long vectors must be divided into 64-element segments before they can be processed in a pipelined fashion.

In an early model of the Fujitsu VP2000, the vector registers are reconfigurable to have variable lengths. The purpose is to dynamically match the register length with the vector length being processed.

As illustrated in Fig. 8.14b, a total of 64 Kbytes in the register file can be configured into 8, 16, 32, 64, 128, and 256 vector registers with 1024, 512, 256, 128, 64, and 32 component registers, respectively. All component registers are 64 bits in length.

In the following Fortran Do loop operations, the three-dimensional vectors are indexed by I with constant values of J and K in the second and third dimensions.

$$
\begin{aligned}
&\textbf{Do } 10 \text{ I} = 0, 31 \\
&\quad \text{ZZ0(I)} = \text{U(I,J,K)} - \text{U(I,J}-1\text{,K)} \\
&\quad \text{ZZ1(I)} = \text{V(I,J,K)} - \text{V(I,J}-1\text{,K)} \\
&\qquad\qquad \vdots \\
&\quad \text{ZZ84(I)} = \text{W(I,J,K)} - \text{W(I,J}-1\text{,K)} \\
&10\quad \textbf{Continue}
\end{aligned}
$$

The program can be vectorized to have 170 input vectors and 85 output vectors with a vector length of 32 elements (I = 0 to 31). Therefore, the optimal partition is to configure the register file as 256 vector registers with 32 components each. ∎

Software support for parallel and vector processing in the above supercomputers will be treated in Part IV. This includes multitasking, macrotasking, microtasking,

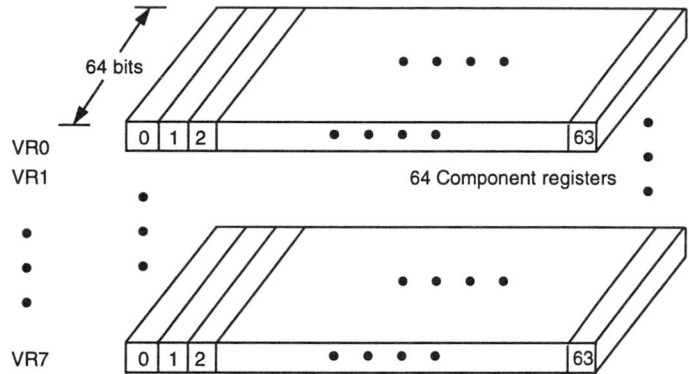

(a) Eight vector registers ($8 \times 64 \times 64$ bits) on Cray machines

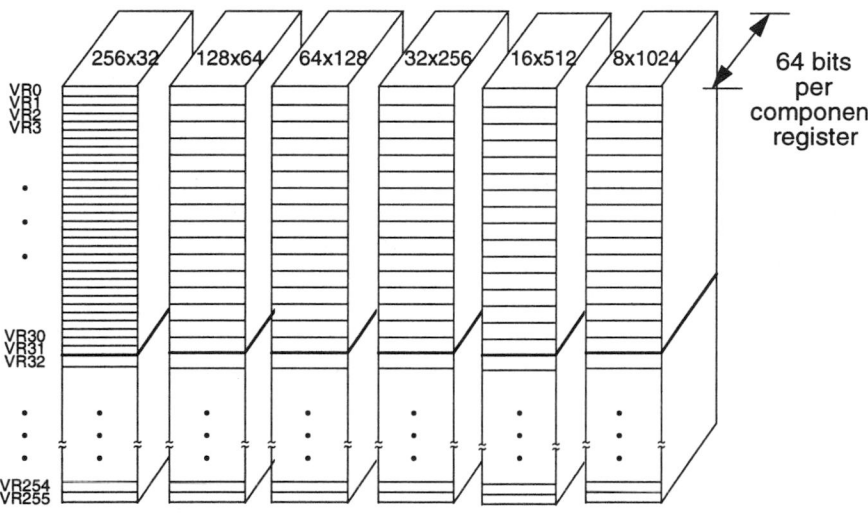

(b) Vector register configurations in the Fujitsu VP2000

Figure 8.14 Vector register file in Cray and Fujitsu supercomputers.

autotasking, and interactive compiler optimization techniques for vectorization or parallelization.

The VPP 500 This is a new supercomputer series from Fujitsu, called *vector parallel processor*. The architecture of the VPP500 is scalable from 7 to 222 PEs, offering a highly parallel MIMD multivector system. The peak performance is targeted for 335 Gflops with the first shipment in September 1993. Figure 8.15 shows the architecture of the VPP500 used as a back-end machine attached to a VP2000 or a VPX 200 host.

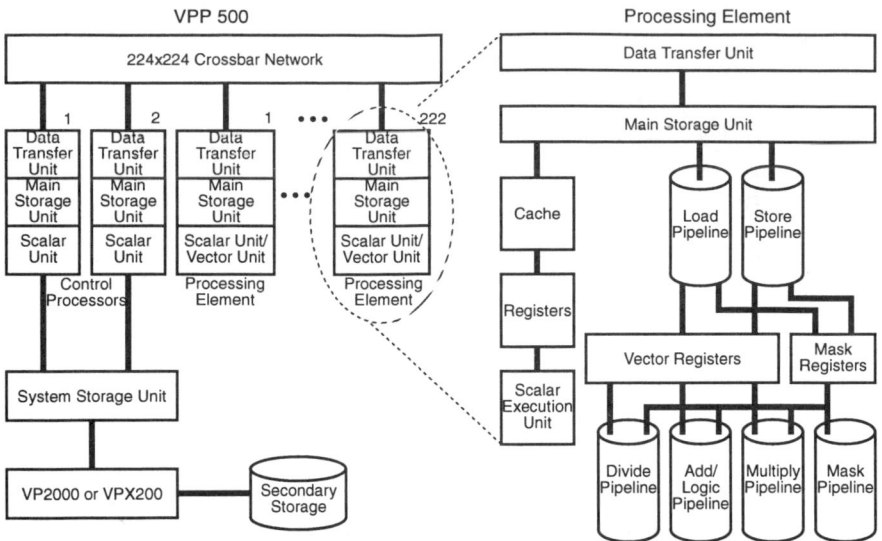

Figure 8.15 The Fujitsu VPP500 architecture. (Courtesy of Fujitsu Ltd., 1992)

Each PE has a peak processing speed of 1.6 Gflops, implemented with 256K-gate GaAs and BiCMOS LSI circuits. Up to two control processors coordinate the activities of the PEs through a crossbar network. The data transfer units in each PE handle inter-PE communications. Each PE has its own memory with up to 256 Mbytes of static RAM. The system applies the global shared virtual memory concept. In other words, the collection of local memories physically distributed over the PEs forms a single address space. The entire system can have up to 55 Gbytes of main memory collectively.

Each PE has a scalar unit and a vector unit operating in parallel. These functional pipelines are very similar to those built into the VP2000 (Fig. 8.13), but the pipeline functions are modified. We have seen the 224×224 crossbar design in Fig. 2.26b. This is by far the largest crossbar built into a commercial MPP system. The crossbar network is conflict-free, and only one crosspoint switch is on in each row or column of the crossbar switch array.

The VPP500 runs jointly with its host the UNIX System V Release 4-based UXP/-VPP operating system that supports closely coupled MIMD operations. The optimization functions of the Fortran 77 compiler work with the parallel scheduling function of the UNIX-based OS to exploit the maximum capability of the vector parallel architecture. The IEEE 754 floating-point standard has been adopted.

The data transfer unit in each PE provides 400 Mbytes/s unidirectional and 800 Mbytes/s bidirectional data exchange among PEs. The unit translates logical addresses to physical addresses to facilitate access to the virtual global memory. The unit is also equipped with special hardware for fast barrier synchronization. We will further review the software environment for the VPP500 in Chapter 11.

The system is scalable with an incremental control structure. A single control processor is sufficient to control up to 9 PEs. Two control processors are used to coordinate a VPP with 30 to 222 PEs. The system performance is expected to scale with the number of PEs spanning a peak performance range from 11 to 335 Gflops and a memory capacity of 1.8 to 55 Gbytes. The real performance of both the Cray/MPP and Fujitsu VPP500 remain to be seen when the systems become operational.

8.2.4 Mainframes and Minisupercomputers

High-end mainframes, minisupercomputers, and supercomputing workstations are studied in this section. Besides summarizing current systems, we examine the architecture designs in the VAX 9000 and Stardent 3000 as case studies. The LINPACK results compiled by Dongarra (1992) are presented to compare a wide range of computers for solving linear systems of equations.

High-End Mainframe Supercomputers   This class of supercomputers have been called *near-supercomputers*. They offer a peak performance of several hundreds of Mflops to 2.4 Gflops as listed in Table 8.4. These machines are not designed entirely for number crunching. Their main applications are still in business and transaction processing. The floating-point capability is only an add-on optional feature of these mainframe machines.

The number of CPUs ranges from one to six in a single system among the IBM ES/9000, VAX 9000, and Cyber 2000 listed in Table 8.4. The main memory is between 32 Mbytes and 1 Gbyte. Extended memory can be as large as 8 Gbytes in the ES/9000.

Vector hardware appears as an optional feature which can be used concurrently with the scalar units. Most vector units consist of an add pipeline and a multiply pipeline. The clock rates are between 9 and 30 ns in these machines. The I/O subsystems are rather sophisticated due to the need to support large database processing applications in a network environment.

DEC VAX 9000   Even though the VAX 9000 cannot provide a Gflops performance, the design represents a typical mainframe approach to high-performance computing. The architecture is shown in Fig. 8.16a.

Multichip packing technology was used to build the VAX 9000. It offers 40 times the VAX/780 performance per processor. With a four-processor configuration, this implies 157 times the 11/780 performance. When used for transaction processing, 70 TPS was reported on a uniprocessor. The peak vector processing rate ranges from 125 to 500 Mflops.

The system control unit is a crossbar switch providing four simultaneous 500-Mbytes/s data transfers. Besides incorporating interconnect logic, the crossbar was designed to monitor the contents of cache memories, tracking the most up-to-date cache coherence.

Up to 512 Mbytes of main memory are available using 1-Mbit DRAMs on 64-Mbyte arrays. Up to 2 Gbytes of extended memory are available using 4-Mbit DRAMs. Various I/O channels provide an aggregate data transfer rate of 320 Mbytes/s. The crossbar has eight ports to four processors, two memory modules, and two I/O controllers. Each port has a maximum transfer rate of 1 Gbyte/s, much higher than in bus-connected

Table 8.4 High-end Mainframe Supercomputers

Machine Characteristics	IBM ES/9000 -900 VF	DEC VAX 9000/440 VP	CDC Cyber 2000V
Number of processors	6 processors each attached to a vector facility	4 processors with vector boxes	2 central processors with vector hardware
Machine cycle time	9 ns	16 ns	9 ns
Maximum memory	1 Gbyte	512 Mbytes	512 Mbytes
Extended memory	8 Gbytes	2 Gbytes	N/A
Processor architecture: vector, scalar, and other functional units	Vector facility (VF) attached to each processor, delivering 4 floating-point results per cycle.	Vector processor (VBOX) connected to a scalar CPU. Two vector pipelines per VBOX. Four functional units in scalar CPU.	FPU for add and multiply, scalar unit with divide and multiply, integer unit and business data handler per processor.
I/O subsystem	256 ESCON fiber optic channels.	4 XMI I/O buses and 14 VAXBI I/O buses.	18 I/O processors with optional 18 additional I/O processors.
Operating system	MVS/ESA, VM/ESA, VSE/ESA	VMS or ULTRIX	NOS/VE
Vectorizing languages / compilers	Fortran V2 with interactive vectorization.	VAX Fortran compiler supporting concurrent scalar and vector processing.	Cyber 2000 Fortran V2.
Peak performance and remarks	2.4 Gflops (predicted).	500 Mflops peak.	210 Mflops per processor.

systems.

Each vector processor (VBOX) is equipped with an add and a multiply pipeline using vector registers and a mask/address generator as shown in Fig. 8.16b. Vector instructions are fetched through the memory unit (MBOX), decoded in the IBOX, and issued to the VBOX by the EBOX. Scalar operations are directly executed in the EBOX.

The vector register file consists of $16 \times 64 \times 64$ bits, divided into sixteen 64-element vector registers. No instruction takes more than five cycles. The vector processor generates two 64-bit results per cycle, and the vector pipelines can be chained for dot-product operations.

The VAX 9000 can run with either a VMS or an ULTRIX operating system. The service processor in Fig. 8.16a uses four Micro VAX processors devoted to system, disk/tape, and user interface control and to monitoring 20,000 scan points throughout the system for reliable operation and fault diagnosis.

Minisupercomputers   These are a class of low-cost supercomputer systems with a performance of about 5 to 15% and a cost of 3 to 10% of that of a full-scale supercomputer. Representative systems include the Convex C Series, Alliant FX Series, Encore

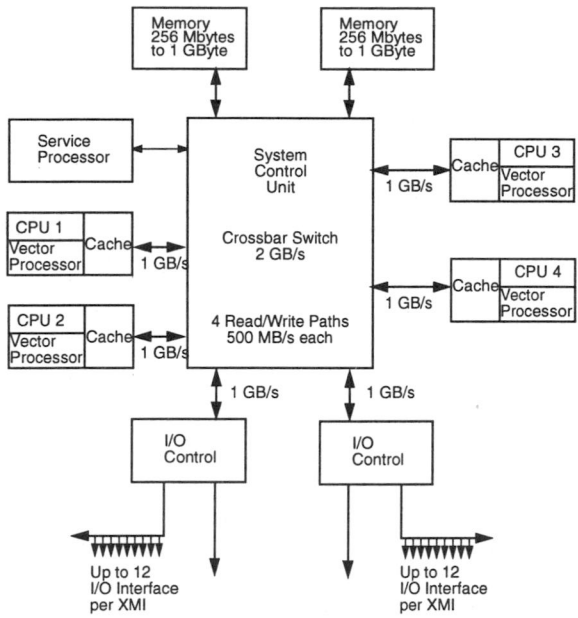

(a) The VAX 9000 multiprocessor system

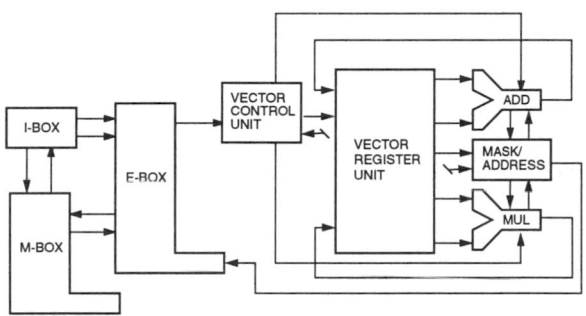

(b) The vector processor (VBOX)

Figure 8.16 The DEC VAX 9000 system architecture and vector processor design.
(Courtesy of Digital Equipment Corporation, 1991)

Multimax Series, and Sequent Symmetry Series.

Some of these minisupercomputers were introduced in Chapters 1 and 7. Most of them have an open architecture using standard off-the-shelf microprocessors and UNIX systems.

Both scalar and vector processing is supported in these multiprocessor systems with shared memory and peripherals. Most of these systems are built with a graphic subsystem for visualization and performance-tuning purposes.

Supercomputing Workstations High-performance workstations are being produced by Sun Microsystems, IBM, DEC, HP, Silicon Graphics, and Stardent using the state-of-the-art superscalar or superpipelined RISC microprocessors introduced in Chapters 4 and 6. Most of these workstations have a uniprocessor configuration with built-in graphics support but no vector hardware.

Silicon Graphics has produced the 4-D Series using four R3000 CPUs in a single workstation without vector hardware. Stardent Computer Systems produces a departmental supercomputer, called the Stardent 3000, with custom-designed vector hardware.

The Stardent 3000 The Stardent 3000 is a multiprocessor workstation that has evolved from the TITAN architecture developed by Ardent Computer Corporation. The architecture and graphics subsystem of the Stardent 3000 are depicted in Fig. 8.17. Two buses are used for communication between the four CPUs, memory, I/O, and graphics subsystems (Fig. 8.17a).

The system features R3000/R3010 processors/floating-point units. The vector processors are custom-designed. A 32-MHz clock is used. There are 128 Kbytes of cache; one half is used for instructions and the other half for data.

The buses carry 32-bit addresses and 64-bit data and operate at 16 MHz. They are rated at 128 Mbytes/s each. The R-bus is dedicated to data transfers from memory to the vector processor, and the S-bus handles all other transfers. The system can support a maximum of 512 Mbytes of memory.

A full graphics subsystem is shown in Fig. 8.17b. It consists of two boards that are tightly coupled to both the CPUs and memory. These boards incorporate rasterizers (pixel and polygon processors), frame buffers, Z-buffers, and additional overlay and control planes.

The Stardent system is designed for numerically intensive computing with two- and three-dimensional rendering graphics. One to two I/O processors are connected to SCSI or VME buses and other peripherals or Ethernet connections. The peak performance is estimated at 32 to 128 MIPS, 16 to 64 scalar Mflops, and 32 to 128 vector Mflops. Scoreboard, crossbar switch, and arithmetic pipelines are implemented in each vector processor.

Gordon Bell, chief architect of the VAX Series and of the TITAN/Stardent architecture, identified 11 rules of minisupercomputer design in 1989. These rules require performance-directed design, balanced scalar/vector operations, avoiding holes in the performance space, achieving peaks in performance even on a single program, providing a decade of addressing space, making a computer easy to use, building on others' work, always looking ahead to the next generation, and expecting the unexpected with slack resources.

The LINPACK Results This is a general-purpose Fortran library of mathematical software for solving dense linear systems of equations of order 100 or higher. LINPACK is very sensitive to vector operations and the degree of vectorization by the compiler. It has been used to predict computer performance in scientific and engineering areas.

Many published Mflops and Gflops results are based on running the LINPACK code with prespecified compilers. LINPACK programs can be characterized as having

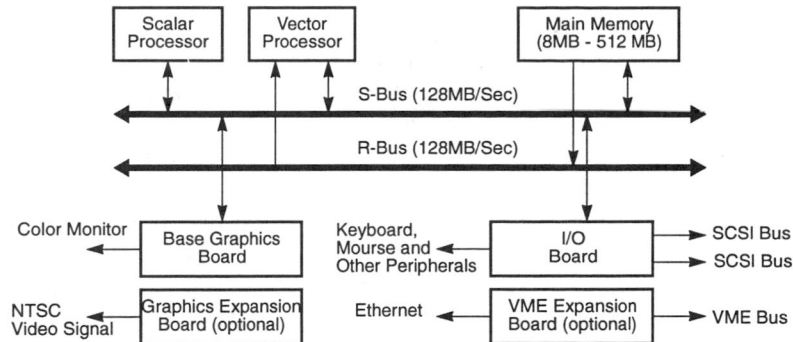

(a) The Stardent 3000 system architecture

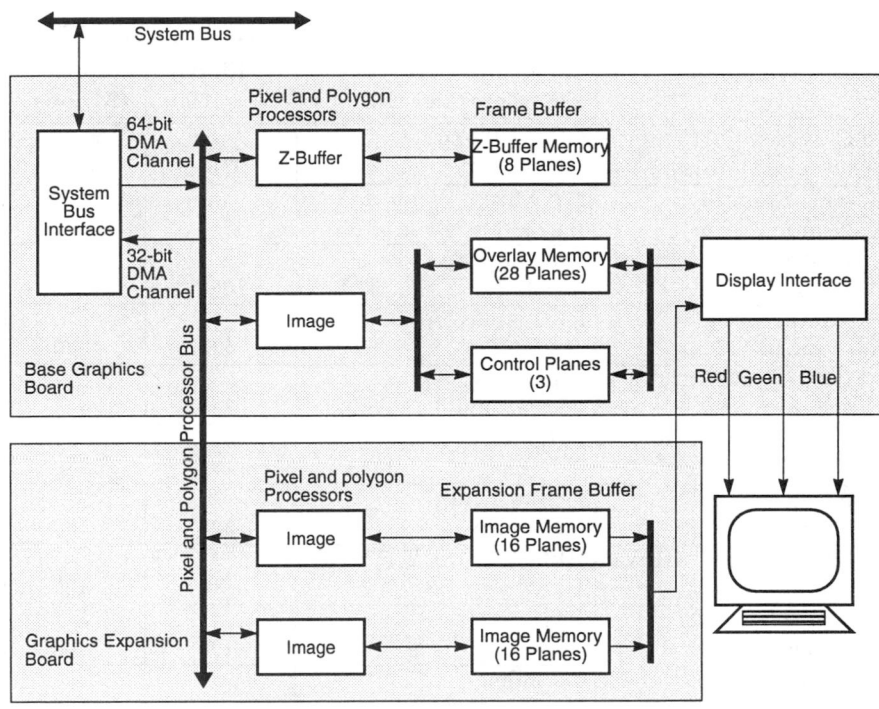

(b) The graphics subsystem architecture

Figure 8.17 The Stardent 3000 visualization departmental supercomputer. (Courtesy of Stardent Computer Inc., 1990)

a high percentage of floating-point arithmetic operations.

In solving a linear system of n equations, the total number of arithmetic operations involved is estimated as $2n^3/3 + 2n^2$, where $n = 1000$ in the LINPACK experiments.

Over the years, Dongarra has compared the performance of various computer systems in solving dense systems of linear equations. His performance experiments have involved about 100 computers.

The timing information presented in this report reflects only one problem area. The results should not be used to judge the overall performance of a computer system. However, the results do reflect the floating-point, parallel, and vector processing capabilities of the machines tested. Since the original reports are quite long, only brief excerpts are quoted in Table 8.5.

The second column reports LINPACK performance results based on a matrix of order $n = 100$ in a Fortran environment. The third column shows the results of solving a system of equations of order $n = 1000$ with no restriction on the method or its implementation. The last column lists the theoretical peak performance of the machines.

Table 8.5 Performance in Solving a System of Linear Equations

Computer Model	LINPACK Benchmark $n = 100$ OS/Compiler, Mflops		Best-effort (Mflops) $n = 1000$	Theoretic Peak (Mflops)
Cray Y-MP C90 (16 proc., 4.2 ns)	CF77 5.0 –Zp –Wd–e68	479	9715	16000
NEC SX-3/14 (1 proc., 2.9 ns)	f77SX020 R1.13 –pi*	314	4511	5500
Fujitsu VP2400/10 (4 ns)	Fortran77 EX/VP V11 L10	170	1688	2000
Convex C3840 (4 proc., 16.7 ns)	fc7.0 –tm c38 –O3 –ep –ds –is	75	425	480
IBM ES/9000-520VF (1 proc., 9 ns)	VAST-2/VS Fortran V2R4	60	338	444
FPS 510S MCP707 (7 proc., 25 ns)	Pgf77 –O4 –Minline	33	184	280
Alliant FX/2800-200 (14 processors)	fortran 1.1.27 –O –inline	31	325	560
DEC VAX9000/410VP (1 proc., 16 ns)	HPO V1.3–163V, DXML	22	89	125
CDC Cyber 205 (4 pipes)	FTN	17	195	400
Stardent 3040	3.0 -inline –nmax = 300	12	77	128
SUN SPARCstation 2	f77 1.4 -O3 –cg89 –dalign	4	N/A	N/A
IBM PC/AT with 80287	Microsoft 3.2	0.0091	N/A N/A	N/A N/A

Source: Jack Dongarra, "Performance of Various Computers Using Standard Linear Equations Software," Computer Science Dept., Univ. of Tennessee, Knoxville, TN 37996-1301, March 11, 1992.

The LINPACK results reported in Table 8.5 were for a small problem size of 100 unknowns. No changes were made in the LINPACK software to exploit vector capabilities on multiple processors in the machines being evaluated. The compilers of some machines may generate optimized code that itself accesses special hardware features.

The third column corresponds to a much larger problem size of 1000 unknowns. All possible optimization means, including user optimizations of the software, were allowed to achieve as high an execution rate as possible, called the *best-effort Mflops*.

The theoretical peak can easily be calculated by counting the maximum number of floating-point additions and multiplications that can be completed during a period of time, usually the cycle time of the machine.

Example 8.6 Peak performance calculation for the Cray Y-MP/8

The Cray Y-MP/8 has a cycle time of 6 ns. During a cycle, the results of both an addition and a multiplication can be completed. Furthermore, there are eight processors operating simultaneously without interference in the best case. Thus, we predict the peak performance of the Cray Y-MP/8 as follows:

$$\frac{2 \text{ operations}}{1 \text{ cycle}} \times \frac{1 \text{ cycle}}{6 \times 10^{-9} \text{ s}} \times 8 \text{ processors} = 2667 \text{ Mflops} = 2.6 \text{ Gflops} \qquad (8.12)$$

The peak performance is often cited by manufacturers. It provides an upper bound on the real performance. Comparing the results in the second and third columns with the peak values, only 2.9 to 86.3% of the peak was achieved in these runs. This implies that the peak performance cannot represent sustained real performance in most cases. Usually 10% of the peak performance is achievable.

8.3 Compound Vector Processing

In this section, we study compound vector operations. Multipipeline chaining and networking techniques are described and design examples given. A graph transformation approach is presented for setting up pipeline networks to implement compound vector functions, which are either specified by the programmer or detected by an intelligent compiler.

8.3.1 Compound Vector Operations

A *compound vector function* (CVF) is defined as a composite function of vector operations converted from a looping structure of linked scalar operations. The following example clarifies the concept.

Example 8.7 A compound vector function called the SAXPY code

Consider the following Fortran loop of a sequence of five scalar operations to be executed N times:

$$
\begin{array}{lll}
\textbf{Do}\ 10\ \ \text{I} = 1,\ \text{N} & & \\
\quad \text{Load} & \text{R1, X(I)} & \\
\quad \text{Load} & \text{R2, Y(I)} & \\
\quad \text{Multiply} & \text{R1, S} & (8.13) \\
\quad \text{Add} & \text{R2, R1} & \\
\quad \text{Store} & \text{Y(I), R2} & \\
10\quad \textbf{Continue} & &
\end{array}
$$

where X(I) and Y(I), I = 1, 2, ..., N, are two source vectors originally residing in the memory. After the computation, the resulting vector is stored back to the memory. S is an immediate constant supplied to the multiply instruction.

After vectorization, the above scalar SAXPY code is converted to a sequence of five vector instructions:

$$
\begin{array}{llll}
M(x : x + N - 1) & \rightarrow & V1 & \textit{Vector load} \\
M(y : y + N - 1) & \rightarrow & V2 & \textit{Vector load} \\
S \times V1 & \rightarrow & V1 & \textit{Vector multiply} \qquad (8.14) \\
V2 + V1 & \rightarrow & V2 & \textit{Vector add} \\
V2 & \rightarrow & M(y : y + N - 1) & \textit{Vector store}
\end{array}
$$

The same vector notation used in Eq. 4.1 is applied here, where x and y are the starting memory addresses of the X and Y vectors, respectively; V1 and V2 are two N-element vector registers in the vector processor.

The vector code in Eq. 8.14 can be expressed as a CVF as follows, using Fortran 90 notation:

$$
Y(1 : N) = S \times X(1 : N) + Y(1 : N) \tag{8.15}
$$

For simplicity, we write the above expression for a CVF as follows:

$$
Y(I) = S \times X(I) + Y(I) \tag{8.16}
$$

where the index I implies that all vector operations involve N elements.

■

Compound Vector Functions   Table 8.6 lists a number of example CVFs involving one-dimensional vectors indexed by *I*. The same concept can be generalized to multidimensional vectors with multiple indexes. For simplicity, we discuss only CVFs defined over one-dimensional vectors. Typical operations appearing in these CVFs include *load, store, multiply, divide, logical,* and *shifting* vector operations. We use "slash" to represent the *divide* operations. All vector operations are defined on a component-wise basis unless otherwise specified.

The purpose of studying CVFs is to explore opportunities for concurrent processing of linked vector operations. The numbers of available vector registers and functional pipelines impose some limitations on how many CVFs can be executed simultaneously.

Table 8.6 Representative Compound Vector Functions

One-dimensional compound vector functions	Maximum chaining degree
$V1(I) = V2(I) + V3(I) \times V4(I)$	2
$V1(I) = B(I) + C(I)$	3
$A(I) = V1(I) \times S + B(I)$	4
$A(I) = V1(I) + B(I) + C(I)$	5
$A(I) = B(I) + S \times C(I)$	5
$A(I) = B(I) + C(I) + D(I)$	6
$A(I) = Q \times V1(I) \times (R \times B(I) + C(I))$	7
$A(I) = B(I) \times C(I) + D(I) \times V1(I)$	7
$A(I) = V1(I) + (1 / A(I) + 1 / B(I)) + Log(V2(I))$	8
$A(I) = \sqrt{V2(I)} + Sin(B(I) + C(I)) + V3(I)$	8
$A(I) = B(I) \times C(I) + D(I) \times E(I) \times S$	9
$A(I) = (A(I) + B(I) \times C(I) + D(I)) \times (I)$	10

Note: Vi(I) are vector registers in the processor. A(I), B(I), C(I), D(I), and E(I) are vectors in memory. Scalars are indicated as Q, R, and S are available from scalar registers in the processor. The chaining degrees include both memory-access and functional pipeline operations.

8.3.2 Vector Loops and Chaining

Vector pipelining and chaining are an integral part of all vector processors built today. Concurrent processing of several vector arithmetic, logic, shift, and memory-access operations require the chaining of multiple pipelines in a linear cascade. The idea of chaining is an extension of the technique of internal data forwarding practiced in a scalar processor (Fig. 5.15).

Chaining affects the speed of vector processors. Each of the CVFs listed in Table 8.6 is potentially a candidate for chaining. However, the implementation may hinge on the particular architecture of a vector machine. Principal concepts and implementations of vector looping, chaining, and recursion are described below.

Vector Loops or Strip-mining   When a vector has a length greater than that of the vector registers, segmentation of the long vector into fixed-length *segments* is necessary. This technique has been called *strip-mining*. One vector segment (one surface of the mine field) is processed at a time. In the case of Cray computers, the vector segment length is 64 elements.

Until all the vector elements in each segment are processed, the vector register cannot be assigned to another vector operation. Strip-mining is restricted by the number of available vector, registers and so is vector chaining. In the Fujitsu VP Series, the vector registers can be reconfigured to match the vector length. This allows strip-mining to be done more dynamically with a different "depth" in different applications.

The program construct for processing long vectors is called a *vector loop*. Vector segmentation is done by machine hardware under software control and should be transparent to the programmer. The loop count is determined at compile time or at run time, depending on the index value. Inside a loop, all vector operands have equal length, equal to that of the vector register.

Functional Units Independence   In order to for vector operations to be linked, they must follow a linear data flow pattern, All functional pipeline units employed must be independent of each other. The same unit cannot be assigned to execute more than one instruction in the same chain.

Furthermore, vector registers must be lined up as interfaces between functional pipelines. The successive output results of a pipeline unit are fed into a vector register, one element per cycle. This vector register is then used as an input register for the next pipeline unit in the chain.

With the requirement of continuous data flow in the successive pipelines, the interface registers must be able to pass one vector element per cycle between adjacent pipelines. There may be extra transition time delays between loading the successive vector segments into the interface registers.

To avoid conflicts among different vector operations, the vector registers and functional pipelines must be reserved before a vector chain can be established. The vector chaining and the timing relationship are illustrated in Figs. 8.18 and 8.19 for executing the vectorized SAXPY code specified in Eq. 8.14.

Example 8.8   Pipeline chaining on Cray Supercomputers and on the Cray X-MP (Courtesy of Cray Research, Inc., 1985)

The Cray 1 has one memory-access pipe for either load or store but not for both at the same time. The Cray X-MP has three memory-access pipes, two for *vector load* and one for *vector store*. These three access pipes can be used simultaneously.

To implement the SAXPY code on the Cray 1, the five vector operations are divided into three chains: The first chain has only one vector operation, *load Y*. The second chain links the *load X* to scalar-vector *multiply* (S×) operations and then to the *vector add* operation. The last chain is for *store Y* as illustrated in Fig. 8.18a.

The same set of vector operations is implemented on the Cray X-MP in a single chain, as shown in Fig. 8.18b, because three memory-access pipes are used simultaneously. The chain links five vector operations in a single connected cascade.

To compare the time required for chaining these pipelines, Fig. 8.19a shows that roughly $5n$ cycles are needed to perform the vector operations sequentially without any overlapping or any chaining. The Cray 1 requires about $3n$ cycles to execute, corresponding to about n cycles for each vector chain. The Cray X-MP requires about n cycles to execute.

In Fig. 8.19, the pipeline flow-through latencies (startup delays) are denoted as s, m, and a for the memory-access pipe, the multiply pipe, and the add pipe, respectively. These latencies equal the lengths of individual pipelines. The exact cycle counts can be slightly greater than the counts of $5n, 3n$, and n due to these extra delays.

■

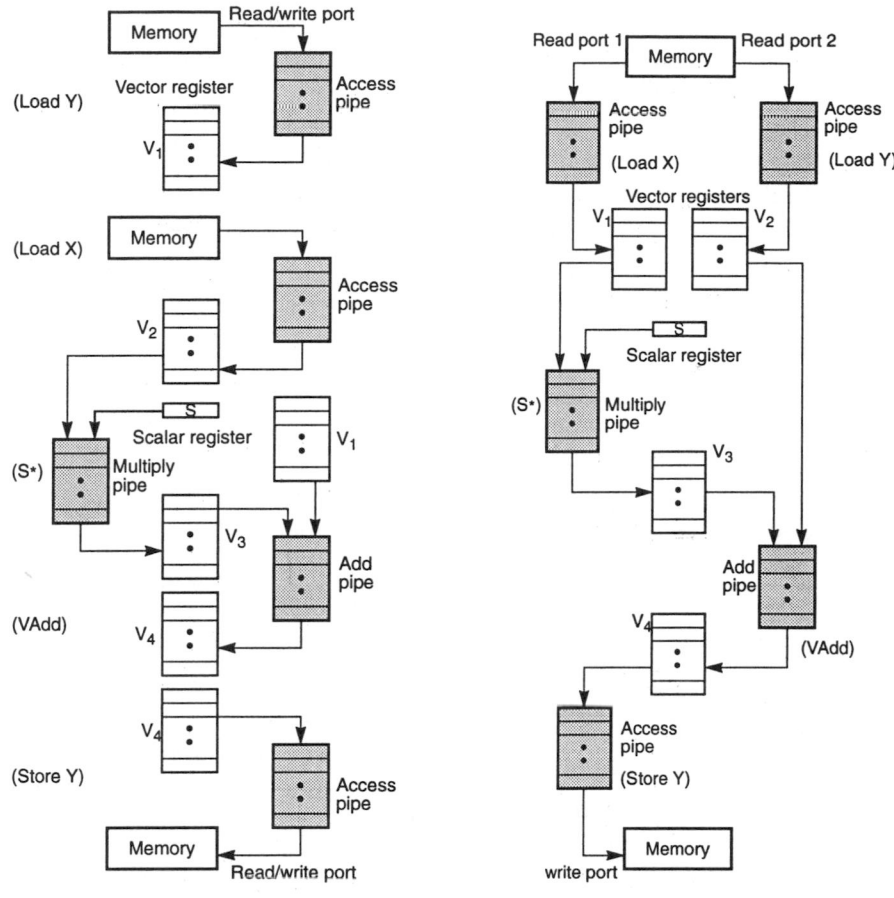

(a) Limited chaining using only one memory-access pipe in the Cray 1

(b) Complete chaining using three memory-access pipes in the Cray X-MP

Figure 8.18 Multipipeline chaining on Cray 1 and Cray X-MP for executing the SAXPY code: $Y(1:N) = S \times X(1:N) + Y(1:N)$. (Courtesy of Cray Research, Inc., 1985)

The above example clearly demonstrates the advantages of vector chaining. A meaningful chain must link two or more pipelines. As far as the amount of time is concerned, the longer the chaining, the better the performance. The *degree of chaining* is indicated by the number of distinct pipeline units that can be linked together.

Vector chaining effectively increases the overall pipeline length by adding the pipeline stages of all functional units in the chain to form a single long pipeline. The potential speedup of this long pipeline is certainly greater according to Eq. 5.5.

Chaining Limitations The number of vector operations in a CVF must be small

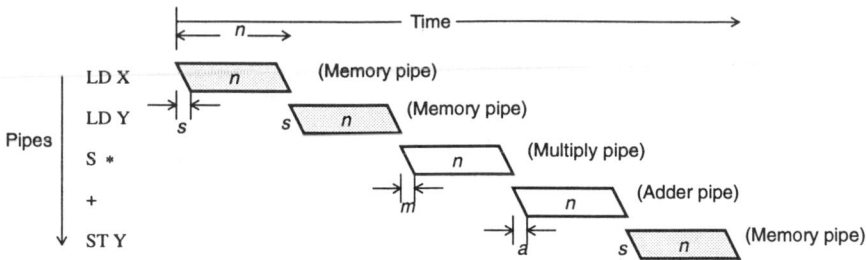

(a) Sequential execution without chaining

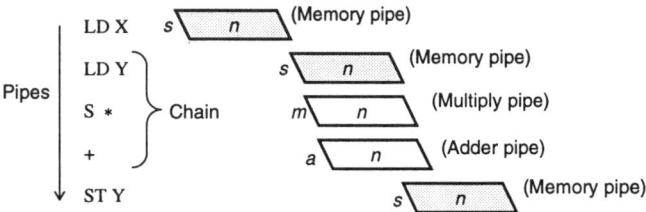

(b) Chaining of arithmetic pipes with only one memory pipe

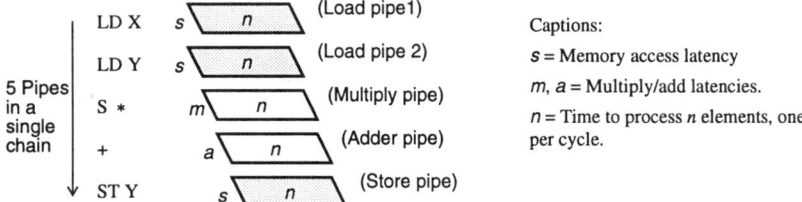

(c) Chaining in two load pipes, two arithmetic pipes, and one store pipe

Figure 8.19 Timing for chaining the SAXPY code Y(1:N) = S × X(1:N) + Y(1:N)
under different memory-access capabilities. (Courtesy of Cray Research,
Inc., 1985)

enough to make chaining possible. Vector chaining is limited by the small number of
functional pipelines available in a vector processor. Furthermore, the limited number
of vector registers also imposes an additional limit on chaining.

For example, the Cray Y-MP has only eight vector registers. Suppose all memory
pipes are used in a vector chain. These require that three vector registers (two for *vector
read* and one for *vector store* be reserved at the beginning and end of the chaining
operations. The remaining five vector registers are used for arithmetic, logic, and shift
operations.

The number of interface registers required between two adjacent pipeline units is at
least one and sometimes two for two source vectors. Thus, the number of non-memory-
access vector operations implementable with the remaining five vector registers cannot

be greater than five. In practice, this number is between two and three.

The actual degree of chaining depends on how many of the vector operations involved are binary or unary and how many use scalar or vector registers. If they are all binary operations, each requiring two source vector registers, then only two or three vector operations can be sandwiched between the memory-access operations. Thus a single chain on the Cray/Y-MP can link at most five or six vector operations including the memory-access operations.

Vector Recurrence These are a special class of vector loops in which the outputs of a functional pipeline may feed back into one of its own source vector registers. In other words, a vector register is used for holding the source operands and the result elements simultaneously.

This has been done on Cray machines using a *component counter* associated with each vector register. In each pipeline cycle, the vector register is used like a shift register at the component level. When a component operand is "shifted" out of the vector register and enters the functional pipeline, a result component can enter the vacated component register during the same cycle. The component counter must keep track of the shifting operations until all 64 components of the result are loaded into the vector register.

Recursive vector summation is often needed in scientific and statistical computations. For example, the dot product of two vectors, $A \cdot B = \sum_{i=1}^{n} a_i \times b_i$, requires the use of recursion. Another example is polynomial evaluation over vector operands.

Summary Our discussion of vector and pipeline chaining is based on a load-store architecture using vector registers in all vector instructions. As the number of functional units increases steadily in modern supercomputers, both the Cray C-90 and the NEC SX-X offer 16-way parallelism within each processor.

The degree of chaining can certainly increase if the vector register file becomes larger and scoreboarding techniques are applied to ensure functional unit independence and to resolve data dependence or resource dependence problems. The use of multiport memory is crucial to enabling large vector chains.

Vector looping, chaining, and recursion represent the state of the art in extending pipelining for vector processing. Furthermore, one can use *masking, scatter,* and *gather* instructions to manipulate sparse vectors or sparse matrices containing a large number of dummy zero entries. A vector processor cannot be considered versatile unless it is designed to handle both dense and sparse vectors effectively.

8.3.3 Multipipeline Networking

The idea of linking vector operations can be generalized to a multipipeline networking concept. Instead of linking vector operations into a linear chain, one can build a *pipenet* by introducing multiple functional pipelines with inserted delays to achieve *systolic computation* of CVFs.

In 1978, Kung and Leiserson introduced systolic arrays for special-purpose computing. Their idea was to map a specific algorithm into a fixed architecture. A *systolic array* is formed with a network of functional units which are locally connected and operates synchronously with multidimensional pipelining. We explain below how a pipeline net can be extended from the systolic concept to build a dynamic vector processor for efficient execution of various CVFs.

Pipeline Net (Pipenet)   Systolic arrays are built with fixed connectivity among the processing cells. This restriction is removed in a pipeline net. A pipenet has programmable connectivity as illustrated in Fig. 8.20. It is constructed from interconnecting multiple *functional pipelines* (FPs) through two *buffered crossbar networks* (BCNs) which are themselves pipelined.

A two-level pipeline architecture is seen in a pipeline net. The lower level corresponds to pipelining within each functional unit. The higher level is the pipelining of FPs through the BCNs. A generic model of a pipeline net is shown in Fig. 8.20d. The register file includes scalar and vector registers, as are often found in a modern vector processor.

The set of functional pipelines should be able to handle important vector arithmetic, logic, shifting, and masking operations. Each FP_i is pipelined with k_i stages. The output terminals of each BCN are buffered with programmable delays. The BCN1 is used to establish the dynamic connections between the register file and the FPs. The BCN2 sets up the dynamic connections among the FPs.

For simplicity, we call a pipeline network a *pipenet*. Conventional pipelines or pipeline chains are special cases of pipenets. Note that a pipenet is programmable with dynamic connectivity. This represents the fundamental difference between a static systolic array and a dynamic pipenet. In a way, one can visualize pipenets as programmable systolic arrays. The programmability sets up the dynamic connections, as well as the number of delays along some connection paths.

Setup of the Pipenet   Figures 8.20a through 8.20d show how to convert from a program graph to a pipenet. Whenever a CVF is to be evaluated, the crossbar networks are programmed to set up a connectivity pattern among the FPs that best matches the data flow pattern in the CVF.

The *program graph* represents the data flow pattern in a given CVF. Nodes on the graph correspond to vector operators, and edges show the data dependence, with delays properly labeled, among the operators.

The program graph in Fig. 8.20a corresponds to the following CVF:

$$E(I) = [A(I) \times B(I) + B(I) \times C(I)] \, / \, [B(I) \times C(I) \times [C(I) + D(I)]] \qquad (8.17)$$

for $I = 1, 2, \ldots, n$. This CVF has four input vectors $A(I)$, $B(I)$, $C(I)$, and $D(I)$ and one output vector $E(I)$ which demand five memory-access operations. In addition, there are seven vector arithmetic operations involved.

In other words, the above CVF demands a chaining degree of 11 if one considers implementing it with a chain of memory-access and arithmetic pipelines. This high degree of chaining is very difficult to implement with a limited number of FPs and vector registers. However, the CVF can be easily implemented with a pipenet as shown in Fig. 8.20b.

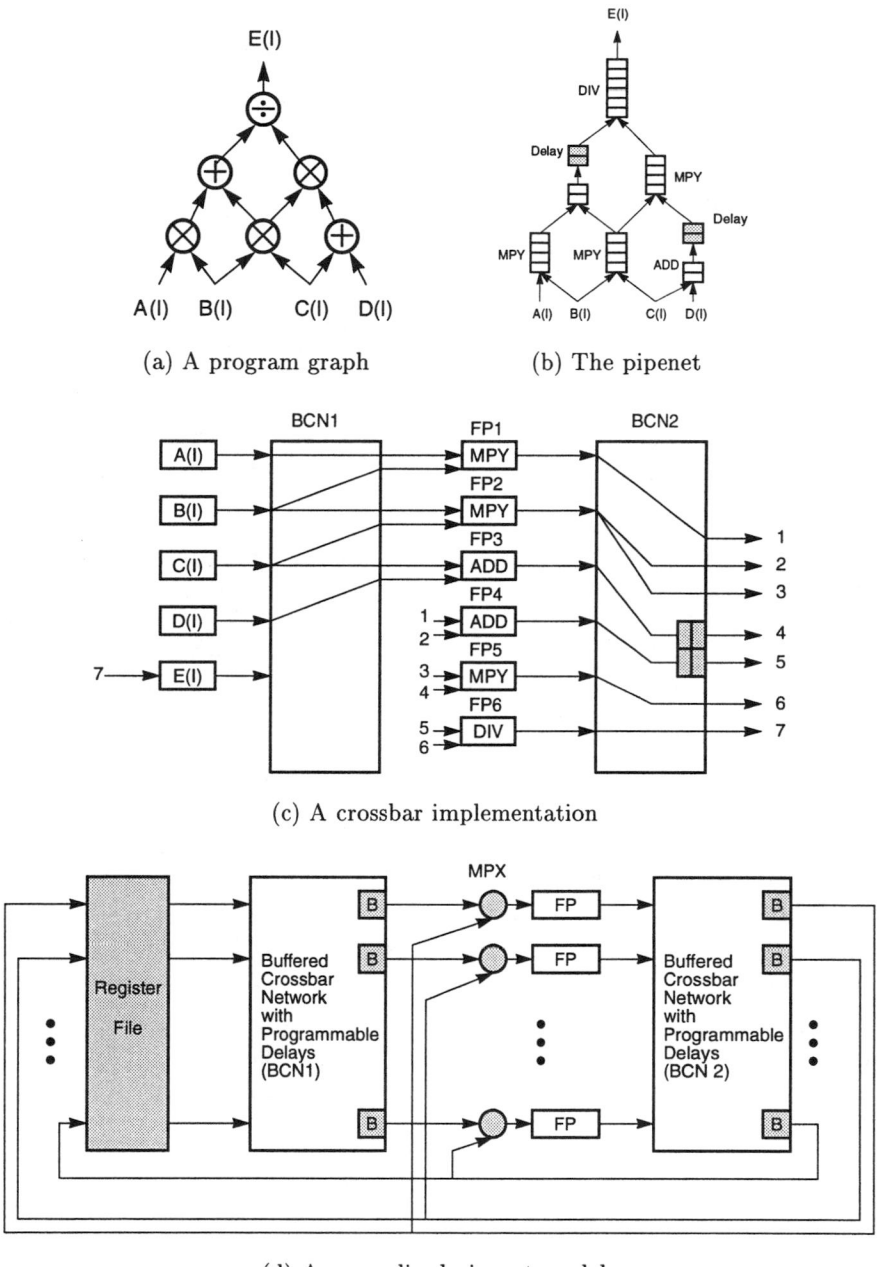

(a) A program graph   (b) The pipenet

(c) A crossbar implementation

(d) A generalized pipenet model

Figure 8.20 The concept of a pipenet and its implementation model. (Reprinted from Hwang and Xu, *IEEE Transactions on Computers*, Jan. 1988)

Six FPs are employed to implement the seven vector operations because the product vector $B(I) \times C(I)$, once generated, can be used in both the denominator and the numerator. We assumes two, four, and six pipeline stages in the ADD, MPY, and DIV units, respectively. Two noncompute delays are being inserted, each with two clock delays, along two of the connecting paths. The purpose is to equalize all the path delays from the input end to the output end.

The connections among the FPs and the two inserted delays are shown in Fig. 8.20c for a crossbar-connected vector processor. The feedback connections are identified by numbers. The delays are set up in the appropriate buffers at the output terminals identified as 4 and 5. Usually, these buffers allow a range of delays to be set up at the time the resources are scheduled.

The program graph can be specified either by the programmer or by a compiler. The pipeline can be set up by using a sequence of microinstructions or by using hardwired control. Various connection patterns in the crossbar networks can be prestored for implementing each CVF type. Once the CVF is decoded, the connect pattern is enabled for setup dynamically.

Program Graph Transformations   The program in Fig. 8.20a is acyclic or loop-free without feedback connections. An almost trivial mapping is used to establish the pipenet (Fig. 8.20b). In general, the mapping cannot be obtained directly without some graph transformations. We describe these transformations below with a concrete example CVF, corresponding to a cyclic graph shown in Fig. 8.21a.

On a directed program graph, nodal delays correspond to the appropriate FPs, and edge delays are the signal flow delays along the connecting path between FPs. For simplicity, each delay is counted as one pipeline cycle.

A *cycle* is a sequence of nodes and edges which starts and ends with the same node. We will concentrate on *synchronous* program graphs in which all cycles have positive delays. In particular, we consider a k-graph, a synchronous program graph in which all nodes have a delay of k cycles. A 0-graph is called a *systolic program graph*.

The following two lemmas provide basic tools for converting a given program graph into an equivalent graph. The equivalence is defined up to graph isomorphism and with the same input/output behaviors.

Lemma 1:   Adding k delays to any node in a systolic program graph and then subtracting k delays from all incoming edges to that node will produce an equivalent program graph.

Lemma 2:   An equivalent program graph is generated if all nodal and edge delays are multiplied by the same positive integer, called the *scaling constant*.

To implement a CVF by setting up a pipenet in a vector processor, one needs first to represent the CVF as a systolic graph with zero delays and positive edge delays. Only a systolic graph can be converted to a pipenet as exemplified below.

Example 8.9 Program graph transformation to set up a pipenet (Hwang

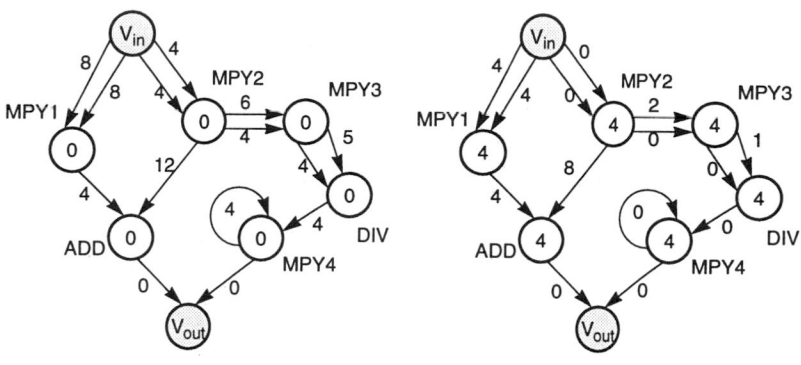

(a) A systolic program graph   (b) After graph transformation

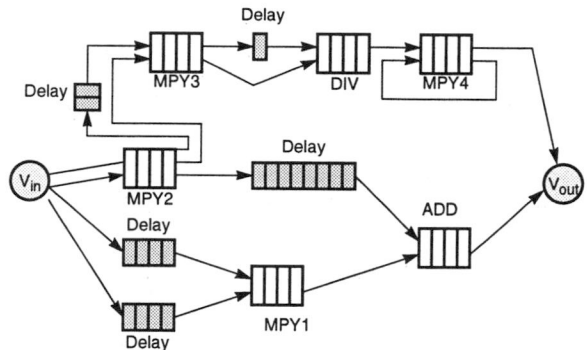

(c) Pipenet implementation with inserted delays between pipelines

Figure 8.21 From synchronous program graph to pipenet implementation. (Reprinted from Hwang and Xu, *IEEE Transactions on Computers*, Jan. 1988)

and Xu, 1988)

Consider the systolic program graph in Fig. 8.21a. This graph represents the following set of CVFs:

$$
\begin{aligned}
E(I) &= [B(I) \times C(I)] + [C(I) \times D(I)] \\
F(I) &= [C(I) \times D(I)] \times [C(I-2) \times D(I-2)] \\
G(I) &= [F(I)/F(I-1)] \times G(I-4)
\end{aligned}
\tag{8.18}
$$

Two multiply operators (MPY1 and MPY2) and one add operator (ADD) are applied to evaluate the vector $E(I)$ from the input end (V_{in}) to the output end (V_{out}) in Fig. 8.21a. The same operator MP2 is applied twice, with different delays (four and six cycles), before it is multiplied by MPY3 to generate the output vector $F(I)$. Finally, the divide (DIV) and multiply (MPY4) operators are applied to generate the output vector G(I).

Applying Lemma 1, we add four-cycle delays to each operator node and subtract four-cycle delays from all incoming edges. The transformed graph is obtained in Fig. 8.21b. This is a 4-graph with all nodal delays equal to four cycles. Therefore, one can construct a pipenet with all FPs having four pipeline stages as shown in Fig. 8.21c. The two graphs shown in Figs. 8.21b and 8.21c are indeed isomorphic.

The inserted delays correspond to the edge delays on the transformed graph. These delays can be implemented with programmable delays in the buffered crossbar networks shown in Fig. 8.20a. Note that the only self-reflecting cycle at node MPY4 represents the recursion defined in the equation for vector G(I). No scaling is applied in this graph transformation.

■

The systolic program graph in Fig. 8.21a can be obtained by intuitive reasoning and delay analysis as shown above. Systematic procedures needed to convert any set of CVFs into systolic program graphs were reported in the original paper by Hwang and Xu (1988).

If the systolic graph so obtained does not have enough edge delays to be transferred into the operator nodes, we have to multiply the edge delays by a scaling constant s, applying Lemma 2. Then the pipenet clock rate must be reduced by s times. This means that successive vector elements entering the pipenet must be separated by s cycles to avoid collisions in the respective pipelines.

Performance Evaluation   The above graph transformation technique has been applied in developing various pipenets for implementing CVFs embedded in Livermore loops. Speedup improvements of between 2 and 12 were obtained, as compared with implementing them on vector hardware without chaining on networking.

In order to build into future vector processors the capabilities of multipipeline networking described above, Fortran and other vector languages must be extended to represent CVFs under various conditions.

Automatic compiler techniques need to be developed to convert from vector expressions to systolic graphs and then to pipeline nets. Therefore, new hardware and software mechanisms are needed to support compound vector processing. This hardware approach can be one or two orders of magnitude faster than the software implementation.

8.4 SIMD Computer Organizations

Vector processing can also be carried out by SIMD computers as introduced in Section 1.3. Implementation models and two example SIMD machines are presented below. We examine their interconnection networks, processing cells, memory, and I/O structures.

8.4.1 Implementation Models

Two SIMD computer models are described below based on the memory distribution and addressing scheme used. Most SIMD computers use a single control unit and distributed memories, except for a few that use associative memories.

The instruction set of an SIMD computer is decoded by the array control unit. The *processing elements* (PEs) in the SIMD array are passive ALUs executing instructions broadcast from the control unit. All PEs must operate in lockstep, synchronized by the same array controller.

Distributed-Memory Model Spatial parallelism is exploited among the PEs in an SIMD computer. A distributed-memory SIMD computer consists of an array of PEs which are controlled by the same array control unit, as shown in Fig. 8.22a. Program and data are loaded into the control memory through the host computer.

An instruction is sent to the control unit for decoding. If it is a scalar or program control operation, it will be directly executed by a scalar processor attached to the control unit. If the decoded instruction is a vector operation, it will be broadcast to all the PEs for parallel execution.

Partitioned data sets are distributed to all the local memories attached to the PEs through a vector data bus. The PEs are interconnected by a data-routing network which performs inter-PE data communications such as shifting, permutation, and other routing operations. The data-routing network is under program control through the control unit. The PEs are synchronized in hardware by the control unit.

In other words, the same instruction is executed by all the PEs in the same cycle. However, masking logic is provided to enable or disable any PE from participation in a given instruction cycle. The Illiac IV was such an SIMD machine consisting of 64 PEs with local memories interconnected by an 8×8 mesh with wraparound connections (Fig. 2.18b).

Almost all SIMD machines built today are based on the distributed-memory model. Various SIMD machines differ mainly in the data-routing network chosen for inter-PE communications. The four-neighbor mesh architecture has been the most popular choice in the past. Besides Illiac IV, the Goodyear MPP and AMT DAP610 were also implemented with the two-dimensional mesh. Variations from the mesh are the hypercube embedded in a mesh, which was implemented in the CM-2, and the X-Net plus a multistage crossbar router implemented in the MasPar MP-1.

Shared-Memory Model In Fig. 8.22b, we show a variation of the SIMD computer using shared memory among the PEs. An alignment network is used as the inter-PE memory communication network. Again this network is controlled by the control unit.

The Burroughs Scientific Processor (BSP) has adopted this architecture, with $n = 16$ PEs updating $m = 17$ shared-memory modules through a 16×17 alignment network. It should be noted that the value m is often chosen to be relatively prime with respect to n, so that parallel memory access can be achieved through skewing without conflicts.

The alignment network must be properly set to avoid access conflicts. Most SIMD computers are built with distributed memories. Some SIMD computers use bit-slice PEs,

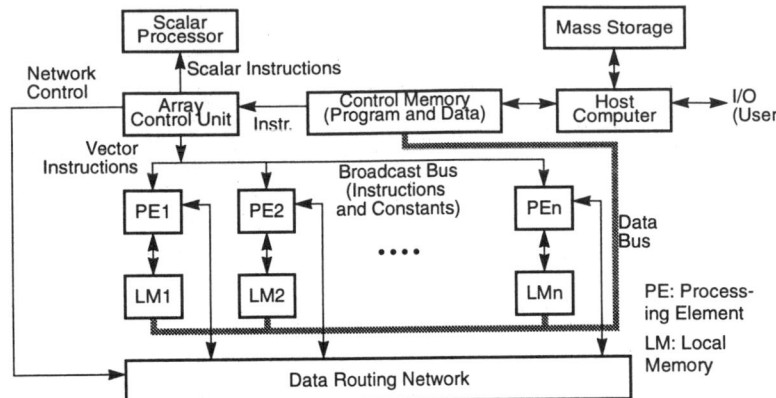

(a) Using distributed local memories (e.g., the Illiac IV)

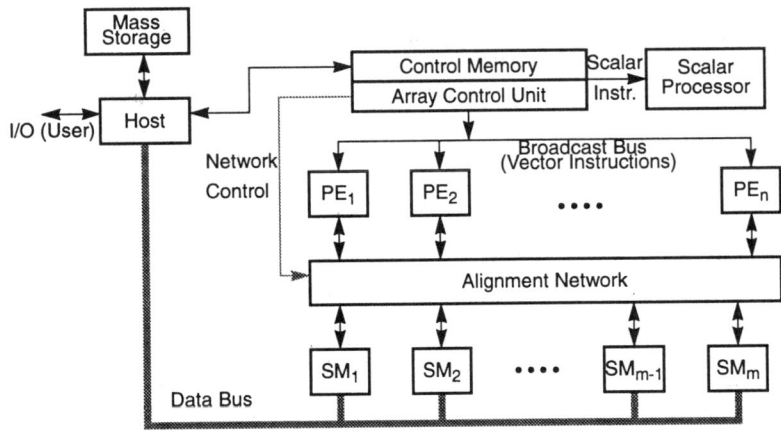

(b) Using shared-memory modules (e.g., the BSP)

Figure 8.22 Two models for constructing SIMD supercomputers.

such as the DAP610 and CM/200. Both bit-slice and word-parallel SIMD computers are studied below.

SIMD Instructions   SIMD computers execute vector instructions for arithmetic, logic, data-routing, and masking operations over vector quantities. In bit-slice SIMD machines, the vectors are nothing but binary vectors. In word-parallel SIMD machines, the vector components are 4- or 8-byte numerical values.

All SIMD instructions must use vector operands of equal length n, where n is the number of PEs. SIMD instructions are similar to those used in pipelined vector processors, except that temporal parallelism in pipelines is replaced by spatial parallelism in multiple PEs.

The data-routing instructions include permutations, broadcasts, multicasts, and various rotate and shift operations. Masking operations are used to enable or disable a subset of PEs in any instruction cycle.

Host and I/O   All I/O activities are handled by the host computer in the above SIMD organizations. A special control memory is used between the host and the array control unit. This is a staging memory for holding programs and data.

Divided data sets are distributed to the local memories (Fig. 8.22a) or to the shared memory modules (Fig. 8.22b) before starting the program execution. The host manages the mass storage or graphics display of computational results. The scalar processor operates concurrently with the PE array under the coordination of the control unit.

8.4.2   The CM-2 Architecture

The Connection Machine CM-2 is a fine-grain MPP computer using thousands of bit-slice PEs in parallel to achieve a peak processing speed of above 10 Gflops. We describe the parallel architecture built into the CM-2. Parallel software developed with the CM-2 will be discussed in Chapter 10.

Program Execution Paradigm   All programs start execution on a *front end*, which issues microinstructions to the back-end processing array when data-parallel operations are desired. The *sequencer* breaks down these microinstructions and broadcasts them to all *data processors* in the array.

Data sets and results can be exchanged between the front end and the processing array in one of three ways: *broadcasting*, *global combining*, and *scalar memory bus* as depicted in Fig. 8.23. Broadcasting is carried out through the broadcast bus to all data processors at once.

Global combining allows the front end to obtain the sum, largest value, logical OR, etc., of values, one from each processor. The scalar bus allows the front end to read or to write one 32-bit value at a time from or to the memories attached to the data processors. Both VAX and Symbolics Machines have been used as the front end and as hosts.

The Processing Array   The CM-2 is a back-end machine for data-parallel computation. The processing array contains from 4K to 64K bit-slice data processors (or PEs), all of which are controlled by a sequencer as shown in Fig. 8.23.

The sequencer decodes microinstructions from the front end and broadcast nanoinstructions to the processors in the array. All processors can access their memories simultaneously. All processors execute the broadcast instructions in a lockstep manner.

The processors exchange data among themselves in parallel through the *router*, *NEWS grids*, or a *scanning* mechanism. These network elements are also connected to I/O interfaces. A mass storage subsystem, called the *data vault*, is connected through the I/O for storing up to 60 Gbytes of data.

Processing Nodes   Figure 8.24 shows the CM-2 processor chips with memory and

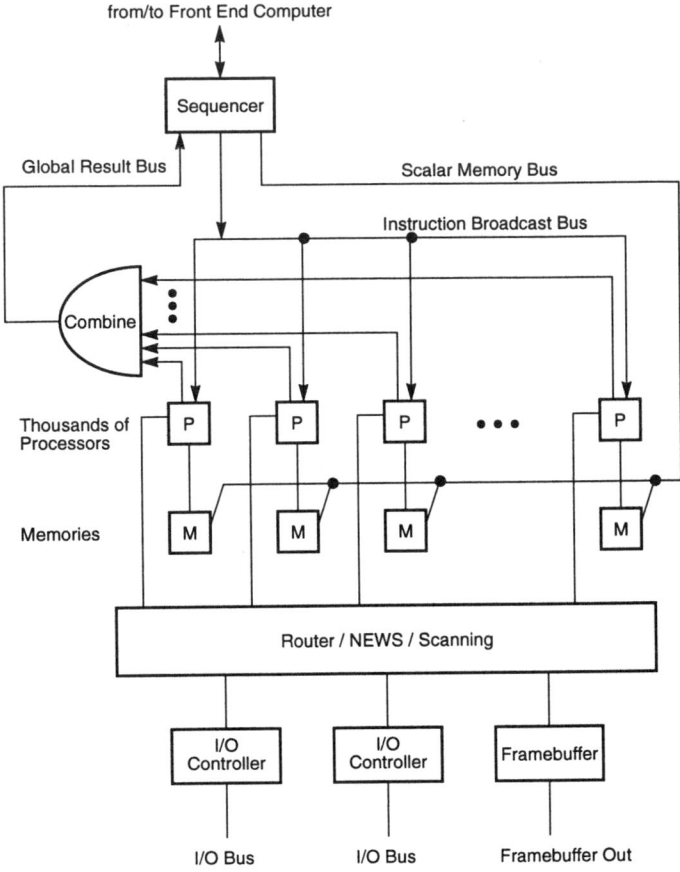

from/to Front End Computer

Sequencer

Global Result Bus

Scalar Memory Bus

Instruction Broadcast Bus

Combine

Thousands of Processors

Memories

Router / NEWS / Scanning

I/O Controller

I/O Controller

Framebuffer

I/O Bus

I/O Bus

Framebuffer Out

Figure 8.23 The architecture of the Connection Machine CM-2. (Courtesy of Thinking Machines Corporation, 1990)

floating-point chips. Each data processing node contains 32 bit-slice data processors, an optional floating-point accelerator, and interfaces for interprocessor communication. Each data processor is implemented with a 3-input and 2-output bit-slice ALU and associated latches and a memory interface. This ALU can perform bit-serial full-adder and Boolean logic operations.

The processor chips are paired in each node sharing a group of memory chips. Each processor chip contains 16 processors. The parallel instruction set, called *Paris*, includes nanoinstructions for memory load and store, arithmetic and logical, and control of the router, NEWS grid, and hypercube interface, floating-point, I/O, and diagnostic operations.

The memory data path is 22 bits (16 data and 6 ECC) per processor chip. The 18-bit memory address allows $2^{18} = 256$K memory words (512 Kbytes of data) shared by

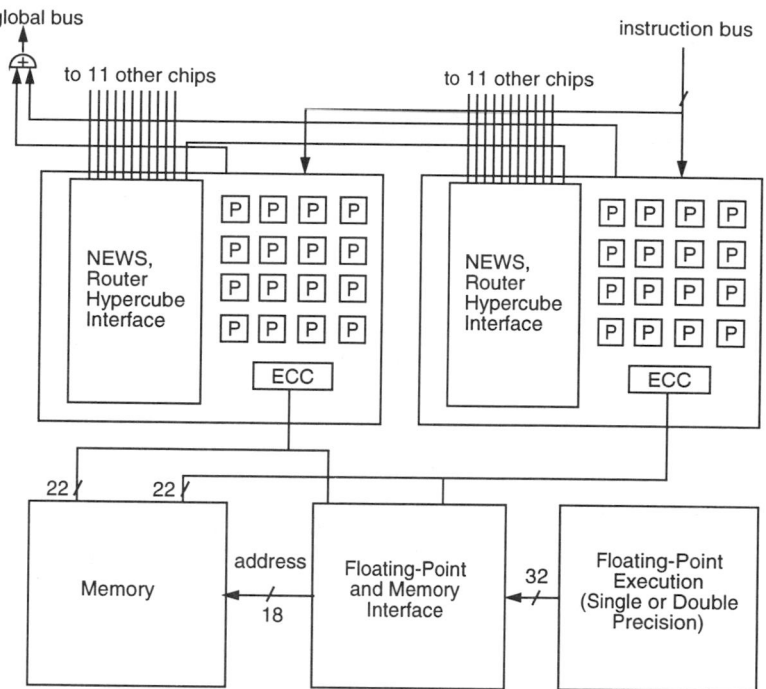

Figure 8.24 A CM-2 processing node consisting of two processor chips and some memory and floating-point chips. (Courtesy of Thinking Machines Corporation, 1990)

32 processors. The floating-point chip handles 32-bit operations at a time. Intermediate computational results can be stored back into the memory for subsequent use. Note that integer arithmetic is carried out directly by the processors in a bit-serial fashion.

Hypercube Routers   Special hardware is built on each processor chip for data routing among the processors. The router nodes on all processor chips are wired together to form a Boolean n-cube. A full configuration of CM-2 has 4096 router nodes on processor chips interconnected as a 12-dimensional hypercube.

Each router node is connected to 12 other router nodes, including its paired node (Fig. 8.24). All 16 processors belonging to the same node are equally capable of sending a message from one vertex to any other processor at another vertex of the 12-cube. The following example clarifies this message-passing concept.

Example 8.10   Message routing on the CM-2 hypercube (Thinking Machines Corporation, 1990)

On each vertex of the 12-cube, the processors are numbered 0 through 15. The hypercube routers are numbered 0 through 4095 at the 4096 vertices. Processor

5 on router node 7 is thus identified as the 117th processor in the entire system because $16 \times 7 + 5 = 117$.

Suppose processor 117 wants to send a message to processor 361, which is located at processor 9 on router node 22 ($16 \times 22 + 9 = 361$). Since router node 7 = $(000000000111)_2$ and router node 22 = $(000000010110)_2$, they differ at dimension 0 and dimension j4.

This message must traverse dimensions 0 and 4 to reach its destination. From router node 7, the message is first directed to router node 6 = $(00000000110)_2$ through dimension 0 and then to router node 22 through dimension 4, if there is no contention for hypercube wires. On the other hand, if router 7 has another message using the dimension 0 wire, the message can be routed first through dimension 4 to router 23 = $(000000010111)_2$ and then to the final destination through dimension 0 to avoid channel conflicts.

∎

The NEWS Grid   Within each processor chip, the 16 physical processors can be arranged as an 8×2, 1×16, 4×4, $4 \times 2 \times 2$, or $2 \times 2 \times 2 \times 2$ grid, and so on. Sixty-four *virtual processors* can be assigned to each physical processor. These 64 virtual processors can be imagined to form a 8×8 grid within the chip.

The "NEWS" grid is based on the fact that each processor has a north, east, west, and south neighbor in the various grid configurations. Furthermore, a subset of the hypercube wires can be chosen to connect the 2^{12} nodes (chips) as a two-dimensional grid of any shape, 64×64 being one of the possible grid configurations.

By coupling the internal grid configuration within each node with the global grid configuration, one can arrange the processors in NEWS grids of any shape involving any number of dimensions. These flexible interconnections among the processors make it very efficient to route data on dedicated grid configurations based on the application requirements.

Scanning and Spread Mechanisms   Besides dynamic reconfiguration in NEWS grids through the hypercube routers, the CM-2 has been built with special hardware support for scanning or spreading across NEWS grids. These are very powerful parallel operations for fast data combining or spreading throughout the entire array.

Scanning on NEWS grids combines communication and computation. The operation can simultaneously scan in every row of a grid along a particular dimension for the partial sum of that row, find the largest or smallest value, or compute bitwise OR, AND, or exclusive OR. Scanning operations can be expanded to cover all elements of an array.

Spreading can send a value to all other processors across the chips. A single-bit value can be spread from one chip to all other chips along the hypercube wires in only 75 steps. Variants of scans and spreads have been built into the Paris instructions for ease of access.

I/O and Data Vault   The Connection Machine emphasizes massive parallelism in computing as well as in visualization of computational results. High-speed I/O channels are available from 2 to 16 channels for data and/or image I/O operations. Peripheral devices attached to I/O channels include a data vault, CM-HIPPI system, CM-IOP system, and VMEbus interface controller as illustrated in Fig. 8.23. The data vault is a disk-based mass storage system for storing program files and large data bases.

Major Applications   The CM-2 has been applied in almost all the MPP and grand challenge applications introduced in Chapter 3. Specifically, the Connection Machine Series has been applied in document retrieval using relevance feedback, in memory-based reasoning as in the medical diagnostic system called QUACK for simulating the diagnosis of a disease, and in bulk processing of natural languages.

Other applications of the CM-2 include the SPICE-like VLSI circuit analysis and layout, computational fluid dynamics, signal/image/vision processing and integration, neural network simulation and connectionist modeling, dynamic programming, context-free parsing, ray tracing graphics, and computational geometry problems. As the CM-2 is upgraded to the CM-5, the applications domain will certainly expand accordingly.

8.4.3   The MasPar MP-1 Architecture

This is a medium-grain SIMD computer, quite different from the CM-2. Parallel architecture and MP-1 hardware design are described below. Special attention is paid to its interprocessor communication mechanisms.

The MasPar MP-1   The MP-1 architecture consists of four subsystems: the *PE array*, the *array control unit* (ACU), a *UNIX subsystem* with standard I/O, and a *high-speed I/O subsystem* as depicted in Fig. 8.25a. The UNIX subsystem handles traditional serial processing. The high-speed I/O, working together with the PE array, handles massively parallel computing.

The MP-1 family includes configurations with 1024, 4096, and up to 16,384 processors. The peak performance of the 16K-processor configuration is 26,000 MIPS in 32-bit RISC integer operations. The system also has a peak floating-point capability of 1.5 Gflops in single-precision and 650 Mflops in double-precision operations.

Array Control Unit   The ACU is a 14-MIPS scalar RISC processor using a demand-paging instruction memory. The ACU fetches and decodes MP-1 instructions, computes addresses and scalar data values, issues control signals to the PE array, and monitors the status of the PE array.

Like the sequencer in CM-2, the ACU is microcoded to achieve horizontal control of the PE array. Most scalar ACU instructions execute in one 70-ns clock. The whole ACU is implemented on one PC board.

An implemented functional unit, called a *memory machine*, is used in parallel with the ACU. The memory machine performs PE array load and store operations, while the ACU broadcasts arithmetic, logic, and routing instructions to the PEs for parallel execution.

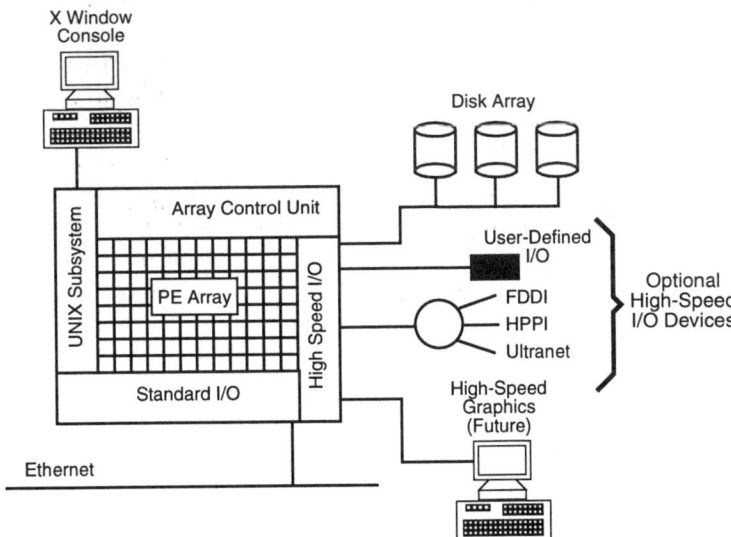

(a) MP-1 System Block Diagram

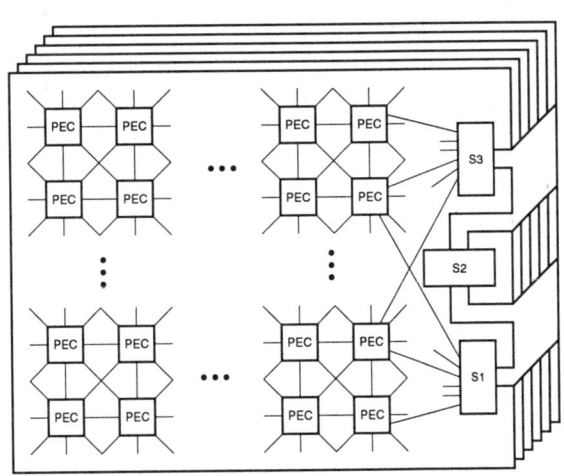

(b) Array of PE clusters

Figure 8.25 The MasPar MP-1 architecture. (Courtesy of MasPar Computer Corporation, 1990)

The PE Array   Each processor board has 1024 PEs and associated memory arranged as 64 *PE clusters* (PEC) with 16 PEs per cluster. Figure 8.25b shows the inter-PEC connections on each processor board. Each PEC chip is connected to eight neighbors via the X-Net mesh and a global multistage crossbar router network, labeled S1, S2, and S3 in Fig. 8.25b.

Each PE cluster (Fig. 8.26a) is composed of 16 PEs and 16 processor memories (PEMs). The PEs are logically arranged as a 4×4 array for the X-Net two-dimensional mesh interconnections. The 16 PEs in a cluster share an access port to the multistage crossbar router. Interprocessor communications are carried out via three mechanisms:

(1) ACU-PE array communications.
(2) X-Net nearest-neighbor communications.
(3) Global crossbar router communications.

The first mechanism supports ACU instruction/data broadcasts to all PEs in the array simultaneously and performs global reductions on parallel data to recover scalar values from the array. The other two IPC mechanisms are described separately below.

X-Net Mesh Interconnect   The X-Net interconnect directly connects each PE with its eight neighbors in the two-dimensional mesh. Each PE has four connections at its diagonal corners, forming an X pattern similar to the BLITZEN X grid network (Davis and Reif, 1986). A tri-state node at each X intersection permits communication with any of eight neighbors using only four wires per PE.

The connections to the PE array edges are wrapped around to form a 2-D torus. The torus structure is symmetric and facilitates several important matrix algorithms and can emulate a one-dimensional ring with two-X-Net steps. The aggregate X-Net communication bandwidth is 18 Gbytes/s in the largest MP-1 configuration.

Multistage Crossbar Interconnect   The network provides global communication between all PEs and forms the basis for the MP-1 I/O system. The three router stages implement the function of a 1024×1024 crossbar switch. Three router chips are used on each processor board.

Each PE cluster shares an originating port connected to router stage S1 and a target port connected to router stage S3. Connections are established from an originating PE through stages S1, S2, and S3 and then to the target PE. The full MP-1 configuration has 1024 PE clusters, so each stage has 1024 router ports. The router supports up to 1024 simultaneous connections with an aggregate bandwidth of 1.3 Gbytes/s.

Processor Elements and Memory   The PE design is mostly data path logic and has no instruction fetch or decode logic. The design is detailed in Fig. 8.26b. Both integer and floating-point computations execute in each PE with a register-based RISC architecture. Load and store instructions move data between the PEM and the register set.

Each PE has forty 32-bit registers available to the programmer and eight 32-bit registers for system use. The registers are bit and byte addressable. Each PE has a 4-bit integer ALU, a 1-bit logic unit, a 64-bit mantissa unit, a 16-bit exponent unit, and a flag unit. The NIBBLE bus is four bits wide and the BIT bus is one bit wide. The PEM can be directly or indirectly addressed with a maximum aggregated memory bandwidth of 12 Gbytes/s.

Most data movement with each PE occurs on the NIBBLE bus and the BIT bus.

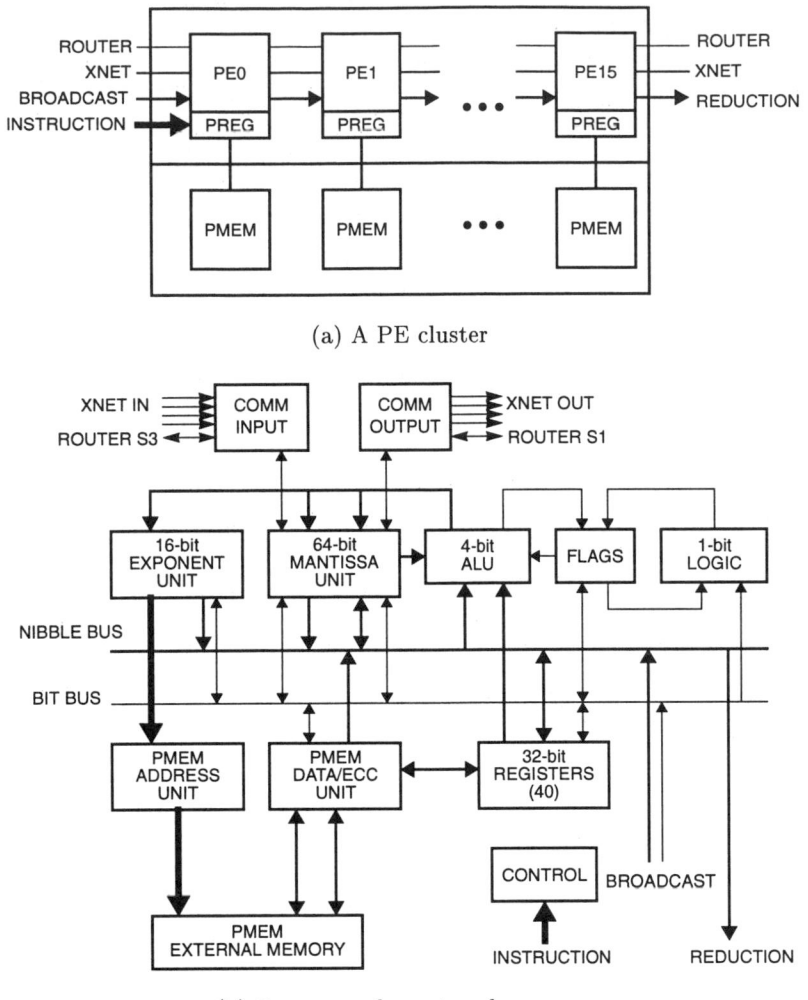

(a) A PE cluster

(b) Processor element and memory

Figure 8.26 Processing element and memory design in the MasPar MP-1. (Courtesy of MasPar Computer Corporation, 1990)

Different functional units within the PE can be simultaneously active during each microstep. In other words, integer, Boolean, and floating-point operations can all perform at the same time. Each PE runs with a slow clock, the speed is obtained through massive parallelism like that implemented in the CM-2.

Parallel Disk Arrays   Another feature worthy of mentioning is the massively parallel I/O architecture implemented in the MP-1. The PE array (Fig. 8.25a) communicates with a parallel disk array through the high-speed I/O subsystem, which is essentially implemented by the 1.3 Gbytes/s global router network.

The disk array provides up to 17.3 Gbytes of formatted capacity with a 9-Mbytes/s sustained disk I/O rate. The parallel disk array is a necessity to support data-parallel computation and provide file system transparency and multilevel fault tolerance.

8.5 The Connection Machine CM-5

The grand challenge applicationswill drive the development of present and future MPP systems to achieve the 3T performance goals. The Connection Machine Model CM-5 is the latest effort of Thinking Machines Corporation toward this end. We describe below the innovations surrounding the CM-5 architectural development, its building blocks, and the application paradigms.

8.5.1 A Synchronized MIMD Machine

The CM-2 and its predecessors were criticized for having a rigid SIMD architecture, limiting general-purpose applications. The CM-5 designers liberated themselves by choosing a universal architecture, which combines the advantages of both SIMD and MIMD machines.

Traditionally, supercomputer programmers were forced to choose between MIMD and SIMD computers. An MIMD machine is good at independent branching but bad at synchronization and communication. On the contrary, an SIMD machine is good at synchronization and communication but poor at branching. The CM-5 was designed with a synchronized MIMD structure to support both styles of parallel computation.

The Building Blocks   The CM-5 architecture is shown in Fig. 8.27. The machine contains from 32 to 16,384 *processing nodes*, each of which can have a 32-MHz SPARC processor, 32-Mbytes of memory, and a 128-Mflops vector processing unit capable of performing 64-bit floating-point and integer operations.

Instead of using a single sequencer (as in the CM-2), the system uses a number of *control processors*, which are Sun Microsystems workstation computers. The number of control processors, varying with different configurations, ranges from one to several tens. Each control processor is configured with memory and disk based on the needs.

Input and output are provided via high-bandwidth *I/O interfaces* to graphics devices, mass secondary storage such as a data vault, and high-performance networks. Additional low-speed I/O is provided by Ethernet connections to the control processors. The largest configuration is expected to occupy a space of 30 m × 30 m, and is shooting for a peak performance over 1 Tflops.

The Network Functions   The building blocks are interconnected by three networks: a *data network*, a *control network*, and a *diagnostic network*. The data network provides high-performance, point-to-point data communications between the processing nodes. The control network provides cooperative operations, including broadcast, synchronization, and scans, as well as system management functions.

The diagnostic network allows "back-door" access to all system hardware to test

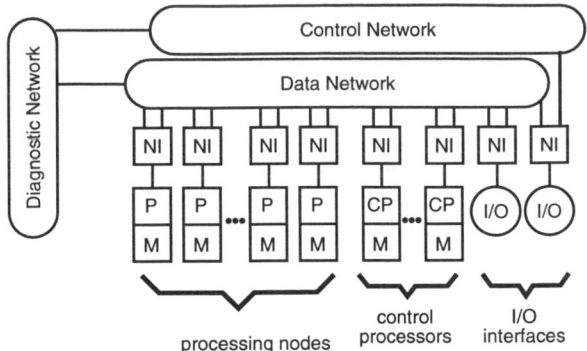

Figure 8.27 The network architecture of the Connection Machine CM-5. (Courtesy of Leiserson et al., Thinking Machines Corporation, 1992)

system integrity and to detect and isolate errors. The data and control networks are connected to processing nodes, control processors, and I/O channels via *network interfaces*.

The CM-5 architecture is considered universal because it is optimized for data-parallel processing of large and complex problems. The data parallelism can be implemented in either SIMD mode, multiple SIMD mode, or synchronized MIMD mode.

The data and control networks are designed to have good *scalability*, making the machine size limited by the affordable cost but not by any architectural or engineering constraint. In other words, the networks depend on no specific types of processors. When new technological advances arrive, they can be easily incorporated into the architecture. The network interfaces are designed to provide an abstract view of the networks.

The System Operations   The system operates one or more user *partitions*. Each partition consists of a control processor, a collection of processing nodes, and dedicated portions of the data and control networks. Figure 8.28 illustrates the distributed control on the CM-5 obtained through the dynamic use of the two interprocessor communication networks. Major system management functions, services, and data distribution are summarized in this diagram.

The partitioning of resources is managed by a system executive. The control processor assigned to each partition behaves like a *partition manager*. Each user process executes on a single partition but may exchange data with processes on other partitions. Since all partitions utilize UNIX time-sharing and security features, each allows multiple users to access the partition, while ensuring no conflicts or interferences.

Access to system functions is classified as either *privileged* or *nonprivileged*. Privileged system functions include access to data and control networks. These accesses can be executed directly by user code without system calls. Thus, OS kernel overhead can be eliminated in network communication within a user task. Access to the diagnostic

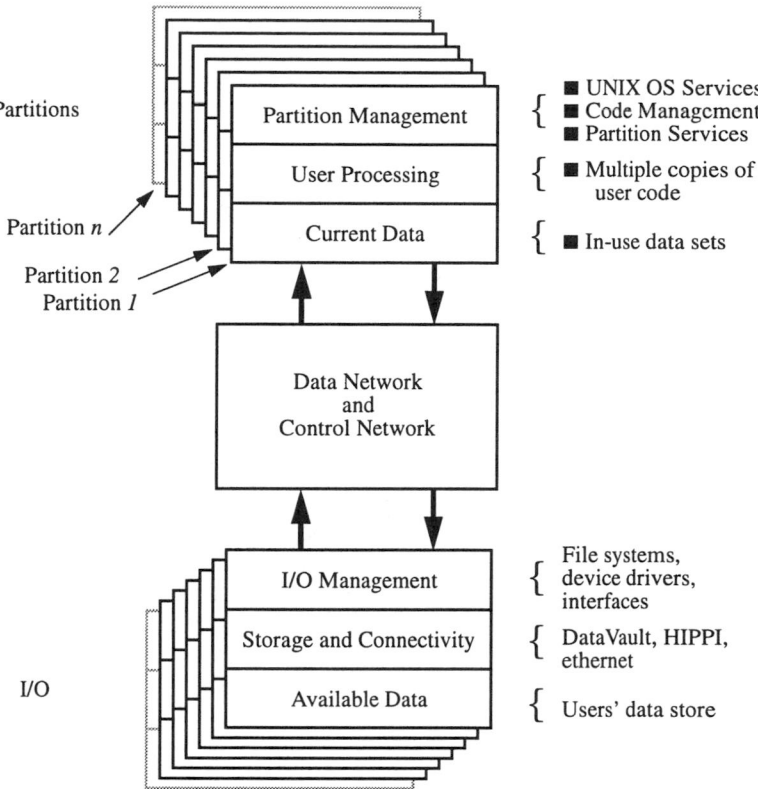

Figure 8.28 Distributed control on the CM-5 with concurrent user partitions and I/O activities. (Courtesy of Thinking Machines Corporation, 1992)

network, to shared I/O resources, and to other partitions is also privileged but must be accomplished via system calls.

Some control processors in the CM-5 are assigned to manage the I/O devices and interfaces. This organization allows a process on any partition to access any I/O device, and ensures that access to one device does not impede access to other devices. Functionally, the system operations, as depicted in Fig. 8.28, are divided into user-oriented partitions, I/O services based upon system calls, dynamic control of the data and control networks, and system management and diagnostics.

The two networks download user code from a control processor to the processing nodes, pass I/O requests, transfer messages of all sorts between control processors, and transfer data among nodes and I/O devices, either in a single partition or among different partitions. The I/O capacity can be scaled with increasing numbers of processing nodes or of control partitions. The CM-5 has embodied the features of hardware modularity, distributed control, latency tolerance, and user abstraction; all are needed for *scalable computing*.

8.5.2 The CM-5 Network Architecture

The data network is based on the *fat-tree* concept introduced by Leiserson (1985). We explain below how it is applied in CM-5 construction. Then we describe the major operations on the control network. Finally, the structure of the diagnostic network is discussed.

Fat Trees A fat tree is more like a real tree in that it becomes thicker as it acquires more leaves. Processing nodes, control processors, and I/O channels are located at the leaves of a fat tree. A *binary fat tree* was illustrated in Fig. 2.17c. The internal nodes are switches. Unlike an ordinary binary tree, the channel capacities of a fat tree increase as we ascend from leaves to root.

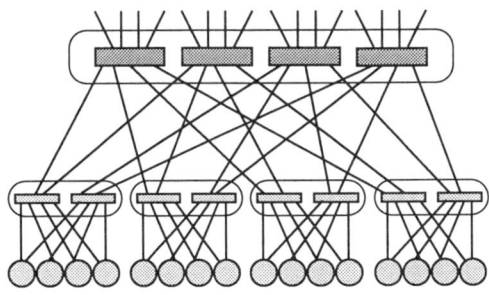

Figure 8.29 CM-5 data network implemented with a 4-ary fat tree (Courtesy of Leiserson et al., Thinking Machines Corporation, 1992)

The hierarchical nature of a fat tree can be exploited to give each user partition a dedicated subtree, which cannot be interfaced with by any other partition's message traffic. The CM-5 data network is actually implemented with a 4-ary fat tree as shown in Fig. 8.29. Each of the internal switch nodes is made up of several router chips. Each router chip is connected to four child chips and either two or four parent chips.

To implement the partitions, one can allocate different subtrees to handle different partitions. The size of the subtrees varies with different partition demands. The I/O channels are assigned to another subtree, which is not devoted to any user partition. The I/O subtree is accessed as shared system resources. In many ways, the data network functions like a hierarchical system bus, except that there is no interference among partitioned subtrees. All leave nodes have unique physical addresses.

The Data Network To route a message from one processor node to another, the message is sent up the tree to the least common ancestor of the two processors and then down to the destination.

In the 4-ary fat-tree implementation (Fig. 8.29) of the data network, each connection provides a link to another chip with a raw bandwidth of 20 Mbytes/s in each direction. By selecting at each level of the tree whether two or four parent links are used, the bandwidths between nodes in the fat tree can be adjusted. Flow control is provided on each link.

Each processor has two connections to the data network, corresponding to a raw bandwidth of 40 Mbytes/s in and out of cach leaf node. In the first two levels, each router chip uses only two parent connections to the next higher level, yielding an aggregate bandwidth of 160 Mbytes/s out of a subtree with 16 leaf nodes. All router chips higher than the second level use four parent connections, which, for example, yield an aggregate bandwidth of 10 Gbytes/s in each direction, from one half of a 2K-node system to the other.

The bandwidth continues to scale linearly up to 16,384 nodes, the largest CM-5 configuration being built. In larger machines, transmission-line techniques are used to pipeline bits across long wires, thereby overcoming the bandwidth limitation that would otherwise be imposed by wire latency.

As a message goes up the tree, it may have several choices as to which parent connection to take. The decision is resolved by pseudo-randomly selecting from among those links that are unobstructed by other messages. After reaching the least common ancestor of the source and destination nodes, the message takes a single available path of links down to the destination. The pseudo-random choice at each level automatically balances the load on the network and avoids undue congestion caused by pathological message sets.

The data network chips are driven by a 40-MHz clock. The first two levels are routed through backplanes. The wires on higher levels are routed through cables, which can be either 9 or 26 ft in length. The cables reliably carry signals in excess of 90% of the speed of light. Message routing is based on the wormhole concept discussed in Section 7.4.

Faulty processor nodes or connection links can be mapped out of the system and quarantined. This allows the system to remain functional while servicing and testing the mapped-out portion. The data network is acyclic from input to output, which precludes deadlock from occurring if the network promises to eventually accept or deliver all messages injected into it and the originating processors promise to eventually remove all messages from the network after they have been successfully delivered.

The Control Network The architecture of the control network is that of a complete binary tree with all system components at the leaves. Each user partition is assigned to a subtree of the network. Processing nodes are located at leaves of the subtree, and a control processor is mapped into the partition at an additional leaf. The control processor executes scalar part of the code, while the processing nodes execute the data-parallel part.

Unlike the variable-length messages transmitted by the data network, control network packets have a fixed length of 65 bits. There are three major types of operations on the control network: *broadcasting*, *combining*, and *global operations*. These operations provide interprocessor communications. Separate FIFOs in the network interface correspond to each type of control operations.

The control network provides the mechanisms allowing data-parallel code to be executed efficiently and supports MIMD execution for general-purpose application. The binary tree architecture makes the control network much simpler to implement than the fat trees used in the data network. The control network has the additional switching capability to map around faults and to connect any of the control processors to any user partition using an off-line routing strategy.

The Diagnostic Network   This network is needed for upgrading system availability. Built-in testability is achieved with scan-based diagnostics. Again, this network is organized as a (not necessarily complete) binary tree for its simplicity in addressing. One or more *diagnostic processors* are at the root. The leaves are *pods*, and each pod is a physical system, such as a board or a backplane. There is a unique path from the root to each pod being tested.

The diagnostic network allows groups of pods to be addressed according to a "hypercube-address" scheme. A special diagnostic interface was designed to form an in-system check of the integrity of all CM-5 chips that support the JTAG (Joint Test Action Group) standard and all networks. It provides scan access to all chips supporting the JTAG standard and programmable ad hoc access to non-JTAG chips. The network itself is completely testable and diagnosable. It is able to map out and ignore faulty or power-down parts of the machine.

8.5.3   Control Processors and Processing Nodes

The functional architectures of the control processors and of the processing nodes are described in this subsection.

Control Processor   As shown in Fig. 8.30, the basic control processor consists of a RISC microprocessor (CPU), memory subsystem, I/O with local disks and Ethernet connections, and a CM-5 network interface. This is equivalent to a standard off-the-shelf workstation-class computer system. The network interface connects the control processor to the rest of the system through the control network and the data network.

Each control processor runs CMOST, a UNIX-based OS with extensions for managing the parallel processing resources of the CM-5. Some control processors manage computational resources in user partitions. Others are used to manage I/O resources. Control processors specialize in managerial functions rather than computational functions. For this reason, high-performance arithmetic accelerators are not needed. Instead, additional I/O connections are more useful in control processors.

Processing Nodes   Figure 8.31 shows the basic structure of a processing node. It is a SPARC-based microprocessor with a memory subsystem, consisting of a memory controller and 8, 16, or 32 Mbytes of DRAM memory. The internal bus is 64 bits wide.

The SPARC processor was chosen for its multiwindow feature to facilitate fast context switching. This is very crucial to the dynamic use of the processing nodes in different user partitions at different times. The network interface connects the node to the rest of the system through the control and data networks . The use of a hardware arithmetic accelerator to augment the microprocessor is optional.

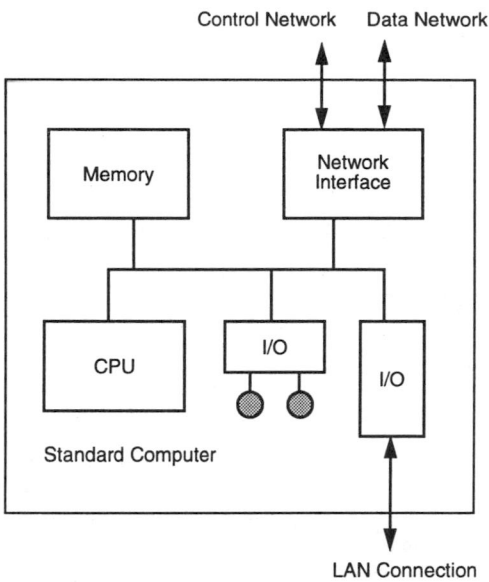

Figure 8.30 The control processor in the CM-5. (Courtesy of Thinking Machines Corporation 1992)

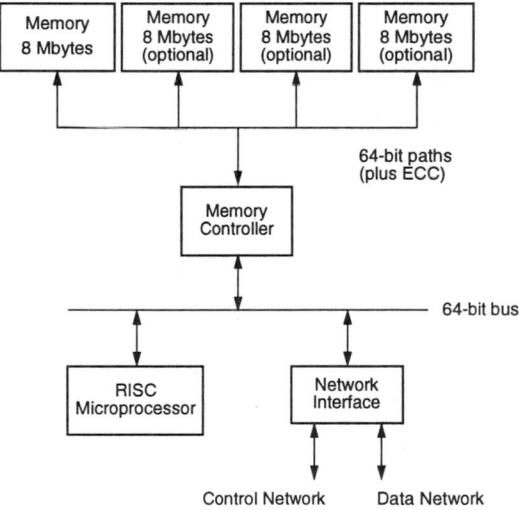

Figure 8.31 The processing node in the CM-5. (Courtesy of Thinking Machines Corporation, 1992)

Vector Units   As illustrated in Fig. 8.32a, vector units can be added between the memory bank and the system bus as an optional feature. The vector units place the memory controller in Fig. 8.31. Each vector unit has a dedicated 72-bit path to its attached memory bank, providing a peak memory bandwidth of 128 Mbytes/s per vector unit.

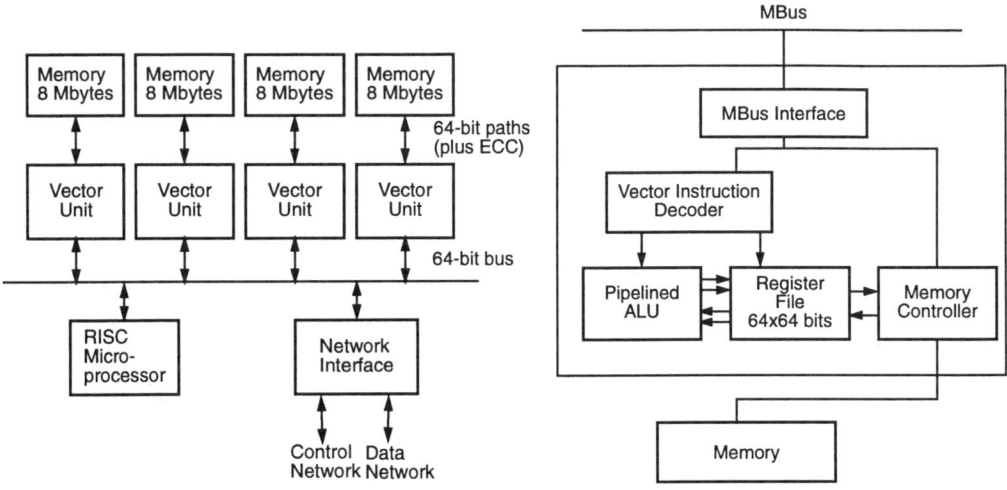

(a) Processing node with vector units   (b) Vector unit functional architecture

Figure 8.32 The processing node with vector units in the CM-5. (Courtesy of Thinking Machines Corporation, 1992)

The vector unit executes vector instructions issued by the scalar microprocessor and performs all functions of a memory controller, including generation and check of ECC (error correcting code) bits. As detailed in Fig. 8.32b, each vector unit has a vector instruction decoder, a pipelined ALU, and sixty-four 64-bit registers like a conventional vector processor.

Each vector instruction may be issued to a specific vector unit or pairs of units or broadcast to all four units at once. The scalar microprocessor takes care of address translation and loop control, overlapping them with vector unit operations. Together, the vector units provide 512 Mbytes/s memory bandwidth and 128 Mflops 64-bit peak performance per node. In this sense, each processing node of the CM-5 is itself a supercomputer. Collectively, 16K processing nodes may yield a peak performance of $2^{14} \times 2^7 = 2^{21}$ Mflops = 2 Tflops.

Initially, SPARC microprocessors are being used in implementing the control processors and processing nodes. As processor technology advances, other new processors may be also combined in the future. The network architectures are designed to be independent of the processors chosen except for the network interfaces which may need some minor modifications when new processors are used.

8.5.4   Interprocessor Communications

We have described the high-speed scanning and spreading mechanisms built into the CM-2. In the CM-5, these mechanisms are being further upgraded into four categories of interprocessor communication : *replication, reduction, permutation, parallel prefix.*

These operations can be applied to regular or irregular data sets including vectors, matrices, multidimensional arrays, variable-length vectors, linked lists, and completely irregular patterns. In this section, we describe the key concepts behind these ICP operations. Data-parallel software has been developed to support these operations on the CM-5. The roles of the control network are identified in these operations.

Replication   Recall the *broadcast* operation, where a single value may be replicated as many copies and distributed to all processors, as illustrated in Fig. 8.33a. Other duplication operations include the *spreading* of a column vector into all the columns of a matrix (Fig. 8.33b), the *expansion* of a short vector into a long vector (Fig. 8.33c), and a completely irregular duplication (Fig. 8.33d).

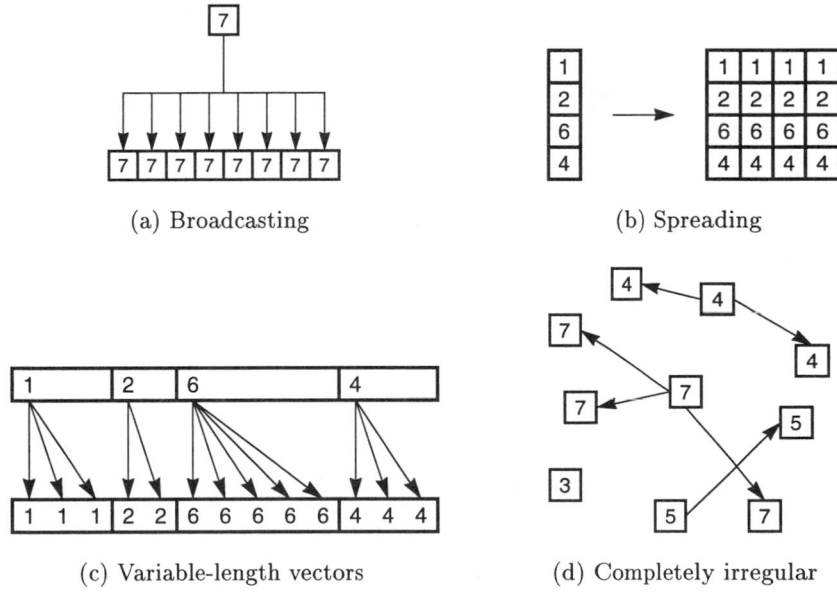

(a) Broadcasting   (b) Spreading

(c) Variable-length vectors   (d) Completely irregular

Figure 8.33 Replication operations for interprocessor communications on CM-5. (Courtesy of Thinking Machines Corporation, 1992)

Replication plays a fundamental role in matrix arithmetic and vector processing, especially on a data-parallel machine. Replication is carried out through the control network in four kinds of broadcasting schemes: *user broadcast, supervisor broadcast, interrupt broadcast,* and *utility broadcast.* These operations can be used to download code and to distribute data, to implement fast barrier synchronization, and to configure partitions through the OS.

Reduction   Vector reduction has been implemented on the CM-2 by fast scanning, and on the CM-5 the mechanism is further generalized as the opposite of replication. As illustrated in Fig. 8.34, *global reduce* produces the sum of vector components (Fig. 8.34a). Similarly, the row/column reductions produce the sums per each row of a matrix (Fig. 8.34b).

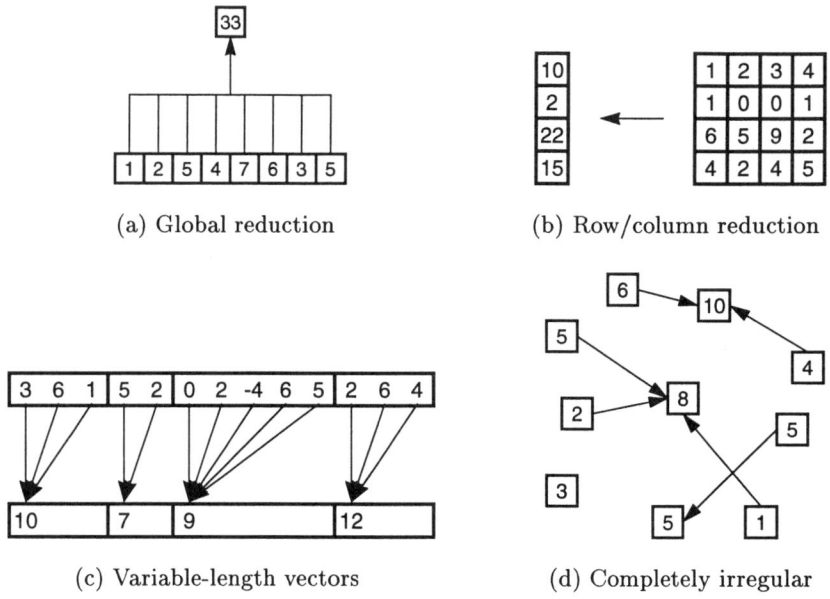

(a) Global reduction   (b) Row/column reduction

(c) Variable-length vectors   (d) Completely irregular

Figure 8.34 Reduction operations on the CM-5. (Courtesy of Thinking Machines Corporation, 1992)

Variable-length vectors are reduced in chunks of a long vector (Fig. 8.34c). The same idea applies to a completely irregular set as well (Fig. 8.34d). In general, reduction functions include the maximum, the minimum, the average, the dot product, the sum, logical AND, logical OR, etc. Fast scanning and combining are necessities in implementing these operations.

Four types of combining operations, *reduction, forward scan* (parallel prefix), *backward scan* (parallel suffix), and *router done*, are supported by the control network. We will describe parallel prefix shortly. Router done refers to the detection of completion of a message-routing cycle, based on Kirchoff's current law, in that the network interfaces keep track of the number of messages entering and leaving the data network. When a round of message sending and acknowledging is complete, the net "current" (messages) in and out of a port should be zero.

Permutation   Data-parallel computing relies on permutation for fast exchange of data among processing nodes. Figure 8.35 illustrates four cases of permutations performed on the CM-5. These permutation operations are often needed in matrix transpose, reversing a vector, shifting a multidimensional grid, and FFT butterfly operations.

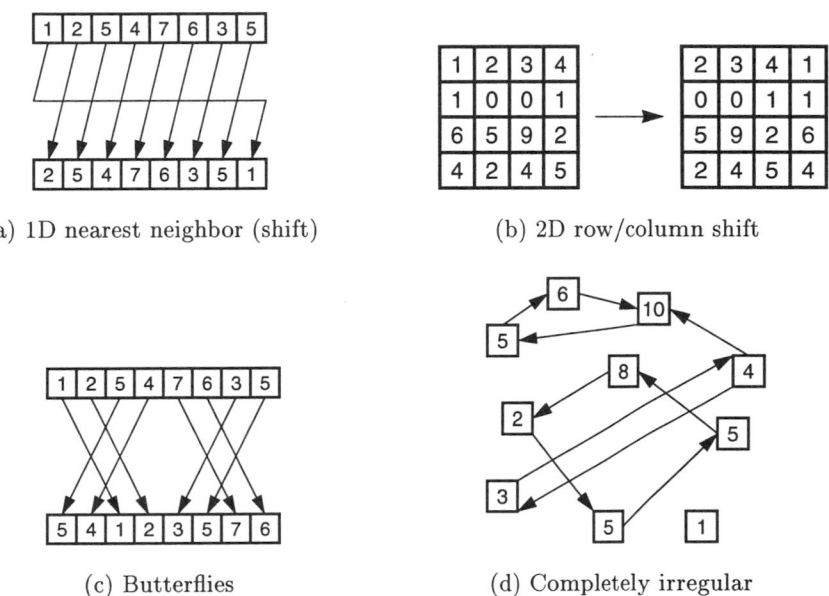

(a) 1D nearest neighbor (shift)   (b) 2D row/column shift

(c) Butterflies   (d) Completely irregular

Figure 8.35 Permutation operations for interprocessor communications on the CM-5. (Courtesy of Thinking Machines Corporation, 1992)

Example 8.11   Discretization of the Laplace equation on the CM-5 using permutation (shifting) operations (Thinking Machines Corporation, 1992)

This example is taken from the *CM-5 Technical Summary*, published by Thinking Machines Corporation in 1992. Consider a finite-difference grid used in the discretization of the Laplace equation. The average of four neighbors is iteratively computed as follows:

$$
\begin{aligned}
C \;=\; 0.25 \;\times\; (&(\text{CSHIFT}(A, 1, +1) \\
&\text{CSHIFT}(A, 1, -1) \\
&\text{CSHIFT}(A, 2, +1) \\
&\text{CSHIFT}(A, 2, -1))
\end{aligned}
$$

The CSHIFT is a Fortran operation that shifts an array with periodic boundary conditions. Elements shifted off one edge are circularly shifted onto the opposite edge, and thus no elements are lost in this permutation.

■

Parallel Prefix   This is a kind of combining operation supported by the control network. A *parallel prefix* operation delivers to the ith processor the result of applying one of the five reduction operators to the values in the preceding $i - 1$ processors, in the linear order given by data address.

The idea is illustrated in Fig. 8.36 with four examples. Figure 8.36a shows the one-dimensional sum-prefix, in which the fourth output 12 is the sum of the first four input elements $(1 + 2 + 5 + 4 = 12)$. The two-dimensional row/column sum-prefix (Fig. 8.36b) can be similarly performed using the forward-scanning mechanism.

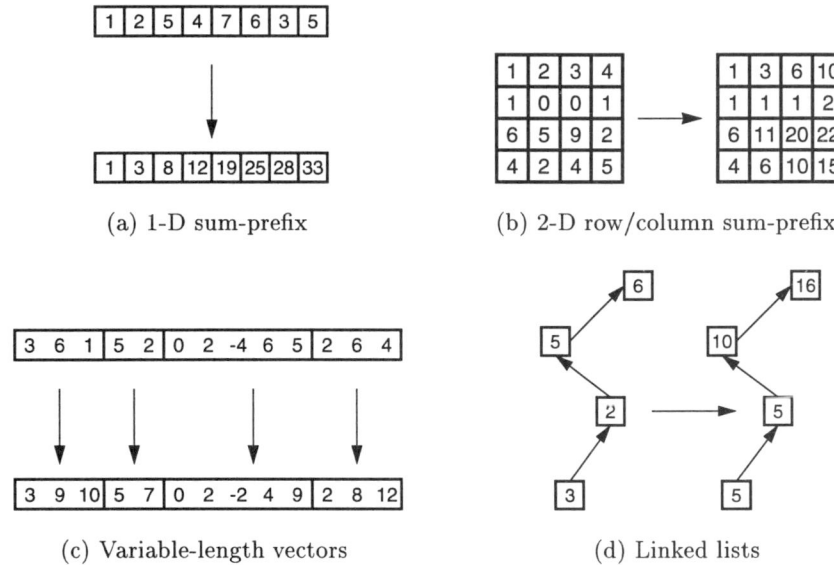

(a) 1-D sum-prefix

(b) 2-D row/column sum-prefix

(c) Variable-length vectors

(d) Linked lists

Figure 8.36 Parallel prefix operations on the CM-5. (Courtesy of Thinking Machines Corporation, 1992)

Figure 8.36c computes the one-dimensional prefix-sum on sections of a long vector independently. Figure 8.36d shows the forward scanning along linked lists to produce the prefix-sums as outputs.

Many prefix and suffix scanning operations appear to be inherently sequential processes. But the scanning and combining mechanisms on the CM-5 can make the process approximately $\log_2 n$ faster, where n is the array length involved. For example, a parallel prefix operation on a vector of 1000 entries can be finished in 10 steps instead of 1000 steps on the CM-5.

8.6 Bibliographic Notes and Exercises

An assessment of future supercomputer architecture was given by [Smith90] , Hsu, and Hsiung. [Dongarra89] and Hinds have compared the Cray X-MP with the Fujitsu VP-200 and Hitachi S-810/20. Recent reports on the Cray Y-MP Series can be found in [Cray91] and [Cray89]. Information on the VP2000 Series is from [Fujitsu90]. The NEC SX-X Series is described in [NEC90]. The VAX 9000 is discussed in [Brunner90] and the Stardent 3000 in [Siewiorek91].

Information on the latest Cray Y-MP C-90 clusters and Cray/MPP development can be obtained from Cray Research, Inc. [Cray92]. The Fujitsu VPP500 has been recently announced [Fujitsu92]. For a performance comparison of supercomputers, consult the article by [Simmons92] et al. The article by [Cybenko92] and Kuck describes the evolution of supercomputer software. The power of massive parallelism in supercomputing was assessed in [Zorpetta92].

Vector chaining benchmark results were reported in [Cheng89]. The discussion of compound vector functions and pipeline nets was based on the work of [Hwang88] and Xu. The vector performance of the IBM 3090 was reported in [Clark86] and Wilson. Vector models for data-parallel computers were described in [Blelloch90]. Reviews of available parallel/vector computers can be found in [Trew91] and Wilson.

A survey of SIMD computers can be found in [Hord90]. The Connection Machine CM-2 was described in [TMC90] and CM-2 applications in [Waltz87] and [Tucker88] and Robertson. The MasPar MP-1 was reported in [Nickolls90] and summarized in [MasPar91]. The BLITZEN architecture was discussed in [Blevins90].

The CM-5 was functionally introduced in [TMC91]. The network architecture of CM-5 was reported in [Leiserson92]. The original paper on the fat tree was in [Leiserson85]. Randomized routing on fat trees was described by [Greenberg89] and Leiserson. Data-parallel algorithms for the Connection Machines are reported in [Hillis86] and Steele.

Exercises

Problem 8.1 Explain the structural and operational differences between register-to-register and memory-to-memory architectures in building multipipelined supercomputers for vector processing. Comment on the advantages and disadvantages in using SIMD computers as compared with the use of pipelined supercomputers for vector processing.

Problem 8.2 Explain the following terms related to vector processing:

(a) Vector and scalar balance point.
(b) Vectorization ratio in user code.
(c) Vectorization compiler or vectorizer.
(d) Vector reduction instructions.
(e) Gather and scatter instructions.

(f) Sparse matrix and masking instruction.

Problem 8.3 Explain the following memory organizations for vector accesses:

(a) S-access memory organization.
(b) C-access memory organization.
(c) C/S-access memory organization.

Problem 8.4 Distinguish among the following vector processing machines in terms of architecture, performance range, and cost-effectiveness:

(a) Full-scale vector supercomputers.
(b) High-end mainframes or near-supercomputers.
(c) Minisupercomputers or supercomputing workstations.

Problem 8.5 Explain the following terms associated with compound vector processing:

(a) Compound vector functions.
(b) Vector loops and pipeline chaining.
(c) Systolic program graphs.
(d) Pipeline network or pipenets.

Problem 8.6 Answer the following questions related to the architecture and operations of the Connection Machine CM-2:

(a) Describe the processing node architecture, including the processor, memory, floating-point unit, and network interface.
(b) Describe the hypercube router and the NEWS grid and explain their uses.
(c) Explain the scanning and spread mechanisms and their applications on the CM-2.
(d) Explain the concepts of broadcasting, global combining, and virtual processors in the use of the CM-2.

Problem 8.7 Answer the following questions about the MasPar MP-1:

(a) Explain the X-Net mesh interconnect (the PE array) built into the MP-1.
(b) Explain how the multistage crossbar router works for global communication between all PEs.
(c) Explain the computing granularity on PEs and how fast I/O is performed on the MP-1.

Problem 8.8 Answer the following questions about the Connection Machine CM-5:

(a) What is a fat tree and its application in constructing the data network in the CM-5?
(b) What are user partitions and their resources requirements?
(c) Explain the functions of the control processors of the control network and of the diagnostic network.
(d) Explain how vector processing is supported in each processing node.

Problem 8.9 Give examples, different from those in Figs. 8.33 through 8.36, to explain the concepts of replication, reduction, permutation, and parallel prefix operations on the CM-5. Check the *Technical Summary of CM-5* published by Thinking Machines Corporation if additional reading is needed.

Problem 8.10 On a Fujitsu VP2000, the vector processing unit is equipped with two load/store pipelines plus five functional pipelines as shown in Fig. 8.13. Consider the execution of the following compound vector function:

$$A(I) = B(I) \times C(I) + D(I) \times E(I) + F(I) \times G(I)$$

for $I = 1, 2, \ldots, N$. Initially, all vector operands are in memory, and the final vector result must be stored in memory.

(a) Show a pipeline-chaining diagram, similar to Fig. 8.18, for executing this CVF.
(b) Show a space-time diagram, similar to Fig. 8.19, for pipelined execution of the CVF. Note that two *vector loads* can be carried out simultaneously on the two vector-access pipes. At the end of computation, one of the two access pipes is used for storing the A array.

Problem 8.11 The following sequence of compound vector function is to be executed on a Cray X-MP vector processor:

$$A(I) = B(I) + s \times C(I)$$
$$D(I) = s \times B(I) \times C(I)$$
$$E(I) = C(I) \times (C(I) - B(I))$$

where $B(I)$ and $C(I)$ are each 64-element vectors originally stored in memory. The resulting vectors $A(I)$, $D(I)$, and $E(I)$ must be stored back into memory after the computation.

(a) Write 11 vector instructions in proper order to execute the above CVFs on a Cray X-MP processor with two vector-load pipes and one vector-store pipe which can be used simultaneously with the remaining functional pipelines.
(b) Show a space-time diagram, similar to Fig. 8.19, for achieving maximally chained vector operations for executing the above CVFs in minimum time.
(c) Show the potential speedup of the above vector chaining operations over the chaining operations on the Cray 1, which has only one memory-access pipe.

Problem 8.12   Consider a vector computer which can operate in one of two execution modes at a time: one is the *vector mode* with an execution rate of $R_v = 10$ Mflops, and the other is the *scalar mode* with an execution rate of $R_s = 1$ Mflops. Let α be the percentage of code that is vectorizable in a typical program mix for this computer.

(a) Derive an expression for the *average execution rate* R_a for this computer.
(b) Plot R_a as a function of α in the range $(0, 1)$.
(c) Determine the vectorization ratio α needed in order to achieve an average execution rate of $R_a = 7.5$ Mflops.
(d) Suppose $R_s = 1$ Mflops and $\alpha = 0.7$. What value of R_v is needed to achieve $R_a = 2$ Mflops?

Problem 8.13   Describe an algorithm using *add, multiply*, and *data-routing* operations to compute the expression $s = A_1 \times B_1 + A_2 \times B_2 + \cdots + A_{32} \times B_{32}$ with minimum time in each of the following two computer systems. It is assumed that add and multiply require two and four time units, respectively. The time required for instruction/data fetches from memory and decoding delays are ignored. All instructions and data are assumed already loaded into the relevant PEs. Determine the minimum compute time in each case.

(a) A serial computer with a processor equipped with one adder and one multiplier, only one of which can be used at a time. No data-routing operation is needed in this uniprocessor machine.
(b) An SIMD computer with eight PEs (PE$_0$, PE$_1$, $\cdots$, PE$_7$), which are connected by a bidirectional circular ring. Each PE can directly *route* its data to its neighbors in one time unit. The operands A$_i$ and B$_i$ are initially stored in PE$_{i \bmod 8}$ for $i = 1, 2, \cdots, 32$. Each PE can *add* or *multiply* at different times.

Problem 8.14   Calculate the peak performance in Gflops with reasoning in each of the following two vector supercomputers.

(a) The Cray Y-MP C-90 with 16 vector processors.
(b) The NEC SX-X with 4 vector processors.
(c) Explain why both machines offer a maximum 64-way parallelism in their vector operations.

Problem 8.15   Devise a minimum-time algorithm to multiply two 64×64 matrices, $A = (a_{ij})$ and $B = (b_{ij})$, on an SIMD machine consisting of 64 PEs with local memory. The 64 PEs are interconnected by a 2D 8×8 torus with bidirectional links.

(a) Show the initial distribution of the input matrix elements (a_{ij}) and (b_{ij}) on the PE memories.
(b) Specify the SIMD instructions needed to carry out the matrix multiplication. Assume that each PE can perform one *multiply*, one *add*, or one *shift* (shifting data to one of its four neighbors) operation per cycle.

You should first compute all the *multiply* and *add* operations on local data before starting to route data to neighboring PEs. The SIMD *shift* operations can be either east, west, south, or north with wraparound connections on the torus.

(c) Estimate the total number of SIMD instruction cycles needed to compute the matrix multiplication. The time includes all arithmetic and data-routing operations. The final product elements $C = A \times B = (c_{ij})$ end up in various PE memories without duplication.

(d) Suppose data duplication is allowed initially by loading the same data element into multiple PE memories. Devise a new algorithm to further reduce the SIMD execution time. The initial data duplication time, using either data *broadcast* instructions or data *routing* (shifting) instructions, must be counted. Again, each result element c_{ij} ends up in only one PE memory.

Problem 8.16   Compare the Connection Machines CM-2 and CM-5 in their architectures, operation modes, functional capabilities, and potential performance based on technical data available in the literature. Write a report summarizing your comparative studies and comment on the improvement made in the CM-5 over the CM-2 from the viewpoints of a computer architect and of a machine programmer.

Problem 8.17   Consider the use of a multivector multiprocessor system for computing the following linear combination of n vectors:

$$\mathbf{y} = \sum_{j=0}^{1023} a_j \times \mathbf{x}_j$$

where $\mathbf{y} = (y_0, y_1, ..., y_{1023})^T$ and $\mathbf{x}_j = (x_{0j}, x_{1j}, ..., x_{1023,j})^T$ for $0 \le j \le 1023$ are column vectors; $\{a_j | 0 \le j \le 1023\}$ are scalar constants. You are asked to implement the above computations on a four-processor system with shared memory. Each processor is equipped with a vector-add pipeline and a vector-multiply pipeline. Assume four pipeline stages in each functional pipeline.

(a) Design a minimum-time parallel algorithm to perform concurrent vector operations on the given multiprocessor, ignoring all memory-access and I/O operations.

(b) Compare the performance of the multiprocessor algorithm with that of a sequential algorithm on a uniprocessor without the pipelined vector hardware.

Problem 8.18   The Burroughs Scientific Processor (BSP) was built as an SIMD computer consisting of 16 PEs accessing 17 shared memory modules. Prove that conflict-free memory access can be achieved on the BSP for vectors of an arbitrary length with a stride which is not a multiple of 17.

Chapter 9

Scalable, Multithreaded, and Dataflow Architectures

This chapter discusses large-scale computers built with scalable, multithreaded, or dataflow architectures. These architectures are candidates for developing massively parallel processing (MPP) systems. Most MPP systems are still in the research and prototyping stages. Therefore, the material is presented with a strong research flavor benefiting mostly researchers, designers, and graduate students.

Major research issues covered include latency-hiding techniques, principles of multithreading, multidimensional scalability, multithreaded architectures, fine-grain multicomputers, dataflow, and hybrid architectures. Example systems studied include the Stanford Dash, Wisconsin Multicube, USC/OMP, KSR-1, Tera, MIT Alewife and J-Machine, Caltech Mosaic C, ETL EM-4, and MIT/Motorola *T.

9.1 Latency-Hiding Techniques

Future scalable systems are most likely to use distributed shared memory. The access of remote memory may significantly increase memory latency. Furthermore, the processor speed is increasing at a much faster rate than memory and interconnection networks. Thus any scalable multiprocessor or large-scale multicomputer must rely on the use of latency-reducing, -tolerating, or -hiding mechanisms. Four latency-hiding mechanisms are studied below for enhancing scalability and programmability.

Latency hiding can be accomplished through four complementary approaches: (i) using *prefetching techniques* which brings instructions or data close to the processor before it is actually needed; (ii) using *coherent caches* supported by hardware to reduce cache misses; (iii) using *relaxed memory consistency* models by allowing buffering and pipelining of memory references; and (iv) using *multiple-contexts* support to allow a processor to switch from one context to another when a long-latency operation is encountered.

The first three mechanisms are described in this section, supported by simulation results obtained by Stanford researchers. Multiple contexts will be treated with multithreaded processors and system architectures in Sections 9.2 and 9.4. However, the effect of multiple contexts is shown here in combination with other latency-hiding mechanisms.

9.1.1 Shared Virtual Memory

Single-address-space multiprocessors/multicomputers must use shared virtual memory. We present a model of such an architectural environment based on the Stanford Dash experience. Then we examine several shared-virtual-memory systems developed at Stanford, Yale, Carnegie-Mellon, and Princeton universities.

The Architecture Environment   The Dash architecture is a large-scale, cache-coherent, NUMA multiprocessor system, as depicted in Fig. 9.1. It consists of multiple multiprocessor clusters connected through a scalable, low-latency interconnection network. Physical memory is distributed among the processing nodes in various clusters. The distributed memory forms a global address space.

Cache coherence is maintained using an invalidating, distributed directory-based protocol (Section 7.2.3). For each memory block, the directory keeps track of remote nodes cacheing it. When a write occurs, point-to-point messages are sent to invalidate remote copies of the block. Acknowledgment messages are used to inform the originating node when an invalidation has been completed.

Two levels of local cache are used per processing node. One can assume a write-through *primary cache* and a write-back *secondary cache*. Loads and writes can be separated with the use of *write buffers* for implementing weaker memory consistency models. The main memory is shared by all processing nodes in the same clusters.

To facilitate prefetching and the directory-based coherence protocol, directory memory and remote-access caches are used for each cluster. The remote-access cache is shared by all processors in the same cluster. To support multiple contexts, each processor must be equipped with additional hardware features to be described in Sections 9.2 and 9.4.

The SVM Concept   Figure 9.2 shows the structure of a distributed shared memory. A global virtual address space is shared among processors residing at a large number of loosely coupled processing nodes. This *shared virtual memory* (SVM) concept was introduced in Section 4.4.1. Implementation and management issues of SVM are discussed below.

Shared virtual memory was first developed in a Ph.D. thesis by Li (1986) at Yale University. The idea is to implement coherent shared memory on a network of processors without physically shared memory. The coherent mapping of SVM on a message-passing multicomputer architecture is shown in Fig. 9.2b. The system uses virtual addresses instead of physical addresses for memory references.

Each virtual address space can be as large as a single node can provide and is shared by all nodes in the system. Li (1988) implemented the first SVM system, IVY, on a network of Apollo workstations. The SVM address space is organized in pages which can be accessed by any node in the system. A memory-mapping manager on each node

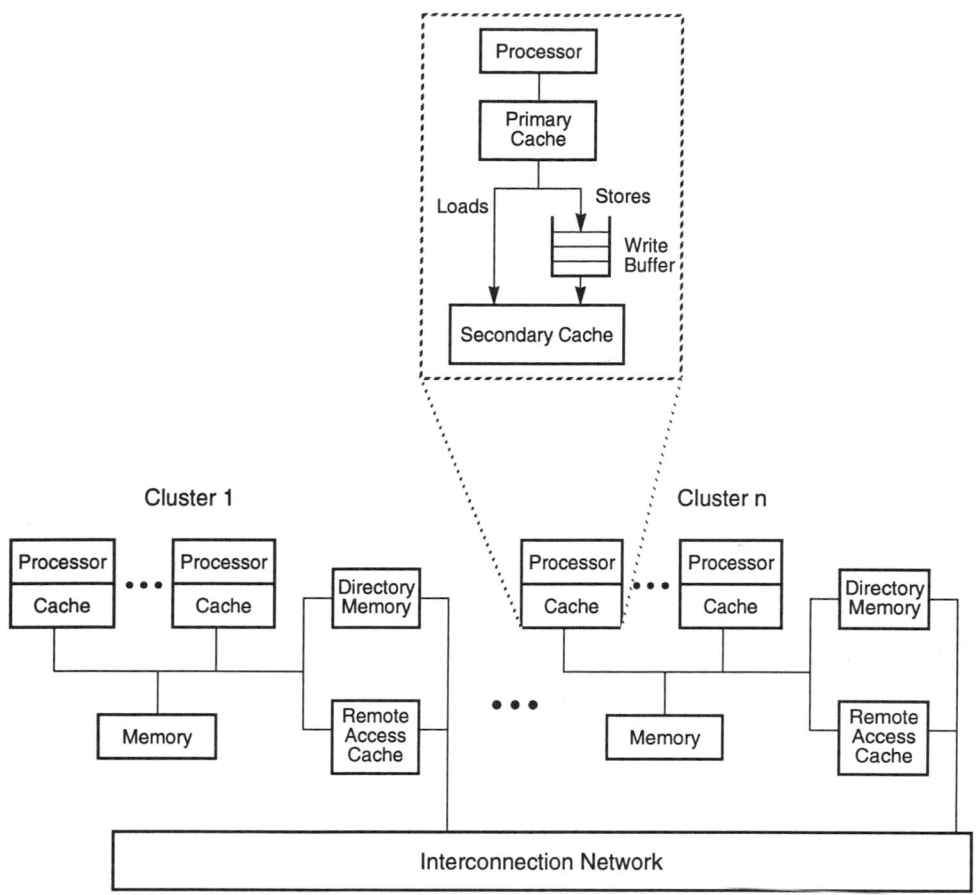

Figure 9.1 A scalable coherent cache multiprocessor with distributed shared memory modeled after the Stanford Dash. (Courtesy of Anoop Gupta et al., *Proc. 1991 Ann. Int. Symp. Computer Arch.*)

views its local memory as a large cache of pages for its associated processor.

Page Swapping   According to Kai Li (1992), pages that are marked read-only can have copies residing in the physical memories of other processors. A page currently being written may reside in only one local memory. When a processor writes a page that is also on other processors, it must update the page and then invalidate all copies on the other processors. Li described the page swapping as follows:

A memory reference causes a page fault when the page containing the memory location is not in a processor's local memory. When a page fault occurs, the memory manager retrieves the missing page from the memory of another processor. If there is a page frame available on the receiving node, the page is moved in. Otherwise, the

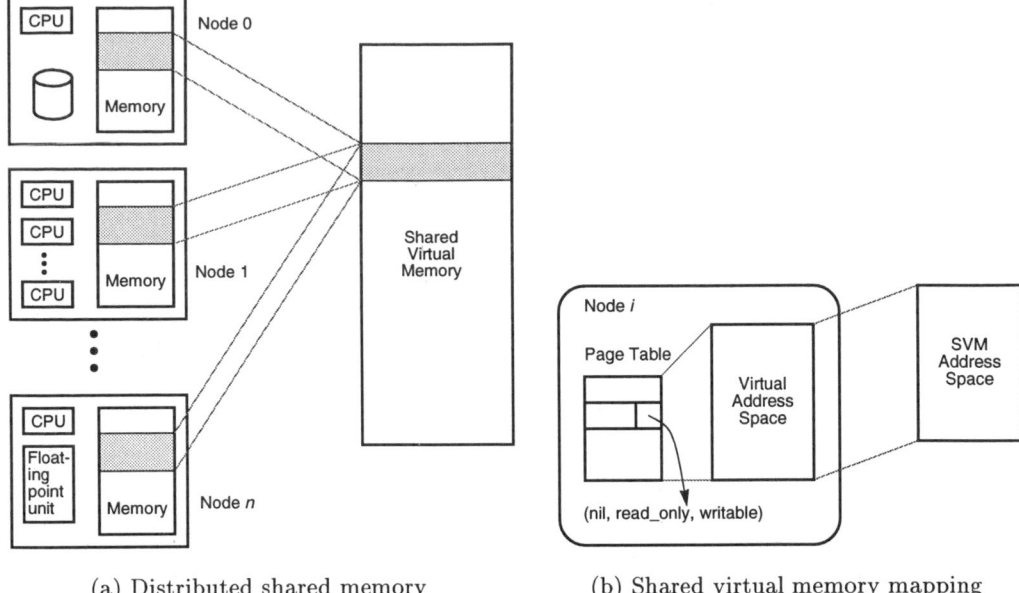

(a) Distributed shared memory   (b) Shared virtual memory mapping

Figure 9.2 The concept of distributed shared memory with a global virtual address
space shared among all processors on loosely coupled processing nodes
in a massively parallel architecture. (Courtesy of Kai Li, 1992)

SVM system uses page replacement policies to find an available page frame, swapping
its contents to the sending node.

A hardware MMU can set the access rights (*nil, read-only, writable*) so that a memory access violating memory coherence will cause a page fault. The memory coherence
problem is solved in IVY through distributed fault handlers and their servers. To client
programs, this mechanism is completely transparent.

The large virtual address space allows programs to be larger in code and data
space than the physical memory on a single node. This SVM approach offers the ease
of shared-variable programming in a message-passing environment. In addition, it improves software portability and enhances system scalability through modular memory
growth.

Example SVM Systems Nitzberg and Lo (1991) have conducted a survey of SVM
systems. Excerpts from their survey, descriptions of four representative SVM systems
are summarized in Table 9.1. The Dash implements SVM with a directory-based coherence protocol. The Linda offers a shared associative object memory with access
functions. The Plus uses a write-update coherence protocol and performs replication
only by program request. The Shiva extends the IVY system for the Intel iPSC/2 hypercube. In using SVM systems, there exists a tendency to use large block (page) sizes

as units of coherence. This tends to increase false-sharing activity.

Table 9.1 Representative Shared-Virtual-Memory Systems (Excerpts from Nitzberg and Lo, *IEEE Comput.*, August 1991)

System and Developer	Implementation and Structure	Coherence Semantics and Protocols	Special mechanics for Performance and Synchronization
Stanford Dash (Lenoski, Laudon, Gharachorloo, Gupta, and Hennessy, 1988–).	Mesh-connected network of Silicon Graphics 4D/340 workstations with added hardware for coherent caches and prefetching.	Release memory consistency with write-invalidate protocol.	Relaxed coherence, prefetching, and use queued locks for synchronization.
Yale Linda (Carriero and Gelernter, 1982–).	Software-implemented system based on the concepts of tuple space with access functions to achieve coherence via virtual memory management.	Coherence varies with environment; hashing is used in associative search; no mutable data.	Linda can be implemented for many languages and machines using C-Linda or Fortran-Linda interfaces.
CMU Plus (Bisiani and Ravishankar, 1988–).	A hardware implementation using MC 88000, Caltech mesh, and Plus kernel.	Uses processor consistency, nondemand write-update coherence, delayed operations.	Pages for sharing, words for coherence, complex synchronization instructions.
Princeton Shiva (Li and Schaefer, 1988).	Software-based system for Intel iPSC/2 with a Shiva/native operating system.	Sequential consistency, write-invalidate protocol, 4-Kbyte page swapping.	Uses data structure compaction, messages for semaphores and signal.wait, distributed memory as backing store.

Scalability issues of SVM architectures include determining the sizes of data structures for maintaining memory coherence and how to take advantage of the fast data transmission among distributed memories in order to implement large SVM address spaces. Data structure compaction and page swapping can simplify the design of a large SVM address space without using disks as backing stores. A number of alternative choices are given in Li (1992).

9.1.2   Prefetching Techniques

Prefetching techniques are studied below. These involve both hardware and software approaches. Some benchmark results for prefetching on the Stanford Dash system are presented to illustrate the benefits.

Prefetching Techniques   Prefetching uses knowledge about the expected misses in a program to move the corresponding data close to the processor before it is actually needed. Prefetching can be classified based on whether it is *binding* or *nonbinding*, and whether it is controlled by *hardware* or *software*.

With binding prefetching, the value of a later reference (e.g., a register load) is bound at the time when the prefetch completes. This places restrictions on when a binding prefetch can be issued, since the value will become stale if another processor modifies the same location during the interval between prefetch and reference. Binding prefetching may result in a significant loss in performance due to such limitations.

In contrast, nonbinding prefetching also brings the data close to the processor, but the data remains visible to the cache coherence protocol and is thus kept consistent until the processor actually reads the value.

Hardware-controlled prefetching includes schemes such as long cache lines and instruction lookahead. The effectiveness of long cache lines is limited by the reduced spatial locality in multiprocessor applications, while instruction lookahead is limited by branches and the finite lookahead buffer size.

With software-controlled prefetching, explicit prefetch instructions are issued. Software control allows the prefetching to be done selectively (thus reducing bandwidth requirements) and extends the possible interval between prefetch issue and actual reference, which is very important when latencies are large.

The disadvantages of software control include the extra instruction overhead required to generate the prefetches, as well as the need for sophisticated software intervention. In our study, we concentrate on *non-binding software-controlled prefetching*.

Benefits of Prefetching   The benefits of prefetching come from several sources. The most obvious benefit occurs when a prefetch is issued early enough in the code so that the line is already in the cache by the time it is referenced. However, prefetching can improve performance even when this is not possible (e.g., when the address of a data structure cannot be determined until immediately before it is referenced). If multiple prefetches are issued back to back to fetch the data structure, the latency of all but the first prefetched references can be hidden due to the pipelining of the memory accesses.

Prefetching offers another benefit in multiprocessors that use an ownership-based cache coherence protocol. If a cache block line is to be modified, prefetching it directly with ownership can significantly reduce the write latencies and the ensuing network traffic for obtaining ownership. Network traffic is reduced in read-modify-write instructions, since prefetching with ownership avoids first fetching a read-shared copy.

Benchmark Results   Stanford researchers (Gupta, Hennessy, Gharachorloo, Mowry, and Weber, 1991) have reported some benchmark results for evaluating various latency-hiding mechanisms. They include a particle-based three-dimensional simulator used in aeronautics (MP3D), an LU-decomposition program (LU), and a digital logic simulation program (PTHOR). The effect of prefetching is illustrated in Fig. 9.3 for running the MP3D code on a simulated Dash multiprocessor (Fig. 9.1).

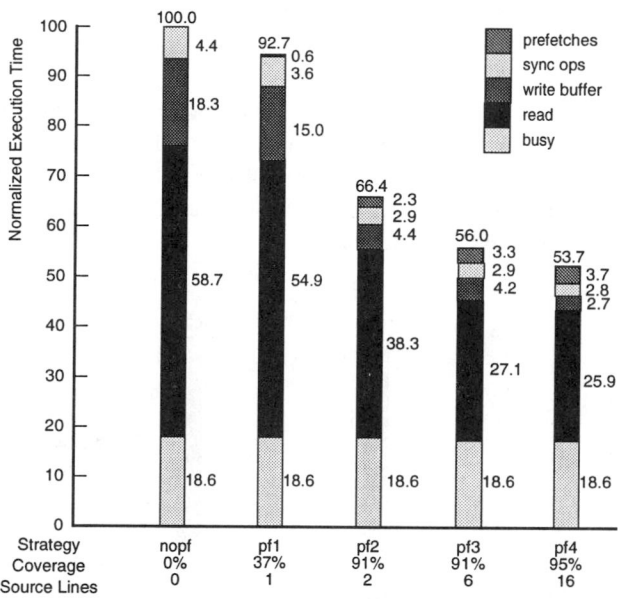

Figure 9.3 Effect of various prefetching strategies for running the MP3D benchmark on a simulated Dash multiprocessor. (Courtesy of Anoop Gupta et al., 1992)

The simulation runs involved 10,000 particles in a $64 \times 8 \times 8$ space array with five time steps. Five prefetching strategies were tested (*nopf, pf1, pf2, pf3,* and *pf4* in Fig. 9.3). These strategies range from no prefetching (*nopf*) to prefetching of the particle record in the same iteration or pipelined across increasing numbers of iterations (*pf1* through *pf4*). The bar diagrams in Fig. 9.3 show the execution times normalized with respect to the *nopf* strategy. Each bar shows a breakdown of the times required for prefetches, synchronization operations, using write buffers, reads, and busy in computing.

The end result is that prefetches are issued for up to 95% of the misses that occur in the case without prefetching (referred to as the *coverage factor* in Fig. 9.3). The prefetching causes significant time reduction in synchronization operations, using write buffers, and performing read operations. The best speedup achieved in Fig. 9.3 is 1.86, when the *pf4* prefetching strategy is compared with the *nopf* strategy. Still the prefetching benefits will be application-dependent. To introduce the prefetches in the MP3D code, only 16 lines of extra code were added to the source code.

9.1.3 Distributed Coherent Caches

While the coherence problem is easily solved for small bus-based multiprocessors through the use of snoopy cache coherence protocols, the problem is much more complicated for large-scale multiprocessors that use general interconnection networks. As a result, some existing large-scale multiprocessors do not provide caches (e.g., BBN Butterfly), others provide caches that must be kept coherent by software (e.g., IBM RP3), and still others provide full hardware support for coherent caches (e.g., Stanford Dash).

Dash Experience   We evaluate the benefits when both private and shared read-write data are cacheable, as allowed by the Dash hardware coherent caches, versus the case where only private data are cacheable. Figure 9.4 presents a breakdown of the normalized execution times with and without cacheing of shared data for each of the applications. Private data are cached in both caches.

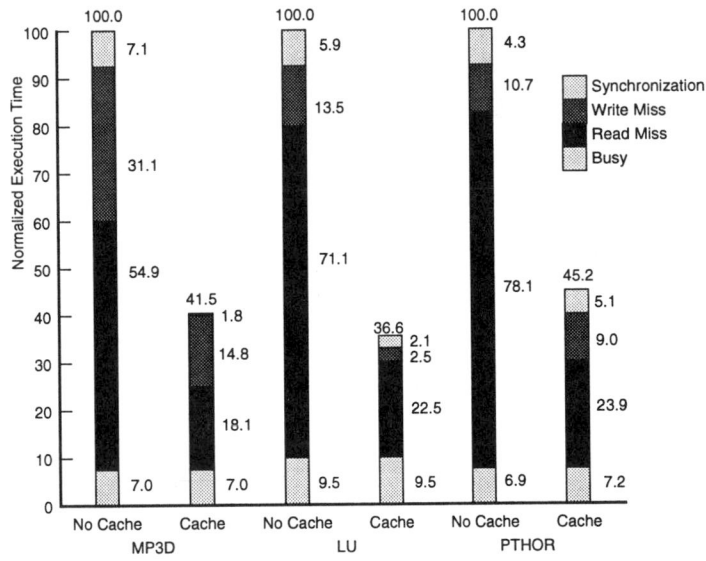

Figure 9.4 Effect of cacheing shared data in simulated Dash benchmark experiments. (Courtesy of Gupta et al., *Proc. Int. Symp. Comput. Archit.*, Toronto, Canada, May 1991)

The execution time of each application is normalized to the execution time of the case where shared data is not cached. The bottom section of each bar represents the busy time or useful cycles executed by the processor. The section above it represents the time that the processor is stalled waiting for reads. The section above that is the amount of time the processor is stalled waiting for writes to be completed. The top section, labeled "synchronization," accounts for the time the processor is stalled due to locks and barriers.

Benefits of Cacheing   As expected, the cacheing of shared read-write data provides substantial gains in performance, with benefits ranging from 2.2- to 2.7-fold improvement for the three Stanford benchmark programs. The largest benefit comes from a reduction in the number of cycles wasted due to read misses. The cycles wasted due to write misses are also reduced, although the magnitude of the benefits varies across the three programs due to different write-hit ratios.

The cache-hit ratios achieved by MP3D, LU, and PTHOR are 80, 66, and 77%, respectively, for shared-read references, and 75, 97, and 47% for shared-write references. It is interesting to note that these hit ratios are substantially lower than the usual uniprocessor hit ratios.

The low hit ratios arise from several factors: The data set size for engineering applications is large, parallelism decreases spatial locality in the application, and communication among processors results in invalidation misses. Still, hardware cache coherence is an effective technique for substantially increasing the performance with no assistance from the compiler or programmer.

9.1.4   Scalable Coherence Interface

A special coherence interconnect structure is needed to extend from conventional bused backplanes to a fully duplex, point-to-point interface specification. The *scalable coherence interface* (SCI) is specified in IEEE Standard 1596-1992. The current SCI supports unidirectional point-to-point connections.

Once fully developed, up to 64K processors, memory modules, or I/O nodes can effectively interface with a shared SCI interconnect. The cache coherence protocols used in the SCI are directory-based. A sharing list is used to chain the distributed directories together for reference purposes.

SCI Interconnect Models   The SCI defines the interface between nodes and the external interconnect. The initial goal was set to use 16-bit links with a bandwidth of 1 Gbyte/s per link. As a result, backplane buses have been replaced by unidirectional point-to-point links. A typical SCI configuration is shown in Fig. 9.5a. Each SCI node can be a processor with attached memory and I/O devices. The SCI interconnect can assume a ring structure or a crossbar switch as depicted in Figs. 9.5b and 9.5c, respectively, among other configurations.

Each node has an input link and an output link which are connected from or to the SCI ring or crossbar. The bandwidth of SCI links depends on the physical standard chosen to implement the links and interfaces.

In such an environment, the concept of broadcast bus-based transactions has been abandoned. Coherence protocols are based on point-to-point transactions initiated by a requester and completed by a responder. A ring interconnect provides the simplest feedback connections among the nodes.

The converters in Fig. 9.5a are used to bridge the SCI ring to the VME or Futurebus as shown. A mesh of rings can also be considered using some bridging modules. The bandwidth, arbitration, and addressing mechanisms of a SCI ring are expected to significantly outperform backplane buses. By eliminating the snoopy cache controllers,

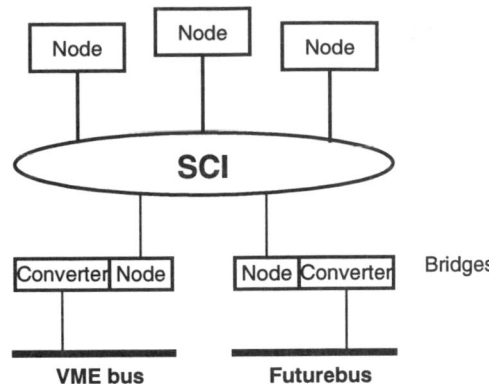

(a) Typical SCI configuration with bridges to other buses

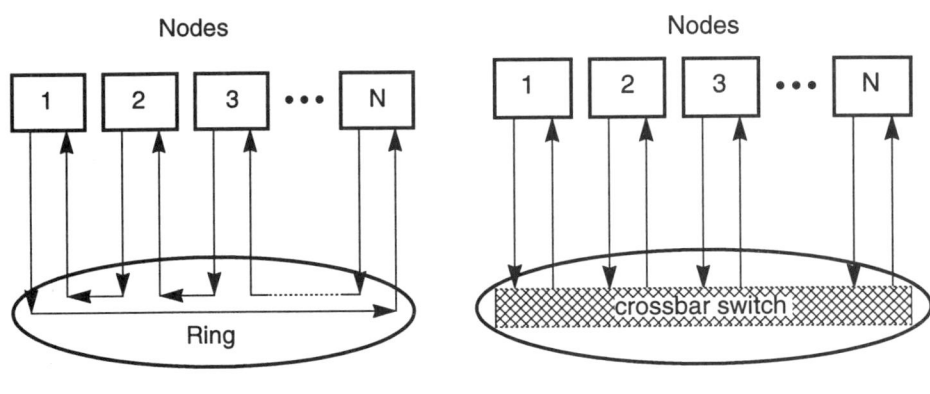

(b) A ring for point-to-point transactions   (c) A crossbar multiprocessor

Figure 9.5 SCI interconnection configurations. (Reprinted with permission from the IEEE Standard 1596-1992, copyright © 1992 by IEEE, Inc.)

the SCI should also be cheaper. But the main advantage lies in its scalability.

Although the SCI is scalable, the amount of memory used in the cache directories also scales up well. The performance of the SCI protocol does not scale, since when the sharing list is long, invalidations will take too much time. This problem is still unsolved by the SCI designers.

Sharing-List Structures   Sharing lists are used in the SCI to build chained directories for cache coherence use. The length of the sharing lists is effectively unbounded. Sharing lists are dynamically created, pruned, and destroyed. Each coherently cached block is entered onto a list of processors sharing the block.

Processors have the option of bypassing the coherence protocols for locally cached data. Cache blocks of 64 bytes are assumed. By distributing the directories among

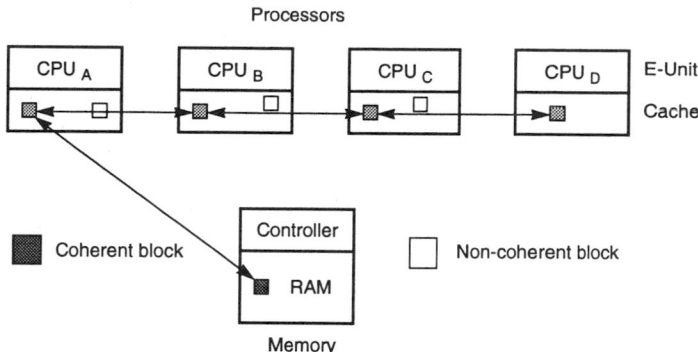

Figure 9.6 SCI cache coherence protocol with distributed directories. (Courtesy of D. V. James et al., *IEEE Computer*, 1990)

the sharing processors, the SCI avoids scaling limitations imposed by using a central directory. Communications among sharing processors are supported by heavily shared memory controllers, as shown in Fig. 9.6.

Other blocks may be locally cached and are not visible to the coherence protocols. For every block address, the memory and cache entries have additional tag bits which are used to identify the first processor (head) in the sharing list and to link the previous and following nodes.

Doubly linked lists are maintained between processors in the sharing list, with forward and backward pointers as shown by the double arrows in each link. Non-coherent copies may also be made coherent by page-level control. However, such higher-level software coherence protocols are beyond the scope of the published SCI standard. The backward pointers support independent deletions of entries in the middle of the linked list.

Sharing-List Creation The states of the sharing list are defined by the state of the memory and the states of list entries. Normally, the shared memory is either in a home (uncached) or a cached (sharing-list) state. The sharing-list entries specify the location of the entry in a multiple-entry sharing list, identify the only entry in the list, or specify the entry's cache properties, such as clean, dirty, valid, or stale.

The head processor is always responsible for list management. The stable and legal combinations of the memory and entry states can specify uncached data, clean or dirty data at various locations, and cached writable or stale data.

The memory is initially in the home state (uncached), and all cache copies are invalid. Sharing-list creation begins at the cache where an entry is changed from an invalid to a pending state. When a read-cache transaction is directed from a processor to the memory controller, the memory state is changed from uncached to cached and the requested data is returned.

The requester's cache entry state is then changed from a pending state to an only-clean state. Sharing-list creation is illustrated in Fig. 9.7a. Multiple requests can be simultaneously generated, but they are processed sequentially by the memory controller.

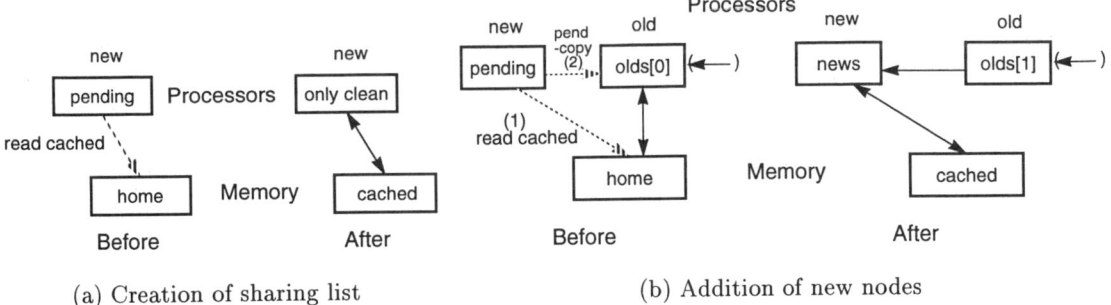

(a) Creation of sharing list   (b) Addition of new nodes

Figure 9.7 Sharing-list creation and update examples. (Courtesy of D. V. James et al., *IEEE Computer*, 1990)

Sharing-List Updates   For subsequent memory access, the memory state is cached, and the cache head of the sharing list has possibly dirty data. As illustrated in Fig. 9.7b, a new requester (cache A) first directs its read-cache transaction to memory but receives a pointer to cache B instead of the requested data.

A second cache-to-cache transaction, called *prepend*, is directed from cache A to cache B. Cache B then sets its backward pointer to point to cache A and returns the requested data. The dashed lines correspond to transactions between a processor and memory or another processor. The solid lines are sharing-list pointers.

After the transaction, the inserted cache A becomes the new head, and the old head, cache B, is in the middle as shown by the new sharing list on the right in Fig. 9.7b.

Any sharing-list entry may delete itself from the list. Details of entry deletions are left as an exercise for the reader. Simultaneous deletions never generate deadlocks or starvation. However, the addition of new sharing-list entries must be performed in first-in-first-out order in order to avoid potential deadlocking dependences.

The head of the sharing list has the authority to purge other entries from the list to obtain an exclusive entry. Others may reenter as a new list head. Purges are performed sequentially. The chained-directory coherence protocols are fault-tolerant in that dirty data is never lost when transactions are discarded.

Implementation Issues   The SCI was developed to support multiprocessor systems with thousands of processors by providing a coherent distributed-cache image of distributed shared memory and bridges that interface with existing or future buses. It can support various multiprocessor topologies using Omega or crossbar networks.

By using combining switches, combining of requests can be achieved to overcome memory hot spots. For effective applications of the SCI scheme, one must consider some implementation issues. Differential emitter coupled logic (ECL) signaling works well at SCI clock rates. The proposed SCI implementation uses a 16-bit data path at 2 ns per word (250-MHz clock rate).

The interface is synchronously clocked. Several models of clock distribution are supported. SCI fiber optic implementation may be useful for moving data out of detectors. The technological trends are to build the Futurebus based on VME/Fastbus experiences and to build the SCI based on Futurebus+ to make scalability a reality in the world of multiprocessors. With distributed shared-memory and distributed cache coherence protocols, the boundary between multiprocessors and multicomputers will become further blurred in the future.

9.1.5 Relaxed Memory Consistency

We have studied *weak consistency* (WC) (Sindhu et al., 1992) and *sequential consistency* (SC) in Section 5.4. Two additional memory models are introduced below for building scalable multiprocessors with distributed shared memory.

Processor Consistency   Goodman (1989) introduced the *processor consistency* (PC) model in which writes issued by each individual processor are always in program order. However, the order of writes from two different processors can be out of program order. In other words, consistency in writes is observed in each processor, but the order of reads from each processor is not restricted as long as they do not involve other processors.

The PC model relaxes from the SC model by removing some restrictions on writes from different processors. This opens up more opportunities for write buffering and pipelining. Two conditions related to other processors are required for ensuring processor consistency:

(1) Before a *read* is allowed to perform with respect to any other processor, all previous *read* accesses must be performed.

(2) Before a write is allowed to perform with respect to any other processor, all previous *read* or *write* accesses must be performed.

These conditions allow *reads* following a *write* to bypass the *write*. To avoid deadlock, the implementation should guarantee that a *write* that appears previously in program order will eventually be performed.

Release Consistency   One of the most relaxed memory models is the *release consistency* (RC) model introduced by Gharachorloo et al. (1990). Release consistency requires that synchronization accesses in the program be identified and classified as either *acquires* (e.g., locks) or *releases* (e.g., unlocks). An acquire is a read operation (which can be part of a read-modify-write) that gains permission to access a set of data, while a release is a write operation that gives away such permission. This information is used to provide flexibility in buffering and pipelining of accesses between synchronization points.

The main advantage of the relaxed models is the potential for increased performance by hiding as much write latency as possible. The main disadvantage is increased hardware complexity and a more complex programming model. Three conditions ensure release consistency:

(1) Before an ordinary *read* or *write* access is allowed to perform with respect to any other processor, all previous *acquire* accesses must be performed.

(2) Before a *release* access is allowed to perform with respect to any other processor, all previous ordinary *read* and *store* accesses must be performed.

(3) *Special accesses* are processor-consistent with one another. The ordering restrictions imposed by weak consistency are not present in release consistency. Instead, release consistency requires processor consistency and not sequential consistency.

Release consistency can be satisfied by (i) stalling the processor on an acquire access until it completes, and (ii) delaying the completion of release access until all previous memory accesses complete. Intuitive definitions of the four memory consistency models, the SC, WC, PC, and RC, are summarized in Fig. 9.8.

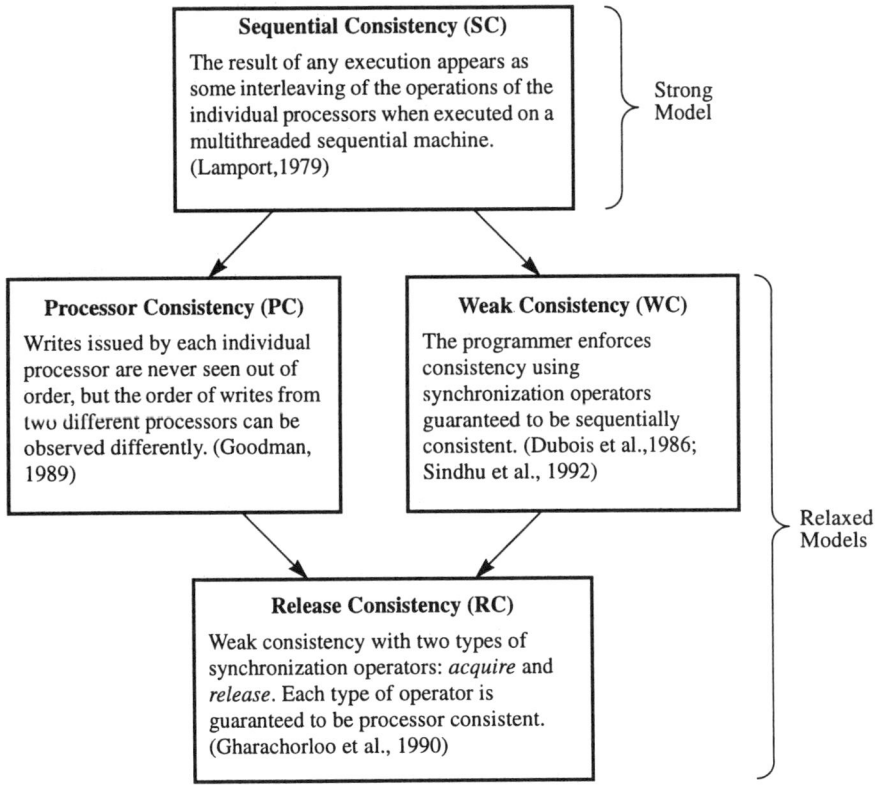

Figure 9.8 Intuitive definitions of four memory consistency models. The arrows point from strong to relaxed consistencies. (Courtesy of Nitzberg and Lo, *IEEE Computer*, August 1991)

The cost of implementing RC over that for SC arises from the extra hardware cost of providing a lockup-free cache and keeping track of multiple outstanding requests. Although this cost is not negligible, the same hardware features are also required to support prefetching and multiple contexts.

Effect of Release Consistency Figure 9.9 presents the breakdown of execution times under SC and RC for the three applications. The execution times are normalized to those shown in Fig. 9.3 with shared data cached. As can be seen from the results, RC removes all idle time due to write-miss latency.

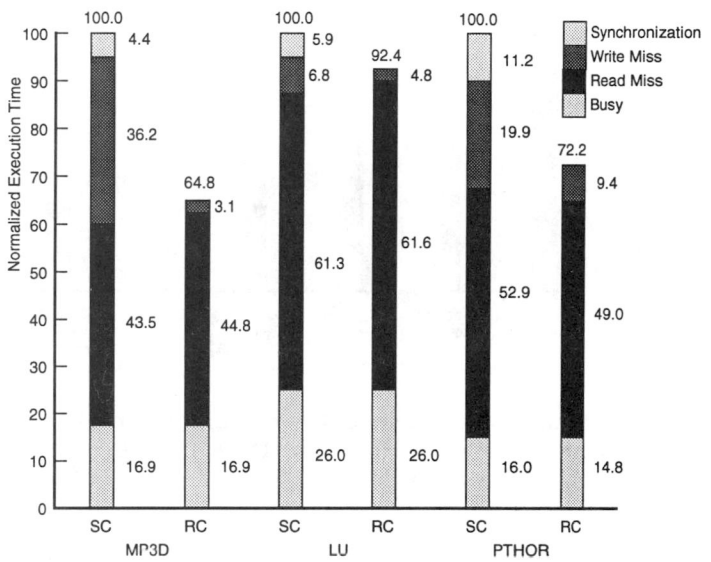

Figure 9.9 Effect of relaxing the shared-memory model from sequential consistency (SC) to release consistency (RC). (Courtesy of Gupta et al., *Proc. Int. Symp. Comput. Archit.*, Toronto, Canada, May 1991)

The gains are large in MP3D and PTHOR since the write-miss time constitutes a large portion of the execution time under SC (35 and 20%, respectively), while the gain is small in LU due to the relatively small write-miss time under SC (7%).

Effect of Combining Mechanisms The effect of combining various latency-hiding mechanisms is illustrated by Fig. 9.10 based on the MP3D benchmark results obtained at Stanford University. The idea of using *multiple-context* processors will be described in Section 9.2. However, the effect of integrating MC with other latency-hiding mechanisms is presented below.

The busy parts of the execution times in Fig. 9.10 are equal in all combinations. This is the CPU busy time for executing the MP3D program. The idle part in the

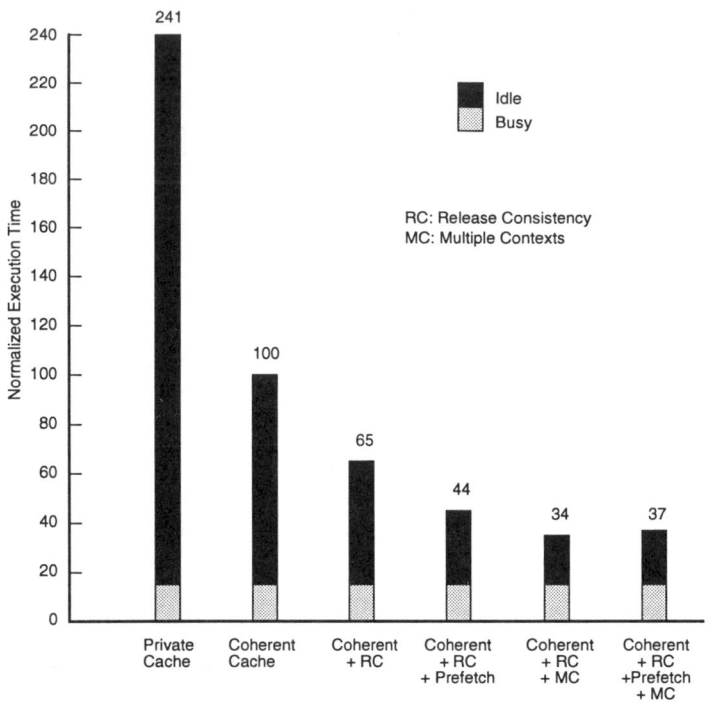

Figure 9.10 Effect of combining various latency-hiding mechanisms from the MP3D benchmark on a simulated Dash multiprocessor. (Courtesy of Gupta, 1992)

bar diagram corresponds to memory latency and includes all cache-miss penalties. All the times are normalized with respect to the execution time (100 units) required in a *cache-coherent* system. The leftmost time bar (with 241 units) corresponds to the worst case of using a private cache exclusively without shared reads or writes. Long overhead is experienced in this case due to excessive cache misses. The use of a cache-coherent system shows a 2.41-fold improvement over the private case. All the remaining cases are assumed to use hardware coherent caches.

The use of *release consistency* shows a 35% further improvement over the coherent system. The adding of prefetching reduces the time further to 44 units. The best case is the combination of using coherent caches, RC, and *multiple contexts* (MC). The rightmost time bar is obtained from applying all four mechanisms. The combined results show an overall speedup of 4 to 7 over the case of using private caches.

The above and other uncited benchmark results reported at Stanford suggest that a coherent cache and relaxed consistency uniformly improve performance. The improvements due to prefetching and multiple contexts are sizable but are much more application-dependent. Combinations of the various latency-hiding mechanisms generally attain a better performance than each one on its own.

9.2 Principles of Multithreading

This section considers multidimensional and/or multithreaded processors and system architectures. Only control-flow approaches are described here. Fine-grain machines are studied in Section 9.3, von Neumann multithreading in Section 9.4, and dataflow multithreading in Section 9.5.

9.2.1 Multithreading Issues and Solutions

Multithreading demands that the processor be designed to handle multiple contexts simultaneously on a context-switching basis. We first specify the typical architecture environment using multiple-context processors. Next we present a multithreaded computation model. Then we look further into the latency and synchronization problems and discuss their solutions in this environment.

Architecture Environment A multithreaded MPP system is modeled by a network of processor (P) and memory (M) nodes as depicted in Fig. 9.11a. The distributed memories form a global address space. Four machine parameters are defined below to analyze the performance of this network:

(1) *The latency* (L): This is the communication latency on a remote memory access. The value of L includes the network delays, cache-miss penalty, and delays caused by contentions in split transactions.

(2) *The number of threads* (N): This is the number of threads that can be interleaved in each processor. A *thread* is represented by a *context* consisting of a program counter, a register set, and the required context status words.

(3) *The context-switching overhead* (C): This refers to the cycles lost in performing context switching in a processor. This time depends on the switch mechanism and the amount of processor states devoted to maintaining active threads.

(4) *The interval between switches* (R): This refers to the cycles between switches triggered by remote reference. The inverse $p = 1/R$ is called the *rate of requests* for remote accesses. This reflects a combination of program behavior and memory system design.

In order to increase efficiency, one approach is to reduce the rate of requests by using distributed coherent caches. Another is to eliminate processor waiting through multithreading. The basic concept of multithreading is described below.

Multithreaded Computations Bell (1992) has described the structure of the multithreaded parallel computations model shown in Fig. 9.11b. The computation starts with a sequential thread (1), followed by supervisory scheduling (2) where the processors begin threads of computation (3), by intercomputer messages that update variables among the nodes when the computer has a distributed memory (4), and finally by synchronization prior to beginning the next unit of parallel work (5).

The communication overhead period (3) inherent in distributed memory structures is usually distributed throughout the computation and is possibly completely over-

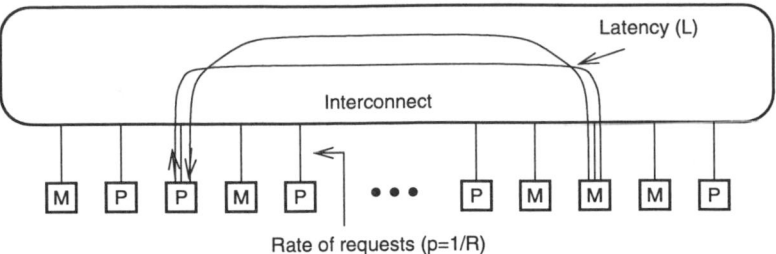

(a) The architecture environment. (Courtesy of Rafael Saavedra, 1992)

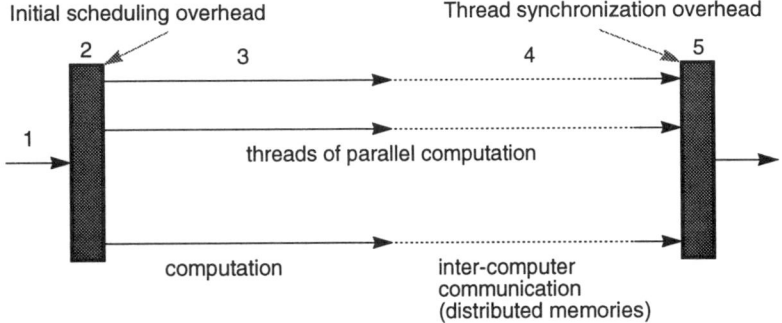

(b) Multithreaded computation model. (Courtesy of Gordon Bell, *Commun. ACM*, August 1992)

Figure 9.11 Multithreaded architecture and its computation model for a massively parallel processing system.

lapped. Message-passing overhead (send and receive calls) in multicomputers can be reduced by specialized hardware operating in parallel with computation.

Communication bandwidth limits granularity, since a certain amount of data has to be transferred with other nodes in order to complete a computational grain. Message-passing calls (4) and synchronization (5) are nonproductive. Fast mechanisms to reduce or to hide these delays are very much needed. Multithreading is not capable of speedup in the execution of single threads, while weak ordering or relaxed consistency models are capable of doing this.

Problems of Asynchrony   Massively parallel processors operate asynchronously in a network environment. The asynchrony triggers two fundamental latency problems: *remote loads* and *synchronizing loads*, as observed by Nikhil (1992). These two problems are explained by the following example:

Example 9.1   Latency problems for remote loads or synchronizing loads (Rishiyun Nikhil, 1992).

The remote load situation is illustrated in Fig. 9.12a. Variables A and B are located on nodes N2 and N3, respectively. They need to be brought to node N1 to compute the difference $A - B$ in variable C. The basic computation demands the execution of two remote loads (rload) and then the subtraction.

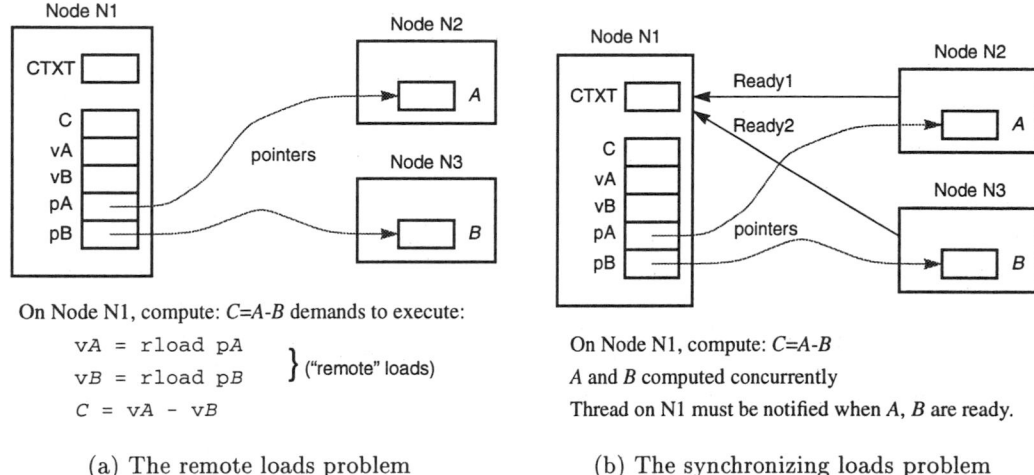

On Node N1, compute: $C=A-B$ demands to execute:

$$vA = rload\ pA$$
$$vB = rload\ pB \quad \Big\} \text{ ("remote" loads)}$$
$$C = vA - vB$$

(a) The remote loads problem

On Node N1, compute: $C=A-B$

A and B computed concurrently

Thread on N1 must be notified when A, B are ready.

(b) The synchronizing loads problem

Figure 9.12 Two common problems caused by asynchrony and communication latency in massively parallel processors. (Courtesy of R.S. Nikhil, Digital Equipment Corporation, 1992)

Let pA and pB be the pointers to A and B, respectively. The two rloads can be issued from the same thread or from two different threads. The *context* of the computation on N1 is represented by the variable CTXT. It can be a stack pointer, a frame pointer, a current-object pointer, a process identifier, etc. In general, variable names like vA, vB, and C are interpreted relative to CTXT.

In Fig. 9.12b, the idling due to synchronizing loads is illustrated. In this case, A and B are computed by concurrent processes, and we are not sure exactly when they will be ready for node N1 to read. The ready signals (ready1 and ready2) may reach node N1 asynchronously. This is a typical situation in the producer-consumer problem. Busy-waiting may result. ∎

The key issue involved in remote loads is how to avoid idling in node N1 during the load operations. The latency caused by remote loads is an architectural property. The latency caused by synchronizing loads also depends on scheduling and the time it takes to compute A and B, which may be much longer than the transit latency. The synchronization latency is often unpredictable, while the remote-load latencies are often predictable.

Multithreading Solutions   This solution to asynchrony problems is to multiplex among many threads: When one thread issues a remote-load request, the processor begins work on another thread, and so on (Fig. 9.13a). Clearly, the cost of thread switching should be much smaller than that of the latency of the remote load, or else the processor might as well wait for the remote load's response.

As the internode latency increases, more threads are needed to hide it effectively. Another concern is to make sure that messages carry continuations. Suppose, after issuing a remote load from thread T_1 (Fig. 9.13a), we switch to thread T_2, which also issues a remote load. The responses may not return in the same order. This may be caused by requests traveling different distances, through varying degrees of congestion, to destination nodes whose loads differ greatly, etc.

One way to cope with the problem is to associate each remote load and response with an identifier for the appropriate thread, so that it can be reenabled on the arrival of a response. These thread identifiers are referred to as *continuations* on messages. A large *continuation name space* should be provided to name an adequate number of threads waiting for remote responses.

The size of the hardware-supported continuation in a name space varies greatly in existing designs: from 1 in the Dash, 4 in the Alewife, 64 in the HEP, and 1024 in the Tera (Section 9.4) to the local memory address space in the Monsoon, Hybrid Dataflow/von Neumann, MDP (Section 9.3), and *T (Section 9.5). Of course, if the hardware-supported name space is small, one can always virtualize it by multiplexing in software, but this has an associated overhead.

Distributed Cacheing   The concept of distributed cacheing is shown in Fig. 9.13b. Every memory location has an owner node. For example, N1 owns B and N2 owns A. The directories are used to contain import-export lists and state whether the data is *shared* (for reads, many caches may hold copies) or *exclusive* (for writes, one cache holds the current value).

The directories multiplex among a small number of contexts to cover the cache loading effects. The MIT Alewife, KSR-1, and Stanford Dash have implemented directory-based coherence protocols. It should be noted that distributed cacheing offers a solution for the remote-loads problem, but not for the synchronizing-loads problem. Multithreading offers a solution for remote loads and possibly for synchronizing loads. However, the two approaches can be combined to solve both types of remote-access problems.

9.2.2   Multiple-Context Processors

Multithreaded systems are constructed with *multiple-context* (or *multithreaded*) processors. In this section, we study an abstract model based on the work of Saavedra et al. (1990). We then present an example of this type of processor. We discuss the processor efficiency issue as a function of memory latency (L), the number of contexts (N), and context-switching overhead (C).

The Enhanced Processor Model   A conventional single-thread processor will *wait* during a remote reference, so we may say it is idle for a period of time L. A multithreaded

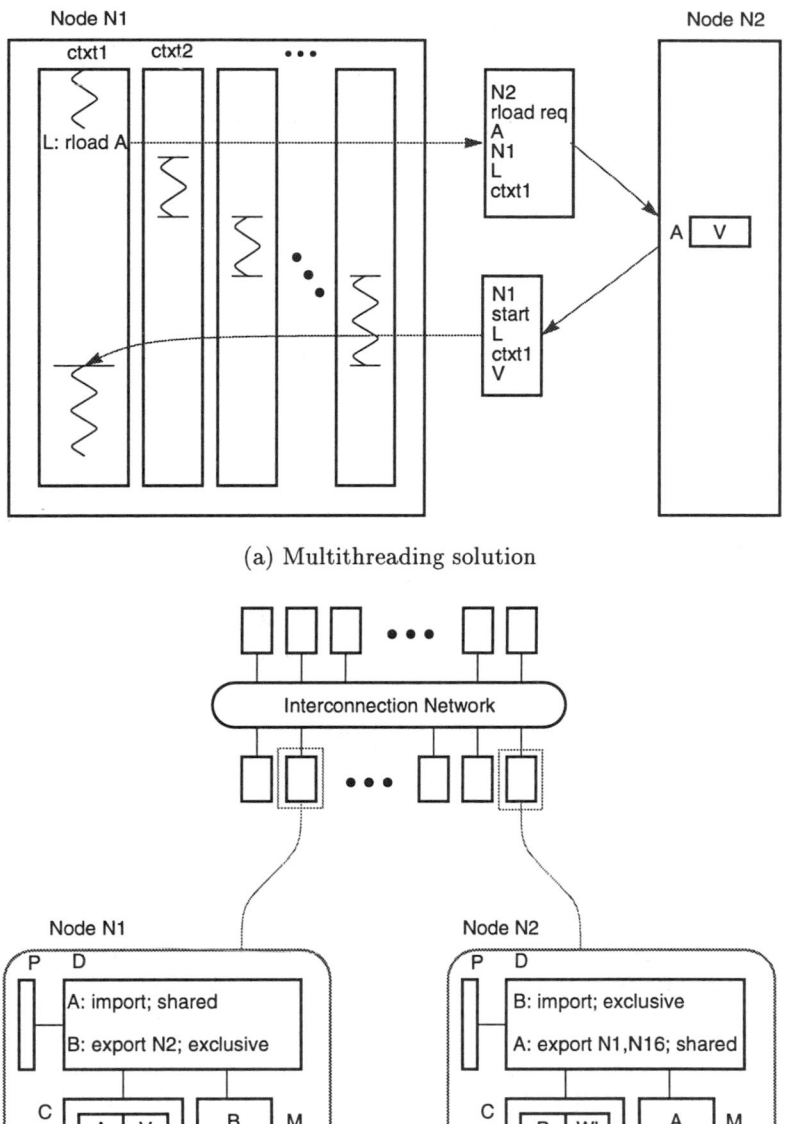

(a) Multithreading solution

(b) Distributed cacheing

Figure 9.13 Two solutions for overcoming the asynchrony problems. (Courtesy of R.S. Nikhil, Digital Equipment Corporation, 1992)

processor, as modeled in Fig. 9.14a, will suspend the current context and switch to another, so after some fixed number of cycles it will again be busy doing useful work, even though the remote reference is outstanding. Only if all the contexts are suspended (blocked) will the processor be idle.

Clearly, the objective is to maximize the fraction of time that the processor is busy, so we will use the *efficiency* of the processor as our performance index, given by

$$Efficiency = \frac{busy}{busy + switching + idle} \qquad (9.1)$$

where *busy*, *switching*, and *idle* represent the amount of time, measured over some large interval, that the processor is in the corresponding state. The basic idea behind a multithreaded machine is to interleave the execution of several contexts in order to dramatically reduce the value of *idle*, but without overly increasing the magnitude of *switching*.

The state of a processor is determined by the disposition of the various contexts on the processor. During its lifetime, a context cycles through the following states: *ready*, *running*, *leaving*, and *blocked*. There can be at most one context running or leaving. A processor is *busy* if there is a context in the running state; it is *switching* while making the transition from one context to another, i.e., when a context is leaving. Otherwise, all contexts are blocked and we say the processor is *idle*.

A running context keeps the processor busy until it issues an operation that requires a context switch. The context then spends C cycles in the *leaving* state, then goes into the *blocked* state for L cycles, and finally reenters the *ready* state. Eventually the processor will choose it and the cycle will start again.

The abstract model shown in Fig. 9.14a assumes one thread per context, and each context is represented by its own program counter (PC), register set, and process status word (PSW). An example multithreaded processor in which three thread slots ($N = 3$) are provided is shown in Fig. 9.14b.

Example 9.2 A multithreaded processor with three thread slots (Hiroaki Hirata et al., 1992).

As shown in Fig. 9.14b, the processor is provided with several instruction queue unit and decode unit pairs, called *thread slots*. Each thread slot, associated with a program counter, makes up a *logical processor*, while an instruction fetch unit and all functional units are physically shared among logical processors.

An instruction queue unit has a buffer which saves some instructions succeeding the instruction indicated by the program counter. The buffer size needs to be at least $B = N \times C$ words, where N is the number of thread slots and C is the number of cycles required to access the instruction cache.

An instruction fetch unit fetches at most B instructions for one thread every C cycles from the instruction cache and attempts to fill the buffers in the instruction queue unit. This fetching operation is done in an interleaved fashion for multiple threads. So, on the average, the buffer in one instruction queue unit is filled once in B cycles.

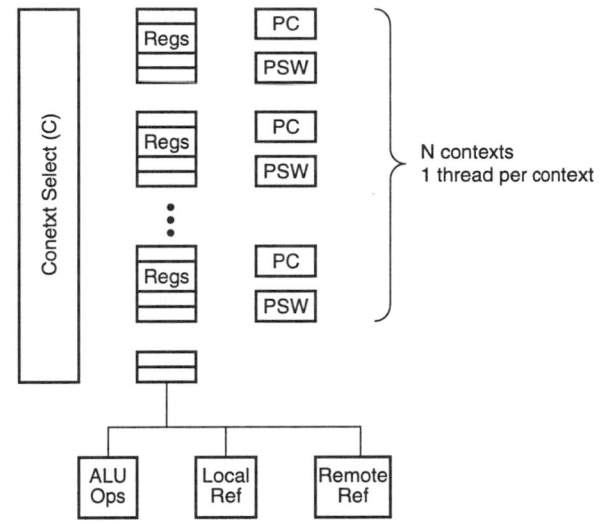

(a) Multithreaded model. (Courtesy of Rafael Saavedra, 1992)

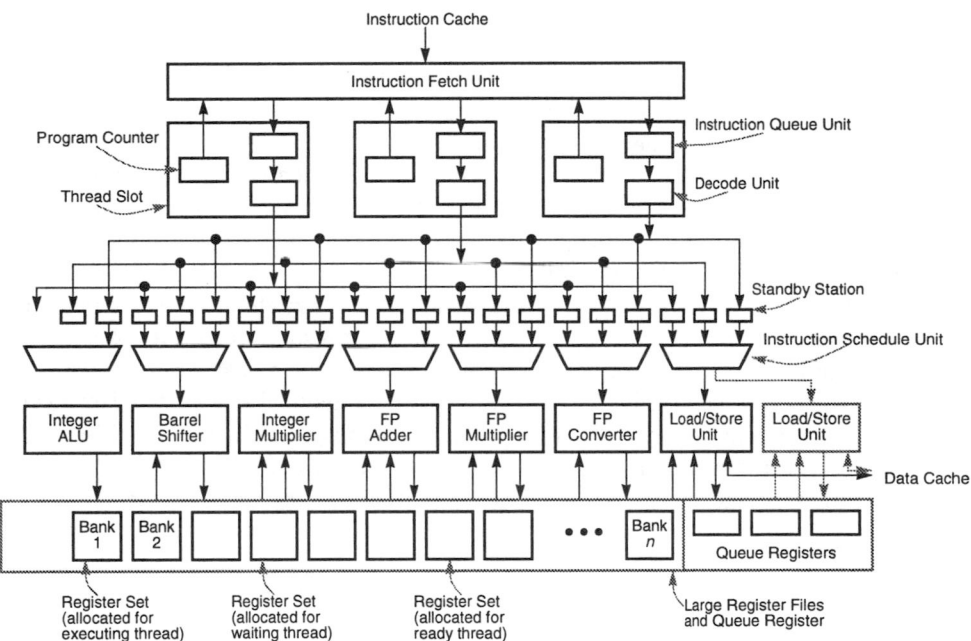

(b) A three-thread processor example. (Courtesy of H. Hirata et al., *Proc. 19th Int. Symp. Comput. Archit.*, Australia, May 1992)

Figure 9.14 Multiple-context processor model and an example design.

When one of the threads encounters a branch instruction, however, that thread can preempt the prefetching operation. The instruction cache and fetch unit might become a bottleneck for a processor with many thread slots. In such cases, another cache and fetch unit would be needed.

Simulation results show that by executing two and four threads in parallel on a nine-functional-unit processor, a 2.02- and a 3.72-fold speedup, respectively, can be achieved over a conventional single-thread processor.

■

Context-Switching Policies   Different multithreaded architectures are distinguished by the context-switching policies adopted. Specified below are four switching policies:

(1) *Switch on cache miss* — This policy corresponds to the case where a context is preempted when it causes a cache miss. In this case, R is taken to be the average interval between misses (in cycles), and L the time required to satisfy the miss. Here, the processor switches contexts only when it is certain that the current one will be delayed for a significant number of cycles.

(2) *Switch on every load* — This policy allows switching on every load, independent of whether it will cause a miss or not. In this case, R represents the average interval between loads. A general multithreading model assumes that a context is blocked for L cycles after every switch; but in the case of a switch-on-load processor, this happens only if the load causes a cache miss.

The general model can be employed if it is postulated that there are two sources of latency (L_1 and L_2), each having a particular probability (p_1 and p_2) of occurring on every switch. If L_1 represents the latency on a cache miss, then p_1 corresponds to what is normally referred to as the miss ratio. L_2 is a zero-cycle memory latency with probability p_2.

(3) *Switch on every instruction* — This policy allows switching on every instruction, independent of whether it is a load or not. In other words, it interleaves the instructions from different threads on a cycle-by-cycle basis. Successive instructions become independent, which will benefit pipelined execution. However, the cache miss may increase due to breaking of locality. It has been verified by some trace-driven experiments at Stanford that cycle-by-cycle interleaving of contexts provides a performance advantage over switching at a cache miss in that the context interleaving could hide pipeline dependences and reduce the context switch cost.

(4) *Switch on block of instruction* — Blocks of instructions from different threads are interleaved. This will improve the cache-hit ratio due to locality. It will also benefit single-context performance.

Processor Efficiencies   A single-thread processor executes a context until a remote reference is issued (R cycles) and then is idle until the reference completes (L cycles). There is no context switch and obviously no switch overhead. We can model this behavior as an alternating renewal process having a cycle of $R + L$. In terms of Eq. 9.1, R and L correspond to the amount of time during a cycle that the processor is *busy* and *idle*, respectively. Thus the efficiency of a single-threaded machine is given by

$$E_1 = \frac{R}{R + L} = \frac{1}{1 + L/R} \tag{9.2}$$

This shows clearly the performance degradation of such a processor in a parallel system with a large memory latency.

With multiple contexts, memory latency can be hidden by switching to a new context, but we assume that the switch takes C cycles of overhead. Assuming the run length between switches is constant with a sufficient number of contexts, there is always a context ready to execute when a switch occurs, so the processor is never idle. The processor efficiency is analyzed below under two different conditions as illustrated in Fig. 9.15.

(1) *Saturation region* — In this saturated region, the processor operates with maximum utilization. The cycle of the renewal process in this case is $R + C$, and the efficiency is simply

$$E_{\text{sat}} = \frac{R}{R + C} = \frac{1}{1 + C/R} \tag{9.3}$$

Observe that the efficiency in saturation is independent of the latency and also does not change with a further increase in the number of contexts.

Saturation is achieved when the time the processor spends servicing the other threads exceeds the time required to process a request, i.e., when $(N-1)(R+C) >$ L. This gives the saturation point, under constant run length, as

$$N_d = \frac{L}{R + C} + 1 \tag{9.4}$$

(2) *Linear region* — When the number of contexts is below the saturation point, there may be no ready contexts after a context switch, so the processor will experience idle cycles. The time required to switch to a ready context, execute it until a remote reference is issued, and process the reference is equal to $R + C + L$. Assuming N is below the saturation point, during this time all the other contexts have a turn in the processor. Thus, the efficiency is given by

$$E_{\text{lin}} = \frac{NR}{R + C + L} \tag{9.5}$$

Observe that the efficiency increases linearly with the number of contexts until the saturation point is reached and beyond that remains constant. The equation for E_{sat} gives the fundamental limit on the efficiency of a multithreaded processor and underlines the importance of the ratio C/R. Unless the context switch is extremely cheap, the remote reference rate must be kept low.

Figures 9.15a and 9.15b show snapshots of context switching in the saturation and linear regions, respectively. The processor efficiency is plotted as a function of the number of contexts in Fig. 9.15c.

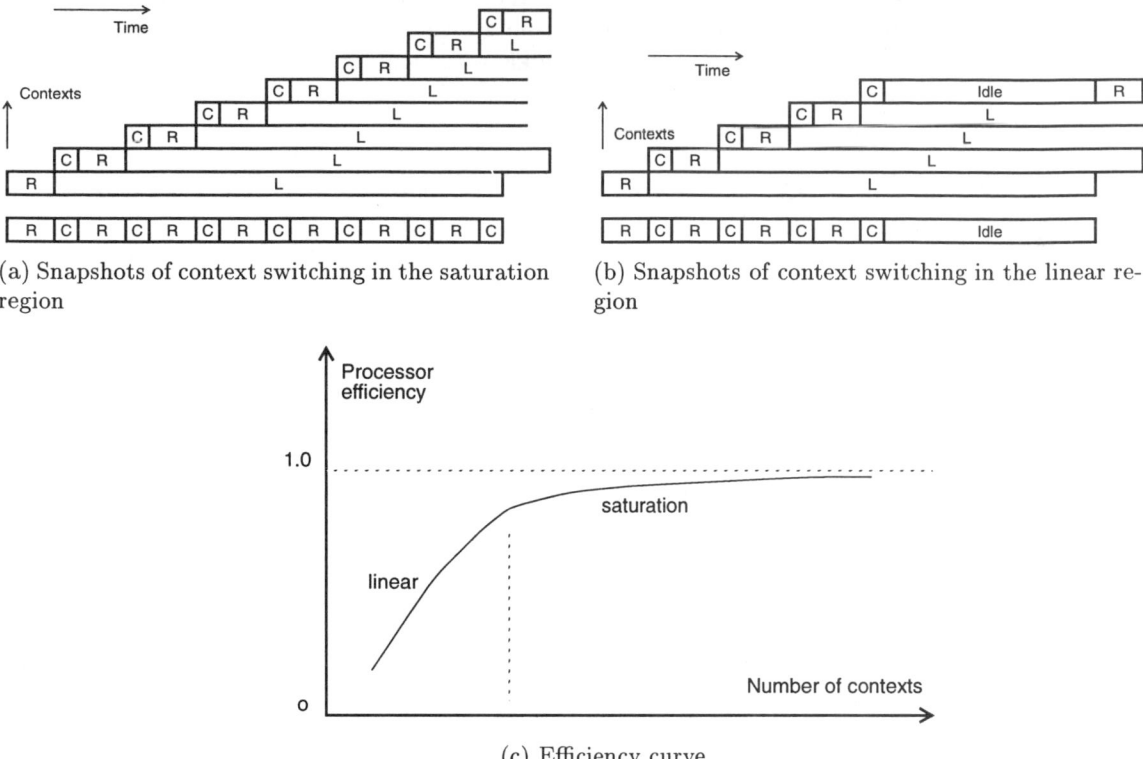

(a) Snapshots of context switching in the saturation region

(b) Snapshots of context switching in the linear region

(c) Efficiency curve

Figure 9.15 Context switching and processor efficiency as a function of the number of contexts. (Courtesy of Rafael Saavedra, 1992)

In Fig. 9.16, the processor efficiency is plotted as a function of the memory latency L with an average run length $R = 16$ cycles. The $C = 0$ curve corresponds to zero switching overhead. With $C = 16$ cycles, about 50% efficiency can be achieved. These results are based on a Markov model of multithreaded architecture by Saavedra (1992). It should be noted that multithreading increases both processor efficiency and network traffic. Tradeoffs do exist between these two opposing goals, and this has been discussed in a recent paper by Agarwal (1992).

9.2.3 Multidimensional Architectures

In order to enhance the scalability of multiprocessor systems, many research groups have explored economical and multidimensional architectures that support fast communication, coherence extension, distributed shared memory, and modular packaging.

The architecture of massively parallel processors has evolved from one-dimensional *rings* to two-dimensional and three-dimensional meshes or tori as illustrated in Fig. 9.17. The Maryland Zmob experimented on a *slotted token ring* for building a multiproces-

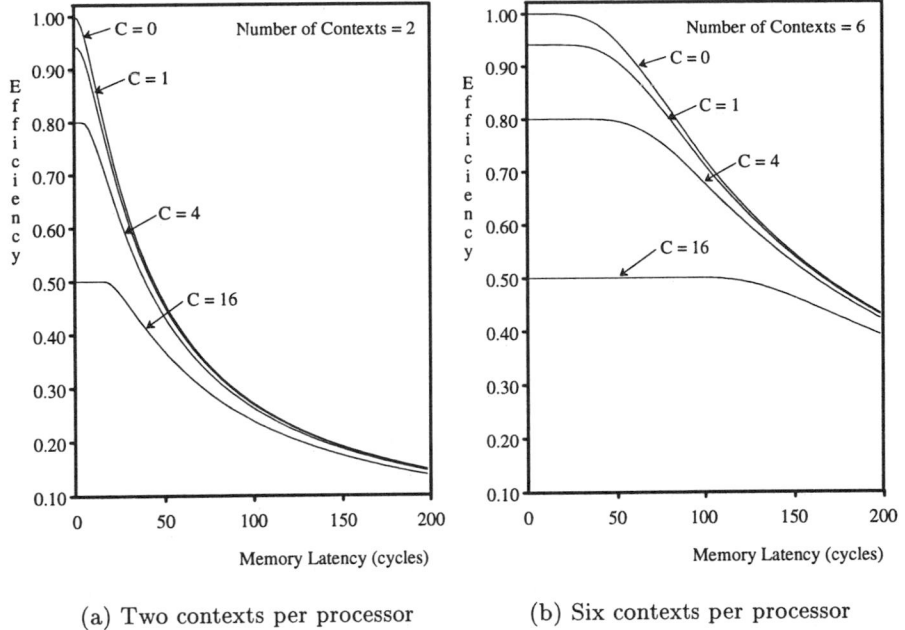

(a) Two contexts per processor   (b) Six contexts per processor

Figure 9.16 Processor efficiency of a multithreaded architecture. (Courtesy of R. Saavedra, D. E. Culler, and T. von Eicken, 1992)

sor. Both the CDC Cyberplus and KSR-1 have used hierarchical (two-level) ring architectures. The ring is the simplest architecture to implement from the viewpoint of backplane packaging.

Two-dimensional meshes have been adopted in the Stanford Dash, the MIT Alewife, the Wisconsin Multicube, the Intel Paragon, and the Caltech Mosaic C. A three-dimensional mesh/torus is being implemented in the MIT J-Machine, the Tera computer, and in the Cray/MPP architecture, called T3D. The USC *orthogonal multiprocessor* (OMP) can be extended to higher dimensions. In the foreseeable future, three-dimensional mesh architectures may become popular in building MPPs. It becomes more difficult to build even higher-dimensional architectures with conventional two-dimensional circuit boards.

Instead of using hierarchical buses or switched network architectures in one dimension, multiprocessor architectures can be extended to a higher *dimensionality* or *multiplicity* along each dimension. The concepts are described below for two- and three-dimensional meshes proposed for the Multicube and OMP architectures, respectively.

The Wisconsin Multicube   This architecture was proposed by Goodman and Woest (1988) at the University of Wisconsin. It employs a snooping cache system over a grid of buses, as shown in Fig. 9.18a. Each processor is connected to a multilevel cache.

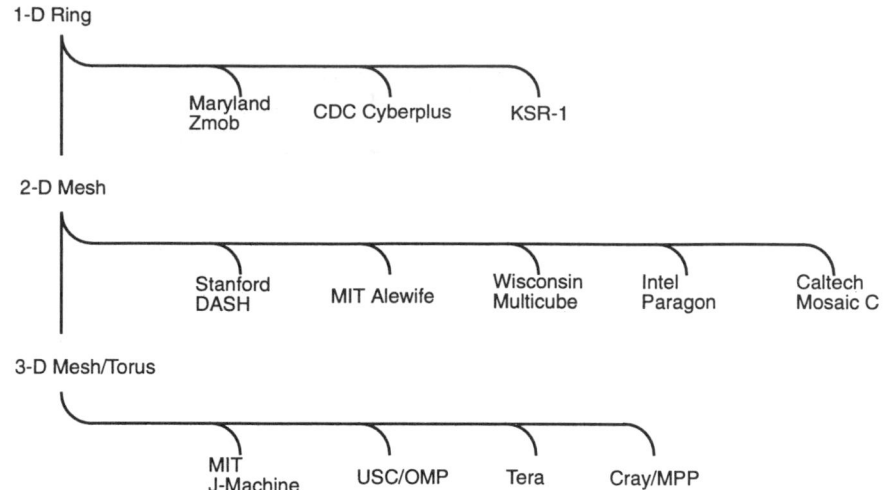

Figure 9.17 The evolution from one-dimensional ring to two-dimensional mesh and then to three-dimensional mesh/torus architecture for building massively parallel processors.

The first-level cache, called the *processor cache*, is a high-performance (SRAM) cache designed with the traditional goal of minimizing memory latency. A second-level cache, referred to as the *snooping cache*, is a very large cache designed to minimize bus traffic.

Each snooping cache monitors two buses, a row bus and a column bus, in order to maintain data consistency among the snooping caches. Consistency between the two cache levels is maintained by using a write-through strategy to ensure that the processor cache is always a strict subset of the snooping cache. The main memory is divided up among the column buses. All processors tied to the same column share the same home memory. The row buses are used for intercolumn communication and cache coherence control.

The proposed architecture is an example of a new class of interconnection topologies, the *multicube*, which consists of $N = n^k$ processors, where each processor is connected to k buses and each bus is connected to n processors. The hypercube is a special case where $n = 2$. The Wisconsin Multicube is a two-dimensional multicube ($k = 2$), where n scales to about 32, resulting in a proposed system of over 1000 processors.

The Orthogonal Multiprocessor   In the proposed OMP architecture (Fig. 9.18b), n processors simultaneously access n rows or n columns of interleaved memory modules. The $n \times n$ *memory mesh* is interleaved in both dimensions. In other words, each row is n-way interleaved and so is each column of memory modules. There are $2n$ logical buses spanning in two orthogonal directions.

The synchronized row access or column access must be performed exclusively. In fact, the row bus R_i and the column bus C_i can be the same physical bus because only one of the two will be used at a time. The memory controller (MC) in Fig. 9.18b

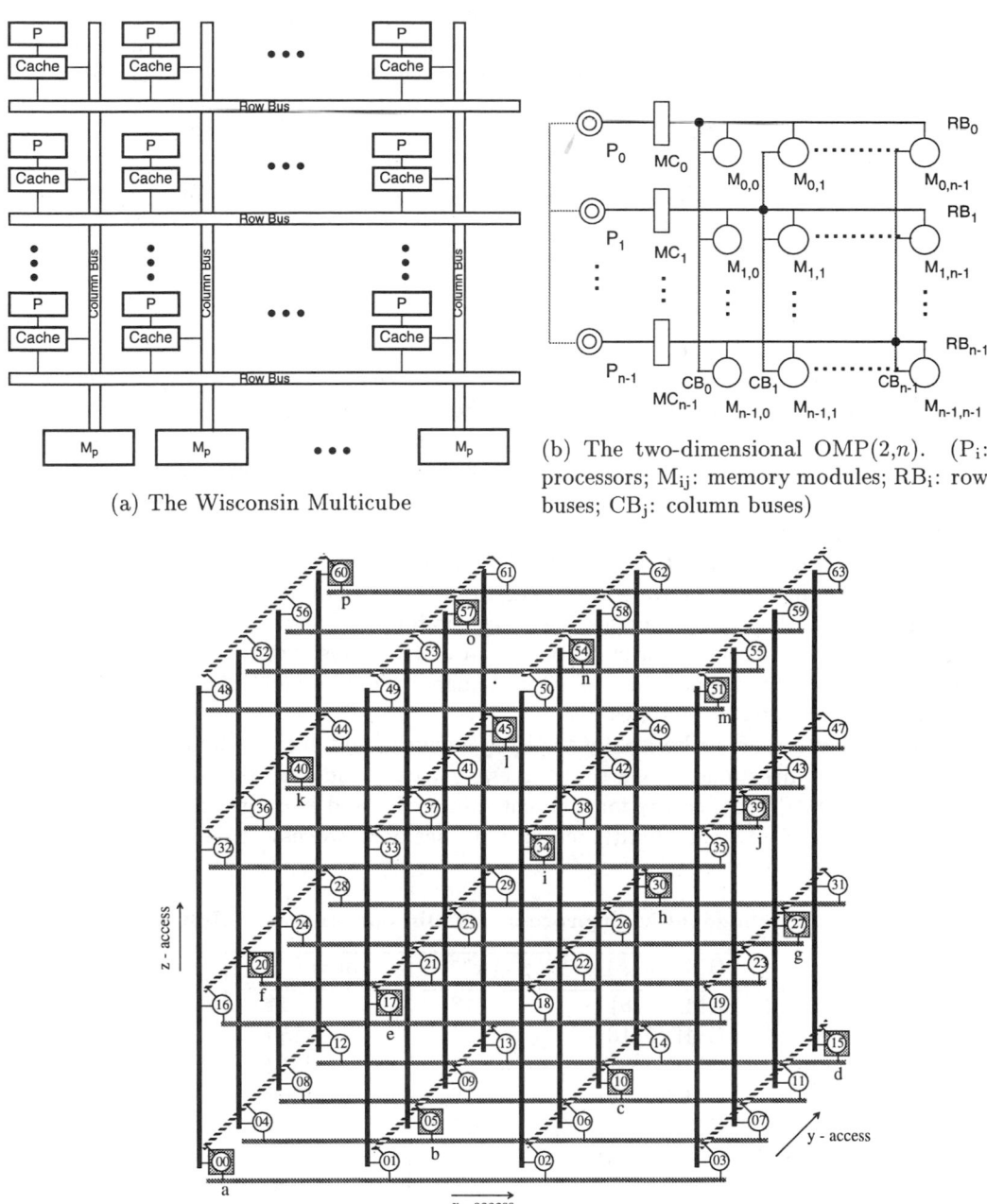

(a) The Wisconsin Multicube

(b) The two-dimensional OMP(2,n). (P_i: processors; M_{ij}: memory modules; RB_i: row buses; CB_j: column buses)

(c) The 3-D OMP(3,4) architecture. (Processors are labeled a, b, ..., p; memory modules are labeled 00, 01, ..., 63)

Figure 9.18 The Multicube and orthogonal multiprocessor architectures. (Courtesy of Goodman and Woest, 1988, and of Hwang et al., 1989)

synchronizes the row access and column access of the shared memory.

The OMP architecture supports special-purpose computations in which data sets can be regularly arranged as matrices. Simulated performance results obtained at USC have verified the effectiveness of using an OMP in matrix algebraic computations or in image processing operations.

In Fig. 9.18b, each of the memory modules M_{ij} is shared by two processors P_i and P_j. In other words, the physical address space of processor P_i covers only the ith row or the ith column of the memory mesh. The OMP is well suited for SPMD operations, in which n processors are synchronized at the memory-access level when data sets are vectorized in matrix format.

Multidimensional Extensions   The above OMP architecture can be generalized to higher dimensions. A generalized orthogonal multiprocessor is denoted as an OMP(n, k), where n is the *dimension* and k is the *multiplicity*. There are $p = k^{n-1}$ processors and $m = k^n$ memory modules in the system, where $p \gg n$ and $p \gg k$.

The system uses p memory buses, each spanning into n dimensions. But only one dimension is used in a given memory cycle. There are k memory modules attached to each spanning bus.

Each module is connected to n out of p buses through an n-way switch. It should be noted that the dimension n corresponds to the number of accessible ports that each memory module has. This implies that each module is shared by n out of $p = k^{n-1}$ processors. For example, the architecture of an OMP(3,4) is shown in Fig. 9.18c, where the circles represent memory modules, the squares processor modules, and the circles inside squares computer modules.

The 16 processors orthogonally access 64 memory modules via 16 buses, each spanning into three directions, called the *x-access*, *y-access*, and *z-access*, respectively. Various sizes of OMP architecture for different values of n and k are given in Table 9.2. A five-dimensional OMP with multiplicity $k = 16$ has 64K processors.

Table 9.2 Orthogonal Multiprocessor of Dimension n and Multiplicity k

OMP(n, k)	$p = k^{n-1}$	$m = k^n$
OMP$(2, 8)$	8	64
OMP$(2, 16)$	16	256
OMP$(3, 8)$	64	512
OMP$(3, 16)$	256	4096
OMP$(4, 8)$	512	4096
OMP$(4, 16)$	4096	65,536
OMP$(5, 16)$	65,536	1,048,576

Note: p = number of processors; m = number of memory modules.

9.3 Fine-Grain Multicomputers

Traditionally, shared-memory multiprocessors like the Cray Y-MP are used to perform coarse-grain computations in which each processor executes programs having a few tasks of 20 s or longer. Message-passing multicomputers are used to execute medium-grain programs with approximately 10-ms task size as in the iPSC/1. In order to build MPP systems, we have to explore a higher degree of parallelism by making the task grain size even smaller.

Fine-grain parallelism appears in SIMD or data-parallel computers like the CM-2 or on the message-driven J-Machine and Mosaic C to be described below. We characterize first fine-grain parallelism and discuss the network architectures proposed in recent years. Special attention is paid to the efficient hardware or software mechanisms developed for achieving fine-grain MIMD computation.

9.3.1 Fine-Grain Parallelism

We compare below the grain sizes, communication latencies, and concurrency in four classes of parallel computers. This comparison leads to the rationales for developing fine-grain multicomputers.

Latency Analysis   The computing granularity and communication latency of multiprocessors, data-parallel computers, and medium- and fine-grain multicomputers are summarized in Table 9.3. These table entries summarize what we have learned in Chapters 7 and 8. Four attributes are identified to characterize these machines. Only typical values for a typical program mix are shown. The intention is to show the order of magnitude in these entries.

The *communication latency* T_c measures the data or message transfer time on a system interconnect. This corresponds to the shared-memory access time on the Cray Y-MP, the time required to send a 32-bit value across the hypercube network in the CM-2, and the network latency on the iPSC/1 or J-Machine. The *synchronization overhead* T_s is the processing time required on a processor, or by a PE, or on a processing node of a multicomputer for the purpose of synchronization.

The sum $T_c + T_s$ gives the total time required for IPC. The shared-memory Cray Y-MP has a short T_c but a long T_s. The SIMD machine CM-2 has a short T_s but a long T_c. The long latency of the iPSC/1 makes it unattractive based on today's standards. The MIT J-Machine is designed to make a major improvement in both communication delays.

Fine-Grain Parallelism   The *grain size* T_g is measured by the execution time of a typical program, including both computing time and communication time involved. Supercomputers handle large grain. Both the CM-2 and the J-Machine are designed as fine-grain machines. The iPSC/1 is a relatively medium-grain machine compared with the rest.

Large grain implies lower concurrency or a lower DOP (degree of parallelism). Fine grain leads to a much higher DOP and also to higher communication overhead.

Table 9.3 Fine-Grain, Medium-Grain, and Coarse-Grain Machine Characteristics

Characteristics	Machine			
	Cray Y-MP	Connection Machine CM-2	Intel iPSC/1	MIT J-Machine
Communication latency, T_c	40 ns via shared memory	600 μs per 32-bit *send* operation	5 ms	2 μs
Synchronization overhead, T_s	20 μs	125 ns per bit-slice operation in lock step	500 μs	1 μs
Grain size, T_g	20 s	4 μs per 32-bit result per PE instruction	10 ms	5 μs
Concurrency (DOP)	2–16	8K–64K	8–128	1K–64K
Remark	Coarse-grain supercomputer	Fine-grain data parallelism	Medium-grain multicomputer	Fine-grain multicomputer

SIMD machines use hardwired synchronization and massive parallelism to overcome the problems of long network latency and slow processor speed. *Fine-grain multicomputers*, like the J-Machine and Caltech Mosaic, are being designed to lower both the grain size and the communication overhead compared to those of traditional multicomputers.

9.3.2   The MIT J-Machine

The architecture and building block of the MIT J-Machine, its instruction set, and system design considerations are described below based on the paper by Dally et al. (1992). The building block is the *message-driven processor* (MDP), which is a 36-bit microprocessor custom-designed for a fine-grain multicomputer.

The J-Machine Architecture   The k-ary n-cube networks have been applied in the MIT J-Machine. The initial prototype J-Machine uses a 1024-node network $(8 \times 8 \times 16)$, which is a reduced 16-ary 3-cube with 8 nodes along the x- and y-dimensions and 16 nodes along the z-dimension. A 4096-node J-Machine will use a full 16-ary 3-cube with $16 \times 16 \times 16$ nodes. The J-Machine designers have called their network a three-dimensional mesh.

The addressing limits the size of the J-Machine to having a maximum configuration of 65,536 nodes, corresponding to a three-dimensional mesh with $32 \times 32 \times 64$ nodes. The architecture of the three-dimensional mesh or a general k-ary n-cube was shown in Fig. 2.20 for the case of $k = 4$. All hidden parts (nodes and links) are not shown for purposes of clarity. Clearly, every node has a constant node degree of 6, and there are three rings crossing each node along the three dimensions. The end-around connections can be folded (Fig. 2.21b) to balance the wire length on all channels.

The MDP Design   The MDP chip includes a processor, a 4096-word by 36-bit memory, and a built-in router with network ports as shown in Fig. 9.19. An on-chip memory controller with error checking and correction (ECC) capability permit local memory to be expanded to 1 million words by adding external DRAM chips. The processor is message-driven in the sense that it processes in response to messages, via the dispatch mechanism. No receive instruction is needed.

The MDP creates a task to handle each arriving message. Messages carrying these tasks advance or drive each computation. It is a general-purpose multicomputer processing node that provides the communication, synchronization, and global naming mechanisms required to efficiently support fine-grain, concurrent programming models. The grain size is as small as 8-word objects or 20-instruction tasks. Fine-grain programs typically execute from 10 to 100 instructions between communication and synchronization actions.

MDP chips provide inexpensive processing nodes with plentiful VLSI commodity parts to construct the Jellybean Machine (J-Machine) multicomputer. As shown in Fig. 9.19a, the MDP appears as a component with a memory port, six two-way network ports, and a diagnostic port.

The memory port provides a direct interface to up to 1M words of ECC DRAM, consisting of 11 multiplexed address lines, a 12-bit data bus, and 3 control signals. Prototype J-Machines use three 1M × 4 static-column DRAMs to form a four-chip processing node with 262,144 words of memory. The DRAMs cycle three times to access a 36-bit data word and a fourth time to check or update the ECC check bits.

The network ports connect MDPs together in a three-dimensional mesh network. Each of the six ports corresponds to one of the six cardinal directions (+x, −x, +y, −y, +z, −z) and consists of nine data and six control lines. Each port connects directly to the opposite port on an adjacent MDP.

The diagnostic port issues supervisory commands and reads and writes MDP memory from a console processor (host). Using this port, a host can read or write at any location in the MDP's address space, as well as reset, interrupt, halt, or single-step the processor. The MDP chip floor plan is shown Fig. 9.19b.

Figure 9.19c shows the components built inside the MDP chip. The chip includes a conventional microprocessor with prefetch, control, register file and ALU (RALU), and memory blocks. The network communication subsystem comprises the routers and network input and output interfaces. The *address arithmetic unit* (AAU) provides addressing functions. The MDP also includes a DRAM interface, control clock, and diagnostic interface.

Instruction-Set Architecture   The MDP extends a conventional microprocessor instruction-set architecture with instructions to support parallel processing. The instruction set contains fixed-format, three-address instructions. Two 17-bit instructions fit into each 36-bit word with 2 bits reserved for type checking.

Separate register sets are provided to support rapid switching among three execution levels: background, priority 0 (P0), and priority 1 (P1). The MDP executes at the background level when no message creates a task and initiates execution at the P0 or P1 level depending on the message's priority.

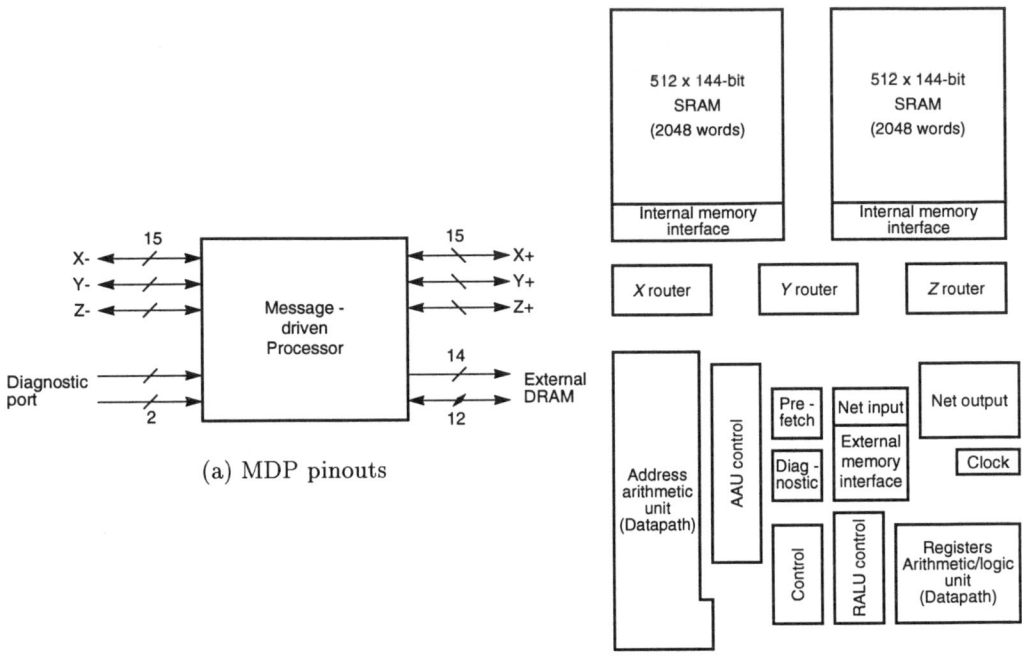

(a) MDP pinouts

(b) MDP chip floor plan

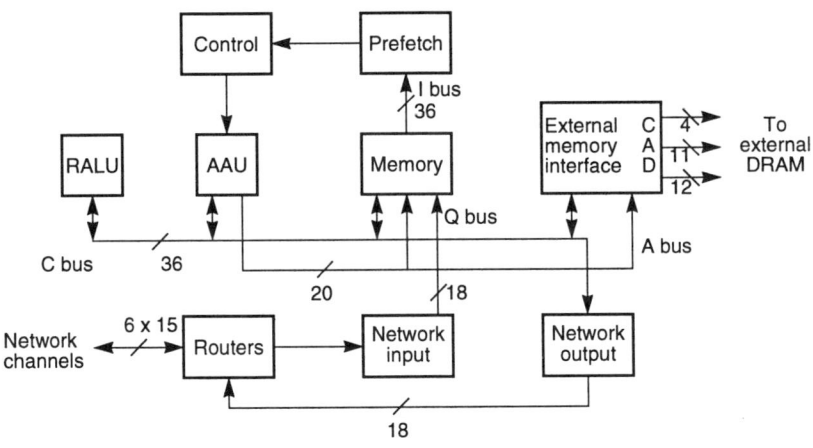

(c) Schematic block diagram

Figure 9.19 The message-driven processor (MDP) architecture. (Courtesy of W. Dally et al.; reprinted with permission from *IEEE Micro*, April 1992)

The P1 level has higher priority than the P0 level. The register set at each priority level includes four GPRs, four address registers, four ID registers, and one instruction pointer (IP). The ID registers are not used in the background register set.

Communication Support   The MDP provides hardware support for end-to-end message delivery including formatting, injection, delivery, buffer allocation, buffering, and task scheduling. A MDP transmits a message using a series of SEND instructions, each of which injects one or two words into the network at either priority 0 or 1.

Consider the following MDP assembly code for sending a four-word message using three variants of the SEND instruction.

```
SEND    R0,0        ;   send net address (priority 0)
SEND2   R1,R2,0     ;   header and receiver (priority 0)
SEND2E  R3,[3,A3],0 ;   selector and continuation end message (priority 0)
```

The first SEND instruction reads the absolute address of the destination node in $< X, Y, Z >$ format from R0 and forwards it to the network hardware. The SEND2 instruction reads the first two words of the message out of registers R1 and R2 and enqueues them for transmission. The final instruction enqueues two additional words of data, one from R3 and one from memory. The use of the SEND2E instruction marks the end of the message and causes it to be transmitted into the network. This sequence executes in four clock cycles.

The J-Machine is a three-dimensional mesh with two-way channels, dimension-order routing, and blocking flow control (Fig. 9.20). The faces of the network cube are open for use as I/O ports to the machine. Each channel can sustain a data rate of 288 Mbps (million bits per second). All three dimensions may operate simultaneously for an aggregate data rate of 864 Mbps per node.

Message Format and Routing   The J-Machine uses deterministic dimension-order E-cube routing. As shown in Fig. 9.20, all messages route first in the x-dimension, then in the y-direction, and then in the z-direction. Since messages route in dimension order and messages running in opposite directions along the same dimension do not block, resource cycles are thus avoided and leave the network provably deadlock-free.

Example 9.3 A typical message in the MIT J-Machine. (W. Dally et al., 1992)

The following message consists of nine flits. The first three flits of the message contain the x-, y-, and z-addresses. Each node along the path compares the address in the head flit of the message. If the two indices match, the node routes the rest to the next dimension. The final flit in the message is marked as the tail.

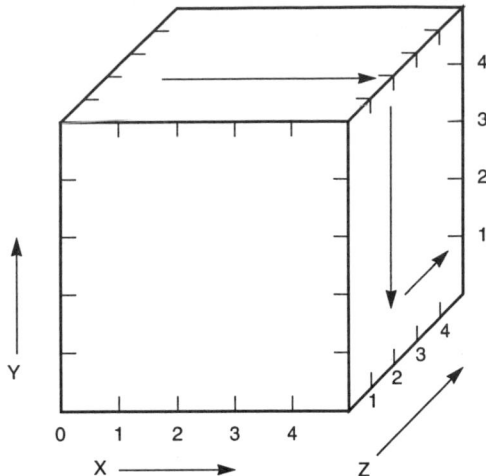

Figure 9.20 E-cube routing from node (1,5,2) to node (5,1,4) on a 6-ary 3-cube.

Flit	Contents		Remark
1	5:+		x-address
2	1:−		y-address
3	4:+		z-address
4	Msg: 00		Method to call
5	00440		
6	INT: 00		Argument to method
7	0023		
8	INT: 0 0		Reply address
9	< 1 : 5 : 2 >	T	

The MDP's network output node formats the address flits of the message. It also precomputes the direction (positive or negative) the message must travel along each dimension, setting additional bits in the address flits. This reduces the latency and complexity of the router nodes.   ■

The MDP supports a broad range of parallel programming models, including shared-memory, data-parallel, dataflow, actor, and explicit message passing, by providing a low-overhead primitive mechanism for communication, synchronization, and naming.

Its communication mechanisms permit a user-level task on one node to send a message to any other node in a 4096-node machine in less than 2 μs. This process doesn't consume any processing resources on intermediate nodes, and it automatically allocates buffer memory on the receiving node. On message arrival, the receiving node creates and dispatches a task in less than 1 μs.

Presence tags provide synchronization on all storage locations. Three separate register sets allow fast context switching. A translation mechanism maintains bindings between arbitrary names and values and supports a global virtual address space. These mechanisms were selected to be general and amenable to efficient hardware implementation. The J-Machine uses wormhole routing and blocking flow control. A combining-tree approach is used for synchronization.

The Router Design   The routers form the switches in a J-Machine network and deliver messages to their destinations. As shown in Fig. 9.21a, the MDP contains three independent routers, one for each bidirectional dimension of the network.

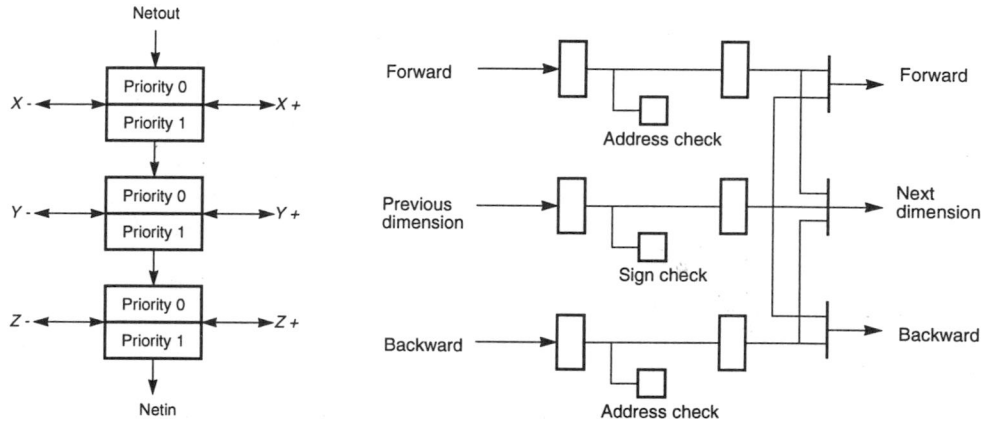

(a) Dual-priority levels per dimension in the router.

(b) Each priority with forward, reverse, and previous data paths to the next dimension.

Figure 9.21 Priority control and dimension-order router design in the MDP chip.
(Courtesy of W. Dally et al.; reprinted with permission from *IEEE Micro*, April 1992)

Each router contains two separate virtual networks with different priorities that share the same physical channels. The priority-1 network can preempt the wires even if the priority-0 network is congested or jammed. The priority levels support multi-threaded operations.

Each of the 18 router paths contains buffers, comparators, and output arbitration (Fig. 9.21). On each data path, a comparator compares the lead flit, which contains the destination's address in this dimension, to the node coordinate. If the head flit does not match, the message continues in the current direction. Otherwise the message is routed to the next dimension.

A message entering the dimension competes with messages continuing in the dimension at a two-to-one switch. Once a message is granted this switch, all other input is locked out for the duration of the message. Once the head flit of the message has set up the route, subsequent flits follow directly behind it.

Blocking Flow Control   The flow control in the J-Machine is illustrated in Fig. 9.22. When a message arrives at a router path already in use by a message of the same priority, it is blocked. The blocked message compresses into routers along its path, occupying one node per word (two flits) of the message.

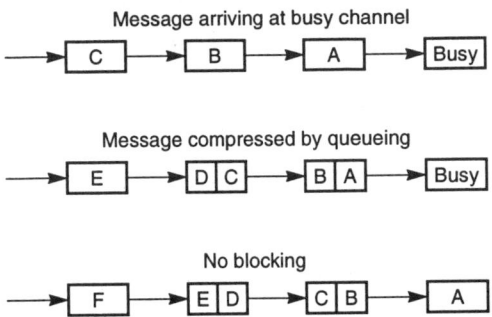

Figure 9.22 Blocking flow control with two stages of queueing per node. (Courtesy of W. Dally et al., 1992)

Two priorities of messages share the physical wires but use completely separate buffers and routing logic. This allows priority-1 messages to proceed through blockages at priority 0. Without this ability, the system could not redistribute data that has caused hot spots in the network.

Global Naming   The AAU, the largest logic block in the MDP, performs all functions associated with memory addressing. To support naming and relocation, the AAU contains the address and ID registers. It protects memory accesses and implements the translation instructions. Each memory reference is offset by the selected address register's base field and checked against its length field.

An attempt to access through an invalid address register (which may occur when an object relocates) or to access beyond the end of an object raises an exception. A translation base/mask register defines an area of memory to be a two-way, set-associative translation buffer.

Synchronization   The MDP synchronizes using message dispatch and presence tags on all states. Because each message arrival dispatches a process, messages can signal events on remote nodes. For example, in the following combining-tree example, each COMBINE message signals its own arrival and initiates the COMBINE routine.

In response to an arriving message, the processor may set presence tags for task synchronization. For example, access to the value produced by the combining tree may be synchronized by initially tagging as empty the location that will hold this value. An attempt to read this location before the combining tree has written it will raise an exception and suspend the reading task until the root of the tree writes the value.

Example 9.4 Using a combining tree for synchronization of events (W. Dally et al., 1992)

A combining tree is shown in Fig. 9.23. This tree sums results produced by a distributed computation. Each node sums the input values as they arrive and then passes a result message to its parent.

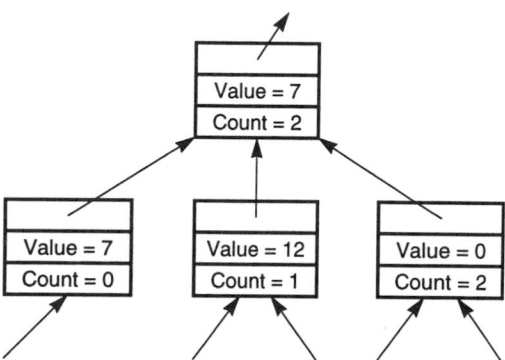

Figure 9.23 A combining tree for internode communication or synchronization. (Courtesy of W. Dally et al., 1992)

A pair of SEND instructions is used to send the COMBINE message to a node. Upon message arrival, the MDP buffers the message and creates a task to execute the following COMBINE routine written in MDP assembly code:

```
COMBINE:   MOVE     [1,A3],COMB                    ; get node pointer from message
           MOVE     [2,A3], R1                     ; get value from message
           ADD      R1, COMB.VALUE, R1
           MOVE     R1, COMB.VALUE                 ; store result
           MOVE     COMB.COUNBT, R2                ; get Count
           ADD      R2, -1, R2
           MOVE     R2, COMB.COUNT                 ; store decremented Count
           BNZ      R2, DONE
           MOVE     HEADER, R0                     ; get message header
           SEND2    COMB.PARENT_NODE, R0           ; send message to parent
           SEND2E   COMB.PARENT, R1                ; with value
DONE:      SUSPEND
```

If the node is idle, execution of this routine begins three cycles after message arrival. The routine loads the combining-node pointer and value from the message, performs the required add and decrement, and, if Count reaches zero, sends a message to its parent.

Research Issues   The J-Machine is an exploratory research project. Rather than being specialized for a single model of computation, the MDP incorporates primitive mechanisms for efficient communication, synchronization, and naming. The machine is being used as a platform for software experiments in fine-grain parallel programming.

Reducing the grain size of a program increases both the potential speedup due to parallel execution and the potential overhead associated with parallelism. Special hardware mechanisms for reducing the overhead due to communication, process switching, synchronization, and multithreading are therefore central to the design of the MDP. Software issues such as load balancing, scheduling, and locality remain open questions.

The MIT research group led by Dally has implemented two languages on the J-Machine: the actor language Concurrent Smalltalk and the dataflow language Id. The machine's mechanism also supports dataflow and object-oriented programming models using a global name space. The use of a few simple mechanisms provides orders of magnitude lower communication and synchronization overhead than is possible with multicomputers built from off-the-shelf microprocessors.

9.3.3   The Caltech Mosaic C

The Caltech Mosaic C is an experimental fine-grain multicomputer that employs single-chip nodes and advanced packaging technology to demonstrate the performance/-cost advantages of fine-grain multicomputer architecture. We describe below the architecture of the Mosaic C and review its application potentials, based on a recent report by Seitz (1992), the project leader at Caltech.

From Cosmic Cube to Mosaic C   The evolution from the Cosmic cube to the Mosaic is an example of one type of *scaling track* in which advances in technology are employed to reimplement nodes of a similar logical complexity but which are faster and smaller, have lower power, and are less expensive. The progress in microelectronics over the past decade has been such that the Mosaic nodes are ≈ 60 times faster, use ≈ 20 times less power, are ≈ 100 times smaller, and are (in constant dollars) ≈ 25 times less expensive to manufacture than the Cosmic Cube nodes.

Each Mosaic node includes 64 Mbytes of memory and an 11-MIPS processor, a packet interface, and a router. The nodes are tied together with a 60-Mbytes/s, two-dimensional routing-mesh network (Fig. 9.24). The compilation-based programming system allows fine-grain reactive-process message-passing programs to be expressed in C+–, an extension of C++, and the run-time system performs automatic distributed management of system resources.

Mosaic C Node   The Mosaic C multicomputer node is a single 9.25 mm × 10.00 mm chip fabricated in a 1.2-μm-feature-size, two-level-metal CMOS process. At 5-V operation, the synchronous parts of the chip operate with large margins at a 30-MHz clock rate, and the chip dissipates ≈ 0.5 W. The tested but unpackaged cost of the silicon at the 40 to 50% yields typical of this chip is approximately \$30 per chip.

The processor also includes two program counters and two sets of general-purpose registers to allow zero-time context switching between user programs and message han-

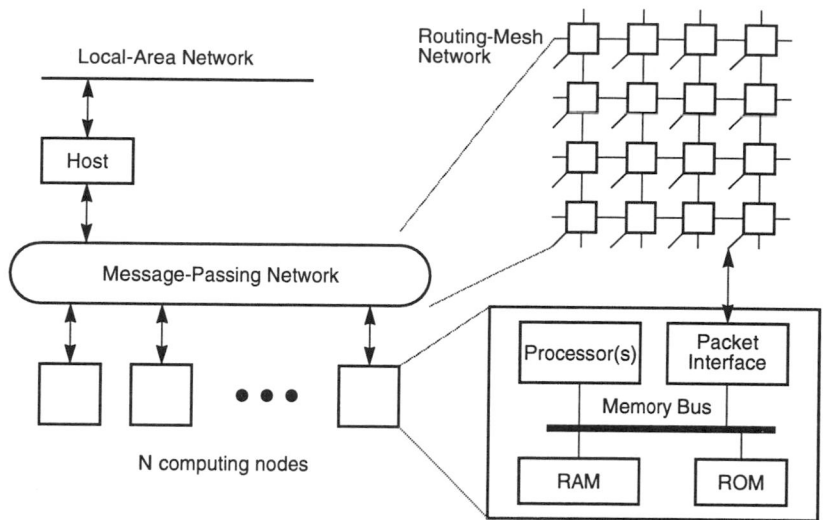

Figure 9.24 The Caltech Mosaic architecture. (Courtesy of C. Seitz, 1992)

dling. Thus, when the packet interface has received a complete packet, received the header of a packet, completed the sending of a packet, exhausted the allocated space for receiving packets, or any of several other events that could be selected, it can interrupt the processor by switching it instantly to the message-handling context.

Instead of several hundred instructions for handling a packet, the Mosaic typically requires only about 10 instructions. The number of clock cycles for the message-handling routines could be reduced to insignificance by placing them in hardware, but the Caltech group chose the more flexible software mechanism so that they could experiment with different message-handling strategies.

Mosaic C 8×8 **Mesh Boards**   The choice of a two-dimensional mesh for the Mosaic was based on a 1989 engineering analysis; originally, a three-dimensional mesh network was planned. But the mutual fit of the two-dimensional mesh network and the circuit board medium provides high packaging density and allows the high-speed signals between the routers to be conveyed on shorter wires.

Sixty-four Mosaic chips are packaged by tape-automated bonding (TAB) in an 8×8 array on a circuit board. These boards allow the construction of arbitrarily large, two-dimensional arrays of nodes using stacking connectors. Thus style of packaging was meant to demonstrate some of the density, scaling, and testing advantages of mesh-connected systems. Host-interface boards are also used to connect the Mosaic arrays and workstations.

Applications and Future Trends   Charles Seitz determined that the most profitable niche and scaling track for the multicomputer, a highly scalable and economical

MIMD architecture, is the fine-grain multicomputer. The Mosaic C demonstrates many of the advantages of this architecture, but the major part of the Mosaic experiment is to explore the programmability and application span of this class of machine.

The Mosaic may be taken as the origin of two scaling tracks: (1) Single-chip nodes are a technologically attractive point in the design space of multicomputers. Constant-node-size scaling will result in single-chip nodes of increasing memory size, processing capability, and communication bandwidth in larger systems than centralized shared-memory multiprocessors. (2) Constant-node-complexity scaling will allow a Mosaic 8×8 board to be implemented as a single chip, with about 20 times the performance per node, within 10 years.

A 16K-node machine is under construction at Caltech to explore the programmability and application span of the Mosaic C architecture for large-scale computing problems. For the loosely coupled computations in which they excel, a multicomputer will be more economically implemented as a network of high-performance workstations connected by a high-bandwidth local-area network . In fact, the Mosaic components and programming tools have already been used by a USC Information Science Institute project (led by Danny Cohen, 1992) to implement a 400-Mbits/s ATOMIC local-area network for this purpose.

9.4   Scalable and Multithreaded Architectures

Three scalable multiprocessor systems are discussed in this section. The Stanford Dash combines several latency-hiding mechanisms. The Kendall Square KSR-1 offers the first commercial attempt to produce a multiprocessor with cache-only memory. The Tera computer has evolved from the HEP/Horizon series developed by Burton Smith.

The Dash and Tera are still under development. Although the KSR-1 is now on the market, very few application data have become available. Therefore, only some architectural features are described below. Real performance data are yet to be released. All three systems are extensions of the traditional von Neumann model. By far, the Tera system represents the most aggressive attempt to build a multithreaded multiprocessor.

9.4.1   The Stanford Dash Multiprocessor

This is an experimental multiprocessor system being developed by John Hennessy and coworkers at Stanford University since 1988. The name Dash is an abbreviation for *Directory Architecture for Shared Memory*. The fundamental premise behind Dash is that it is possible to build a scalable high-performance machine with a single address space, coherent caches, and distributed memories. The directory-based coherence gives Dash the ease of use of shared-memory architectures, while maintaining the scalability of message-passing machines.

The Prototype Architecture   A high-level organization of the Dash architecture was illustrated in Fig. 9.1 when we studied the various latency-hiding techniques. The Dash prototype is illustrated in Fig. 9.25. It incorporates up to 64 MIPS R3000/R3010 microprocessors with 16 clusters of 4 PEs each. The cluster hardware is modified from

Silicon Graphics 4D/340 nodes with new directory and reply controller boards as depicted in Fig. 9.25a.

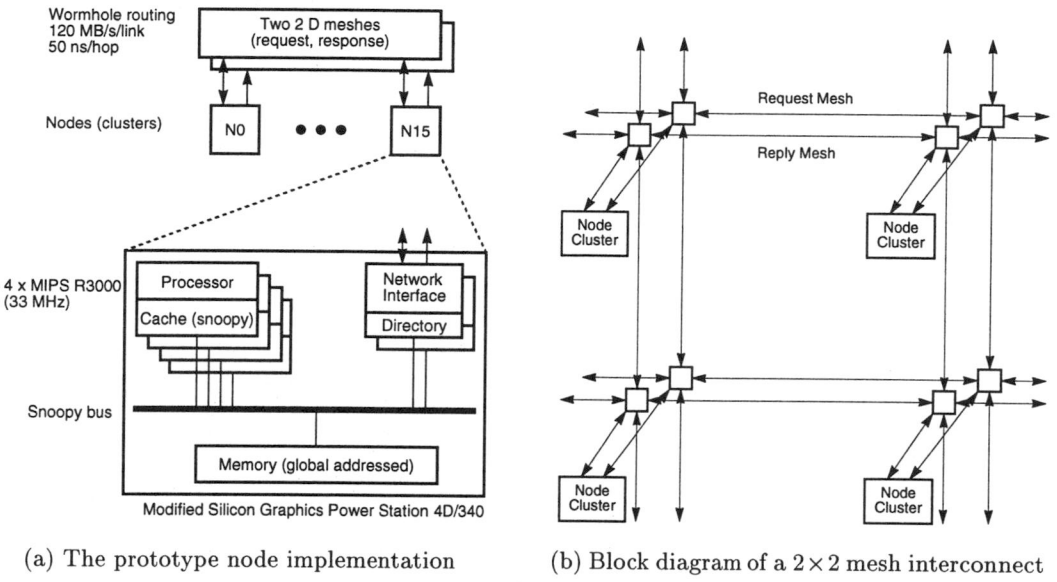

(a) The prototype node implementation   (b) Block diagram of a 2×2 mesh interconnect

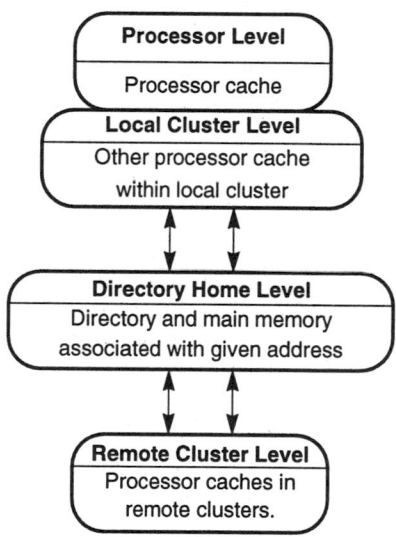

(c) Logic memory hierarchy

Figure 9.25 The Stanford Dash prototype system. (Courtesy of D. Lenoski et al., *Proc. 19th Int. Symp. Comput. Archit.*, Australia, May 1992)

The interconnection network among the 16 multiprocessor clusters is a pair of wormhole-routed mesh networks. The channel width is 16 bits with a 50-ns fall-through time and a 35-ns cycle time. One mesh network is used to *request* remote memory, and the other is a *reply* mesh as depicted in Fig. 9.25b, where the small squares at mesh intersections are the 5 × 5 mesh routers.

The Dash designers claimed scalability for the Dash approach. Although the prototype is limited to at most 16 clusters (a 4 × 4 mesh), due to the limited physical memory addressability (256 Mbytes) of the 4D/340 system, the system is scalable to support hundreds to thousands of processors in the future.

To use the 4D/340 in the Dash, the Stanford team made minor modifications to the existing system boards and designed a pair of new boards to support the directory memory and intercluster interface. The main modification to the existing boards was to add a bus retry signal that is used when a request requires service from a remote cluster.

The central bus arbiter has also been modified to accept a mask from the directory. The mask holds off a processor's retry until the remote request has been serviced. This effectively creates a split-transaction bus protocol for requests requiring remote service.

The new directory controller boards contain the directory memory, the intercluster coherence state machines and buffers, and a local section of the global interconnection network. The directory logic is split between the two logic boards along the lines of the logic used for outbound and inbound portions of intercluster transactions.

The mesh networks support a scalable local and global memory bandwidth, which rewards locality. The single-address space with coherent caches makes it easier for compilers and programmers, permits incremental porting or tuning of applications, and exploits temporal and spatial locality. Other factors contributing to improved performance include mechanisms for reducing and tolerating latency and current I/O capabilities.

Dash Memory Hierarchy   Dash implements an invalidation-based cache coherence protocol. A memory location may be in one of three states:

- *Uncached* — not cached by any cluster;
- *Shared* — in an unmodified state in the caches of one or more clusters; or
- *Dirty* — modified in a single cache of some cluster.

The directory keeps the summary information for each memory block, specifying its state and the clusters that are cacheing it. The Dash memory system can be logically broken into four levels of hierarchy, as illustrated in Fig. 9.25c.

The first level is the processor's cache which is designed to match the processor speed and support snooping from the bus. It takes only one clock to access the processor's cache. A request that cannot be serviced by the processor's cache is sent to the *local cluster*. The prototype allows 30 processor clocks to access the local cluster. This level includes the other processors' caches within the requesting processor's cluster.

Otherwise, the request is sent to the *home cluster* level. The home level consists of the cluster that contains the directory and physical memory for a given memory address. It takes 100 processor clocks to access the directory at the home level. For many accesses (for instance, most private data references), the local and home cluster are the same, and the hierarchy collapses to three levels. In general, however, a request will travel through the interconnection network to the home cluster.

The home cluster can usually satisfy the request immediately, but if the directory entry is in a dirty state, or in a shared state when the requesting processor requests exclusive access, the fourth level must also be accessed. The *remote cluster* level for a memory block consists of the clusters marked by the directory as holding a copy of the block. It takes 135 processor clocks to access processor caches in remote clusters in the prototype design.

The Directory Protocol The directory memory relieves the processor caches of snooping on memory requests by keeping track of which caches hold each memory block. In the home node, there is a directory entry per block frame. Each entry contains one *presence bit* per processor cache. In addition, a *state bit* indicates whether the block is uncached, shared in multiple caches, or held exclusively by one cache (i.e., whether the block is dirty).

Using the state and presence bits, the memory can tell which caches need to be invalidated when a location is written. Likewise, the directory indicates whether memory's copy of the block is up-to-date or which cache holds the most recent copy. If the memory and directory are partitioned into independent units and connected to the processors by a scalable interconnect, the memory system can provide a scalable memory bandwidth.

By using the directory memory, a node writing a location can send point-to-point invalidation or update messages to those processors that are actually cacheing that block. This is in contrast to the invalidating broadcast required by the snoopy protocol. The scalability of the Dash depends on this ability to avoid broadcasts.

Another important attribute of the directory-based protocol is that it does not depend on any specific interconnection network topology. As a result, one can readily use any of the low-latency scalable networks, such as meshes or hypercubes, that were originally developed for message-passing machines.

Example 9.5 Cache coherence protocol using distributed directories in the Dash multiprocessor (Daniel Lenoski and John Hennessy et al., 1992.)

Figure 9.26a illustrates the flow of a read request to remote memory with the directory in a dirty remote state. The read request is forwarded to the owning dirty cluster. The owning cluster sends out two messages in response to the read. A message containing the data is sent directly to the requesting cluster, and a sharing writeback request is sent to the home cluster. The sharing writeback request writes the cache block back to memory and also updates the directory.

This protocol reduces latency by permitting the dirty cluster to respond directly to the requesting cluster. In addition, this forwarding strategy allows the directory controller to simultaneously process many requests (i.e., to be multithreaded) without the added complexity of maintaining the state of outstanding requests. Serialization is reduced to the time of a single intercluster bus transaction. The only resource held while intercluster messages are being sent is a single entry in the

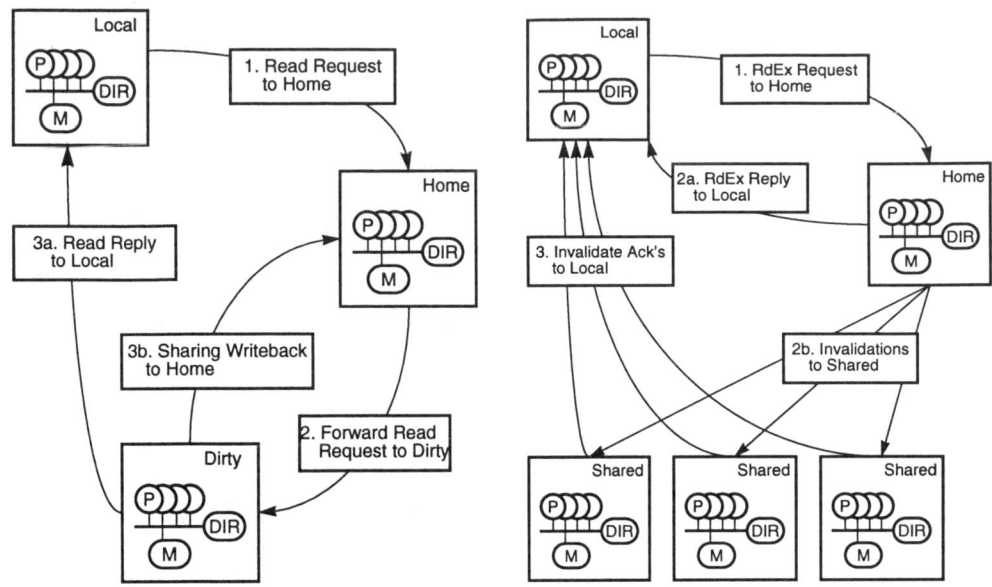

(a) Read of dirty remote cache block   (b) Write to shared remote cache block

Figure 9.26 Two examples of a directory-based cache coherence protocol in the Dash. (Courtesy of Lenoski and Hennessy, 1992)

originating cluster's remote-access cache.

Figure 9.26b shows the corresponding sequence for a write operation that requires remote service. The invalidation-based protocol requires the processor (actually the write buffer) to acquire exclusive ownership of the cache block before completing the store. Thus, if a write is made to a block that the processor does not have cached, or only has cached in a shared state, the processor issues a read-exclusive request on the local bus.

In this case, no other cache holds the block entry dirty in the local cluster, so a RdEx Request (message 1) is sent to the home cluster. As before, a remote-access cache entry is allocated in the local cluster. At the home cluster, the pseudo-CPU issues the read-exclusive request to the bus. The directory indicates that the line is in the shared state. This results in the directory controller sending a RdEx Reply (message 2a) to the local cluster and invalidation requests (Inv-Req, message 2b) to the sharing cluster.

The home cluster owns the block, so it can immediately update the directory to the dirty state, indicating that the local cluster now holds an exclusive copy of the memory line. The RdEx Reply message is received in the local cluster by the reply controller, which can then satisfy the read-exclusive request.

To ensure consistency at release points, however, the remote-access cache entry is deallocated only when it receives the number of invalidate acknowledgments (Inv-

Ack, message 3) equal to an invalidation count sent in the original reply message.

■

An important feature of the coherence protocol is its forwarding strategy. If a cluster cannot reply directly to a given request, it forwards responsibility for the request to a cluster that should be able to respond. This technique also minimizes the serialization of requests since no requests are blocked while intercluster messages are being sent. Forwarding allows the directory controller to work on multiple requests concurrently (i.e., makes it multithreaded) without having to retain any additional state about forwarded requests.

The Dash prototype as it is is rather small in size. A future plan is to extend it to 512 clusters with a total of $512 \times 4 = 2048$ processors. If each processor assumes a five-issue superscalar operation with a 100-MHz clock, the extended machine then has the potential to become a system with 1 tera operations per second. This future system has been named *TeraDash*.

The TeraDash demands an integrated implementation with lower overhead in the scalable directory structure. A three-dimensional torus network ($8 \times 8 \times 8$) is being considered with 16-bit data paths, a 20-ns fall-through delay, and a 4-ns cycle time. The access time ratio among the four levels of memory hierarchy will be approximately 1:5:16:80:120, where 1 corresponds to one processor clock. Hopefully, the address limitations of the current Dash prototype will be removed and the scalability of the Dash architecture will be confirmed for a large number (≥ 1000) of processors.

9.4.2 The Kendall Square Research KSR-1

This is the first commercial attempt to build a scalable multiprocessor with *cache-only memory architecture* (COMA). The Kendall Square Research KSR-1 is a size-and-generation-scalable shared-memory multiprocessor computer. It is formed as a hierarchy of "ring multis" as depicted in Fig. 9.27.

The KSR-1 Architecture   Scalability in the KSR-1 is achieved by connecting 32 processors to form a ring multi (search engine 0 in Fig. 9.27) operating at 1 Gbyte/s (128 million accesses per second). Interconnection bandwidth within a ring scales linearly, since every ring slot has roughly the capacity of a typical crosspoint switch found in a supercomputer room that interconnects eight to sixteen 100-Mbytes/s HIPPI channels.

The KSR-1 uses a two-level hierarchy to interconnect 34 Ring:0s by a top-level Ring:1 (1088 processors) and is therefore massive. The ring design supports an arbitrary number of levels, permitting ultras to be built (Fig. 9.28).

Each node comprises a primary cache, acting as a 32-Mbyte primary memory, and a 64-bit superscalar processor with roughly the same performance as an IBM RS/6000 operating at the same clock rate. The superscalar processors containing 64 floating-point and 32 fixed-point registers of 64 bits are designed for both scalar and vector operations.

For example, 16 elements can be prefetched at one time. A processor also has a 0.5-Mbyte subcache supplying 20 million accesses per second to the processor (a

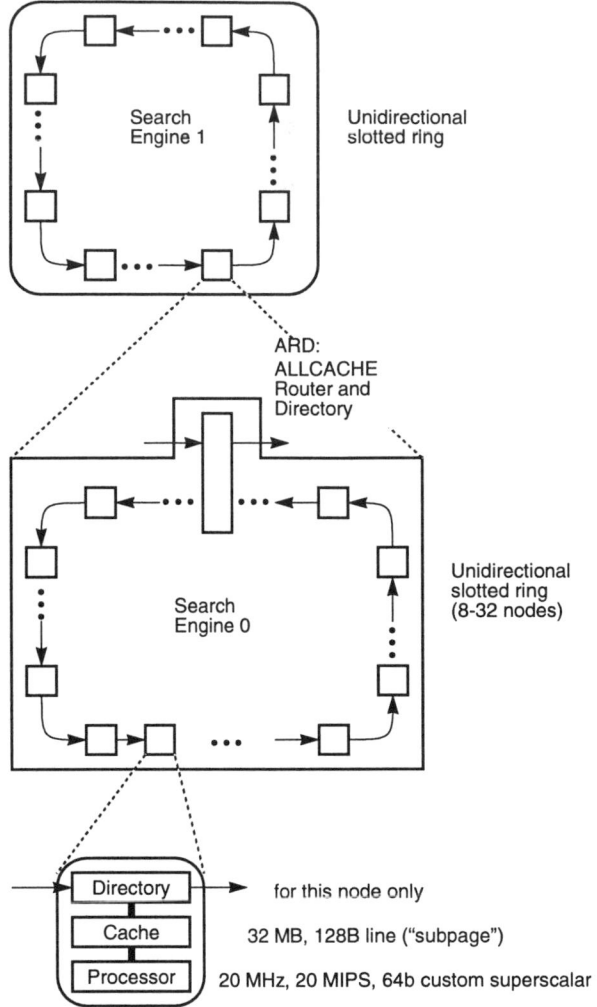

Figure 9.27 The KSR-1 architecture with a slotted ring for communication. (Courtesy of Kendall Square Research Corporation, 1991)

computational efficiency of 0.5). A processor operates at 20 MHz and is fabricated in 1.2-μm CMOS.

The processor, without caches, contains 3.9 million transistors on 6 types of 12 custom chips. Three-quarters of each processor consists of the search engine responsible for migrating data to and from other nodes, for maintaining memory coherence throughout the system using distributed directories, and for ring control.

The ALLCACHE Memory   The KSR-1 eliminates the memory hierarchy found in

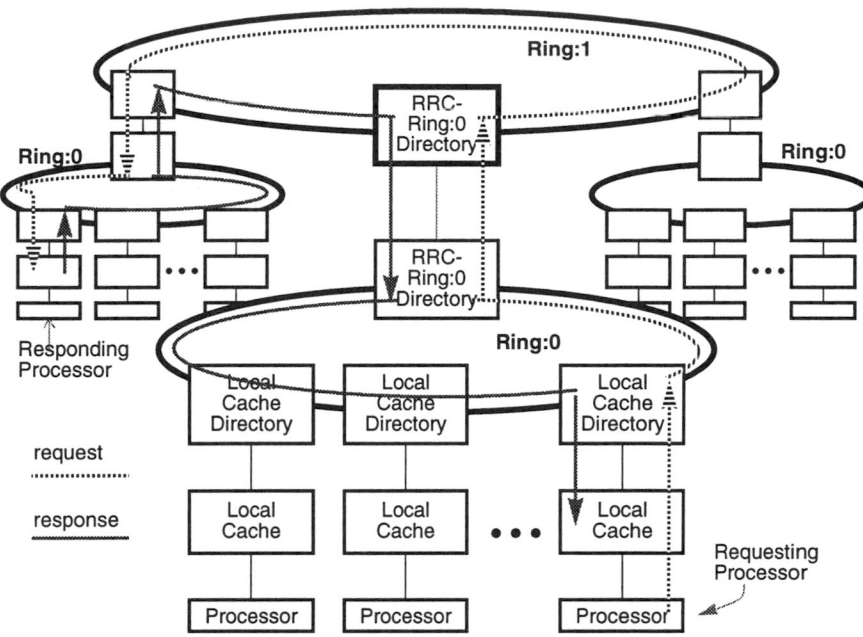

Figure 9.28 Remote cache (memory) access through two levels of communication rings in the KSR-1. (Courtesy of Kendall Square Research Corporation, 1991)

conventional computers and the corresponding physical memory addressing overhead. Instead, it offers a single-level memory, called *ALLCACHE* by KSR designers. This ALLCACHE design represents the confluence of cache and shared virtual memory concepts that exploit locality required by scalable distributed computing. Each local cache has a capacity of 32 Mbytes (2^{25} bytes). The global virtual address space has 2^{40} bytes.

Bell (1992) considers the KSR machine the most likely blueprint for future scalable MPP systems. This is a revolutional architecture and thus was more controversial when it was first introduced in 1991. The architecture provides size (including I/O) and generation scalability in that every node is identical, and it offers an efficient environment for both arbitrary workloads and sequential to parallel processing through a large hardware-supported address space with an unlimited number of processors.

Programming Model   The KSR machine provides a strictly sequential consistent programming model and dynamic management of memory through hardware migration and replication of data throughout the distributed processor memory nodes using its ALLCACHE mechanism.

With sequential consistency, every processor returns the latest value of a written value, and results of an execution on multiple processors appear as some interleaving of operations of individual nodes when executed on a multithreaded machine. With ALL-CACHE, an address becomes a name, and this name automatically migrates throughout the system and is associated with a processor in a cachelike fashion as needed.

Copies of a given cell are made by the hardware and sent to other nodes to reduce access time. A processor can prefetch data into a local cache and poststore data for other cells. The hardware is designed to exploit spatial and temporal locality.

For example, in the SPMD programming model, copies of the program move dynamically and are cached in each of the operating nodes' primary and processor caches. Data such as elements of a matrix move to the nodes as required simply by accessing the data, and the processor has instructions that prefetch data to the processor's register. When a processor writes to an address, all cells are updated and memory coherence is maintained. Data movement occurs in subpages of 128 bytes (16 words) of its 16K pages.

Example 9.6 Multiring searching with requesting and responding processors on different Ring:0s (Courtesy of Kendall Square Research Corporation, 1991).

The internode communication for remote memory access is achieved through a searching process. When the requester and responder are in the same Ring:0, the searching is restricted to a single connected Ring:0. Local cache directories show what addresses can be found in the connected local cache. Each Ring:0 is a unidirectional slotted ring for pipelined searching until the destination is reached.

Figure 9.28 illustrates the situation when the requester and responder reside in different Ring:0s. The top level, Ring:1, consists entirely of *ring routing cells* (RRCs), each containing a directory for the Ring:0 to which it is connected. Each RRC directory on Ring:1 is essentially a duplicate of the RRC directory on the corresponding Ring:0.

When a packet reaches an RRC on Ring:1, it will be moved to the next RRC on the ring if the RRC directory indicates that the requested data is not on the corresponding ring. Otherwise, the packet is routed down to the RRC on Ring:0. The packet-passing speed of a Ring:0 is 8 million packets per second. Ring:1 can be configured to handle 8, 16, 32, or 64 million packets per second.

∎

Environment and Performance   Every known form of parallelism is supported via the KSR's Mach-based operating system. Multiple users may run multiple sessions comprising multiple applications or multiple processes (each with independent address spaces), each of which may consist of multiple threads of control running and simultaneously sharing a common address space. Message passing is supported by pointer passing in the shared memory to avoid data copying and enhance performance.

The KSR also provides a commercial programming environment for transaction processing that accesses relational databases in parallel with unlimited scalability as an alternative to multicomputers formed from multiprocessor mainframes. A 1K-node system provides almost two orders of magnitude more processing power, primary memory, I/O bandwidth, and mass storage capacity than a multiprocessor mainframe.

For example, unlike the typical tera candidates, a 1088-node system can be configured with 15.3 terabytes of disk memory, providing 500 times the capacity of its main memory. The 32- and 320-node systems are projected to deliver over 1000 and 10,000 transactions per second, respectively, giving them over 100 times the throughput of a multiprocessor mainframe.

9.4.3 The Tera Multiprocessor System

Multithreaded von Neumann architecture can be traced back to the CDC 6600 manufactured in the mid-1960s. Multiple functional units in the 6600 CPU can execute different operations simultaneously using a scoreboarding control. The very first multithreaded multiprocessor was the Denelcor HEP designed by Burton Smith in 1978. The HEP was built with 16 processors driven by a 10-MHz clock, and each processor can execute 128 *threads* (called *processes* in HEP terminology) simultaneously.

The HEP failed to survive due to inadequate software and compiler support. The Tera is very much a HEP descendant but is implemented with modern VLSI circuits and packaging technology. A 400-MHz clock is proposed for use in the Tera system, again with a maximum of 128 threads (*i-streams* in Tera terminology) per processor.

In this section, we describe the Tera architecture, its processors and thread state, and the tagged memory/registers. The unique features of the Tera include not only the high degree of multithreading but also the explicit-dependence lookahead and the high degree of superpipelining in its processor-network-memory operations. These advanced features are mutually supportive. The first Tera machines are expected to appear in late 1993.

The Tera Design Goals The Tera architecture was designed with several major goals in mind. First, it needed to be suitable for very high-speed implementations, i.e., have a short clock period and be scalable to many processors. A maximum configuration of the first implementation of the architecture (Fig. 9.29a) will have 256 processors, 512 memory units, 256 I/O cache units, 256 I/O processors, 4096 interconnection network nodes, and a clock period of less than 3 ns.

Second, it was important that the architecture be applicable to a wide spectrum of problems. Programs that do not vectorize well, perhaps because of a preponderance of scalar operations or too frequent conditional branches, will execute efficiently as long as there is sufficient parallelism to keep the processors busy. Virtually any parallelism applicable in the total computational workload can be turned into speed, from operation-level parallelism within program basic blocks to multiuser time and space sharing.

A third goal was ease of compiler implementation. Although the instruction set does have a few unusual features, they do not seem to pose unduly difficult problems for the code generator. There are no register or memory addressing constraints and only three addressing modes. Condition code setting is consistent and orthogonal. Although the richness of the instruction set often allows several ways to do something, the variation in their relative costs as the execution environment changes tends to be small.

Because the architecture permits the free exchange of spatial and temporal locality for parallelism, a highly optimizing compiler may work hard improving locality and

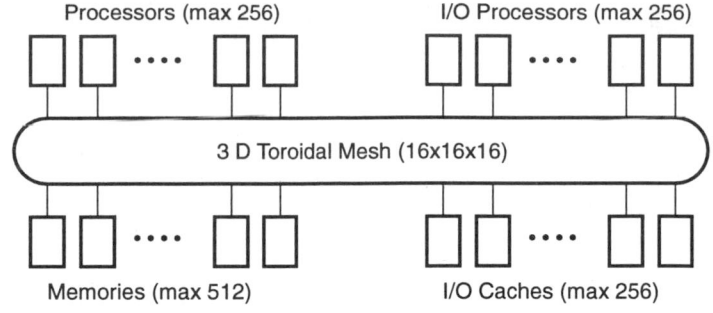

(a) The Tera computer system

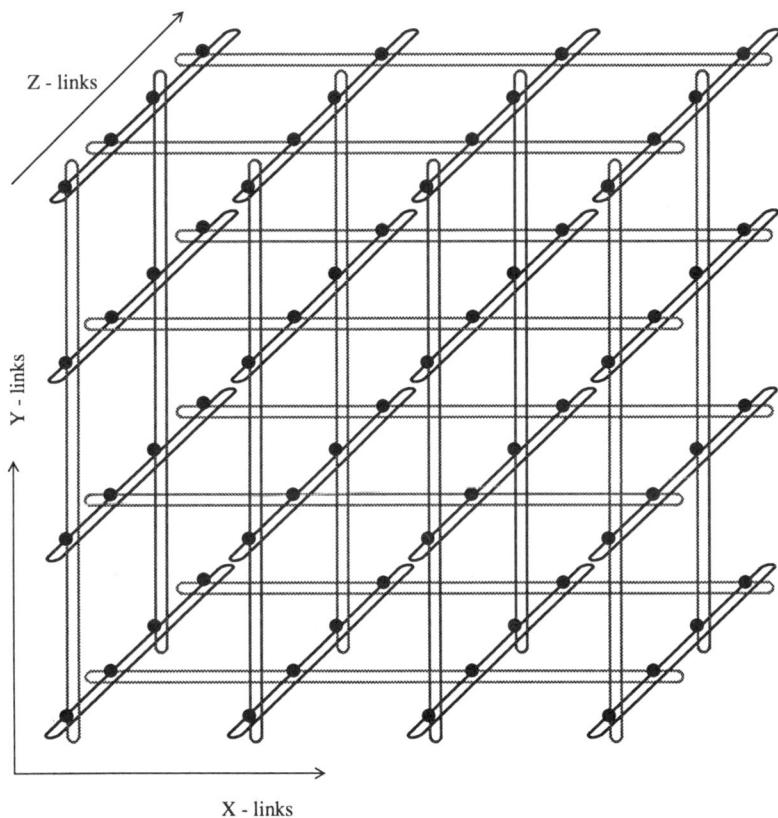

(b) A sparse 4 × 4 × 4 torus with X-links and Y-links missing on alternate Z-layers, respectively

Figure 9.29 The Tera multiprocessor and its three-dimensional sparse torus architecture shown with a 4 × 4 × 4 configuration. (Courtesy of Tera Computer Company, 1992)

trade the parallelism thereby saved for more speed. On the other hand, if there is sufficient parallelism, the compiler has a relatively easy job.

The Sparse Three-Dimensional Torus The interconnection network is a three-dimensional sparsely populated torus (Fig. 9.29b) of pipelined packet-switching nodes, each of which is linked to some of its neighbors. Each link can transport a packet containing source and destination addresses, an operation, and 64 data bits in both directions simultaneously on every clock tick. Some of the nodes are also linked to *resources*, i.e., processors, data memory units, I/O processors, and I/O cache units.

Instead of locating the processors on one side of the network and the memories on the other (a "dance hall" configuration), the resources are distributed more-or-less uniformly throughout the network. This permits data to be placed in memory units near the appropriate processor when possible and otherwise generally maximizes the distance between possibly interfering resources.

The interconnection network of one 256-processor Tera system contains 4096 nodes arranged in a $16 \times 16 \times 16$ toroidal mesh; i.e., the mesh "wraps around" in all three dimensions. Of the 4096 nodes, 1280 are attached to the resources comprising 256 cache units and 256 I/O processors. The 2816 remaining nodes do not have resources attached but still provide message bandwidth.

To increase node performance, some of the links are missing. If the three directions are named x, y, and z, then x-links and y-links are missing on alternate z-layers (Fig. 9.29b). This reduces the node degree from 6 to 4, or from 7 to 5, counting the resource link. In spite of its missing links, the bandwidth of the network is very large.

Any plane bisecting the network crosses at least 256 links, giving the network a data bisection bandwidth of one 64-bit data word per processor per tick in each direction. This bandwidth is needed to support shared-memory addressing in the event that all 256 processors are addressing memory on the other side of some bisecting plane simultaneously.

As the Tera architecture scales to larger numbers of processors p, the number of network nodes grows as $p^{3/2}$ rather than as the $p \log p$ associated with the more commonly used multistage networks. For example, a 1024-processor system would have 32,768 nodes. The reason for the overhead per processor of $p^{1/2}$ instead of $\log p$ stems from the fact that the system is limited by the speed of light.

One can argue that memory latency is fully masked by parallelism only when the number of messages being routed by the network is at least $p \times l$, where l is the (round-trip) latency. Since messages occupy volume, the network must have a volume proportional to $p \times l$; since the speed of light is finite, the volume is also proportional to l^3 and therefore l is proportional to $p^{1/2}$ rather than $\log p$.

Superpipelined Support Each processor in a Tera computer can execute multiple instruction streams (threads) simultaneously. In the current implementation, as few as 1 or as many as 128 program counters may be active at once. On every tick of the clock, the processor logic selects a thread that is ready to execute and allows it to issue its next instruction. Since instruction interpretation is completely pipelined by the processor and by the network and memories as well (Fig. 9.30), a new instruction from a

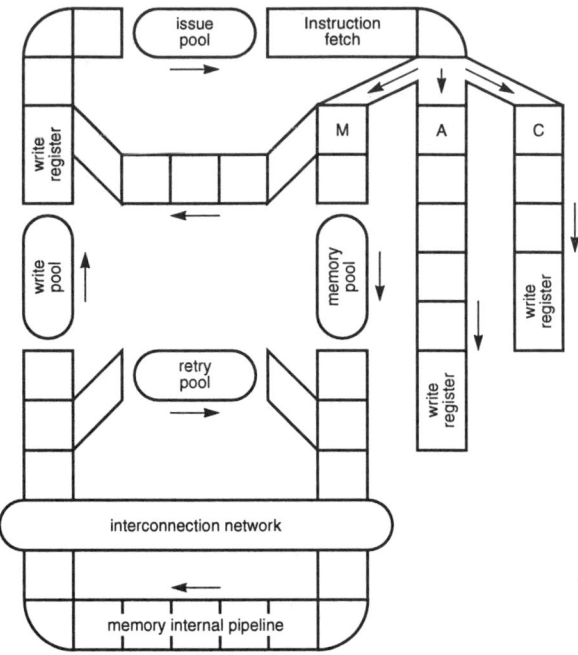

Figure 9.30 Superpipelined processor-network-memory structure. (Courtesy of Tera Computer Company, 1992)

different thread may be issued during each tick without interfering with its predecessors.

When an instruction finishes, the thread to which it belongs becomes ready to execute the next instruction. As long as there are enough threads in the processor so that the average instruction latency is filled with instructions from other threads, the processor is being fully utilized. Thus, it is only necessary to have enough threads to hide the expected latency (perhaps 70 ticks on average); once latency is hidden, the processor is running at peak performance and additional threads do not speed the result.

If a thread were not allowed to issue its next instruction until the previous instruction is completed, then approximately 70 different threads would be required on each processor to hide the expected latency. The lookahead described later allows threads to issue multiple instructions in parallel, thereby reducing the number of threads needed to achieve peak performance.

In Fig. 9.30, three operations can be executed simultaneously per instruction per processor. The *M-pipeline* is for memory-access operations, the *A-pipeline* is for arithmetic operations, and the *C-pipeline* is for control or arithmetic operations. The instructions are 64 bits wide. If more than one operation in an instruction specifies the same register or setting of condition codes, the priority is $M > A > C$.

It has been estimated that a peak speed of 1G operations per second can be achieved per processor if driven by a 333-MHz clock. However, a particular thread will not exceed about 100M operations per second because of interleaved execution. The processor pipeline is rather deep, about 70 ticks, as compared with 8 ticks in the HEP pipeline.

Thread State and Management Figure 9.31 shows that each thread has the following states associated with it:

- One 64-bit stream status word (SSW);
- Thirty-two 64-bit general-purpose registers (R0–R31);
- Eight 64-bit target registers (T0–T7).

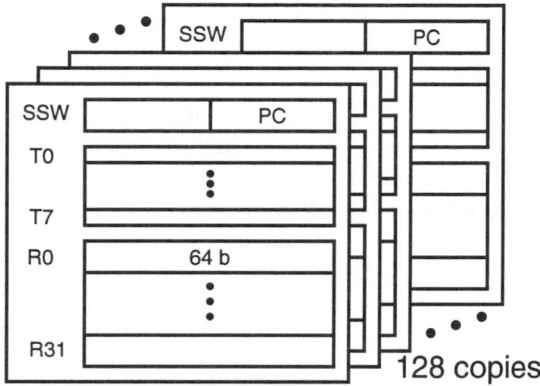

Stream Status Word (SSW)

- 32 bit PC (Program Counter)
- Modes (e.g., rounding, lookahead disable)
- Trap disable mask (e.g., data alignment, overflow)
- Condition codes (last four emitted)

No synchronization bits on R0-R31

Target Registers (T0-T7) look like SSWs

Figure 9.31 The thread management scheme used in the Tera computer. (Courtesy of Tera Computer Company, 1992)

Context switching is so rapid that the processor has no time to swap the processor-resident thread state. Instead, it has 128 of everything, i.e., 128 SSWs, 4096 general-purpose registers, and 1024 target registers. It is appropriate to compare these registers in both quantity and function to vector registers or words of caches in other architectures. In all three cases, the objective is to improve locality and avoid reloading data.

Program addresses are 32 bits in length. Each thread's current program counter is located in the lower half of its SSW. The upper half describes various modes (e.g., floating-point rounding, lookahead disable), the trap disable mask (e.g., data alignment, floating overflow), and the four most recently generated condition codes.

Most operations have a _TEST variant which emits a condition code, and branch operations can examine any subset of the last four condition codes emitted and branch appropriately. Also associated with each thread are thirty-two 64-bit general-purpose registers. Register R0 is special in that it reads as 0 and output to it is discarded. Otherwise, all general-purpose registers are identical.

The target registers are used as branch targets. The format of the target registers is identical to that of the SSW, though most control transfer operations use only the low 32 bits to determine a new PC. Separating the determination of the branch target address from the decision to branch allows the hardware to prefetch instructions at the branch targets, thus avoiding delay when the branch decision is made. Using target registers also makes branch operations smaller, resulting in tighter loops. There are also skip operations which obviate the need to set targets for short forward branches.

One target register (T0) points to the trap handler which is nominally an unprivileged program. When a trap occurs, the effect is as if a coroutine call to a T0 had been executed. This makes trap handling extremely lightweight and independent of the operating system. Trap handlers can be changed by the user to achieve specific trap capabilities and priorities without loss of efficiency.

Explicit-Dependence Lookahead   If there are enough threads executing on each processor to hide the pipeline latency (about 70 ticks), then the machine is running at peak performance. However, if each thread can execute some of its instructions in parallel (e.g., two successive loads), then fewer threads and parallel activities are required to achieve peak performance.

The obvious solution is to introduce instruction lookahead; the only difficulty is controlling it. The traditional register reservation approach requires far too much scoreboard bandwidth in this kind of architecture. Either multithreading or horizontal instruction alone would preclude scoreboarding. The traditional alternative, exposing the pipeline, is also impractical because multithreading and unpredictable memory operation latency make it impossible to generate code that is both efficient and safe.

The Tera architecture uses a new technique called *explicit-dependence lookahead.* Each instruction contains a 3-bit lookahead field that explicitly specifies how many instructions from this thread will be issued before encountering an instruction that depends on the current one. Since seven is the maximum possible lookahead value, at most 8 instructions and 24 operations can be concurrently executing from each thread.

A thread is ready to issue a new instruction when all instructions with lookahead values referring to the new instruction have completed. Thus, if each thread maintains a lookahead of seven, then nine threads are needed to hide 72 ticks of latency.

Lookahead across one or more branch operations is handled by specifying the minimum of all distances involved. The variant branch operations JUMP_OFTEN and JUMP_SELDOM, for high- and low-probability branches, respectively, facilitate optimization by providing a barrier to lookahead along the less likely path. There are also SKIP_OFTEN and SKIP_SELDOM operations. The overall approach is philosophically similar to exposed-pipeline lookahead except that the quanta are instructions instead of ticks.

Advantages and Drawbacks   The Tera uses multiple contexts to hide latency. The machine performs a context switch every clock cycle. Both pipeline latency (eight cycles) and memory latency are hidden in the HEP/Tera approach. The major focus is on latency tolerance rather than latency reduction.

With 128 contexts per processor, a large number (2K) of registers must be shared finely between threads. The thread creation must be very cheap (a few clock cycles). Tagged memory and registers with full/empty bits are used for synchronization. As long as there is plenty of parallelism in user programs to hide latency and plenty of compiler support, the performance is potentially very high.

However, these Tera advantages may be embedded in a number of potential drawbacks. The performance must be bad for limited parallelism, such as guaranteed low single-context performance. A large number of contexts (threads) demands lots of registers and other hardware resources which in turn implies higher cost and complexity. Finally, the limited focus on latency reduction and cacheing entails lots of slack parallelism to hide latency as well as lots of memory bandwidth; both require a higher cost for building the machine.

9.5   Dataflow and Hybrid Architectures

Multithreaded architectures can be designed with a pure dataflow approach or with a hybrid approach combining von Neumann and data-driven mechanisms. In this final section, we briefly review the historical development of dataflow computers. Then we consider the design of the ETL/EM-4 in Japan and the prototype design of the MIT/Motorola *T project.

9.5.1   The Evolution of Dataflow Computers

As introduced in Section 2.3, dataflow computers have the potential for exploiting all the parallelism available in a program. Since execution is driven only by the availability of operands at the inputs to the functional units, there is no need for a program counter in this architecture, and its parallelism is limited only by the actual data dependences in the application program. While the dataflow concept offers the potential of high performance, the performance of an actual dataflow implementation can be restricted by a limited number of functional units, limited memory bandwidth, and the need to associatively match pending operations with available functional units.

Arvind and Iannucci (1987) have identified *memory latency* and *synchronization overhead* as two fundamental issues in multiprocessing. Scalable multiprocessors must address the loss in processor efficiency in these cases. Using various latency-hiding mechanisms and multiple contexts per processor makes the conventional von Neumann architecture quite expensive to implement, and only certain types of parallelism can be exploited efficiently.

HEP/Tera computers offer an evolutionary step beyond the von Neumann architectures. Dataflow architectures represent a radical alternative to von Neumann architectures because they use dataflow graphs as their machine languages. Dataflow graphs, as opposed to conventional machine languages, specify only a partial order for the execution of instructions and thus provide opportunities for parallel and pipelined execution at the level of individual instructions.

Dataflow Graphs   We have seen a dataflow graph in Fig. 2.13. Dataflow graphs can be used as a machine language in dataflow computers. Another example of a dataflow graph (Fig. 9.32a) is given below.

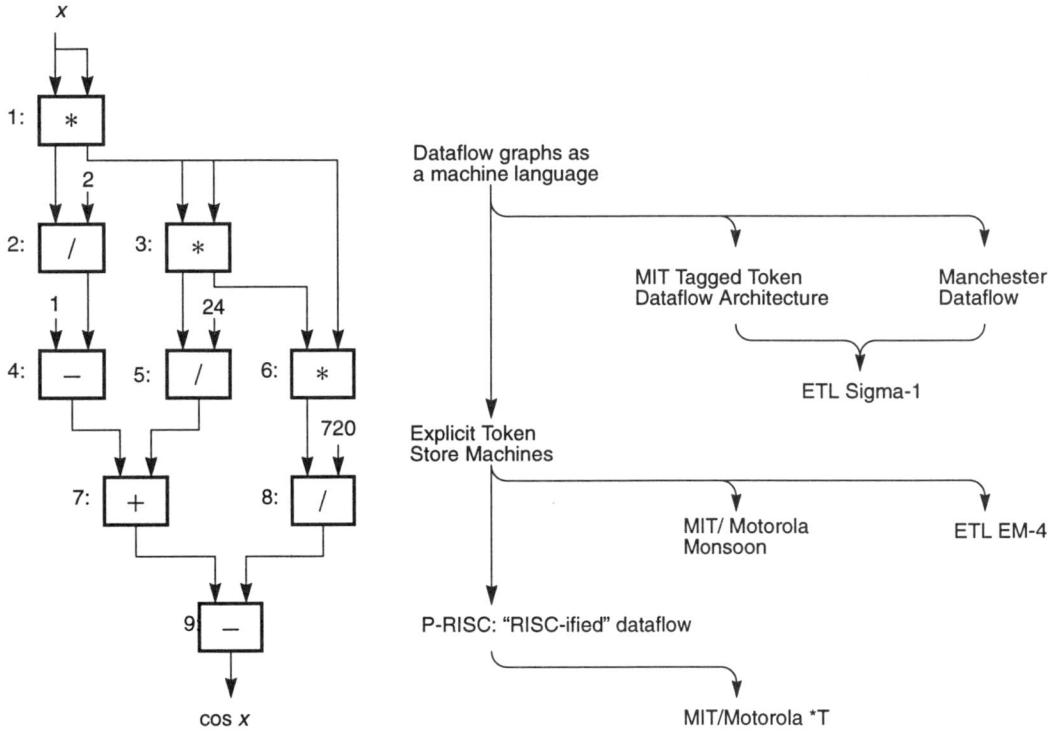

(a) Dataflow graph for computing $\cos x$. (Courtesy of Arvind)

(b) Evolution tree of dynamic dataflow machines (Courtesy of R. Nikhil)

Figure 9.32 An example dataflow graph and dataflow machine projects.

Example 9.7   The dataflow graph for the calculation of cos x (Arvind, 1991).

This dataflow graph shows how to obtain an approximation of $\cos x$ by the following power series computation:

$$\cos x \;\simeq\; 1 - \frac{x^2}{2!} + \frac{x^4}{4!} - \frac{x^6}{6!}$$

$$= 1 - \frac{x^2}{2} + \frac{x^4}{24} - \frac{x^6}{720} \qquad (9.6)$$

The corresponding dataflow graph consists of nine operators (actors or nodes). The edges in the graph interconnect the operator nodes. The successive powers of x are obtained by repeated multiplications. The constants (divisors) are fed into the nodes directly. All intermediate results are forwarded among the nodes. ∎

Static versus Dynamic Dataflow   *Static dataflow computers* simply disallow more than one token to reside on any one arc, which is enforced by the firing rule: A node is enabled as soon as tokens are present on all input arcs and there is no token on any of its output arcs. Jack Dennis proposed the very first static dataflow computer in 1974.

The static firing rule is difficult to implement in hardware. Special feedback *acknowledge signals* are needed to secure the correct token passing between producing nodes and consuming nodes. Also, the static rule makes it very inefficient to process arrays of data. The number of acknowledge signals may grow too fast to be supported by hardware.

However, static dataflow has inspired the development of *dynamic dataflow computers*, which are being researched vigorously at MIT and in Japan. In a dynamic architecture, each data token is tagged with a context descriptor, called a *tagged token*. The firing rule of tagged-token dataflow is changed to: A node is enabled as soon as tokens with identical tags are present at each of its input arcs.

With tagged tokens, tag matching becomes necessary. Special hardware mechanisms are needed to achieve this. In the rest of this section, we discuss only dynamic dataflow computers. Arvind of MIT has pioneered the development of tagged-token architecture for dynamic dataflow computers.

Although data dependence does exist in dataflow graphs, they do not force unnecessary sequentialization, and dataflow computers schedule instructions according to the availability of the operands. Conceptually, "token"-carrying values flow along the edges of the graph. Values or tokens may be memory locations.

Each instruction waits for tokens on all inputs, consumes input tokens, computes output values based on input values, and produces tokens on outputs. No further restriction on instruction ordering is imposed. No side effects are produced with the execution of instructions in a dataflow computer. Both dataflow graphs and machines implement only functional languages.

Pure Dataflow Machines   Figure 9.32b shows the evolution of dataflow computers. The MIT *tagged-token dataflow architecture* (TTDA) (Arvind et al., 1983), the Manchester Dataflow Computer (Gurd and Watson, 1982), and the ETL Sigma-1 (Hiraki and Shimada, 1987) are all pure dataflow computers. The TTDA was simulated but never built. The Manchester machine was actually built and became operational in mid-1982. It operates asynchronously using a separate clock for each processing element with a performance comparable to that of the VAX/780.

The ETL Sigma-1 was developed at the Electrotechnical Laboratory, Tsukuba,

Japan. It consists of 128 PEs fully synchronous with a 10-MHz clock. It implemented the I-structure memory proposed by Arvind. The full configuration became operational in 1987 and achieved a 170-Mflops performance. The major problem in using the Sigma-1 was lack of high-level language for users.

Explicit Token Store Machines   These are successors to the pure dataflow machines. The basic idea is to eliminate associative token matching. The waiting token memory is directly addressed, plus the use of full/empty bits. This idea was used in the MIT/Motorola Monsoon (Papadopoulos and Culler, 1988) and in the ETL EM-4 system (Sakai et al., 1989).

Multithreading is supported in the Monsoon using multiple register sets. Thread-based programming was conceptually introduced in the Monsoon. The maximum configuration built consists of eight processors and eight I-structure memory modules using an 8 × 8 crossbar network. It became operational in 1991.

The EM-4 is an extension of the Sigma-1. It was designed for 1024 nodes, but only an 80-node prototype became operational in 1990. The prototype achieved 815 MIPS in an 80 × 80 matrix multiplication benchmark. We will study the details of the EM-4 in Section 9.5.2.

Hybrid and Unified Architectures   These are architectures combining positive features from the von Neumann and dataflow architectures. The best examples include the MIT P-RISC (Nikhil and Arvind, 1988), the IBM Empire (Iannucci et al., 1991), and the MIT/Motorola *T (Nikhil, Papadopoulos, Arvind, and Greiner, 1991).

The P-RISC is a "RISC-ified" dataflow architecture. It allows tighter encodings of the dataflow graphs and produces longer threads for better performance. This was achieved by splitting "complex" dataflow instructions into separate "simple" component instructions that can be composed by the compiler. It uses traditional instruction sequencing. It performs all intraprocessor communication via memory and implements "joins" explicitly using memory locations.

The P-RISC replaces some of the dataflow synchronization with conventional program counter-based synchronization. The Empire is a von Neumann/dataflow hybrid architecture under development at IBM based on the thesis of Iannucci (1988). The *T is the latest effort at MIT joining both the dataflow and von Neumann ideas, to be discussed in Section 9.5.3.

9.5.2   The ETL/EM-4 in Japan

The EM-4 has a global organization as shown in Fig. 9.33a. Each EMC-R node is a single-chip processor without floating-point hardware but including a switch of the network. Each node plays the role of I-structure memory and has 1.31 Mbytes of static RAM. An Omega network is built to provide the interconnections among the nodes.

The Node Architecture   The internal design of the processor chip and of the node memory are shown in Fig. 9.33b. The processor chip communicates with the network through a 3 × 3 crossbar *switch unit*. The processor and its memory are interfaced with

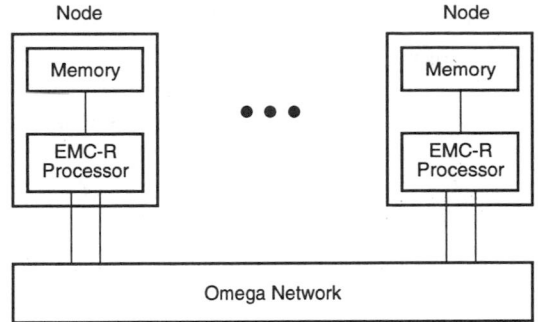

(a) Global organization

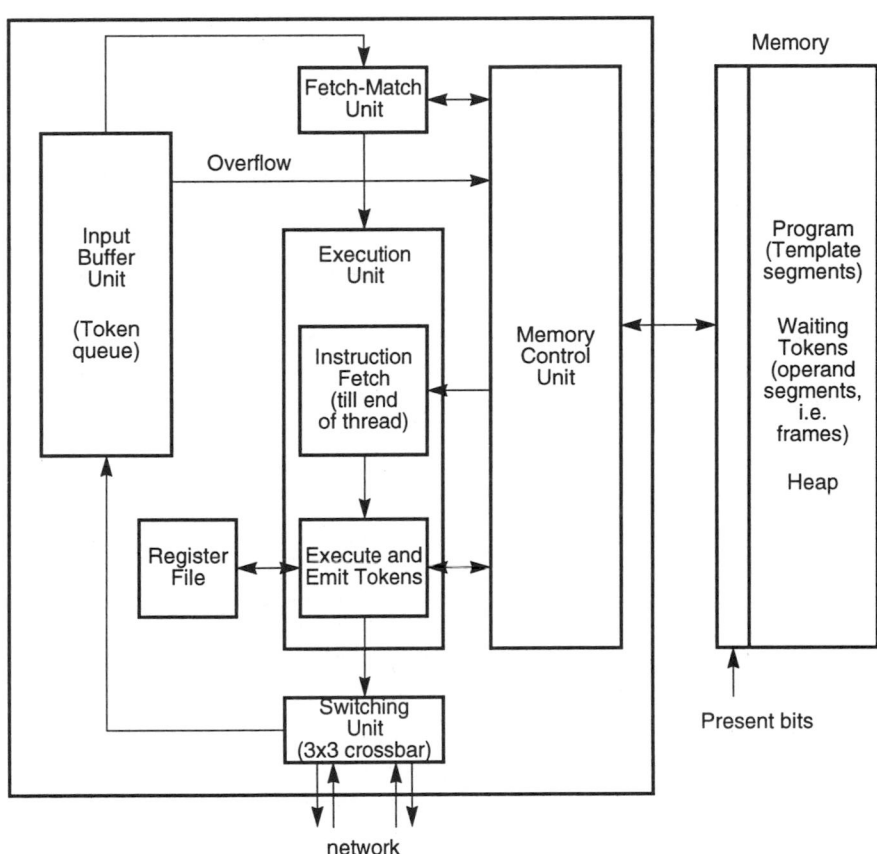

(b) The EMC-R processor design

Figure 9.33 The ETL EM-4 dataflow architecture. (Courtesy of Sakai, Yamaguchi et al., Electrotechnical Laboratory, Tsukuba, Japan, 1991)

a *memory control unit*. The memory is used to hold programs (template segments) as well as tokens (operand segments, heaps, or frames) waiting to be fetched.

The processor consists of six component units. The *input buffer* is used as a token store with a capacity of 32 words. The *fetch-match unit* fetches tokens from the memory and performs the tag-matching operations among the tokens fetched in. Instructions are directly fetched from the memory through the memory controller.

The heart of the processor is the *execution unit*, which fetches instructions until the end of a thread. Instructions with matching tokens are executed. Instructions can emit tokens or write to registers. Instructions are fetched continually using traditional sequencing (PC + 1 or branch) until a "stop" flag is raised to indicate the end of a thread. Then another pair of tokens is accepted. Each instruction in a thread specifies the two sources for the next instruction in the thread.

The same idea is used as in the Monsoon for token matching, but with different encoding. All data tokens are 32 bits, and instruction words are 38 bits. The EM-4 supports remote loads and synchronizing loads. The requesting node behaves like the Monsoon; the responding node behaves like the J-machine. The *full/empty* bits present in memory words are used to synchronize remote loads associated with different threads.

Future Effort   The EM-4 really represents the cumulative efforts of its predecessors. EM-4+, an EM-4 upgrade, is under development. The EM-4+ will include floating-point hardware and is expected to double the EM-4 performance. At present, a C compiler exists for the EM-4 using explicit forks and joins.

The future EM-5 is possibly the centerpiece of the NIPT, the new Japanese national computer project. It has been speculated that the EM-5 will have 16K nodes, yielding a performance of 655 GIPS or 1.3 Tflops with 64-bit words. All of this is still in the planning stage.

9.5.3   The MIT/Motorola *T Prototype

The *T project is a direct descendant of a series of MIT dynamic dataflow architectures unifying with the von Neumann architectures. In this final section, we describe the *T, a prototype multithreaded MPP system based on the work of Nikhil, Papadopoulos, and Arvind of MIT in collaboration with Greiner and Traub of Motorola. Finally, we compare the dataflow and von Neumann perspectives in building fine-grain, massively parallel systems.

The Prototype Architecture   The proposed *T prototype is a single-address-space system. A "brick" of 16 nodes is packaged in a 9-in cube (Fig. 9.34a). The local network is built with 8×8 crossbar switching chips. A brick has the potential to achieve 3200 MIPS or 3.2 Gflops. The memory is distributed to the nodes. One gigabyte of RAM is used per brick. With 200-Mbytes/s links, the I/O bandwidth is 6.4 Gbytes/s per brick.

A 256-node machine can be built with 16 bricks as illustrated in Fig. 9.34b. The 16 bricks are interconnected by four switching boards. Each board implements a 16×16 crossbar switch. The entire system can be packaged into a 1.5-m cube. No cables are used between the boards. The package is limited by connector-pin density. The 256-

node machine has the potential to achieve 50,000 MIPS or 50 Gflops. The bisection bandwidth is 50 Gbytes/s.

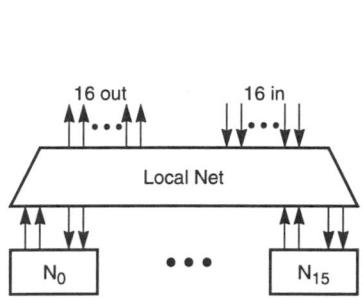

(a) A brick of 16 nodes with 3.2-Gflops and 3200-MIPS peak performance, packaged in a 9-in cube

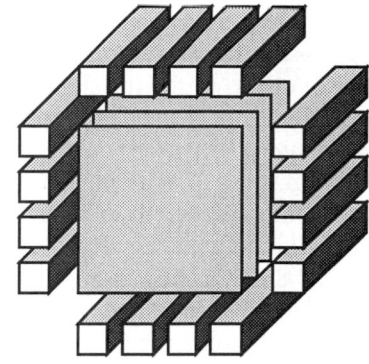

(b) A 256-node machine consisting of 16 bricks interconnected by 4 boards of 16×16 switches and packaged in a 1.5-m cube

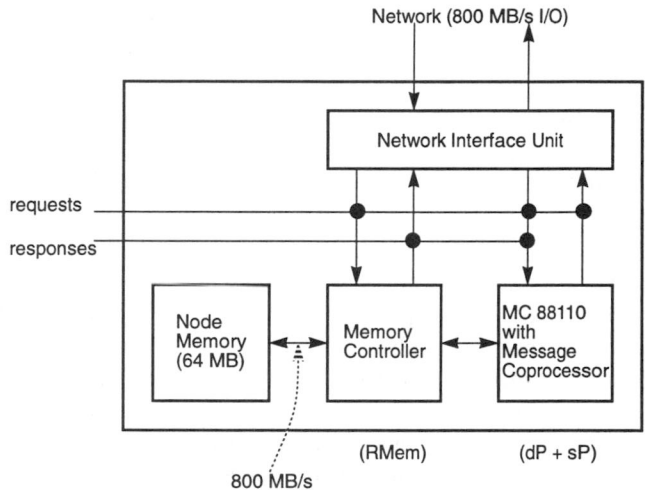

(c) Interior node architecture with data processor (MC 88110) and synchronized coprocessor (SP)

Figure 9.34 The MIT/Motorola *T prototype multithreaded architecture. (Courtesy of Nikhil, Papadopoulos, and Arvind, *Proc. 19th Int. Symp. Computer Arch.*, Australia, May 1992

The *T Node Design   Each node is proposed to be implemented with four component units. A Motorola superscalar RISC microprocessor (MC88110) is modified as a *data processor* (dP). This dP is optimized for long threads. Concurrent integer and floating-point operations are performed within each dP.

A *synchronization coprocessor* (sP) is implemented as an 88000 special-function unit (SFU), which is optimized for simple, short threads. Both the dP and the SP can handle fast loads. The dP handles incoming continuation, while the sP handles incoming messages, rload/rstore responses, and joins for messaging or synchronization purposes. In other words, the sP off-loads simple message-handling tasks from the main processor (the dP). Thus the dP will not be disrupted by short messages.

The *memory controller* handles requests for remote memory load or store, as well as the management of node memory (64 Mbytes). The *network interface unit* receives or transmits messages from or to the network, respectively, as illustrated in Fig. 9.34c. It should be noted that the sP is built as an on-chip SFU of the dP.

The MC 88110 family allows additional on-chip SFUs, with reserved opcode space, common instruction-issue logic and caches, etc., and direct access to processor registers. Example SFUs include the floating-point unit, graphics unit, coprocessor, etc. The MC 88110 is itself a two-way superscalar processor driven by a 50-MHz clock.

New SFUs are added into the MC 88110 to provide 16 buffers for incoming messages and 4 buffers for outgoing messages. Other SFUs include a *continuation stack* with 64 entries and a *microthreaded scheduler*, which supplies continuations from messages and the continuation stack, etc. Special instructions are available for packing or unpacking continuations.

Research Experiments   The *T prototype, once completed, will be used to test the effectiveness of the unified architecture in supporting multithreading operations. The development of the *T has been influenced by other multithreaded architectures, including the Tera, the Alewife, and the J-Machine.

The I-structure semantics is also implemented in the *T. Full/empty bits are used on producer-consumer variables. The *T treats messages as virtual continuations. Thus busy-waiting is eliminated. Other optimizations in the *T include speculative avoidance of the extra loads and stores through multithreading and coherent cacheing.

The *T designers want to provide a superset of the capabilities of the Tera, J-Machine, and EM-4. Compiler techniques developed for these machines should be applicable to the *T. To achieve these goals, a promising approach is to start with declarative languages while the compiler can effortlessly extract a large amount of fine-grain parallelism. Remaining research issues include dynamic storage management and load balancing.

Multithreading: A Perspective   The Dash, KSR-1, and Alewife leverage existing processor technology. The advantages of these directory-based cacheing systems include compatibility with existing hardware and software. But they offer a less aggressive pursuit of parallelism and depend heavily on compilers to obtain locality. The synchronizing loads are still problematic in these distributed cacheing solutions.

In von Neumann multithreading approaches, the HEP/Tera replicates the conventional instruction stream. Synchronizing-loads problems are solved by a hardware trap and software. Hybrid architectures, such as the Empire, still replicate conventional instruction streams, but they do not preserve registers across threads. The synchronizing loads are entirely supported in hardware. The J-Machine supports three instruction streams (priorities). It grows out of message-passing machines but adds support for global addressing. Remote synchronizing loads are supported by software convention.

In the dataflow approaches, the system-level view has stayed constant from the Tagged-Token Dataflow Architecture to the *T. The various designs differ in internal node architecture, with trends toward the removal of intranode synchronization, using larger threads, using longer threads, using high-speed registers, and compatibility with existing machine codes. The *T designers claimed that the unification of dataflow and von Neumann ideas will support a scalable shared-memory programming model using existing SIMD/SPMD codes.

9.6   Bibliographic Notes and Exercises

Latency tolerance through software-controlled prefetching was studied by [Mowry91] and Gupta. [Gupta91] et al. have compared various latency reducing and tolerating techniques through simulated benchmark experiments on the Stanford Dash. Shared virtual memory was studied by [Li86], [Li88], [Bisiani88] and Ravishankar, [Li89] and Schaefer, [Nitzberg91] and Lo, and [Li92]. The weak consistency model was pioneered by [Dubois86] et al. and [Sindhu90] et al. Processor consistency was developed by [Goodman89]. Release consistency was originally defined in [Gharachorloo90] et al. A follow-up paper by [Gharachorloo91] et al. evaluated various memory consistency models.

The Stanford Dash multiprocessor was reported in [Lenoski92] et al. The directory-based cache coherence protocol for Dash was first reported in [Lenoski90] et al. The IEEE scalable coherent interface was described in [James90b] et al. [Censier78] and Feautrier proposed the first directory structure for distributed cacheing systems. A delayed consistency model was described in [Dubois92b].

An excellent survey of multithreaded architectures was given in a tutorial by [Nikhil92a]. [Saavedra90], Culler, and Eicken have modeled multithreaded architectures using Markov chains. Multiple-context processors are discussed in [Laudon92] et al., in [Hirata92] et al., and in [Agarwal92a]. [Bell92] also modeled multithreaded computations. The description of the MIT J-Machine is based on the work of [Dally92] et al. The material on the Caltech Mosaic C was contributed by [Seitz92].

The Maryland Zmob was reported in [Weiser85] et al. The CDC Cyberplus was described in [Ferrante87]. The KSR-1 was described in [KSR91], [Burkhardt92], and [Rothnie91]. A subjective evaluation of KSR-1 can also be found in [Bell92]. The MIT Alewife was described in [Agarwal91] et al. The Wisconsin Multicube was reported by [Goodman88] and Woest. The USC orthogonal multiprocessor was described in [Hwang89b], Tseng, and Kim and in [Hwang91] et al. The Tera was reported in [Alverson90] et al.

Arvind and Iannucci [Arvind87] have studied latency and synchronization problems for multiprocessors. [Chaiken90] et al. and [Agarwal90] et al. have evaluated directory-based cacheing in multiprocessors. The HEP was reported in [Jordan83] and [Smith85]. [Kowalik85] edited a book on HEP applications. Gaudiot and Bic have edited a book on dataflow computers [Gaudiot91]. Static dataflow was reported in [Dennis80] and [Dennis91]. The dynamic dataflow and MIT tagged-token architectures are described in [Arvind84] et al. and [Arvind90] and Nikhil.

The Manchester dataflow machine was discussed in [Gurd85] et al. The ETL Sigma 1 was reported in [Hiraki87] et al. The EM-4 was reported in [Kodama90] et al., [Sakai91] et al., and [Yamaguchi91] et al. The MIT P-RISC was described in [Nikhil89] and Arvind. The *T was reported in [Nikhil92b] et al. [Arvind91] et al. have assessed the evaluation of dataflow computers.

Exercises

Problem 9.1   Consider a scalable multiprocessor with p processing nodes and distributed shared memory. Let R be the rate of each processing node generating a request to access remote memory through the interconnection network. Let L be the average latency for remote memory access. Derive expressions for the processor efficiency E under each of the following conditions:

(a) The processor is single-threaded, uses only a private cache, and has no other latency-hiding mechanisms. Express E as a function of R and L.

(b) Suppose a coherent cache is supported by hardware with proper data sharing and h is the probability that a remote request can be satisfied by a local cache. Express E as a function of R, L, and h.

(c) Now assume each processor is multithreaded to handle N contexts simultaneously. Assume a context-switching overhead of C. Express E as a function of N, R, L, h, and C.

(d) Now consider the use of a 2-D $r \times r$ torus with $r^2 = p$ and bidirectional links. Let t_d be the time delay between adjacent nodes and t_m be the local memory-access time. Assume that the network is fast enough to respond to each request without buffering. Express the latency L as a function of p, t_d, and t_m. Then express the efficiency E as a function of N, R, h, C, p, t_d, and t_m.

Problem 9.2   The following two questions are related to the effect of prefetching on latency tolerance:

(a) Perform an analytical study of the effects of data prefetching on the performance (efficiency) of processors in a scalable multiprocessor system without multithreading.

(b) Repeat part (a) for a multithreaded multiprocessor system under reasonable assumptions.

Problem 9.3   The following questions are related to the effects of memory consistency models:

(a) Perform an analytical study of the effects of using a relaxed consistency memory model in a scalable multiprocessor without multithreading.

(b) Repeat part (a) for a multithreaded multiprocessor system under reasonable assumptions.

(c) Can you derive an efficiency expression for a multiple-context processor supported by both prefetching and release memory consistency?

Problem 9.4 Consider a two-dimensional multicube architecture with m row buses and m column buses (Fig. 9.18a). Each bus has a bandwidth of B bits/s. The bus is considered active when it is actually in progress. The bus utilization rate a ($0 < a \leq 1$) is defined as the number of active bus cycles over the total cycles elapsed. The per-processor request rate r is defined as the number of requests that a processor sends on either of the two buses (for the purpose of memory access, cache coherence, synchronization, etc.) per second.

(a) Consider a single-column bus with associated processors and memory module and express the bus bandwidth as a function of m, a, and r.

(b) What is the total bus bandwidth available in the entire system?

(c) If r is kept constant as the number of processors increases, how many requests can be sent to the system without exceeding the limit?

(d) Each request goes through at a maximum of two buses in the multicube. What bus bandwidth will be needed to satisfy all the requests?

(e) In parts (b) and (d), does the multicube provide enough bus bandwidth? Justify the answer with reasoning.

Problem 9.5 Consider the use of an orthogonal multiprocessor consisting of 4 processors and 16 orthogonally shared memory modules (Fig. 9.18b) to perform an unfolded multiplication of two 8×8 matrices in a partitioned SPMD mode.

(a) Show how to distribute the 2×2 submatrices of the input matrix $A = (a_{ij})$ and $B = (b_{ij})$ to the 16 orthogonally shared memory modules.

(b) Specify the SPMD algorithm by involving all four processors in a synchronized manner to access either the row memories or the column memories. Synchronization is handled at the loop level.

   You can assume the use of a pipeline-read to fetch either one column or one row vector of the input matrix A or B at a time, and a pipeline-write to store the product matrix, $C = A \times B = (c_{ij})$, elements in a similar fashion. Assume that sufficient large register windows are available within each processor to hold all 2×2 submatrix elements. Each processor can perform inner product operations.

(c) Let $N \times N$ be the matrix size and $k = N/n$ the partitioned block size in mapping a large matrix in the orthogonal memory. Estimate the number of orthogonal memory accesses and the number of synchronizations needed in an SPMD algorithm for multiplying two $N \times N$ matrices on an n-processor OMP.

(d) Repeat the above for a two-dimensional fast Fourier transform over $N \times N$ sample points on an n-processor OMP, where $N = n \cdot k$ for some integer $k \geq 2$. The idea of performing a two-dimensional FFT on an OMP is to perform a one-dimensional FFT along one dimension in a row-access mode.

All n processors then synchronize, switch to a column-access mode, and perform another one-dimensional FFT along the second dimension. First try the case where $N = 8$, $n = 4$, and $k = 2$ and then work out the general case for large $N \gg n$.

Problem 9.6   The following questions are related to shared virtual memory:

(a) Why has shared virtual memory (SVM) become a necessity in building a scalable system with memories physically distributed over a large number of processing nodes?

(b) What are the major differences in implementing SVM at the cache block level and the page level?

Problem 9.7   The release consistency (RC) model has combined the advantages of both the processor consistency (PC) and the weak consistency (WC) models. Answer the following questions related to these consistency models:

(a) Compare the implementation requirements in the three consistency models.

(b) Comment on the advantages and shortcomings of each consistency model.

Problem 9.8   Answer the following questions involving the MIT J-Machine:

(a) What are the unique features of the message-driven processors (MDP) making it suitable for building fine-grain multicomputers?

(b) Explain the E-cube routing mechanism built into the MDP.

(c) What are the functions of the address arithmetic unit (AAU)?

(d) Explain the concept of using a combining tree for synchronization of events on various nodes in the J-Machine.

Problem 9.9   Explain the following architectural trends in building multicomputers or scalable MPP systems:

(a) Why are hypercube networks (binary n-cube networks), which were very popular in first-generation multicomputers, being replaced by 2D or 3D meshes or tori in the second and third generations of multicomputers?

(b) Can you relate first-level wormhole routing to the asynchronous superpipelining practiced in processor designs? Identify their similarities and differences.

Problem 9.10   Answer the following questions on the SCI standards:

(a) Explain the sharing-list creation and update methods used in the IEEE Scalable Coherence Interface (SCI) standard.

(b) Comment on the advantages and disadvantages of chained directories for cache coherence control in large-scale multiprocessor systems.

Problem 9.11 Compare the four context-switching policies: switch on cache miss, switch on every load, switch on every instruction (cycle by cycle), and switch on block of instructions.

(a) What are the advantages and shortcomings of each policy?

(b) What additional research is needed to make an optimal choice among the four policies?

Problem 9.12 After studying the Dash memory hierarchy and directory protocol, answer the following questions with an analysis of potential performance:

(a) Define the cache states used in the Dash.

(b) How are the cache directories implemented in the memory hierarchy?

(c) Explain the Dash directory-based coherence protocol when reading a remote cache block that is dirty in a remote cluster.

(d) Repeat part (c) for the case of writing to a shared remote cache block.

Problem 9.13 Answer the following questions on multiprocessors:

(a) Describe the ALLCACHE architecture implemented in the Kendall Square Research KSR-1.

(b) Explain how to maintain cache coherence in the KSR-1.

(c) Study the paper on COMA architectures by Stenström et al. (1992) and Hagersten et al. (1990). Compare the differences between the KSR-1 and the Data Diffusion Machine (DDM) architecture.

Problem 9.14 Answer the following questions on the development of the Tera computer.

(a) What are the design goals of the Tera computer?

(b) Explain the sparse 3D torus used in the Tera. What are the advantages of the sparse structure?

(c) Explain how superpipelining is applied in supporting the multithreaded operations in each Tera processor.

(d) Explain the thread state and management scheme used in the Tera.

(e) Explain the idea of explicit-dependence lookahead and its effects on multithreading in the Tera.

(f) What are the contributions of the Tera architecture and software development? Compare the advantages and potential drawbacks of the Tera computer.

Problem 9.15 Answer the following questions related to dataflow computers:

(a) Distinguish between static dataflow computers and dynamic dataflow computers.

(b) Draw a dataflow graph showing the computations of the roots of a sequence of quadratic equations $A_i x_i^2 + B_i x_i + C_i = 0$ for $i = 1, 2, ..., N$.

(c) Consider the parallel execution of the successive root computations 25 with a four-PE tagged-token dataflow computer (Fig. 2.12). Show a minimum-time schedule for using the four PEs to compute the N pairs of roots.

Problem 9.16   Consider the mapping of a one-dimensional circular convolution computation on a multiprocessor with 4 processors and 32 memory modules which are 32-way interleaved for pipelined access of vector data. Assume no contention between processors and memories in the interconnection network. The one-dimensional convolution is defined over a $1 \times n$ image and a $1 \times m$ kernel as follows:

$$Y(i) = \sum_{j=0}^{m-1} W(j) \cdot X((i - j) \bmod n) \qquad \text{for} \quad 0 \le i \le n - 1$$

(a) How many multiplications and additions are involved in the above computations? Map the image pixels $X(i)$ to memory module M_j if $j = i \pmod{32}$ and assume $n = 256$. The output image $Y(i)$ is also stored in module M_j if $j = i \pmod{32}$ for $0 \le i \le 255$. The kernel is also stored in a similar manner. Assume $m = 4$ and each processor handles the computation of one output image.

(b) Show how to partition the computations among the four processors such that minimum time is spent in both memory-access and CPU executions. Assume no memory conflicts and up to four fetch or store operations (but not mixed) can be performed at the same time. The interleaved memory can be accessed by one or more processors at the same time.

(c) What is the minimum execution time (including both memory and CPU operations) if each multiply and add and each interleaved memory access is considered one time unit. Assume enough working registers are available in each CPU.

(d) What is the speedup factor of the above multiprocessor solution over a uniprocessor solution? You can make similar assumptions about use of the 32-way interleaved memory for both uniprocessor and multiprocessor configurations.

Problem 9.17   Answer the following questions on fine-grain multicomputer and massive parallelism:

(a) Why are fine-grain processors chosen for future multiprocessors over medium-grain processors used in the past?

(b) Why is a single global addressing space desired over distributed address spaces?

(c) From scalability point of view, why is fine-grain parallelism more appealing than medium-grain or coarse-grain parallelism for building MPP systems?

Part IV

Software for Parallel Programming

Chapter 10 **Parallel Models, Languages, and Compilers** **Chapter 11** **Parallel Program Development and Environments** **Chapter 12** **UNIX, Mach, and OSF/1 for Parallel Computers**

Summary

Part IV discusses software and programming requirements of parallel/vector computers. We begin with a characterization of parallel programming models: shared-variable, message-passing, data-parallel, object-oriented, functional, logic, and heterogeneous. Then we evaluate parallel languages and compiler technologies for parallel programming. This includes the study of language features, programming environments, compilers for parallelization and vectorization, and performance tuning. We will describe locks, semaphores, monitors, synchronization, multitasking, and various program decomposition techniques.

In the final chapter, we identify multiprocessor UNIX design goals and study three kernel models for extending UNIX on multiprocessors or multicomputers. Finally, we consider the developmental experiences reported for the CMU Mach/OS kernel and its porting into the Open Software Foundation's OSF/1, which have evolved from UNIX BSD, System V, and Mach for general-purpose parallel computers.

Chapter 10

Parallel Models, Languages, and Compilers

This chapter is devoted to programming and compiler aspects of parallel and vector computers. To study beyond architectural capabilities, we must learn about the basic models for parallel programming and how to design optimizing compilers for parallelism. Models studied include those for shared-variable, message-passing, object-oriented, data-parallel, functional, and logic programming. We examine language extensions, parallelizing, vectorizing, and trace-driven compilers designed to support parallel programming.

10.1 Parallel Programming Models

A programming model is a collection of program abstractions providing a programmer a simplified and transparent view of the computer hardware/software system. Parallel programming models are specifically designed for multiprocessors, multicomputers, or vector/SIMD computers. Five models are characterized below for these computers that exploit parallelism with different execution paradigms.

10.1.1 Shared-Variable Model

In all programming systems, we consider processors active resources and memory and I/O devices passive resources. The basic computational units in a parallel program are *processes* corresponding to operations performed by related code segments. The granularity of a process may vary in different programming models and applications.

A *program* is a collection of processes. Parallelism depends on how interprocess communication (IPC) is implemented. Fundamental issues in parallel programming are centered around the *specification, creation, suspension, reactivation, migration, termination*, and *synchronization* of concurrent processes residing in the same or different processors.

By limiting the scope and access rights, the process address space may be shared or restricted. To ensure orderly IPC, a mutual exclusion property requires the exclusive access of a shared object by one process at a time. We address these issues and explore their solutions below.

Shared-Variable Communication   Multiprocessor programming is based on the use of shared variables in a common memory for IPC. As depicted in Fig. 10.1a, shared-variable IPC demands the use of shared memory and mutual exclusion among multiple processes accessing the same set of variables.

Fine-grain MIMD parallelism is exploited in tightly coupled multiprocessors. Interprocessor synchronization can be implemented either unconditionally or conditionally, depending on the mechanisms used.

The main issues in using this model include protected access of critical sections, memory consistency, atomicity of memory operations, fast synchronization, shared data structures, and fast data movement techniques to be studied in Section 10.2.

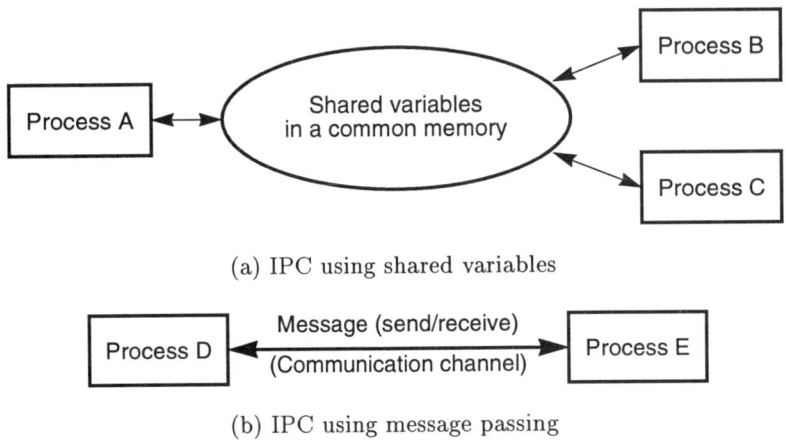

(a) IPC using shared variables

(b) IPC using message passing

Figure 10.1 Two basic mechanisms for interprocess communication (IPC).

Critical Section   A *critical section* (CS) is a code segment accessing shared variables, which must be executed by only one process at a time and which, once started, must be completed without interruption. In other words, a CS operation is indivisible and satisfies the following requirements:

- *Mutual exclusion* — At most one process executing the CS at a time.
- *No deadlock in waiting* — No circular wait by two or more processes trying to enter the CS; at least one will succeed.
- *Nonpreemption* — No interrupt until completion, once entered the CS.
- *Eventual entry* — A process attempting to enter its CS will eventually succeed.

Protected Access   The main problem associated with the use of a CS is avoiding race conditions where concurrent processes executing in different orders produce different results. The granularity of a CS affects the performance. If the boundary of a CS is too large, it may limit parallelism due to excessive waiting by competing processes.

When the CS is too small, it may add unnecessary code complexity or software overhead. The trick is to shorten a heavy-duty CS or to use conditional CSs to maintain a balanced performance.

In Chapter 11, we will study shared variables in the form of *locks* for implementing mutual exclusion in CSs. *Binary* and *counting semaphores* are used to implement CSs and to avoid system deadlocks. *Monitors* are suitable for structured programming.

Shared-variable programming requires special atomic operations for IPC, new language constructs for expressing parallelism, compilation support for exploiting parallelism, and OS support for scheduling parallel events and avoiding resource conflicts. Of course, all of these depend on the memory consistency model used.

Shared-memory multiprocessors use shared variables for interprocessor communications. Multiprocessing takes various forms, depending on the number of users and the granularity of divided computations. Four operational modes used in programming multiprocessor systems are specified below:

Multiprogramming   Traditionally, *multiprogramming* is defined as multiple independent programs running on a single processor or on a multiprocessor by time-sharing use of the system resources. A multiprocessor can be used in solving a single large problem or in running multiple programs across the processors.

A multiprogrammed multiprocessor allows multiple programs to run concurrently through time sharing of all the processors in the system. Multiple programs are interleaved in their CPU and I/O activities. When a program enters I/O mode, the processor switches to another program. Therefore, multiprogramming is not restricted to a multiprocessor. Even on a single processor, multiprogramming is often implemented.

Multiprocessing   When multiprogramming is implemented at the process level on a multiprocessor, it is called *multiprocessing*. Two types of multiprocessing are specified below. If interprocessor communications are handled at the instruction level, the multiprocessor operates in MIMD mode. If interprocessor communications are handled at the program, subroutine, or procedural level, the machine operates in MPMD (*multiple programs over multiple data streams*) mode.

In other words, we define MIMD multiprocessing with fine-grain instruction-level parallelism. MPMD multiprocessing exploits coarse-grain procedure-level parallelism. In both multiprocessing modes, shared variables are used to achieve interprocessor communication. This is quite different from the operations implemented on a message-passing system.

Multitasking   A single program can be partitioned into multiple interrelated tasks concurrently executed on a multiprocessor. This has been implemented as *multitasking* on Cray multiprocessors. Thus multitasking provides the parallel execution of two or more parts of a single program. A job efficiently multitasked requires less execution time. Multitasking is achieved with added codes in the original program in order to provide proper linkage and synchronization of divided tasks.

Tradeoffs do exist between multitasking and not multitasking. Only when overhead is short should multitasking be practiced. Sometimes, not all parts of a program can be divided into parallel tasks. Therefore, multitasking tradeoffs must be analyzed before implementation. Section 11.2 will treat this issue.

Multithreading The traditional UNIX/OS has a single-threaded kernel in which only one process can receive OS kernel service at a time. In a multiprocessor as studied in Chapter 9, we want to extend the single kernel to be multithreaded. The purpose is to allow multiple *threads* of lightweight processes to share the same address space and to be executed by the same or different processors simultaneously.

The concept of *multithreading* is an extension of the concepts of multitasking and multiprocessing. The purpose is to exploit fine-grain parallelism in modern multiprocessors built with multiple-context processors or superscalar processors with multiple-instruction issues. Each thread may use a separate program counter. Resource conflicts are the major problem to be resolved in a multithreaded architecture.

The levels of sophistication in securing data coherence and in preserving event order increase from monoprogramming to multitasking, to multiprogramming, to multiprocessing, and to multithreading in that order. Memory management and special protection mechanisms must be developed to ensure correctness and data integrity in parallel thread operations.

Partitioning and Replication The goal of parallel processing is to exploit parallelism as much as possible with the lowest overhead. *Program partitioning* is a technique for decomposing a large program and data set into many small pieces for parallel execution by multiple processors.

Program partitioning involves both programmers and the compiler. Parallelism detection by users is often explicitly expressed with parallel language constructs. Program restructuring techniques can be used to transform sequential programs into a parallel form more suitable for multiprocessors. Ideally, this transformation should be carried out automatically by a compiler.

Program replication refers to duplication of the same program code for parallel execution on multiple processors over different data sets. Partitioning is often practiced on a shared-memory multiprocessor system, while replication is more suitable for distributed-memory message-passing multicomputers.

So far, only special program constructs, such as independent loops and independent scalar operations, have been successfully parallelized. Clustering of independent scalar operations into vector or VLIW instructions is another approach toward this end.

Scheduling and Synchronization Scheduling of divided program modules on parallel processors is much more complicated than scheduling of sequential programs on a uniprocessor. *Static scheduling* is conducted at post compile time. Its advantage is low overhead but the shortcoming is a possible mismatch with the run-time profile of each task and therefore potentially poor resource utilization.

Dynamic scheduling catches the run-time conditions. However, dynamic scheduling requires fast context switching, preemption, and much more OS support. The advantages of dynamic scheduling include better resource utilization at the expense of higher scheduling overhead. Static and dynamic methods can be jointly used in a sophisticated multiprocessor system demanding higher efficiency.

In a conventional UNIX system, *interprocessor communication* (IPC) is conducted at the process level. Processes can be created by any processor. All processes asynchronously accessing the shared data must be protected so that only one is allowed to access the shared writable data at a time. This *mutual exclusion* property is enforced with the use of locks, semaphores, and monitors to be described in Chapter 11.

At the control level, virtual program counters can be assigned to different processes or threads. Counting semaphores or barrier counters can be used to indicate the completion of parallel branch activities. One can also use atomic memory operations such as *Test&Set* and *Fetch&Add* to achieve synchronization. Software-implemented synchronization may require longer overhead. Hardware barriers or combining networks can be used to reduce the synchronization time.

Cache Coherence and Protection   Besides maintaining data coherence in a memory hierarchy, multiprocessors must assume data consistency between private caches and the shared memory. The multicache coherence problem demands an invalidation or update after each write operation. These coherence control operations require special bus or network protocols for implementation as noted in previous chapters. A memory system is said to be *coherent* if the value returned on a read instruction is always the value written by the latest write instruction on the same memory location. The access order to the caches and to the main memory makes a big difference in computational results.

The shared memory of a multiprocessor can be used in various consistency models as discussed in Chapters 4 and 9. Sequential consistency demands that all memory accesses be strongly ordered on a global basis. A processor cannot issue an access until the most recently shared writable memory access has been globally performed. A weak consistency model enforces ordering and coherence at explicit synchronization points only. Programming with the processor consistency or release consistency may be more restricted, but memory performance is expected to improve.

10.1.2 Message-Passing Model

Multicomputer programming is depicted in Fig. 10.1b. Two processes D and E residing at different processor nodes may communicate with each other by passing messages through a direct network. The messages may be instructions, data, synchronization, or interrupt signals, etc. The communication delay caused by message passing is much longer than that caused by accessing shared variables in a common memory. Multicomputers are considered loosely coupled multiprocessors. Two message-passing programming models are introduced below. Techniques for message-passing programming are treated in Sections 11.4 and 11.5.

Synchronous Message Passing   Since there is no shared memory, there is no need for mutual exclusion. *Synchronous message* passing must synchronize the sender process and the receiver process in time and space, just like a telephone call using circuit-switched lines. In general, no buffers are used in the communication channels. That is why synchronous communication can be blocked by channels being busy or in error since only one message is allowed to be transmitted via a channel at a time.

In a *synchronous* paradigm, the passing of a message must synchronize the sending process and the receiving process in time and space. Besides having a time connection, the sender and receiver must also be linked by physical communication channels in space. A path of channels must be ready to enable the message passing between them.

In other words, the sender and receiver must be coupled in both time and space synchronously. If one process is ready to communicate and the other is not, the one that is ready must be blocked (or wait). In this sense, synchronous communication has been also called a blocking communication scheme.

Asynchronous Message Passing   Asynchronous communication does not require that message sending and receiving be synchronized in time and space. Buffers are often used in channels, which results in nonblocking in message passing provided sufficiently large buffers are used or the network traffic is not saturated.

However, arbitrary communication delays may be experienced because the sender may not know if and when the message has been received until acknowledgment is received from the receiver. This scheme is like a postal service using mailboxes (channel buffers) with no synchronization between senders and receivers.

Nonblocking can be achieved by *asynchronous message passing* in which two processes do not have to be synchronized either in time or in space. The sender is allowed to send a message without blocking, regardless of whether the receiver is ready or not.

Asynchronous communication requires the use of buffers to hold the messages along the path of the connecting channels. Since channel buffers are finite, the sender will eventually be blocked. In a synchronous multicomputer, buffers are not needed because only one message is allowed to pass through a channel at a time.

The critical issue in programming this model is how to distribute or duplicate the program codes and data sets over the processing nodes. Tradeoffs between computation time and communication overhead must be considered.

Message-passing programming is gradually changing, once the virtual memory spaces from all nodes are combined. As explained in Chapter 9, fine-grain concurrent programming with global naming merges the shared-variable and message-passing mechanisms for heterogeneous processing.

Distributing the Computations   Program replication and data distribution are used in multicomputers. The processors in a multicomputer (or a NORMA machine) are loosely coupled in the sense that they do not share memory. Message passing in a multicomputer is handled at the subprogram level rather than at the instructional or fine-grain process level as in a tightly coupled multiprocessor. That is why explicit parallelism is more attractive for multicomputers.

Example 10.1   A concurrent program for distributed computing on a multicomputer (Justin Rattner, Intel Scientific Computers, 1990)

The computation involved is the evaluation of π as the area under the curve $f(x)$ between 0 and 1 as shown in Fig. 10.2. Using a rectangle rule, we write the integral in discrete form:

$$\pi = \int_0^1 f(x)\ dx = \int_0^1 \frac{4}{1+x^2}\ dx \doteq h \sum_{i=1}^n f(x_i)$$

where $h = 1/n$ is the panel width, $x_i = h(i - 0.5)$ are the midpoints, and n is the number of panels (rectangles) to be computed.

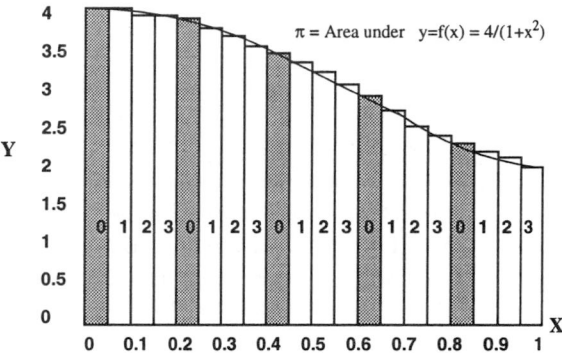

Figure 10.2 Domain decomposition for concurrent programming on a multicomputer with four processors.

Assume a four-node multicomputer with four processors labeled 0, 1, 2, and 3. The rectangle rule decomposition is shown with $n = 20$ and $h = 1/20 = 0.05$. Each processor node is assigned to compute the areas of five rectangular panels. Therefore, the computational load of all four nodes is balanced.

Host program	Node program
input(n)	p = numnodes()
send(n,allnodes)	me = mynode()
recv(Pi)	recv(n)
output(Pi)	h = 1.0 / n
	sum = 0
	Do i = me + 1, n, p
	x = h × (i − 0.5)
	sum = sum + f(x)
	End Do
	pi = h × sum
	gop('+', Pi, host)

Each node executes a separate copy of the node program. Several system calls are used to achieve message passing between the host and the nodes. The host program *sends* the panel number n as a message to all the nodes, which receive it accordingly in the node program. The commands *numnodes* and *mynode* specify how big the system is and which node it is, respectively.

The software for the iPSC system offers a global summing operation *gop*('+', pi, host) which iteratively pairs nodes that exchange their current partial sums. Each partial sum received from another node is added to the sum at the receiving node, and the new sum is sent out in the next round of message exchange.

Eventually, all the nodes accumulate the global sum multiplied by the height (pi = $h\times$ sum) which will be returned to the host for printout. Not all pairs of node communications need to be carried out. Only $\log_2 N$ rounds of message exchanges are required to compute the adder-tree operations, where N is the number of nodes in the system.

■

10.1.3   Data-Parallel Model

With the lockstep operations in SIMD computers, the data-parallel code is easier to write and to debug because parallelism is explicitly handled by hardware synchronization and flow control. Data-parallel languages are modified directly from standard serial programming languages. For example, Fortran 90 is specially tailored for data parallelism. Thinking Machines' C* is specially designed for programming the Connection Machines.

Data-parallel programs require the use of predistributed data sets. Thus the choice of parallel data structures makes a big difference in data-parallel programming. Interconnected data structures are also needed to facilitate data exchange operations. In summary, data-parallel programming emphasizes local computations and data routing operations (such as permutation, replication, reduction, and parallel prefix). It is applied to fine-grain problems using regular grids, stencils, and multidimensional signal/image data sets.

Data parallelism can be implemented either on SIMD computers or on SPMD multicomputers, depending on the grain size and operation mode adopted. In this section, we consider mainly parallel programming on SIMD computers that emphasize fine-grain data parallelism under synchronous control. Data parallelism often leads to a high degree of parallelism involving thousands of data operations concurrently. This is rather different from control parallelism which offers a much lower degree of parallelism at the instruction level.

Synchronization of data-parallel operations is done at compile time rather than at run time. Hardware synchronization is enforced by the control unit to carry out the lockstep execution of SIMD programs. We address below instruction/data broadcast, masking, and data-routing operations separately. Languages, compilers, and the conversion of SIMD programs to run on MIMD multicomputers are also discussed.

Data Parallelism   Ever since the introduction of the Illiac IV computer, programming

SIMD array processors has been a challenge for computational scientists. The main difficulty in using the Illiac IV has been to match the problem size with the fixed machine size. In other words, large arrays or matrices must be partitioned into 64-element segments before they can be effectively processed by the 64 processing elements (PEs) in the Illiac IV machine.

Today's SIMD computers, like the Connection Machine CM-2, offer bit-slice fine-grain data parallelism using 16,384 PEs concurrently in a single-array configuration. This demands a lower degree of array segmentation and thus higher flexibility in programming.

Synchronous SIMD programming differs from asynchronous MIMD programming in that all PEs in an SIMD computer operate in a lockstep fashion, whereas all processors in an MIMD computer execute different instructions asynchronously. As a result, SIMD computers do not have the mutual exclusion or synchronization problems associated with multiprocessors or multicomputers.

Instead, inter-PE communications are directly controlled by hardware. Besides lockstep in computing operations among all PEs, inter-PE data communication is also carried out in lockstep. These synchronized instruction executions and data-routing operations make SIMD computers rather efficient in exploring spatial parallelism in large arrays, grids, or meshes of data.

In an SIMD program, scalar instructions are directly executed by the control unit. Vector instructions are broadcast to all processing elements. Vector operands are loaded into the PEs from local memories simultaneously using a global address with different offsets in local index registers. Vector stores can be executed in a similar manner. Constant data can be broadcast to all PEs simultaneously.

The masking pattern (binary vector) can be set under program control so that PEs can be enabled or disabled dynamically in any instruction cycle. Masking instructions are directly supported by hardware. Data-routing vector operations are supported by an inter-PE routing network, which is also under program control on a dynamic basis.

Array Language Extensions   Array extensions in data-parallel languages are represented by high-level data types. We will specify Fortran 90 array notations in Section 10.2.2. The array syntax enables the removal of some nested loops in the code and should reflect the architecture of the array processor.

Examples of array processing languages are CFD for the Illiac IV, DAP Fortran for the AMT/Distributed Array Processor, C* for the TMC/Connection Machine, and MPF for the MasPar family of massively parallel computers.

An SIMD programming language should have a global address space, which obviates the need for explicit data routing between PEs. The array extensions should have the ability to make the number of PEs a function of the problem size rather than a function of the target machine.

Connection Machine C* language fits these requirements nicely. A Pascal-based language, *Actus*, has been developed by R.H. Perrott for problem-oriented SIMD programming. Actus offers hardware transparency, application flexibility, and explicit control structures in both program structuring and data typing operations.

Compiler Support To support data-parallel programming, the array language expressions and their optimizing compilers must be embedded in familiar standards such as Fortran 77, Fortran 90, and C. The idea is to unify the program execution model, facilitate precise control of massively parallel hardware, and enable incremental migration to data-parallel execution.

Compiler-optimized control of SIMD machine hardware allows the programmer to drive the PE array transparently. The compiler must separate the program into scalar and parallel components and integrate with the UNIX environment.

The compiler technology must allow array extensions to optimize data placement, minimize data movement, and virtualize the dimensions of the PE array. The compiler generates data-parallel machine code to perform operations on arrays.

Array sectioning allows a programmer to reference a section or a region of a multi-dimensional array. Array sections are designated by specifying a start index, a bound, and a stride. *Vector-valued subscripts* are often used to construct arrays from arbitrary permutations of another array. These expressions are vectors that map the desired elements into the target array. They facilitate the implementation of *gather* and *scatter* operations on a vector of indices.

SIMD programs can be recompiled for MIMD architecture. The idea is to develop a source-to-source precompiler to convert, for example, from Connection Machine C* programs to C programs running on an nCUBE message-passing multicomputer in SPMD mode.

In fact, SPMD programs are a special class of SIMD programs which emphasize medium-grain parallelism and synchronization at the subprogram level rather than at the instruction level. In this sense, the data-parallel programming model applies to both synchronous SIMD and loosely coupled MIMD computers. Program conversion between different machine architectures is very much needed to broaden software portability.

10.1.4 Object-Oriented Model

If one considers special language features and their implications, additional models for parallel programming can be introduced. An object-oriented programming model is characterized below.

In this model, *objects* are dynamically created and manipulated. Processing is performed by sending and receiving messages among objects. Concurrent programming models are built up from low-level objects such as processes, queues, and semaphores into high-level objects like monitors and program modules.

Concurrent OOP The popularity of *object-oriented programming* (OOP) is attributed to three application demands: First, there is increased use of interacting processes by individual users, such as the use of multiple X windows. Second, workstation networks have become a cost-effective mechanism for resource sharing and distributed problem solving. Third, multiprocessor technology has advanced to the point of providing supercomputing power at a fraction of the traditional cost.

As a matter of fact, program abstraction leads to program modularity and software reusability as is often found in OOP. Other areas that have encouraged the growth of

OOP include the development of CAD (computer-aided design) tools and word processors with graphics capabilities.

Objects are program entities which encapsulate data and operations into single computational units. It turns out that concurrency is a natural consequence of the concept of objects. In fact, the concurrent use of coroutines in conventional programming is very similar to the concurrent manipulation of objects in OOP.

The development of *concurrent object-oriented programming* (COOP) provides an alternative model for concurrent computing on multiprocessors or on multicomputers. Various object models differ in the internal behavior of objects and in how they interact with each other.

An Actor Model COOP must support patterns of reuse and classification, for example, through the use of inheritance which allows all instances of a particular class to share the same property. An *actor model* developed at MIT is presented as one framework for COOP.

Actors are self-contained, interactive, independent components of a computing system that communicate by asynchronous message passing. In an actor model, message passing is attached with semantics. Basic actor primitives include:

(1) *Create*: Creating an actor from a behavior description and a set of parameters.
(2) *Send-to*: Sending a message to another actor.
(3) *Become*: An actor replacing its own behavior by a new behavior.

State changes are specified by behavior replacement. The replacement mechanism allows one to aggregate changes and to avoid unnecessary control-flow dependences. Concurrent computations are visualized in terms of concurrent actor creations, simultaneous communication events, and behavior replacements. Each message may cause an object (actor) to modify its state, create new objects, and send new messages.

Concurrency control structures represent particular patterns of message passing. The actor primitives provide a low-level description of concurrent systems. High-level constructs are also needed for raising the granularity of descriptions and for encapsulating faults. The actor model is particularly suitable for multicomputer implementations.

Parallelism in COOP Three common patterns of parallelism have been found in the practice of COOP. First, *pipeline concurrency* involves the overlapped enumeration of successive solutions and concurrent testing of the solutions as they emerge from an evaluation pipeline.

Second, *divide-and-conquer concurrency* involves the concurrent elaboration of different subprograms and the combining of their solutions to produce a solution to the overall problem. In this case, there is no interaction between the procedures solving the subproblems. These two patterns are illustrated by the following examples taken from the paper by Agha (1990).

Example 10.2 Concurrency in object-oriented programming (Gul Agha, 1990)

A prime-number generation pipeline is shown Fig. 10.3a. Integer numbers are generated and successively tested for divisibility by previously generated primes in a linear pipeline of primes. The circled numbers represent those being generated.

A number enters the pipeline from the left end and is eliminated if it is divisible by the prime number tested at a pipeline stage. All the numbers being forwarded to the right of a pipeline stage are those indivisible by all the prime numbers tested on the left of that stage.

Figure 10.3b shows the multiplication of a list of numbers $(10, 7, -2, 3, 4, -11, -3)$ using a divide-and-conquer approach. The numbers are represented as leaves of a tree. The problem can be recursively subdivided into subproblems of multiplying two sublists, each of which is concurrently evaluated and the results multiplied at the upper node.

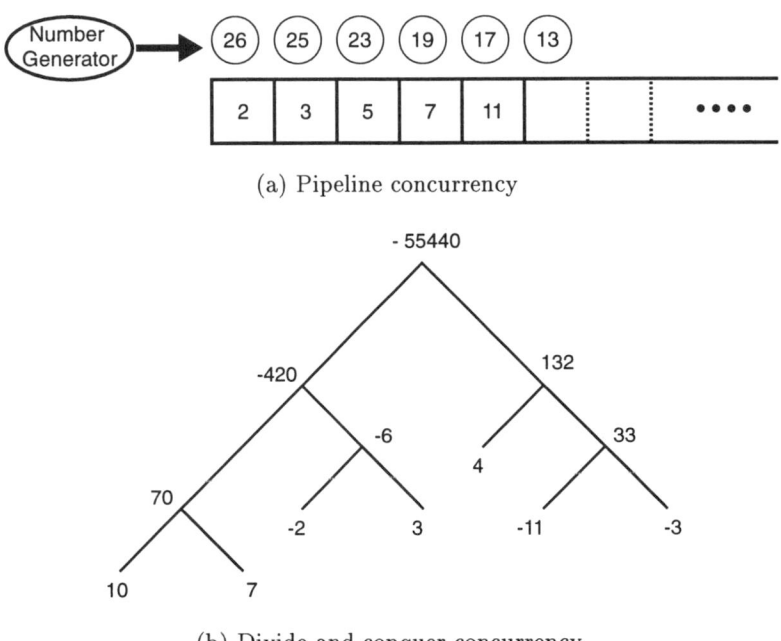

(a) Pipeline concurrency

(b) Divide-and-conquer concurrency

Figure 10.3 Two concurrency types in object-oriented programming. (Courtesy of G. Agha, *Commun. ACM*, September 1990)

A third pattern is called *cooperative problem solving*. A simple example is the dynamic path evaluation (computational objects) of many physical bodies (objects) under the mutual influence of gravitational fields. In this case, all objects must interact with each other; intermediate results are stored in objects and shared by passing messages between them. Interested readers should refer to the book on actors by Agha (1986).

10.1.5   Functional and Logic Models

Two language-oriented programming models for parallel processing are described below. The first model is based on using functional programming languages such as pure *Lisp*, *SISAL*, and *Strand 88*. The second model is based on logic programming languages such as *Concurrent Prolog* and *Parlog*. We reveal opportunities for parallelism in these two models and discuss their potential in AI applications.

Functional Programming Model   A functional programming language emphasizes the *functionality* of a program and should not produce side effects after execution. There is no concept of storage, assignment, and branching in functional programs. In other words, the history of any computation performed prior to the evaluation of a functional expression should be irrelevant to the meaning of the expression.

The lack of side effects opens up much more opportunity for parallelism. Precedence restrictions occur only as a result of function application. The evaluation of a function produces the same value regardless of the order in which its arguments are evaluated. This implies that all arguments in a dynamically created structure of a functional program can be evaluated in parallel. All *single-assignment* and *dataflow languages* are functional in nature. This implies that functional programming models can be easily applied to data-driven multiprocessors. The functional model emphasizes fine-grain MIMD parallelism and is referentially transparent.

The majority of parallel computers designed to support the functional model were oriented toward Lisp, such as the Multilisp developed at MIT. Other dataflow computers have been used to execute functional programs, including SISAL used in the Manchester dataflow machine.

Logic Programming Model   Based on predicate logic, *logic programming* is suitable for knowledge processing dealing with large databases. This model adopts an implicit search strategy and supports parallelism in the logic inference process. A question is answered if the matching facts are found in the database. Two facts match if their predicates and associated arguments are the same. The process of matching and unification can be parallelized under certain conditions. Clauses in logic programming can be transformed into dataflow graphs. Parallel unification has been attempted on some dataflow computers built in Japan.

Concurrent Prolog, developed by Shapiro (1986), and *Parlog*, introduced by Clark (1987), are two parallel logic programming languages. Both languages can implement relational language features such as AND-parallel execution of conjunctive goals, IPC by shared variables, and OR-parallel reduction.

In *Parlog*, the resolution tree has one chain at AND levels, and OR levels are partially or fully generated. In *Concurrent Prolog*, the search strategy follows multiple paths or depth first. Stream parallelism is also possible in these logic programming systems.

Both functional and logic programming models have been used in artificial intelligence applications where parallel processing is very much in demand. Japan's *Fifth-Generation Computing System* (FGCS) project has attempted to develop parallel logic systems for problem solving, machine inference, and intelligent human-machine interfacing.

In many ways, the FGCS project is a marriage of parallel processing hardware and AI software. The *Parallel Inference Machine* (PIM-I) in this project was designed to perform 10 *million logic inferences per second* (MLIPS). The actual performance result is yet to be verified with real benchmark experiments.

10.2   Parallel Languages and Compilers

The environment for parallel computers is much more demanding than that for sequential computers. A programming environment is a collection of software tools and system software support. Software for driving parallel computers is still in the early developmental stage. Users are still forced to spend a lot of time programming hardware details instead of concentrating on program parallelism using high-level abstraction.

To break this hardware/software barrier, we need a parallel software environment which provides better tools for users to implement parallelism and to debug programs. Most of the recently developed software tools are still in the research and testing stage, and a few have become commercially available.

10.2.1   Language Features for Parallelism

Chang and Smith (1990) have classified the language features for parallel programming into six categories according to functionality. These features are idealized for general-purpose applications. In practice, the real languages developed or accepted by the user community might have some or no features in some of the categories. Some of the features are identified with existing language/compiler development. The listed features set guidelines for developing a user-friendly programming environment.

Optimization Features   These features are used for program restructuring and compilation directives in converting sequentially coded programs into parallel forms. The purpose is to match the software parallelism with the hardware parallelism in the target machine.

- Automated parallelizer — The Express C automated parallelizer and the Alliant FX Fortran compiler.
- Semiautomated parallelizer — Needs compiler directives or programmer's interaction, such as DINO.
- Interactive restructure support — Static analyzer, run-time statistics, dataflow graph, and code translator for restructuring Fortran code, such as the MIMDizer from Pacific Sierra.

Availability Features   These are features that enhance the user-friendliness, make the language portable to a large class of parallel computers, and expand the applicability of software libraries.

- Scalability — The language is scalable to the number of processors available and independent of hardware topology.
- Compatibility — The language is compatible with an established sequential language.
- Portability — The language is portable to shared-memory multiprocessors, message-passing multicomputers, or both.

Synchronization/Communication Features   Listed below are desirable language features for synchronization or for communication purposes:

- Single-assignment languages
- Shared variables (locks) for IPC
- Logically shared memory such as the tuple space in Linda
- Send/receive for message passing
- Rendezvous in Ada
- Remote procedure call
- Dataflow languages such as Id
- Barriers, mailbox, semaphores, monitors

Control of Parallelism   Listed below are features involving control constructs for specifying parallelism in various forms:

- Coarse, medium, or fine grain
- Explicit versus implicit parallelism
- Global parallelism in the entire program
- Loop parallelism in iterations
- Task-split parallelism
- Shared task queue
- Divide-and-conquer paradigm
- Shared abstract data types
- Task dependency specification

Data Parallelism Features   Data parallelism is used to specify how data are accessed and distributed in either SIMD or MIMD computers.

- Run-time automatic decomposition — Data are automatically distributed with no user intervention as in Express.
- Mapping specification — Provides a facility for users to specify communication patterns or how data and processes are mapped onto the hardware, as in DINO.
- Virtual processor support — The compiler maps the virtual processors dynamically or statically onto the physical processors, as in PISCES 2 and DINO.
- Direct access to shared data — Shared data can be directly accessed without monitor control, as in Linda.

- SPMD (single program multiple data) support — SPMD programming, as in DINO and Hypertasking.

Process Management Features   These features are needed to support the efficient creation of parallel processes, implementation of multithreading or multitasking, program partitioning and replication, and dynamic load balancing at run time.

- Dynamic process creation at run time
- Lightweight processes (threads) — Compared to UNIX (heavyweight) processes
- Replicated workers — Same program on every node with different data (SPMD mode)
- Partitioned networks — Each processor node might have more than one process and all process nodes might run different processes
- Automatic load balancing — The workload is dynamically migrated among busy and idle nodes to achieve the same amount of work at various processor nodes

The above language features cannot be implemented without compiler support, operating system assistance, and integration with an existing environment. Software assets based on conventional languages form the basis for building an efficient parallel programming environment.

The optimization features emphasize code parallelization and vectorization at compile time. The availability features widen the application domains and make the languages machine-independent.

The synchronization features must be supported by efficient hardware and software mechanisms for their implementation. The control features often depend on tradeoffs among grain size, memory demand, and communication and scheduling overhead. Data parallelism exploits fine-grain computations on SIMD machines and medium-grain on MIMD computers.

The process management features are closely tied to the OS functions provided. Therefore, the languages, compilers, and OS must be developed jointly in an integrated fashion.

10.2.2   Parallel Language Constructs

Special language constructs and data array expressions are presented below for exploiting parallelism in programs. We first specify Fortran 90 array notations. Then we describe commonly used parallel constructs for program flow control.

Fortran 90 Array Notations   A multidimensional data array is represented by an array name indexed by a sequence of subscript triplets, one for each dimension. Triplets for different dimensions are separated by commas as specified below:

$$e_1 : e_2 : e_3$$
$$e_1 : e_2$$
$$e_1 : * : e_3 \tag{10.1}$$

$$e_1 : *$$

$$e_1$$

$$*$$

where each e_i is an arithmetic expression that must produce a scalar integer value. The first expression e_1 is a *lower bound*, the second e_2 an *upper bound*, and the third e_3 an *increment (stride)*. For example, $B(1 : 4 : 3, 6 : 8 : 2, 3)$ represents four elements $B(1, 6, 3)$, $B(4, 6, 3)$, $B(1, 8, 3)$, and $B(4, 8, 3)$ of a three-dimensional array.

When the third expression in a triplet is missing, a unit stride is assumed. The $*$ notation in the second expression indicates all elements in that dimension starting from e_1, or the entire dimension if e_1 is also omitted. When both e_2 and e_3 are omitted, the e_1 alone represents a single element in that dimension. For example, $A(5)$ represents the fifth element in the array $A(3 : 7 : 2)$. This notation allows us to select array sections or particular array elements.

Array assignments are permitted under the following constraints: The array expression on the right must have the same shape and the same number of elements as the array on the left. For example, the assignment $A(2 : 4, 5 : 8) = A(3 : 5, 1 : 4)$ is valid, but the assignment $A(1 : 4, 1 : 3) = A(1 : 2, 1 : 6)$ is not valid, even though each side has 12 elements. When a scalar is assigned to an array, the value of the scalar is assigned to every element of the array. For instance, the statement $B(3 : 4, 5) = 0$ sets $B(3, 5)$ and $B(4, 5)$ to 0.

Parallel Flow Control   The conventional Fortran Do loop declares that all scalar instructions within the (**Do, Enddo**) pair are executed sequentially, and so are the successive iterations. To declare parallel activities, we use the (**Doall, Endall**) pair. All iterations in the Doall loop are totally independent of each other. This implies that they can be executed in parallel if there are sufficient processors to handle different iterations. However, the computations within each iteration are still executed serially in program order.

When the successive iterations of a loop nest depend on each other, we use the (**Doacross, Endacross**) pair to declare parallelism with loop-carried dependences. Synchronizations must be performed between the iterations that depend on each other. For example, dependence along the J-dimension exists in the following program. We use **Doacross** to declare parallelism along the I-dimension, but synchronization between iterations is required. The (**Forall, Endall**) and (**Pardo, Parend**) commands can be interpreted either as a Doall loop or as a Doacross loop.

```
Doacross I=2, N
    Do J=2, N
S₁:     A(I,J)=(A(I,(J−1))+A(I,J+1))/2
    Enddo
Endacross
```

Another program construct is the (**Cobegin, Coend**) pair. All computations specified within the block could be executed in parallel. But parallel processes may be created with a slight time difference in real implementations. This is quite different from the semantics of the Doall loop or Doacross loop structures. Synchronizations among concurrent processes created within the pair are implied. Formally, the command

Cobegin
P_1
P_2
$\vdots$
P_n
Coend

causes processes P_1, P_2, ..., P_n to start simultaneously and to proceed concurrently until they have all ended. The command (**Parbegin, Parend**) has equivalent meaning.

Finally, we introduce the **Fork** and **Join** commands in the following example. During the execution of a process P, we can use a **Fork** Q command to spawn a new process Q:

Process P	Process Q
$\vdots$	$\vdots$
Fork Q	$\vdots$
$\vdots$	**End**
Join Q	

The **Join** Q command recombines the two processes into one process. Execution of Q is initialized when the **Fork** Q statement in P is executed. Programs P and Q are executed concurrently until either P executes the **Join** Q statement or Q terminates. Whichever one finishes first must wait for the other to complete execution, before they can be rejoined.

In a UNIX environment, the **Fork-Join** statements provide a direct mechanism for dynamic process creation including multiple activations of the same process. The **Cobegin-Coend** statements provide a structured single-entry, single-exit control command which is not as dynamic as the **Fork-Join**. The (**Parbegin, Parend**) command is equivalent to the (**Cobegin, Coend**) command.

10.2.3 Optimizing Compilers for Parallelism

Using high-level languages in source code, compilers have become a necessity in modern computers. The role of a compiler is to remove the burden of program optimization and code generation from the programmer. A parallelizing compiler consists of the following three major phases: *flow analysis, optimizations,* and *code generation,* as depicted in Fig. 10.4.

Flow Analysis   This phase reveals the program flow patterns in order to determine data and control dependences in the source code. We have discussed data dependence relations among scalar-type instructions in previous chapters. Scalar dependence analysis is extended below to structured data arrays or matrices. Depending on the machine

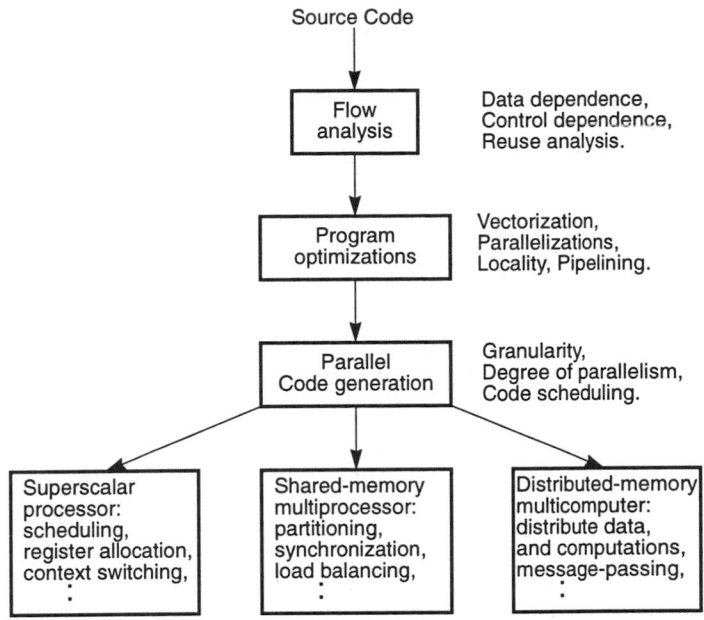

Figure 10.4 Compilation phases in parallel code generation.

structure, the granularities of parallelism to be exploited are quite different. Thus the flow analysis is conducted at different execution levels on different parallel computers.

Generally speaking, instruction-level parallelism is exploited in superscalar or VLSI processors; loop level in SIMD, vector, or systolic computers; and task level in multiprocessors, multicomputers, or a network of workstations. Of course, exceptions do exist. For example, fine-grain parallelism is being pushed down to the next generation of multicomputers with a globally shared address space. The flow analysis must also reveal some code/data reuse and memory-access patterns.

Program Optimizations This refers to the transformation of user programs in order to explore the hardware capabilities as much as possible. Transformation can be conducted at the loop level, locality level, or prefetching level with the ultimate goal of reaching global optimization. The optimization often transforms a code into an equivalent but "better" form in the same representation language. These transformations should be machine-independent.

In reality, most transformations are constrained by the machine architecture. This is the main reason why many compilers are machine-dependent. At the least, we want to design a compiler which can run on most machines with only minor modifications. One can also conduct certain transformations preceding the global optimization. This may require a source-to-source optimization (sometimes called a *precompiler*), which transforms the program from one high-level language to another before using a dedicated compiler for the second language on a target machine.

The ultimate goal of program optimization is to maximize the speed of code execution. This involves the minimization of code length and of memory accesses and the exploitation of parallelism in programs. The optimization techniques include vectorization using pipelined hardware and parallelization using multiple processors simultaneously. The compiler should be designed to reduce the running time with minimum resource binding. Other optimizations demand the expansion of routines or procedure integration with inlining. Both local and global optimizations are needed in most programs. Sometimes the optimization should be conducted at the algorithmic level and must involve the programmer.

Machine-dependent transformations are meant to achieve more efficient allocation of machine resources, such as memory, registers, and functional units. Replacement of complex operations by cheaper ones is often practiced. Other optimizations include elimination of unnecessary branches or common expressions. Instruction scheduling can be used to eliminate pipeline or memory delays in executing consecutive instructions.

Parallel Code Generation Code generation usually involves transformation from one representation to another, called an *intermediate form*. A code model must be chosen as an intermediate form. Parallel code is even more demanding because parallel constructs must be included. Code generation is closely tied to the instruction scheduling policies used. Basic blocks linked by control-flow commands are often optimized to encourage a high degree of parallelism . Special data structures are needed to represent instruction blocks.

Parallel code generation is very different for different computer classes. For example, a superscalar processor may be software-scheduled or hardware-scheduled. How to optimize the register allocation on a RISC or superscalar processor, how to reduce the synchronization overhead when codes are partitioned for multiprocessor execution, and how to implement message-passing commands when codes/data are distributed (or replicated) on a multicomputer are added difficulties in parallel code generation. Compiler directives can be used to help generate parallel code when automated code generation cannot be implemented easily.

Two exploratory optimizing compilers have been developed over the last decade: one is Parafrase at the University of Illinois, and the other is the PFC (Parallel Fortran Converter) at Rice University. These systems are briefly introduced below.

Parafrase and Parafrase 2 This system, developed by David Kuck and coworkers at Illinois, is a source-to-source program restructurer (or compiler preprocessor) which transforms sequential Fortran 77 programs into forms suitable for vectorization or parallelization. Parafrase contains more than 100 program transformations which are encoded as passed. A *pass list* is used to identify the particular sequence of transformations needed for restructuring a given sequential program. The output of Parafrase is the converted concurrent program.

Different programs use different pass list and thus go through different sequences of transformations. The pass lists can be optimized for specific machine architectures and specific program constructs. A Parafrase 2 is under development for handling programs written in C and Pascal, in addition to converting Fortran codes. Information on Parafrase can be found in [Kuck84] et al. and on Parafrase 2 in [Polychronopoulos89] et al.

Parafrase is retargetable to produce code for different classes of parallel/vector computers. The program transformed by Parafrase still needs a conventional optimizing compiler to produce the object code for the target machine. The Parafrase technology has been transferred to implement the KAP vectorizer by Kuck and Associates, Inc.

The PFC and ParaScope   Ken Kennedy and his associates at Rice University developed the PFC as an automatic source-to-source vectorizer. It translates Fortran 77 codes into Fortran 90 codes. A categorized dependence testing scheme was developed in the PFC for revealing opportunities for loop vectorization. The PFC package is also being extended to a PFC+ for parallel code generation on shared-memory multiprocessors. The PFC and PFC+ also support the ParaScope programming environment.

The PFC (Allen and Kennedy, 1984) performs syntax analysis, including the following four steps:

(1) Interprocedural flow analysis using call graphs.
(2) Standard transformations such as Do-loop normalization, subscript categorization, deletion of dead codes, etc.
(3) Dependence analysis which applies the separability, GCD, and Banerjee tests jointly.
(4) Vector code generation. The PFC+ further implements a parallel code generation algorithm (Callahan et al., 1988).

Commercial Compilers   Optimizing compilers have also been developed in a number of commercial parallel/vector computers, including the Alliant FX/Fortran compiler, the Convex parallelizing/vectorizing compiler, the Cray CFT compiler, the IBM vectorizing Fortran compiler, the VAST vectorizer by Pacific Sierra, Inc., and Intel iPSC-VX compiler. IBM is also developing a PTRAN (Parallel Fortran) system based on control dependence with interprocedural analysis. Interested readers may refer to the bibliographic notes for additional information.

10.3   Dependence Analysis of Data Arrays

Dependence testing of successive iterations in multidimensional data arrays is described in this section. This provides a theoretical foundation for the development of vectorizing or parallelizing compilers.

10.3.1   Iteration Space and Dependence Analysis

Flow dependence, antidependence, and output dependence were defined for scalar data in Section 2.1.2. They can be summarized by the existence of a dynamic references of R_1 and R_2, if and only if either R_1 or R_2 is a write operation, R_1 executes before R_2, or R_1 and R_2 both write the same variable. When the referenced object is a data array indexed by a multidimensional subscript, the dependence may become very difficult to determine at compile time.

Precise and efficient dependence tests are essential to the effectiveness of a parallelizing compiler. The process of computing all the data dependences in a program is called *dependence analysis*. The testing scheme presented below is based on the work of Goff, Kennedy, and Tseng (1991). These dependence tests are being implemented at Rice University in the PFC with the parallel ParaScope programming environment.

Dependence Testing   Calculating data dependence for arrays is complicated by the fact that two array references may not access the same memory location. Dependence testing is the method used to determine whether dependences exist between two subscripted references to the same array in a loop nest. For the purpose of this explication, we ignore any control flow except for the loops themselves. Suppose we wish to test whether or not there exists a dependence from statement S_1 to S_2 in the following model loop nest of n levels, represented by n integer indices $i_1, i_2, ..., i_n$.

$$
\begin{aligned}
&\textbf{Do } i_1 = L_1, U_1 \\
&\quad \textbf{Do } i_2 = L_2, U_2 \\
&\qquad \cdots \\
&\quad \textbf{Do } i_n = L_n, U_n \\
&S_1: \qquad \text{A}(f_1(i_1, ..., i_n), ..., f_m(I_1, ..., i_n)) = \cdots \\
&S_2: \qquad \cdots = \text{A}(g_1(i_1, ..., i_n), ..., g_m(I_1, ..., i_n)) \\
&\quad \textbf{Enddo} \\
&\qquad \cdots \\
&\quad \textbf{Enddo} \\
&\textbf{Enddo}
\end{aligned}
$$

Iteration Space   The n-dimensional discrete Cartesian space for n-deep loops is called an *iteration space*. The *iteration* is represented as coordinates in the iteration space. The following example clarifies the concept of *lexicographic order* for the successive iterations in a loop nest.

Example 10.3 Lexicographic order for sequential execution of successive iterations in a loop structure (Monica Lam, 1992)

Consider a two-dimensional iteration space (Fig. 10.5) representing the following two-level loop nest in unit-increment steps:

$$
\begin{aligned}
&\textbf{Do } i = 0, 5 \\
&\quad \textbf{Do } j = i, 7 \\
&\qquad f(i, j) = \cdots \\
&\quad \textbf{Enddo} \\
&\textbf{Enddo}
\end{aligned}
$$

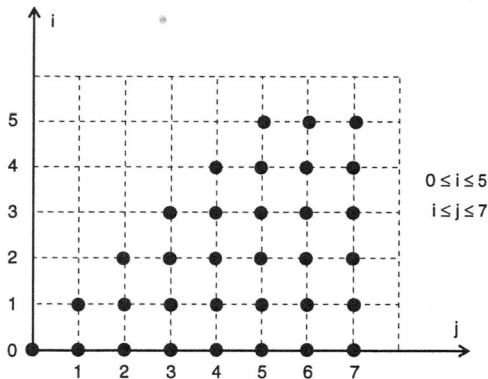

Figure 10.5 A two-dimensional iteration space for the loop nest in Example 10.3.

The following sequential order of iteration is a lexicographic order:

$$(0,0), (0,1), (0,2), (0,3), (0,4), (0,5), (0,6), (0,7)$$
$$(1,1), (1,2), (1,3), (1,4), (1,5), (1,6), (1,7)$$
$$(2,2), (2,3), (2,4), (2,5), (2,6), (2,7)$$
$$(3,3), (3,4), (3,5), (3,6), (3,7)$$
$$(4,4), (4,5), (4,6), (4,7)$$
$$(5,5), (5,6), (5,7)$$

The lexicographic order is important to performing matrix transformation, which can be applied for loop optimization. We will apply lexicographic orders for loop parallelization in Section 10.5.

Dependence Equations   Let α and β be vectors of n integer indices within the ranges of the upper and lower bounds of the n loops. There is a *dependence* from S_1 to S_2 if and only if there exist α and β such that α is lexicographically less than or equal to β and the following system of *dependence equations* is satisfied:

$$f_i(\alpha) = g_i(\beta) \qquad \forall i, 1 \le i \le m \tag{10.2}$$

Otherwise the two references are *independent*.

The dependence equations in Eq. 10.2 are linear expressions of the loop index variables. Dependence testing is thus equivalent to the problem of linear Diophantine equations, which is an NP-complete problem. *Exact tests* are dependence tests that will detect dependences if and only if they exist. In practice, exact tests are not performed due to the excessive overhead involved. Only approximate solutions (which are efficient to implement) are sought.

Parallelizing compilers have traditionally relied on two dependence tests to detect data dependences between pairs of array references: *Banerjee's inequalities* (Banerjee, 1988) and *GCD tests* (Wolfe, 1989). However, these tests are usually more general than necessary.

In Section 10.3.2, we present a practical testing algorithm developed by Rice University researchers led by Ken Kennedy. The test algorithm is based on partitioning the subscripts in a pair of array references. A suite of simple tests is developed to reduce the cost of performing dependence analysis, making it more practical for most compilers.

Distance and Direction Vectors Suppose there exists a data dependence for $\alpha = (\alpha_1, \alpha_2, ..., \alpha_n)$ and $\beta = (\beta_1, \beta_2, ..., \beta_n)$. Then the *distance vector* $\mathbf{D} = (D_1, ..., D_n)$ is defined as $\beta - \alpha$. The *direction vector* $\mathbf{d} = (d_1, d_2, ..., d_n)$ of the dependence is defined by

$$d_i = \begin{cases} < & \text{if } \alpha_i < \beta_i \\ = & \text{if } \alpha_i = \beta_i \\ > & \text{if } \alpha_i > \beta_i \end{cases} \tag{10.3}$$

The elements are always displayed in order from left to right and from the outermost to the innermost loop in the nest.

For example, consider the following loop nest:

> **Do** $i = L_1, U_1$
>   **Do** $j = L_2, U_2$
>   **Do** $k = L_3, U_3$
> $\text{A}(i+1, j, k-1) = \text{A}(i, j, k) + C$
>   **Enddo**
>   **Enddo**
> **Enddo**

The distance and direction vectors for the dependence between iterations along three dimensions of the array A are $(1, 0, -1)$ and $(<, =, >)$, respectively. Since several different values of α and β may satisfy the dependence equations, a set of distance and direction vectors may be needed to completely describe the dependence.

Direction vectors are useful for calculating the level of *loop-carried* dependences. A dependence is *carried* by the outermost loop for which the direction in the direction vector is not "=". For instance, the direction vector $(<, =, >)$ for the dependence above shows the dependence is carried on the i-loop.

Carried dependences are important because they determine which loops cannot be executed in parallel without synchronization. Direction vectors are also useful in determining whether loop interchange is legal and profitable. Distance vectors are more precise versions of direction vectors that specify the actual distance in loop iterations between two accesses to the same memory location. They may be used to guide optimizations to exploit parallelism or the memory hierarchy.

10.3.2 Subscript Separability and Partitioning

Dependence testing thus has two goals. It tries to disprove the dependence between pairs of subscripted references to the same array variable. If dependences exist, it tries to characterize them in some manner, usually as a minimum complete set of distance and direction vectors. Dependence testing must also be *conservative* and assume the existence of any dependence it cannot disprove. Otherwise the validity of any optimizations based on dependence information is not guaranteed.

Subscript Categories   The term *subscript* refers to one of the subscripted positions in a pair of array references, i.e., the pair of subscripts in some dimension of the two array references. When testing for dependence, we classify subscript positions by the total number of distinct loop indices they contain.

A subscript is said to be *zero index variable* (ZIV) if the subscript position contains no index in either reference. A subscript is said to be *single index variable* (SIV) if only one index occurs in that position. Any subscript with more than one index is said to be *multiple index variable* (MIV).

Example 10.4 Subscript types in a loop computation

Consider the following loop nest of three levels, identified by indices i, j, and k.

> **Do** $i_1 = L_1, U_1$
> > **Do** $j = L_2, U_2$
> > > **Do** $k = L_n, U_n$
> > > > $A(5, i + 1, j) = A(N, i, k) + C$
> > > **Enddo**
> > **Enddo**
> **Enddo**

When testing for a flow dependence between the two references to A in the code, the first subscript is ZIV because 5 and N are both constants, the second is SIV because only index i appears in this dimension, and the third is MIV because both indices j and k appear in the third dimension. For simplicity, we have ignored the output dependence in this example.

∎

Subscript Separability   When testing multidimensional arrays, we say that a subscript position is *separable* if its indices do not occur in the other subscripts. If two different subscripts contain the same index, we say they are *coupled*. Separability is important because multidimensional array references can cause imprecision in dependence testing.

If all the subscripts are separable, we may compute the direction vector for each subscript independently and merge the direction vectors on a positional basis with full precision. The following examples clarify these concepts.

Example 10.5 From separability to direction vector and distance vector

Consider the following loop nest:

$$\textbf{Do } i_1 = L_1, U_1$$
$$\textbf{Do } j = L_2, U_2$$
$$\textbf{Do } k = L_3, U_3$$
$$\text{A}(i, j, j) = \text{A}(i, j, k) + C$$
$$\textbf{Enddo}$$
$$\textbf{Enddo}$$
$$\textbf{Enddo}$$

The first subscript is separable because index i does not appear in the other dimensions, but the second and third are coupled because they both contain the index j. ZIV subscripts are separable because they contain no indices.

Consider another loop nest:

$$\textbf{Do } i_1 = L_1, U_1$$
$$\textbf{Do } j = L_2, U_2$$
$$\textbf{Do } k = L_n, U_n$$
$$\text{A}(i + 1, j, k - 1) = \text{A}(i, j, k) + C$$
$$\textbf{Enddo}$$
$$\textbf{Enddo}$$
$$\textbf{Enddo}$$

The leftmost direction in the direction vector is determined by testing the first subscript, the middle direction by testing the second subscript, and the rightmost direction by testing the third subscript.

The resulting direction vector $(<, =, >)$ is precise. The same approach applied to distances allows us to calculate the exact distance vector $(1, 0, -1)$.

∎

Subscript Partitioning We need to classify all the subscripts in a pair of array references as separable or as part of some minimal coupled group. A coupled group is *minimal* if it cannot be partitioned into two nonempty subgroups with distinct sets of indices. Once a partition is achieved, each separable subscript and each coupled group has completely disjoint sets of indices.

Each partition may then be tested in isolation and the resulting distance or direction vectors merged without any loss of precision. Since each variable and coupled subscript group contains a unique subset of indices, a merge may be thought of as a Cartesian product.

In the following loop nest, the first subscript yields the direction vector $(<)$ for the i-loop. The second subscript yields the direction vector $(=)$ for the j-loop. The resulting Cartesian product is the single vector $(<, =)$.

$$\textbf{Do } i = L_1, U_1$$
$$\textbf{Do } j = L_2, U_2$$
$$\text{A}(i + 1, j) = \text{A}(i, j) + C$$
$$\textbf{Enddo}$$
$$\textbf{Enddo}$$

Consider another loop nest where the first subscript yields the direction vector $(<)$ for the i-loop.

> **Do** $i = L_1, U_1$
>   **Do** $j = L_2, U_2$
> $A(i+1,5) = A(i, N) + C$
>   **Enddo**
> **Enddo**

Since j does not appear in any subscript, we must assume the full set of direction vectors for the j-loop: $\{(<),(=),(>)\}$. Thus, a merge yields the following set of direction vectors for both dimensions:

$$\{(<,<),(<,=),(<,>)\}$$

10.3.3 Categorized Dependence Tests

The goal of dependence testing is to construct the complete set of distance and direction vectors representing potential dependences between an arbitrary pair of subscripted references to the same array variable. Since distance vectors may be treated as precise direction vectors, we will simply refer to direction vectors.

The Testing Algorithm   The following procedure is for dependence testing based on a partitioning approach, which can isolate unrelated indices and localize the computation involved and thus is easier to implement.

(1) Partition the subscripts into separable and minimal coupled groups using the following algorithm:

> **Subscript Partitioning Algorithm** (Goff, Kennedy, and Tseng, 1991)
>   **Input**: A pair of m-dimensional array references
>   containing subscripts $S_1...S_m$ enclosed in n loops
>   with indices $I_1...I_n$.
>   **Output**: A set of partitions $P_1...P_{n'}$, $n' \leq n$, each
>   containing a separable or minimal coupled group.
>   **For** each i, $1 \leq i \leq n$ **Do**
> $P_i \leftarrow \{S_i\}$
>   **Endfor**
>   **For** each index I_i, $1 \leq i \leq n$ **Do**
> $k \leftarrow \{none\}$
>   **For** each remaining partition P_j **Do**
>   **if** $\exists S_1 \in P_j$ such that S_1 contains I_i, **then**
>   **if** $k = \{none\}$ **then**
> $k \leftarrow j$
>   **else**
> $P_k \leftarrow P_k \cup P_j$
>   Discard P_j
>   **Endif**

Endif

Endfor

Endfor

(2) Label each subscript as ZIV, SIV, or MIV.

(3) For each separable subscript, apply the appropriate single subscript test (ZIV, SIV, MIV) based on the complexity of the subscript. This will produce independence or direction vectors for the indices occurring in that subscript.

(4) For each coupled group, apply a multiple subscript test to produce a set of direction vectors for the indices occurring within that group.

(5) If any test yields independence, no dependences exist.

(6) Otherwise merge all the direction vectors computed in the previous steps into a single set of direction vectors for the two references.

Test Categories   Dependence test results for ZIV subscripts are treated specially. If a ZIV subscript proves independence, the dependence test algorithm halts immediately. If independence is not proved, the ZIV test does not produce direction vectors, and so no merge is necessary. For the implementation of the above algorithm, we have specified how to perform the single subscript tests (ZIV, SIV, MIV) separately. We consider below the trivial case of ZIV first, then SIV, and finally MIV which is more involved.

We first consider dependence tests for single separable subscripts. All tests presented assume that the subscript being tested contains expressions that are linear in the loop index variables. A subscript expression is linear if it has the form $a_1 i_1 + a_2 i_2 + \cdots + a_n i_n + e$, where i_k is the index for the loop at nesting level k; all $a_k, 1 \le k \le n$, are integer constants; and e is an expression possibly containing loop-invariant symbolic expressions.

The ZIV Test   The ZIV test is a dependence test performed on two loop-invariant expressions. If the system determines that the two expressions cannot be equal, it has proved independence. Otherwise the subscript does not contribute any direction vectors and may be ignored. The ZIV test can be easily extended for symbolic expressions. Simply form the expression representing the difference between the two subscript expressions. If the difference simplifies to a nonzero constant, we have proved independence.

The SIV Test   An SIV subscript for index i is said to be *strong* if it has the form $(ai + c_1, ai' + c_2)$, i.e., if it is linear and the coefficients of the two occurrences of the index i are constant and equal. For strong SIV subscripts, define the *dependence distance* as

$$d = i' - i = \frac{c_1 - c_2}{a} \tag{10.4}$$

A dependence exists if and only if d is an integer and $|d| \le U - L$, where U and L are the loop upper and lower bounds. For dependences that do exist, the *dependence direction* is given by

$$\text{Direction} = \begin{cases} < & \text{if } d > 0 \\ = & \text{if } d = 0 \\ > & \text{if } d < 0 \end{cases} \tag{10.5}$$

The strong SIV test is thus an exact test that can be implemented very efficiently in a few operations. A bounded iteration space is shown in Fig. 10.6a. The case of a strong SIV test is shown in Fig. 10.6b.

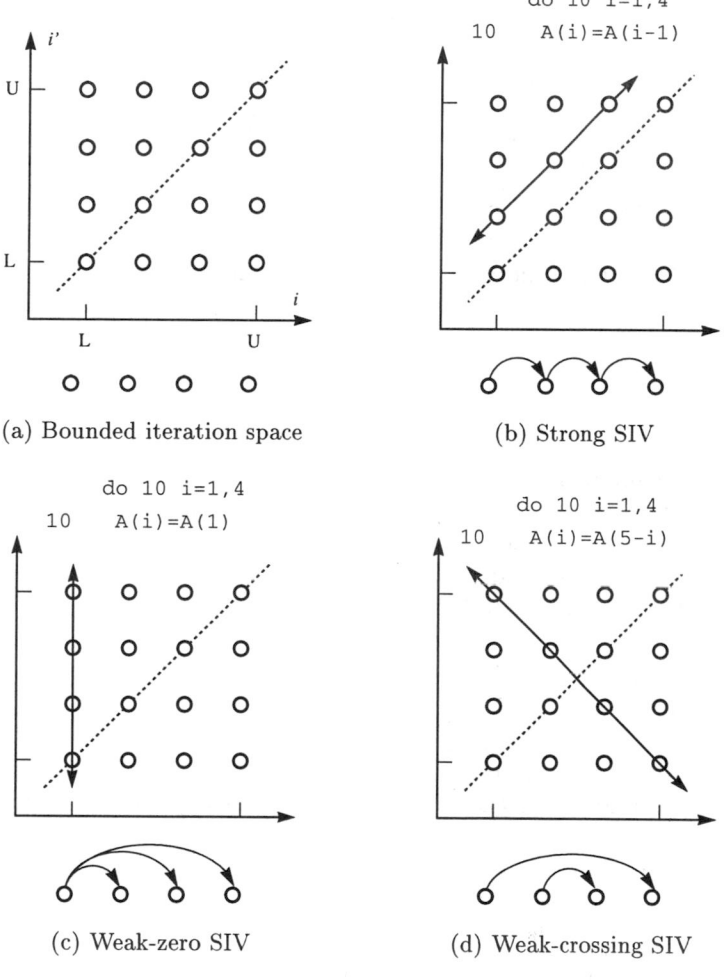

(a) Bounded iteration space

(b) Strong SIV

(c) Weak-zero SIV

(d) Weak-crossing SIV

Figure 10.6 Geometric view of SIV tests in four cases. (Courtesy of Goff et al., 1991; reprinted from ACM SIGPLAN *Conf. Programming Language Design and Implementation*, Toronto, Canada, 1991).

Another advantage of the strong SIV test is that it can be easily extended to handle loop-invariant symbolic expressions. The trick is to first evaluate the dependence distance d symbolically. If the result is a constant, then the test may be performed as above. Otherwise calculate the difference between the loop bounds and compare the result with d symbolically.

A *weak* SIV subscript has the form $\langle a_1 i + c_1, a_2 i' + c_2 \rangle$, where the coefficients of the two occurrences of index i have different constant values. As stated previously, weak SIV subscripts may be solved using the single-index exact test. However, we also find it helpful to view the problem geometrically, where the dependence equation

$$a_1 i + c_1 = a_2 i' + c_2$$

describes a line in the two-dimensional plane with i and i' as axes.

The weak SIV test can then be formulated as determining whether the line derived from the dependence equation intersects with any integer points in the space bounded by the loop upper and lower bounds, as shown in Fig. 10.5.

Two special cases should be studied separately.

Weak-Zero SIV Test   The case in which $a_1 = 0$ or $a_2 = 0$ is called a *weak-zero SIV* subscript, as illustrated in Fig. 10.6c. If $a_2 = 0$, the dependence equation reduces to

$$i = \frac{c_2 - c_1}{a_1}$$

It is only necessary to check that the resulting value for i is an integer and within the loop bounds. A similar check applies when $a_1 = 0$.

The weak-zero SIV test finds dependences caused by a particular iteration i. In scientific codes, i is usually the first or last iteration of the loop, eliminating one possible direction vector for the dependence.

Consider the following loop for a strong SIV test:

> **Do** $i = 1, \text{N}$
> $\text{A}(i + 2\text{N}) = \text{A}(i + \text{N})$
> **Enddo**

The strong SIV test can evaluate the dependence distance d as $2N - N$, which simplifies to N. This is compared with the loop bounds symbolically, proving independence since $N > N - 1$.

Consider the following simplified loop in the program *tomcatv* from the SPEC benchmark suite (Uniejewski, 1989):

> **Do** $i = 1, \text{N}$
> $\text{Y}(i + \text{N}) = \text{Y}(i + \text{N}) + \text{Y}(\text{N}, \text{N})$
> **Enddo**

The weak-zero SIV test can determine that the use of $\text{Y}(1, \text{N})$ causes a loop-carried true dependence from the first iteration to all the other iterations. Similarly, with aid from symbolic analysis, the weak-zero SIV test can discover that the use of $\text{Y}(\text{N}, \text{N})$

causes a loop-carried antidependence from all iterations to the last iteration. By identifying the first and last iterations as the only cause of dependences, the weak-zero SIV test advises the user or compiler to peel the first and last iterations of the loop, resulting in the following parallel loop:

$$Y(1,N) = Y(1,N) + Y(N,N)$$
Do $i = 2, N - 1$
$$\qquad Y(i, N) = Y(i, N) + Y(N, N)$$
Enddo
$$Y(N,N) = Y(1,N) + Y(N,N)$$

Weak-Crossing SIV Test   All subscripts where $a_2 = -a_1$ are *weak-crossing SIV*. These subscripts typically occur as part of Cholesky decomposition, also illustrated in Fig. 10.6d. In these cases we set $i = i'$ and derive the dependence equation

$$i = \frac{c_2 - c_1}{2a_1}$$

This corresponds to the intersection of the dependence equation with the line $i = i'$. To determine whether dependences exist, we simply need to check that the resulting value i is within the loop bounds and is either an integer or has a noninteger part equal to $1/2$.

Weak-crossing SIV subscripts cause *crossing* dependences, loop-carried dependences whose end points all cross iteration i. These dependences may be eliminated using a *loop-splitting* transformation (Kennedy et al. 1991) as described below.

Consider the following loop from the Callahan-Dongarra-Levine vector test (Callahan et al., 1988):

Do $i = 1, N$
$$\qquad A(i) = A(N{-}i + 1) + C$$
Enddo

The weak-crossing SIV test determines that dependences exist between the definition and use of A and that they all cross iteration $(N + 1)/2$. Splitting the loop at that iteration results in two parallel loops:

Do $i = 1, (N + 1)/2$
$$\qquad A(i) = A(N - i + 1) + C$$
Enddo
Do $i = (N + 1)/2 + 1, N$
$$\qquad A(i) = A(N - i + 1) + C$$
Enddo

The MIV Tests   SIV tests can be extended to handle complex iteration spaces where loop bounds may be functions of other loop indices, e.g., triangular or trapezoidal loops. We need to compute the minimum and maximum loop bounds for each loop index.

Starting at the outermost loop nest and working inward, we replace each index in a loop upper bound with its maximum value (or minimal if it is a neighbor term). We do the opposite in the lower bound, replacing each index with its minimal value (or maximal if it is a negative term).

We evaluate the resulting expressions to calculate the minimal and maximal values for the loop index and then repeat for the next inner loop. This algorithm returns the maximal range for each index, all that is needed for SIV tests.

The Banerjee-GCD test may be employed to construct all legal direction vectors for linear subscripts containing multiple indices. In most cases the test can also determine the minimal dependence distance for the carrier loop.

A special case of MIV subscripts, called RDIV (restricted double-index variable) subscripts, have the form $(a_1 i + c_1, a_2 j + c_2)$. They are similar to SIV subscripts except that i and j are distinct indices. By observing different loop bounds for i and j, SIV tests may also be extended to test RDIV subscripts exactly.

A large body of work has been performed in the field of dependence testing at Rice University, the University of Illinois, and Oregon Graduate Institute. What was described above is only one of the many dependence testing algorithms being proposed. Experimental results are reported from these research centers. Readers are advised to read published material on Banerjee's test and the GCD test, which provide other inexact and conservative solutions to the problem.

The development of a parallelizing compiler is limited by the difficulty of having to deal with many nonperfectly nested loops. The equivalence of subscript conditions and complex numbers have not been addressed by compiler researchers. The lack of dataflow information is often the ultimate limit on automatic compilation of parallel code.

10.4 Code Optimization and Scheduling

In this section, we describe the roles of compilers in code optimization and code generation for parallel computers. In no case should one expect production of a true optimal code which matches the hardware behavior perfectly. Compilation is a software technique which transforms the source program to generate better object code, which can reduce the running time and memory requirement. On a parallel computer, program optimization often demands an effort from both the programmer and the compiler.

10.4.1 Scalar Optimization with Basic Blocks

Instruction scheduling is often supported by both compiler techniques and dynamic scheduling hardware. In order to exploit *instruction-level parallelism* (ILP), we need to optimize the code generation and scheduling process under both machine and program constraints. Machine constrains are caused by mutually exclusive use of functional units, registers, and memory. program constraints are caused by data and control dependences. Some processors, like those with VLIW architecture, explicitly specify ILP in their instructions. Others may use hardware interlock, out-of-order execution, or speculative execution. Even machines with dynamic scheduling hardware can benefit from compiler scheduling techniques.

There are two alternative approaches to supporting code scheduling. One is to provide a set of nontrapping instructions so that the compiler can perform aggressive *static code scheduling*. This approach requires an extension of the instruction set of existing processors. The second approach is to support out-of-order execution in the microarchitecture so that the hardware can perform aggressive *dynamic code scheduling*. This approach usually does not require the instruction set to be modified but requires complex hardware support.

In general, code scheduling methods ensure that control dependences, data dependences, and resource limitations are properly handled during concurrent execution. The goal is to produce a code schedule that minimizes the execution time or the memory demand, in addition to enforcing correctness of execution. Static scheduling at compile time requires intelligent compilation support, whereas dynamic scheduling at run time requires sophisticated hardware support. In practice, dynamic scheduling is assisted by static scheduling in improving performance and reducing hardware cost. On the other hand, static scheduling is assisted by hardware interlocking in enforcing correctness of execution.

Precedence Constraints   A hardware scheduler is needed to direct the flow of instructions after a branch. A software scheduler is used to handle only conditionally executed instructions. Speculative execution requires the use of program profiling to estimate effectiveness. Speculative exceptions must not terminate execution. In other words, precise exception handling is desired to alleviate the control dependence problem. The data dependence problem involves instruction ordering and register allocation issues.

If a flow dependence is detected, the *write* must proceed ahead of the *read* operation involved. Similarly, output dependence produces different results if two *writes* to the same location are executed in a different order. Antidependence enforces a *read* to be ahead of the *write* operation involved. We need to analyze the memory variables. Scalar data dependence is much easier to detect. Dependence among arrays of data elements is much more involved, as shown in Section 10.3. Other difficulties lie in interprocedural analysis, pointer analysis, and register allocations interacting with code scheduling.

Basic Block Scheduling   A *basic block* (or just a *block*) is a sequence of statements satisfying two properties: (1) No statement but the first can be reached from outside the block; i.e., there are no branches in from the middle of the block. (2) All statements are executed consecutively if the first one is. Therefore, no branches out or halts are allowed until the end of the block. All blocks are required to be *maximal* in the sense that they cannot be extended up or down without violating these properties.

For local optimization only, an *extended basic block* is defined as a sequence of code statements in which the first statement is the only entry point. Thus an extended block may have branches out in the middle of the code but no branches into it. The basic steps for constructing basic blocks are summarized below:

(1) Find the *leaders*, which are the first statements in a block. Leaders are identified as being one or more of the following:

   (a) The first statement of the code.
   (b) The target of a conditional or unconditional branch.
   (c) A statement following a conditional branch.

(2) For a leader a basic block consists of the leader and all statements following up to but excluding the next leader. Note that the beginning of inaccessible code (dead code) is not considered a leader. In fact, dead code should be eliminated.

Example 10.6 Basic block construction in a bubble sort program (S. Graham, J. L. Hennessy, and J. D. Ullman, 1992)

A bubble sort program sorts an array $A[j]$ with statically allocated storage. Each element of A requires 4 bytes of byte-addressable machine. The elements of $A[j]$ are numbered $j = 1, 2, ..., n$, where n is a variable. To be specific, $A[j]$ is stored in location $addr(A) + 4 * (j - 1)$, where $addr(A)$ produces the starting address of the array A. The following source code is for bubble sort:

> **For** $i := n - 1$ **downto** 1 **do**
>   **For** $j := 1$ **to** i **do**
>   **If** $A[j] > A[j + 1]$ **then**
>   **Begin**
> $temp := A[j]$
> $A[j] := A[j + 1]$
> $A[j + 1] := temp$
>   **End**
>   **End** of j-loop
> **End** of i-loop

If a three-address machine is assumed, the above code is translated into the following assembly language code. Variable names on the right of := stand for values, and on the left for addresses.

```
          i := n - 1
    s5:   if i < 1 goto s1
          j := 1
    s4:   if j > i goto s2
          t1 := j - 1
          t2 := 4 * t1
          t3 := A[t2]              /A[j]/
          t4 := j + 1
          t5 := t4 - 1
          t6 := 4 * t5
          t7 := A[t6]              /A[j+1]/
          if t3 <= t7 goto s3      /if A[j] > A[j+1] then begin .../
          t8 := j - 1
```

```
        t9 := 4 * t8
        temp := A[t9]           /temp := A[j]/
        t10 := j + 1
        t11 := t10 - 1
        t12 := 4 * t11
        t13 := A[t12]           /A[j+1]/
        t14 := j - 1
        t15 := 4 * t14
        A[t15] := t13           /A[j] := A[j+1]/
        t16 := j + 1
        t17 := t16 - 1
        t17 := t16 - 1
        t18 := 4 * t17
        A[t18] := temp          /A[j+1] := temp/
    s3: j := j + 1
        goto s4
    s2: i := i - 1
        goto s5
    s1: halt
```

The above 31 statements are divided into 8 basic blocks as shown in Fig. 10.7. A program *flow graph* is drawn to show the precedence relationship among the basic blocks. Each node in the flow graph corresponding to one basic block may contain different numbers of statements. The entry node is B1, and the exit node is B2.

■

Basic blocks are created and destroyed as the program is modified. Space must be made available for growing collections of blocks. Block records should keep pointers to predecessors and successors. Storage reclamation techniques or linked list structures can be used to represent blocks. Sources of program optimization include algorithm redesign such as the use of heap sort in place of the bubble sort, algebraic optimization to eliminate redundant operations, and other optimizations conducted within the basic blocks locally or on a global basis.

10.4.2 Local and Global Optimizations

We first describe local code optimization within basic blocks. Then we study global optimizations among basic blocks. Both intraprocedural and interprocedural optimizations are discussed. Finally, we identify some machine-dependent optimizations. Readers will realize the limitations and potentials of these code optimization methods.

Local Optimizations   These are code optimizations performed only within basic blocks. The information needed for optimization is gathered entirely from a single basic block, not from an extended basic block. No control-flow information between blocks is considered. Listed below are some local optimizations often performed:

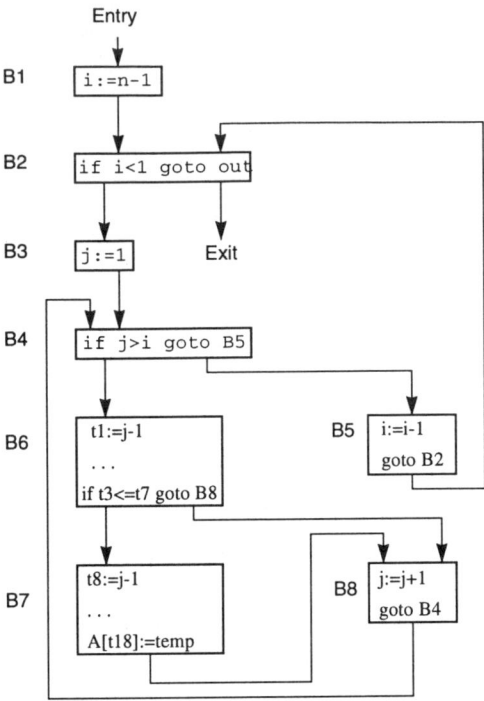

Figure 10.7 A flow graph showing the precedence relationship among basic blocks in the bubble sort program. (Courtesy of S. Graham, J. L. Hennessy, and J. D. Ullman, *Course on Code Optimization and Code Generation*, Western Institute of Computer Science, Stanford University, 1992)

(1) **Local common subexpression elimination**: If a subexpression is to be evaluated more than once within a single block, it can be replaced by a single evaluation. For Example 10.6, in block B7, t9 and t15 each compute $4 * (j - 1)$, and t12 and t18 each compute 4*j. Replacing t15 by t9, and t18 by t12, we obtain the following revised code for B7, which is shorter to execute.

$$t8 := j - 1$$
$$t9 := 4 * t8$$
$$temp := A[t9]$$
$$t12 := 4 * j$$
$$t13 := A[t12]$$
$$A[t9] := t13$$
$$A[t12] := temp$$

(2) **Local constant folding or propagation**: Sometimes some constants used in instructions can be computed at compile time. This often takes place in the initialization blocks. The compile-time generated constants are then folded to eliminate unnecessary calculations at run time. In other cases, a local copy may be propagated to eliminate unnecessary calculations.

(3) **Algebraic optimization to simplify expressions**: For example, one can replace the *identity statement* $A := B + 0$ or $A := B * 1$ by $A := B$ and later even replace references to this A by references to B. Or one can use the *commutative law* to combine expressions $C := A + B$ and $D := B + A$. The *associative* and *distributive laws* can also be applied on equal-priority operators, such as replacing $(a - b) + c$ by $a - (b - c)$ if $(b - c)$ has already been evaluated earlier.

(4) **Instruction reordering**: Code reordering is often practiced to maximize the pipeline utilization or to enable overlapped memory accesses. Some orders yield better code than others. Reordered instructions lead to better scheduling, preventing pipeline or memory delays. In the following example, instruction I3 may be delayed in memory accesses:

```
I1:    Load    R1, A
I2:    Load    R2, B
I3:    Add     R2, R1, R2 — delayed
I4:    Load    R3, C
```

With reordering, the instruction I3 may experience no delay:

```
I1:    Load    R1, A
I2:    Load    R2, B
I4:    Load    R3, C
I3:    Add     R2, R1, R2 — not delayed
```

(5) **Elimination of dead code or unary operators**: Code segments or even basic blocks which are not accessible or will never be referenced can be eliminated to save compile time, run time, and space requirements. Sometimes, this code cannot be determined until run time. In this situation, test runs are needed to reveal the dead codes.

Unary operators, such as arithmetic negation and logical complement, can often be eliminated by applying algebraic laws, such as $x + (-y) = x - y$, $-(x - y) = y - x$, $(-x) * (-y) = x * y$, Not(Not A) = A, etc. Boolean expression evaluation can often be optimized after some form of minimization.

Global Optimizations   These are code optimizations performed across basic block boundaries. Control-flow information among basic blocks is needed. John Hennessy (1992) has classified intraprocedural global optimizations into three types:

(1) **Global versions of local optimizations**: These include global common subexpression elimination, global constant propagation, dead code elimination, etc. The following example further optimizes the code in Example 10.6 if some global optimizations are performed.

Example 10.7   Global optimizations in the bubble sort program (S. Graham, J. L. Hennessy, and J. D. Ullman, 1992)

In Example 10.6, block B7 needs to compute $A[t9] = A[4 * (j-1)]$, which was computed in block B6. To reach B7, B6 must be executed first and the value of j never changes between the two nodes. Thus the first three statements of B7 can be replaced by temp := t3. Similarly, t9 computes the same value as t2, t12 computes the same value as t6, and t13 computes the same value as t7. The entire block B7 may be replaced by

$$temp := t3$$
$$A[t2] := t7$$
$$A[t6] := temp$$

and, substituting for temp by

$$A[t2] := t7$$
$$A[t6] := t3$$

The revised program, after both local and global optimizations, is obtained as follows:

```
B1:  i := n - 1
B2:  If i < 1 goto out
B3:  j := 1
B4:  If j > i goto B5
B6:  t1 := j - 1
     t2 := 4 * t1
     t3 := A[t2]              /A[j]/
     t6 := 4 * j
     t7 := A[t6]              /A[j+1]/
     If t3 <= t7 goto B8
B7:  A[t2] := t7
     A[t6] := t3
B8:  j := j + 1
     goto B4
B5:  i := i - 1
     goto B2
out:
```

■

(2) **Loop optimizations**: These include various loop transformations to be described in subsequent sections for the purpose of vectorization, parallelization, or both. Very often *code motion* and *induction variable elimination* can simplify loop structures. For example, one can replace the calculation of an induction variable involving a multiplication by an addition to its former value. The addition takes less time to perform and thus results in a shorter execution time.

In other cases, loop-invariant variables or codes can be moved out of the loop to simplify the loop nest. One can also lower the loop control overhead using loop unrolling to reduce iteration or loop fusion to merge loops. Loops can be exchanged to facilitate pipelining of long vectors. Many loop optimization examples will be given in subsequent sections.

(3) **Control-flow optimization**: These are other global optimizations dealing with control structure but not directly with loops. A good example is *code hoisting*, which eliminates copies of identical code on parallel paths in a flow graph. This can often save space significantly.

Interprocedural global optimizations are much more difficult to perform due to historical sensitivity and global dependence relationships. Sometimes, procedure integration can be performed to replace a call to a procedure by an instantiation of the procedure body. This may also trigger the discovery of other optimization opportunities. Interprocedure dependence analysis must be performed in order to reveal these opportunities.

Machine-Dependent Optimizations   With a finite number of registers, memory cells, and functional units in a machine, the efficient allocation of machine resources affects both space and time optimization of programs. For example, *strength reduction* replaces complex operations by cheaper operations, such as replacing $2a$ by $a + a$, a^2 by $a * a$, and $length(S1 + S2)$ by $length(S1) + length(S2)$. We will address register and memory allocation problems in code generation methods in the next section.

Other flow control optimizations can be conducted at the machine level. Good examples include the elimination of unnecessary branches, the replacement of instruction sequences by simpler equivalent code sequences, instruction reordering supported by hardware interlocking, and multiple instruction issues in superscalar and/or superpipelined processors. Machine-level parallelism is often exploited at the fine-grain instruction level. System-level parallelism is often exploited in coarse-grain computations.

10.4.3   Vectorization and Parallelization Methods

Besides scalar optimizations, we need to perform vector and/or parallel optimizations. The purpose is to improve the performance of programs that manipulate large data arrays or can be partitioned for parallel execution. *Vectorization* is the process of converting scalar looping operations into equivalent vector instruction execution. *Parallelization* aims at converting sequential code into parallel form, which can enable parallel execution by multiple processors.

An optimizing compiler that does vectorization automatically or semiautomatically with directives from programmers is called a *vectorizing compiler* or simply a *vectorizer*. Similarly, a *parallelizing compiler* should be designed to generate parallel code from sequential code automatically or semiautomatically. We introduce below various methods suggested for vectorization and parallelization. Vector hardware must be provided to speed up vector operations. Multiprocessors or multicomputers must be used to execute parallelized codes. Inhibitors of vectorization and parallelization are also identified in some program constructs in order to avoid unrewarding attempts.

Vectorization Methods   We describe below several methods for vectorization. Many other methods can be found in the references cited in the bibliographic notes. We use Fortran 90 notations, for example, successive iterations in the following loop are totally independent:

Do 20 I = 8, 120, 2
20   A(I) = B(I+3)+C(I+1)

This scalar loop can be converted into one *vector-add* instruction defined by the following array assignment:

$$A(8{:}120{:}2) = B(11{:}123{:}2)+C(9{:}121{:}2)$$

(1) **Use of temporary storage**: Consider the following Do loop:

Do 20 I = 1, N
          A(I) = B(I) + C(I)
20   B(I) = 2 $*$ A(I+1)

This loop represents the following sequence of scalar operations:

A(1) = B(1) + C(1)
B(1) = 2 $*$ A(2)
A(2) = B(2) + C(2)
B(2) = 2 $*$ A(3)
$\vdots$

In order to enable pipelined execution by vector hardware, we need to introduce a temporary array TEMP(1:N) to produce the following vector code:

TEMP(1:N) = A(2:N+1)
A(1:N) = B(1:N) + C(1:N)
B(1:N) = 2 $*$ TEMP(1:N)

Without the TEMP array, the second array assignment B(1:N) may use the modified A(1:N) array which was intended in the original code.

(2) **Loop interchanging**: Vectorization is often performed in the inner loop rather than in the outer loop. Sometimes we can interchange the loops to enable the vectorization. The general rules for loop interchanges are to make the most profitable vectorizable loop the innermost loop, to make the most profitable parallelizable loop the outermost loop, to enable memory accesses to consecutive elements in arrays, and to bring a loop with longer vector length (iteration count) to the innermost loop for vectorization. The profitability is defined by improvement in execution time. Consider the following loop nest with two levels:

Do 20 I = 2, N
          Do 10 J = 2, N
S_1:   A(I,J) = (A(I,J–1) + A(I,J+1))/2   (10.6)
10   **Continue**
20 **Continue**

The statement S_1 is both flow-dependent and antidependent on itself with the direction vectors $(=, <)$ and $(=, >)$, or more precisely the distance vectors $(0, -1)$ and $(0, 1)$, respectively. This implies that the loop cannot be vectorized along the J-dimension. Therefore, we have to interchange the two loops:

> **Do** 20 J = 2, N
>   **Do** 20 I = 2, N
>   A(I,J) = (A(I,J-1) + A(I,J+1))/2
> 20   **Continue**

Now, the inner loop (I-dimension) can be vectorized with the zero distance vector. The vectorized code is

> **Do** 20 J = 2, N
>   A(2:N,J) = (A(2:N,J–1) + A(2:N,J+1))/2
> 20   **Continue**

In general, an innermost loop cannot be vectorized if forward dependence direction ($<$) and backward direction ($>$) coexist. The ($=$) direction does not prevent vectorization.

(3) **Loop distribution**: Nested loops can be vectorized by distributing the outermost loop and vectorizing each of the resulting loops or loop nests.

> **Do** 10 I = 1, N
>   B(I,1) = 0
>   **Do** 20 J = 1, M
>   A(I) = A(I) + B(I,J) * C(I,J)
> 20   **Continue**
>   D(I) = E(I) + A(I)
> 10   **Continue**

The I-loop is distributed to three copies, separated by the nested J-loop from the assignment to array B and D, and vectorized as follows:

> B(1:N,1) = 0 (a zero vector)
> **Do** 30 I = 1, N
>   A(I) = A(I) + B(I, 1:M) * C(I, 1:M)
> 30   **Continue**
> D(1:N) = E(1:N) + A(1:N)

(4) **Vector reduction**: We have defined vector reduction instructions in Eqs. 8.6 and 8.7. A number of reductions are defined in Fortran 90. In general, a vector reduction produces a scalar value from one or two data arrays. Examples include the *sum*, *product*, *maximum*, and *minimum* of all the elements in a single array. The *dot product* produces a scalar $S = \sum_{i=1}^{n} A_i \times B_i$ from two arrays A(1:n) and B(1:n). The loop

> **Do** 40 I = 1,N
> S_1:   A(I) = B(I) + C(I)
> S_2:   S = S + A(I)
> S_3:   AMAX = MAX(AMAX, A(I))
> 40   **Continue**

has the following dependence relations: $S_1(=)S_2$, $S_1(=)S_3$, $S_2(<)S_2$, $S_3 < S_3$. Although statements S_2 and S_3 each form a dependence cycle, each statement is recognized as a reduction operation and can be vectorized as follows:

S_1: A(1:N) = B(1:N) + C(1:N)
S_2: S = S + SUM(A(1:N))
S_3: AMAX = MAX(AMAX, MAXVAL(A(1:N)))

where SUM, MAX, and MAXVAL are all vector operations.

(5) **Node splitting:** The data dependence cycle can sometimes be broken by node splitting. Consider the following loop:

Do 50 I = 2, N
S_1:   T(I) = A(I − 1) + A(I + 1)
S_2:   A(I) = B(I) + C(I)
50 **Continue**

Now we have the dependence cycle $S_1(<)S_2$ and $S_1(>)S_2$, which seems to prevent vectorization. However, we can split statement S_1 into two parts and apply statement reordering:

Do 50 I=2,N
S_{1a}:   X(I) = A(I + 1)
S_2:   A(I) = B(I) + C(I)
S_{1b}:   T(I) = A(I − 1) + X(I)
50 **Continue**

The new loop structure has no dependence cycle and thus can be vectorized:

S_{1a}: X(2:N) = A(3:N+1)
S_2:   A(2:N) = B(2:N) + C(2:N)
S_{1b}: T(2:N) = A(1:N − 1) + X(2:N)

It should be noted that node splitting cannot resolve all dependence cycles.

(6) **Other vector optimizations:** There are many other methods for vectorization, and we do not intend to exhaust them all. For example, *scalar variables* in a loop can sometimes be expanded into dimensional arrays to enable vectorization. *Subexpressions* in a complex expression can be vectorized separately. Loops can be *peeled*, *unrolled*, *rerolled*, or *tiled* (blocking) for vectorization.

Some machine-dependent optimizations can be performed, such as *strip mining* (loop sectioning) and *pipeline chaining*, introduced in Chapter 8. Sometimes a vector register can be used as an accumulator, making it possible for the compiler to move loads and stores of the register outside the vector loop. The movement of an operation out of a loop to a basic block preceding the loop is called *hoisting*, and the inverse is called *sinking*. Vector loads and stores can be hoisted or sunk only when the array reference and assignment have the same subscripts and all the subscripts are the induction variable of a vectorized loop or loop constants.

Vectorization Inhibitors   Listed below are some conditions inhibiting or preventing vectorization:

(1) Computed conditional statements such as IF statements which depend on runtime conditions.

(2) Multiple loop entries or exits (not basic blocks).
(3) Function or subroutine calls.
(4) Input/output statements.
(5) Recurrences and their variations.

A *recurrence* exists when a value calculated in one iteration of a loop might be referenced in another iteration. This occurs when dependence cycles exist between loop iterations. In other words, there must be at least one *loop-carried dependence* for a recurrence to exist. Any number of *loop-independent dependences* can occur in a loop, but a recurrence does not exist unless that loop contains at least one loop-carried dependency.

Code Parallelization Parallel code optimization spreads a single program into many threads for parallel execution by multiple processors. The purpose is to reduce the total execution time. Each *thread* is a sequence of instructions that must execute on a single processor. The most often parallelized code structure is performed over the outermost loop if dependence can be properly controlled and synchronized.

Consider the two-deep loop nest in Eq. 10.6. Because all the dependence relations have a "=" direction in the I-loop, this outer loop can be parallelized with no need for synchronization between the loop iterations. The parallelized code is as follows:

$$
\begin{aligned}
&\textbf{Doall } I = 2, N \\
&\quad\textbf{Do } J = 2, N \\
S_1: &\quad\quad A(I, J) = (A(I, J - 1) + A(I, J + 1)) \,/\, 2 \\
&\quad\textbf{Enddo} \\
&\textbf{Endall}
\end{aligned}
$$

Each of the $N-1$ iterations in the outer loop can be scheduled for a single processor to execute. Each newly created thread consists of one entire J-loop with a constant index value for I. If dependence does exist between the iterations, the Doacross construct can be used with proper synchronization among the iterations. The following example shows five execution modes of a serial loop execution to various combinations of parallelization and vectorization of the same program.

Example 10.8 Five execution modes of a FX/Fortran loop on the Alliant FX/80 multiprocessor (Alliant Computer Systems Corporation, 1989)

FX/Fortran generates code to execute a simple Do loop in *scalar, vector, scalar-concurrent, vector-concurrent,* and *concurrent outer/vector inner* (COVI) modes. The computations involved are performed either over a one-dimensional data array A(1:2048) or over a two-dimensional data array B(1:256, 1:8), where A(K)=B(I,J) for K=8(I–1)+J.

By using array A, the computations involved are expressed by a pure scalar loop:

Do K = 1, 2048
    A(K) = A(K) + S
Enddo

where S is a scalar constant. Figure 10.8a shows the scalar (serial) execution on a single processor in 30,616 clock cycles.

A(1) = A(1) + S, A(2) = A(2) + S, ..., A(2048) = A(2048) + S

(a) Scalar execution on one processor in 30,616 cycles

A(1 : 256) = A(1 : 256) + S, A(257 : 512) = A(257 : 512) + S, ..., A(1793 : 2048) = A(1793 : 2048) + S

(b) Vector execution on one processor sequentially in 6048 cycles

P_1: B(1, 1) = B(1, 1) + S, B(1, 2) = B(1, 2) + S, ..., B(1, 256) = B(1, 256) + S P_2: B(2, 1) = B(2, 1) + S, B(2, 2) = B(2, 2) + S, ..., B(2, 256) = B(2, 256) + S $\vdots$ P_8: B(8, 1) = B(8, 1) + S, B(8, 2) = B(8, 2) + S, ..., B(8, 256) = B(8, 256) + S

(c) Scalar-concurrent execution on eight processors in 3992 cycles

P_1: A(1 : 2041 : 8) = A(1 : 2041 : 8) + S P_2: A(2 : 2042 : 8) = A(2 : 2042 : 8) + S $\vdots$ P_8: A(8 : 2048 : 8) = A(8 : 2048 : 8) + S

(d) Vector-concurrent execution on eight processors in 960 cycles

P_1: B(1, 1 : 256) = B(1, 1 : 256) + S P_2: B(2, 1 : 256) = B(2, 1 : 256) + S $\vdots$ P_8: B(8, 1 : 256) = B(8, 1 : 256) + S

(e) COVI execution on eight processors in 756 cycles

Figure 10.8 Five execution modes of a FX/Fortran loop on the Alliant Multiprocessor. (Courtesy of Alliant Computer Systems Corporation, 1989)

The same code can be vectorized into eight vector instructions and executed serially on a single processor equipped with vector hardware. Each vector instruction works on 256 iterations of the following loop:

$$A(1:2048:256) = A(1:2048:256) + S$$

The total execution is reduced to 6048 clock cycles as shown in Fig. 10.8b.

The scalar-concurrent mode is shown in Fig. 10.8c. Eight processors are used in parallel, performing the following scalar computations:

Doall J = 1, 8
    Do I = 1, 256
        B(I,J) = B(I,J) + S
    Enddo
Endall

Now, the total execution time is further reduced to 3992 clock cycles.

Figure 10.8d shows the vector-concurrent mode on eight processors, all equipped with vector hardware. The following vector codes are executed in parallel:

Doall J = 1, 8
    A(K:2040+K:8) = A(K:2040+K:8) + S
Endall

This vectorized execution by eight processors results in a total time of 960 clock cycles.

Finally, the same program can be executed in COVI mode. The inner loop executes in vector mode, and the outer loop is executed in parallel mode. In Fortran 90 notation, we have:

$$B(1:8,1:256) = B(1:8,1:256) + S$$

The total execution time is now reduced to 756 clock cycles, the shortest among the five execution modes.

                                                                 ■

Inhibitors of Parallelization Most inhibitors of vectorization also prevent parallelization. Listed below are some inhibitors of parallelization:

(1) Multiple entries or exits.
(2) Function or subroutine calls.
(3) Input/output statements.
(4) Nondeterminism of parallel execution.
(5) Loop-carried dependences.

While only backward dependences interfere with vectorization, forward and backward dependences both affect parallelization. The overhead of synchronization code can outweigh performance gains from parallelization. We will illustrate this tradeoff analysis for multitasking on the Cray X-MP in Chapter 11. Most code parallelization is conducted at the loop level. To reduce or increase grain size, one must consider the tradeoffs between computations and communications. None of the existing compilers for parallelism has this capability yet. In most cases, the tradeoff studies are done by programmers. However, compiler directives can be used to guide the code optimization process.

10.4.4   Code Generation and Scheduling

Code generation for parallel computers is still confined within each processor. Issues involved in code generation include order of execution, instruction selection, register allocation, branch handling, postoptimizations, etc. We describe the concepts of basic blocks and instruction scheduling schemes for basic blocks. Then we consider register allocation, pattern matching, and other table-driven methods for advanced code generation. How to expand code generation methods for multiple processors systematically is still a wide-open area. It must integrate with algorithm designers, programmers, and machine architects vertically.

Directed Acyclic Graphs   Because instructions within each basic block are sequenced without any backtracks, computations performed can thus be represented by a *directed acyclic graph* (DAG). A DAG can be built in one pass through a basic block. The nodes in a DAG represent *values*. Each interior node is labeled by the operator that produces its value. Edges on the DAG show the data dependence constraints. The children of a node are the nodes producing the operand values. The leaf nodes carry the initial values or constants existing on entry to a basic block.

DAG construction repeats the following steps from node to node. Consider the statement $A := B + C$ in a basic block. We first find nodes representing the values of B and C. If B and C are not computed in the block, they must be retrieved from leaf nodes. Otherwise, B and C should come from interior nodes of the DAG. Then we create a node labeled "+". Children of this node are the nodes for values of B and C. If there is already an identical node (same label and same child nodes), node creation can be skipped. The node for "+" becomes the current node for A. In the case of a data transfer operation $A := B$, find the node representing the value of B. Then the node representing B becomes the current node for A. Exceptions do exist. A procedure call must assume all variable values have changed. If a variable could possibly point to another variable, then that variable could now have a new value. Assignment to elements of an array must be assumed to alter the entire array.

Example 10.9 Construction of a DAG for the inner loop kernel of the bubble sort program (S. Graham, J. L. Hennessy, and J. D. Ullman, 1992)

Listed below are the statements contained in the basic block B7 of the bubble sort program in Fig. 10.7.

$$
\left.\begin{array}{rcl}
t8 & := & j - 1 \\
t9 & := & 4 * t8 \\
temp & := & A[t9]
\end{array}\right\} temp := A[j]
$$

$$
\left.\begin{array}{rcl}
t10 & := & j + 1 \\
t11 & := & t10 - 1 \\
t12 & := & 4 * t11 \\
t13 & := & A[t12]
\end{array}\right\} A[j+1]
$$

$$
\left.\begin{array}{rcl}
\text{t14} & := & j-1 \\
\text{t15} & := & 4*\text{t14} \\
A[\text{t15}] & := & \text{t13}
\end{array}\right\} A[j] := A[j+1]
$$

$$
\left.\begin{array}{rcl}
\text{t16} & := & j+1 \\
\text{t17} & := & \text{t16}-1 \\
\text{t18} & := & 4*\text{t17} \\
A[\text{t18}] & := & \text{temp}
\end{array}\right\} A[j+1] := \text{temp}
$$

The corresponding DAG representation of block B7 is shown in Fig. 10.9. For nodes with the same operator, one or more names are labeled provided they consume the same operands (although they may use different values at different times). The initial value of any variable x is denoted by x_0, such as the A_0, j_0, and $temp_0$ at the leaf nodes.

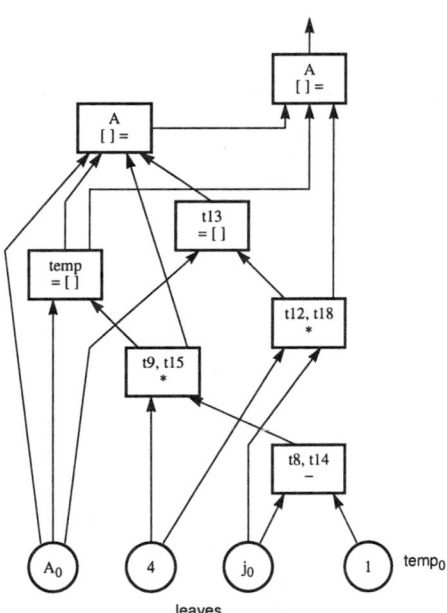

Figure 10.9 Directed acyclic graph representation of the basic block B7, the inner loop of the bubble sort program in Examples 10.6 and 10.7. (Courtesy of S. Graham, J. L. Hennessy, and J. D. Ullman, *Course on Code Optimization and Code Generation*, Western Institute of Computer Science, Stanford University, 1992)

In order to construct a DAG systematically, an auxiliary table can be used to keep track of variables and temporaries. The DAG construction process automatically detects common subexpressions and eliminates them accordingly. *Copy propagation* can be used to compute only one of the variables in each class. The construction process can easily discover the variables used or assigned and the nodes whose values can be computed at compile time. Any node whose children are constants is itself a constant. One can also label the edges in a DAG with the delays.

List Scheduling A DAG represents the flow of instructions in a basic block. A *topological sort* can be used to schedule the operations. Let READY be a buffer holding all nodes which are ready to execute. Initially, the READY buffer holds all leaf nodes with zero predecessors. Schedule each node in READY as early as possible, until it becomes empty. After all the predecessor (children) nodes are scheduled, the successor (parent) node should be immediately inserted into the READY buffer.

With list scheduling, each interior node is scheduled after its children. Additional ordering constraints are needed for a procedure call or assignment through a pointer. When the root nodes are reached, the schedule is produced. The length of the schedule equals the critical path on the DAG. To determine the critical path, both edge delays and nodal delays must be counted.

Some priority scheme can be used in selecting instructions from the READY buffer for scheduling. For example, the seven interior nodes of the DAG in Fig. 10.9 can be scheduled as follows, based on the topological order. In the case of two temporaries using the same node, we select the lower-numbered one. The following sequential code results:

$$t12 := 4 * j$$
$$t8 := j - 1$$
$$t13 := A[t12]$$
$$t9 := 4 * t8$$
$$temp := A[t9]$$
$$A[t9] := t13$$
$$A[t12] := temp$$

List scheduling schedules operations in topological order. There are no backtracks in the schedule. It is considered the best among critical-path, branch-and-bound for microinstruction scheduling (Joseph Fisher, 1979). Variations of topological list scheduling do exist such as introducing a *priority function* for ready nodes, using *top-down* versus *bottom-up* direction, and using cycle scheduling as explained below. Whenever possible, parallel scheduling of nodes should be exploited, of course, subject to data, control, and resource dependence constraints.

Cycle Scheduling List scheduling is operation-based, which has the advantage that the highest-priority operation is scheduled first. Another scheduling method for instructions in basic blocks is based on a *cycle scheduling* concept in which "cycles" rather "operations" are scheduled in order. Let READY be a buffer holding nodes with zero unscheduled predecessors ready to execute in a current cycle. Let LEADER be a buffer holding nodes with zero unscheduled predecessors but not ready in a current cycle (e.g., due to some latency unfulfilled). The following cycle scheduling algorithm is modified from the list scheduling algorithm:

> **Current-cycle** $= 0$
> **Loop until** READY and LEADER are empty
>   **For** each node n in READY (in decreasing priority order)
>   Try to schedule n in current cycle
>   If successful, update READY and LEADER
>   **Increment Current-cycle** by 1
> **end of loop**

The advantages of cycle scheduling include simplicity in implementation for single-cycle resources, such as in a superpipelined superscalar processor. There is no need to keep records of source usage and it is also easier to keep track of register lifetimes. It can be considered an improvement over the list scheduling scheme, which may result in more idle cycles. LEADER provides another level of buffering. Nodes in LEADER that have become ready should be immediately loaded into the READY queue.

Register Allocation   Traditional instruction scheduling methods minimize the number of registers used, which also reduces the degree of parallelism exploited. To optimize the code generated from a DAG, one can convert it to a sequence of *expression trees* and then study optimization for the trees. Often minimum number of registers is used for expression trees. The registers can be allocated with instructions in the scheduling scheme.

In general, more registers would allow more parallelism. The above bottom-up scheduling methods shorten register lifetimes for expression trees. A round-robin scheme can be used to allocate registers while the schedule is being generated. Or one can assume an infinite number of registers to produce a schedule first and then allocate registers and add spill code later. Another approach is to integrate register allocation with scheduling by keeping track of the liveness of registers. When the remaining registers are greater than a given threshold, one should maximize parallelism. Otherwise, one should reduce the number of registers allocated.

Register allocation can be optimized by register descriptors (or tags) to distinguish among constant, variable, indexed variable, frame pointer, etc. This tagged register may enable some additional local or global code optimizations. Another advanced feature is special branch handling, such as delayed branches or using shorter delay slots if possible.

Code generation can be improved with a better instruction selection scheme. We can first generate code for expression trees needing no register spills. One can also select instructions by recursively matching templates to parts of expression trees. A match causes code to be generated and the subtree to be rewritten with a subtree for the result. The process ends when the subtree is reduced to a single node. When template matching fails, heuristics can be used to generate subgoals for another matching. The key ideas of instruction selection by pattern matching include:

(1) Convert code generation to primarily a systematic process.
(2) Use tree-structured patterns describing instructions and use a tree-structured intermediate form.

(3) Select instructions by covering input to instruction patterns.

Major issues in pattern-based instruction selection include development of pattern matching algorithms, design of intermediate form and target machine descriptions, and interaction with low-level optimization. Extensive table descriptions are needed. Therefore table compression techniques are needed to handle large numbers of patterns to be matched.

Advanced code generation needs to be directed toward exploitation of parallelism. Therefore special compilation support is very much in demand for superscalar and multithreaded processors. These are still wide-open research problems. Partial solutions to these problems can be found in subsequent sections. There is still a long way to go in developing really "intelligent" compilers for parallel computers, Vertical integration among algorithm designers, programmers, and system architects is expected. Readers will realize in subsequent sections that existing compilers are rather primitive and have limited capabilities.

10.4.5 Trace Scheduling Compilation

Branch prediction has been used in a software scheduling technique called *trace scheduling*. The idea was originally developed for scheduling and packing operations into *horizontal microinstructions*. Trace scheduling was proposed for use in a VLIW architecture designed for scientific computation without vectorization.

The concept of trace scheduling is illustrated in Fig. 10.10. A *trace* is formed by a sequence of basic blocks, separated by assuming a particular outcome for every branch encountered in the sequence. The code example shows the first trace involving three basic blocks (A, B, and C). There are many traces for different combinations of branch outcomes. The second trace corresponds to another branch combination. Each trace is scheduled for parallel execution by a VLIW processor.

Code Compaction Independent instructions in the trace are compacted into VLIW instructions. Each VLIW word can be packed with multiple short instructions which can be independently executed in parallel. The decoding of these independent instructions is carried out simultaneously. Multiple function units (such as the memory-access unit, arithmetic unit, branch unit, etc.) are employed to carry out the parallel execution. Only independent operations are packed into a VLIW instruction.

Branch prediction is based on software heuristics or on using profiles of previous program executions. Each trace should correspond to the most likely execution path. The first trace should be the most likely one, the second trace should be the second most likely one, and so on. In fact, reduction in the execution time of an earlier trace is obtained at the expense of that of later traces. In other words, the execution time of likely traces is reduced at the expense of that of unlikely traces.

Compensation Code The effectiveness of trace scheduling depends on correct predictions at successive branches in a program. To cope with the problem of incorrect predictions, compensation codes are added to off-trace paths to provide correct linkage with the rest of program. Because code compaction may move short instructions up

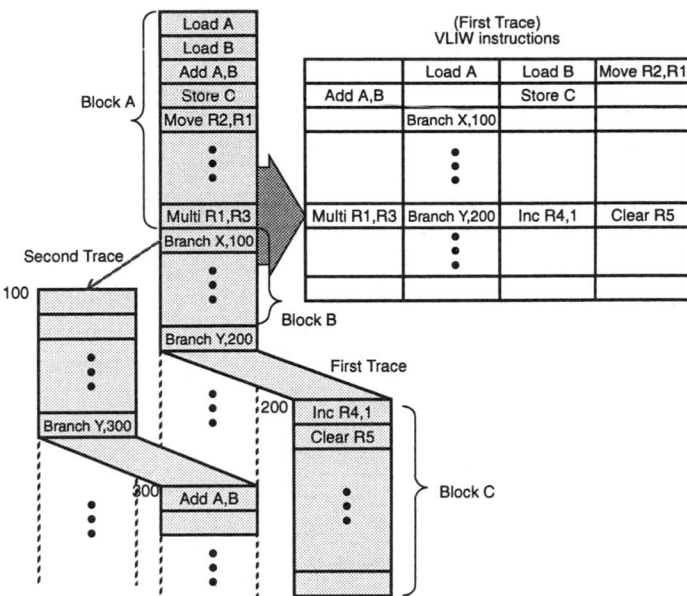

Figure 10.10 Code compaction for VLIW processor based on trace scheduling developed by Joseph Fisher (1981).

or down in the program, different compensation codes must be inserted to restore the original code distribution in the code blocks involved.

Example 10.10 Trace scheduling with code compaction and compensation

Consider an example program consisting of five basic blocks in Fig. 10.11a. The initial trace contains blocks A, B, and D, after it is predicted that execution of I4 and I9 will lead to the left path. In Fig. 10.11b, instruction I3 has been moved from block A to block B, and I7 moved to block D, to form the new blocks A′, B′ and D′.

Therefore, we need to insert instruction I3 in block C′ also and modify the target address to 201 in branch instruction I9. Similarly, some instructions have been moved up to preceding blocks (Fig. 10.11c). Compensation codes, Undo I5 and I10, must be inserted in block C′ to restore the original program semantics. ■

Compensation codes (in shaded boxes in Fig. 10.11) are needed on off-trace paths. This will make the second trace correctly executed, without being affected by the first trace due to code movement. The compensation code added should be a small portion of the code blocks. Sometimes the Undo operation cannot be used due to the lack of

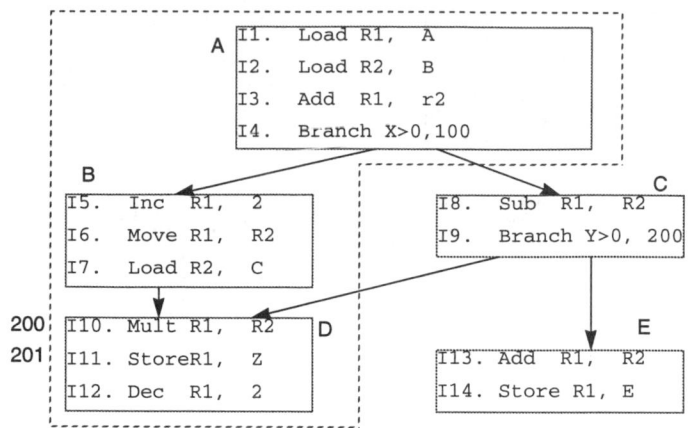

(a) The first trace on the flow graph.

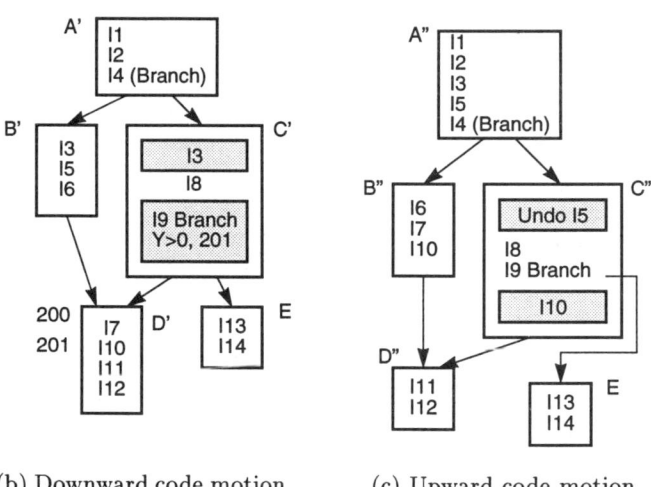

(b) Downward code motion.   (c) Upward code motion

Figure 10.11 Code motions for trace scheduling compaction in Example 10.10.

an inverse operation in the instruction set. By adding compensation code, software is performing the function of the branch history buffer described in Chapter 6.

The efficiency of trace scheduling depends on the degree of correct prediction of branches. With accurate branch predictions, the compensation code added may not be executed at all. The fewer the number of most likely traces to be executed, the better the performance will be.

By all means, branches should be used as infrequently as possible in any program. Spare processors can be used to add compensation code without causing additional overhead. Trace scheduling was mainly designed for VLIW processors. For superscalar processors, similar techniques can exploit parallelism in a program whose branch behavior is relatively easier to predict, such as in some scientific applications.

10.5 Loop Parallelization and Pipelining

This section describes the theory and application of loop transformations for vectorization or parallelization purposes. At the end, we address software pipelining techniques.

10.5.1 Loop Transformation Theory

Parallelizing loop nests is one of the most fundamental program optimization techniques demanded in a vectorizing and parallelizing compiler. In this section, we study a loop transformation theory and loop transformation algorithms derived from this theory. The bulk of the material is based on the work by Wolf and Lam (1991).

The theory unifies all combinations of loop interchange, skewing, reversal, and tilting as *unimodular transformations*. The goal is to maximize the degree of parallelism or data locality in a loop nest. These transformations also support efficient use of the memory hierarchy on a parallel machine.

Elementary Transformations   A loop transformation rearranges the execution order of the iterations in a loop nest. Three elementary loop transformations are introduced below.

(1) *Permutation* — A permutation δ on a loop nest transforms iteration $(p_1, \ldots, p_n)$ to $(p_{\delta_1}, \ldots, p_{\delta_n})$. This transformation can be expressed in matrix form as I_δ, the $n \times n$ identity matrix with rows permuted by δ. For $n = 2$, a loop interchange transformation maps iteration (i, j) to iteration (j, i). In matrix notation, we can write this as

$$\begin{bmatrix} 0 & 1 \\ 1 & 0 \end{bmatrix} \begin{bmatrix} i \\ j \end{bmatrix} = \begin{bmatrix} j \\ i \end{bmatrix}$$

The following two-deep loop nest is being transformed using the permutation matrix $\begin{bmatrix} 0 & 1 \\ 1 & 0 \end{bmatrix}$ as depicted in Fig. 10.12a.

Do $i = 1, \mathrm{N}$	**Do** $j = 1, \mathrm{N}$
Do $j = 1, \mathrm{N}$	**Do** $i = 1, \mathrm{N}$
$\mathrm{A}(j) = \mathrm{A}(j) + \mathrm{C}(i, j)$ $\Longrightarrow$	$\mathrm{A}(j) = \mathrm{A}(j) + \mathrm{C}(i, j)$
Enddo	**Enddo**
Enddo	**Enddo**

(2) *Reversal* — Reversal of the ith loop is represented by the identity matrix with the ith element on the diagonal equal to -1.

The following loop nest is being reversed using the transformation matrix $\begin{bmatrix} 1 & 0 \\ 0 & -1 \end{bmatrix}$ as depicted in Fig. 10.12b.

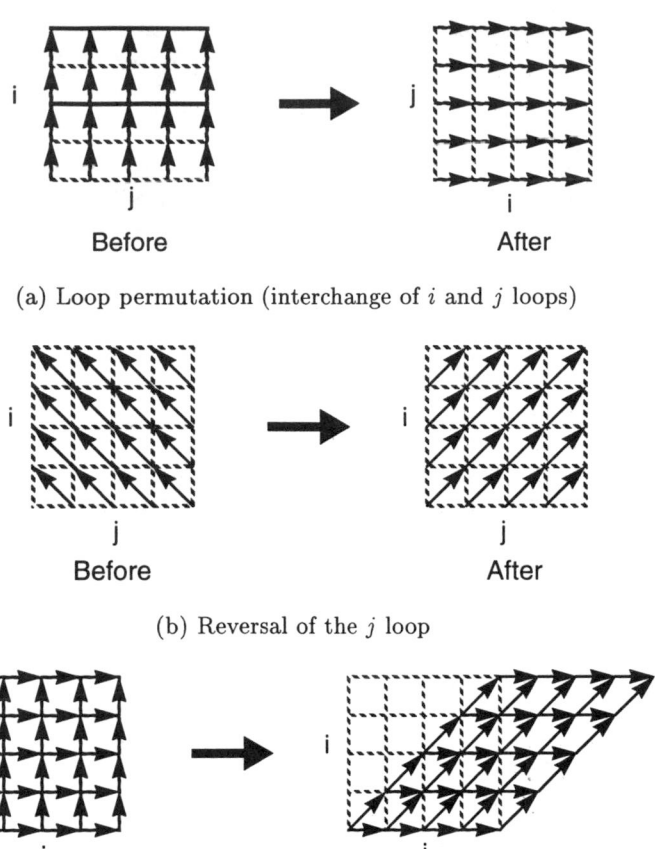

(a) Loop permutation (interchange of i and j loops)

(b) Reversal of the j loop

(c) Skewing of the inner loop by a factor of 1

Figure 10.12 Elementary loop transformations. (Courtesy of Monica Lam, *WISC Tutorial Notes*, Stanford University, 1992)

$$
\begin{array}{ll}
\textbf{Do } i = 1, \text{N} & \textbf{Do } i = 1, \text{N} \\
\quad \textbf{Do } j = 1, \text{N} & \quad \textbf{Do } j = -\text{N}, -1 \\
\quad\quad \text{A}(i,j) = \text{A}(i-1,j+1) \quad \Longrightarrow & \quad\quad \text{A}(i,-j) = \text{A}(i-1,-j+1) \\
\quad \textbf{Enddo} & \quad \textbf{Enddo} \\
\textbf{Enddo} & \textbf{Enddo}
\end{array}
$$

(3) *Skewing* — Skewing loop I_j by an integer factor f with respect to loop I_i maps iteration $(p_1, \ldots, p_{i-1}, p_i, p_{i+1}, \ldots, p_{j-1}, p_j, p_{j+1}, \ldots, p_n)$ to $(p_1, \ldots, p_{i-1}, p_i, p_{i+1}, \ldots, p_{j-1}, p_j + f p_i, p_{j+1}, \ldots, p_n)$.

In the following loop nest, the transformation performed is a skew of the inner loop with respect to the outer loop by a factor of 1, represented by the transformation matrix $\begin{bmatrix} 1 & 0 \\ 1 & 1 \end{bmatrix}$ as depicted in Fig. 10.12c.

Do $i = 1, N$	**Do** $j = 1, N$
Do $j = 1, N$	**Do** $j = 1, N$
$A(i, j) = A(i, j - 1) +$ $\implies$	$A(i, j - i) = A(i, j - i - 1) +$
$A(i - 1, j)$	$A(i - 1, j - i)$
Enddo	**Enddo**
Enddo	**Enddo**

Various combinations of the above elementary loop transformations can be defined. Wolf and Lam have called these *unimodular transformations*. The optimization problem is thus to find the unimodular transformation that maximizes an objective function given a set of schedule constraints.

Transformation Matrices   Unimodular transformations are defined by *unimodular matrices*. A unimodular matrix has three important properties. First, it is square, meaning that it maps an n-dimensional iteration space into an n-dimensional iteration space. Second, it has all integer components, so it maps integer vectors to integer vectors. Third, the absolute value of its determinant is 1.

Because of these properties, the product of two unimodular matrices is unimodular, and the inverse of a unimodular matrix is unimodular, so that combinations of unimodular loop transformations and the inverse of unimodular loop transformations are also unimodular loop transformations. A loop transformation is said to be *legal* if the transformed dependence vectors are all lexicographically positive.

A compound loop transformation can be synthesized from a sequence of primitive transformations, and the effect of the loop transformation is represented by the products of the various transformation matrices for each primitive transformation. The major issues for loop transformation include how to apply a transform, correctness or locality, and desirability or advantages of applying a transform. Wolf and Lam (1991) have stated the following conditions for unimodular transformations:

(1) Let D be the set of distance vectors of a loop nest. A unimodular transformation (matrix) T is *legal*, if and only if $\forall\ d \in D$,

$$T \cdot d \geq 0 \tag{10.7}$$

(2) Loops i through j of a nested computation with dependence vectors D are *fully permutable*, if $\forall d \in D$,

$$((d_1, d_2, ..., d_{i-1}) > 0 \quad \text{or} \quad (\forall i \leq k \leq j : d_k \geq 0)). \tag{10.8}$$

Proofs of these two conditions are left as exercises for the reader. The following example shows how to determine the legality of unimodular transformations.

Do $i = 1, N$
Do $j = 1, N$
$A(i, j) = f(A(i, j), A(i + 1, j - 1))$

Enddo
       Enddo

This code has the dependence vector $d = (1, -1)$. The loop interchange transformation is represented by the matrix

$$T = \begin{bmatrix} 0 & 1 \\ 1 & 0 \end{bmatrix}$$

The transformation is illegal since $T \cdot d = (-1, 1)$ is lexicographically negative. However, compounding the interchange with a reversal represented by the transformation matrix

$$T' = \begin{bmatrix} -1 & 0 \\ 0 & 1 \end{bmatrix} \begin{bmatrix} 0 & 1 \\ 1 & 0 \end{bmatrix} = \begin{bmatrix} 0 & -1 \\ 1 & 0 \end{bmatrix}$$

is legal because $T' \cdot d = (1, 1)$ is lexicographically positive.

10.5.2   Parallelization and Wavefronting

The theory of loop transformation can be applied to execute loop iterations in parallel. In this section, we describe loop parallelization procedures. A wavefronting approach is presented for fine-grain parallelization. Tiling is applied to reduce synchronization costs in coarse-grain computations.

Parallelization Conditions   The purpose of loop parallelization is to maximize the number of parallelizable loops. For n-deep loops whose dependences can be represented with distance vectors, at least $(n - 1)$ degrees of parallelism can be exploited in both fine-grain and coarse-grain computations.

The algorithm for loop parallelization consists of two steps: It first transforms the original loop nest into a canonical form, namely, a *fully permutable* loop nest. It then transforms the fully permutable loop nest to exploit coarse- and/or fine-grain parallelism according to the target architecture.

(1) *Canonical form.* Loops with distance vectors have the special property that they can always be transformed into a fully permutable nest via skewing. It is easy to determine how much to skew an inner loop with respect to an outer one to make these loops fully permutable. For example, if a doubly nested loop has dependences $\{(0, 1), (1, -2), (1, -1)\}$, then skewing the inner loop by a factor of 2 with respect to the outer loop produces $\{(0, 1), (1, 0), (1, 1)\}$.

(2) *Parallelization process.* Iterations of a loop can execute in parallel if and only if no dependences are carried by that loop. Such a loop is called a Doall loop. To maximize the degree of parallelism is to transform the loop nest to maximize the number of Doall loops.

Let $(I_1, ..., I_n)$ be a loop nest with lexicographically positive dependences $d \in D$. I_i is parallelizable if and only if $\forall d \in D, (d_1, ..., d_{i-1}) > (0, ..., 0)$, the zero vector, or $d_i = 0$. Once the loops are made fully permutable, the steps to generate Doall parallelism are simple. In the following discussion, we show that the loops in canonical form can be trivially transformed to produce both fine- and coarse-grain parallelism.

Fine-Grain Wavefronting A nest of n fully permutable loops can be transformed into code containing at least $(n-1)$ degrees of parallelism. In the degenerate case where no dependences are carried by these n loops, the degree of parallelism is n. Otherwise, $(n-1)$ parallel loops can be obtained by skewing the innermost loop in the fully permutable nest by each of the other loops and moving the innermost loop to the outermost position.

This transformation, called *wavefront transformation*, is represented by the following matrix:

$$T = \begin{bmatrix} 1 & 1 & \cdots & 1 & 1 \\ 1 & 0 & \cdots & 0 & 0 \\ 0 & 1 & \cdots & 0 & 0 \\ \vdots & & \ddots & & \vdots \\ 0 & 0 & \cdots & 1 & 0 \end{bmatrix} \tag{10.9}$$

Fine-grain parallelism is often exploited on vector machines, superscalar processors, and systolic arrays. The following example shows the entire process of loop parallelization exploiting fine-grain parallelism. The process includes skewing and wavefront transformation.

Example 10.11 Loop skewing and wavefront transformation (Michael Wolf and Monica Lam, 1991)

Figure 10.13a shows the iteration space and dependence of a source loop nest. The skewed loop nest is shown in Fig. 10.13b after applying the matrix $T = \begin{bmatrix} 1 & 0 \\ 1 & 1 \end{bmatrix}$. Figure 10.13c shows the result of applying wavefront transformation to the skewed loop code, which is a result of first skewing the innermost loop to make the two-dimensional loop nest fully permutable and then applying the wavefront transformation to create one degree of parallelism.

There are no dependences between iterations within the innermost loop nest. The transform is a wavefront transformation because it causes iterations along the diagonal of the original loop nest to execute in parallel.

∎

This wavefront transformation automatically places the maximum Doall loops in the innermost loops, maximizing fine-grain parallelism. This is the appropriate transformation for superscalar or VLIW machines. Although these machines have a low degree of parallelism, finding multiple parallelizable loops is still useful. Coalescing multiple Doall loops prevents the pitfall of parallelizing only a loop with a small iteration count.

Coarse-Grain Parallelism For MIMD coarse-grain multiprocessors, having as many outermost Doall as possible reduces the synchronization overhead. A wavefront transformation produces the maximum degree of parallelism but makes the outermost loop sequential if any are. For example, consider the following loop nest:

```
For I₁:=0 to 5 do
   For I₂:=0 to 6 do
      A[I₂+1]:=1/3*(A[I₂]+A[I₂+1]+A[I₂+2]
```

$$D = \{ (0,1), (1,0), (1,-1) \}$$

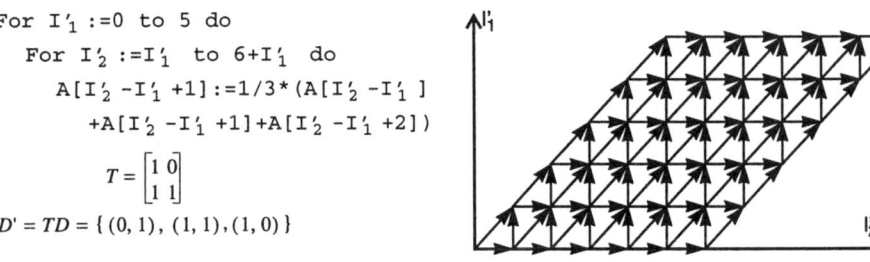

(a) Extract dependence information from source loop nest

```
For I'₁ :=0 to 5 do
   For I'₂ :=I'₁ to 6+I'₁ do
      A[I'₂ -I'₁ +1]:=1/3*(A[I'₂ -I'₁ ]
            +A[I'₂ -I'₁ +1]+A[I'₂ -I'₁ +2])
```

$$T = \begin{bmatrix} 1 & 0 \\ 1 & 1 \end{bmatrix}$$

$$D' = TD = \{ (0,1), (1,1), (1,0) \}$$

(b) Skew to make inner loop nest fully permutable

```
For I'₁ :=0 to 16 do
   Doall I'₂ :=max(0,⌈(I'₁ -6)/2⌉) to min(5,⌊I'₁ /2⌋) do
      A[I'₁ -2I'₂ +1]:=1/3*A[I'₁ -2I'₂ ]+A[I'₁ ,-2I'₂ +1]
            +A[I'₁ -2I'₂ +2]
```

$$T = \begin{bmatrix} 1 & 1 \\ 1 & 0 \end{bmatrix} \begin{bmatrix} 1 & 0 \\ 1 & 1 \end{bmatrix} = \begin{bmatrix} 2 & 1 \\ 1 & 0 \end{bmatrix}$$

$$D' = TD = \{ (1,0), (2,1), (1,1) \}$$

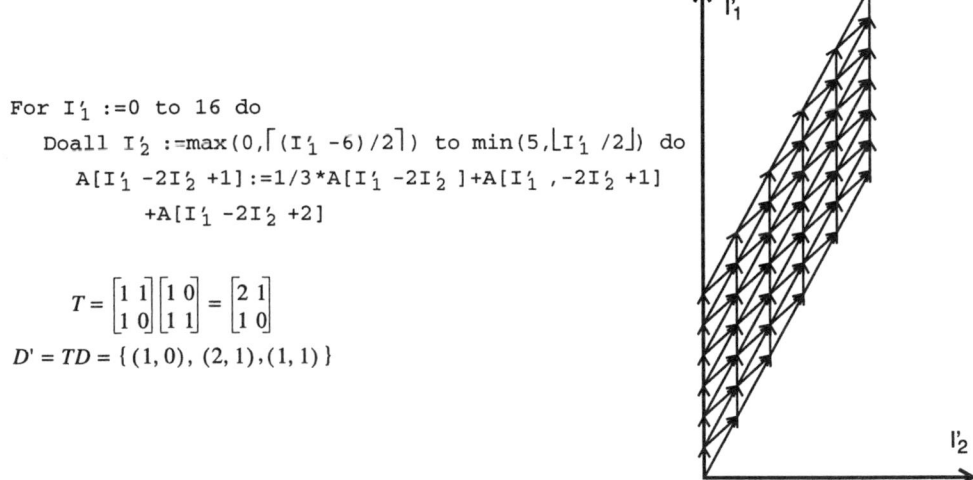

(c) Wavefront transformation on the skewed loop nest

Figure 10.13 Fine-grain parallelization by loop skewing and wavefront transformation in Example 10.11. (Courtesy of Wolf and Lam; reprinted from *IEEE Trans. Parallel Distributed Systems*, 1991)

Do $i = 1, N$
    Do $j = 1, N$
        $A(i, j) = f(A(i - 1, j - 1))$
    Enddo
Enddo

This loop nest has the dependence $(1, 1)$, and so the outermost loop is sequential and the innermost loop is a Doall. The wavefront transformation $\begin{bmatrix} 1 & 1 \\ 1 & 0 \end{bmatrix}$ does not change this. In contrast, the unimodular transformation $\begin{bmatrix} 1 & -1 \\ 0 & 1 \end{bmatrix}$ transforms the dependence to $(0, 1)$, making the outer loop a Doall and the inner loop sequential.

In this example, the dimensionality of the iteration space is two, but the dimensionality of the space spanned by the dependence vectors is only one. When the dependence vectors do not span the entire iteration space, it is possible to perform a transformation that makes outermost Doall loops.

A heuristic though nonoptimal approach for making loops Doall is simply to identify loops I_i such that all d_i are zero. Those loops can be made outermost Doall. The remaining loops in the tile can be wavefronted to obtain the remaining parallelism.

Loop parallelization can be achieved through unimodular transformations as well as tiling. For loops with distance vectors, n-deep loops have at least $(n - 1)$ degrees of parallelism. The loop parallelization algorithm has a common step for fine- and coarse-grain parallelism in creating an n-deep fully permutable loop nest by skewing. The algorithm can be tailored for different machines based on the following guidelines:

- Move Doall loop innermost (if one exists) for fine-grain machines. Apply a wavefront transformation to create up to $(n - 1)$ Doall loops.
- Create outermost Doall loops for coarse-grain machines. Apply tiling to a fully permutable loop nest.
- Use tiling to create loops for both fine- and coarse-grain machines.

10.5.3 Tiling and Localization

Tiling and locality optimization techniques are studied in this section. The ultimate purpose is to reduce synchronization overhead and to enhance multiprocessor efficiency when loops are distributed for parallel execution.

Tiling to Reduce Synchronization   It is possible to reduce the synchronization cost and improve the data locality of parallelized loops via an optimization known as *tiling* (Wolfe, 1989). Tiling is not a unimodular transformation. In general, tiling maps an n-deep loop nest into a $2n$-deep loop nest where the inner n loops include only a small fixed number of iterations. Figure 10.14a shows the code after tiling the example in Fig. 10.13b using a tile size of 2×2. The two innermost loops execute the iterations within each tile, represented as 2×2 squares. The two outer loops, represented by the two axes, execute the 3×4 tiles. The outer loops of the tiled code control the execution of the tiles.

```
For II'₁ :=0 to 5 by 2 do
   For II'₂ :=0 to 11 by 2 do
      For I'₁ :=II'₁  to min(I'₁ +1,5) do
         For I'₂ :=max(II'₁ ,II'₂ ) to min(6+I'₁ ,II'₂ +1) do
            A[I'₂ ]:=1/3*(A[I'₂ -1]+A[I'₂ ]+A[I'₂ +1])
```

(a) Tiled code from the skewed code in Fig. 10.13b

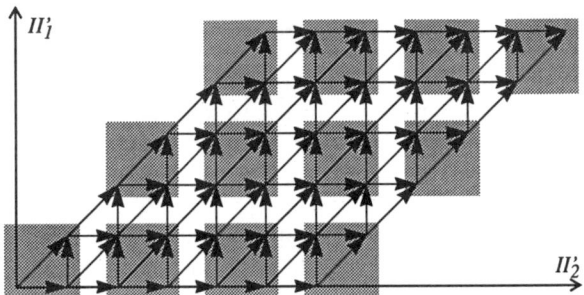

(b) Iteration space and dependences of the tiled code

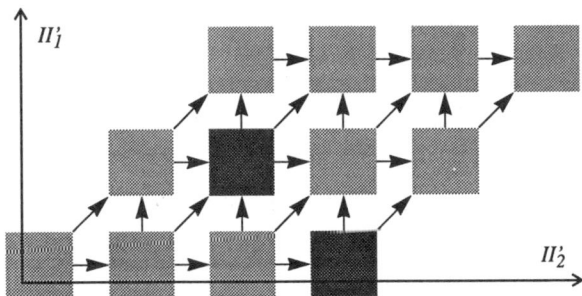

(c) Parallel execution of the tiled loops

Figure 10.14 Tiling of the skewed loops for parallel execution on a coarse-grain multiprocessor. (Courtesy of Wolf and Lam; reprinted from *IEEE Trans. Parallel Distributed Systems*, 1991)

The same property that supports parallelization, full permutability, is also the key to tiling: Loops I_i through I_j of a legal computation can be tiled if they are fully permutable.

Thus, loops in the canonical form of the parallelization algorithm can also be tiled. Moreover, the characteristics of the controlling loops resemble those of the original set of loops. An abstract view giving the dependences of the tiles is shown in Fig. 10.14b. These controlling loops are themselves permutable and so are easily parallelizable. However, each iteration of the outer n loops is a tile of iterations instead of an individual iteration. Parallel execution of the tiled loops is shown in Fig. 10.14c.

Tiling can therefore increase the granularity of synchronization and data are often reused within a tile. Without tiling, when a Doall loop is nested within a non-Doall loop, all processors must be synchronized with a barrier at the end of each Doall loop.

Using tiling, we can reduce the synchronization cost in the following two ways. First, instead of applying wavefront transformation to the loops in canonical form, we first tile the loops and then apply a wavefront transformation to the controlling loops of the tiles. In this way, the synchronization cost is reduced by the size of the tile. Certain loops cannot be represented as distances. Direction vectors can be used to represent these loops. The idea is to represent direction vectors as an infinite set of distance vectors.

Locality optimization in user programs is meant to reduce memory-access penalties. Software pipelining can reduce the execution time. Both are desired improvements in the performance of parallel computers. Program locality can be enhanced with loop interchange, reversal, tiling, and prefetching techniques. The effort requires a reuse analysis of a "localized" iteration space. Software pipelining relies heavily on sufficient support from a compiler working effectively with the scheduler.

The fetch of successive elements of a data array is pipelined for an interleaved memory. In order to reduce the access latency, the loop nest can interchange its indices so that a long vector is moved into the innermost loop, which can be more effective with pipelined loads. Loop transformations are performed to reuse the data as soon as possible or to improve the effectiveness of data caches.

Prefetching is often practiced to hide the latency of memory operations. This includes the insertion of special instructions to prefetch data from memory to cache. This will enhance the cache hit ratio and reduce the register occupation rate with a small prefetch overhead. In scientific codes, data prefetching is rather important. Instruction prefetching, as studied in Chapter 6, is often practiced in modern processors using prefetch buffers and an instruction cache. In the following discussion, we concentrate on data prefetching techniques.

Tiling for Locality   Blocking or tiling is a well-known technique that improves the data locality of numerical algorithms. Tiling can be used for different levels of memory hierarchy such as physical memory, caches, and registers; multilevel tiling can be used to achieve locality at multiple levels of the memory hierarchy simultaneously.

To illustrate the importance of tiling, consider the example of matrix multiplication:

```
Do i = 1, N
    Do j = 1, N
        Do k = 1, N
            C(i, k) = C(i, k) + A(i, j) × B(j, k)
        Enddo
    Enddo
Enddo
```

In this code, although the same rows of C and B are reused in the next iteration of the middle and outer loops, respectively, the large volume of data used in the intervening iterations may replace the data from the register file or the cache before they can be reused. Tiling reorders the execution sequence such that iterations from loops of the outer dimensions are executed before all the iterations of the inner loop are completed. The tiled matrix multiplication is

$$
\begin{aligned}
&\textbf{Do } \ell = 1, \text{N}, s\\
&\quad \textbf{Do } m = 1, \text{N}, s\\
&\quad\quad \textbf{Do } i = 1, \text{N}\\
&\quad\quad\quad \textbf{Do } j = \ell, \min(\ell + s - 1, \text{N})\\
&\quad\quad\quad\quad \textbf{Do } k = m, \min(m + s - 1, \text{N})\\
&\quad\quad\quad\quad\quad \text{C}(i, k) = \text{C}(i, k) + \text{A}(i, j) \times \text{B}(j, k)\\
&\quad\quad\quad\quad \textbf{Enddo}\\
&\quad\quad\quad \textbf{Enddo}\\
&\quad\quad \textbf{Enddo}\\
&\quad \textbf{Enddo}\\
&\textbf{Enddo}
\end{aligned}
$$

Tiling reduces the number of intervening iterations and thus data fetched between data reuses. This allows reused data to still be in the cache or register file and hence reduces memory accesses. The tile size s can be chosen to allow the maximum reuse for a specific level of memory hierarchy. For example, the tile size is relevant to the cache size used or the register file size used.

The improvement obtained from tiling can be far greater than that obtained from traditional compiler optimizations. Figure 10.15 shows the performance of 500×500 matrix multiplication on an SGI 4D/380 machine consisting of eight MIPS R3000 processors running at 33 MHz. Each processor has a 64-Kbyte direct-mapped first-level cache and a 256-Kbyte direct-mapped second-level cache. Results from four different experiments are reported: without tiling, tiling to reuse data in caches, tiling to reuse data in registers, and tiling for both register and caches. For cache tiling, the data are copied into consecutive locations to avoid cache interference.

Tiling improves the performance on a single processor by a factor of 2.75. The effect of tiling on multiple processors is even more significant since it reduces not only the average data-access latency but also the required memory bandwidth. Without cache tiling, contention over the memory bus limits the speedup to about 4.5 times. Cache tiling permits speedups of over 7 for eight processors, achieving an overall speed of 64 Mflops when combined with register tiling.

Localized Iteration Space   Reusing vector spacing offers opportunities for locality optimization. However, reuse does not imply locality. For example, if reuse does not occur soon enough, it may miss the temporal locality. Therefore, the idea is to make reuse happen soon enough. A *localized iteration space* contains the iterations that can exploit reuse. In fact, tiling increases the number of dimensions in which reuse can be exploited.

Consider the following two-deep nest. Reuse is exploited for loops i and j only, which form the localized iteration space:

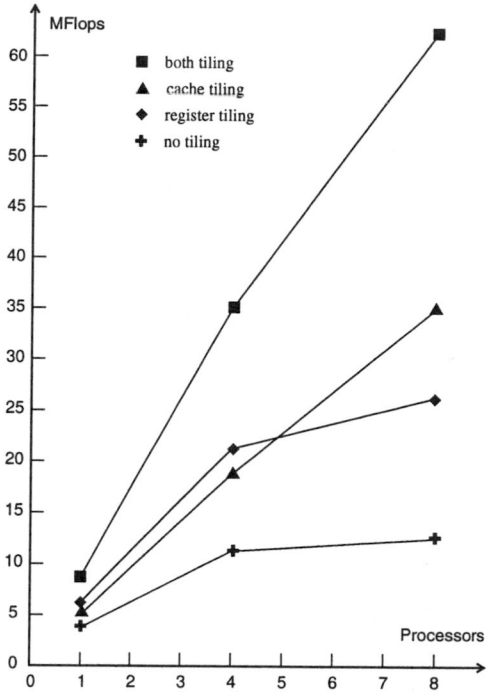

Figure 10.15 **Performance of a 500 × 500 double precision matrix multiplication on the SGI 4D/380. Cache tiles are 64 × 64 iterations and register tiles are 4 × 2.** (Courtesy of Wolf and Lam; reprinted from ACM SIGPLAN *Conf. Programming Language Design and Implementation*, Toronto, Canada, 1991)

```
Do i = 1, N
    Do j = 1, N
        B(i, j) = f(A(i), A(j))
    Enddo
Enddo
```

Reference $A(j)$ touches different data within the inner loop but reuses the same elements across the outer loop. More precisely, the same data $A(j)$ is used in iterations $(i, j), 1 \leq i \leq N$. There is reuse, but the reuse is separated by accesses to $N - 1$ other data. When N is large, the data is removed from the cache before it can be reused, and there is no locality. Therefore, a reuse does not guarantee locality. The tiled code is shown below:

```
Do ℓ = 1, N, s
    Do i = 1, N
        Do j = ℓ, max(ℓ + s - 1, N)
            B(i, j) = f(A(i), A(j))
```

> **Enddo**
> **Enddo**
> **Enddo**

We choose the tile size such that the data used within the tile can be held within the cache. For this example, as long as s is smaller than the cache size, $A(j)$ will still be present in the cache when it is reused. Thus, reuse is exploited for loops i and j only, and so the localized iteration space includes only these loops.

In general, if n is the first loop with a large bound, counting from innermost to outermost, then reuse occurring within the inner n loops can be exploited. Therefore the localized vector space of a tiled loop is simply that of the innermost tile, whether the tile is coalesced or not.

Obviously, memory optimizations are important. Locality can be upheld by intersecting the localized vector space with the reuse vector space. In other words, reuse directs the search for unimodular and tiling transformations. One should use locality information to eliminate unnecessary prefetches.

10.5.4 Software Pipelining

This refers to the pipelining of successive iterations of a loop in the source programs. The advantage of software pipelining is to reduce the execution time with compact object code. The idea has been validated by implementation of a compiler for the Warp, a systolic array of 10 processors built at CMU (Lam, 1988). Obviously, software pipelining is more effective for deep hardware pipelines. The concept is illustrated with an example taken from Lam's Tutorial Notes on Compilers for Parallel Machines (1992).

Pipelining of Loop Iterations   Successive iterations of the following loop nest are to be executed on a two-issue processor first without software pipelining and then with pipelining.

> **Do** I = 1, N
>   A(I) = A(I) × B + C
> **Enddo**

This is an example of Doall loops in which all iterations are independent. It is assumed that each memory access (*Read* or *Write*) takes one cycle and each arithmetic operation (*Mul* and *Add*) requires two cycles. Without pipelining, one iteration requires six cycles to execute as listed below:

Cycle	Instruction	Comment
1	Read	/Fetch A[I]/
2	Mul	/Multiply by B/
3		
4	Add	/Add to C/
5		
6	Write	/Store A[I]/

Therefore, N iterations require 6N cycles to complete, ignoring the loop control overhead. Listed below is the execution of the same code on an 8-deep instruction pipeline:

Cycle	Iteration			
	1	2	3	4
1	Read			
2	Mul			
3		Read		
4		Mul		
5	Add		Read	
6			Mul	
7		Add		Read
8	Write			Mul
9			Add	
10		Write		
11				Add
12			Write	
13				
14				Write

Four iterations of the software-pipelined code are shown. Although each iteration requires 8 cycles to flow through the pipeline, the four overlapped iterations require only 14 clock cycles to execute. Compared with the nonpipelined execution, a speedup factor of $24/14 = 1.7$ is achieved with the pipelining of four iterations.

N iterations require $2N + 6$ cycles to execute with the pipeline. Thus, a speedup factor of $6N/(2N + 6)$ is achieved. As N approaches infinity, a speedup factor of 3 is expected. This shows the advantage of software pipelining, if other overhead is ignored.

Doacross loops Unlike unrolling, software pipelining can give optimal results. Locally compacted code may not be globally optimal. The Doall loops can fill arbitrarily long pipelines with infinite iterations. In the following Doacross loop with dependence between iterations, software pipelining can still be done but is harder to implement.

Do I = 1, N		Read
	A(I) = A(I) × B	Mul
	Sum = Sum + A(I);	Add
Enddo		Write

The software-pipelined code is shown below:

1	Read	
2	Mul	
3	Add	Read
4	Write	Mul
5		Add
6		Write

It is assumed that one memory access and one arithmetic operation can be concurrently executed on the two-issue superscalar processor in each cycle. Thus recurrences can also be parallelized with software pipelining.

As in the hardware pipeline scheduling in Chapter 6, the objective of software pipelining is to minimize the interval at which iterations are initiated; i.e., the *initiation latency* determines the throughput for the loop. The basic units of scheduling are minimally indivisible sequences of microinstructions. In the above Doall loop example, the initiation latency is two cycles per iteration, and the Doacross loop is also pipelined with an initiation latency of two cycles.

To summarize, software pipelining demands maximizing the throughput by reducing the initiation latency, as well as by producing small, compacted code size. The trick is to find an identical schedule for every iteration with a constant initiation latency. The scheduling problem is tractable, if every operation takes unit execution time and no cyclic dependences exist in the loop.

10.6 Bibliographic Notes and Exercises

Parallel programming models have been treated in books [Ben-Ari90], [Perrott87], and [Andrews91]. [Brawer89] introduced multiprocessor programming with Fortran 77 and C in a UNIX environment. [Babb88] described programming techniques for a number of commercially available parallel processors.

The book by [Holt78] et al. introduced concurrent programming with operating systems applications. In [Hwang89a] and DeGroot, software techniques for programming supercomputers and AI machines are discussed.

The material on concurrent object-oriented programming is based on [Agha90]. Actors were described in [Agha86]. Data-parallel programming was treated in [Perrott87], [Christy90], and [Quinn90] and Hatcher. *Concurrent Prolog* [Shapiro86] and *Parlog* [Clark83] were developed for parallel logic programming.

The use of SISAL for parallel functional programming was treated in [Allan85] and Oldehoeft. Japan's FGCS was described in [Hertzberger84]. Parallel programs for solving the traveling salesperson problem can be found in [Felten85] and in [Kallstrom88] and Thakkar.

Compilers for supercomputers are reported in [Padua86] and Wolfe [Kuck84] et al. and in [Allen87] and Kennedy. PFC Fortran extensions are described in [Leasure90]. The Parafrase has been extended to Parafrase 2 [Polychronopoulos89] et al. for handling C and Pascal programs besides Fortran codes. PFC+ and ParaScope information can be found in [Callahan88] et al. The book by [Zima90] and Chapman provides a comprehensive coverage of compilers developed for parallel and vector computers. Material on trace scheduling compilers can be found in [Fisher81].

Dependence analysis of data arrays started with the pioneering work of [Banerjee88]. The GCD test was reported in [Banerjee79] and in [Wolfe82]. The categorical dependence testing was invented by [Goff91], Kennedy, and Tseng. The material on code optimization and generation was based on the tutorials by [Graham92], Hennessy, and Ullman. Vectorization and parallelization methods have been studied in [Hwang84] and Briggs, [Wolfe89], [Zima90] and Chapman, [Convex90], and [Alliant89].

Trace scheduling compilation was proposed by [Fisher81]. The theory of loop transformation was developed by [Wolf91b] and Lam. The material on list scheduling and cycle scheduling is based on the tutorial by [Lam92]. Locality optimization with tiling was proposed in [Wolf91a] and Lam. Software pipelining is based on the work of [Lam88]. A *Molecule* language construct for layered development of parallel programs was proposed by [Xu89] and Hwang. Programming with Fortran 90 can be found in the book by [Brainerd90] et al. Parallel languages and compilers are also treated by contributing authors in the monograph edited by [Gelernter90], Nicolau, and Padua.

Exercises

Problem 10.1 Explain the following terms associated with message-passing programming of multicomputers:

(a) Synchronous versus asynchronous message-passing schemes.
(b) Blocking versus nonblocking communications.
(c) The *rendezvous* concept introduced in the Ada programming system.
(d) Name-addressing versus channel-addressing schemes for message passing.
(e) Uncoupling between sender and receiver using buffers or mailboxes.
(f) Lost-message handling and interrupt-message handling.

Problem 10.2 Concurrent object-oriented programming was introduced in Section 10.1.4. Chain multiplication of a list of numbers was illustrated in Example 10.2 based on a *divide-and-conquer* strategy. A fragment of a Lisp-like code for multiplying the sequence of numbers is given below:

```
(define tree-product
    (lambda [tree]
        (if (number ? tree)
            tree
            (*(tree-product(left-tree)   tree))
            (*(tree-product(right-tree)   tree)))))
```

In this code, a *tree* is passed to *tree-product*, which tests to see if the *tree* is a number (i.e., a singleton at the leaf node). If so, it returns the *tree*; otherwise it subdivides the problem into two recursive calls. The *left-tree* and *right-tree* are functions which pick off the left and right branches of the tree. Note that the argument to * may be evaluated concurrently.

Write a Lisp code to implement this divide-and-conquer algorithm for chain multiplication on a multiprocessor or a multicomputer system. Compare the execution time by running the same program sequentially on a uniprocessor system. The chain should be sufficiently long to see the difference.

Problem 10.3 *Gaussian elimination* with partial pivoting has been implemented by

[Quinn90] and Hatcher in C* code on the Connection Machine, as well as in a concurrent C code on an nCUBE 3200 multicomputer.

(a) Discuss the translation/compiler effort from C* to C in the two machines after a careful reading of the paper by Quinn and Hatcher.

(b) Comment on SPMD (single program and multiple data streams) programming style as opposed to SIMD programming style on the Connection Machine in terms of synchronization implementation and related performance issues.

(c) Repeat the program conversion experiments for a *fast Fourier transform* (FFT) algorithm if both machines and their programming manuals are available to you. Otherwise, perform the program conversion at least manually at the algorithm level using pseudo codes with parallel constructs.

Problem 10.4   Explain the following terms related to shared-variable programming on multiprocessors.

(a) Multiprogramming.   (d) Multitasking.
(b) Multiprocessing in MIMD mode.   (e) Multithreading.
(c) Multiprocessing in MPMD mode.   (f) Program partitioning.

Problem 10.5   The following arrays are declared in Fortran 90.

        REAL A(10,10,5)
        REAL B(9,9)
        REAL C(3,4,5)

(a) List array elements specified by the following array expressions: A(5,8:*,*), B(3:*:3,5:8), and C(*,3,4).

(b) Can you make the following array assignments?

        A(3:5,7,4:6) = C(*,3,3:5),
        B(1:2,7:9) = B(7:9,4:6),
        C(*,4,4:5) = B(7:9,8:9), and
        A(5,9:10,2:4) = A(7,3:4,3:*) + C(2,4:5,1:3).

Problem 10.6   Determine the dependence relations among the three statements in the following loop nest. The direction vector and distance vector should be specified in all dependence relations.

           Do I = 1, N
                Do J = 2, N
    S_1:   A(I,J) = A(I, J−1) + B(I, J)
    S_2:   C(I,J) = A(I, J) + D(I+1, J)
    S_3:   D(I,J) = 0.1
                Enddo
           Enddo

Problem 10.7 Consider the following loop nest:

$$
\begin{aligned}
&\textbf{Do } I = 1, N \\
S_1:\quad &\qquad A(I) = B(I) \\
S_2:\quad &\qquad C(I) = A(I) + B(I) \\
S_3:\quad &\qquad E(I) = C(I+1) \\
&\textbf{Enddo}
\end{aligned}
$$

(a) Determine the dependence relations among the three statements.
(b) Show how to vectorize the code with Fortran 90 statements.

Problem 10.8 Consider the following loop nest:

$$
\begin{aligned}
&\textbf{Do } I = 1, N \\
&\qquad \textbf{Do } J = 2, N \\
S_1:\quad &\qquad\qquad A(I,J) = B(I, J) + C(I, J) \\
S_2:\quad &\qquad\qquad C(I,J) = D(I, J)/2 \\
S_3:\quad &\qquad\qquad E(I,J) = A(I,J-1)**2 + E(I,J-1) \\
&\qquad \textbf{Enddo} \\
&\textbf{Enddo}
\end{aligned}
$$

(a) Show the data dependences among the statements.
(b) Show how to parallelize the loop scheduling the parallelizable iterations to concurrent processors.

Problem 10.9 Consider the following loop nest:

$$
\begin{aligned}
&\textbf{Do } J = 1, N \\
&\qquad \textbf{Do } I = 1, N \\
S_1:\quad &\qquad\qquad A(I,J+1) = B(I, J) + C(I, J) \\
S_2:\quad &\qquad\qquad D(I,J) = A(I, J)/2 \\
&\qquad \textbf{Enddo} \\
&\textbf{Enddo}
\end{aligned}
$$

(a) Show how to compile the code for vectorization in the I-loop, assuming Fortran column-major storage order.
(b) Show how to compile the code for parallelization in the J-loop using the **Doacross** and **Endacross** commands. You can use a conditional statement or $Signal(J)$ and $Wait(J-1)$ for synchronization in the concurrent loop.
(c) Show how to compile the loop to perform the J-loop in vector mode, while using the **Doall** and **Endall** commands for the outer I-loop.

Problem 10.10 Explain the following loop transformations and discuss how to apply them for loop vectorization or parallelization:

(a) Loop permutation.   (e) Wavefront transformation.
(b) Loop reversal.   (f) Locality optimization.
(c) Loop skewing.   (g) Software pipelining.
(d) Loop tiling.

Problem 10.11   Loop-carried dependence (LCD) exists in the following loops:

(a) Consider the forward LCD in the following loop:

> **Do** I = 1, N
> $\quad$ A(I) = A(I+1) + 3.14159
> **Enddo**

Explain why a forward LCD does not prevent vectorization of a loop.

(b) The following loop contains backward LCDs:

> **Do** I = 1, N − 1
> $\quad$ A(I) = B(I) + C(I)
> $\quad$ B(I+1) = D(I) ∗ 3.14159
> **Enddo**

Show that the loop can be vectorized by statement reordering.

Problem 10.12   Vectorize or parallelize the following loops if possible. Otherwise, explain why it is not possible.

(a)

> **Do** I = 1, N
> $\quad$ A(I+1) = A(I) + 3.14159
> **Enddo**

(b)

> **Do** I = 1, N
> $\quad$ **If** (A(I) .LE. 0.0) **then**
> $\quad\quad$ S = S + B(I) ∗ C(I)
> $\quad\quad$ X = B(I)
> $\quad$ **Endif**
> **Enddo**

Problem 10.13   Tanenbaum and associates have suggested a hybrid parallel programming paradigm using shared objects and broadcasting. Study the paper that appeared in *IEEE Computer* (August 1992) and explain how to apply the software paradigm for either multiprocessors or multicomputers.

Chapter 11

Parallel Program Development and Environments

This chapter describes software environments and program development techniques for parallel computers. We first introduce environments, synchronization, and execution modes. Then we describe methods for shared-variable and message-passing program development. The emphasis is on program modularity, fast communications, load balancing, and performance tuning.

11.1 Parallel Programming Environments

An environment for parallel programming consists of hardware platforms, languages supported, OS and software tools, and application packages. The hardware platforms vary from shared-memory, message-passing, vector processing, and SIMD to dataflow computers. Usually, we directly identify the machine models used, such as the Cray Y-MP, BBN Butterfly, iPSC/860, etc.

11.1.1 Software Tools and Environments

Pure parallel programming languages such as Linda and Strand 88 represent the coarsest parallel programming environment. Others form an integrated environment consisting of an editor, a debugger, performance monitors, and a program visualizer for improving software productivity and the quality of application programs, such as the packages Express and TOPSYS.

Figure 11.1 shows a classification of environment types on the line between pure languages and integrated environments. Integrated environments can be divided into *basic*, *limited*, and *well-developed* classes, depending on the maturity of the tool sets.

A basic environment provides a simple program tracing facility for debugging and performance monitoring or a graphic mechanism for specifying the task dependence graph in SCHEDULE, the process call graph in FAUST, and the process component graph in PIE.

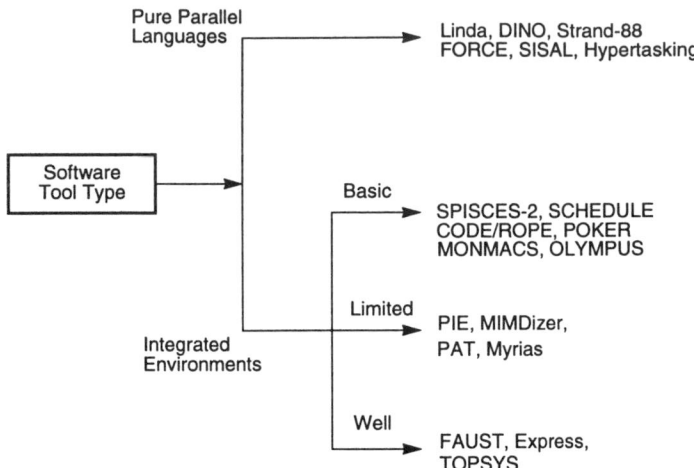

Limited integration provides tools for parallel debugging, performance monitoring, or program visualization beyond the capability of the basic environments listed. Well-developed environments provide intensive tools for debugging programs, interaction of textual/graphical representations of a parallel program, visualization support for performance monitoring, program visualization, parallel I/O, parallel graphics, etc.

The classification of a particular tool will likely change with time. A pure language might become an integrated environment for parallel programming someday when a complete set of tools is added to it. For example, C-Linda and Fortran-Linda were developed to help C and Fortran programmers write parallel programs using the tuple spaces in Linda.

Environment Features   In designing a parallel programming language, one often faces a dilemma involving *compatibility, expressiveness, ease of use, efficiency,* and *portability.* Parallel languages are developed either by introducing new languages such as Linda and Occam or by extending existing sequential languages such as Fortran 90, C*, and Concurrent Pascal.

A new parallel programming language has the advantage of using high-level parallel concepts or constructs for parallelism instead of using imperative (algorithmic) languages which might have a sequential nature or exhibit the von Neumann bottleneck and therefore are inherently sequential. New languages are usually difficult to market for acceptance in a short time period because they are often incompatible with existing sequential languages.

Most parallel computer designers choose the language extension approach to solving the compatibility problem. High-level parallel constructs were added to Fortran, C, Pascal, and Lisp to make them suitable for use on parallel computers. Special optimizing compilers are needed to automatically detect parallelism and transform sequential constructs into parallel ones.

High-level parallel constructs can be implicitly embedded in the syntax or explicitly specified by users. We have identified three compiler approaches: *preprocessors*, *precompilers*, and *parallelizing compilers*, as illustrated in Fig. 11.2.

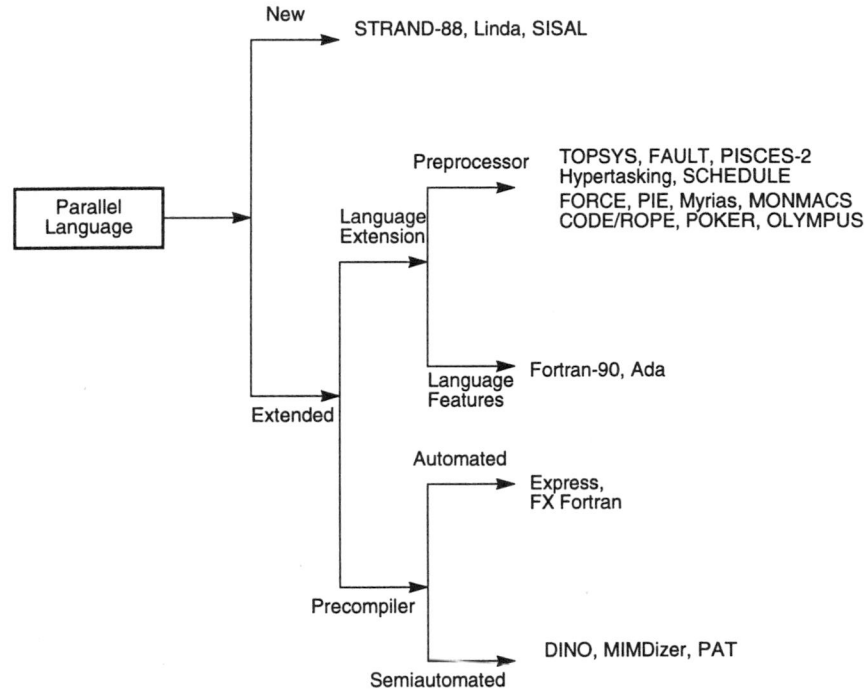

Figure 11.2 Parallel languages and compiler technology for parallel programming at different automation levels.

Preprocessors, such as FORCE and MONMACS, use compiler directives or macroprocessors. Precompilers include automated Alliant FX Fortran compilers, the Express C automatic parallelizer, and semiautomated compilers such as DINO, PAT, and MIMDizer.

Known parallel language approaches are differentiated in Fig. 11.2 based on the degree of compilation support developed. The categorization is by no means fixed. A language tool can be upgraded from the semiautomated category to the automated category if the compiler is upgraded in the future. Several parallel programming tools are summarized in Table 11.1.

In addition to language/compiler development, a parallel programming environment must have supporting tools to facilitate the development and testing of parallel programs. Such an environment should provide software tool sets which can be applied

Table 11.1 Representative Parallel Programming Tools

Tool Name, Computing Model	Language, OS, GUI and Environment Features	Hardware Platform	Applications Remarks and References
MIMDizer shared-memory message-passing systems	Fortran, UNIX, VMS, X-Window, SunView, Do-loop parallelization compiler directives, array decomposition code restructuring, message-passing support.	X-MP, Y-MP, Cray 2, iPSC, Sun, IRIS, NEC.	Scientific numerical computations; Pacific Sierra Research Corp. [Harrison90] and [PSR90].
Express message-passing systems	Fortran 77, Fortran 90, C, C++, UNIX, DOS, X-Window, SunView, code parallelization, domain decomposition, communication monitoring, load balancing.	iPSC, Y-MP, Sun, nCUBE, PC, Macintosh.	Scientific (finite-element, etc.); Parasoft Corp. [Parasoft90].
C-Linda tuple-space protocol	C (Fortran called from C), UNIX, X-Window, new language extension to parallel programming.	iPSC/860, Y-MP, Sun, IBM R6000, Encore, Sequent, Apollo.	Scientific and distributed database processing; Scientific Computing Ass., Inc. [Ahuja86].
SCHEDULE shared memory systems	Fortran, C, UNIX, X-Window, SunView, parallel algorithm development and functional parallelization.	Cray 2, Encore, Sequent, Alliant.	Scientific, numeric; [Dongarra86], University of Tennessee.

GUI = Graphics user interface.

to different phases of the program development cycle such as for editing, debugging, performance monitoring, and tuning.

Recent advances in visualization allow the programmer to visualize parallel computations through dynamic graphical animation of control flow and data flow patterns. Program visualization permits users to identify performance bottlenecks in a parallel program more easily than with a conventional approach.

Summary   The important environment features are summarized below:

(1) Control-flow graph generation.
(2) Integrated textual/graphical map.
(3) Parallel debugger at source code level.
(4) Performance monitoring by either software or hardware means.
(5) Performance prediction model.
(6) Program visualizer for displaying program structures and data flow patterns.
(7) Parallel input/output for fast data movement.

(8) Visualization support for program development and guidance for parallel computations.

(9) OS support for parallelism in front-end or back-end environments.

(10) Communication support in a network environment.

11.1.2 Y-MP, Paragon, and CM-5 Environments

The software and programming environments of the Cray Y-MP, Intel Paragon XP/S, and Connection Machine CM-5 are examined below. Readers can check the bibliographic notes for additional information on these and other parallel computer environments.

Cray Y-MP Software The Cray Y-MP runs with the Cray operating systems UNICOS or COS. Two Fortran compilers, CFT77 and CFT, provide automatic vectorizing, as do C and Pascal compilers. Large library routines, program management utilities, debugging aids, and a Cray assembler are included in the system software. Communications and applications software are also provided by Cray, by a third party, or from the public domain.

UNICOS is written in C and is a time-sharing OS extension of UNIX. UNICOS supports optimizing, vectorizing, concurrentizing Fortran compilers and optimizing and vectorizing Pascal and C compilers. Besides interactive mode, it supports the Network Queueing System (NQS) for batch processing.

COS is a multiprogramming, multiprocessing, and multitasking OS. It offers a batch environment to the user and supports interactive jobs and data transfers through the front-end system. COS programs can run with a maximum of four processors up to 16M words of memory on the Y-MP system. Migration tools are available to ease conversion from COS to LINICOS.

We will describe three multiprocessing/multitasking methods — macrotasking, microtasking, and autotasking — in Section 11.2.3. The Cray Y-MP has implemented all three methods and they can work together in a single program.

CFT77 is a multipass, optimizing, transportable compiler. It does vectorization and scalar optimization of Fortran 77 programs. The Cray assembler, CAL, enables a user to tailor a program to the architecture of the Cray Y-MP.

Subroutine libraries contain various utilities, high-performance I/O subroutines, numerous math and scientific routines, and some special-purpose routines for communications and applications. A directory of applications software for Cray supercomputers is available.

Intel Paragon XP/S Software The Intel Paragon XP/S system is an extension of the Intel iPSC/860 and Delta systems. A summary of the XP/S system is given in Table 11.2. It is claimed to be a scalable, wormhole, mesh-connected multicomputer using distributed memory. The processing nodes used are 50-MHz i860 XP processors.

Paragon runs a distributed UNIX/OS based on OSF technology and is in conformance with POSIX, System V.3, and Berkeley 4.3BSD. The languages supported include C, Fortran, Ada, C++, and data-parallel Fortran and their compilers.

Table 11.2 Intel Paragon XP/S Multicomputer System

Capacity	5–300 Gflops peak 64-bit results, 2.8–160 GIPS peak integer performance, node-to-node message routing at 200 Mbytes/s (full duplex), 1–128 Gbytes main memory, up to 500 Gbytes/s aggregate bandwidth, 6 Gbytes − 1 Tbyte internal disk storage, up to 6.4 Gbyte/s aggregate I/O bandwidth.
Node architecture	Nodes based on Intel's 50 MHz i860 XP processor, 75 Mflops, 42 VAX MIPS peak per processor, 16-128 Mbytes DRAM per node.
Operating system	Distributed UNIX based on OSF technology, conformance with POSIX, System V.3, 4.3 BSD virtual memory, simultaneous batch and interactive operation.
Programming environment	C, Fortran, Ada, C++, data-parallel Fortran, integrated tool suite with a Motif-based GUI, FORGE and CAST parallelization tools, Intel ProSolver parallel equation solvers, BLAS, NAG, SEGlib and other math libraries, interactive parallel debugger (IPD), hardware-aided performance visualization system (PVS).
Visualization	X Window System, PEX, Distributed Graphics Library (DGL) client support, AVS and Explorer interactive visualizers, connectivity to HIPPI frame buffers.

Source: Intel Corporation, Supercomputer Systems Division, Beaverton, Oregon, 1991.

The integrated tool suite includes FORGE and CAST parallelization tools, Intel ProSolver parallel equation solvers, and BLAS, NAG, SEGlib, and other math libraries. The programming environment provides an interactive parallel debugger (IPD) with a hardware-aided performance visualization system (PVS).

CM-5 Software   The software support and programming environment for the CM-5 are introduced here. The CM-5 designers claim independent scalability of processing, communication, and I/O. This claim must be supported by extensive software, languages, and application libraries.

The software layers of the Connection Machine systems are shown in Fig. 11.3. The operating system used is CMOST, an enhanced version of UNIX/OS which supports time sharing and batch processing. The low-level languages and libraries match the hardware instruction-set architectures.

CMOST provides not only standard UNIX services but also supports fast IPC capabilities and data-parallel and other parallel programming models. It also extends the UNIX I/O environment to support parallel reads and writes and manages large files on data vaults.

The Prism programming environment is an integrated Motif-based graphical environment. Users can develop, execute, debug, and analyze the performance of programs with Prism, which can operate on terminals or workstations running the X-Window system.

High-level languages supported on the CM-5 include CM Fortran, C++, and *Lisp. CM Fortran is based on standard Fortran 77 with array processing extensions of standard Fortran 90. CM Fortran also offers several extensions beyond Fortran 90, such as

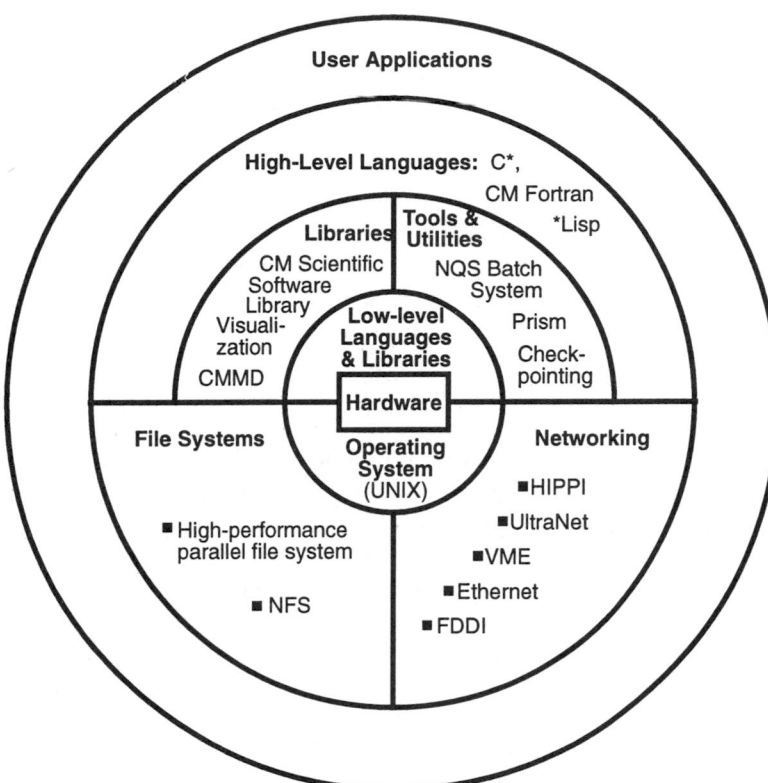

Figure 11.3 Software layers of the Connection Machine system. (Courtesy of Thinking Machines Corporation, 1992)

a FORALL statement and some additional intrinsic functions.

C* is an extension of the C programming language which supports data-parallel programming. Similarly, the *Lisp is an extension of the Common Lisp programming language for data-parallel programming. Both can be used to structure parallel data and to compute, communicate, and transfer data in parallel.

The CM Scientific Software Library includes linear algebra routines, fast Fourier transforms, random number generators, and some statistical analysis packages. Data visualization is aided by an integrated environment including the CMX11 library. The CMMD is a CM message-passing library permitting concurrent processing in which synchronization occurs only between matched sending and receiving nodes.

11.1.3 Visualization and Performance Tuning

The performance of a parallel computer can be enhanced at five processing stages: At the machine design stage, the architecture and OS should be optimized to yield high resource utilization and maximum system throughput. At the algorithm design/data structuring stage, the programmer should match the software with the target hardware. At the compiling stage, the source code should be optimized for concurrency, vectorization, or scalar optimization at various granularity levels. Finally, the compiled object code can go through further fine-tuning with the help of program profilers or run-time information.

Visualization Support   We consider below performance tuning at the programming, compiling, and execution stages. To probe further, we add a few performance measures to reflect the resource utilization rate for a wide class of application programs. These measures are useful in predicting the performance and in performing program tuning in an interactive manner. Special program trace methods are needed for event monitoring in the performance tuning process. A special *graphics user interface* is required to support performance monitoring, prediction, and visualization.

In the I/O and program development areas, visualization is also needed. Performance tuning requires extensive software experiments with the help of GUI and utilization support. Tuning an operating system and an application program requires effort from both ends. System tuning involves the tuning of virtual memory and process priorities, such as adjusting the resident set size and scheduling policy.

Even the best computer architect cannot guarantee the performance of a system until the machine is tested by actually running real software programs. During the architecture design stage, simulation may be used to predict performance. However, simulation experiments are often biased or restricted by theoretical load characteristics.

During the compiling and execution stages, user programs can be modified either through programmer guidance or through the use of an intelligent compiler for automatic code transformations toward optimization or vectorization. Program modification can be extended all the way back to the algorithm design stage. Choosing better data structures may make a difference.

Performance Tuning   If special hardware/software mechanisms are available, one can also collect run-time information, such as processor and memory utilization patterns, to guide the code transformation. One can also conduct a critical-path analysis of programs in order to reveal the bottleneck. Bottleneck removal or shortening the critical path through grain packing or other techniques can improve the system performance.

Besides measuring the MIPS, Mflops, or TPS rate, performance tuning may require a check of the CPU utilization rate, cache hit ratio, page fault rate, load index, synchronization frequency, memory-access pattern, OS/compiler overhead, and interprocessor communication or data movement delays. These measures will reflect the degree of matching between software and hardware.

To tune a computer system for a given application, the gap between hardware and software must be closed. *Compiler directives* can be inserted to guide code optimization. Program profilers can be used to modify the object code in multiple passes through the compiler. Program tuning at run time must be assisted by program traces and event monitoring. These may require special hardware/software support. The purpose of these traces is to produce system control parameters for more efficient processor allocation, better memory utilization, higher cache hit ratios, fewer page faults, fewer synchronizations, and lower communication overhead.

Run-time program tuning is much more difficult to implement than compile-time tuning. However, the latter lacks run-time conditions, making it difficult to predict the performance accurately. Thus both compile-time and run-time techniques are needed in the performance tuning process.

11.2 Synchronization and Multiprocessing Modes

Principles of various synchronization mechanisms for interprocess communications are studied first. Then we describe various modes for multiprocessing with shared memory.

11.2.1 Principles of Synchronization

The performance and correctness of a parallel program execution rely heavily on efficient synchronization among concurrent computations in multiple processors. As revealed in Chapter 6, both hardware and software mechanisms are needed to synchronize various granules of parallel operations.

The source of the synchronization problem is the sharing of writable objects (data or structures) among processes. Once a writable object permanently becomes read-only, the synchronization problem vanishes at that point. Synchronization consists of implementing the order of operations in an algorithm by observing the dependences for writable data. Shared object access in an MIMD architecture requires dynamic management at run time, which is much more complex than that of an SIMD architecture using lockstep to achieve synchronization at compile time.

Low-level synchronization primitives are often implemented directly in hardware. Resources such as the CPU, bus or network, and memory units may also be involved in synchronization of parallel computations.

We examine below *atomic operations, wait protocols, fairness policies, access order*, and *sole-access protocols*, based on the work of Bitar (1991), for implementing efficient synchronization schemes.

Atomic Operations   Two classes of shared memory access operations are (1) an individual *read* or *write* such as Register1 $:= x$ and (2) an indivisible *read-modify-write* such as $x := f(x)$ or $y := f(x)$.

From the synchronization point of view, the order of program operations is described by *read-modify-write* operations over shared writable objects called *atoms*. An operation on an atom is called an *atomic operation*.

A hard atom is one whose access races are resolved by hardware such as Test&Set, whereas a soft atom is one whose access races are resolved by software such as a shared data structure protected by a Test&Set bit.

The atomicity of objects must be explicitly implemented by the software (on soft atoms), or the software must explicitly delegate the responsibility to the hardware (for hard atoms).

The execution of operations may be out of program order as long as the execution order preserves the meaning of the code. Three kinds of program dependences are identified below:

- *Data dependences*: WAR, RAW, and WAW as defined in Chapters 2 and 5.
- *Control dependences*: Program flow control statements such as *goto* and *if–then*.
- *Side-effect dependences*: Due to exceptions, traps, I/O accesses, time out, etc.

The correct execution order, as enforced by correct synchronization, must observe the program dependences. The atomicity of operations is maintained by observing dependences. Therefore, synchronized execution order must resolve any race conditions at run time.

Wait Protocols   There are two kinds of *wait protocols* when the sole-access right is denied due to conflicts. In a *busy wait*, the process remains loaded in the processor's context registers and is allowed to continually retry. The reason for using busy wait is that it offers a faster response when the shared object becomes available.

In a *sleep wait*, the process is removed from the processor and put in a wait queue. The process being suspended must be notified of the event it is waiting for. The hardware complexity increases in a multiprocessor using sleep wait as compared with those implementing busy wait.

When locks are used to synchronize processes in a multiprocessor, busy wait is used more often than sleep wait. Busy wait may offer a better performance if it entails less use of processors, memories, or network channels. If sleep-wait queues are managed using lock synchronization, it may be necessary for a process to wait for access to a sleep-wait queue.

Busy wait can be implemented with a self-service protocol by polling across the network, or with a full-service protocol by being notified across the network when the atom becomes available.

Fairness Policies   Busy wait may reduce synchronization delay when the shared object becomes available. However, it wastes processor cycles by continually checking the object state and also may cause hot spots in memory access.

In sleep wait, the resources are better utilized, but a longer synchronization delay may result. For all suspended processes waiting in a queue, a *fairness policy* must be used to revive one of the waiting processes. Three fairness policies are summarized below:

- *FIFO*: The wait queue follows a first-in-first-out policy.
- *Bounded*: The number of turns a waiting process will miss is upper-bounded.
- *Livelock-free*: One waiting process will always proceed; not all will wait forever.

In general, the higher level of fairness corresponds to the more expensive implementation. Although FIFO fairness guarantees that no process will be served out of turn, its associated costs in algorithmic complexity and memory usage may be too high. Another concern is the prevention of deadlock among competing processes.

Sole-Access Protocols   Conflicting atomic operations are serialized by means of *sole-access protocols*. Three synchronization methods are described below based on who updates the atom and whether sole access is granted before or after the atomic operations:

(1) *Lock synchronization*: In this method, the atom is updated by the requester process and sole access is granted before the atomic operation. For this reason, it is also called *presynchronization*. The method can be applied for shared read-only access. The lock granularity is a major issue in lock synchronization.

Most hardware-implemented locking appeals more to finer-grain physical units such as the memory module, cache, memory/cache block, etc. Lock mechanisms are described in Section 11.3.1.

(2) *Optimistic synchronization*: This method also updates the atom by the requester process. But sole access is granted after the atomic operation via abortion. It is also called *postsynchronization*. A process may secure sole access after first completing an atomic operation on a *local* version of the atom and then executing another atomic operation on the *global* version of the atom.

The second atomic operation determines if a concurrent update of the atom has been made since the first operation was begun. If a concurrent update has taken place, the global version is not updated; instead, the first atomic operation is aborted and restarted from the new global version.

The name *optimistic* is due to the fact that the method expects that there will be no concurrent access to the atom during the execution of a single atomic operation. This method was designed to eliminate the bottleneck created by a coarse-grain lock for the object of the first atomic operation.

This idea led to the concept of *optimistic concurrency* developed by Kung and Robinson (1981). Optimistic synchronization requires extra work in order to implement the global update operations, and the method incurs an abortion cost.

(3) *Server synchronization*: This method updates the atom by the server process of the requesting process, as suggested by the name. Compared with lock synchronization and optimistic synchronization, *server synchronization* offers full service.

An atom has a unique update server. A process requesting an atomic operation on the atom sends the request to the atom's update server. The update server may be a specialized *server processor* (SP) associated with the atom's memory module.

Remote procedure call and object-oriented or actor systems, in which shared objects are encapsulated by a process, provide examples of server synchronization. In a shared-memory multiprocessor using hard atoms, all three synchronization methods can be implemented, whereas only server synchronization is used in a message-passing multicomputer.

A soft atom occurs only under lock synchronization or optimistic synchronization. A specialized server processor must be used to implement server synchronization in a multiprocessor with a centralized shared memory.

In a message-passing system, the node processors with local memories can implement server synchronization without using additional server processors.

Synchronization Environment—After learning about the principles of IPC and synchronization methods, we consider several implementation issues in developing parallel programs for multiprocessors.

Parallel program development hinges on how efficient synchronization is implemented with locks, semaphores, and monitors. We assess below various synchronization environments.

It is often desired to move the synchronization logic closer to the shared memory units in order to reduce bus traffic or network contention and CPU usage in the synchronization process. Such a synchronization environment may require the use of a server processor for coordinating the synchronization process and virtual memory management.

The Cedar multiprocessor system has included this feature in globally shared memory units. Most commercial multiprocessors do not include this option.

In general, synchronization controls the granularity of partitioned algorithms, affects the ease of writing correct programs, determines the fairness in selecting among competing processes, and ultimately influences the efficiency of parallel program execution. Special hardware support and sophisticated software development may be needed to create a user-friendly environment.

Due to the dynamic behavior, a poor synchronization environment may cause excessive waste in CPU cycles or network bandwidths, which otherwise could be more effectively used by other active processes.

A poorly written parallel program may result in too many synchronizations that cancel all the advantages of parallelism. Therefore, an ideal synchronization environment should be jointly developed by designers and programmers.

11.2.2 Multiprocessor Execution Modes

Multiprocessor supercomputers are built for vector processing as well as for parallel processing across multiple processors. Multiprocessing modes include parallel execution from the fine-grain process level to the medium-grain task level, and to the coarse-grain program level.

In this section, we examine the programming requirements for intrinsic multiprocessing as well as for multitasking. Experiences using the Cray Y-MP are presented along with a programming example.

Multiprocessing Requirements   Multiprocessing at the process level requires the use of shared memory in a tightly coupled system. Summarized below are special requirements for facilitating efficient multiprocessing:

- Fast context switching among multiple processes resident in processors.
- Multiple register sets to facilitate context switching.
- Fast memory access with conflict-free memory allocations.

- Effective synchronization mechanism among multiple processors.
- Software tools for achieving parallel processing and performance monitoring.
- System and application software for interactive users.

As machine size increases, the problems of communication overhead and effective exploitation of parallelism in a user program become more serious. These are the main reasons why commercial multiprocessor systems are limited to having only a few processors.

The use of a large number of processors for fine-grain multiprocessing is still unrealized today, and much research is needed to accomplish the goal. Meeting the challenges will rely to a great extent on the development of system hardware and software support to free programmers from dealing with tedious program partitioning, parallel scheduling, memory consistency, and latency tolerance.

Multitasking Environments Three generations of multitasking software have been developed by Cray Research, NEC, and other multiprocessor manufacturers. Multitasking exploits parallelism at several levels:

- Functional units are pipelined or chained together.
- Multiple functional units are used concurrently.
- I/O and CPU activities are overlapped.
- Multiple CPUs can cooperate on a single program to achieve minimal execution time.

In a multitasking environment, the tasks and data structures of a job must be properly partitioned to allow parallel execution without conflict. However, the availability of processors, the order of execution, and the completion of tasks are functions of the run-time conditions of the machine. Therefore, multitasking is nondeterministic with respect to time.

On the other hand, tasks themselves must be deterministic with respect to results. To ensure successful multitasking, the user must precisely define and include the necessary communication and synchronization mechanisms and provide protection for shared data in critical sections.

Reentrancy is a useful property allowing one copy of a program module to be used by more than one task in parallel. Nonreentrant code can be used only once during the lifetime of the program. Reentrant code, if residing in the critical section, can be used in a serial fashion and is called *serially reusable code.*

Reentrant code, which is called many times by different tasks, must be assigned with local variables and control indicators stored in independent locations each time the routine is called. A stack mechanism has been employed in Cray multiprocessors to support reentrancy.

11.2.3 Multitasking on Cray Multiprocessors

Three levels of multitasking are described below for parallel execution on Cray X-MP or Y-MP multiprocessors. Multitasking tradeoffs are demonstrated by an example program execution.

Macrotasking   When multitasking is conducted at the level of subroutine calls, it is called *macrotasking* with medium to coarse grains. Macrotasking has been implemented ever since the introduction of Cray X-MP systems. The concept of macrotasking is depicted in Fig. 11.4a.

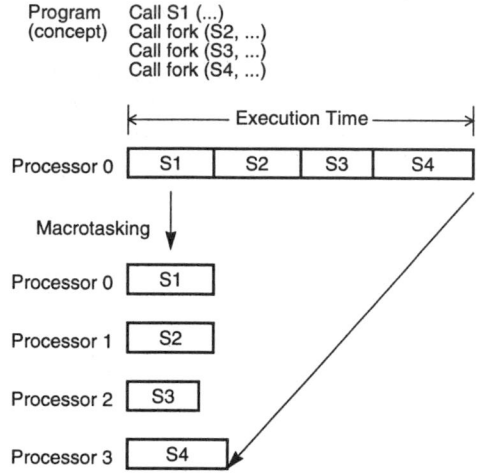

(a) Macrotasking across subroutine calls

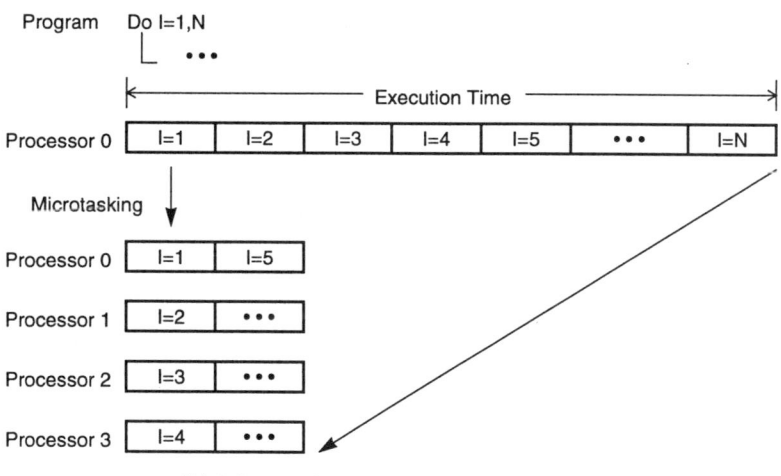

(b) Microtasking across loop iterations

Figure 11.4 Multitasking at two different processing levels.

A main program calls a subroutine S1 and then forks out three additional calls of the same subroutine. Macrotasking is best suited to programs with larger, longer-running tasks. The user interface with the Cray Y-MP system's macrotasking capability is a set of Fortran-callable subroutines that explicitly define and synchronize tasks at the subroutine level.

The execution of these four subroutine calls on a uniprocessor (processor 0) is done sequentially. Macrotasking can spread the calls across four processors. Note that each processor may have a different execution time. However, the overall execution time is reduced due to parallel processing.

Microtasking This corresponds to multitasking at the loop control level with finer granularity. Compiler directives are often used to declare parallel execution of independent or dependent iterations of a looping program construct.

This technique was implemented in Alliant FX multiprocessors using the **Doacross** directive. Figure 11.4b illustrates the spread of every four instructions of a Do loop to four processors simultaneously through microtasking.

When the iterations are independent of each other, microtasking is easier to implement. When dependence does exist between the iterations, the system must resolve the dependence before parallel execution can be continued. Interprocessor communications are needed to resolve the dependences.

In addition to working efficiently on parts of programs where the granularity is small, microtasking works well when the number of processors available for the job is unknown or may vary during the program's execution. Additionally, in a batch environment where processors may become available for short periods, the microtasked job can dynamically adjust to the number of available processors.

Autotasking The autotasking feature automatically divides a program into discrete tasks for parallel execution on a multiprocessor. In the past, macrotasking was achieved by direct programmer intervention. Microtasking was aided by an interactive compiler.

Autotasking demands much higher degrees of automation. Only recent Cray multiprocessors, like the Cray Y-MP and C-90, have been implemented with autotasking software. Some compiler or assembler directives are provided to allow programmers to fine-tune their code for even better performance, especially in production environments.

It should be noted that through program fine-tuning, plus the use of autotasking software support, the above multitasking feature can improve both individual job performance and overall system throughput in a production environment.

At its highest capability, autotasking should be able to achieve fully automatic multiprocessing. It allows user programs to be automatically partitioned over multiple CPUs (without user intervention). Autotasking is based on the microtasking design and shares several advantages with microtasking: very low overhead synchronization cost, excellent dynamic performance independent of the number of CPUs available, both large and small granularity parallelism, and so on. In addition to being fully automatic, autotasking exceeds microtasking in overall performance and in the various levels of parallelism that can be employed.

Mutlitasking Tradeoffs Speedup from multitasking may occur only when the time saved in executing parallel tasks outweighs the overhead penalty. The overhead is very sensitive to task granularity; it includes the initiation, management, and interaction of tasks. These are often accomplished by adding code to the original code, as exemplified below.

Example 11.1   Macrotasking on a Cray X-MP dual-processor system
(Courtesy of Cray Research, Inc., 1987)

This program can benefit from multitasking depending on the execution time of the subroutine SUB(J) and the overhead introduced in service routines. Before one attempts to convert serial code into multitasked code, the expected performance should be predicted to ensure a net gain.

Consider the sequential execution of the following code on an X-MP uniprocessor:

```
        Program Main
          ⋮
        Do 100 I = 1, 50
              ⋮
            Do 10 J = 1, 2
                  CALL SUB(J)
 10         Continue
              ⋮
100     Continue
          ⋮
        STOP
        END
```

The 100 loop has dependent iterations which cannot be executed in parallel. The 10 loop has independent iterations which are being attempted for multitasking. The following multitasked code is written for a dual-processor X-MP.

```
    Program Main                            SUBROUTINE T
    Common /MT/ IST, IDN, JOB               Common /MT/ IST, IDN, JOB
    CALL TSKSTART (IDTASK, T)          101  CALL EVWAIT(IST)
    JOB = 1                                 CALL EVCLEAR(IST)
    Do 100 I = 1, 50                        IF (JOB .NE. 1) GOTO 102
        CALL EVPOST(IST)                    CALL SUB(2)
        CALL SUB(1)                         CALL EVPOST(IDN)
        CALL EVWAIT(IDN)                    GOTO 101
        CALL EVCLEAR(IDN)             102    RETURN
100 Continue                                END
    JOB = 2
    CALL EVPOST(IST)
    CALL TSKWAIT(IDTASK)
    STOP
    END
```

The execution of the sequential program on one CPU consists of two parts:

$$\text{Time(1 CPU)} = \text{time(Seq)} + \text{time(SUB)} = (0.04 + 0.96) \times (20.83 \text{ s}) = 20.83 \text{ s}$$

where time(SUB) accounts for 96% of the time spent in subroutine SUB, and time(Seq) for 4% spent on the remaining portion of the program. The total run time measured on one CPU was 20.83 s.

To execute the multitasked program on two CPUs requires

$$\text{Time(2 CPUs)} = \text{time(Seq)} + \frac{1}{2}\text{time(SUB)} + \text{overhead}$$

The subroutine SUB was equally divided between two CPUs. This reduces time(SUB) by one-half. The overhead is estimated below with some approximation of the delays caused by workload imbalance and memory contention. The service routines TSKSTART, TSKWAIT, EVPOST, EVCLEAR, and EVWAIT have been used in the Cray X-MP to establish the multitasking structure.

$$
\begin{aligned}
\text{Overhead} \ =\ & \text{time(TSKSTART)} + \text{time(TSKWAIT)} + 51 \times \text{time(EVPOST)} +\\
& (\text{workload imbalance delay}) + (\text{memory contention delay})\\
=\ & 1500 \text{ CP} + 1500 \text{ CP} + 51 \times 1500 \text{ CP} + 50 \times 200 \text{ CP} +\\
& 50 \times 1500 \text{ CP} + (0.02 \times 50 \times 0.2 \text{ s}) = 0.216 \text{ s}
\end{aligned}
$$

where the CP (clock period) is equal to 9.5 ns. Therefore,

$$\text{Time(2 CPUs)} = (0.4 \times 20.83) + \frac{1}{2} \times (0.96 \times 20.83) + 0.216 = 11.05 \text{ s}.$$

We thus project the following speedup:

$$\text{Speedup} = \frac{\text{time(1 CPU)}}{\text{time(2 CPUs)}} = \frac{20.83}{11.02} = 1.88$$

This speedup helps decide whether multitasking is worthwhile. The actual speedup of this program as measured by Cray programmers was 1.86. This indicates that the above prediction is indeed very close.

■

Factors affecting performance include task granularity, frequency of calls, balanced partitioning of work, and programming skill in the choice of multitasking mechanisms.

Multitasking offers a speedup which is upper-bounded by the number of processors in a system. Because vector processing offers a greater speedup potential over scalar processing (in the Cray X-MP, vectorization offers a speedup in the range of 10 to 20), multitasking should not be employed at the expense of vectorization.

In the case of a short vector length, scalar processing may outperform vector processing. In the case of a small task size, vector processing (or even scalar processing) may outperform multitasking.

Both scalar and vector codes may be multitasked, depending on the granularity and the overhead. For coarse-grain computations with reasonably low overhead (as in the above example), multitasking is appropriate and advantageous.

11.3   Shared-Variable Program Structures

We describe below the use of spin locks, suspend locks, binary and counting semaphores, and monitors for shared-variable programming. These mechanisms can be used to implement various synchronization methods among concurrent processes. Shared-variable constructs are also used in OS kernel development for protected access of certain kernel areas.

11.3.1   Locks for Protected Access

Lock and unlock mechanisms are described below using shared variables among multiple processes. *Binary locks* are used globally among multiple processes. *Spin locks* are based on a time-slot concept. *Dekker's locks* are based on using distributed requests jointly with a spin lock. Special multiprocessor instructions are needed to implement these locking mechanisms.

Spin Locks   The entrance and exit of a CS can be controlled by a binary *spin lock* mechanism in which the gate is protected by a single binary variable x, which is shared by all processes attempting to enter the CS.

Example 11.2   Definition of a binary spin lock

The gate variable x is initially set to 0, corresponding to the *open* status. Each process P_i is allowed to test the value of x until it becomes 0. Then it can enter the CS. The gate must be *closed* by setting $x = 1$ after entering.

```
Shared var x: (0,1)          /The spin lock/
    x := 0                   /The CS is open initially/
Process Pi for all i
    Repeat
        ⋮                    /Spinning with busy wait/
    Until x = 0
    x := 1                   /Close gate after entry/
        ⋮                    /The critical section/
    x := 0                   /Open gate after done and exit/
```

After the CS is completed, the gate is reopened. A busy-wait protocol is used in spin locks. Precautions must be exercised to prevent simultaneous entries into the CS by multiple processes.   ∎

Example 11.3   Definition of a generalized spin lock with n possible values

One way to guarantee mutually exclusive entry is to use a *generalized spin lock* with n processes as defined below. The gate variable x is allowed to assume n

integer values $(1, 2, \ldots, n)$. The fact that $x = i$ implies that the gate is open only to process P_i.

```
Shared var x: (1, 2, 3, ..., n)   /The spin lock/
     x := 1                       /The CS is initially open to process P1/
Process Pi for all i = 1, 2, ..., n−1
     Repeat
         ⋮                       /Spinning with busy wait/
     Until x = i                  /Entry/
         ⋮                       /The critical section/
     x := i + 1                   /Exit/

Process Pn                        /The last process/
     Repeat
         ⋮                       /Spinning with busy wait/
     Until x = n                  /Entry/
         ⋮                       /The critical section/
     x := 1                       /Exit and return to P1/
```

Initially, the lock is open to P_1. After P_1 finishes with the CS, the gate is open to P_2, and so on. The last process P_n will reset the lock to $x = 1$. ∎

This solution guarantees mutual exclusion at the expense of longer waiting times, while the lock is spinning among n circularly ordered processes. The processes must wait even if there are no conflicting requests at the same time.

Dekker's Protocol   To guarantee mutual exclusion without unnecessary waiting, Dekker has suggested the use of separate request variables by different processes along with the use of a spin lock.

Example 11.4   Dekker's protocol for protected access to a critical section by two processes

The following program shows Dekker's solution for two processes. Each process uses a request variable (p_i) to indicate if it wishes to be *inside* (1) or *outside* (0) the CS. When both processes indicate the same wish to be inside, a spin lock ($x = 1$ or 2) is used to resolve the conflict.

```
var p1, p2: (inside, outside)
Shared var x: (1, 2)
    x := 1   /The CS is initially open to process P1/
    p1 := outside; p2 := outside
Process 1                          Process 2
```

p1 := inside
if p2 = inside then
begin
    if x = 2 then
    begin
        p1 := outside
        Repeat until x = 1
        p1 := inside
    end
    Repeat
    until p2 = **outside**
end

  ⋮   /The critical section/
x := 2 /Open to process 2/
p1 := outside;

p2 := inside
if p1 = inside then
begin
    if x = 1 then
    begin
        p2 := outside
        Repeat until x = 2
        p2 := inside
    end
    Repeat
    until p1 = **outside**
end

  ⋮   /The critical section/
x := 1 /Open to process 1/
p2 := outside

■

This scheme avoids unnecessary waiting delays whenever there is no conflicting request. The generalization of Dekker's method to a large number of processes is very cumbersome and also expensive to implement (see Problem 11.2).

Suspend Locks   These types of locks use the sleep-wait protocol. A process blocked from entering a CS is removed from the processor's ready-to-run queue. Instead, the suspended process is put in a wait queue. When the suspend lock is opened, one of the suspended processes in the wait queue will be reactivated using one of the fairness policies. Suspend locks are used less often due to the added hardware cost and increased waiting delays.

Spin locks or suspend locks can be implemented in a multiprocessor system using special instructions such as Test&Set, Fetch&Add, and Compare&Swap, depending on the atomic operations supported by the hardware in a given computer.

The Test&Set(x, y_i) instruction operates on two boolean variables: The spin lock x is shared by multiple processes, and the local condition variable y_i indicates the *outside* (0) and *inside* (1) wishes of process P_i. As an atomic action, this instruction sets y_i to the old value of the lock x and closes the lock (1) as follows:

Test&Set(x, y_i):
   $< y_i := x;\quad x := 1 >$

If both processes simultaneously wish to enter the CS, only one can succeed in *closing* the spin lock x and thus mutual exclusion is guaranteed.

Example 11.5 Test&Set implementation of Dekker's protocol for accessing a critical section

```
Shared var x: (0, 1)
var y1, y2: (0, 1)
Process 1
   if y1 = 1 then Test&Set(x, y1)
      if y1 = 1 then Test&Set(x, y2)

         ⋮                          /The critical section/
      x := 0                        /Exit/

         ⋮                          /Noncritical section/


Process 2
   if y2 = 1 then Test&Set(x, y2)
      if y2 = 1 then Test&Set(x, y1)

         ⋮                          /The critical section/
      x := 0                        /Exit/

         ⋮                          /Noncritical section/
```

The Test&Set instruction can be used to implement Dekker's protocol, as shown in the above program in which y_i corresponds to the request variable p_i from process i for $i = 1$ and 2, respectively.

∎

11.3.2 Semaphores and Applications

Spin locks using the busy-wait protocol may cause excessive waiting in a multi-programmed multiprocessor system. Eliminating busy waiting in processes would make better use of the resources if a process blocked from entering a CS goes to sleep and is awakened when the CS is opened.

This may improve processor utilization. Instead of using processors to execute a spinning process, they could be used more productively in executing other nonblocking processes. *Semaphores* were developed based on the sleep-wait protocol.

Binary Semaphores Critical sections can be viewed as sections of railroad track. *Semaphores* are control signals for avoiding collisions between trains (processes) on the same track section (CS). Dijkstra (1968) introduced the use of binary semaphores for the management of concurrent processes seeking to access CSs.

A *binary semaphore s* is a boolean variable taking the value of 0 or 1. Each shared resource or CS can be associated with a dedicated semaphore. Only two atomic operations (primitives), P and V, are used to access the CS represented by the semaphore apart from initialization by setting $s = 1$.

- The $P(s)$ operation causes the value of the semaphore s to be decreased by 1 if s is not already 0.
- The $V(s)$ operation causes the value of the semaphore s to be increased by 1 if s is not already 1.

The physical meaning of $s = 1$ is the availability of a single copy of the resource represented by s. On the other hand, $s = 0$ means the resource is being occupied. When the resource is a CS, a binary semaphore corresponds essentially to a gate or lock variable using the sleep-wait protocol.

Counting Semaphores   A *counting semaphore* s is a nonnegative integer taking $(n+1)$ possible values $0, 1, 2, ..., n$ for some integer $n \geq 1$. Therefore a binary semaphore corresponds to the special case of $n = 1$. A counting semaphore with a maximum value of n acts as a collection of n permits corresponding to the use of n copies of a shared resource.

A shared resource can be a program segment, a data table, or any passive device such as memory or I/O resources. A permit is issued upon each request until all copies are taken. Formally, we define $P(s)$ and $V(s)$ on counting semaphores as follows:

- $P(s)$: If $s > 0$, then $s := s - 1$; else suspend the execution of the current process and put it in a wait queue.
- $V(s)$: If the wait queue is not empty, then wake up one of them; else $s := s + 1$.

Intuitively, the operation $P(s)$ corresponds to the submission of a request for a permit. $V(s)$ corresponds to the return of a permit after a process finishes using a copy of the resource. Note that both operators are atomic. Operations defined between a pair of P and V operators are indivisible and cannot be interrupted by other processes.

However, nested levels of (P, V) pairs can be used when multiple resource types are used. If the maximum value of a counting semaphore $(s = n)$ has already been reached, the $V(s)$ operation will not increase the value of s beyond the upper bound n.

System Deadlock   *System deadlock* refers to the situation in a multiprocessor when concurrent processes are holding resources and preventing each other from completing their execution.

In general, a deadlock can be prevented if one or more of the following four necessary conditions are removed:

(1) *Mutual exclusion* — Each process has exclusive control of its allocated resources.
(2) *Nonpreemption* — A process cannot release its allocated resources until completion.
(3) *Wait for* — Processes can hold resources while waiting for additional resources.
(4) *Circular wait* — Multiple processes wait for each other's resources in a circular dependence situation.

Shared-Resource Allocation   *P-V* operators and semaphores can be used to allocate and deallocate shared resources in a multiprocessor to avoid deadlock among concurrent processes.

Example 11.6 Resource allocation using P-V operators to prevent deadlock

Four processes are sharing six resources in the process declaration shown in Fig. 11.5a. The six resource types are represented by binary semaphores S_i for $i = 1, 2, ..., 6$. Allocation of the resource S_i to process P_j is requested by $P(S_i)$, and the release of resource S_i is requested by $V(S_i)$. The resource request-release pattern of each process P_j for $j = 1, 2, 3, 4$ is shown by a column of such P-V pairs.

All four processes can submit their requests asynchronously. It is the request ordering that leads to deadlock. The release ordering does not make any difference.

Process 1	Process 2	Process 3	Process 4	Process 4 (Modified)
•	P(S2)	•	•	•
P(S1)	•	P(S3)	•	P(S1)
•	P(S4)	•	P(S5)	•
P(S2)	•	P(S5)	•	•
•	P(S5)	•	P(S6)	P(S6)
P(S3)	•	•	•	•
•	•	•	P(S1)	P(S5)
•	V(S5)	•	•	•
•	•	V(S5)	•	•
•	•	•	V(S6)	V(S6)
V(S1)	V(S4)	•	V(S1)	V(S5)
•	•	V(S3)	V(S5)	V(S1)
V(S2)	•	•	•	•
V(S3)	V(S2)	•	•	•
•	•	•	•	•

(a) Four concurrent processes

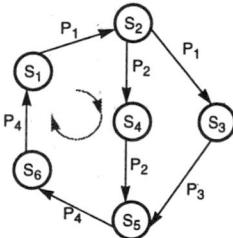
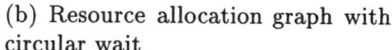

(b) Resource allocation graph with circular wait

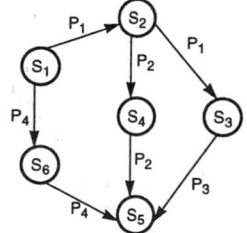

(c) Modified resource allocation graph without circular wait

Figure 11.5 Shared resource allocation using P-V operators to prevent system deadlock. (Reprinted from Hwang, *Proc. IEEE*, 1987)

The *resource allocation graph* shown in Fig. 11.5b is a directed graph where the nodes correspond to the six resource types. An edge with the label P_k from node S_i to node S_j means process P_k is requesting resource S_j while holding resource S_i.

A cycle, $S_1 \rightarrow S_2 \rightarrow S_4 \rightarrow S_5 \rightarrow S_6 \rightarrow S_1$, in Fig. 11.5b implies the possibility of a *circular wait* among processes P_1, P_2, and P_4. To break the deadlock possibility, one can modify the resource request pattern in process 4 as shown in the fifth column.

A new allocation graph results in Fig. 11.5c, where no cycle exists after reversing the edge between S_1 and S_6. Of course, this reversing should not invalidate the demand by process 4.

∎

Deadlock Avoidance   *Static prevention* as outlined above may result in poor resource utilization. *Dynamic deadlock avoidance* depends on the run-time conditions, which may introduce a heavy overhead in detecting the potential existence of a deadlock.

Although dynamic detection may lead to better resource utilization, the tradeoff in detection and recovery costs must be considered before choosing between static and dynamic methods.

Most parallel computers have a static prevention method due to its simplicity of implementation. Sophisticated dynamic avoidance or a recovery scheme for the deadlock problem requires one to minimize the incurred costs to justify the net gains. Breaking a deadlock problem by aborting some noncritical processes should result in a minimum recovery cost.

A meaningful analysis of the recovery costs associated with various options is very time-consuming. This is the main reason why sophisticated deadlock recovery mechanisms have not been built into most multiprocessors.

The P and V operators are usually implemented by the underlying operating system kernel. They can also be implemented by special hardware or by software traps. Spin locks and P-V operators are both low-level, fine-grain, atomic operations which are more often used in system programming than in user programming.

The main problem is that such low-level operations are error-prone and not appealing to ordinary programmers who enjoy the simplicity of using high-level languages.

Semaphores can be used in extending UNIX/OS for multiprocessor systems. Only binary semaphores are used in controlling a CS. Counting semaphores are used in the deadlock-free allocation of shared resources with multiple copies each.

11.3.3   Monitors and Applications

In using locks or semaphores to define critical sections, shared variables are global to all processes. CSs are embodied within processes, which may be scattered throughout the entire program. This may pose some problems in program modularization or debugging, which in turn limits parallelism.

Hoare (1974) proposed a *monitor* construct for structuring an operating system. We describe below the structure of monitors for parallel programming applications.

Monitor Structure A *monitor* is a high-level program construct for structured programming that emphasizes modularity and encapsulation. As shown in Fig. 11.6, a monitor collects shared variables and associated procedures into a single construct which allows only one process access at a time.

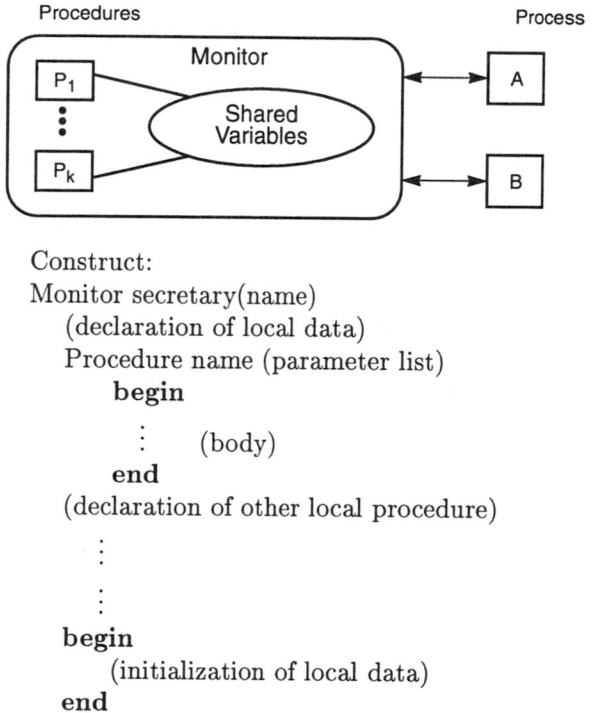

Construct:
Monitor secretary(name)
    (declaration of local data)
    Procedure name (parameter list)
      begin
        ⋮   (body)
      end
    (declaration of other local procedure)
      ⋮

      ⋮

begin
    (initialization of local data)
end

Figure 11.6 The structure of a monitor for structured programming.

A typical monitor consists of the following three parts: The first part defines the monitor name and declares all local variables to be used. The second part is a collection of procedures using the variables declared. The third part is for the initialization of all local variables.

In a monitor, the CSs are removed from the bodies of processes and become procedures or functions defined over variables confined within the boundary of the monitor. A process invokes the appropriate procedure within the monitor when it wishes to enter the desired CS. Instead of providing individual process management, a monitor behaves like a secretary in an office who provides services to a number of persons (processes). Each process can declare as many monitors as the program requires.

Producer-Consumer Implementation The monitor procedures should not access any nonlocal variables outside the monitor. An example monitor for implementing the *producer-consumer* problem is shown below:

Example 11.7 Monitor for a producer process and a consumer process

In this monitor, two procedures are defined over the declared local variables. The producer process sends messages that are received by the consumer process. The communication between producers and consumers is handled by the *deposit* and *fetch* procedures defined below:

```
Monitor Producer-Consumer
    Buffer[0:n−1]: integer
    Inpointer, Outpointer: integer
    Notfull, Notempty: Buffer conditions
    Count: integer index

    Procedure Deposit(I)
        begin
            if Count = n then wait(Notfull)
            Buffer(Inpointer) := I
            Inpointer := (Inpointer + 1) mod n
            signal(Notempty)
        end

    Procedure Fetch(I)
        begin
            if Count = 0 then wait(Notempty)
            I := Buffer(Outpointer)
            Outpointer := (Outpointer + 1) mod n
            signal(Notfull)
        end
    begin
        Inpointer := 0
        Outpointer := 0
        Count := 0
    end
```

The deposit procedure appends to or inserts messages into the communication buffer. The fetch procedure takes or fetches messages for the consumers. The producer process calls the deposit procedure for service, and the consumer process calls the fetch procedure for service.

Only one process (either producer or consumer) can call the desired procedure at a time. However, priorities can be set to ensure that deposit calls will be executed ahead of fetch calls. A monitor is not itself a process. A monitor is a static module of data and procedure declarations. The producer and consumer are active processes which must be programmed separately as shown below:

```
Producer
    I: integer
    begin
        Produce(I)
        Deposit(I)
    end

Consumer
    I: integer
    begin
        Fetch(I)
        Consume(I)
    end
```

∎

Monitor Applications   All the operations of semaphores can be simulated by monitors, and vice versa. Monitors have been suggested for use in a number of parallel programming applications.

Brinch-Hansen (1977) has used monitors for Concurrent Pascal programming. In general, a parallel program may contain two different kinds of program modules: *active processes* and *passive monitors*. As in the producer-consumer example, all shared variables are defined within the monitors; interprocess communications are handled by calling procedures in the same monitor. Different groups of processes may use different monitors (secretaries) for different types of services (procedures) which may require special shared resources (local variables).

Several distinct advantages are observed in using monitors to support IPC or interprocessor synchronization. First, all dedicated services (procedures) are lumped together within each monitor. Therefore, a calling process can be freed from worrying about these procedure details. Second, the monitor designer does not have to worry about how many user processes may access it. Third, monitors provide modularity in program debugging and maintenance.

For example, one can program a disk scheduler as a dedicated monitor. All user processes can submit requests to the disk scheduler, and only one is serviced by a driver procedure at a time. Monitors hide information within their boundaries, which implies the potential for concurrent object-oriented programming and efficient handling of abstract data types. Programs using monitors are not only easier to debug but also are easier to use in exploiting parallelism. The localization of shared variables is an important asset in concurrent programming.

Monitors can be implemented directly with data and procedure declarations, or indirectly through the use of semaphores such as using *P-V* primitives to declare the entry and exit of a monitor. Monitors can also be implemented with the help of a kernel of data structures and routines. A monitor kernel may include special primitives for *entry*, *wait*, and *signal* operations. Wirth (1977) has proposed a Modula kernel for implementing monitors when using the language for parallel programming.

11.4   Message-Passing Program Development

Multicomputer programming demands the distribution of computational load and data structures to various node processors for balanced parallel processing. Message-passing paradigms are needed for internode communications.

Three program/data decomposition techniques, namely, *domain, control,* and *object decomposition,* are presented for programming a multicomputer. In each case, example problems are given to illustrate the decomposition technique involved.

11.4.1   Distributing the Computation

The key to using a multicomputer system is to distribute the computations among an ensemble of computer (processor-memory) nodes. Each node executes its own program, and all nodes are interconnected by a direct network. Concurrent processes created at different nodes communicate by passing messages.

In this section, we assess the basic programming environment for multicomputers. A distributed program is used to explain the interaction between host and node programs. We study program tuning to achieve load balancing directed toward higher performance.

Host and Node Environments   The programming environment of a multicomputer includes a host run-time system and resident operating systems (kernels) in all node computers. The host system provides uniform communication between processes independently of the nodes, the host workstation, and the network connection.

Host processes are those located outside the nodes, either in the host machine, which provides a message path between the multicomputer and a network, or in other host machines connected via the network.

For example, a Cosmic Environment has been developed for the hypercube computer at Caltech. The host environment is a UNIX processor and uses UNIX and language processor utilities to communicate with the node processes and other host processes through messages.

A separate OS kernel is located in each node computer that supports multiprogramming, with an address space confined by local memory. Many node processes can be created at each node. In fact, the total number of node processes may be far greater than the number of node computers in the system.

All node processes execute concurrently in different physical nodes or interleaved through multiprogramming within the same node. Node processes communicate with each other by sending or receiving messages.

The *Cosmic kernel* (later modified to the Reactive Kernel) at Caltech is one such node operating system that supports this process-model programming.

There are no shared variables between node processes in the Cosmic kernel, even if they reside in the same node. The node processes do not have access to input/output devices, and all I/O activities must be handled by the hosts.

The Cosmic Cube programming environment does not use new programming languages. Existing sequential programming languages such as C, Pascal, Fortran, Lisp,

Assembly, etc., are used to write process codes. A library of C functions or procedures has been developed to control message-passing and process-spawning operations.

This approach uses explicit parallelism built on top of existing compiler technology. Many interesting Cosmic features have been built into or modified in commercial systems such as the Intel iPSC and nCUBE computers. Various multicomputer programming environments differ mainly in the languages and message-passing paradigms used. We use an example to illustrate how to distribute computations to the nodes of an Intel iPSC system.

In programming a multicomputer, the process involves separation of the user interface from the computational kernel, leaving the user interface on the host, moving the kernel to each single node, and adding a message interface between host and nodes.

In order to distribute the computation, the programmer chooses a decomposition method and then maps the decomposition to all the nodes. A node-to-node communication protocol must be established. Finally, one needs to balance the load and reduce the communication/computation ratio.

Message Types and Parameters   Example 10.1 demonstrated the need for message passing. In a multicomputer program, a process must distinguish between a number of different message types. A particular field of all messages can be reserved to carry a *message type* (identified by an integer).

Different message types may demand different actions by the sending or receiving processes. Messages of different types are handled in a specific order. For example, the Cosmic Environment/Reactive Kernel dispatches messages according to the types received, allows the process to exercise discretion, and supports customized message functions on top of the X-window primitives.

Let us examine the basic message parameters associated with *send* and *receive* system calls. The message-passing primitives are specified by:

send (type, buffer, length, node, process)
receive (type, buffer, length)

where *type* identifies the message type, *buffer* indicates the location of the message, *length* specifies the length of the message (in bytes), *node* designates the destination node, and *process* is the process ID at the destination node.

The *send* and *receive* primitives are used by the sending and receiving processes, respectively. Therefore, the buffer field in *send* specifies the memory location of the message to be retrieved from. On the other hand, the buffer field in *receive* specifies where the arriving message will be stored.

Once stored in a local memory, a message can be retrieved only by the local processor. No remote memory access is allowed in a pure message-passing multicomputer. The implementation requirements of the two message-passing models are studied below.

11.4.2   Synchronous Message Passing

The message-passing process involves a sender and one or more receiver(s). When a process sends a message, the system must decide a number of issues: first, whether the receiver should cooperate or be ready to receive it; second, whether the communication path has been established or not; and third, whether one or more messages can be sent to the same destination node or to multiple destinations.

In a synchronous communication network, the sender process and receiver process must be synchronized in time and space. Time synchronization means both processes must be ready before message transmission can take place. Space synchronization demands the availability of a communication path, i.e., a sequence of connected channels from source to destination.

The simplest implementation allows only one message on a communication channel at a time. No buffers are used in such a communication network. Therefore, blocking is possible if the channel requested is busy or in error. For this reason, synchronous message passing belongs to the class of blocking communication systems. The correct protocol must be adopted to ensure the coupling of the sender and receiver in time and space. Blocking may take place very often in such a network. How to minimize the delays caused by blocking is a major issue to be considered.

Synchronous message handling simply ignores blocked messages, assuming no buffers are used with the communication channels. This scheme has been implemented as one of two possible message-handling modes in the Intel iPSC systems. The idea is to halt further execution of instructions in a process until the desired message is sent or received.

When a process issues a request to receive a message, the process, being blocked, executes no further instructions until the expected message has been received. If the message arrives ahead of the request, the receiver will wait only until the message is copied into the local memory from the system buffer in which it was temporarily stored.

Similarly, when a sender initiates a message transfer, the sending process is blocked until the message is copied from local memory into the message-passing network by the node kernel. Synchronous message handling is easier to implement but may not result in the highest performance possible in a given network.

The Ada Experience   Example synchronous message-passing programming systems include the unbuffered Ada system, which uses a *rendezvous* concept to synchronize the sender with the receiver. In such a system, the early arrival at the rendezvous must wait for the late arrival. Ada uses a *name-addressing* scheme in which (Node, PID) is used to identify a process residing at a node.

The Ada system allows two-way data flow during a single rendezvous. A process calls another process by name without divulging its own identity. Ada implements a *select* primitive, which allows a process (task) to conditionally select an entry call to execute among multiple entries.

Programming with the rendezvous concept can easily implement a *remote procedure call* (RPC) besides the *select* option. It also supports dynamic task creation and priorities in task selection.

The Occam Experience   The CSP (communicating sequential processes) described by Hoare employs a selective synchronous scheme based on a tightly synchronized form of message passing. In 1988 the CSP was modified and called the Occam system by

INMOS/transputer developers, based on a *channel-addressing* scheme. This scheme establishes a communication path between sender and receiver by directional channel tracing. The Occam system uses one-way data flow.

In a synchronous message-passing system, the number of completed *receive* operations is identical to the number of completed *send* operations at the other end. The blocking problem can be avoided or alleviated using buffers in the message-passing network.

Buffering is like a telephone system with an answering machine or like a fax machine. The sender and receiver do not have to be available at the same time, and yet they can communicate with each other.

11.4.3 Asynchronous Message Passing

This programming paradigm requires the use of buffers on communication channels or the use of a global mailbox. Message sending and receiving do not have to be synchronized in time or space. Nonblocking communication is possible if infinite-size buffers are used along the communication channels.

Blocked messages are buffered for later transmission. This system is like a postal service system using many mailboxes as buffers. No synchronization is needed because the sender does not have to know if and when a message is received. The receiver expecting a message will not be suspended from regular execution.

Even when we consider a buffered asynchronous system nonblocking, the system will eventually be blocked due to the use of limited-size channel buffers. If the network traffic is not heavy or saturated, moderate-size buffers will lead to essentially nonblocking communications.

Most advanced message-passing networks prefer using asynchronous communications for the sake of better resource utilization and shorter communication delays. In such systems, the number of posted *receive* operations is greater than the number of completed *send* operations directed to that process. An arbitrary delay may result in a buffered communication scheme.

Caltech's Cosmic programming environment supports asynchronous message passing. Intel's iPSC system also supports asynchronous message handling. Asynchronous *receives* allow processes to alert the node kernel that certain messages are expected and should be delivered as soon as they arrive.

In the meantime, the receiving process can continue its work if needed. An asynchronous *send* allow a sender to alert the kernel that it wants to send a message, but the process does not wait until the message is sent. Obviously, asynchronous communication is more efficient and sometimes faster.

The Linda Experience The *Linda* programming system also operates asynchronously. Uncoupling of the sender and receiver in time and space is expected. Linda is based on a global mailbox or bulletin-board concept for achieving asynchronous communications. This is done through the use of a *tuple space*, which is logically shared by all concurrent processes.

The tuple space consists of *tuples*, which are typed data sequences. Any process

can add a tuple to the shared tuple space or remove a matching tuple from the space. If no matching is found, the process suspends until a match is found later. One can also read a copy of a matching tuple without removing it from the space.

In many ways, tuple space behaves like a bulletin board where anybody can add a notice (tuple), read a notice, or remove a notice. The tuple space, therefore, becomes a global mailbox which can be accessed by all processes for the purpose of either sending, broadcasting, or receiving messages (tuples).

Concurrent programming works in Linda by pattern matching on the tuple signatures (tag fields). The concept of tuple space can be implemented on either shared-memory multiprocessors or distributed-memory multicomputers.

Interrupt and Lost Messages   Interrupt messages are a special form of asynchronous message handling. Instead of continuing working while waiting for a regular message to arrive, interrupt handling is carried out immediately by the receiver without having the receiving process post a *receive* when it is ready to receive. After the interrupt is serviced, the interrupted process may resume its original work.

Messages are often lost in a message-passing system. When an incoming message is not expected or needed, it may be ignored and thus lost. Messages directed to the wrong process or wrong node will not be found or retrieved by the intended process, and the node kernel may not be able to cope with the problem.

These messages may end up in the system message buffers and eventually be lost. This may not be the fault of a programmer. Special detection aid or debugging tools are needed to inspect message buffers and to redirect lost messages to the right place.

11.5   Mapping Programs onto Multicomputers

This section describes program decomposition techniques based on data domains, control structures, functionality, and object-oriented concepts. In all cases, we aim at performance tuning and enhancement of message-passing programming. At the end, we characterize the environment for heterogeneous programming.

11.5.1   Domain Decomposition Techniques

Programming a multicomputer requires three major steps: *decomposition*, *mapping*, and *tuning*. The goals are to balance the load, minimize communication overhead, reduce sequential bottlenecks, and make the program scalable. Decomposition of the π calculation is a perfect example of domain decomposition.

In general, if a calculation is based on a large, static data structure and the amount of work is about the same for each data element, then one should partition the data structure evenly.

The resulting programming technique is called *domain decomposition*. This technique can be used with a wide range of applications, including physical modeling, matrix computations, and data/knowledge base management.

Perfect Decomposition   To choose the best decomposition method for a given ap-

plication, one needs to understand the mathematical formulation, the data domain, the algorithm used, and the flow of control (communication pattern). Certain applications fall naturally into the *perfect decomposition* category. Such parallel applications can be divided evenly into a set of processes that require little or no communication with each other.

This category is the easiest to decompose. The π computation is a good example of a perfect decomposition in which only the partial sums need to be communicated. The order in which these summations are performed is immaterial, and thus interprocess synchronization is unnecessary.

Perfect decomposition often leads to SPMD operational mode. In other words, the same node program is replicated at all the nodes. Real applications of this kind range from the modeling of proteins with thousands of atoms and millions of possible configurations to financial speculation involving investments subject to parallel evaluation of multiple hypothetical portfolio cases simultaneously.

Perfect decomposition requires very little communication overhead, and the balanced computations often result in nearly 100% efficiency. This kind of decomposition requires the least amount of programmer effort.

Often the same sequential program running on a single processor will be running on all node processors, each with a different data set. Once an interprocess dependence relationship exists, the conditions for perfect decomposition may be destroyed.

Domain Decomposition The key to domain decomposition is the *regularity* of the data structures involved. Three kinds of problem domains are identified below as natural candidates for domain decomposition:

(1) *Static data structure* — For example, matrix factorization for solving a large finite-difference problem on a system with a regular network topology.

(2) *Dynamic data structures tied to a single entity* — For example, in a many-body problem, subsets of the bodies can be distributed to different nodes. Through gravitational forces, the bodies may be interacting with each other and moving in space. The calculation for each body can stay on the original node assigned.

(3) *Fixed domain with dynamic computations with various regions of the domain* — For example, a program that models fluid vortices, where the domain stays fixed but the whirlpools move around.

Three major steps are specified below to decompose the domain of a given application:

(1) Distribute the subdomain of data to various nodes.

(2) Restrict the computation so that each node program updates its own subdomain of data.

(3) Put the communication in node programs.

The easiest way to implement the above steps is first to port sequential programs to various nodes. The porting procedure involves the following operations: Compile and test an existing serial program on one node and then run multiple copies of the same program on many nodes at once. Put in the communication commands and run it on all the nodes. Localize the node program execution over its own data. Finally, tune the program to enhance the performance.

Example 11.8 LINPACK matrix factorization using domain decomposition (Justin Rattner, Intel Scientific Computers, 1990)

Gaussian elimination is often performed in factoring a square matrix in the LINPACK linear equations package. The regular nature of this algorithm and the regularity of the domain make it inherently parallel and suitable for domain decomposition.

Distributing the domain only requires partitioning the characteristic matrix into sections and distributing them among the processor nodes. We will first examine the sequential LINPACK Gaussian elimination algorithm. After determining the matrix distribution, we then distribute the computation, (mapping) result in a parallel matrix factorization algorithm.

The LU factorization is a triply nested loop. The outer loop controls how much of the matrix remains to be factorized. At each iteration, the remaining part of the matrix is a smaller submatrix in the lower right-hand corner. The elimination process requires that information from one node be broadcast to all the other nodes. A sequential factor algorithm in LINPACK is as follows:

Sequential Factor Algorithm

for $i = 1$ to $n - 1$ **do**

    Find $\max[A(i, i)$ to $A(n, i)]$ in ith column,

    Swap rows to make $A(i, i)$ the pivot,

    Divide $A(i + 1, i)$ to $A(n, i)$ by $A(i, i)$,

    for $j = i + 1$ to n **do**

      for $k = i + 1$ to n **do**

        $A(k, j) \leftarrow A(k, j) - A(k, i) \times A(i, j)$

      end of k-loop

    end of j-loop

end of i-loop

This sequential code can be executed directly on a single node. The next step is to distribute the matrix elements. The matrix is distributed by columns among the processor nodes. The matrix domain is mapped to the nodes so that all the processors own approximately the same number of columns of the matrix.

Figure 11.7 shows a column-wrapped mapping of a 12×12 matrix onto four processors. Once a column becomes the pivot column, it requires no further computation. By using this column mapping, all processors can be kept busy during most of the computation, and a new processor owns the pivot column at each step.

The *parallel factor algorithm* is designed to make the pivoting processor control the computation at each outer-loop iteration. The processor owning the first column finds the pivot element, swaps the pivot row with the first row, and divides the pivot column by the pivot, just as in the sequential factor algorithm.

Column Number of Matrix

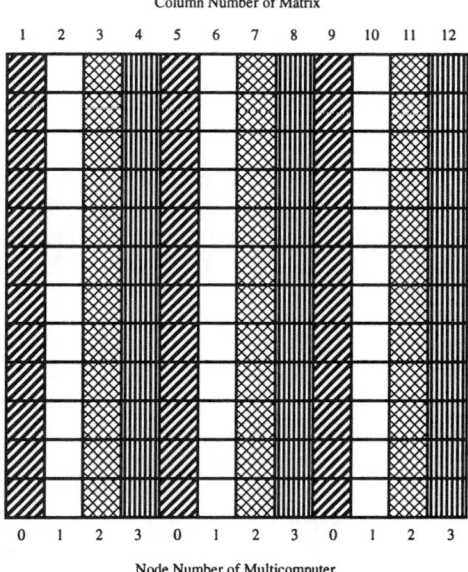

Node Number of Multicomputer

Figure 11.7 Column-wrapped mapping of a 12×12 **matrix onto four nodes of a multicomputer.** (Courtesy of Justin Rattner, Intel Scientific Computers, 1990)

After the first iteration, the pivot processor broadcasts the pivot column and the pivot row number as a message to all the remaining processors. Any processor receiving this message swaps its pivot row with its first row. Then all processors perform the remaining subtraction and scaling from each remaining column simultaneously.

This parallel factorization algorithm is specified below, where n is the order of the matrix, N is the total number of nodes, i is the global index, h is a local index with an initial value of 1, and p is the index of the processor. The value of p is different for each processor and is in the range $0 \leq p < N$. In addition, all processors use the same node program as stated below.

Parallel Factor Algorithm

$p \leftarrow$ index of this processor

for $i = 1$ to $n - 1$ **do**

    if $(i - 1) \bmod N = p$ **then**

        Find max in the hth column

        Swap row i and h to make $A(i, i)$ the pivot

        Divide $A(i + 1, i)$ to $A(n, i)$ by $A(i, i)$

        Broadcast pivot column, $A(i + 1, i)$ to $A(n, i)$,

          and pivot row number h

        $h \leftarrow h + 1$

        else

           Receive the pivot package

           Swap rows

           Scale and **subtract** the pivot column from each remaining column

      end of i-loop

The above parallel algorithm is modified from the serial algorithm by the addition of a *test* to determine the pivot processor and system calls to *send* (broadcast) the pivot information to all the processor nodes. This simple change may result in a maximum speedup proportional to the number N of processors used in the multicomputer.

Performance Tuning   From a successive pivoting point of view, the above algorithm is essentially sequential. While a pivot column is being determined, all the remaining processors are waiting for it. When the pivot is finally broadcast, all the processors can subtract in parallel and then proceed with the next pivot node, etc.

From a performance point of view, the load is not fully balanced across the nodes. The sequential bottleneck is in the successive pivoting processors. The mapping of the matrix columns in a modulo fashion has already provided a better load balance than assigning adjacent columns to the same processor.

The performance can be further tuned by reordering the algorithm to speed up the pivot and broadcast process, or by dividing the row in parallel instead of the column. Furthermore, one can change the granularity by using a *block algorithm* to obtain fewer and longer messages, which will improve the communication/computation ratio.

In other applications, such as seismic processing, finite-element analysis, vision integration, and multidimensional complex FFT, the performance bottleneck is in different areas such as extensive synchronization delays or dynamically changing data structures. In each case, performance tuning will be focused differently. However, the general idea of keeping more of the processors busy more of the time can always improve efficiency.

11.5.2   Control Decomposition Techniques

When the domain and data structure are irregular or unpredictable, we cannot apply domain decomposition. One alternative is to focus on distributing the flow of control of the computation rather than on distributing the domain.

An example of a domain that is not suitable for domain decomposition is the irregular search space created by a game tree where the branching factor varies from node to node. The search tree must be dynamically assigned to maintain a balanced load.

In general, *control decomposition* is used for symbolic processing problems in artificial intelligence applications. In this section, we will study several control decomposition strategies, including *functional decomposition* and a *manager-worker approach*.

Functional Decomposition   An algorithm can be visualized as a set of interconnected functional modules. The flow of control is indicated by directed edges in the diagram. For small problems, these functional modules tend to be executed sequentially, one after another. However, large problems may have significant overlap between the modules.

Example 11.9 Image understanding with functional decomposition

Figure 11.8 shows an example of functional decomposition for image understanding.

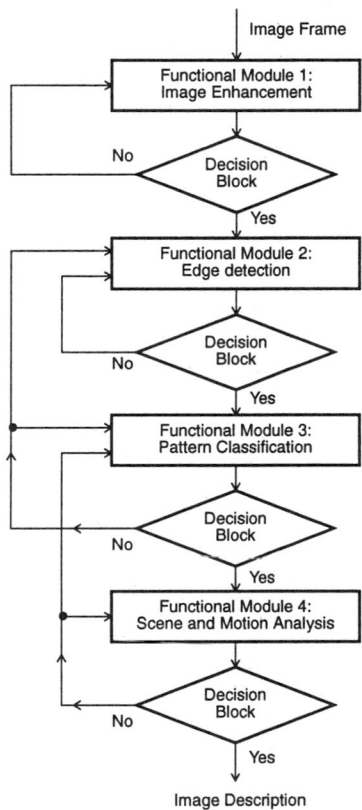

Figure 11.8 Functional decomposition for image understanding.

Image enhancement, edge detection, pattern recognition, and scene and motion analysis are processed by four functional modules in a pipelined fashion (with a feedback branch, if adaptively done) over a sequence of image frames. ∎

The functionality requires the application of different decomposition techniques, depending on the data structures and computations involved. One needs to add special message interfaces between different functional modules.

Usually different functional modules are assigned to different processor nodes. Therefore, hybrid nodes are needed to deal with functional needs. Some nodes will be floating-point-intensive, some for symbolic manipulation, some for input/output activities, etc.

Manager-Worker Approach   This is a divide-and-conquer technique. The idea is to divide the application into tasks, not necessarily having the same size, and use one of the processes to serve as a manager node and the rest as worker nodes.

As illustrated in Fig. 11.9, the manager is responsible for dispatching tasks out as worker nodes become available. The manager must also communicate with the host for input/output operations.

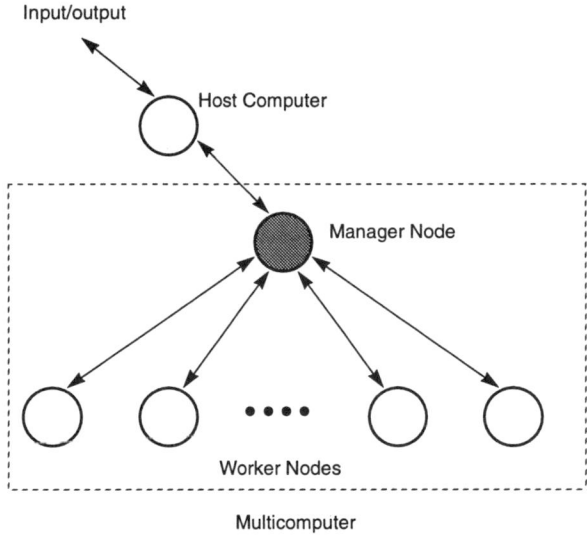

Figure 11.9 Manager-worker mapping for control decomposition. (Courtesy of Justin Rattner, Intel Scientific Computers, 1990)

The manager functions include management of global data structures, maintaining a list of subprograms (tasks), and assigning problems/tasks to workers. The worker node, once it becomes available, should request the job, receive the job, and perform the assigned task. The manager must perform dynamic load balancing among the workers in order to enhance the performance of the entire system.

Example 11.10 Solving the N-queens problem on a multicomputer

A typical application of the manager-worker decomposition is to solve the N-

queens problem on a multicomputer. The problem is to find all possible solutions to placing N queens on an $N \times N$ chessboard so that no more than one queen is on each row, each column, or each diagonal.

In other words, each row, column, or diagonal must have at most one queen and no queen can attack another, as illustrated in Fig. 11.10. The way the N-queen problem is solved is to generate a search tree using the workers to solve each leaf node of the search tree.

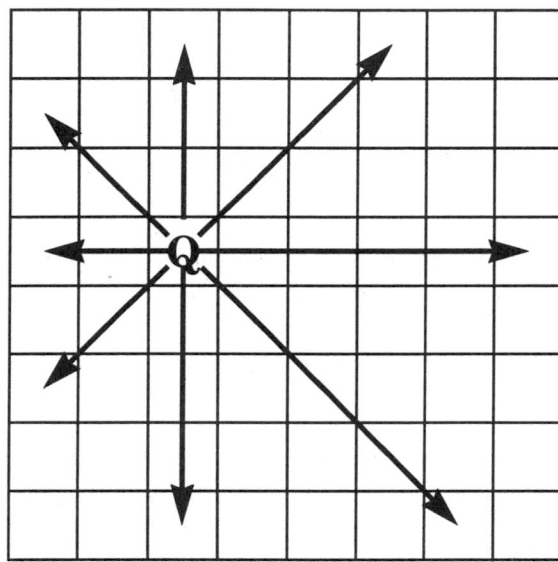

Problem: Find all solutions to placing N queens on an $N \times N$ chessboard so that there is exactly one queen on each row, or each column, or each diagonal.

Figure 11.10 N**-queens problem: An example of manager-worker decomposition.** (Courtesy of Justin Rattner, Intel Scientific Computers, 1990)

The manager builds and maintains the top level of the search tree, assigns workers to build different branches of the tree, and keeps track of the total number of solutions (chessboard patterns) generated. Each worker should be able to split the problem into subproblems and to solve a subproblem.

Once a branch is too heavily extended, the worker can report to the manager which will off-load subtree operations to other available nodes. Close communication between the manager and the workers is necessary to keep up a well-balanced processing of the search tree on a dynamic basis. ∎

Performance Tuning   One technique for tuning the performance is to use double-buffer messages in a manager-worker decomposition. The idea is to send a worker two pieces of work the first time. As soon as the worker is finished working on the first piece, another piece is readily available.

Once the communication/computation issues have been well balanced, the results from the first piece of work are returned to the manager who can send over another piece of work before the worker finishes the second piece. There is always one job waiting in the worker's queue, and thus workers are kept busy all the time.

A potential problem with the manager-worker approach is that the manager may become the bottleneck. According to Intel iPSC experience, up to 50 workers managed by a single manager do not create a serious bottleneck problem as long as a good communication/computation ratio can be maintained.

Clearly, for a multicomputer consisting of thousands of processors in the future, the manager bottleneck problem will become more serious. One can always consider a hierarchy of managers.

Another possible solution is to use floating managers or multitasking processors, which can execute both a worker process or a manager process on the same processor. These options must be carefully experimented with before they can be adopted in real applications.

11.5.3   Heterogeneous Processing

In this section, we learn how to combine object-oriented programming with message-passing techniques for distributed computing applications. We first characterize objects in relation to parallelism. Then we illustrate the *object decomposition* concept using an air traffic control simulation example.

Finally, we present the concept of *layered parallelism* using a seismic monitoring example which involves all the decomposition techniques we have learned to solve very large-scale problems.

Objects and Parallelism   The object-oriented approach to parallel programming offers a formal basis for decomposing the data structures and threads of control in user programs. In what follows, we define objects and reveal the relationship between objects and parallel processing.

The idea of objects comes from *data abstraction*, in order to hide low-level details from programmers for large-scale problems, such as in archive data/knowledge processing. An object encompasses a set of logically related data and a set of procedures which operate on the object's data, as illustrated by the example in Fig. 11.11a.

The example shows that temporary storage in the form of a stack can be treated as an object consisting of a last-in-first-out queue of data which can be pushed down or popped up in its management. It should be noted that an object type (class) is conceptually different from instances of the object type.

Those instances, called objects, are the ones used in program execution. Only the object's procedures have access to the object's data. A programmer can be freed from knowing the detailed implementation of the objects.

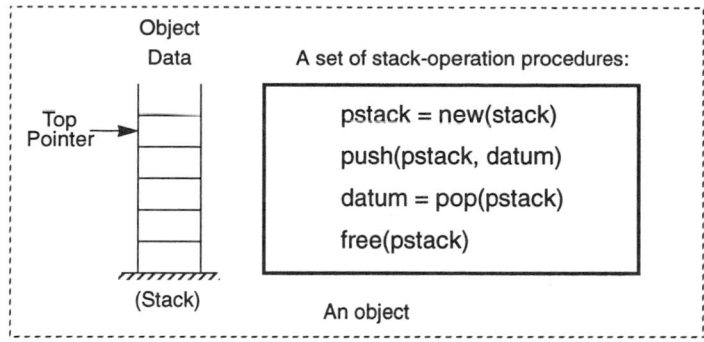

(a) Defining a stack as an object

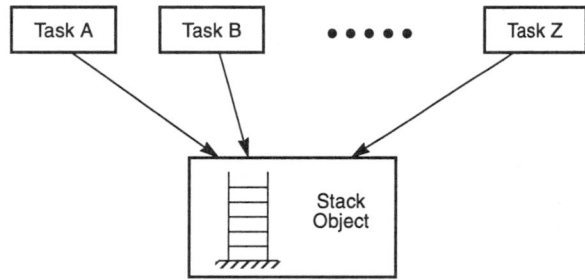

(b) Shared access to a stack object by multiple tasks

Figure 11.11 A stack object and its shared access by multiple tasks.

This will simplify program debugging and testing efforts and offer modularity in program development, which are all desired features for parallel programming.

Shared data structures can be organized as objects. Shared access to a stack object by multiple tasks is shown in Fig. 11.11b. A parallel program may be composed of multiple threads of execution that access both private and shared data.

Threads of execution can be allocated as instances of a task type. If tasks must communicate by exchanging messages, the shared object is accessed via message passing, which must be synchronized by a system-defined class of queue objects.

Object Decomposition This technique offers natural advantages for parallel computers. It avoids the use of global variables, simplifies the program/data partitioning process, and provides higher granularity of interaction among objects through the use of predefined procedures for accessing objects.

Ada and C++ both support data abstraction in object-oriented programming on message-passing systems. The following air traffic simulation system explains the key concepts behind object decomposition for parallel programming. The goal of the simulation is to measure the effects of scheduling, weather, etc., on air traffic flow control.

Example 11.11 Air traffic simulation on a multicomputer

Three fundamental object types are identified in Fig. 11.12. First, the airports contain information regarding their location, runways, etc. Second, airplanes contain information related to position, velocity, fuel capacity, etc. Third, air space sectors are prespecified by the American Flight Agency.

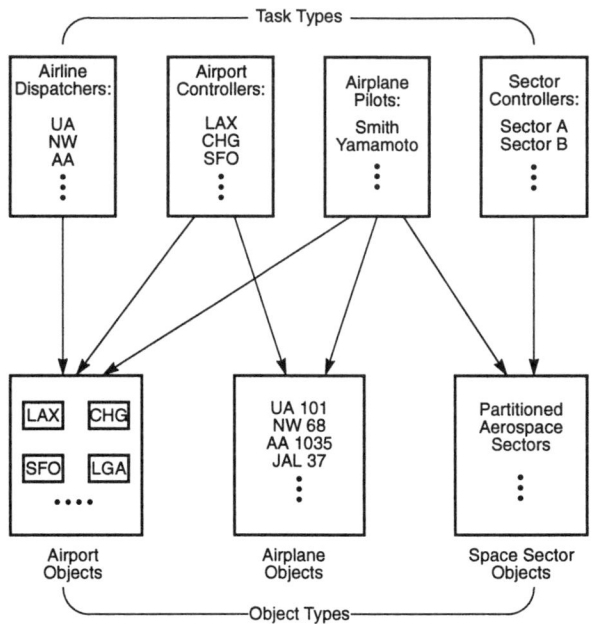

Figure 11.12 Air traffic simulation using decomposition techniques

Objects of these types are manipulated by four task types: The air dispatchers allocate airplanes and pilots (scheduling). The pilots operate the airplanes. The air traffic controllers manage the safe use of airports and airspace sectors to avoid collisions.

The tasks interact by invoking procedures on shared airplane, airport, and airspace objects. The communications among the tasks and between tasks and objects can be simulated by a message-passing multicomputer. Appropriate protocols must be established in these message-passing operations.

The use of multiple tasks with separate functions in this application is also an example of control decomposition. The computational complexity of this air traffic simulation problem is controlled by partitioning the airspace into sectors.

The air traffic controller manages the separation of airplanes within each sector. When an airplane crosses between two sectors, the controller must pass the duty to the next sector controller on the route of the flight.

Sector partitioning of the airspace is indeed a domain decomposition of the problem. Different sector controllers are simulated at different nodes of a multicomputer. In order to conduct real-time simulations, all the object information must be retrieved from the local memory without page faults from the backup store.

For example, distribution of the airports should be made to associate the airports and their controllers within the same sector node. As airplanes fly from one airport to another, the objects are sent as messages among nodes. Load balance can be achieved by defining the section boundaries, interleaving sector objects, or allowing sector migration dynamically as traffic becomes dense.

Heterogeneous Processing   A growing trend in computing is to use a network of resources for *heterogeneous processing* (HP). A large-scale complex problem can be solved by combining a number of computers of various kinds in a network environment.

On a multicomputer-based network, HP can be practiced using a combination of decomposition techniques. In what follows, we describe computations with embedded parallelism at various processing layers.

Layered Decomposition   In solving large and complex problems, very often we have to employ a programatic layered approach to extracting parallelism using different decomposition techniques at different levels. Such an approach can be called *layered decomposition.*

The functional approach to image understanding (Fig. 11.8) can be considered a four-layered decomposition. As machine size may reach thousands of nodes in the future, we have to develop programming techniques on that scale.

The layered approach certainly addresses the scalability problem with foresight. Example applications include weather simulation, fluid flow, structural analysis, molecular dynamics, and seismic monitoring.

Example 11.12 Heterogeneous processing based on layered parallelism

The problem of seismic monitoring requires the processing of a large amount of seismic data using a layered approach consisting of three levels, as illustrated in Fig. 11.13.

At the highest level, perfect parallelism is expected because real-time signals come in from monitors located at various sites using separate processors mostly running with Fortran seismic code.

Once the seismic signals are cleaned up by the distributed processors, we need to use a middle level of functional decomposition corresponding to a sequence of AI decisions.

Finally, the interpretation of real-time data may require domain decomposition for each signal domain.

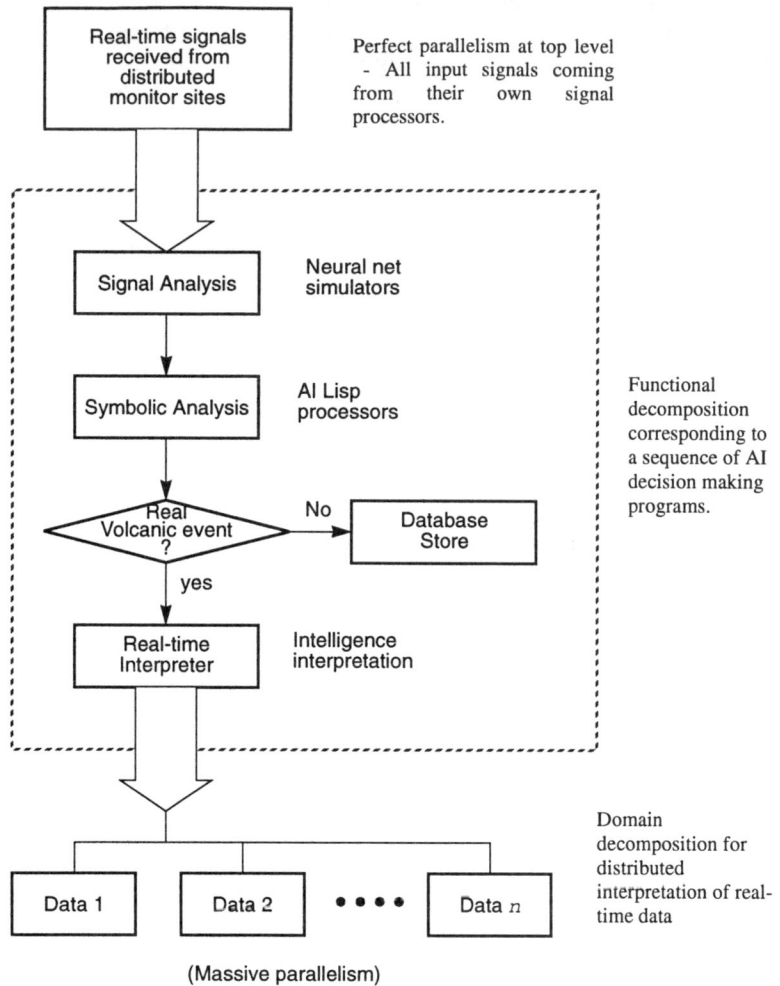

Figure 11.13 Layered parallelism in decomposing the seismic monitoring problem for heterogeneous processing.

Different types of computers are used at different levels in the layered approach. The monitor processors used at the top level are nothing but separated signal processors with numeric, Fortran, and vector processing capabilities.

The functional decomposition may use neural simulators, Lisp processors, or other symbolic machines with extended memory and intelligence interpretation capabilities.

The bottom level of domain decomposition demands a higher degree of parallelism if the survey sites are numerous. Graphics/visualization capabilities are needed to interpret the final results. The above example presents a typical case of heterogeneous processing in a network environment.

11.6   Bibliographic Notes and Exercises

The material on shared-variable synchronization and access order is based on [Bitar91] and [Bitar92]. Various synchronization methods were surveyed in [Dinning89]. [Graunke90] and Thakkar have compared synchronization algorithms on shared-memory multiprocessors.

Binary semaphores were originally proposed by [Dijkstra68]. [Hoare74] developed the use of *monitors* for operating system structuring. [Wirth77] introduced *Modula* for modular multiprogramming. [Perrott79] developed the *Actus* language for array/vector processing. [Brinch Hansen75] introduced the *Concurrent Pascal* language for shared-variable programming.

For message-passing programming, readers are referred to the *Cosmic C Programming Manual* by [Seitz89] et al. and the *Intel Parallel Programming Primer* edited by [Ragsdale90]. [Hoare74] developed CSP, which was later modified to *Occam* by [Pountain87] and May.

[Gelernter85a] proposed the *Linda* programming system for asynchronous communications. The material on domain, control, functional, and object decomposition techniques is based on Intel's iPSC experience. Additional programming examples using these techniques can be found in the Intel iPSC publication [Ragsdale90]. A parallel program for solving the *N*-queens problem can be found in [Athas88] and Seitz.

Surveys of parallel programming tools were reported in [Chang90] and Smith and in [Cheng91]. The MIMDizer was introduced in [Harrison90]. A user's guide for MIMDizer is also directly available from Pacific-Sierra Research Corporation [PSR90]. The Express is documented in [Parasoft90]. Linda was first introduced by [Gelernter85b] et al. C-Linda was discussed in [Ahuja86] et al., and SCHEDULE was reported by [Dongarra86] and Sorensen.

Cray Y-MP software overview can be found in [Cray89]. The Paragon programming environment is summarized in [Intel91]. The CM-5 software discussion is based on the technical summary [TMC91]. Additional information on CM-2 software can be found in [TMC90]. The supercompilers for iPSC/860 are described in [Intel90].

Exercises

Problem 11.1   Explain the following terms associated with fast and efficient synchronization schemes on a shared-memory multiprocessor:

(a) Busy-wait versus sleep-wait protocols for sole access of a critical section.
(b) Fairness policies for reviving one of the suspended processes waiting in a queue.
(c) Lock mechanisms for presynchronization to achieve sole access to a critical section.
(d) Optimistic concurrency or the postsynchronization method.
(e) Server synchronization and the corresponding synchronization environment.

Problem 11.2   Distinguish between spin locks and suspend locks for sole access to a critical section. Generalize Dekker's protocol from two procedures to three or more procedures sharing critical sections. Also implement the generalized Dekker's protocol using the Test&Set atomic operation.

Problem 11.3 There are many ways to solve the mutual exclusion problem based on different implementation schemes, such as the use of spin locks or Dekker's protocol. Describe the following schemes:

(a) Implementing mutually exclusive access to a critical section using *binary semaphores*.

(b) Implementing mutual exclusion using a *monitor* called by processes competing for access to a critical section.

Problem 11.4 Dijkstra [Dijkstra68] has defined a dining philosophers problem: There are five philosophers dining around a table as shown.

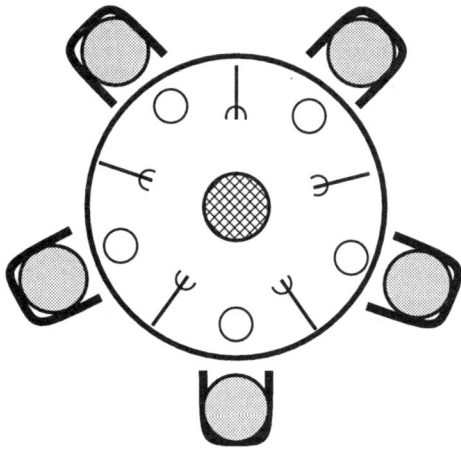

Each of the philosophers engages in only two activities, *thinking* or *eating*, as characterized below as a process P_i for $i = 1, 2, \ldots, 5$.

```
P_i:  begin
          loop
              Think
              Fetch protocol
              Eat
              Release protocol
          endloop
      end
```

A bowl with an endless supply of spaghetti is placed in the center of the table. Each philosopher is given a plate. There are five forks on the table, one between adjacent philosophers. Each philosopher enters a thinking period while not eating. In order to be able to eat, the philosopher must get hold of two neighboring forks from his left and right sides. The *fetch protocol* specifies how each of the forks will be picked up. The *release protocol* specifies how the forks will be released after eating. Each philosopher is allowed to pick up one fork at a time from his left or from his right side. Each fork can be used by only one philosopher at a time. No fork can be passed around the table, and a fork must be put back where it was picked up.

The problem is to design the fetch protocol and the release protocol for the philosophers so that no deadlock will occur. A deadlock means a circular wait situation in which each philosopher holds one fork and refuses to release it. A deadlock means starvation, so we should avoid it in the solution. Note that different protocols can be specified for different philosophers. Not all philosophers must follow the same protocol.

(a) Use the *P* and *V* operators and binary semaphores to specify the fetch and release protocols. One semaphore can be used to represent the fork on the right, and another semaphore to represent the fork on the left. The purpose of the protocols is to prevent deadlock from occurring so that no individual starvation will occur. Initially, all the forks are placed on the table corresponding to an initial value of 1 for the semaphores. When a fork is picked up, its semaphore is changed to a value of 0.

(b) Design a *monitor* to control the fetch and release of the forks. In this case, the fetch and release are each specified by a procedure in the monitor. Also, specify the philosophers as user processes calling the monitor to claim forks. Again, no deadlock or starvation is allowed under the same assumption made in part (a).

Problem 11.5 Explain why *mutual exclusion, nonpreemption, wait for,* and *circular wait* are necessary conditions but not sufficient conditions for a system deadlock to occur. Also distinguish among the *deadlock prevention, avoidance, detection,* and *recovery* schemes. Comment on their implementation costs and expected performances.

Problem 11.6 Five concurrent processes are specified below using four resource types represented by four semaphores. Answer the following questions with reasoning and justification.

> **Begin**
> > **shared record**
> > > **begin**
> > > > **var** S_1, S_2, S_3, S_4: **semaphore;**
> > > > **var** blocked, unblocked: **integer;**
> > > **end**
> > > **initial** blocked = 0, unblocked = 1;
> > > **initial** $S_1 = S_2 = S_3 =$ unblocked; $S_4 =$ blocked;
> > > **cobegin**
> > > > A: **begin** P(S_1); V(S_1); P(S_2); V(S_2); **end;**
> > > > B: **begin** P(S_1); P(S_2); V(S_4); V(S_2); V(S_1); **end;**
> > > > C: **begin** P(S_2); P(S_3); V(S_2); V(S_3); **end;**
> > > > D: **begin** P(S_4); P(S_2); P(S_1); V(S_1); V(S_2); **end;**

E: **begin** $P(S_3)$; $P(S_2)$; $V(S_2)$; $V(S_3)$; **end**;

     coend

End

(a) Is a deadlock possible among the five code segments represented by A, B, C, D, and E? Which subset of code segments may enter a deadlock on what resources?

(b) If the deadlock situation does occur in part (a), what additional code segments could be indefinitely blocked?

(c) Is a deadlock inevitable or does it depend on race conditions? Justify your answer with reasoning using a resource allocation graph.

(d) Make a minor change in one program segment to prevent a deadlock from occurring. Justify the claim with a resource allocation graph similar to Fig. 11.5.

Problem 11.7 Scheduling access to a moving-head disk can be implemented by a monitor. The implementation consists of three components: *user processes* which request, access, and release the disk service; a *disk scheduler* which performs the scheduling of disk data to be accessed by one user at a time; and *driver procedures* that perform actual data transfer.

(a) Write a *monitor* to implement the disk scheduler. The monitor should consist of two procedures, one for a request for and one for a release from disk access.

(b) Specify how a user process can call the monitor for disk access. The disk driver procedures are considered given.

Problem 11.8 Write a monitor as a barrier counter for the synchronization of n concurrent processes. The barrier counter should be resettable, and a user process should be specified to call the monitor when it reaches the barrier. Note that local and shared variables must be declared and initialization of local data must be given.

Problem 11.9 Answer the following questions on decomposition techniques for message-passing programming on multicomputer nodes:

(a) What is a perfect decomposition? Explain the advantages and discuss the differences in program replication techniques on multicomputers as opposed to program partitioning on multiprocessors.

(b) Based on data domain, algorithm used, and flow of control in applications, distinguish the opportunities for applying domain, control, and object decomposition techniques in distributed computing on multicomputers.

Problem 11.10 The *N-queens problem* (Fig. 11.10) was introduced along with the manager-worker approach to control decomposition in programming a multicomputer. Suppose $N = 8$. There are 92 possible solutions to the 8-queens problem.

(a) Write a program that searches for a solution. First run the program on a sequential computer (such as on a workstation or even a personal computer). Record the time required to conduct the sequential search. A sequential search involves backtracking once an impossible configuration is exposed. The backtracking step systematically removes the current emplacements of the queens and then continues with a new emplacement.

(b) Develop a parallel program to run on a message-passing multicomputer if one is available. For a concurrent search for solutions to the N-queens problem, backtracking is not necessary because all solutions are equally pursued. A detected impossible configuration is simply discarded by the node. Observe the dynamics of the concurrent search activities and record the total execution time. Compare the measured execution time data and comment on speedup gain and other performance issues.

Note that an 8-queens program written in an object-oriented programming language Cantor is given in the paper by Athas and Seitz (1988). Interested readers may want to compare their results with those reported by the Caltech researchers.

Problem 11.11 The *traveling salesperson problem* is to find the shortest route connecting a set of cities, visiting each city only once. The difficulty is that as the number of cities grows, the number of possible paths connecting them grows exponentially. In fact, $(n-1)!/2$ paths are possible for n cities. A parallel program, based on *simulated annealing*, has been developed by Caltech researchers Felten, Karlin, and Otto [Felten85] for solving the problem for 64 cities grouped in 4 clusters of 16 each on a multicomputer.

(a) Kallstrom and Thakkar (1988) have implemented the Caltech program in C language on an iPSC/1 hypercube computer with 32 nodes. Study this C program for solving the traveling salesperson problem using a simulated annealing technique. Describe the concurrency opportunities in performing the large number of iterations (such as 60,000) per temperature drop.

(b) Rerun the code on a modern message-passing multicomputer. Check the execution time and performance results and compare them with those reported by Kallstrom and Thakkar. Contact these authors for original codes. You may need to modify the code in order to run on a different machine.

Problem 11.12 Choose an example program to demonstrate the concepts of macrotasking, microtasking, and autotasking on a Cray-like multiprocessor supercomputer. Perform a tradeoff study on the relative performance of the three multitasking schemes based on the example program execution.

Problem 11.13 Write a multitasked vectorized code in Fortran 90 for matrix multiplication using four processors with a shared memory. Assume square matrices of order $n = 4k$. The entire data set is available from the shared memory.

Problem 11.14 Design a message-passing program for performing fast Fourier transform (FFT) over 1024 sample points on a 32-node hypercube computer. Both host and node programs should be specified, including all communication commands. Initially each node holds 32 sample points without duplicating the data set. The results should be sent to the host for output.

Problem 11.15 A typical two-dimensional image is represented by a rectangular array of pixels (picture elements). Each pixel (i, j) is represented by a $\log_2 b$-bit integer corresponding to the gray level (between 0 and $b - 1$) at coordinate (i, j) of a black-and-white picture. *Histogramming* is a process to count the frequency of occurrences of each gray level. Let $\mathrm{histog}(0 : b - 1)$ be the array of a histogram of b gray levels.

The following serial code is written for histogramming on a uniprocessor system:

```
Var pixel(0 : m − 1, 0 : n − 1);
Var histog(0 : b − 1): integer;
histog(0 : b − 1) = 0;
for i = 0 to m − 1 do
        for j = 0 to n − 1 do
                histog(pixel(i, j)) = histog(pixel(i, j)) + 1
```

The time complexity (number of counts) of this serial program is $O(mn)$, where mn corresponds to the image size.

Partition the image, $\mathrm{pixel}(i, j)$ for $0 \le i \le m - 1$ and $0 \le j \le n - 1$ into p disjoint segments, where each segment has $m/p = s$ rows of the image. Develop a parallel program which can spawn a set of p processes to histogram the entire image. The p concurrent processes share the same $\mathrm{histog}(0 : b - 1)$ array.

(a) Use the **Doall** and **Endall** statements to specify a parallel program for counting the histogram simultaneously on a p-processor system with shared memory.

(b) What is the potential speedup of the parallel program over the above serial program? You can ignore the image I/O overhead by assuming the entire image database is in the main memory.

Chapter 12

UNIX, Mach, and OSF/1
for Parallel Computers

After specifying multiprocessor UNIX design goals, we present three OS kernel models for extending UNIX to meet the multiprocessor demand. The three operating system models are called *master-slave*, *floating-executive*, and *multithreaded kernels*. These UNIX kernels can be implemented with locks, semaphores, and monitors.

In addition, implementation and performance issues are addressed. The complexities in implementing, debugging, and using UNIX on multiprocessors are compared. UNIX extensions are also developed for supporting loosely coupled message-passing multicomputers or for supporting a network of heterogeneous computers.

Case studies of UNIX extensions include the Purdue dual-processor UNIX, the IBM IX/370, the Caltech Cosmic Environment and Reactive Kernel, the Intel NX/2 node kernel, the CMU Mach, the NeXT Mach, and the OSF/1 system.

12.1 Multiprocessor UNIX Design Goals

Porting the UNIX environment to a parallel computer is a complex task. Depending on the hardware platform and operational modes desired, there are abundant options for a system designer to choose from. The original UNIX developed by Brian Kernighan and Dennis Ritchie was explicitly written as a portable, general-purpose, time-sharing operating system for uniprocessor computers.

With the rapid progress made in processor, memory, and network technologies, a modern parallel architecture places even greater responsibility on the operating system to manage resources. Major OS functions include processor scheduling, virtual memory management, I/O device and network-access control, etc.

These OS functions are implemented with a large amount of system software, much too large to fit in its entirety in the main memory. The portion of the OS that resides continuously in the main memory is called the *kernel*. The remaining OS components, such as the file management system and compilers for various languages, are stored on disk and brought into the main memory only when needed.

Impact of RISC   Due to the popularity of using RISC and superscalar processors in modern computer systems, the reduction in hardware functions increases the burden on the operating system kernel. In other words, the kernel must exercise more software control on RISC hardware as compared with firmware control on CISC hardware.

For example, the MIPS R3000 OS kernel must save registers and read special registers during trap handling and context switching. It must also generate a TLB miss trap, instead of relying on a hardware MMU to handle the TLB miss, as compared with direct hardware control in the M68000.

On the other hand, the increasing OS role provides more flexibility in customizing the kernel control in dedicated environments for real-time and transaction processing applications. In these environments, the OS functions must be specially tuned to match the processing characteristics.

12.1.1 Conventional UNIX Limitations

Reviewed below are several limitations of conventional UNIX for parallel processing. Methods for coping with these problems are given in subsequent sections.

UNIX Kernel Structure   The structure of a conventional UNIX kernel is shown in Fig. 12.1, where the kernel is built around a *file subsystem* and a *process control subsystem*, in addition to some *device drivers*.

The kernel guarantees atomic operations like *read* and *write* system calls, protects access to shared kernel data structures and shared devices, manages files, schedules CPU time, handles interrupts, coordinates IPC, performs memory management, etc.

The interrupt handler may access data structures shared by user process codes. Interested readers are referred to the book by Back (1986) for details on designing UNIX/OS.

Single Thread of Control   Conventional UNIX running on a uniprocessor system has a single thread of control. This implies that only one system call from a user process can be serviced at a time in the kernel. In other words, the kernel execution (except interrupts) is done within the context of a single process in a mutually exclusive manner.

The kernel is nonpreemptive in this sense. Processes executing in kernel mode must be serialized through explicit context switching. No time slicing is practiced while a process is in kernel mode.

Since the kernel provides global services, such as in *init, sched,* and *vhand* system calls, and mutual exclusion is a necessity in sharing kernel data structures, many user processes may be blocked from entering the kernel while they sleep wait for I/O, locks, synchronizations, etc. This serialization of kernel services may significantly lower the uniprocessor performance.

Multitasking Bottleneck   UNIX supports *multitasking* in which multiple processes (tasks) within a program can run at the same time. Like a banking service, multitasking is achieved in a uniprocessor system by having only one processor (a teller) running among several service windows where there are long lines of user processes (customers)

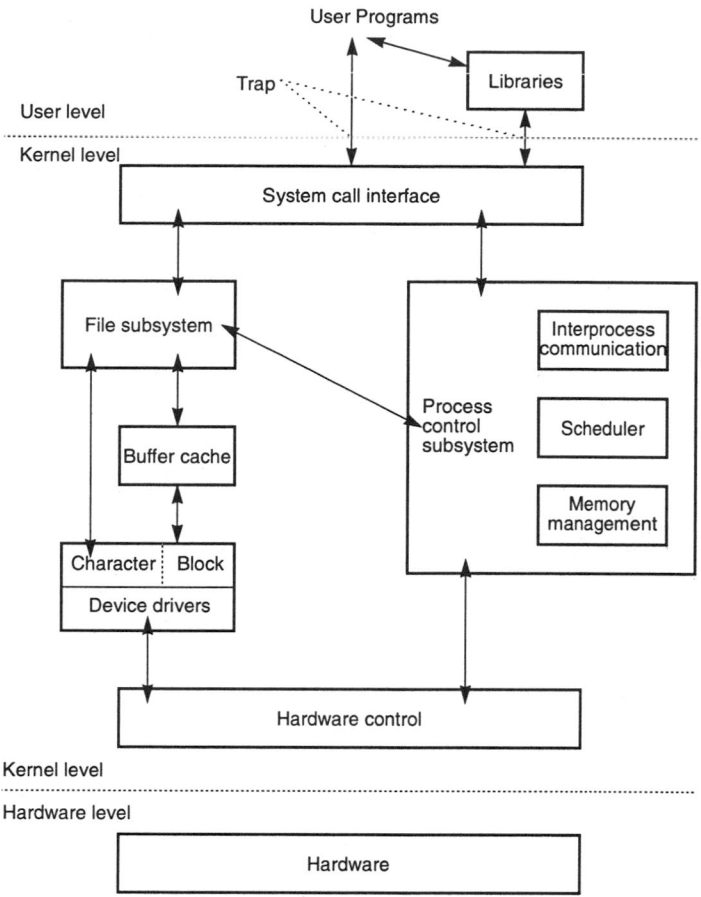

Figure 12.1 Architecture of the conventional UNIX kernel for a uniprocessor system. (Reprinted with permission from *The Design of the UNIX Operating System* by Maurice Back, Prentice-Hall, Englewood Cliffs, NJ, 1986)

waiting in line for service. Obviously, the kernel becomes the bottleneck in the system when multitasking is implemented on a uniprocessor computer.

Furthermore, a *parent process* must copy its entire address space and give the space to a newly created *child process* when a *fork* system call is made. This may result in excessive *copy-on-write* operations in a single-threaded kernel. Of course, multitasking between foreground and background tasks can be supported by using a *file server* tied to several uniprocessor systems as depicted in Fig. 12.2.

The foreground tasks are executed at user processors through interactive keyboard/display operations performed by the users. The background tasks, such as *mail-a-message* and *print-a-file*, are handled simultaneously, overlapping with the foreground tasks. In this case, the user processors may be running DOS or UNIX, but the file server

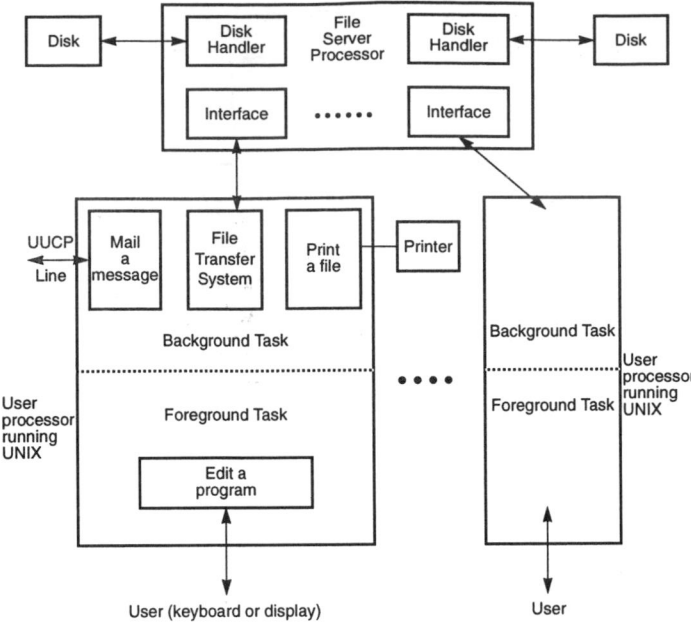

Figure 12.2 Multitasking in a network of workstations connected to a common file server in a UNIX environment. (Courtesy of Tom Jermoluk, 1990; reprinted with permission)

must run UNIX in order to achieve this kind of foreground and background multitasking.

Limited Resources Sharing   In a tightly coupled system, two or more processors share peripherals, memory, and process space. Either the child process address space can be copied by the parent process, or the parent and child processes can share the same address space. In conventional UNIX with a single thread of control, address-space copying is an expensive operation.

The shortcomings of conventional UNIX on sequential machines led to the desire for a multiprocessor UNIX kernel. Some of the design goals have been set high in order to enhance compatibility and portability across many machines.

12.1.2   Compatibility and Portability

Multiprocessor UNIX should be designed to control and to coordinate the access to system resources and kernel data structures. These functions should be independent of machine size and transparent from machine architecture. In other words, the presence of multiple processors should be hidden to provide portability and compatibility.

These are two very important issues in porting UNIX on computer systems for parallel or distributed processing. The kernel functions should be designed to be architecture-independent with the help of special hardware and user interfaces. UNIX for parallel computers should be compatible with some existing UNIX system such as Berkeley BSD or Bell Laboratory's System V, among others.

Besides retaining uniprocessor UNIX system call interfaces, commands, and applications, multiprocessor UNIX should be designed to explore all opportunities for parallelism. These include using multiple threads of control in the kernel, supporting fast interprocessor synchronizations and self-scheduling, and allowing multitasking or multithreading at various MIMD grain levels.

12.1.3   Address Space and Load Balancing

The address space of modern computer systems has doubled in size every 2 years during the last two decades. Based on this trend, a 64-bit address space will be needed by 1997. UNIX extensions for future parallel computers must satisfy this demand.

Forming a single address space for all related processes is a necessity for a multiprocessor under the control of an integrated UNIX system. The address spaces for processes created in unrelated tasks can still be disjoint. This will enable multitasking in a heterogeneous system.

Codes originally written for uniprocessor systems must be enhanced to prevent races and deadlocks in a multiprocessor system. Different approaches in adapting UNIX for a multiprocessor may trade off performance and complexity in implementing shared address space.

The problems of deadlock prevention, avoidance of race conditions, and achieving dynamic load balancing are much more difficult to solve on a multiprocessor system than on a uniprocessor system. System deadlocks can be prevented, avoided, or recovered.

Load balancing requires monitoring of the load index on each processor, applies process migration, and achieves an evenly distributed or executed workload. Load balancing affects the effective or sustained performance of parallel computers.

At present, most load balancing is done statically at algorithm design time, at program time, or at compile time. Dynamic load balancing through OS intervention has a long way to go, and UNIX designers cannot ignore this issue.

12.1.4   Parallel I/O and Network Services

Parallel I/O and ease of network access must be further upgraded in future UNIX systems. Heterogeneous processing is an unavoidable trend. High-end supercomputers or MPP systems can be accessed only through a network for ordinary users. This has been set as one of the goals of the HPCC program.

A multiprocessor UNIX should be able to support parallel I/O activities, share virtual memory in addition to sharing physical memory, facilitate better network messaging, allow distributed program debugging, and provide a large body of UNIX libraries and utilities for multiprocessing or multicomputing applications.

These design goals will be reexamined as we proceed with the study of various UNIX extensions for multiprocessors and multicomputers. For special application domains, such as real-time transaction processing and fault-tolerant computing, one may require additional OS features and eliminate some of the unnecessary features.

The extension of UNIX has been focused on general-purpose parallel computers. The trend toward customization of UNIX for dedicated applications on parallel computers is increasing in business and military areas, in addition to the strong interest in the academic and scientific communities.

12.2   Master-Slave and Multithreaded UNIX

Three multiprocessor UNIX variations are studied below. In order of increasing implementation complexity and performance capability, we present a *master-slave kernel* using a fixed processor to handle all system calls, and then a *floating-executive kernel* using giant locks to switch the master among several processors.

A *multithreaded kernel* is the most difficult one to implement. A higher degree of parallelism and a better performance are expected in a multiprocessor running multithreaded UNIX/OS. Both fixed-master and floating-master kernels have been developed in existing computers and the demand for multithreaded kernels is on the rise.

12.2.1   Master-Slave Kernels

This is the simplest approach to adapting UNIX for a multiprocessor system. Only one processor is designated as the *master*, which runs the kernel code and handles all system calls and interrupts. The remaining processors in the system are *slaves* running only user codes. When a slave process requests a system call, the process is context-switched to the master.

Single Fixed Master   This is practically a uniprocessor model from the viewpoint of the OS kernel, as illustrated in Fig. 12.3. A *single master kernel* is the most economical OS to implement and to debug on top of the conventional UNIX porting base.

A process in the conventional UNIX system switches among four possible states. When executing in *user mode*, a process can be preempted for various reasons. It enters the *kernel mode* upon a system call or an interrupt. However, the UNIX kernel can handle only one process at a time. In other words, a process running in kernel mode cannot be preempted by another process.

Thus, the kernel is nonpreemptive and maintains the consistency of its data structures through mutual exclusion. When a system call cannot be serviced, the kernel allows a *context switch* to put the requesting process into a sleep wait in memory. Sleeping processes do not consume CPU cycles. A suspended process is later woken up and enters a *ready-to-run* state in which it is not executing but is ready to be selected by the scheduler and to reenter the kernel mode.

Implementation Issues   In implementing a master-slave kernel for a multiprocessor, only critical resources are tied to the *run queue*. In this case, the master selects and runs any selected process. A traditional sleep/wakeup mechanism is used by the master. Slaves can run processes only in user mode. They return processes to the run queue when they execute a system call, take a trap, or are interrupted, or when the time slice expires.

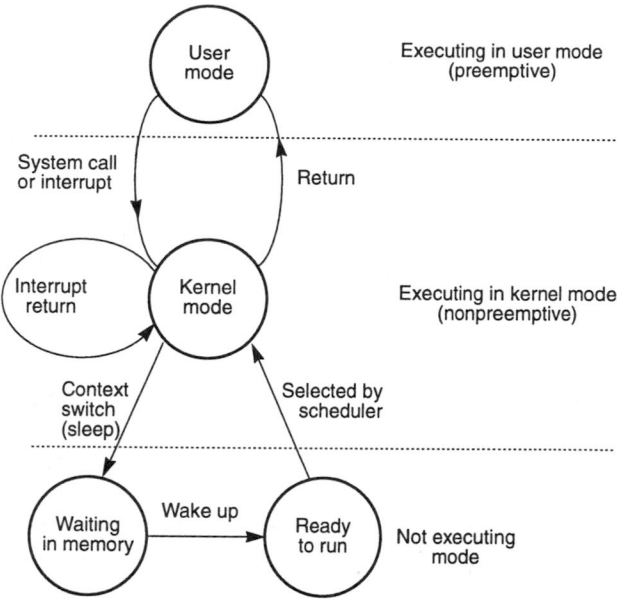

Figure 12.3 The life cycle of a UNIX process managed by a single master operating system kernel.

The run queue must be protected (locked) to prevent simultaneous updates. Often spin locks are used to achieve protected access for the run queue. In a multiprocessor with a master-slave kernel, the kernel should avoid races among multiple slaves in the sleep/wakeup process.

In protecting the run queue, the system cannot sleep or context-switch while holding a spin lock. Spin locks are efficient only when resources are held briefly. The entire system can be brought to a standstill if a critical spin lock is held for too long. The run queue is usually implemented as an unordered, linked list of ready-to-run processes.

The scheduler should select the highest-priority process for execution. The run queue is held locked during the search process in order to prevent races caused by simultaneous updates. The lock is also held while new processes are added to the list. When a user process running on a slave processor executes a system call such as

```
systrap()
{
  setrq(u.u_proc);
  context-switch();
}
```

each processor handles its own interrupts using its own clock hardware. The slave clock interrupt handlers keep track of the time slice for the current process and any pending signals. The master clock interrupt does the rest of the UNIX chores. All I/O interrupts are handled by the master, not by the slaves.

Example 12.1   The Purdue master-slave UNIX kernel running on a VAX/780 dual-processor system

An example of master-slave UNIX was developed for a dual-processor VAX/780 system by George Goble (1979) at Purdue University. In this case, all UNIX kernel code ran on the master while user code ran on the master or the slave. The master simply treated the slave as a schedulable resource.

System throughput increased to the amount of user code the system had available to run. Dynamic load balancing (Hwang et al., 1982b) was also implemented at Purdue University on a UNIX-based network of VAX computers.   ∎

Performance Issues   Although the master-slave kernel is simple and economic to implement, the main problem comes from the fact that the master becomes the bottleneck for the entire multiprocessor. Most UNIX programs spend half the time in kernel mode.

Since there is only one processor designated as the master, the bottleneck problem prevents the system from having a scalable performance as the number of slaves increases. Diminishing returns are expected as the multiprocessor system adds more processors. However, the basic master-slave model can be modified to alleviate the bottleneck problem to some extent, as revealed in the next two sections.

Having a single master simplifies the control table conflict problem. The entire system is subject to catastrophic failure that requires operator intervention to restart when the master-designate fails or commits an irrecoverable error. Idle time on the slaves can build up, and slaves may be underutilized if the master cannot cope with the bottleneck problem in handling all system calls.

For special applications where the workload is well defined, or for an asymmetric multiprocessor in which the slaves are much less capable, the single-master model is the logical choice. In fact, most commercial multiprocessors start with the construction of a master-slave OS configuration before the kernel is upgraded to yield a better performance or higher reliability.

12.2.2   Floating-Executive Kernels

The basic master-slave model can be enhanced by allowing the master to float among several processors capable of executing kernel functions. *Giant locks* must be used to implement the floating master at three ascending levels of sophistication.

The first approach implements a *single floating master*. The second approach, called *single master with assistants*, locks only critical sections and leaves some noncritical system calls unlocked. The third approach allows *multiple masters* to coexist using multiple giant locks. These floating-executive approaches are illustrated in Fig. 12.4 and described below separately.

Single Floating Master   A giant lock is used to protect the entire OS kernel, resulting in a single master floating among several master-capable processors (Fig. 12.4a). Upon a system call, an executive-capable processor must first acquire the lock on entry to the kernel. If the lock is already taken, the processor can run another user-mode process. Once the lock is acquired, this processor becomes the *master*.

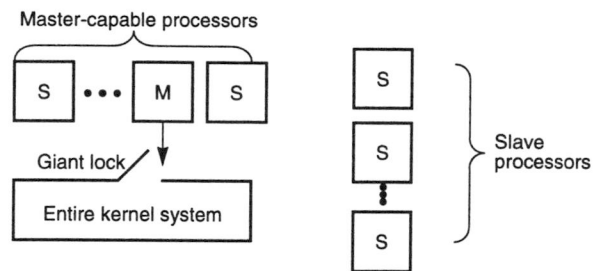

(a) Single floating master

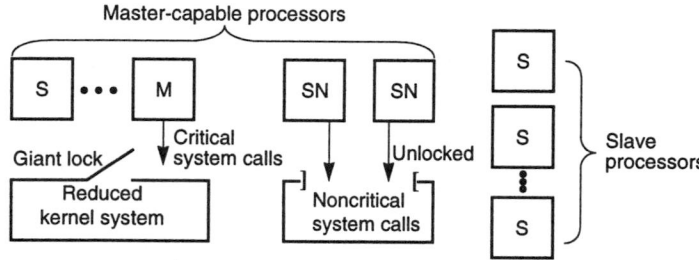

(b) Single master with assistants

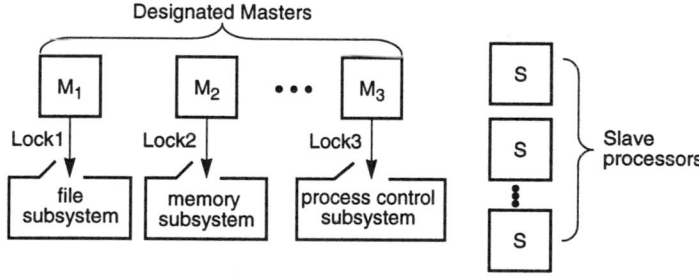

(c) Multiple masters executing different subsystem kernel functions

Figure 12.4 Three floating-executive models for extending UNIX for multiprocessors (M: master-designate; S: slave processor running user code; SN: slave running noncritical system calls; M$_i$: master for the ith kernel subsystem).

Other processors continue executing user-mode codes as slaves. The present master releases the lock on exit from the kernel. Note that not every processor can be the master; only those connected with I/O and having access to the kernel code can serve in this capacity. The idea of moving the master among several processors results in nearly the same poor performance as in the fixed master-slave model. Moving the bottleneck around does not solve the problem.

The major advantage of this approach is achieving better reliability because the executive can be switched around. Whenever the present master fails to function properly, the system can remove it from being in charge. Thus floating-executive UNIX may enhance the fault-tolerance capability of a multiprocessor system.

The kernel lock prevents context switching to the master. If the lock is taken, then context switching may not take place. This may also degrade the system performance if it is not properly designed.

Single Master with Assistants   Instead of locking the entire kernel on one processor, multiprocessor UNIX can be designed to lock critical system calls to the master and leave the remaining noncritical system calls, such as *getid*, *time*, *alarm*, *nice*, etc., unlocked and executed on the slaves (Fig. 12.4b).

Still, one giant lock is used to install the master on a floating basis. This approach will eliminate part of the bottleneck and alleviate the burden on the master to some extent. However, heavy-duty system calls must still be locked to the master. The bottleneck problem cannot be fully resolved in this model.

Multiple Cooperative Masters   In this OS model, multiple giant locks are used to lock major kernel subsystems separately on different processors (Fig. 12.4c). Noncritical system calls run on slaves, as in the assisted master model. The subsystems separately handle different system calls related to file management, memory management, process management, etc.

A lock hierarchy must be established to cope with an ordering problem for system calls affecting different subsystems. This model allows some parallelism by using multiple masters, which will enhance the system performance.

This approach essentially replaces one bottleneck by a handful of smaller bottlenecks. Therefore, the improvement in performance is still limited.

The above floating-executive kernels provide enhanced reliability and flexibility in supporting general-purpose applications. The workload is better balanced among multiple processors. Conflicts in servicing system calls must be resolved with sophisticated and dynamic control of the lock hierarchy.

Most of the kernel code and shared data structures must be reentrant since multiple masters may coexist. Table-access conflicts and lock-out delays may take place in a floating-executive system. The divided kernel subsystems must be protected to maintain the integrity of the shared objects.

Performance Issues   Floating-executive kernel should result in a better performance than fixed-master kernel. The switching of kernel functions from one processor to another may create some overhead, which can degrade the expected performance.

A multiprocessor running a floating-executive UNIX kernel has the advantage of providing graceful degradation and better availability in a reduced-capacity system in case of processor failure.

Most multiprocessor OS designers started with implementing a fixed master-slave kernel to gain experience before meeting the challenge of implementing the floating-executive kernel in a UNIX environment. Examples include the UNIX/370 developed by Bell Laboratories to run UNIX on some multiprocessor models of the IBM 370 mainframe.

Example 12.2   The IX/370 OS developed for an IBM multiprocessor

UNIX/370 was the first system to protect kernel data structures using semaphores. This allows the kernel to run on more than one processor, thus reducing the system bottleneck problem.

The first version of UNIX/370 was created as a port of UNIX layered on the kernel of an IBM OS called TSS. That kernel, called SSS, was used to interface UNIX with the hardware. This was the basis for the system shipped by IBM known as IX/370 (Fig. 12.5). Some of the IX/370 features were also built into some multiprocessor kernels for IBM/390 mainframes.

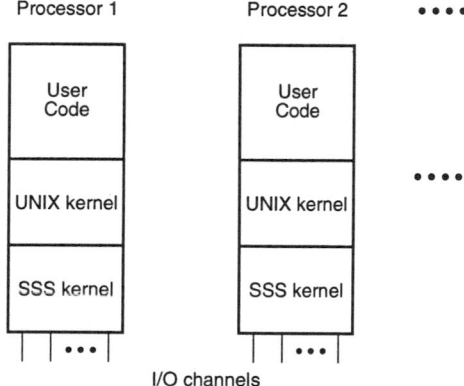

Figure 12.5 Layered development of the IX/370, a UNIX extension for some IBM 370 multiprocessor models.

By alleviating the kernel bottleneck problem, enhancement of system performance was also observed in multiprocessors built by DEC, CDC, Sequent, and Alliant.

12.2.3   Multithreaded UNIX Kernel

A *thread* is a computational entity capable of being executed alone. Threads are lightweight processes with a necessary minimal state. The minimal state includes the processor state and the contents of relevant registers.

On a multiprocessor, multiple threads can execute in parallel on separate processors. This implies that threads assigned to different processors should be unrelated and that all threads assigned to the same processor may share resources.

The scheduling of multiple threads on a set of available processors is called *gang-scheduling*. To implement a *multithreaded kernel*, spin locks or semaphores are used to separate multiple threads with different sets of kernel resources.

The gang-scheduling concept and multithreading implementation techniques are studied below. A case study on multithreading in the Mach and OSF/1 kernels will be discussed in Section 12.4.

Advantages and Challenges   Multithreading is a further refinement of the floating-executive concept. It is meant to maximize the degree of parallelism in kernel operations. Unrelated threads can run on different processors without resource contention, and CPU utilization should be greatly enhanced.

System performance is expected to scale up when a multiprocessor increases in size. A multithreaded kernel should optimize the cost/performance ratio of processor resources. Creating all these advantages will require extensive developmental effort.

It may take many years of team effort to develop an efficient multithreaded kernel. The main difficulties lie in determining how to separate multiple threads of control and how to debug them through program trace and performance tuning.

At present, the computer community has only limited experience with these challenges. The degree of multithreading in an extended UNIX kernel is limited by UNIX semantics and by the ability to map virtual processes to physical processors.

The concept of multithreading is tied to multitasking in Mach/OS kernel development. We study multithreading for a single task in this section. Multithreaded multitasking will be discussed later. Multithreading must be developed with low overhead.

Some superscalar processors allow multiple instruction threads within the same CPU. For clarity, we restrict this study to assigning one thread to each processor at a time. Multiple threads within a single superscalar processor demand extensive compilation support.

Gang Scheduling   The idea of using giant locks to separate the functions of kernel subsystems can be continually expanded until the kernel is fully multithreaded to handle many lightweight processes simultaneously.

Understanding the concept of gang- scheduling will help readers appreciate the mechanisms used for resource locking. For simplicity in illustration, we concentrate on a single task consisting of multiple threads corresponding to the demands of a single application, as demonstrated in Fig. 12.6. Each thread is assigned a priority and a scheduling policy.

The primary data structure used by the scheduler is the *run queue*, a priority queue of runnable threads implemented by an array of doubly linked queues. Programming models for applications can introduce concurrency beyond the hardware parallelism at two levels as shown in Fig. 12.6b.

Independently schedulable threads are formed from an application supported by a

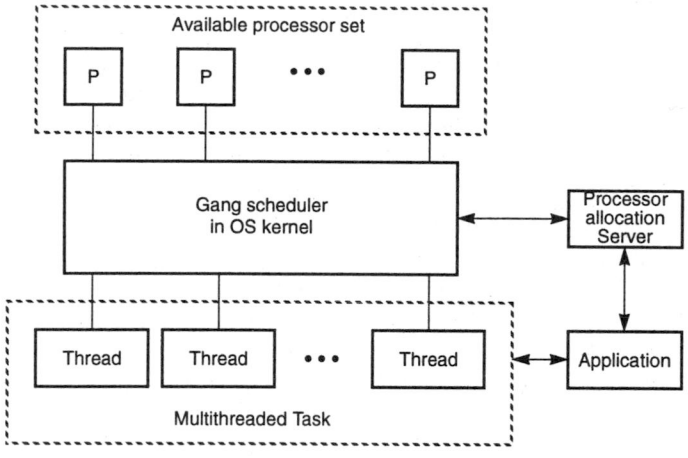

(a) Gang-scheduling facility

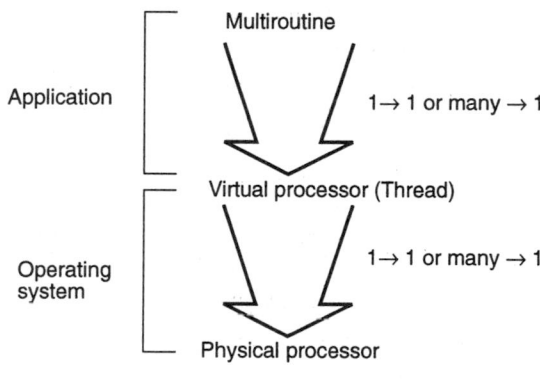

(b) Two levels of parallelism

Figure 12.6 Gang-scheduling of multiple threads using virtual processors.

user library or language directives. These threads are generally known as *virtual processors*. The OS kernel multiplexes the threads onto *physical processors*. The processor allocation in a gang-scheduling kernel is supported by the facility shown in Fig. 12.6a.

The overhead involved in gang-scheduling can be significant, so processors are often dedicated for periods longer than the typical time-sharing quantum to amortize this overhead.

The *processor allocation server* is a facility for implementing an allocation policy. The server has direct control over the processors required by the application. The *available processor set* is assigned to multiple threads dynamically. The application creates the processor set, and a program will execute only threads belonging to the same processor set.

Detailed thread management will be treated later along with the Mach kernel. A kernel performance monitor may be needed to tune the system performance. The gang-scheduler can run on any available processor or be distributed on several processors.

Multithreaded Kernel   Some UNIX kernel data structures and process regions have well-defined locking needs. An example is the internal representation of a file given by an *inode*, which contains a description of the disk layout for the file data, file owner, access permissions, and access times. These restrictions must be properly locked.

Other examples include *buffers* and *regions*. On the other hand, many UNIX system calls, such as *signal* interaction with *pause* and *sleep* or parent-child synchronization during *wait-exit*, do not have well-defined locking needs and thus are inappropriate for multithreading.

A multithreaded kernel uses implicit mutual exclusion based on the nonpreemptive rule used in a uniprocessor kernel. These protected accesses to critical sections must be properly locked. Furthermore, *sleep-wakeup* is inappropriate for a multiprocessor which does not allow for ordered wakeup.

In order to eliminate races with sleep-wakeup, a multithreaded kernel demands *spin-locked sleeps*. Any *sleep routine* must be enhanced to release a lock once a thread is asleep and to reacquire the lock when it is awakened. Spin locks should be used to protect sleep queues. This will result in multithreading with minimal code changes.

Example 12.3 Locking needs in the UNIX system calls *plock* **and** *prele*

We show the use of the system calls *plock* and *prele* in implementing spin-locked sleeps. Not all the locking needs of the UNIX kernel are as apparent as in *plock-prele* operations. The plock and prele operations are shown below:

```
Plock(ip)
struct inode *ip;
{
    acquire(ip -> i_lock);
    while (ip -> i_flags & ILOCK)
    {
        ip -> i_flags |= IWANT;
        sleep (ip, PID);
    }
    ip -> i_flags |= ILOCK;
    release(ip -> i_lock);
}

Prele(ip)
struct inode *ip;
{
    acquire(ip -> i_lock);
```

```
            ip -> i_flags &= ~ILOCK;
            if (ip -> i_flags & IWANT)
            {
                  ip -> i_flags &= ~IWANT;
                  wakeup (ip);
            }
            release(ip -> i_lock);
      }
```

The uniprocessor kernel thinks it is nonpreemptive. More locks and/or more sleeps must be added. Waking up all sleeping processes is not a problem with uniprocessors, since the lock is generally free by the time others are ready to run.

On a multiprocessor, all awakened processes may run at once, or one is locked and others go back to sleep. *System thrashing* is a concern in implementing a multithreaded kernel using spin-locked sleeps.

Lock Granularity and Convoys   The granularity of locks will affect the performance. Coarse granularity occurs with the use of giant locks in a master-slave OS kernel. Giant locks are not efficient in handling lightweight threads.

Finer granularity incurs extra overhead for locking primitives. It also requires more memory space for locks. The problem of *convoys* may demand the parallelization of critical sections and the use of sequential locking.

Example 12.4   Sequential locking in solving the convoy problem

When threads are formed behind a coarse-grained lock, each thread must follow the others sequentially through the lock as illustrated in Fig. 12.7.

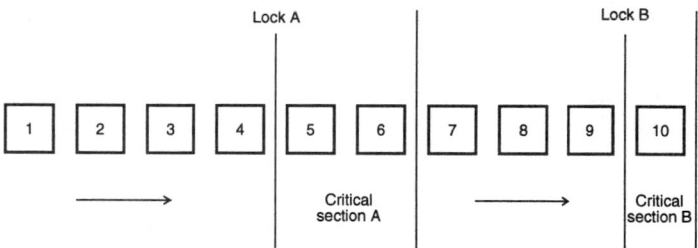

Figure 12.7 Convoys flowing through sequential locking from user mode to critical sections in the kernel. (Courtesy of Curt Schimmel, 1990)

The convoys are facing a bottleneck problem, which must be eliminated to guarantee the throughput rate. Adding more sequential locks certainly cannot help alleviate the bottleneck before the critical section. One way to resolve the problem is to parallelize the critical section, which is in keeping with the spirit of multithreading.

The convoys can extend all the way back from the user mode in order to improve the performance.

■

In general, the performance must be measured. Statistics should be gathered and analyzed. Code modification may be needed in optimizing the performance of a multithreaded kernel. Remeasurement of performance is then required. This performance tuning process may be very time-consuming, but the results will make a big difference in performance. Process migration is another way to achieve balanced multithreading.

Semaphored Kernels Instead of using spin locks, one can also use *semaphores* to implement multithreading. Spin locks are usually cheaper to apply and perform well in cases where resources are held for short durations. Semaphores are more expensive than spin locks.

Semaphores work for any number of processes or processors. In building a semaphored kernel, one must watch out for deadlocks and lock ordering problems. Again, performance tuning is required during the development period.

A semaphored kernel essentially replaces all occurrences of sleep-wakeup with appropriate *P-V* operations. The granularity of locks must be decided. For example, the *inode* requires individual data structure locks, and giant locks apply to much larger kernel areas.

Semaphored locking for multithreading offers better control of lock granularity. This may lead to a higher performance than when using spin-locked sleeps. The semaphore approach is also better if an event may take longer than the context-switching time to occur. For this reason, one cannot use semaphores in interrupt handlers.

Summary The discussion of master-slave, floating-executive, and multithreaded kernels in the above can be summarized as follows. The type of multiprocessor UNIX kernel chosen is determined by considering the performance level and the implementation cost compared to conventional UNIX.

Multithreaded kernels are a stripped-down version of floating-master kernels. The amount of work put into kernel implementation and tuning experiments determines the real performance. A patched-up uniprocessor kernel implementation may run on a multiprocessor but may not be able to deliver a high performance.

Standard UNIX interface must work with multiprocessor UNIX. This compatibility is a necessity; otherwise, all the advantages may disappear. The extended UNIX kernel for multiprocessors should hide the implementation from the user. In conclusion, we have to build on top of an existing UNIX system instead of away from it. This will become evident as we proceed in subsequent sections.

12.3 Multicomputer UNIX Extensions

In this section, we study a multicomputer OS for building future UNIX-compatible systems. These UNIX extensions are based on message passing for IPC. We review first the host/OS and then the node/OS kernels based on different message-passing schemes and programming environments.

In a multicomputer with distributed memory, the OS functions are distributed between the host computer at the front end and the node processors at the back end. Message passing must be supported by both front-end and back-end environments.

12.3.1 Message-Passing OS Models

From the perspective of a distributed operating system, message passing can be implemented with three distinct approaches as specified below.

(1) *Object-oriented model.* This model treats an arbitrary number of tasks (processes) concurrently, each task using its own address space (memory object). Multitasking is practiced on each node.

Intertask communication across the nodes is carried out by either a synchronous or an asynchronous message-passing protocol. The user is not aware of the number of nodes or even of their existence as separable entities. Message passing supports IPC as well as process migration for load balancing purposes. The Mach/OS kernel is based on this approach.

(2) *Node-addressed model.* In this case, only one task is running on each node at a time. Message passing is often implemented with an asynchronous scheme. The actual topology of the machine is hidden from the user. Most commercial multicomputers use this approach.

The early models used store-and-forward message routing, and more recent models use wormhole routing, which provides rapid IPC without interrupting intermediate node processors.

Caltech's *Cosmic Environment* and *Reactive Kernel* support multiple-process message-passing C programs on a UNIX host, based on a node-addressed model.

(3) *Channel-addressed model.* Again, this model runs one task on each node. Data are communicated by synchronous message passing on communication channels. The user must be aware of hardware communication topology.

The *Occam environment* is based on channel addressing. Caltech's CrOS III is a high-performance channel-addressed communication environment. It is also available commercially as *Express* from the Parasoft Corporation.

12.3.2 Cosmic Environment and Reactive Kernel

The *Cosmic Environment* (CE) and *Reactive Kernel* (RK) were developed at Caltech to support a message-passing C programming environment on network hosts and multicomputer nodes, respectively.

Network UNIX Hosts   The CE system runs on a network host. Its major function is to support uniform communication between the UNIX host and node processes. A set of daemon processes, utility programs, and libraries constitute the CE system.

Besides being able to interface with a single multicomputer, the CE can be used as a stand-alone system for running message-passing C programs on collections of network-connected UNIX hosts. It can also be used to handle the allocation of, and interfaces with, more than one multicomputer as depicted in Fig. 12.8.

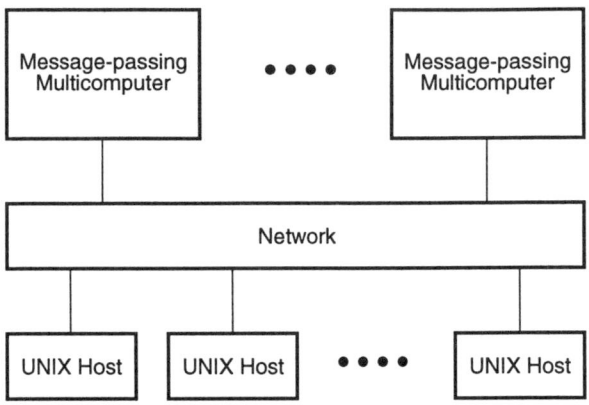

Figure 12.8 A network of UNIX hosts and several multicomputers.

Node Operating System   The RK is a node operating system for a multicomputer. It was originally developed as the *Cosmic kernel* for first-generation Cosmic Cubes and later modified and ported to second-generation multicomputers like the Symult Series 2010. The same C programming environment is also supported on Intel iPSC/1 and iPSC/2 multicomputers with an NX node OS through the use of a compatibility library.

The difference between running message-passing C programs on UNIX hosts under the CE system RK and on multicomputer nodes is in their process creation functions. Every node runs a copy of the RK system. It supports multiprogramming within a single node. Based on 1991 standards, message passing is about 100 times faster in a multicomputer network than in a local area network.

The CE provides the same message-passing functions for UNIX processes that the RK provides for node processes. Thus the CE and the RK together provide an integrated programming environment in which uniform communication between host and node processes can be achieved independently of the physical allocation of these processes.

The entire set of processes created for a given computation is called a *process group*. The allocation of a process group and a set of multicomputer nodes is controlled by the user with a host facility called *getcube* in first-generation binary n-cube multicomputers.

The n-cube nodes are shared by recursive partitioning into subcubes of lower dimensions. A subcube is specified by its "cube" size as characterized by its dimension, $n = \log_2 N$, where N is the total number of nodes involved in subcube operations.

Space-Sharing Systems   A multicomputer is normally *space-shared* rather than *time-shared* as in a multiprocessor. Space sharing allows a user to select the number of nodes that is appropriate to the size of a process group matching the load balancing requirement of a given computation. For example, a 128-node iPSC 7-cube can be assigned as follows: 64 nodes (a 6-cube) to one user running a computation, 32 nodes (a 5-cube) to another user, and the remaining 32 nodes (a 5-cube) free.

Mesh-connected multicomputers like the Symult Series 2010 indicate the number of nodes needed rather than the dimensions of a binary cube. Every node process in a process group is identified with a unique ID = (node, pid). A host process is identified by the HOST constant and a unique ID maintained by the host run-time system.

The CE/RK environment supports dynamic process spawning, placement and control, and the sending and receiving of messages in either a blocking or nonblocking fashion. Other functions supported include node allocation, process identification, message broadcast, performance measurement, server interfacing, and process termination.

These functions provide a low-level portable environment which can be used as a compilation target for higher-level concurrent programming languages. Other host utilities include those for process kill, process status, load average, C compiler, archiver, debugger, etc.

A reactive handler supports message sending, receiving, queueing, discarding, and acknowledgment, as well as code or data creation, initialization, and spawning of new handlers. Standard CE/RK Fortran interface routines have also been developed in the environment.

12.3.3   Intel NX/2 Kernel and Extensions

The iPSC/860 is a hypercube multicomputer using i860 microprocessors as node processors. The front-end host runs UNIX V with a Sun remote host and a VMS link. The node OS is the NX/2 kernel with a message-passing library.

The NX/2 Kernel   NX/2 message-passing services include *synchronous* (*csend*, *crecv*), *asynchronous* (*isend*, *irecv*, *msgdone*), and *interrupt-driven* (*hrecv*) operations.

Synchronous operations are used in simple and initial application development, which demands explicit process synchronization. Asynchronous operations allow computation to overlap communication, help keep system message buffers clear, and speed up communications.

Interrupt-driven operations avoid blocking or polling of incoming messages and allow them to be processed as soon as they arrive. The interrupt feature is especially useful in applications where message arrival time is unpredictable.

The NX/2 supports full end-to-end message routing with flow control and buffer management. Message types are distinguishable with receive buffers. The NX/2 features greater utilization, a better I/O performance, and higher reliability due to simplicity and lock-up-free, easy-to-use utilities, compared with the earlier hypercube systems iPSC/1 and iPSC/2.

A UNIX-compatible application interface is available to support a concurrent file system using standard Fortran and C I/O statements. The I/O subsystem is faster, with parallel access to multiple disk drives. Disk caches are available on both computer nodes and I/O nodes.

OS in the Delta and Paragon Systems   The Intel-Darpa Touchstone Project is targeted to achieve multicomputing with more than 100-Gflops performance. The Paragon demands an even higher performance. We have reviewed the software environment for the Paragon in Section 11.1.2. Let us further examine the OS requirements in these message-based systems.

Both the Delta and Paragon feature the use of flat interconnects, fat nodes, stock processors, and commodity disks. The system supports MIMD and SPMD programming models with *send-receive* messages but no shared resources. The software design philosophy of the Delta and Paragon systems emphasizes the use of lightweight operating systems with layered support for specific programming models, applications, and tools.

The purpose is to minimize overhead for messages and processes. High-level services are distributed via service nodes or across a LAN. The emphasis is on UNIX compatibility with low space or time overhead. Several alternative OS kernels are being considered, including the Mach, Reactive Kernel, and OSF/1. The tradeoffs lie in efficiency versus functional capability. A scalable and reliable parallel file system is very much in demand.

The OS also considers shared virtual memory using a paged or object-oriented approach. Fortran or C/C++ languages, interactive parallelization tools, multilanguage debugging, real-time parallel program analysis, and environment integration are other goals set for Paragon software development.

12.4   Mach/OS Kernel Architecture

The Mach/OS kernel provides task/thread management, multithreaded multitasking, and IPC with messages via access ports. We study below Mach scheduling techniques, including the Mach scheduler, context switching, thread priorities, and concurrency support. Mach virtual memory implementation and management are also described.

Throughout the discussion, we consider the factors contributing to the portability of Mach on various parallel architectures. Development of the Mach/OS is still in progress. Performance statistics are gradually being released as more computers are being ported with Mach or with variations of Mach such as the OSF/1.

Besides studying the original design specifications of the Mach/OS kernel developed by the CMU design team, we also examine some of the Mach implementation features on the NeXT Computer and on BBN Butterfly processors.

UNIX compatibility makes Mach attractive to a wide audience. This enables it to transcend its role as a research project and emerge as a viable commercial OS for future UNIX-based computers consisting of multiple processors either tightly or loosely coupled.

12.4.1 Mach/OS Kernel Functions

While the computer hardware has changed radically since it was introduced 20 years ago, the single thread of control in the UNIX kernel has remained the same. Mach is a multiprocessor OS kernel developed at Carnegie-Mellon University. It was designed with the intention of integrating both distributed and multiprocessor functionality.

Successive versions of UNIX from computer manufacturers have increased its size to exceed 2 Mbytes. Mach provides a stripped-down approach to modifying UNIX for use in the network of uniprocessors and multiprocessors depicted in Fig. 12.9.

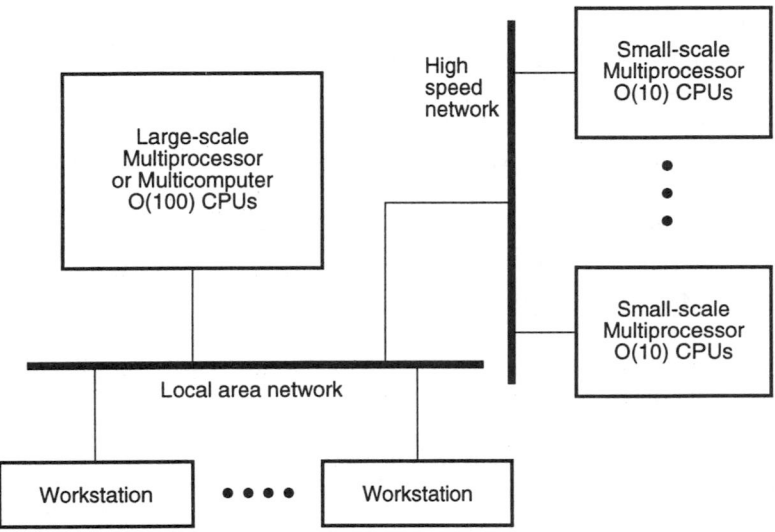

Figure 12.9 A typical Mach computing environment consisting of a network of uniprocessor workstations and multiprocessors of various sizes. (Courtesy of Accetta et al., Carnegie-Mellon University, 1986)

Mach designers introduced five program abstractions, *task*, *thread*, *port*, *message*, and *memory objects*, to be defined shortly. These have been improved from existing UNIX for centralized processing to parallel and distributed computing. The Mach kernel contains essentially three basic service functions: *processor scheduling*, *interprocess communication*, and *virtual memory management*.

All the others are *service tasks* which are treated as independent user-level programs. We introduce below the concept of multithreaded multitasking. The Mach implementations are presented based on the experience of CMU and computer manufacturers who have adapted Mach.

Conventional UNIX has been upgraded in Mach to support parallel processing on multiprocessors and multicomputers. The CMU design team included in Mach code-level compatibility with Berkeley UNIX 4.3 BSD.

Mach Design Goals   Mach has already been used as the basis of NeXT/OS and the UNIX-compatible OSF/1. Summarized below are some of the evolving goals for the Mach design.

- An object-oriented kernel foundation for building UNIX facilities and for development of a UNIX-like system in new computer architectures.
- Support for both tightly coupled multiprocessors and loosely coupled multicomputers by separating the process abstraction into *tasks* and *threads*, with the ability to execute multiple threads within a task simultaneously.
- A new virtual memory design which provides large, sparse virtual address spaces, *copy-on-write* virtual copy operations, *copy-on-write* and *read-write* memory sharing between tasks, memory-mapped files, and user-provided backing-store objects and pagers.
- A capability-based interprocess communication facility that transparently extends across network boundaries and integrates with a virtual memory system to allow the transfer of large amounts of data via *copy-on-write* techniques.
- Easy portability and compatibility with Berkeley UNIX 4.3 BSD plus a number of basic system support facilities for internal kernel debugging, remote file access, and language support for *remote procedure call* style interfaces between tasks written in C, Pascal, and Common Lisp.

The CMU Mach team has reduced the size of the kernel by moving many kernel operations up to the user level where they utilize the basic services provided by the kernel. With this design, the Mach kernel becomes modular, extensible, and easier to debug as shown in the following example.

Example 12.5   The CMU Mach/OS kernel and system architecture

Figure 12.10 shows the Mach system architecture. The kernel retains only the functions of port and message management, virtual memory management, and task/thread scheduling. Many other UNIX functions are handled by dedicated servers specified at the user level.

As illustrated, network messaging, external memory, processor allocation, and distributed memory are handled by separate servers. The system is UNIX-compatible with user processes serviced by both kernel and dedicated servers jointly and simultaneously.

■

12.4.2   Multithreaded Multitasking

We formally define below those program abstractions introduced with the Mach/OS. After the basic definitions, we describe parallelism exploited in a Mach-based system. Then we present a Mach scheduler from a NeXT computer implementation. Thread management and concurrency support issues are then discussed.

Tasks   A *task* defines the resources environment in which threads run. Thus the task

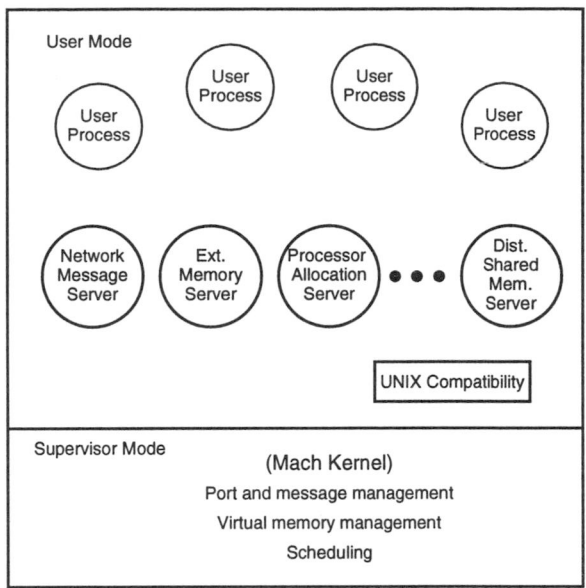

Figure 12.10 The Mach/OS kernel and various user-level servers with UNIX compatibility.

is a basic unit of resource allocation. A task includes protected access and control of all system resources, including CPUs, physical I/O ports, and either virtual or real memory.

A task address space uses a structured map of memory objects. The UNIX notion of a process is a task with a single thread of control. The task itself performs no computation; rather, it is a framework for running threads.

Threads A *thread* is the basic unit of CPU utilization. It is equivalent to a program stream with an independent program counter operating within a task. Usually, threads are lightweight processes that run within the environment defined by a task. A task may have multiple threads with all threads under the same task-sharing capabilities and resources.

As a matter of fact, the Mach kernel itself can be thought of as a task spawning multiple threads to do its work. Spawning of a new thread is much faster and more efficient than forking a traditional UNIX process because processes copy the parent's address space, whereas threads simply share the address space of the parent task. A traditional UNIX process can be thought of as a task with a single thread.

A thread can also be treated as an *object* capable of performing computations with low overhead and minimal state representation. In this sense, a task is generally a high-overhead object (much like a traditional UNIX process), whereas a thread is a relatively low-overhead object.

Thread Management   On a tightly coupled multiprocessor with shared memory, multiple threads with the same task may execute in parallel. Operations on tasks and threads are invoked by sending a message to a port representing the task or thread. Threads may be dynamically created, destroyed, suspended, and resumed.

The suspend and resume operations, when applied to a task, affect threads within that task. Furthermore, tasks may be effectively forked (created) and destroyed depending on the resource mapping strategy used.

Threads are independently scheduled by the Mach kernel while incurring only a modest overhead on the OS. Mach does not provide any explicit mechanisms for synchronizing threads that access shared data structures. However, user-level packages for synchronization may be constructed to exploit hardware synchronization features (e.g., interlocked memory access).

Tasks are related to one another in a tree structure by task creation operations. Regions of virtual memory may be marked as inheritable *read-write* or *copy-on-write* or not marked at all by future child tasks.

Parallelism Opportunities   The application of parallelism in Mach is achieved in any of three ways:

- *Shared-resource multithreading* — The creation of a single task with many threads of control executing in a shared-address space, using shared memory for communication and synchronization.

- *Partially shared-memory multitasking* — The creation of many tasks related by task creation, which share restricted regions of memory.

- *Message-passing multitasking* — The creation of many tasks (most likely on distributed processors) communicating via messages.

These alternatives in Mach's abstraction of tasks and threads make it possible to support various multiprocessor architectures (UMA, NUMA, and NORMA) — tightly coupled, loosely coupled, or networks connected through message passing.

Time-Sharing Schedulers   To understand the Mach scheduler, it is important to review some useful characteristics of a time-sharing scheduler in conventional UNIX. Such a scheduler allocates system resources so that competing processes receive approximately equal portions of CPU time. Each process is allocated a maximum execution time. The scheduler continually decreases a process's priority as it consumes CPU time and always favors higher-priority processes.

A drawback of this system is that on a heavily loaded system with many jobs with short execution times, the priority of a lengthy job can decrease until little or no further CPU time becomes available. To cope with this problem, it is necessary to periodically elevate permanently depressed priorities.

An *event-based elevation* technique favors interactive processes over computation-bound ones. A process' priority is increased through certain interactive events such as I/O completion.

Another elevation technique is based on expected usage functions that define the goals for processor usage based on the *elapsed time* of a job in the system. The goal function determines the time it takes for the job's current processor usage to reach the *critical time*. Each job's priority is the difference between the elapsed time and the critical time. The larger the difference, the higher the priority.

The Mach/OS implements both the time-sharing scheduler and other scheduling techniques which cater to parallelism. The scheduler described below is based on the experience of NeXT Computer, Inc.

Example 12.6 The Mach scheduler in the NeXT computer

With multiple threads in a task, the NeXT scheduler optimizes the context switch time. Figure 12.11 shows 14 possible scheduling states. There are three nested squares in this state transition diagram. The components of state names correspond to the state of the thread: *running* (R), *waiting* (W), *suspended* (S), and *swapped out* (O). The suffix *-proc* is added to states in which the thread is executing on a processor, and the suffix *-runq* indicates that the thread is queued to run.

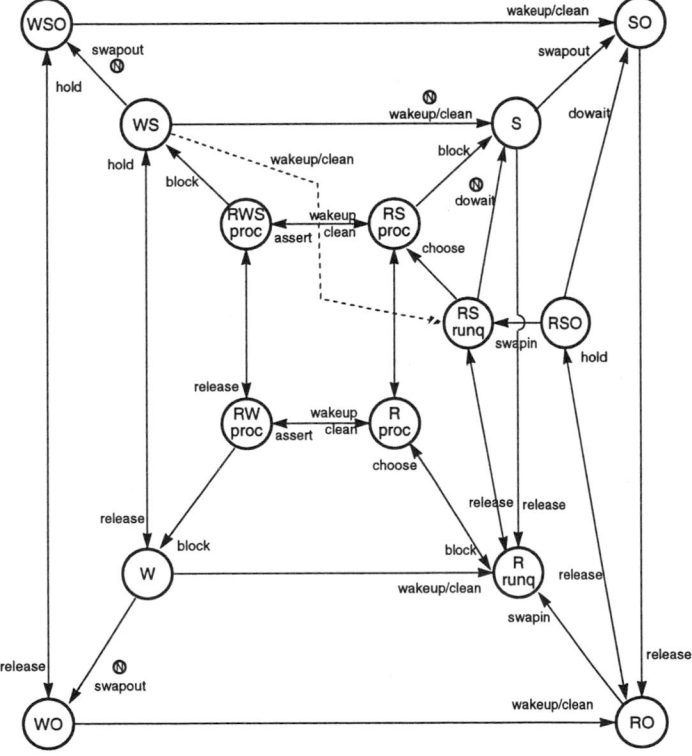

Figure 12.11 State transition diagram for a Mach scheduler implemented in a NeXT computer. (Courtesy of NeXT Computers, Inc., 1991)

The innermost square contains four states in which a thread is executing on a processor. The outermost square contains five states in which a thread is swapped out. In this state, the thread's kernel stack is not in the memory and must be forced into the memory before the thread can be executed.

The middle square contains five states in which a thread is swapped in (the kernel stack is memory-resident) but not actually running; such a thread may be waiting for an event, suspended by the kernel, or queued to run on a processor. Table 12.1 summarizes the scheduler states.

Table 12.1 The Mach Scheduler States

State	Description
R-proc	Running on a processor.
RS-proc	Running on a processor, suspend pending.
RW-proc	Running on a processor, event wait pending.
RWS-proc	Running on a processor, suspend and event wait pending.
R-runq	Queued to run.
RS-runq	Queued to run, suspend pending.
S	Suspended.
W	Waiting for an event.
WS	Waiting for an event and suspended if interruptible; Waiting with suspend pending if not interruptible.
RO	Swapped out with pending swapin.
SO	Suspended and swapped out.
WO	Waiting and swapped out.
WSO	Waiting, suspended, and swapped out.

Context switching is implemented in the Mach scheduler using the notion of a time quantum. A thread is assigned a time quantum at the beginning of its execution. Periodic clock interrupts decrement this time quantum.

A rescheduling check is done when the quantum expires. If there are other executable threads having higher priorities, then a context switch is made to the thread with the highest priority. Next, we discuss scheduling policies based on different thread priority schemes.

Thread Priorities   The run queue in the NeXT Mach scheduler is implemented with 32 queues corresponding to priority levels 0 to 31, where lower priorities have higher numbers. Each thread is assigned a priority and a policy.

The priority indicates the probability of the thread running, and a thread with a lower priority number is more likely to run. Whenever a thread times out or blocks, the thread with the highest priority will be run. In order to prevent starvation, priorities will be reduced when they age, thereby preventing a high-priority thread from keeping the processor(s) from a lower-priority thread. Gang-scheduling should be based on the fairness policy in a time-sharing scheduler.

Figure 12.12 shows a run queue with three threads. The kernel maintains two queues — a local run queue and a global run queue. The local queue is used by threads that have been bound to a specific processor.

When a new thread is needed for execution, each processor examines its local queue. If it is empty, the processor examines the global queue. Then it dequeues and runs the highest-priority thread.

If both queues are empty, the processor becomes idle. An idle processor is assigned a dummy *idle thread*. Idle processors are dispatched by a mechanism that bypasses the run queues.

Threads that become executable are matched with idle processors. This context-switching mechanism increases the performance by preventing an idle processor from polling the queue for new threads.

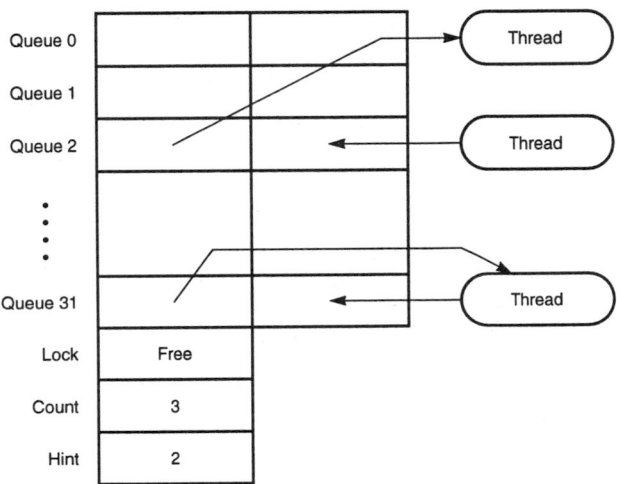

Figure 12.12 A run queue structure with 32 priority levels in the Mach scheduler. (Courtesy of David Black; reprinted with permission from *IEEE Computer*, May 1990)

Interactive or Fixed Scheduling   Besides time-sharing scheduling, there is another method called *interactive scheduling* which is optimized for threads requiring interactive responses. It is optimized to where it will be scheduled more often, so the priorities are probably set higher. This feature is yet to be implemented in real systems.

A third scheduling policy employs a *fixed priority* in which threads do not age like normal time-sharing threads. This feature may cause the starvation of lower-priority threads. Usually, a fixed-priority thread runs until one of the following events occurs:

- A higher-priority thread becomes ready.
- The allocated time quantum expires.
- The thread blocks.

Concurrency Support   Applications using more virtual than physical processors can benefit from user input to scheduling decisions. The user may have application-specific information (hints) about which virtual processors should or should not be running.

The NeXT Mach scheduler accepts user hints. These hints usually contain local information about a particular thread so that users can avoid maintaining overall status information on their applications. Mach recognizes two classes of hints:

- *Discouragement hints* — These hints suggest that the current thread should not run by giving up the processor to any other thread if possible, by temporarily depressing the priority of the current thread, or by blocking the current thread for a specified period.
- *Handoff hints* — These hints indicate that a specific thread should run instead of the current one.

A useful application of discouragement hints is in the implementation of shared-memory synchronization. The second class of hints hands off the processor to the specified thread, bypassing the internal scheduler mechanisms.

When a low-priority thread holds a lock needed by a higher-priority thread, handoff hints become very useful. The higher-priority thread can detect this situation and hand off the processor to the other thread so that the lock is released in due time. Scheduling hints benefit the performance in multiprocessor systems.

12.4.3   Message-Based Communications

Mach IPC abstractions include *messages*, *ports*, and *port sets*. These terms are defined below. Then we describe their use in achieving IPC with message passing in a Mach system.

Ports   A *port* is a communication channel — logically a queue for messages protected by the kernel. Ports are the reference objects for tasks, threads, and other objects. *Send* and *receive* are the fundamental primitive operations on ports. Application programs communicate with objects managed by the kernel and server tasks through the objects' ports.

This is the software counterpart of the communication ports on the hardware. An object is said to have *access right* to a port if it has dealings with that port. A port can be moved from object to object, just as a PC board and the cables connected to it can be moved from one machine to another.

Messages   *Messages* are used to communicate between objects or among many objects. A message typically consists of two parts: a fixed-size header and a variable-size body containing the message. One can also view a message as a typed collection of data objects used in communication between threads or tasks. Messages may be of any size and may contain pointers and typed capabilities for ports. A message can be large enough to transfer the entire address space of a task.

Port Sets   Port access rights can also be delivered through messages. A task can hold multiple access rights, *send* and/or *receive*, on ports. Multiple tasks can hold *send access* to a single port. But only one task can hold *receive access* at a time.

All threads within a task will have the same send and receive access rights to ports. It is also possible to have *port sets*, where a task can have either all or none of the access rights to a group of ports. Port sets must be mutually exclusive — a port can be in only one port set. Port sets cannot be sent through messages.

Protocols   Communication can take place only if the correct port access rights are granted. Each port is actually a message queue which holds a sent message until a thread dequeues it. When a port queue is filled, the initiating thread can do one of three things:

- Sender can block until the message is sent successfully.
- Sender can ask the kernel to hold it temporarily.
- Sender can give an error message.

There are two properties of the Mach communication facility that make it so extensible: *architecture independence* and *network transparency*. These two features enable the porting of Mach on all types of parallel computer architectures without modifying the kernel. Mach provides a consistent, object-oriented interface with all services through message passing.

Interprocess Communication   In traditional UNIX, IPC is done by *pipes*, *signals*, and *sockets*, while network communication is implemented by *interdomain sockets*. This has the disadvantage of using global machine-specific names with no location independence and no protection. Data is passed uninterpreted by the kernel as streams of bytes.

The Mach kernel provides a richer IPC interface which allows both synchronous and asynchronous communication. Synchronous IPC is implemented by *streams*, and asynchronous IPC by *remote procedure calls* (RPCs). The basic communication abstraction provides location independence, security, and data-type tagging.

When *send rights* are sent, they are duplicated at the sender and recipient. When a *receive right* is sent in a message, it is removed from the sender; when the message is received, the right is given to the receiver.

Example 12.7 A typical message exchange in Mach using access ports

Figure 12.13 shows a typical message exchange sequence. Task A sends a message to port P_2. Task A has send rights to P_2 and receive rights to P_1. Task B, having receive rights to P_2, receives the message sent by task A. Task B optionally replies by sending a message to P_1.

If port P_2 had been full, Task A would have had the option at the point of

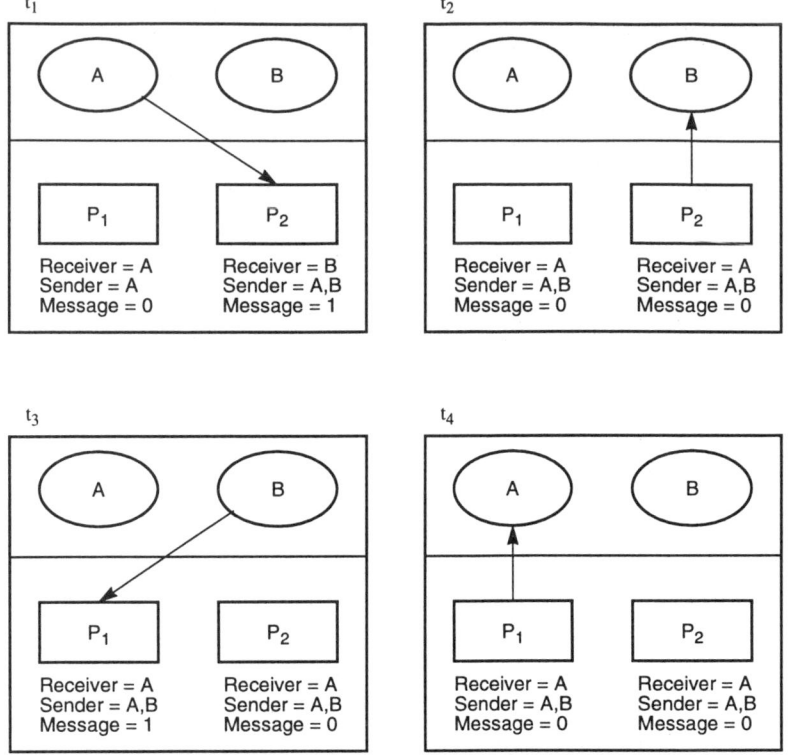

Legends: Tasks: A, B;   Ports: P_1, P_2,   Time instants: t_i.

Figure 12.13 A typical message exchange sequence using access ports (P_1 and P_2) between two tasks (A and B). (Courtesy of Accetta et al., Carnegie-Mellon University, 1986)

sending the message to: (1) be suspended until the port was no longer full; (2) have the message send operation return a port-full error code; or (3) have the kernel accept the message for future transmission to P_2.   ∎

The Mach kernel does not support IPC over the network; however, IPC can be transparently extended over the network by user-level tasks called *network servers*. A network server is essentially a representative of tasks on remote nodes.

When a network server receives a message from a task trying to send a message to a remote destination port, it maps the destination port into a destination network port identifier. The network server then obtains the destination address from the identifier and sends the message over the network. When the destination network server receives the message, it maps the network port identifier into a local destination port and forwards the message to its destination.

An IPC interface generates RPC code for communication between a client and a server process. The IPC interface is language-independent. A *Mach interface generator* is used to generate procedures in C to *pack* and *send* or *receive* and *unpack* the IPC messages.

12.4.4   Virtual Memory Management

The Mach virtual memory design provides functions for managing a large, sparse virtual address space. Each Mach task can use up to 4 Gbytes of virtual memory for the execution of its threads. A restriction imposed by Mach is that memory regions must be aligned on system page boundaries. The size of a system page is defined at boot time and can be any multiple of the hardware page size.

Task Memory Allocation   The *task address space* consists of a series of mappings of memory addresses to the task and its associated *memory objects*. The virtual memory design allows tasks to:

- Allocate or deallocate regions of virtual memory,
- Set protections on regions of virtual memory, and
- Specify the inheritance of regions of virtual memory.

Virtual Memory Operations   Table 12.2 lists some typical virtual memory operations that can be performed on a task. Mach supports *copy-on-write* virtual copy operations, and *copy-on-write* and *read-write* memory sharing between tasks. It also supports *memory-mapped files* and user-provided *backing-store objects and pagers*.

An example of Mach's virtual memory operations is the *fork* command in UNIX. When a *fork* is invoked, the child's address map is created based on the parent's address map's inheritance values. Inheritance may be specified as *shared, copy,* or *none* and may be specified on a per-page basis.

Shared pages are for *read* and *write* access by both the parent and the child. Copy pages are marked for copying onto the child's map. When a *write* request is made to the marked address, it is copied into the child's address space. A none page is not passed to the child at all.

Memory Mapping Mechanisms   The Mach virtual memory management hardware has been implemented with minimal machine architecture dependence. The machine-dependent portion of the virtual memory subsystem consists of a single code module and its related header file.

The machine-dependent data structures contain mappings necessary for the execution of a specific program on a particular machine. Most other virtual memory management functions are maintained in machine-independent data structures. Several important data structures used are specified below.

- *Address maps* — A doubly linked list of map entries, each of which describes a mapping from a range of addresses to a region of virtual memory. There is a single address map associated with each task.

Table 12.2 Virtual Memory Operations in Mach/OS

---

vm_allocate(task_address,size,anywhere)
Allocate and fill with zeros new virtual memory either anywhere or at a specified address.

vm_copy(target_task,source_address,count,dest_address)
Virtually copy a range of memory from one address to another.

vm_deallocate(target_task,address,size)
Deallocate a range of address — make them no longer valid.

vm_inherit(target_task,address,size,new_inheritance)
Set the inheritance attributes of an address range.

vm_protect(target_task,address,size,set_maximum,new_protection)
Set the protection attributes of an address range.

vm_read(target_task,address,size,data,data_count)
Read the contents of a region of a task's address space.

vm_regions(target_task,address,size,elements,elements_count)
Return description of a specified region of a task's address space.

vm_statistics(target_task,vm_state)
Return statistics about the use of memory by target_task.

vm_write(target_task,address,count,data,data_count)
Write the contents of a region of a task's address space.

- *Sharing maps* — A special address map describing a region of virtual memory that is shared between tasks. A sharing map provides a level of indirection from address maps to backing-store objects.
- *Memory objects* — A unit of secondary storage managed by the kernel or the user task. A virtual memory object specifies resident pages as well as where to find nonresident pages. Memory objects are pointed to by address maps. *Shadow objects* are used to hold pages that have been copied after a *copy-on-write* fault.
- *Resident page table* — A table used to keep track of information about machine-independent pages residing in the main memory.

Figure 12.14a illustrates the mappings between address maps associated with two tasks and the indirect mappings between address maps and a sharing map, all to some virtual memory objects in the backing store.

Figure 12.14b shows the *machine-dependent* page tables and the *machine-independent* resident memory, address maps (for three tasks), and their mappings to the virtual memory objects.

The machine-independent portion of the virtual memory implementation has full knowledge of all memory information. The machine-dependent portion, on the other hand, has a single-page validate/invalidate/protect interface. The page sizes for different sections of the implementation need not be the same. For example, a VAX has 512 bytes

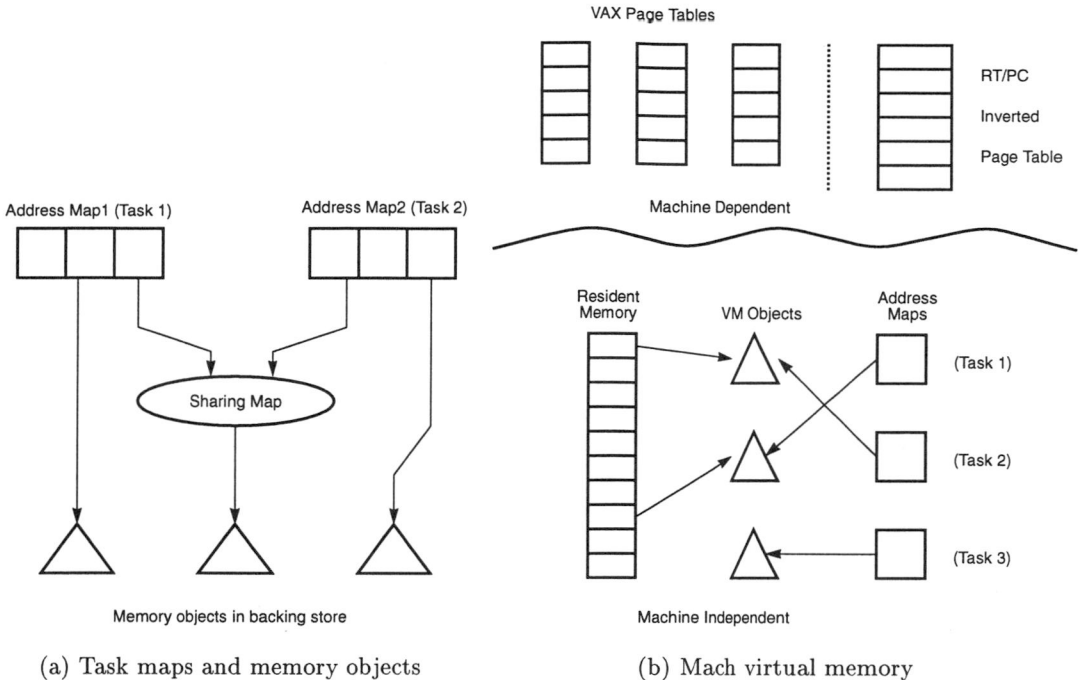

(a) Task maps and memory objects (b) Mach virtual memory

Figure 12.14 Task address mapping in the Mach virtual memory system. (Courtesy of Accetta et al., Carnegie-Mellon University, 1986)

per page. The machine-independent page size is a boot-time variable which can be a power of 2 of the 512-byte page size.

The actual data structures used in the machine-dependent portion depend on the target machine. As shown in Fig. 12.14b, a VAX implementation maintains VAX page tables, whereas an IBM RT/PC maintains an inverted page table. Mach's virtual memory implementation allows not only normal demand paging of tasks but also portions of the kernel to be paged.

Physical memory in Mach is treated as the cache for the virtual memory. An address map entry also contains information about the inheritance and protection attributes of the region of the virtual memory. A memory object is used to store data, indexed by byte, upon which various *read* or *write* operations can be performed.

Memory Objects and Sharing   Each memory object is associated with a reference counter which keeps track of the number of references made to the object. The object is garbage-collected when all mapped references to it are removed.

If a memory object is referenced frequently, then it is worthwhile to retain information about it even after the last mapped reference to it is removed. By doing so, its subsequent reuse can be made very inexpensive. Hence, Mach maintains a cache of such frequently used memory objects. A pager may use domain-specific knowledge to request that an object be kept in the cache after it is no longer referenced.

Mach's virtual memory handles page faults and page-out requests outside the kernel. This is accomplished by associating with each memory object a managing task called a *pager*. For example, to implement a memory-mapped file, virtual memory is created with its pager specified as the file system.

When a page fault occurs, the kernel translates the fault into a request for data from the file system. In fact, a pager may be either internal to the Mach kernel or an external user task. When a *copy-on-write* copy is performed, the various address maps containing copies point to the same memory object, which is being shared by multiple tasks.

When a shared memory object is modified by one of the tasks, a new page accessible only to the writing task must be allocated. Mach manages this type of memory sharing by creating *shadow objects*, which are just memory objects used for holding modified pages originally belonging to another object. New shadow objects often completely overlap the object they are sharing.

Mach garbage-collects shadow objects when it realizes that the immediate shadow objects are redundant and no longer required. Address map entries are allowed to point to a sharing map as well as to a memory object (Fig. 12.14b). The sharing points to shared memory objects.

Mach Portability   Mach's virtual memory system makes few assumptions about the underlying machine architecture, which has facilitated porting of the Mach system to various machines. Mach needs no in-memory hardware-defined data structure to manage the virtual memory. The Mach virtual memory code was originally implemented on a VAX uniprocessor machine.

VAX allows 2 Gbytes of address space to be allocated in the user state to a VAX process. VAX allows paging of the page table within the kernel's virtual address space. VAX pages in Mach may be created or destroyed as necessary to conserve space and improve run-time performance.

Porting Mach to multiprocessors encounters a major problem because facilities to ensure TLB consistency are lacking. Listed below are three approaches to coping with this problem:

- Forcibly interrupt all CPUs, which may be using a shared portion of an address map, so that their address translation buffers can be flushed.
- Postpone the use of a change mapping until all CPUs have taken a timer interrupt and had a chance to flush.
- Allow temporary inconsistency acceptance only when the operation being performed does not require simultaneity.

The multiprocessor system also adds to the difficulty of connecting a universal model of virtual memory. In addition to address translation hardware, multiprocessors also differ in the kind of shared-memory access they provide for individual CPUs. Three possible strategies are introduced below based on porting Mach on different machine architectures:

- With a fully shared memory with uniform access time (Encore Multimax and Sequent Balance).
- With a shared memory with nonuniform access (BBN Butterfly and IBM RP3).
- With message-based, nonshared memory systems (Intel iPSC).

By 1991, Mach had been ported only to multiprocessors with uniform or nonuniform shared memory. By taking advantage of its elaborate message-passing facility, Mach is able to integrate loosely coupled multicomputer systems. Mach has demonstrated its potential to become the model for future UNIX.

Besides VAX processors, Mach has been ported on Sun-3 workstations, the IBM RT/PC and RP3, the BBN Butterfly, the Encore Multimax, the Sequent Balance 21000, the NeXT Computer, and various 80386 machines.

The impact of Mach on UNIX systems is increasing steadily in the parallel processing community. For example, the BBN Butterfly multiprocessor is a NUMA machine. A pSOS^{+m} real-time executive has been developed for this machine. This OS provides dynamic clustering of its processors based on the concept of space sharing rather than time sharing.

Multitasked Clustering   Clusters allow the user to determine how many processors are needed to work on an application. They permit the separation of interactive and computational tasks. The dedication of processors also reduces overhead due to context switching and scheduling.

This multitasked clustering as illustrated in Fig. 12.15 permits the Butterfly system to be space-shared rather than time-shared. The system always reserves some processors for the public cluster. A superuser may change the size of the public cluster dynamically. Users can create and destroy private clusters on a command-to-command basis.

The dedicated processors are allocated from a systemwide free list. Private clusters can expand and contract as needed. Other systems like the IBM OS/2 also support threads, virtual memory, and message-passing mechanisms. However, the OS/2 treats some threads specially, manages the virtual memory in segments, imposes more restrictions, and lacks multiuser operations. This implies a need to standardize the Mach/OS in future UNIX systems.

12.5   OSF/1 Architecture and Applications

In this final section, we introduce the Open Software Foundation's OSF/1 operating system. The bulk of the material is based on the Technical Seminar on OSF/1 [OSF90]. We start with a description of the OSF/1 architecture and its evolutional relationship with other UNIX systems. Then we describe the OSF/1 programming environment and address the issues of performance with multithreading and advanced application development.

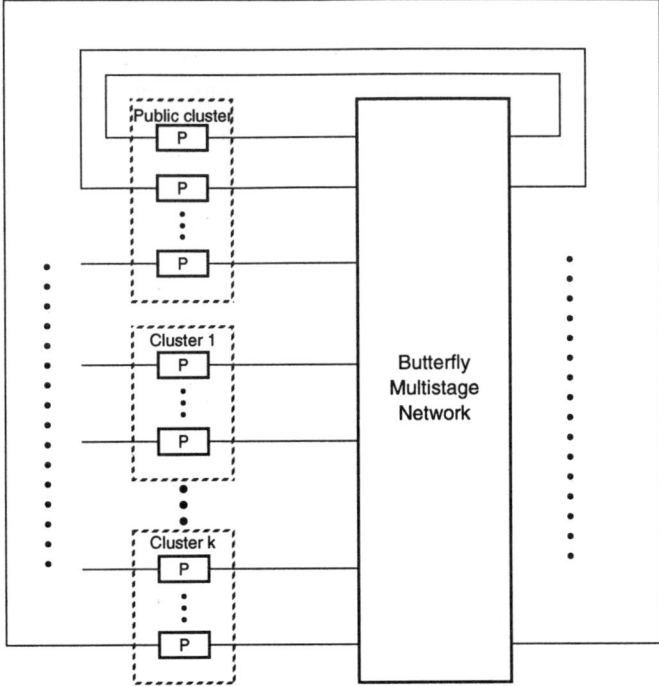

Figure 12.15 Space sharing through multitasked clustering on the BBN TC2000 multiprocessor system (P: processor and local memory).

12.5.1   The OSF/1 Architecture

The OSF was established in 1988 as a nonprofit organization. Its chartered goal is to promote worldwide research and design of open systems environments across the computer industry, as well as user groups. With over 200 participating members, the technology areas pursued by OSF include the operating system (OSF/1), user environment (Motif), distributed computing environment (DCE), architecture-neutral distribution format (ANDF), and distributed management environment (DME). We describe mainly OSF/1 architecture and its interactions with other system components.

Innovations in OSF/1   OSF/1 is an operating system compliant with standards, is compatible with other versions of UNIX, and contains some innovative features. As shown in Fig. 12.16a, OSF/1 has evolved from CMU/Mach, Berkeley UNIX 4.4 BSD, and UNIX System V. OSF/1's innovation involving the kernel comes primarily from its Mach-based technology and the provision of loadable device drivers and streams modules while the operating system is running.

OSF/1 further upgraded the Mach kernel in order to support symmetric multiprocessing, advanced virtual memory management, secure interprocess communication,

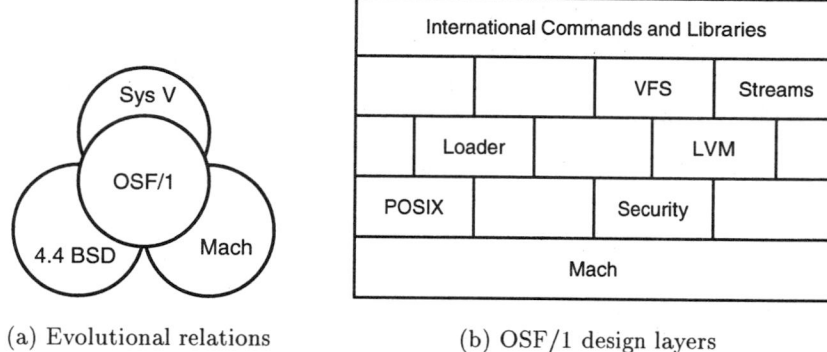

| (a) Evolutional relations | (b) OSF/1 design layers |

Figure 12.16 OSF/1 architecture. (Courtesy of Open Software Foundation, Inc., 1990)

and the use of multiple threads for performance improvement. These features will be described in subsequent subsections. Other innovations in OSF/1 include an execution environment with an extensible loader, shared libraries, and dynamic linking and loading; a *virtual file system* (VFS); a *logical volume manager* (LVM); a C2/B1 security system; and POSIX compliance as summarized in Fig. 12.16b. A set of internationalized commands and libraries is at the top layer of OSF/1.

Multithreaded Mach Kernel The OSF/1 kernel is a highly multithreaded version of the Mach/OS from Carnegie Mellon University. The long-term goal was to return to UNIX's original small, compact kernel. This "microkernel" would support multiple "envelopes" that provide specific variations of OS for specific application environments. Besides portability, scalability, and maturity, multiprocessor support was a key requirement for OSF/1. The basic programming abstractions were determined with *symmetric multiprocessing* capabilities for general-purpose parallel processing applications.

The Mach kernel handles program scheduling, memory management, and interprocess communication as described in previous sections. The symmetric UNIX multiprocessor support is achieved through Mach kernel abstractions of tasks, threads, ports, messages, and memory objects. The OSF implemented the IEEE POSIX Pthreads. Essentially, the task is a "container" for resources such as virtual memory, access rights, file descriptors, etc. The thread should contain a minimal state and be kept small in order to minimize context-switching overhead.

The ports and messages provide a secure, transparent interprocess communications facility optimized for performance. Examples include remote procedure calls and memory allocation to users. Each port provides a communications channel for messages. Ports are implemented as a protected message queue in the kernel. The messages are typed collections of data objects. Memory objects provide a handle for a chunk of virtual memory to construct pagers in a modular, extensible way and to page over the network.

Multiple File Systems Component technologies that comprise OSF/1 include multiple file systems, POSIX compliance, a program loader, a logical volume manager, streams, and C2/B1 security. The OSF/1 kernel includes an internal file system switch (Fig. 12.17a) to facilitate using multiple file systems: the local file system from Berkeley UNIX 4.4 BSD UFS, an NFS-compatible file system from 4.4 BSD, and a UNIX System V file system.

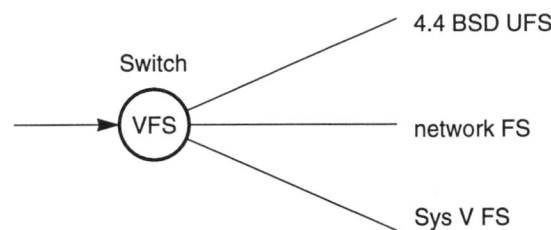

(a) Multiple file systems through a VFS switch in a Mach kernel

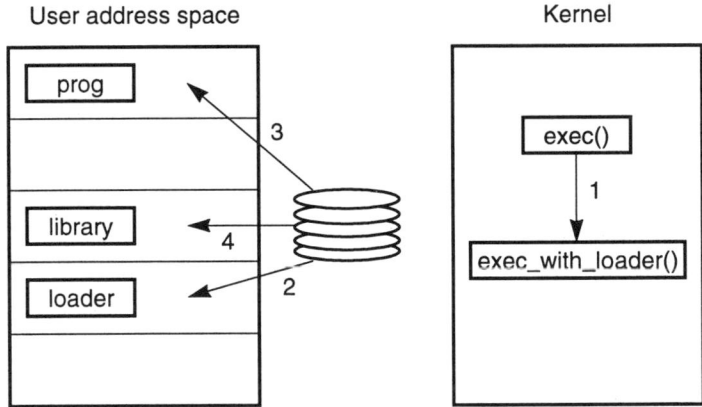

(b) Operations in a program loader

Figure 12.17 Multiple file systems and program loading in OSF/1. (Courtesy of Open Software Foundation, Inc., 1990)

Program Loader The program loader is a new feature developed in OSF/1. It extends from the object file format, has a format-independent design, and supports a shared library. The loader actually runs primarily in user space. An **exec** call will call a special kernel routine (step 1 in Fig. 12.17b), which takes as input the specification for a user program. The loader is mapped into the processor's virtual space (step 2). The loader then maps the application program (step 3) into the virtual memory. It is also responsible for bringing any shared library into the address space (step 4) to satisfy the applications references.

POSIX Compliance The IEEE 1003.1 POSIX standard is described below. It requires the OSF/1 kernel to work in 4.4 BSD *signal handling, tty driver*, and *session and job control*. The following example shows the POSIX compliance for session control among concurrent processes created in multiple programs.

Example 12.8 Session control in OSF/1 (Open Software Foundation, 1990)

Session control is an extension of the process hierarchy in UNIX. The concept is illustrated in Fig. 12.18. Session A contains a program pg9 running a current process p9. Session B handles three programs (pg10, pg11, and pg14), each creating some individual processes or process groups (p10 through p15). The session control features new calls such as setting the session identification and creating a session leader.

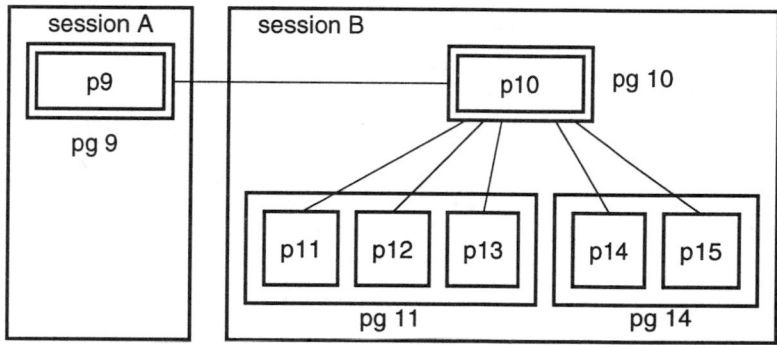

Figure 12.18 OSF/1 session control in compliance with the IEEE POSIX standard. (Courtesy of Open Software Foundation, Inc., 1990)

Besides a session leader, the concepts of a process group, a process, and some traditional UNIX ideas are used in OSF/1. The controlling terminal for a process or a process group becomes an attribute of the session leader, as opposed to the process group in conventional UNIX.

∎

Besides POSIX, the OSF/1 is also in compliance with other key standards, such as ANSI C, FIPS 151-1, and X/Open Portability Guide Issue 3 including base and transport interface (XTI).

Streams I/O Streams is a mechanism that allows customization of the character I/O processing inside the OS kernel. An application handles the I/O streams by inserting modules into the processing sequence.

The kernel software for handling a stream consists of a driver and one or more modules. The application creates a specific driver for a stream via the open system call and returns with a standard UNIX file descriptor. Data is passed between streams modules in messages. The OSF/1 uses a thread for a compute-intensive module. Multiple streams are handled by multiple threads on the symmetric multiprocessor system.

Example 12.9 Streams I/O in the OSF/1 (Open Software Foundation, 1990)

Figure 12.19 shows an example of how the stream steps are handled by the OSF/1 kernel. A program issues a *write*() system call. The stream head executes a *put*() and adds the message queue for the first module (step 2).

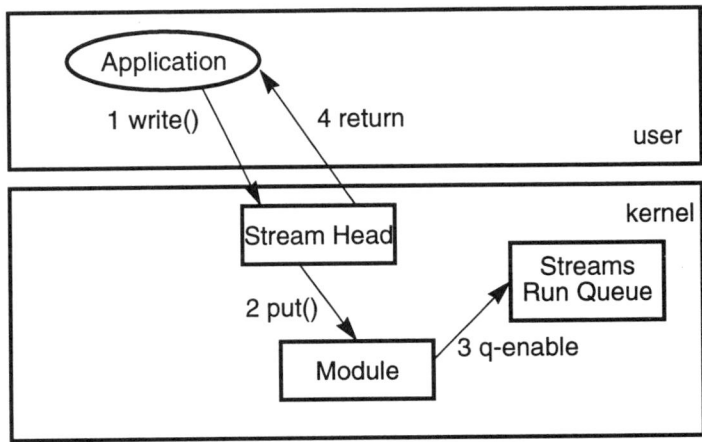

Figure 12.19 OSF/1 kernel operations for stream I/O. (Courtesy of Open Software Foundation, Inc., 1990)

If the module is idle, a *q-enable* operation is called, which adds the module to streams RunQ and awakens the streams thread (step 3). After flow control and memory exhaustion, the I/O process returns the called file descriptor to the application program. ∎

When multiple streams modules are involved, messages are exchanged in message queues. A set of queues can be associated with every module. An awakened streams thread calls the module's service procedure, which processes the message. It then calls the next module's *put* call. The last module called returns to the streams head and then returns to the application program.

Logical Volume Manager The OSF/1 *logical volume manager* (LVM) is a mechanism not only for addressing a problem with traditional UNIX files, which can only span the size of a single disk volume but also addressing files much larger than the physical disk. Also, the LVM monitors the disk for increased data availability. The data can be replicated for greater availability through disk mirroring.

OSF/1 Security Subsystem   Security features are implemented in OSF/1 in order to protect file accesses. A modular approach is adopted. Internally, the kernel invokes a security policy module for handling the application's system call. This module interrogates a policy database on a user-mode program. The policy program determines if the requested operation should be granted. Two levels of security features are implemented in OSF/1: the C2-level for ordinary security control and the B-level for maximum control.

12.5.2   The OSF/1 Programming Environment

The programming environment supported by OSF/1 consists of two basic parts: the software used by program developers and that used by applications as illustrated in Fig. 12.20a. We first introduce these environment features, and then we describe how to construct applications and how to port applications to OSF/1.

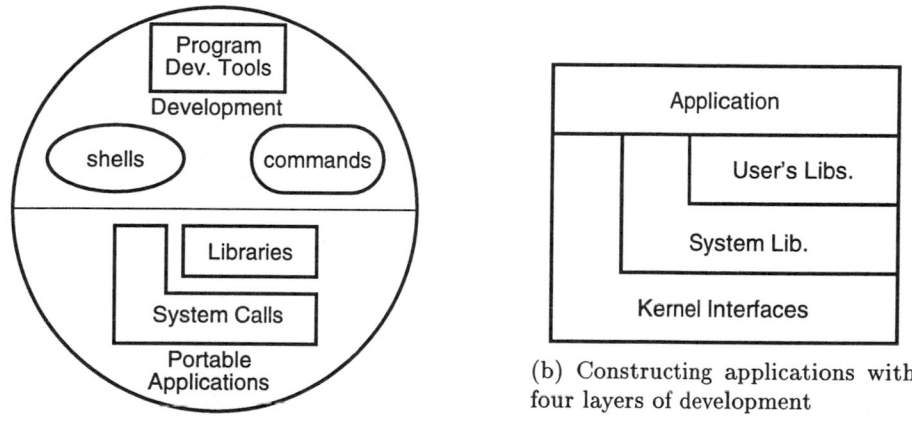

(a) Development tools and portable applications

(b) Constructing applications with four layers of development

Figure 12.20 Multiple file systems and program loading in OSF/1.   (Courtesy of Open Software Foundation, Inc., 1990)

Program Development Environment   The environment for program development contains a number of *program development tools*, such as the compiler, linker, and debugger. The *shells* provide the environment seen by both end users and developers. The *commands* are the basic UNIX commands expanding the shells to satisfy user demands.

The GCC compiler, GAS assembler, GLD loader, and GDB debugger are provided with OSF/1 through modified Free Software Foundation (FSF) tools for application programmers. OSF/1 provides all common UNIX shells, including the Bourne, C, and Korn shells. Any compiler can run OSF/1 as long as it is ANSI C compliant. The

OSF/1 administrative commands come primarily from Berkeley BSD. Other commands are internationalized, including those from System V and 4.3/4.3 BSD.

Runtime Support   This consists of the libraries, kernel, loader, and other subsystems depicted below the dotted line in Fig. 12.20a. Subsystems can be thought of as the higher-level layers of the OS, such as the file system and streams. The libraries and kernel provide entry points that are available to applications while they are executing. The loader enables shell-like applications to start program execution, as explained in Fig. 12.17b.

Developers are allowed to build their own subsystems to suit their needs. The kernel provides support for subsystems, which can be either statically configured via a BSD model or dynamically configured. Kernel-supported subsystems include physical and pseudo devices, file systems, and network protocols. The configuration at run time allows the user to add new subsystems while the program continues executing. OSF supplies kernel documentation to make the subsystem implementation easy.

Porting to OSF/1   Use of the OSF/1 programming environment requires standards and procedures for porting applications. These include porting existing applications, writing new applications, and modifying applications to run on OSF/1. Standards applied in OSF/1 include the AES (Application Environment Specification), ANSI-C XJ311, FIPS 151-1, and POSIX 1003.1. Running programs already written to run on System V or BSD systems demands source compatibility as well as binary compatibility which depends on machine-specific features.

In writing portable applications, one needs to follow the AES standards, conform to both system and language standards, and use implementation-independent features only. Changes are needed when there is a need for standards compliance, entry-point conflicts, architecture-dependence optimizations, or specific data structures. The OSF/1 uses the System V.4 directory tree layouts for these changes. The OSF/1 kernel interfaces are well-defined to make the programming environment friendly to both System V and BSD developers.

Constructing Applications   The application environment, which is built out of four layers of software, shown in Fig. 12.20b. This environment provides the ability to link application libraries and to load applications easily. Thus, code sharing, placement, and linking are the key facilities needed. Shared libraries provide a single copy of the object for main memory as needed, reduce disk space requirements, and improve performance with reduced development overhead.

Dynamic linking allows external symbols to be located dynamically at run time rather than statically at link time. The benefits include name lookup performed at load time, on-the-fly updates, piece-by-piece application construction, and faster software development. Position-independent code contains only "relative" offsets. This enables the system to place other code objects in a relative address space without overwriting the original code. Thus code objects can be cached, increasing performance.

Position-independent code placement brings the benefits of using dynamically loaded libraries, finding the best spot for code, avoiding programs bumping into each other, and making code sharing easier. In summary, OSF/1 was designed to construct new applications that are sharable, dynamically linkable, position-independent, high performing, and fully compatible with existing applications. OSF/1 encourages distributing applications through networking resources and activities.

12.5.3 Improving Performance with Threads

Both Mach and OSF/1 are highly motivated with multithreading or multitasking, as seen in previous sections. In this final subsection, we further elaborate on the use of threads to improve application performance. A thread implies a single sequential flow of control. Therefore, the conventional program execution follows a single thread. On a parallel computer or in a distributed system, an application may create multiple threads, all of which execute simultaneously.

Creating threads is relatively inexpensive in terms of execution time and OS resources. Therefore, modern systems encourage the use of a high degree of multithreading. One must realize that all threads within a task (or process) share the same memory-resident variables or use a common virtual address space. Threads are created to cooperate with each other in order to solve a large problem. Furthermore, data access must be synchronized. IEEE POSIX Standard P1003.4a is a standard for the behavior and programming interfaces relevant to threads, often referred to as the *Pthreads* for POSIX threads.

Basic Concept of Pthreads   As noted in Chapter 9, even on a uniprocessor system, multiple threads may bring benefits in performance. Pthreads can cross-develop multiprocessor applications even on a uniprocessor. Many applications can be more efficiently executed with threads. In its simplest form, one can think of threads as the ability to apply many programs to solve a common problem. Each program has a small, self-contained job to perform. A more sophisticated approach is a single program being divided into many threads for parallel processing.

One example of the use of threads is to build a *multithreaded server*, which contains a simple main program which creates a thread upon each request. The threads can block. Each thread can wait for input on a separate input channel and then handle each service request. While some threads are blocked, others continue to execute. The requests are serviced simultaneously on multiprocessors. Other advantages of multithreading come from parallel processing, which overlaps computations, communications, and I/O activities.

A third motivation for threads is to handle asynchronous events more efficiently. This prevents inconvenient interrupts and avoids complex control flows. The following example shows some Pthreads codes in the main program and in a typical thread handling a request.

Example 12.10 Pthreads code for a multithreaded server (Open Software Foundation, 1990).

The Multithreaded server operates as an appointment manager keeping data in an in-memory database. Multiple threads created share the in-memory database. A sample main program code is given below:

```
void doarequest();
do while (1) {
/Wait for a new request/
      . . .
/Create and start a thread to handle the request/
ret = Pthread_creat(&newthread, Pthread_attr_default, doarequest, chan);
}
```

The application requests are sent out over some unspecified type of communications channel to store or to retrieve data. The **doarequest** code supplies the address of an initial function which creates a new thread to handle a new request. The thread begins executing in an initial function when a **Pthread_creat**() call returns successfully. The example shows the default attributes specified for the thread characteristics. On completing the initial function specified in **Pthread_creat**(), the thread terminates.

The following sample code is for request processing using Pthreads:

```
doarequest(channel)
{    op = read(channel);
     key = read(channel);
     lock(database);
     switch (op) {
          get: write(channel, database[key]);
          set: database[key] = read(channel);
     }
     unlock(database);
}
```

After reading an operation from the communication channel, we look to check which record in the database is being requested. The **lock** is used to provide exclusive update of the database by one request at a time. The **get** and **set** are for writing into the database and reading from it and then returning the data to the communication channel, respectively. The **unlock** then releases the control of the database once the request is serviced.

■

Pthread Synchronization   Multiple threads must be serialized to access shared data using some synchronization objects, such as *mutex* (for mutual exclusive) in OSF/1. Only one thread can hold a *mutex* at a time. Threads trying to acquire a *mutex* are blocked if it is already held by another thread. Thus thread synchronization involves waiting and signaling. Other related issues include threads cancellation and functions used in doing so. Reentrant libraries are called through POSIX Standard 1003.4a. The

Pthreads are used to improve performance with implicit multiprocessor support, cross development, and standard interfaces.

Advanced Applications The useful features of Mach with extensions in OSF/1 facilitate efficient interprocess communication, shared virtual memory, memory-mapped files, and exception handling. Mach supports message passing through ports and security control. A network message server can be used to pass IPC message between machines or for networkwide port registration and lookup. Two applications are exemplified below and illustrated in Fig. 12.21.

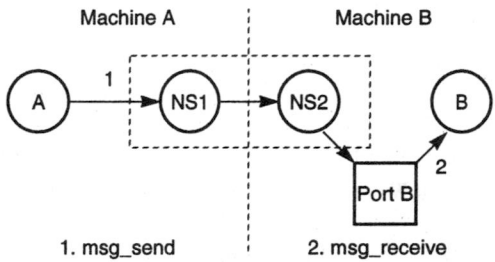

(a) Network interprocess communication

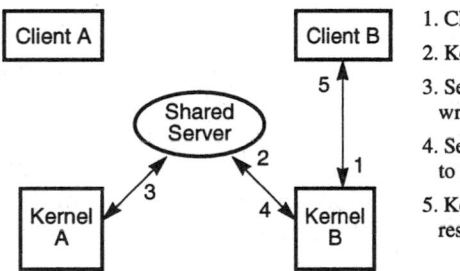

1. Client B reads faults page

2. Kernel B requests data

3. Server asks kernel A to write-protect page

4. Server provides data via IPC to kernel B

5. Kernel B fixes page map and resumes client B

(b) Shared access of the same memory object

Figure 12.21 Network IPC and shared access of common memory object. (Courtesy of Open Software Foundation, Inc., 1990)

Example 12.11 Interprocess communication across the network (Open Software Foundation, 1990).

The network IPC is depicted in Fig. 12.21a, where a message is sent from machine A to machine B. The network servers (NSs) extend IPC across the network. Messages for ports with remote servers must first go to a local server before being forwarded to the destination.

A thread in task A (on machine A) directs a message to port B, but port B is on machine B. Port B has been registered with the network name server NS2. The sender cannot tell if the message is destined for the same or a different machine until the two servers establish exchanged port information.

∎

Example 12.12 Shared-memory manager in OSF/1 (Open Software Foundation, 1990).

This example shows how multiple clients residing in different machines can perform orderly access of the same shared memory object. Figure 12.21b shows the case of two machines identified by kernel A and kernel B, respectively. The shared server may be managed by a third machine. Both the client-server and kernel-server protocols use Mach IPC, which can be transparently extended across a network. It is assumed that clients A and B are on machines A and B, respectively.

In step 1, client B reads faults the same page that client A has. A page fault is delivered to kernel B. In step 2, kernel B sends a message to the shared server asking for the data. In step 3, the shared server sends a message to kernel A asking it to write-protect the page in A's address space. In step 4, the shared server returns the data to kernel B via the IPC. In step 5, kernel B updates B's page table and resumes client B.

∎

Other advanced applications may demand shared-memory management, arbitrating among read/write accesses, memory-mapped files, and exception handling. Interested readers are referred to the *OSF/1 System Programmer's Reference* and the *OSF/1 System Execution Guide* for details on advanced application development. The intention of OSF is to develop applications for an international, cross-cultural user community. At the time this book was completed, IBM, Digital, Hitachi, HP/Apollo, Intel, Siemens, etc., had ported OSF/1 on their machines.

12.6 Bibliographic Notes and Exercises

Multiprocessor UNIX has been introduced in [Bach86] and [Jermoluk90]. An excellent survey of UNIX extensions on multiprocessors or multicomputers was given by [Schimmel90]. The original work on the UNIX system for a time-sharing uniprocessor was reported in [Ritchie74] and Thompson.

The master-slave kernel was first developed by [Goble81] and Marsh at Purdue University.   Dynamic load balancing on a network of UNIX/VAX computers was reported in [Hwang82] et al. Studies on floating-executive and multithread kernels can be found in [Bach86], [Ousterhout88] et al., [Caswell90] and Black, and [Tevanian87] et al.

The Mach/OS was reported in [Accetta86] et al., [Rashid86], [Young87] et al., and [Black90] from CMU. Object-oriented distributed OS development can be traced back to [Jones86] and Rashid, and [Rashid81] and Robertson.

NeXT Computer, Inc. has released a description of the Mach/OS implemented on NeXT computers [NcXT90]. [Russell87] and Waterman and [Tevanian89] and Smith have assessed Mach as a model for future UNIX-based systems.

The Caltech CE/RK environment was described in [Seitz90]. Information on the nX and pSOS^{+m} operating systems used in the TC-2000 can be found in [BBN89]. The OSF/1 is described in [OSF90].

Exercises

Problem 12.1   A multiprocessor OS can introduce concurrency by multiplexing UNIX processes, Mach threads, or virtual processors (VP) onto physical processors (PP). The *multiroutines* (MR), mentioned in Fig. 12.6, are a multiprocessor generalization of coroutines which can be introduced by a user-level library to enhance concurrency through the multiplexing concept.

Parallel programming has been classified by [Black90] into four models as shown in the following table, depending on the relationships among the MRs, VPs, and PPs they support:

Relationship	Model
MR = VP = PP	Pure parallelism
MR $\geq$ VP = PP	User concurrency
MR = VP $\geq$ PP	System concurrency
MR $\geq$ VP $\geq$ PP	Dual concurrency

MR = multiroutines, VP = virtual processors, PP = physical processors.

The *pure parallelism model* corresponds to a perfect match between software and hardware parallelism. The *user concurrency model* introduces additional concurrency at the user level. The *system concurrency model* introduces additional concurrency only at the system level. Finally, the *dual-concurrency model* introduces concurrency at both user and system levels.

Readers are referred to the paper by [Black90] for experimental performance data in managing different numbers of threads using discouragement hints or handoff hints, as introduced in Section 12.4.

(a) Give example applications where each of the four programming models can be used. Intuitive reasoning must be provided in each case.

(b) The choice of programming model depends not only on the applications but also on the language, library, multithreading, and debugging support provided. Discuss the scheduling support for concurrency in the Mach/OS.

Problem 12.2   A special system call, *ptrace*, has been described in [Caswell90] and Black for implementing a Mach debugger for multithreaded applications. Read the paper and discuss how multiple threads and debugging are handled by the improved version of *ptrace*. You should address the issues of accessing memory and registers, thread execution control, exception handling, user interfaces, UNIX compatibility, deadlock avoidance, and processing of application events.

Problem 12.3 Examine the relationship between memory and communication requirements in implementation of the Mach/OS. The paper by [Young87] et al. would be a good reference for this evaluation study. Address the issues related to overall performance, applicability of Mach to new multiprocessor architectures (such as NUMA or NORMA, besides UMA models), and the structure of application programs.

Problem 12.4

(a) Explain the limitations of the conventional UNIX kernel for parallel processing.
(b) Explain the multiprocessor UNIX design goals in the areas of compatibility, portability, address space, load balancing, parallel I/O, and network services.

Problem 12.5 Explain the operations of each of the following master-slave kernel configurations. Also, comment on the advantages and disadvantages in each case.

(a) Single, fixed master-slave kernel.
(b) Single, floating master-slave kernel.
(c) Single master kernel with assistants.
(d) Multiple cooperative masters for executing different subsystem functions.
(e) A multithreaded UNIX kernel.

Problem 12.6 Explain the differences and operations in the following three message-passing OS models:

(a) Object-oriented model.
(b) Node-addressed model.
(c) Channel-addressed model.

Problem 12.7

(a) Explain the differences between time-sharing and space-sharing operating systems.
(b) Describe the Cosmic Environment and the Reactive Kernel developed at Caltech to support message-passing C programming on a multicomputer system.

Problem 12.8 Answer the following questions associated with the development of the CMU Mach/OS kernel:

(a) Define threads, task, messages, port, and port sets.
(b) What are the opportunities for exploiting parallelism in the Mach kernel?

Problem 12.9 Explain the thread management problem and characterize various thread scheduling policies.

Problem 12.10

(a) Explain the message-based communications in a Mach-based system.
(b) Explain how virtual memory is managed in a Mach-based system.
(c) Explain the concept of multithreaded clustering as implemented on the BBN TC2000 system.

Problem 12.11 Answer the following questions on the architecture and applications of a multiprocessor running OSF/1:

(a) Explain the relationships among UNIX System V, UNIX 4.4 BSD, Mach, and OSF/1.
(b) What are the IEEE POSIX standards and the OSF/1 compliance with these standards?
(c) Explain the development and application environments supported by OSF/1.

Problem 12.12 Answer the following questions on using threads to improve performance on an OSF/1-based multiprocessor system:

(a) Define POSIX threads (Pthreads) and comment on the motivation for Pthreads.
(b) Analyze the following Pthreads sample code for database locking. Explain how to lock or unlock your database by calling the program. The use of db-mutex ensures that only one thread uses the in-memory database at a time.

```
doarequest(channel)
{    op = read(channel);
     key = read(channel);
     Pthread_mutex_lock (db_mutex);
     switch (op) {
         get: write(channel, database[key]);
         set: database[key] = read(channel);
     }
     Pthread_mutex_unlock (db_mutex);
}
```

Problem 12.13 Examine the LINPACK benchmark program for solving a linear system of equations and answer the following questions on executing the LINPACK program on a shared-memory multiprocessor system:

(a) Explain the algorithm being implemented in the LINPACK.
(b) What compiler support is provided for vectorization, parallelization, and other optimizations of the LINPACK code on a multiprocessor of your choice?
(c) What operating system support do you wish to have for parallel I/O and paral-

lel execution of a LINPACK program with a very large data set which must be retrieved from a disk?

Problem 12.14   Repeat the questions of Problem 12.13 on a message-passing multicomputer. You can choose any target multicomputer that is available for the study. Both algorithm and software support for executing the LINPACK code in the distributed computing environment should be explained.

Problem 12.15   A parallel computer system has been attached with 20 tape drives. All jobs running on the system requires a maximum of four tape drives to complete each job execution. Assume that each job starts running with only three tape drives for a long period before requesting the fourth tape drive for a short period toward the end of its operation. Also assume an endless supply of jobs to the system.

(a) If the job scheduler in the multiprocessor OS operates with the policy that it will not start a job unless there are four tape drives available. Once four drives are assigned to a job, they will not be released during the entire job period. What is the maximum number of jobs that can be in progress at once? What are the maximum and minimum numbers of tape drives that may be left idling as a result of this policy?

(b) Figure out a better scheduling policy to improve the tape drive utilization rate and at the same time avoid a system deadlock. What is the maximum number of jobs that can be in progress simultaneously using the new scheduling policy? What are the bounds on the number of idling tape drives?

Bibliography

[Accetta86] M. Accetta, R. Baron, W. Bolosky, D. Golub, R. Rashid, A. Tevanian, and M. Young, "Mach: A New Kernel Foundation for UNIX Development," *Proc. Summer 1986 USENIX Conf.*, pp. 93–113, Atlanta, GA, June 1986.

[ACM91] ACM, *Resources in Parallel and Concurrent Systems with an Introduction by Charles Seitz*, ACM Press, New York, 1991.

[Acosta86] R. D. Acosta, J. Kjelstrup, and H. C. Torng, "An Instruction Issuing Approach to Enhancing Performance in Multiple Functional Unit Processors," *IEEE Trans. Computers*, pp. 815–825, Sept. 1986.

[Adam74] T. L. Adam, K. M. Chandy, and J. R. Dickson, "A Comparison of List Schedules for Parallel Processing Systems," *Commun. ACM*, 17(12):685–690, 1974.

[Adve90] S. V. Adve and M. D. Hill, "Weak Ordering: A New Definition," *Proc. 17th Annu. Int. Symp. Computer Arch.*, 1990.

[Adve91] S. V. Adve, V. S. Adve, M. D. Hill, and M. Vernon, "Comparison of Hardware and Software Cache and Coherence Schemes," *Proc. 18th Annu. Int. Symp. Computer Arch.*, pp. 298–308, 1991.

[Agarwal88] A. Agarwal, R. Simoni, J. Hennessy, and M. Horowitz, "An Evaluation of Directory Schemes for Cache Coherence," *Proc. 15th Annu. Int. Symp. Computer Arch.*, 1988.

[Agarwal90] A. Agarwal, B. H. Lim, D. Kranz, and J. Kubiatowicz, "APRIL: A Processor Architecture for Multiprocessing," *Proc. 17th Annu. Int. Symp. Computer Arch.*, pp. 104–114, 1990.

[Agarwal91] A. Agarwal, D. Chaiken, G. D'Souza, K. Johnson, D. Kranz, J. Kubiatowicz, K. Kurihara, B. H. Lim, G. Maa, D. Nussbaum, M. Parkin, and D. A. Yeung, "The MIT Alewife Machine: A Large-Scale Distributed-Memory Multiprocessor," *Proc. Workshop Multithreaded Computers, Supercomputing 91*, 1991.

[Agarwal92a] A. Agarwal, "Performance Tradeoffs in Multithread Processors," *IEEE Trans. Parallel Distri. Systems*, 3(5):525–539, 1992.

[Agarwal92b] A. Agarwal, D. Chaiken, K. Johnson, D. Kranz, J. Kobiatowicz, K. Kurihara, B. Lim, G. Maa, and D. Mussbaum, "The MIT Alewife Machine : A Large-Scale Distributed-Memory Multiprocessor," in Dubois and Thakkar (eds.), *Scalable Shared-Memory Multiprocessors*, Kluwer Academic Publishers, Boston, MA, 1992.

[Agha86] G. Agha, *Actors: A Model of Concurrent Computation in Distributed Systems*, MIT Press, Cambridge, MA, 1986.

[Agha90] G. Agha, "Concurrent Object-Oriented Programming," *Commun. ACM*, 33(9):125–141, Sept. 1990.

[Aho74] A. Aho, J. E. Hopcroft, and J. D. Ullman, *The Design and Analysis of Computer Algorithms*, Addison-Wesley, Reading, MA, 1974.

[Ahuja86] S. Ahuja, N. Carriero, and D. Gelernter, "Linda and Friends," *IEEE Computer*, 19(8):16–34, 1986.

[Allan85] S. J. Allan and R. Oldehoeft, "HEP SISAL: Parallel Functional Programming," in Kowalik (ed.), *Parallel MIMD Computation: HEP Supercomputers and Applications*, MIT Press, Cambridge, MA, 1985.

[Allen84] J. R. Allen and K. Kennedy, "PFC: A Program to Convert Fortran to Parallel Fortran," in Hwang (ed.), *Supercomputers: Design and Applications*, IEEE Computer Society Press, Los Alamitos, CA, 1984.

[Allen87] R. Allen and K. Kennedy, "Automatic Translation of Fortran Programs to Vector Form," *ACM Trans. Prog. Lang. and Systems*, pp. 491–542, Oct. 1987.

[Alliant89] Alliant, *Alliant Product Summary*, Alliant Computer Systems Corporation, Littleton, MA, 1989.

[Almasi89] G. S. Almasi and A. Gottlieb, *Highly Parallel Computing*, Benjamin/Cummings, Redwood, CA, 1989.

[Alverson90] R. Alverson, D. Callahan, D. Cummings, B. Koblenz, A. Porterfield, and B. Smith, "The Tera Computer System," *Proc. ACM Int. Conf. Supercomputing*, pp. 1–6, Amsterdam, The Netherlands, June 1990.

[Amdahl67] G. M. Amdahl, "Validity of Single-Processor Approach to Achieving Large-Scale Computing Capability," *Proc. AFIPS Conf.*, pp. 483–485, Reston, VA., 1967.

[Anaratone86] M. Anaratone, E. Arnould, T. Gross, H. T. Kung, M. S. Lam, O. Menzilcioglu, K. Sarocky, and J. A. Webb, "Warp Architecture and Implementation," *Proc. 13th Annu. Int. Symp. Computer Arch.*, pp. 346–356, Tokyo, June 1986.

[Anderson67] D. W. Anderson, J. G. Earle, R. E. Goldschmidt, and D. M. Powere, "The IBM System/360 Model 91: Floating-Point Execution Unit," *IBM Systems Journal*, pp. 34–53, Feb. 1967.

[Andrews91] G. R. Andrews, *Concurrent Programming: Principles and Practice*, Benjamin/Cummings, Redwood, CA, 1991.

[Archibald86] J. Archibald and J. L. Baer, "Cache Coherence Protocols: Evaluation Using a Multiprocessor Simulation Model," *ACM Trans. Computer Systems*, 4(4):273–298, Nov. 1986.

[Arden81] B. W. Arden and H. Lee, "Analysis of Chordal Ring Network," *IEEE Trans. Computers*, 30(4):291–295, 1981.

[Arvind83] Arvind and R. A. Iannucci, "A Critique of Multiprocessing von Neumann Style," *Proc. 10th Symp. Computer Arch.*, Stockholm, Sweden, 1983.

[Arvind84] Arvind, D. E. Culler, R. A. Iannucci, V. Kathail, K. Pingali, and R. E. Thomas, "The Tagged Token Dataflow Architecture," Technical report, MIT Laboratory for Computer Science, 545 Technology Square, Cambridge, MA 02139, 1984.

[Arvind87] Arvind and R. A. Iannucci, "Two Fundamental Issues in Multiprocessing," *Proc. Parallel Processing on Science and Engineering*, 1987.

[Arvind90] Arvind and R. S. Nikhil, "Executing a Program on the MIT Tagged-Token Dataflow Architecture," *IEEE Trans. Computers*, 39(3):300–318, 1990.

[Arvind91] Arvind, L. Bic, and T. Ungerer, "Evolution of Dataflow Computers," in Gaudiot and Bic (eds.), *Advanced Topics in Dataflow Computing*, pp. 3–34, Prentice-Hall, Englewood Cliffs, NJ, 1991.

[Athas88] W. C. Athas and C. L. Seitz, "Multicomputers: Message-Passing Concurrent Computers," *IEEE Computer*, 21(8):9–24, Aug. 1988.

[Axelrod86] T. S. Axelrod, "Effects of Synchronization Barriers on Multiprocessor Performance," *Parallel Computing*, 3(2):129–140, May 1986.

[Babb88] R. G. Babb, *Programming Parallel Processors*, Addison-Wesley, Reading, MA, 1988.

[Bach84] M. J. Bach and S. J. Buroff, "Multiprocessor UNIX Operating Systems," *AT&T Bell Lab. Tech. Journal*, 63(8), Oct. 1984.

[Bach86] M. J. Bach, *The Design of the UNIX Operating System*, Prentice-Hall, Englewood Cliffs, NJ, 1986.

[Baer80] J. L. Baer, *Computer Systems Architecture*, Computer Science Press, Rockville, MD, 1980.

[Banerjee79] U. Banerjee, *Speedup of Ordinary Programs*, Ph.D. thesis, University of Illinois, 1979.

[Banerjee88] U. Banerjee, *Dependence Analysis for Supercomputing*, Kluwer Academic Press, Boston, MA, 1988.

[Barnes68] G. H. Barnes, R. M. Brown, M. Kato, D. J. Kuck, D. L. Slotnick, and R. A. Stokes, "The ILLIAC IV Computer," *IEEE Trans. Computers*, pp. 746–757, Aug. 1968.

[Batcher76] K. Batcher, "The Flip Network in STARAN," *Proc. Int. Conf. Parallel Processing*, pp. 65–71, 1976.

[Batcher80] K. E. Batcher, "Design of a Massively Parallel Processor," *IEEE Trans. Computers*, pp. 836–840, Sept. 1980.

[BBN89] BBN Advanced Computers Inc., Cambridge, MA., *TC2000 Technical Product Summary*, Nov. 1989.

[Beetem85] J. Beetem, M. Denneau, and D. Weingarten, "The GF11 Supercomputer," *Proc. 12th Annu. Int. Symp. Computer Arch.*, pp. 363–376, Boston, MA, May 1985.

[Bell92] G. Bell, "Ultracomputer: A Teraflop Before Its Time," *Commun. ACM*, 35(8):27–47, 1992.

[Ben-Ari90] M. Ben-Ari, *Principles of Concurrent and Distributed Programming*, Prentice-Hall, Englewood Cliffs, NJ, 1990.

[Bernstein66] A. J. Bernstein, "Analysis of Programs for Parallel Processing," *IEEE Trans. Computers*, pp. 746–757, Oct. 1966.

[Berntsen90] J. Berntsen, "Communication-Efficient Matrix Multiplication on Hypercubes," *Parallel Computing*, pp. 335–342, 1990.

[Bhuyan83] L. N. Bhuyan and D. P. Agrawal, "Design and Performance of Generalized Interconnection Networks," *IEEE Trans. Computers*, pp. 1081–1090, Dec. 1983.

[Bisiani88] R. Bisiani and M. Ravishankar, "Plus: A Distributed Shared-Memory System," *Proc. 17th Annu. Int. Symp. Computer Arch.*, pp. 115–124, 1988.

[Bitar86] P. Bitar and A. M. Despain, "Multiprocessor Cache Synchronization: Issues, Innovations, and Evolution," *Proc. 13th Annu. Int. Symp. Computer Arch.*, 1986.

[Bitar91] P. Bitar, "MIMD Synchronization and Coherence," Technical Report UCB/CSD 90/605, University of California, Berkeley, May 1991.

[Bitar92] P. Bitar, "The Weakest Memory-Access Order," *J. Para. Distri. Computing*, 15:305–331, 1992.

[Black90] D. L. Black, "Scheduling Support for Concurrency and Parallelism in the Mach Operating System," *IEEE Computer*, 23(5):35–42, May 1990.

[Blelloch90] G. E. Blelloch, *Vector Models for Data-Parallel Computing*, MIT Press, Cambridge, MA, 1990.

[Blevins90] D. W. Blevins, E. W. Davis, R. Heaton, and J. H. Reif, "BLITZEN: A Highly Integrated Massively Parallel Machine," *J. Para. Distri. Computing*, pp. 150–160, 1990.

[Boothe92] B. Boothe and A. Ranade, "Improved Multithreading Techniques for Hiding Communication Latency in Multiprocessor," *Proc. 19th Annu. Int. Symp. Computer Arch.*, Australia, May 1992.

[Borkar90] S. Borkar, R. Cohn, G. Fox, T. Gross, H. Hung, M. Lam, M. Levine, B. Moore, W. Moore, C. Peterson, J. Susman, J. Sutton, J. Urbanski, and J. Webb, "Supporting Systolic and Memory Communication in iWARP," *Proc. 17th Annu. Int. Symp. Computer Arch.*, pp. 70–81, May 1990.

[Brainerd90] W. S. Brainerd, C. H. Golberg, and J. C. Adams, *Programmer's Guide to Fortran 90*, McGraw-Hill, New York, 1990.

[Brawer89] S. Brawer, *Introduction to Parallel Programming*, Academic Press, New York, 1989.

[Briggs82] F. A. Briggs, K. S. Fu, K. Hwang, and B. W. Wah, "PUMPS Architecture for Pattern Analysis and Image Database Management," *IEEE Trans. Computers*, pp. 969–982, Oct. 1982.

[Brinch Hansen75] P. Brinch Hansen, "The Programming Language Concurrent Pascal," *IEEE Trans. Software Engineering*, SE-1(2):199–206, June 1975.

[Brunner90] R. A. Brunner, D. P Bhandarkar, F. X. McKeen, B. Patel, W. J. Rogers, and G. L. Yoder, "Vector Processing on the VAX 9000 System," *Digital Technical Journal*, 2(4):61–79, 1990.

[Burkhardt92] H. Burkhardt, *Technical Summary of KSR-1*, Kendall Square Research Corporation, 170 Tracer Lane, Waltham, MA 02154, 1992.

[Butler91] M. Butler, T. Y. Yeh, Y. Patt, M. Alsup, H. Scales, and M. Shebanow, "Single-Instruction-Stream Parallelism is Greater Than Two," *Proc. 18th Annu. Int. Symp. Computer Arch.*, pp. 276–286, 1991.

[Buzbee83] B. L. Buzbee et al., "Supercomputing Value and Trends," unpublished slide presentation, Los Alamos National Laboratory, NM, July 1983.

[Callahan85] D. Callahan, "Task Granularity Studies on a Many-Processor CRAY X-MP," *Parallel Computing*, pp. 109–118, June 1985.

[Callahan88] D. Callahan, K. Cooper, R. Hood, K. Kennedy, and L. Torczon, "ParaScope: A Parallel Programming Environment," *Int. J. Supercomputer Appl.*, 2(4), 1988.

[Caswell90] D. Caswell and D. Black, "Implementing a Mach Debugger for Multithread Applications," *Proc. Winter 1990 USENIX Conf.*, Washington, DC, Jan. 1990.

[CDC80] CDC, *Cyber 200/Model 205 Technical Description*, Control Data Corporation, Nov. 1980.

[CDC90] CDC, "Introduction to Cyber 2000 Architecture," Technical Report 60000457, Control Data Corporation, St. Paul, MN, 1990.

[Cekleov90] M. Cekleov, M. Dubois, J. C. Wang, and F. A. Briggs, "Virtual-Address Cache in Multiprocessors," in Dubois and Thakkar (eds.), *Cache and Interconnect Architectures in Multiprocessors*, Kluwer Academic Press, Boston, MA, 1990.

[Censier78] L. M. Censier and P. Feautrier, "A New Solution to Coherence Problems in Multicache Systems," *IEEE Trans. Computers*, C-27(12):1112–1118, Dec. 1978.

[Chaiken90] D. Chaiken, C. Fields, K. Kwihara, and A. Agrawal, "Directory-Based Cache Coherence in Large-Scale Multiprocessor," *IEEE Computer*, 23(6):49–59, 1990.

[Chang88] A. Chang and M. F. Mergen, "801 Storage: Architecture and Programming," *ACM Trans. Computer Systems*, 6(1):28–50, 1988.

[Chang90] L. C. Chang and B. T. Smith, "Classification and Evaluation of Parallel Programming Tools," Technical Report CS 90-22, University of New Mexico, Albuquerque, NM 87131, 1990.

[Chang91] P. P. Chang, W. Y. Chen, S. A. Mahlke, and W. M Hwu, "Computing Static and Dynamic Code Scheduling for Multiple-Instruction-Issue Processors," *Proc. 24th Int. Symp. Microarch.*, 1991.

[Cheng89] H. Cheng, "Vector Pipelining, Chaining, and Speed on the IBM 3090 and Cray X/MP," *IEEE Computer*, 22(9):31–44, 1989.

[Cheng91] D. Y. Cheng, "A Survey of Parallel Programming Tools," Technical Report RND-91-005, NASA Ames Research Center, Moffett Field, CA, 1991.

[Chin84] C. Y. Chin and K. Hwang, "Packet Switching Networks for Multiprocessors and Dataflow Computers," *IEEE Trans. Computers*, pp. 991–1003, Nov. 1984.

[Chow74] C. K. Chow, "On Optimization for Storage Hierarchies," *IBM J. Res. and Develop.*, pp. 194–203, 1974.

[Christy90] P. Christy, "Software to Support Massively Parallel Computing on the MasPar MP-1," *Digest of Papers Spring Compcon*, San Francisco, CA, Feb. 1990.

[Clark83] K. L. Clark and S. Gregory, "Parlog: A Parallel Logic Programming Language," Technical Report DOC 83-5, Dept. of Computing, Imperial College, London, May 1983.

[Clark86] R. S. Clark and T. L. Wilson, "Vector System Performance of the IBM 3090," *IBM Systems Journal*, 25(1), 1986.

[Cocke90] J. Cocke and V. Markstein, "The Evolution of RISC Technology at IBM," *IBM J. Res. and Develop.*, 34(1):4–11, 1990.

[Convex90] Convex, *Fortran Optimization Guide*, Convex Computer Corporation, Richardson, TX, 1990.

[Cormen90] T. H. Cormen, C. E. Leiserson, and R. L. Rivest, *Introduction to Algorithms*, MIT Press, Cambridge, MA, 1990.

[Cragon89] H. G. Cragon and W. J. Watson, "The TI Advanced Scientific Computer," *IEEE Computer*, 22(1):55–64, 1989.

[Cragon92a] H. G. Cragon, *Branch Strategy Taxonomy and Performance Models*, IEEE Computer Society Press, Los Alamitos, CA, 1992.

[Cragon92b] H. G. Cragon, "Memory Systems and Pipeline Processors," Class notes, Department of ECE, University of Texas, Austin, 1992.

[Crawford90] J. H. Crawford, "The i486 CPU: Executing Instructions in One Clock Cycle," *IEEE Micro*, 10(1):27–36, Feb. 1990.

[Cray77] Cray, *CRAY-1 Computer System Hardware Reference Manual*, Cray Research Institute, 1977.

[Cray89] Cray, *The Cray Y/MP Functional Description Manual*, Cray Research Inc., Eagan, MN, 1989.

[Cray91] Cray, *The Cray Y/MP C-90 Supercomputer System*, Cray Research Inc., Eagan, MN, 1991.

[Cray92] Cray, *Cray/MPP Announcement*, Cray Research, Inc., Eagan, MN, 1992.

[CSRD91] CSRD, "Perfect Club Benchmark Evaluation Package," Technical report, Center for Supercomputer Research and Development, University of Illinois, Urbana, 1991.

[Cybenko92] G. Cybenko and D. J. Kuck, "Revolution or Evolution," *IEEE Spectrum*, 29(9):39–41, 1992.

[Dally86] W. J. Dally and C. L. Seitz, "The Torus Routing Chip," *Journal of Distributed Computing*, 1(3):187–196, 1986.

[Dally87a] W. J. Dally et al., "Architecture of a Message-Driven Processor," *Proc. 14th Annu. Int. Symp. Computer Arch.*, pp. 189–205, IEEE CS Press, June 1987.

[Dally87b] W. J. Dally and C. L. Seitz, "Deadlock-Free Message Routing in Multiprocessor Interconnection Network," *IEEE Trans. Computers*, C-36(5):547–553, May 1987.

[Dally87c] W. J. Dally and P. Song, "Design of a Self-Timed VLSI Multicomputer Communication Controller," *Proc. Int. Conf. Computer Design*, pp. 230–234, IEEE CS Press, Oct. 1987.

[Dally90a] W. J. Dally, "Network and Processor Architecture for Message-Driven Computers," in Suaya and Birtwistle (eds.), *VLSI and Parallel Computation*, Chapter 3, Morgan Kaufmann, San Mateo, CA, 1990.

[Dally90b] W. J. Dally, "Performance Analysis of k-ary n-Cube Interconnection Networks," *IEEE Trans. Computers*, 39(6):775–785, 1990.

[Dally90c] W. J. Dally, "Virtual Channel Flow Control," *Proc. 17th Annu. Int. Symp. Computer Arch.*, pp. 60–68, May 1990.

[Dally92] W. J. Dally, J. Fiske, J. Keen, R. Lethin, M. Noakes, P. Nuth, R. Davison, and G. Fyler, "The Message-Driven Processor: A Multicomputer Processing Node with Efficient Mechanisms," *IEEE Micro*, 12(2):23–39, Apr. 1992.

[Davidson71] E. S. Davidson, "The Design and Control of Pipelined Function Generators," *Proc. Int. IEEE Conf. System Networks and Computers*, pp. 19–21, 1971.

[Davidson75] E. S. Davidson, D. P. Thomas, L. E. Shar, and J. H. Patel, "Effective Control for Pipelined Computers," *Proc. COMPCON*, pp. 181–184, 1975.

[DEC86] DEC, "The Whetstone Performance," Technical report, Digital Equipment Corporation, Bedford, MA, 1986.

[DEC92] DEC, "Alpha Architecture Handbook," Technical report, Digital Equipment Corporation, Boxboro, MA, 1992.

[DeCegamma89] A. L. DeCegamma, *The Technology of Parallel Processing:* Architectures and VLSI Hardware, Prentice-Hall, Englewood Cliffs, NJ, 1989.

[Dekel81] E. Dekel, D. Nassimi, and S. Sahni, "Parallel Matrix and Graph Algorithms," *SIAM J. Computing*, pp. 657–673, 1981.

[Denning68] P. J. Denning, "Working Set Model for Program Behavior," *Commun. ACM*, 11(6):323–333, 1968.

[Dennis80] J. B. Dennis, "Data Flow Supercomputers," *IEEE Computer*, pp. 48–56, Nov. 1980.

[Dennis91] J. Dennis, "The Evolution of "Static" Dataflow Architecture," in Gaudiot and Bic (eds.), *Advanced Topics in Dataflow Computing*, pp. 35–91, Prentice-Hall, Englewood Cliffs, NJ, 1991.

[Diefendorff92] K. Diefendorff and M. Allen, "Organization of the Motorola 88110 Superscalar RISC Microprocessor," *IEEE Micro*, 12(2):40–63, Apr. 1992.

[Dijkstra68] E. W. Dijkstra, "Cooperating Sequential Processes," in Genuys (ed.), *Programming Languages*, Academic Press, New York, 1968.

[Dinning89] A. Dinning, "A Survey of Synchronization Methods for Parallel Computers," *IEEE Computer*, 22(7), 1989.

[Dongarra86] J. J. Dongarra and D. C. Sorensen, "SCHEDULE: Tools for Developing and Analyzing Parallel Fortran Programs," Technical Memo 86, Argonne National Laboratory, 1986.

[Dongarra89] J. Dongarra and A. Hinds, "Comparison of the Cray X/MP-4, Fujitsu VP-200, and Hitachi S-810/20," in Hwang and DeGroot (eds.), *Parallel Processing for Supercomputing and Artificial Intelligence*, pp. 289–323, McGraw-Hill, New York, 1989.

[Dongarra92] J. Dongarra, "Performance of Various Computers Using Standard Linear Equations Software," Technical report, Computer Science Department, University of Tennessee, Knoxville, TN, 1992.

[Dubois86] M. Dubois, C. Scheurich, and F. A. Briggs, "Memory Access Buffering in Multiprocessors," *Proc. 13th Annu. Int. Symp. Computer Arch.*, pp. 434–442, 1986.

[Dubois88] M. Dubois, C. Scheurich, and F. A. Briggs, "Synchronization, Coherence and Event Ordering in Multiprocessors," *IEEE Computer*, 21(2), 1988.

[Dubois90a] M. Dubois and F. A. Briggs, "Tutorial Notes on Shared-Memory Architectures for Multiprocessors," *Proc. 17th Symp. Computer Arch.*, Seattle, WA, 1990.

[Dubois90b] M. Dubois and S. Thakkar (eds.), *Cache and Interconnect Architectures in Multiprocessors*, Kluwer Academic Publishers, Boston, MA, 1990.

[Dubois92a] M. Dubois, "Delayed Consistency," in Dubois and Thakkar (eds.), *Scalable Shared-Memory Multiprocessors*, Kluwer Academic Publishers, Boston, MA, 1992.

[Dubois92b] M. Dubois and S. Thakkar (eds.), *Scalable Shared-Memory Multiprocessors*, Kluwer Academic Publishers, Boston, MA, 1992.

[Eager89] D. L. Eager, J. Zahorjan, and E. D. Lazowska, "Speedup Versus Efficiency in Parallel Systems," *IEEE Trans. Computers*, 38(3):408–423, Mar. 1989.

[Edenfield90] R. Edenfield, M. Gallup, W. Ledbetter, R. McGarity, E. Quintana, and R. Reininger, "The 68040 Processor: Part I. Design and Implementation," *IEEE Micro*, 10(1):66–78, Feb. 1990.

[Emma87] P. G. Emma and E. S. Davidson, "Characterization of Branch and Data Dependences in Programs for Evaluating Pipeline Performance," *IEEE Trans. Computers*, 36:859–875, 1987.

[Encore87] Encore, *Multimax Technical Summary*, Encore Computer Corporation, Ft. Lauderdale, FL, Mar. 1987.

[Enslow74] P. H. Enslow (ed.), *Multiprocessors and Parallel Processing*, Wiley, New York, 1974.

[Felten85] E. Felten, S. Karlin, and S. W. Otto, "The Traveling Salesman on a Hypercube MIMD Computer," *Proc. Int. Conf. Parallel Processing*, St. Charles, IL, Aug. 1985.

[Feng81] T. Y. Feng, "A Survey of Interconnection Networks," *IEEE Computer*, 14(12):12–27, 1981.

[Ferrante87] M. W. Ferrante, "Cyberplus and Map V Interprocessor Communications for Parallel and Array Processor Systems," in Karplus (ed.), *Multiprocessors and Array Processors*, Simulation Councils, Inc., San Diego, CA, 1987.

[Fisher81] J. A. Fisher, "Trace Scheduling: A Technique for Global Microcode Compaction," *IEEE Trans. Computers*, 30(7):478–490, 1981.

[Fisher83] J. A. Fisher, "Very Long Instruction Word Architectures and the ELI-512," *Proc. 10th Symp. Computer Arch.*, pp. 140–150, ACM Press, New York, 1983.

[Flynn72] M. J. Flynn, "Some Computer Organizations and Their Effectiveness," *IEEE Trans. Computers*, 21(9):948–960, 1972.

[Fortune78] S. Fortune and J. Wyllie, "Parallelism in Random Access Machines," *Proc. ACM Symp. Theory of Computing*, pp. 114–118, 1978.

[Fox87] G. C. Fox, S. W. Otto, and A. J. Hey, "Matrix Algorithms on Hypercube (I): Matrix Multiplication," *Parallel Computing*, pp. 17–31, 1987.

[Fujitsu90] Fujitsu, *VP2000 Series Supercomputers*, Fujitsu Ltd., Japan, 1990.

[Fujitsu92] Fujitsu, *VPP500 Vector Parallel Processor*, Fujitsu America, Inc., San Jose, CA, 1992.

[Gajski82] D. D. Gajski, D. Padua, D. J. Kuck, and R. H. Kuhn, "A Second Opinion on Dataflow Machines and Languages," *IEEE Computer*, 15(2), 1982.

[Gajski85] D. D. Gajski and J. K. Peir, "Essential Issues in Multiprocessor Systems," *IEEE Computer*, 18(6), 1985.

[Gaudiot91] J.-L. Gaudiot and L. Bic, *Advanced Topics in Dataflow Computing*, Prentice-Hall, Englewood Cliffs, NJ, 1991.

[Gelernter85a] D. Gelernter, "Generative Communication in Linda," *ACM Trans. Prog. Lang. and Systems*, 7(1):80–112, Jan. 1985.

[Gelernter85b] D. Gelernter, N. Carriero, S. Chandran, and S. Chang, "Parallel Programming in Linda," *Proc. Int. Conf. Parallel Processing*, pp. 255–263, 1985.

[Gelernter90] D. Gelernter, A. Nicolau, and D. Padua, *Languages and Compilers for Parallel Computing*, MIT Press, Cambridge, MA, 1990.

[Gharachorloo90] K. Gharachorloo, D. Lenoski, J. Laudon, P. Gibbons, and J. Hennessy, "Memory Consistency and Event Ordering in Scalable Shared-Memory Multiprocessors," *Proc. 17th Annu. Int. Symp. Computer Arch.*, June 1990.

[Gharachorloo91] G. M. Gharachorloo and K. R. Traub, "Multithreading: A Revisionist View of Dataflow Architecture," *Proc. 18th Annu. Int. Symp. Computer Arch.*, May 1991.

[Gharachorloo92a] K. Gharachorloo, S. Adve, A. Gupta, J. L. Hennessy, and M. Hill, "Programming for Different Memory Consistency Models," *J. Para. Distri. Computing*, Aug. 1992.

[Gharachorloo92b] K. Gharachorloo, A. Gupta, and J. L. Hennessy, "Hiding Memory Latency Using Dynamic Scheduling in Shared-Memory Multiprocessors," *Proc. 19th Annu. Int. Symp. Computer Arch.*, Gold Coast, Australia, May 1992.

[Gharachorloo92c] K. Gharachorloo, A. Gupta, and J. L. Hennessy, "Performance Evaluation of Memory Consistency Models for Shared-Memory Multiprocessors," *Proc. Fourth Int. Conf. Arch. Support for Prog. Lang. and OS*, 1992.

[Gjessing92] S. Gjessing, G. B. Gustavson, J. R. James, and E. H. Kristiansen, "The SCI Cache Coherence Protocol," in Dubois and Thakkar (eds.), *Scalable Shared-Memory Multiprocessors*, Kluwer Academic Publishers, Boston, MA, 1992.

[Glass92] C. J. Glass and L. M. Ni, "The TURN Model for Adaptive Routing," *Proc. 19th Annu. Int. Symp. Computer Arch.*, 1992.

[Goble81] G. H. Goble and M. H. Marsh, "A Dual-Processor UNIX VAX 11/780," Technical report, Dept. of Electrical Engineering, Purdue University, West Lafayette, IN, Sept. 1981.

[Goff91] G. Goff, K. Kennedy, and C. W. Tseng, "Practical Dependence Testing," *Proc. ACM SIGPLAN Conf. Prog. Lang. Design and Implementation*, 1991.

[Goodman83] J. R. Goodman, "Using Cache Memory to Reduce Processor-Memory Traffic," *Proc. 10th Symp. Computer Arch.*, pp. 124–131, June 1983.

[Goodman88] J. R. Goodman and P. Woest, "The Wisconsin Multicube: A New Large-Scale Cache-Coherent Multiprocessor," *Proc. 15th Annu. Int. Symp. Computer Arch.*, pp. 422–431, 1988.

[Goodman89] J. R. Goodman, M. K. Vernon, and P. J. Woest, "Efficient Synchronization Primitives for Large-Scale Cache-Coherent Multiprocessors," *Proc. Third Int. Conf. Arch. Support for Prog. Lang. and OS*, pp. 64–73, 1989.

[Goodman90] J. R. Goodman, "Cache Consistency and Sequential Consistency," Technical Report 61, IEEE SCI Committee, 1990.

[Goor89] A. J. van de Goor, *Computer Architecture and Design*, Addison-Wesley, Reading, MA, 1989.

[Gottlieb83] A. Gottlieb, R. Grishman, C. P. Kruskal, K. P. McAuliffe, L. Rudolph, and M. Snir, "The NYU Ultracomputer—Designing an MIMD Shared Memory Parallel Computer," *IEEE Trans. Computers*, C-32(2):175–189, February 1983.

[Goyal84] A. Goyal and T. Agerwala, "Performance Analysis of Future Shared-Storage Systems," *IBM J. Res. and Develop.*, pp. 95–98, Jan. 1984.

[Graham92] S. Graham, J. L. Hennessy, and J. D. Ullman, "Course on Code Optimization and Code Generation," in *Tutorial on Code Optimization and Generation*, Western Institute of Computer Science, Stanford University, Aug. 1992.

[Graunke90] G. Graunke and S. Thakkar, "Synchronization Algorithms for Shared-Memory Multiprocessors," *IEEE Computer*, 23(6):60–69, 1990.

[Greenberg89] R. I. Greenberg and C. E. Leiserson, "Randomized Routing on Fat Trees," *Advances in Computing Research*, 7:345–374, 1989.

[Gross83] T. R. Gross, "Code Optimization Techniques for Pipelined Architectures," *Proc. IEEE Computer Society Spring Int. Conf.*, pp. 278–285, 1983.

[Gupta90] A. Gupta, W. D. Weber, and T. Mowry, "Reducing Memory and Traffic Requirements for Scalable Directory-Based Cache coherence Schemes," *Proc. Int. Conf. Parallel Processing*, pp. 312–321, 1990.

[Gupta91] A. Gupta, J. L. Hennessy, K. Gharachorloo, T. Mowry, and W. D. Weber, "Computative Evaluation of Latency Reducing and Tolerating Techniques," *Proc. 18th Annu. Int. Symp. Computer Arch.*, pp. 254–263, Toronto, May 1991.

[Gupta92] A. Gupta and V. Kumar, "Scalability of Parallel Algorithms for Matrix Multiplication," Technical report, University of Minnesota, 1992.

[Gupta93] A. Gupta and V. Kumar, "The Scalability of FFT on Parallel Computers," *IEEE Trans. Parallel Distri. Systems*, to appear in 1993.

[Gurd85] J. R. Gurd, C. Kirkham, and J. Watson, "The Manchester Prototype Dataflow Computer," *Commun. ACM*, 28(1):36–45, 1985.

[Gustafson86] J. L. Gustafson, S. Hawkinson, and K. Scott, "The Architecture of a Homogeneous Vector Supercomputer," *Proc. Int. Conf. Parallel Processing*, pp. 649–652, 1986.

[Gustafson88] J. L. Gustafson, "Reevaluating Amdahl's Law," *Commun. ACM*, 31(5):532–533, May 1988.

[Gustafson91] J. Gustafson, D. Rover, S. Elbert, and M. Carter, "The Design of a Scalable, Fixed-Time Computer Benchmark," *J. Para. Distri. Computing*, 11, Aug. 1991.

[Gustavson86] D. B. Gustavson, "Introduction to the Fastbus," *Microprocessors and Microsystem*, 10(2):77–85, 1986.

[Hagersten92] E. Hagersten, A. Landin, and S. Haridi, "Multiprocessor Consistency and Synchronization Through Transient Cache States," in Dubois and Thakkar (eds.), *Scalable Shared-Memory Multiprocessors*, Kluwer Academic Publishers, Boston, MA, 1992.

[Halstead85] R. H. Halstead, Jr., "Multilisp: A Language for Concurrent Symbolic Computation," *ACM Trans. Prog. Lang. and Systems*, 7(4):501–538, Oct. 1985.

[Harrison90] W. Harrison, B. Kramer, W. Rudd, S. Shatz, C. Chang, Z. Segall, D. Clemmer, J. Williamson, B. Peek, B. Appelbe, K. Smith, and A. Kolawa, "Tools for Multiple-CPU Environments," *IEEE Software*, 7(3):45–51, May 1990.

[Hayes86] J. P. Hayes, T. N. Mudge, Q. F. Stout, S. Colley, and J. Palmer, "Architecture of a Hypercube Supercomputer," *Proc. Int. Conf. Parallel Processing*, pp. 653–660, 1986.

[Heath87] M. T. Heath, "Hypercube Applications at Oak Ridge National Laboratory," Heath (ed.), *Hypercube Multiprocessors*, SIAM, Philadelphia, 1987.

[Hellerman67] H. Hellerman, *Digital Computer System Principles*, McGraw-Hill, New York, 1967.

[Hennessy90] J. L. Hennessy and D. A. Patterson, *Computer Architecture: A Quantitative Approach*, Morgan Kaufmann, San Mateo, CA, 1990.

[Hennessy92] J. L. Hennessy, "Introduction to a Tutorial on Code Optimization and Generation," WICS, Stanford University, 1992.

[Hertzberger84] L. O. Hertzberger, "The Architecture of the Fifth-Generation Inference Computers," *Future Generation Computer Systems*, 1(1):19–21, July 1984.

[Hill92] M. D. Hill, "What Is Scalability?," in Dubois and Thakkar (eds.), *Scalable Shared-Memory Multiprocessors*, Kluwer Academic Press, Boston, MA, 1992.

[Hillis86] W. D. Hillis and G. L. Steele, "Data Parallel Algorithms," *Commun. ACM*, 29(12):1170–1183, 1986.

[Hiraki87] K. Hiraki, K. Nishida, S. Sekiguchi, T. Shimada, and T. Yiba, "The SIGMA-1 Dataflow Supercomputer: A Challenge for New Generation Supercomputing Systems," *J. Info. Processing*, 10(4):219–226, 1987.

[Hirata92] H. Hirata, K. Kimura, S. Nagamine, Mochizuki, A. Nishimura, and Y. Nakase, "An Elementary Processor Architecture with Simultaneous Instruction Issuing from Multiple Threads," *Proc. 19th Annu. Int. Symp. Computer Arch.*, 1992.

[Hoare74] C. A. R. Hoare, "Monitors: An Operating System Structuring Concept," *Commun. ACM*, 17(10):549–557, Oct. 1974.

[Holt78] R. C. Holt, G. S. Graham, E. D. Lazowska, and M. A. Scott, *Structured Concurrent Programming with Operating Systems Applications*, Addison-Wesley, Reading, MA, 1978.

[Homewood87] M. Homewood, D. May, D. Shepherd, and R. Shepherd, "THE IMS T800 Transputer," *IEEE Micro*, 7(5):10–26, Oct. 1987.

[Hord90] R. M. Hord, *Parallel Supercomputing in SIMD Architectures*, CRC Press, Boca Raton, FL, 1990.

[Horwat89] W. Horwat, "Concurrent Smalltalk on the Message-Driven Processor," Master's thesis, Laboratory for Computer Science, MIT, 1989.

[Hotchips91] Hotchips, *Proc. Hot Chips III Symp. on High-Performance Chips*, Stanford University, Palo Alto, CA, 1991.

[Hwang77] K. Hwang and S. B. Yao, "Optimal Batched Searching of Tree-Structured Files in Multiprocessor System," *J. ACM*, pp. 441–454, July 1977.

[Hwang78] K. Hwang, *Computer Arithmetic: Principles, Architecture and Design*, Wiley, New York, 1978.

[Hwang82a] K. Hwang and Y. H. Cheng, "Partitioned Matrix Algorithms for VLSI Arithmetic Systems," *IEEE Trans. Computers*, pp. 1215–1224, Dec. 1982.

[Hwang82b] K. Hwang, W. Croft, G. H. Goble, B. W. Wah, F. A. Briggs, W. Simmons, and C. L. Coates, "A UNIX-Based Local Area Network with Load Balancing," *IEEE Computer*, 15(4):55–66, 1982.

[Hwang84] K. Hwang and F. A. Briggs, *Computer Architecture and Parallel Processing*, McGraw-Hill, New York, 1984.

[Hwang87a] K. Hwang, "Advanced Parallel Processing with Supercomputer Architectures," *Proc. IEEE*, vol. 75, Oct. 1987.

[Hwang87b] K. Hwang and J. Ghosh, "Hypernet: A Communication-Efficient Architecture for Constructing Massively Parallel Computers," *IEEE Trans. Computers*, 36:1450–1466, Dec. 1987.

[Hwang87c] K. Hwang, J. Ghosh, and R. Chowkwanyun, "Computer Architectures for AI Processing," *IEEE Computer*, pp. 19–29, Jan. 1987.

[Hwang88] K. Hwang and Z. Xu, "Multipipeline Networking for Compound Vector Processing," *IEEE Trans. Computers*, 37(1):33–47, 1988.

[Hwang89a] K. Hwang and D. DeGroot (eds.), *Parallel Processing for Supercomputers and Artificial Intelligence*, McGraw-Hill, New York, 1989.

[Hwang89b] K. Hwang, P. S. Tseng, and D. Kim, "An Orthogonal Multiprocessor for Parallel Scientific Computations," *IEEE Trans. Computers*, C-38(1):47–81, Jan. 1989.

[Hwang90] K. Hwang et al., "OMP: A RISC-based multiprocessor Using Orthogonal Access Memories and Multiple Spanning Buses," *Proc. ACM Int. Conf. Supercomputing*, pp. 7–22, Amsterdam, The Netherlands, June 1990.

[Hwang91] K. Hwang and S. Shang, "Wired-NOR Barrier Synchronization for Designing Shared-Memory Multiprocessor," *Proc. Int. Conf. Parallel Processing*, St. Charles, IL, Aug. 1991.

[Hwu91] W. M. Hwu, "Tutorial Notes on Compiler Support for Superscalar Processors," *Proc. 18th Annu. Int. Symp. Computer Arch.*, Toronto, 1991.

[Iannucci88] R. A. Iannucci, *A Dataflow/von Neumann Hybrid Architecture*, Ph.D. thesis, MIT Laboratory for Computer Science, 545 Technology Square, Cambridge, MA 02139, 1988.

[IBM90a] IBM, *RISC System/6000 Technology*, IBM Advanced Workstations Division, IBM Austin Communications Dept., Austin, TX, 1990.

[IBM90b] IBM, *System/390 Processors, System Functions*, International Business Machines, White Plains, NY, 1990.

[IEEE85] IEEE, *Standard 754, Order No. CN-953*, IEEE Computer Society Press, Los Alamitos, CA, 1985.

[IEEE91] IEEE, *Futurebus+: Logical Layer Specifications, 896.1-1991*, Microprocessor Standards Subcommittee, IEEE Computer Society, 1991.

[Intel84] Intel, *Multibus II Bus Architecture Specification Handbook*, Intel Corporation, Santa Clara, CA, no. 146077-c, 1984.

[Intel90] Intel, *Supercompilers for the iPSC/860, Technical Summary*, Intel Scientific Computers, Beaverton, OR, 1990.

[Intel91] Intel, *Paragon XP/S Product Overview*, Supercomputer Systems Division, Intel Corporation, Beaverton, OR 97006, 1991.

[James90a] D. D. James, A. T. Laundrie, S. Gjessing, and G. S. Sohni, "Scalable Coherence Interface," *IEEE Computer*, 23(6):74–77, 1990.

[James90b] D. V. James, A. T. Laundrie, S. Gjessing, and G. S. Sohi, "Distributed-Directory Scheme: Scalable Coherent Interface," *IEEE Computer*, 23(6):74–77, 1990.

[Jermoluk90] T. Jermoluk, *Multiprocessor UNIX*, Silicon Graphics Inc., Santa Clara, CA, 1990.

[Johnson91] M. Johnson, *Superscalar Microprocessor Design*, Prentice-Hall, Englewood Cliffs, NJ, 1991.

[Johnsson90] L. Johnsson, "Communication in Network Architectures," in Suaya and Birtwistle (eds.), *VLSI and Parallel Computation*, Morgan Kaufmann, San Mateo, CA, 1990.

[Jones80] A. K. Jones and P. Schwarz, "Experience Using Multiprocessor Systems—A Status Report," *ACM Computing Survey*, 12(2):121–165, 1980.

[Jones86] M. B. Jones and R. F. Rashid, "Mach and Matchmaker: Kernel and Language Support for Object-Oriented Distributed Systems," *Proc. OOPSLA 1986*, pp. 67–77, Portland, OR, Sept. 1986.

[Jordan83] H. F. Jordan, "Performance Measurement on HEP—A Pipelined MIMD Computer," *Proc. 10th Symp. Computer Arch.*, pp. 207–212, 1983.

[Jordan86] H. F. Jordan, "Structuring Parallel Algorithms in an MIMD, Shared Memory Environment," *Parallel Computing*, pp. 93–110, May 1986.

[Jouppi89] N. P. Jouppi and D. W. Wall, "Available Instruction-Level Parallelism for Superscalar and Superpipelined Machines," *Proc. Third Int. Conf. Arch. Support for Prog. Lang. and OS*, pp. 272–282, ACM Press, New York, 1989.

[Kallstrom88] M. Kallstrom and S. S. Thakkar, "Programming Three Parallel Computers," *IEEE Software*, pp. 11–22, Jan. 1988.

[Kane88] G. Kane, *MIPS R2000 RISC Architecture*, Prentice-Hall, Englewood Cliffs, NJ, 1988.

[Karp66] R. M. Karp and R. E. Miller, "Properties of a Model for Parallel Computations: Determinacy, Termination, Queueing," *SIAM J. Appl. Math.*, pp. 1390–1411, Nov. 1966.

[Karp88] R. M. Karp and V. Ramachandran, "A Survey of Complexity of Algorithms for Shared-Memory Machines," Technical Report 408, University of California, Berkeley, 1988.

[Katz90] R. H. Katz and J. L. Hennessy, "High-Performance Microprocessor Architectures," *Int. J. High-Speed Electronics*, 1(1), Jan. 1990.

[Kawabe87] S. Kawabe et al., "The Single Vector-Engine Supercomputer S-820," *Nikkei Electronics*, pp. 111–125, 1987.

[Kermani79] P. Kermani and L. Kleinrock, "Virtual Cut-Through: A New Communication Switching Technique," *Computer Networks*, 3(4):267–286, 1979.

[Kodama90] Y. Kodama, S. Sakai, and Y. Yamaguchi, "A Prototype of Highly Parallel Dataflow Machine EM-4 and Its Preliminary Evaluation," *Prof. Info., Japan*, 1990.

[Kogge81] P. M. Kogge, *The Architecture of Pipelined Computers*, McGraw-Hill, New York, 1981.

[Kowalik85] J. S. Kowalik (ed.), *Parallel MIMD Computation: HEP Supercomputer and Applications*, MIT Press, Cambridge, MA, 1985.

[Kruatrachue88] B. Kruatrachue and T. Lewis, "Grain Size Determination for Parallel Processing," *IEEE Software*, 5(1):23–31, Jan. 1988.

[KSR91] KSR, *KSR-1 Overview*, Internal Report, Kendall Square Research Corporation, 170 Tracer Lane, Waltham, MA 02154, 1991.

[Kuck78] D. J. Kuck, *The Structure of Computers and Computations*, Wiley and Sons, New York, 1978.

[Kuck82] D. J. Kuck and R. A. Stokes, "The Burroughs Scientific Processor (BSP)," *IEEE Trans. Computers*, pp. 363–376, May 1982.

[Kuck84] D. J. Kuck, R. H. Kuhn, B. Leasure, and M. Wolfe, "The Structure of an Advanced Retargetable Vectorizer," in Hwang (ed.), *Supercomputers: Design and Applications*, IEEE Computer Society Press, Los Alamitos, CA, 1984.

[Kuck86] D. J. Kuck, E. S. Davidson, D. H. Lawrie, and A. H. Sameh, "Parallel Supercomputing Today—The Cedar Approach," *Science*, 231(2), Feb. 1986.

[Kumar87] V. Kumar and N. Rao, "Parallel Depth-First Search, Part II: Analysis," *Int. J. Para. Programming*, 16(6):501–519, 1987.

[Kumar88] M. Kumar, "Measuring Parallelism in Computation-Intensive Scientific/Engineering Applications," *IEEE Trans. Computers*, 37(9):1088–1098, 1988.

[Kumar90] V. Kumar and V. Singh, "Scalability of Parallel Algorithms for the All-Pairs Shortest Path Problem," *Proc. Int. Conf. Parallel Processing*, pp. 136–140, 1990.

[Kumar92] V. Kumar and A. Gupta, "Analyzing Scalability of Parallel Algorithms and Architectures," Technical Report AHPCRC 92-020, Army High-Performance Computing Research Center, University of Minnesota, Minneapolis, MN, 1992.

[Kung78] H. T. Kung and C. E. Leiserson, "Systolic Arrays (for VLSI)," Duff and Stewart (eds.), *Sparse Matrix Proceedings*, Knoxville, TN, 1978, SIAM, Philadelphia.

[Kung80] H. T. Kung, "The Structure of Parallel Algorithms," in Yovits (ed.), *Advances in Computers*, vol. 19, pp. 65–112, Academic Press, New York, 1980.

[Kung84] S. Y. Kung, "On Supercomputing with Systolic and Wavefront Array Processors," *Proc. IEEE*, pp. 867–884, July 1984.

[Kung88] S. Y. Kung, *VLSI Array Processors*, Prentice-Hall, Englewood Cliffs, NJ, 1988.

[Kung90] H. T. Kung, "How to Move Parallel Processing into the Mainstream," *Proc. First Workshop on Parallel Processing*, Taiwan, China, Dec. 1990.

[Lam88] M. S. Lam, "Software Pipelining: An Effective Scheduling Technique for VLIW Machines," *Proc. ACM SIGPLAN Conf. Prog. Lang. Design and Implementation*, pp. 318–328, 1988.

[Lam92] M. S. Lam, "Tutorial on Compilers for Parallel Machines," Western Institute of Computer Science, Stanford University, 1992.

[Lamport78] L. Lamport, "Time, Clock, and Ordering of Events in a Distributed System," *Commun. ACM*, July 1978.

[Lamport79] L. Lamport, "How to Make a Multiprocessor Computer That Correctly Executes Multiprocess Programs," *IEEE Trans. Computers*, 28(9):241–248, 1979.

[Lan90] Y. Lan, A. H. Esfahanian, and L. M. Ni, "Multicast in Hypercube Multiprocessors," *J. Para. Distri. Computing*, pp. 30–41, Jan. 1990.

[Lang82] T. Lang, M. Valero, and I. Alegre, "Bandwidth Analysis of Crossbar and Multiple-Bus Contentions for Multiprocessors," *IEEE Trans. Computers*, pp. 1227–1233, Jan. 1982.

[Larson73] A. G. Larson, "Cost-Effective Processor Design with an Application to FFT," Technical Report SU-SEL-73-037, Stanford University, Aug. 1973.

[Larson84] J. L. Larson, "Multitasking on the Cray X-MP-2 Multiprocessor," *IEEE Computer*, pp. 62–69, July 1984.

[Laudon92] J. Laudon, A. Gupta, and M. Horowitz, "Architectural and Implementation Tradeoffs in the Design of Multiple-Context Processors," Technical Report CSL-TR-92-523, Computer Systems Laboratory, Stanford University, Stanford, CA 94305-4055, 1992.

[Lawrie75] D. H. Lawrie, "Access and Alignment of Data in a Array Processor," *IEEE Trans. Computers*, Dec. 1975.

[Leasure90] B. Leasure, *PCF Fortran Extension*, Kuck & Associates, Champaign, IL 61820, 1990.

[Lee80] R. B. Lee, "Empirical Results on the Speedup, Efficiency, Redundancy, and Quality of Parallel Computations," *Proc. Int. Conf. Parallel Processing*, pp. 91–96, Aug. 1980.

[Lee84] J. K. Lee and A. Smith, "Branch Prediction Strategies and Branch Target Buffer Design," *IEEE Computer*, 17(1):6–22, 1984.

[Leiserson85] C. E. Leiserson, "Fat-Trees: Universal Networks for Hardware-Efficient Super-computing," *IEEE Trans. Computers*, 34:892–901, 1985.

[Leiserson92] C. E. Leiserson et al., "The Network Architecture of the Connection Machine CM-5," *Proc. ACM Symp. Parallel Algorithms and Architecture*, San Diego, CA, 1992.

[Lenoski90] D. Lenoski, J. Laudon, K. Gharachorloo, A. Gupta, and J. Hennessy, "The Directory-Based Cache Coherence Protocol for the DASH Multiprocessor," *Proc. 17th Annu. Int. Symp. Computer Arch.*, pp. 148–159, 1990.

[Lenoski92] D. Lenoski, J. Laudon, K. Gharachorloo, W. D. Weber, A. Gupta, J. Hennessy, M. Horowitz, and M. Lam, "The Stanford Dash Multiprocessor," *IEEE Computer*, pp. 63–79, Mar. 1992.

[Lewis92] T. G. Lewis and H. El-Rewini, *Introduction to Parallel Computing*, Prentice-Hall, Englewood Cliffs, NJ, 1992.

[Li85] K. C. Li and H. Schwetman, "Vectorizing C: A Vector Processing Language," *J. Para. Distri. Computing*, 2(2):132–169, May 1985.

[Li86] K. Li, "Shared Virtual Memory on Loosely Coupled Multiprocessors," Technical report, Yale University, 1986.

[Li88] K. Li, "IVY: A Shared Virtual Memory System for Parallel Computing," *Proc. Int. Conf. Parallel Processing*, pp. 94–101, 1988.

[Li89] K. Li and P. Hudak, "Memory Coherence in Shared-Memory Systems," *ACM Trans. Computer Systems*, pp. 321–359, Nov. 1989.

[Li91] K. Li and K. Petersen, "Evaluation of Memory System Extensions," *Proc. 18th Annu. Int. Symp. Computer Arch.*, Toronto, 1991.

[Li92] K. Li, "Scalability Issues of Shared Virtual Memory for Multiprocessors," in Dubois and Thakkar (eds.), *Scalable Shared-Memory Multiprocessors*, Kluwer Academic Publishers, Boston, MA, 1992.

[Lilja88] D. J. Lilja, "Reducing the Branch Penalty in Pipelined Processors," *IEEE Computer*, 21(7):47–55, 1988.

[Lilja92] D. J. Lilja, *Architectural Alternatives for Exploiting Parallelism*, IEEE Computer Society Press, Los Alamitos, CA, 1992.

[Lin91a] X. Lin, P. K. McKinley, and L. M. Ni, "Performance Evaluation of Multicast Worm-hole Routing in 2D-Mesh Multicomputers," *Proc. Int. Conf. Parallel Processing*, vol. I, pp. 435–442, 1991.

[Lin91b] X. Lin and L. M. Ni, "Deadlock-Free Multicast Wormhole Routing in Multicomputer Networks," *Proc. 18th Annu. Int. Symp. Computer Arch.*, pp. 116–125, 1991.

[Linder91] D. H. Linder and J. C. Harden, "An Adaptive and Fault Tolerant Wormhole Routing Strategy for k-ary n-Cubes," *IEEE Trans. Computers*, 40(1):2–12, Jan. 1991.

[Margulis90] N. Margulis, *i860 Microprocessor Architecture*, Intel Osborne/McGraw-Hill, Berkeley, CA, 1990.

[Marson88] M. A. Marson, G. Balbo, and G. Conte, *Performance Modules of Multiprocessor Systems*, MIT Press, Cambridge, MA, Cambridge, MA, 1988.

[MasPar91] MasPar, "The MasPar Family Data-Parallel Computer," Technical summary, MasPar Computer Corporation, Sunnyvale, CA, 1991.

[Mirapuri92] S. Mirapuri, M. Woodacre, and N. Vasseghi, "The MIPS R4000 Processor," *IEEE Micro*, 12(2):10–22, Apr. 1992.

[Mowry91] T. Mowry and A. Gupta, "Tolerating Latency Through Software-Controlled Prefetching in Shared-Memory Multiprocessors," *J. Para. Distri. Computing*, 12:87–106, June 1991.

[Muchnick88] S. S. Muchnick, "Optimizing Compilers for SPARC," *Sun Technology*, Summer:64–77, 1988.

[Mudge87] T. N. Mudge, J. P. Hayes, and D. D. Winsor, "Multiple Bus Architectures," *IEEE Computer*, 20(6):42–49, 1987.

[Nassi87] I. R. Nassi, "A Preliminary Report on the Ultramax: A Massively Parallel Shared-Memory Multiprocessor," Technical Report ETR 87-4, Encore Computer Corporation, Fort Lauderdale, FL, 1987.

[nCUBE90] nCUBE, *nCUBE 6400 Processor Manual*, nCUBE Company, Beaverton, OR 97006, 1990.

[NEC90] NEC, "SX-X Series HNSX," Technical report, Nippon Electric Company, Japan, 1990.

[NeXT90] NeXT Computer, Inc., Redwood City, CA, *The Mach Operating System, Chapter 1*, 1990.

[Ni85a] L. M. Ni and K. Hwang, "Optimal Load Balancing in a Multiple Processor System with Many Job Classes," *IEEE Trans. Software Engineering*, pp. 491–496, May 1985.

[Ni85b] L. M. Ni and K. Hwang, "Vector Reduction Techniques for Arithmetic Pipeline," *IEEE Trans. Computers*, pp. 404–411, May 1985.

[Ni91] L. M. Ni, "A Layered Classification of Parallel Computers," *Proc. 1991 Int. Conf. for Young Computer Scientists*, pp. 28–33, Beijing, China, May 1991.

[Nickolls90] J. R. Nickolls, "The Design of the MasPar MP-1: A Cost-Effective Massively Parallel Computer," in *IEEE Digest of Papers – Comcom*, pp. 25–28, IEEE Computer Society Press, Los Alamitos, CA, 1990.

[Nicol88] D. M. Nicol and F. H. Willard, "Problem Size, Parallel Architecture, and Optimal Speedup," *J. Para. Distri. Computing*, 5:404–420, 1988.

[Nicolau84] A Nicolau and J. A. Fisher, "Measuring the Parallelism Available for Very Long Instruction Word Architectures," *IEEE Trans. Computers*, 33(11):968–976, 1984.

[Nikhil89] R. S. Nikhil and Arvind, "Can Dataflow Subsume von Neumann Computing?," *Proc. 16th Annu. Int. Symp. Computer Arch.*, pp. 262–272, 1989.

[Nikhil92a] R. S. Nikhil, "Tutorial Notes on Multithreaded Architectures," *Proc. 19th Annu. Int. Symp. Computer Arch.*, 1992, Contact DEC Cambridge Res. Lab., 1 Kendall Square, Bldg. 700, Cambridge, MA 02139.

[Nikhil92b] R. S. Nikhil and G. M. Papadopoulos, "*T: A Multithreaded Massively Parallel Architecture," *Proc. 19th Annu. Int. Symp. Computer Arch.*, May 1992.

[Nitzberg91] B. Nitzberg and V. Lo, "Distributed Shared Memory: A Survey of Issues and Algorithms," *IEEE Computer*, 24(8):52–60, 1991.

[Noakes90] M. Noakes and W. J. Dally, "System Design of the J-Machine," in Dally (ed.), *Proc. Sixth MIT Conf. Advanced Research in VLSI*, pp. 179–194, MIT Press, Cambridge, MA, 1990.

[NS88] NS, *NS32532 Performance Analysis: A Benchmark Study*, National Semiconductor, 1988.

[NSF92] NSF, "Grand Challenge: High-Performance Computing and Communications," Report, Committee on Physical, Mathematical, and Engineering Sciences, U.S. Office of Science and Technology Policy, National Science Foundation, Washington, DC, 1992.

[Nussbaum91] D. Nussbaum and A. Agarwal, "Scalability of Parallel Machines," *Commun. ACM*, 34(3):57–61, 1991.

[OSF90] OSF, *OSF/1 Technical Seminar*, Open Software Foundation, Inc., Cambridge, MA, 1990.

[Ousterhout88] J. K. Ousterhout, A. R. Cherenson, F. Douglis, M. N. Nelson, and B. B. Welch, "The Sprite Network Operating System," *IEEE Computer*, 21(2):23–36, 1988.

[Padua80] D. A. Padua, D. J. Kuck, and D. H. Lawrie, "High-Speed Multiprocessors and Compilation Techniques," *IEEE Trans. Computers*, pp. 763–776, Sept. 1980.

[Padua86] D. A. Padua and M. J. Wolfe, "Advanced Compiler Optimizations for Supercomputers," *Commun. ACM*, pp. 1184–1201, Dec. 1986.

[Panda91] D. K. Panda and K. Hwang, "Fast Data Manipulation in Multiprocessors Using Parallel Pipelined Memories," *J. Para. Distri. Computing*, 12:130–145, June 1991.

[Parasoft90] Parasoft, *Express User's Guide Version 3.0*, Parasoft Corporation, Pasadena, CA 90025, 1990.

[Parker91] K. Parker, "The Next Generation Furturebus+," *Futurebus+ Design*, (1):12–28, Jan. 1991.

[Patel78] J. H. Patel, "Pipelines with Internal Buffers," *Proc. 5th Symp. Computer Arch.*, pp. 249–254, 1978.

[Patel81] J. H. Patel, "Performance of Processor-Memory Interconnections for Multiprocessors," *IEEE Trans. Computers*, pp. 771–780, Oct. 1981.

[Patel82] J. H. Patel, "Analysis of Multiprocessors with Private Caches," *IEEE Trans. Computers*, C-31(4):296–304, Apr. 1982.

[Patterson82] D. Patterson and C. Sequin, "A VLSI RISC," *IEEE Computer*, 15(9), 1982.

[Perrott79] R. H. Perrott, "A Language for Array and Vector Processors," *ACM Trans. Prog. Lang. and Systems*, 1(2):177–195, Oct. 1979.

[Perrott87] R. H. Perrott, *Parallel Programming*, Addison-Wesley, Reading, MA, 1987.

[Pfister85a] G. F. Pfister, W. C. Brantley, D. A. George, S. L. Harvey, W. J. Kleinfelder, K. P. McAuliffe, E. A. Melton, V. A. Norlton, and J. Weiss, "The IBM Research Parallel Processor Prototype (RP3): Introduction and Architecture," *Proc. Int. Conf. Parallel Processing*, pp. 764–771, Aug. 1985.

[Pfister85b] G. F. Pfister and V. A. Norton, "Hot Spot Contention and Combining in Multistage Interconnection Networks," *Proc. Int. Conf. Parallel Processing*, pp. 790–797, Aug. 1985.

[Polychronopoulos89] C. D. Polychronopoulos, M. Girkar, M. R. Haghighat, C. L. Lee, B. Leung, and D. Schouten, "Parafrase 2: An Environment for Parallelizing, Partitioning, Synchronizing Programs on Multiprocessors," *Proc. Int. Conf. Parallel Processing*, pp. 765–777, 1989.

[Pountain87] D. Pountain and D. May, *A Tutorial Introduction to Occam Programming*, McGraw-Hill, New York, 1987.

[Prasanna Kumar87] V. K. Prasanna Kumar and C. S. Raghavendra, "Array Processor with Multiple Broadcasting," *J. Para. Distri. Computing*, pp. 1202–1206, Apr. 1987.

[Preparata79] F. P. Preparata and J. E. Vuillemin, "The Cube-Connected Cycles: A Versatile Network for Parallel Computation," *Proc. 20th Symp. Foundations Computer Sci.*, pp. 140–147, 1979.

[Przybylski90] S. Przybylski, *Cache and Memory Hierarchy Design*, Morgan Kaufmann, San Mateo, CA, 1990.

[PSR90] PSR, *MIMDizer User's Guide Version 7.01*, Pacific Science Research, Placerville, CA 90025, 1990.

[Quinn87] M. J. Quinn, *Designing Efficient Algorithms for Parallel Commuters*, McGraw-Hill, New York, 1987.

[Quinn90] M. J. Quinn and P. J. Hatcher, "Data-Parallel Programming on Multicomputers," *IEEE Software*, 7(5):69–76, Sept. 1990.

[Ragsdale90] S. Ragsdale (ed.), *Parallel Programming Primer*, Intel Scientific Computers, Beaverton, OR, 1990.

[Ramamoorthy77] C. V. Ramamoorthy and H. F. Li, "Pipeline Architecture," *ACM Computing Survey*, pp. 61–102, Mar. 1977.

[Rashid81] R. F. Rashid and G. G. Robertson, "Accent: A Communication-Oriented Network Operating System Kernel," *Proc. 8th ACM Symp. Operating System Principles*, pp. 64–75, Dec. 1981.

[Rashid86] R. F. Rashid, "From RIG to Accent to Mach: The Evolution of a Network Operating System," *Proc. Fall Joint Computer Conf.*, pp. 1128–1137, Dallas, TX, Nov. 1986.

[Rice85] J. R. Rice, "Problems to Test Parallel and Vector Languages," Technical Report CSD-TR 516, Purdue University, May 1985.

[Ritchie74] D. M. Ritchie and K. Thompson, "The UNIX Time-Sharing System," *Commun. ACM*, 17(7):365–375, July 1974.

[Rothnie91] J. Rothnie, "KSR-1 Memory System," Technical report, Kendall Square Research, Cambridge, MA, 1991.

[Russell87] C. H. Russell and P. J. Waterman, "Variations on UNIX for Parallel Processing Computers," *Commun. ACM*, 30(12):1048–1055, Dec. 1987.

[Saavedra90] R. H. Saavedra, D. E. Culler, and T. von Eicken, "Analysis of Multithreaded Architectures for Parallel Computing," *Proc. ACM Symp. Parallel Algorithms and Architecture*, Greece, July 1990.

[Sakai91] S. Sakai, Y. Kodama, and Y. Yamaguchi, "Prototype Implementation of a Highly Parallel Dataflow Machine EM-4," *Proc. Int. Parallel Processing Symposium*, 1991.

[Sayanarayanan80] M. Sayanarayanan, "Commercial Multiprocessing Systems," *IEEE Computer*, 13(5):75–96, 1980.

[Scheurich89] C. Scheurich, *Access Ordering and Coherence in Shared-Memory Multiprocessors*, Ph.D. thesis, University of Southern California, 1989.

[Schimmel90] C. Schimmel, "UNIX on Modern Architectures," *Proc. Summer 1990 USENIX Conf.*, Anaheim, CA, June 1990.

[Schwartz80] J. T. Schwartz, "Ultra-Computers," *ACM Trans. Prog. Lang. and Systems*, 2(4):484–521, 1980.

[SCSI84] Computer Business Equipment Manufacturers, Wash. DC, *SCSI Small Computer System Interface, ANSC X3*, 1984.

[Seitz85] C. L. Seitz, "The Cosmic Cube," *Commun. ACM*, 28(1), 1985.

[Seitz88] C. L. Seitz, W. C. Athas, C. M. Flaig, A. J. Martin, J. Seizovic, C. S. Steele, and W.-K. Su, "The Architecture and Programming of the Ametek Series 2010 Multicomputer," *Proc. Conf. Hypercube Computers and Concurrent Applications*, pp. 33–36, Pasadena, CA, Jan. 1988.

[Seitz89] C. L. Seitz, J. Seizovic, and W. K. Su, "The C-Programmer's Guide to Multicomputer Programming," Technical Report CS-TR-88-1, California Institute of Technology, Pasadena, CA, 1989.

[Seitz90] C. L. Seitz, "Concurrent Architectures," in Suaya and Birtwistle (eds.), *VLSI and Parallel Computation*, Chapter 3, Morgan Kaufmann, San Mateo, CA, 1990.

[Seitz92] C. L. Seitz, "Mosaic C: An Experimental Fine-Grain Multicomputer," Technical report, California Institute of Technology, Pasadena, CA 91125, 1992.

[Sevcik89] K. Sevcik, "Characterization of Parallelism in Applications and Their Use in Scheduling," *Proc. ACM SIGMETRICS and Performance*, May 1989.

[Shapiro86] E. Shapiro, "Concurrent Prolog," *IEEE Computer*, 19(1):44–58, 1986.

[Shar72] L. E. Shar, "Design and Scheduling of Statistically Configured Pipelines," Lab Report SU-SEL-72-042, Stanford University, 1972.

[Sheperdson63] J. C. Sheperdson and H. E. Sturgis, "Computability of Recursive Functions," *J. ACM*, 10:217–255, 1963.

[Shih89] Y. Shih and J. Fier, "Hypercube Systems and Key Applications," in Hwang and DeGroot (eds.), *Parallel Processing for Supercomputing and Artificial Intelligence*, pp. 203–244, McGraw-Hill, New York, 1989.

[Siegel79] H. J. Siegel, "A Model of SIMD Machines and a Comparison of Various Interconnection Networks," *IEEE Trans. Computers*, 28(12):907–917, 1979.

[Siegel89] H. J. Siegel, *Interconnection Networks for Large-Scale Parallel Processing: Theory and Case Studies*, 2nd ed., McGraw-Hill, New York, 1989.

[Siewiorek91] D. P. Siewiorek and P. J. Koopman, *The Architecture of Supercomputers, TITAN: A Case Study*, Academic Press, New York, 1991.

[Simmons92] M. L. Simmons, H. J. Wasserman, O. M. Lubeck, C. Eoyang, R. Mendez, H. Harada, and M. Ishiguro, "A Performance Comparison of Four Supercomputers," *Commun. ACM*, 35(8):116–124, 1992.

[Sindhu92] P. S. Sindhu, J. M. Frailong, and M. Cekleov, "Formal Specification of Memory Modules," in Dubois and Thakkar (eds.), *Scalable Shared-Memory Multiprocessors*, Kluwer Academic Publishers, Boston, MA, 1992.

[Smith82] A. J. Smith, "Cache Memories," *ACM Computing Survey*, pp. 473–530, Sept. 1982.

[Smith85] B. Smith, "The Architecture of the HEP," in Kowalik (ed.), *Parallel MIMD Computation: HEP Supercomputer and Applications*, MIT Press, Cambridge, MA, 1985.

[Smith88] J. E. Smith, "Characterizing Computer Performance with a Single Number," *Commun. ACM*, 31(10):1202–1206, 1988.

[Smith89] J. E. Smith, "Dynamic Instruction Scheduling and The Astronautics ZS-1," *IEEE Computer*, 22(7):21–35, 1989.

[Smith90] J. E. Smith, W. C. Hsu, and C. Hsiung, "Future General-Purpose Supercomputer Architecture," *Proc. ACM Supercomputing Conf. 1990*, New York, Nov. 1990.

[Snir82] M. Snir, "On Parallel Search," *Proc. Principles of Distributed Computing*, pp. 242–253, 1982.

[Sohi90] G. S. Sohi, "Instruction Issue Logic for High-Performance, Interruptible, Multiple Functional Unit, Pipelined Computers," *IEEE Trans. Computers*, 39(3):349–359, March 1990.

[SPARC90] Sun Microsystems, *SPARC Architecture Reference Manual V8*, Dec. 1990.

[Stallings90] W. Stallings, *Reduced Instruction Set Computers*, 2nd ed., IEEE Computer Society Press, Los Alamitos, CA, 1990.

[Stenström90] P. Stenström, "A Survey of Cache Coherence Schemes for Multiprocessors," *IEEE Computer*, 23(6):12–25, 1990.

[Stenström92] P. Stenström, T. Joe, and A. Gupta, "Comparative Performance Evaluation of Cache-Coherent NUMA and COMA Architectures," *Proc. 19th Annu. Int. Symp. Computer Arch.*, 1992.

[Stone71] H. S. Stone, "Parallel Processing with the Perfect Shuffle," *IEEE Trans. Computers*, C20:153–161, 1971.

[Stone90] H. S. Stone, *High-Performance Computer Architecture*, Addison-Wesley, Reading, MA, 1990.

[Sullivan77] H. Sullivan and T. R. Bashkow, "A Large Scale, Homogeneous, Fully Distributed Parallel Machine," *Proc. 4th Symp. Computer Arch.*, vol. 5, pp. 105–124, Mar. 1977.

[Sun91] X. H. Sun and D. T. Rover, "Scalability of Parallel Algorithm-Machine Combinations," Technical Report IS-5057, UC-32, Ames Laboratory, Iowa State University, Ames, Iowa, 1991.

[Sun93] X. H. Sun and L. M. Ni, "Scalable Problems and Memory-Bound Speedup," *J. Para. Distri. Computing*, 1993, also appeared in *Proc. ACM Supercomputing*, 1990.

[Sweazey89] P. Sweazey, "Cache Coherence on SCI," *IEEE Computer Architecture Workshop*, Elilat, Israel, May 1989.

[Tabak90] D. Tabak, *Multiprocessors*, Prentice-Hall, Englewood Cliffs, NJ, 1990.

[Tabak91] D. Tabak, *Advanced Microprocessors*, McGraw-Hill, New York, 1991.

[Tanenbaum92] A. S. Tanenbaum, M. F. Kaashoek, and H. E. Bal, "Parallel Programming Using Shared Objects and Broadcasting," *IEEE Computer*, 25(8):10–20, 1992.

[Tevanian87] A. Tevanian, R. F. Rashid, M. W. Young, D. B. Golub, D. L. Black, and E. Cooper, "Mach Threads and the UNIX Kernel: The Battle for Control," *Proc. Summer 1987 USENIX Conf.*, pp. 185–197, Phoenix, AZ, June 1987.

[Tevanian89] A. Tevanian and B. Smith, "Mach: The Model for Future UNIX," *Byte*, 14(12):411–416, Nov. 1989.

[Thakkar90] S. S. Thakkar, M. Dubois, A. T. Laundrie, G. S. Sohi, D. V. James, S. Gjessing, M. Thapar, B. Delagi, M. Carlton, and A. Despain, "New Directions in Scalable Shared-Memory Multiprocessor Architectures," *IEEE Computer*, 23(6):71–83, 1990.

[Thompson80] S. D. Thompson, *A Complexity Theory for VLSI*, Ph.D. thesis, Carnegie-Mellon University, 1980.

[Thornton70] J. E. Thornton, *Design of a Computer: The CDC 6600*, Scott and Foresman, Glenview, IL, 1970.

[TI83] Texas Instruments Inc., Dallas, TX, *NuBus Specification*, 1983.

[TMC90] TMC, *The CM-2 Technical Summary*, Thinking Machines Corporation, Cambridge, MA, 1990.

[TMC91] TMC, *The CM-5 Technical Summary*, Thinking Machines Corporation, Cambridge, MA, 1991.

[Tomasulo67] R. M. Tomasulo, "An Efficient Algorithm for Exploiting Multiple Arithmetic Units," *IBM J. Res. and Develop.*, 11(1):25–33, 1967.

[Treleaven85] P. C. Treleaven, "Control-Driven, Data-Driven, and Demand-Driven Computer Architecture," *Parallel Computing*, 2, 1985.

[Trew91] A. Trew and G. Wilson (eds.), *Past, Present, Parallel: A Survey of Available Parallel Computer Systems*, Springer-Verlag, London, 1991.

[Tucker88] L. W. Tucker and G. G. Robertson, "Architecture and Applications of the Connection Machine," *IEEE Computer*, 21(8):26–38, 1988.

[Ullman84] J. D. Ullman, *Computational Aspects of VLSI*, Computer Science Press, Rockville, MD, 1984.

[VITA90] VME Bus International Trade Association and IEEE P1014 Working Group, *64-Bit VMEbus Specification, Edition D*, Jan. 1990.

[Wah90] B. W. Wah and C. V. Ramamoorthy (eds.), *Computers for Artificial Intelligence Processing*, Wiley, New York, 1990.

[Wall91] D. W. Wall, "Limits of Instruction-Level Parallelism," *Proc. Fourth Int. Conf. Arch. Support for Prog. Lang. and OS*, pp. 176–188, 1991.

[Wallace64] C. E. Wallace, "A Suggestion for Fast Multiplier," *IEEE Trans. Computers*, pp. 14–17, Feb. 1964.

[Waltz87] D. L. Waltz, "Applications of the Connection Machine. (New Computer Architecture from Thinking Machines Corporation," *IEEE Computer*, 20(1):85–97, 1987.

[Wang92] H. C. Wang, *Parallelization of Iterative PDE Solvers on Shared-Memory Multiprocessors*, Ph.D. thesis, University of Southern California, 1992.

[Wang93] H. C. Wang and K. Hwang, "Multicoloring Parallelization of Grid-Structured PDE Solvers," Technical report, University of Southern California, Los Angeles, 1993.

[Weicker84] R. P. Weicker, "Dhrystone: A Synthetic Systems Programming Benchmark," *Commun. ACM*, 27(10):1013–1030, 1984.

[Weiser85] W. Weiser et al., "Status and Performance of the Zmob Parallel Processing Systems," *Proc. COMPCON*, pp. 71–73, 1985.

[Weiss84] S. Weiss and J. E. Smith, "Instruction Issue Logic in Pipelined Supercomputers," *IEEE Trans. Computers*, pp. 1013–1022, 1984.

[Wilson87] A. W. Wilson, "Hierarchical Cache/Bus Architecture for Shared-Memory Multi-processors," *Proc. 14th Annu. Int. Symp. Computer Arch.*, pp. 244–252, 1987.

[Wirth77] N. Wirth, "Modula: A Language for Modular Multiprogramming," *Software Practice and Experience*, 7:3–35, Jan. 1977.

[Wolf91a] M. E. Wolf and M. S. Lam, "A Loop Transformation Theory and an Algorithm to Maximize Parallelism," *IEEE Trans. Parallel Distri. Systems*, 3(10):452–471, 1991.

[Wolf91b] M. E. Wolf and M. S. Lam, "A Data Locality Optimization Algorithm," *Proc. ACM SIGPLAN Conf. Prog. Lang. Design and Implementation*, pp. 30–44, 1991.

[Wolfe82] M. J. Wolfe, *Optimizing Supercompilers for Supercomputers*, Ph.D. thesis, University of Illinois, 1982.

[Wolfe89] M. J. Wolfe, "Automatic Vectorization, Data Dependence, and Optimizations for Parallel Computers," in Hwang and DeGroot (eds.), *Parallel Processing for Supercomputing and Artificial Intelligence*, Chapter 11, McGraw-Hill, New York, 1989.

[Worlton84] J. Worlton, "Understanding Supercomputer Benchmarks," *Automation*, pp. 121–130, Sept. 1984.

[Wu80] C. L. Wu and T. Y. Feng, "On a Class of Multistage Interconnection Networks," *IEEE Trans. Computers*, pp. 696–702, Aug. 1980.

[Wulf72] W. A. Wulf and C. G. Bell, "C.mmp—A Multi-Miniprocessor," *Proc. Fall Joint Computer Conf.*, pp. 765–777, 1972.

[Xu89] Z. Xu and K. Hwang, "Molecule: A Language Construct for Layered Development of Parallel Programs," *IEEE Trans. Computers*, 38(5):587–599, 1989.

[Xu91] J. Xu and K. Hwang, "Mapping Rule-Based Systems onto Multicomputers using Simulated Annealing," *J. Para. Distri. Computing*, pp. 442–455, Dec. 1991.

[Yamaguchi91] Y. Yamaguchi, S. Sakai, and Y. Kodama, "Synchronization Mechanisms of a Highly Parallel Dataflow Machine EM-4," *IEICE Trans.*, 74(1):204–213, 1991.

[Yew87] P. C. Yew, N. F. Tseng, and D. H. Lawrie, "Distributing Hot-Spot Addressing in Large-Scale Multiprocessors," *IEEE Trans. Computers*, pp. 388–395, Apr. 1987.

[Yew91] P. C. Yew and B. W. Wah (eds.), Special Issue on Shared-Memory Multiprocessors, *J. Para. Distri. Computing*, June 1991.

[Young87] M. W. Young, A. Tevanian, R. F. Rashid, D. B. Golub, J. Eppinger, J. Chew, W. Bolosky, D. L. Black, and R. Baron, "The Duality of Memory and Communication in the Implementation of a Multiprocessor Operating System," *Proc. 11th ACM Symp. Operating System Principles*, pp. 63–76, 1987.

[Zima90] H. Zima and B. Chapman, *Supercompiler for Parallel and Vector Computers*, Addison-Wesley, Reading, MA, 1990.

[Zorpetta92] G. Zorpetta, "The Power of Parallelism," *IEEE Spectrum*, 29(9):28–33, 1992.

Index

Copyright Statement

Three IEEE Standards are introduced in this book. The information presented in the following pages and figures was written within the context of the cited IEEE Standards. This information is copyrighted by the Institute of Electrical and Electronics Engineers, Inc.

1. The floating-point standard specified in Example 6.9 on pages 298 and 299 is based on IEEE Standard 754-1985, copyright © 1985 by the IEEE, Inc.

2. The Futurebus+ standard described in Section 5.1.4 on pages 221 to 224 including Fig. 5.6, with applications illustrated in Fig. 7.5 on page 337 and in Fig. 7.17 on page 356, is based on IEEE Standard 896.1-1991, copyright © 1991 by the IEEE, Inc.

3. The scalable coherence interface (SCI) described in Section 9.1.4 on pages 483 to 487 including Fig. 9.5 is based on the IEEE Standard 1596-1992, copyright © 1992 by the IEEE, Inc.

The IEEE takes no responsibility or liability for and will assume no liability for any damages resulting from the reader's misinterpretation of the said information resulting from the placement and context in this publication. Information was reproduced with the permission of the IEEE on March 8, 1993.

Answers to Selected Problems

Provided below are brief or partial answers to a few selected exercise problems. These answers are meant for readers to verify the correctness of their answers. Derivations or detailed computational steps in obtaining these answers are left for readers.

Problem 1.1 Average CPI=1.55 cycles per instruction. Effective processor performance = 25.8 MIPS. Execution time = 3.875 ms.

Problem 1.4 (a) Average CPI = 2.24. (b) MIPS rate = 17.86.

Problem 1.8 (a) Sequential execution time = 1664 CPU cycles. (b) SIMD execution time = 26 machine cycles. (c) Speedup factor = 64.

Problem 2.5

(a)

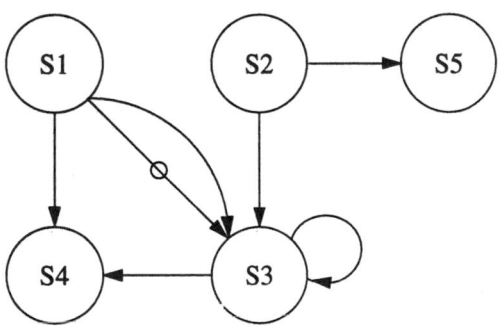

(b) S_4 and S_5 need to use the same Store Unit in accessing the memory. Therefore they are potentially storage-dependent.

Problem 2.11

(a)

Network	d	D	l	$(d \times D \times l)^{-1}$	Rank
Torus	6	6	192	1/6912	2
6-cube	6	6	192	1/6912	2
CCC	3	9	96	1/2592	1

(b)

Network	Mean Internode Distance	Rank
Torus	2.67	1
6-cube	2.67	1
CCC	3.67	2

Problem 2.14   (a) A 4×4 switch has 256 legitimate input-output connections, 24 of which are permutations. (b) A 64-input Omega network requires the use of 48 4×4 switches in 3 stages, with 16 switches per stage. Interstage connections are 4-way shuffles. 24^{48} permutations can be implemented in one pass through the network. (c) The fraction of one-pass permutations equals $24^{48}/64! \approx 1.4 \times 10^{-23}$.

Problem 3.2   Effective speedup $= 3$. Vectorization ratio $\alpha = 0.75$. (b) New speedup $= 3.43$ with vector/scalar speed ratio $= 18$. (c) α must be improved to 0.8.

Problem 3.3   (a) MIPS rate $= \dfrac{nx}{n - (n-1)\alpha}$. (b) $\alpha = 0.96$.

Problem 3.7   Sequential execution time $= 1{,}051{,}648$ cycles. (b) The speedup $= 16.28$. (c) Each processor is assigned 32 iterations balanced between the beginning and ending of the I-loop. (d) The ideal speedup 32 is achieved.

Problem 3.9

Machine	Arithmetic Mean Execution Time	Harmonic Mean Execution Rate	Rank
A	4.00 μs per instruction	0.25 MIPS	2
B	4.78 μs per instruction	0.21 MIPS	3
C	0.48 μs per instruction	2.10 MIPS	1

Problem 4.11

(a) Average cost $c = (c_1 s_1 + c_2 s_2)/(s_1 + s_2)$. $c \longrightarrow c_2$ when $s_2 \gg s_1$ and $c_2 s_2 \gg c_1 s_1$.

(b) $t_a = h t_1 + (1 - h) t_2$.

(c) $E = \dfrac{1}{h + (1 - h)r}$.

(e) $h = 0.99$.

Problem 4.15 (a) Hit ratio $h = 16/33$ for LRU policy. (b) Hit ratio $h = 16/33$ for the circular FIFO policy. (c) These two policies are equally effective for this particular page trace.

Problem 4.17 (a) $t_{eff} = 0.95t_1 + 0.05t_2$. (b) Total cost $c = c_1s_1 + c_2s_2$. (c) s_2 cannot exceed 18.6 Mbytes, $t_2 = 420ns$.

Problem 5.12
(a) $t_a = f_i(h_ic + (1-h_i)(b+c)) + (1-f_i)(h_dc + (1-h_d)((b+c)(1-f_{dir}) + (2b+c)f_{dir}))$.
(b) $t'_a = t_a + (1-f_i)f_{inv}i$.

Problem 5.14 (a) MIPS rate $= px/(1 - mtx)$. (b) $x = 5.83$ MIPS. (c) Effective MIPS rate equals 15.24 MIPS.

Problem 5.17 (a) There are 20 program orders: abcdef, abdcef, abdecf, abdefc, adbcef, adbecf, adbefc, adebcf, adebfc, adefbc, dabcef, dabecf, dabefc, daebcf, daebfc, daefbc, deabcf, deabfc, deafbc, defabc.
(b) Possible output patterns: 0111, 1011, and 1111. (c) Possible outputs are: 1001, 1011, 1101, 0110, 0111, 1110, and 1111.

Problem 6.1 (a) Speedup $= 4.99$. (b) Efficiency $= 0.99$. (c) Throughput $= 24.9$ MIPS.

Problem 6.9 (a) Forbidden latency is 3 with a collision vector (100). (b) State transition diagram is shown below:

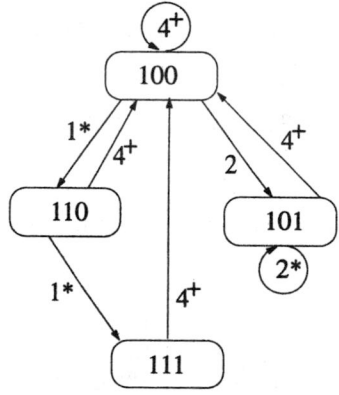

(c) Simple cycle: (2), (4), (1,4), (1,1,4), and (2,4). (d) Optimal constant latency cycle: (2), MAL=2. (e) Throughput $= 25$ MIPS.

Problem 6.15 (a) Speedup factor $= 3.08$. (b) 6.25 MIPS for the X processor and

19.2 MIPS for the Y processor.

Problem 6.17   (a) The four stages perform Exponent subtract, Align, Fraction Add, and Normalize, respectively. (b) 111 cycles to add 100 floating-point numbers.

Problem 7.1   (a) Memory bandwidth $= mc/[(c+m)\tau] = 32$ million words per second. (b) Memory utilization $= cm/(c+m) = 3.2$ requests per memory cycle.

Problem 7.4   A minimum of 21 time steps are needed to schedule the 24 code segments on the 4 processors, which are updating the 6 memory modules simultaneously without conflicts. The average memory bandwidth is thus equal to 70/21=3.33 words per time step, where 70 accounts for 70 memory accesses by four processors in 21 time steps without conflicts.

Problem 7.14   (a) (101101) → (101100) → (101110) → (101010) → (111010) → (011010). (b) Use either a route with a minimum of 20 channels and distance 9 or another route with a minimum distance of 8 and 22 channels. (c) The following tree shows multicast route on the hypercube:

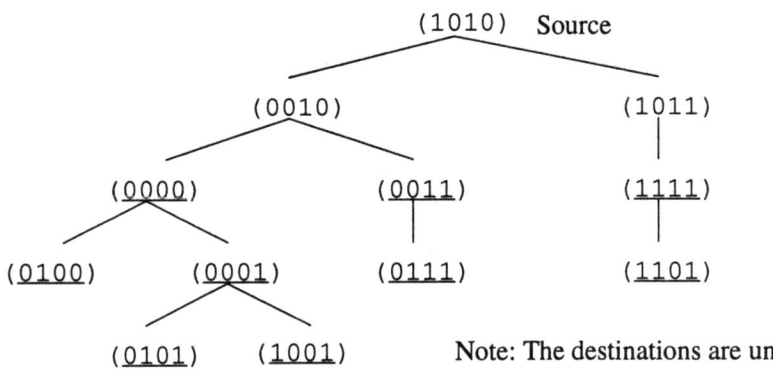

Problem 8.12   (a) $R_a = 10/(10 - 9\alpha)$ (in Mflops). (c) Vectorization ratio $\alpha = 26/27 = 0.963$. (d) $R_v = 3.5$ Mflops.

Problem 8.13   (a) Serial execution time =190 time units. (b) SIMD execution time = 35 time units.

Problem 8.14   (a) C-90 can execute 64 floating-point operations per cycle of 4.2ns, resulting in a peak performance $= 64/(4.2 \times 10^{-9}) = 15.2$ Gflops. (b) Similarly, NEC SX-X has a peak performance $= 64/(2.9 \times 10^{-9}) = 22$ Gflops.

Problem 9.1
(a) $E = 1/(1 + RL)$.
(b) $R' = (1 - h)R$, $E = 1/(1 + R'L) = 1/[1 + RL(1 - h)]$.

(c) If $N \geq N_d = \dfrac{L}{1/R' + C} + 1$, $E_{\text{sat}} = \dfrac{1}{1 + CR'} = \dfrac{1}{1 + (1-h)CR}$.

If $N \leq N_d$, $E_{\text{lin}} = \dfrac{N/R'}{1/R' + C + L} = \dfrac{N}{1 + R'C + R'L} = \dfrac{N}{1 + (1-h)(L+C)R}$.

(d) The mean internode distance $\overline{D} = (r+4)/3$.

Thus $L = 2\overline{D}t_d + t_m = \dfrac{2(r+4)}{3}t_d + t_m = \dfrac{2(\sqrt{p}+4)}{3}t_d + t_m$.

$E_{\text{sat}} = \dfrac{1/R'}{1/R' + C} = \dfrac{1}{1 + (1-h)CR}$.

$E_{\text{lin}} = \dfrac{\dot{N}}{1 + (1-h)R(L+C)} = \dfrac{N}{1 + (1-h)R([\frac{2(\sqrt{p}+4)}{3}t_d + t_m] + C)}$.

Problem 10.5
(a) A(5,8:*,*) declares A(5,8,1), A(5,9,1), A(5,10,1), A(5,8,2), A(5,9,2), A(5,10,2), A(5,8,3), A(5,9,3), A(5,10,3), A(5,8,4), A(5,9,4), A(5,10,4), A(5,8,5), A(5,9,5), A(5,10,5). B(3:*:3,5:8) corresponds to B(3,5), B(3,6), B(3,7), B(3,8), B(6,5), B(6,6), B(6,7), B(6,8), B(9,5), B(9,6), B(9,7), B(9,8). C(*,3,4) stands for C(1,3,4), C(2,3,4), C(3,3,4).
(b) Yes, no, no, and yes, respectively, for the four array assignments.

Problem 10.7
(a) $S_1 \longrightarrow S_2 \dashv\vdash S_3$
(b)

```
S₁:   A(1:N) = B(1:N)
S₃:   E(1:N) = C(2:N+1)
S₂:   C(1:N) = A(1:N) + B(1:N)
```

Problem 10.12
(a) Vectorized code:

```
TEMP(1:N) = A(1:N)
A(2:N+1) = TEMP(1:N) + 3.14159
```

(b) If intermediate vectors are introduced to store the value of $B(I) * C(I)$, then vector or parallel processing can be achieved. This is illustrated in the following code for performing the conditional inner-product operation:

```
D(1:N) = 0
where (A(1:N) .LE. 0.0) do
    D(1:N) = B(1:N) * C(1:N)
endwhere
```

The *where* operator serves as a binary control vector to mask off the operation on some vector elements. See [Wolfe89] for more details. The elements of D can then be summed up in parallel using a binary tree computing structure to obtain S. Alternatively, S can be obtained by a vector reduction operation as follows:

$$S = \text{sum}(D(1{:}N))$$

Similarly, the determination of X in the original loop can be vectorized. Let vector P be initialized so that $P(I) = I$ for $I = 1..N$, and Q is a zero vector. The following vector code yields the desired result:

```
where (A(1:N) .LE. 0.0) do
    Q(1:N) = P(1:N)
endwhere
K = max(Q(1:N))
X = B(K)
```

In the above, *max* is a vector reduction function which finds the maximum value of a vector.

Problem 11.15   (a) Suppose the image is partitioned into p segments, each consisting of $s = m/p$ rows. Vector *histog* is shared among the processors. Therefore its update has to be performed in a critical section to avoid race conditions. Assume it is possible to protect each element of vector *histog* by a separate semaphore. The following program performs parallel histogramming:

```
Var pixel(0 : m − 1, 0 : n − 1);
Var histog(0 : b − 1): integer;
Var lock(0 : b − 1): [0,1];
histog(0 : b − 1) = 0;
lock(0 : b − 1) = 1;
for k = 0 until  p − 1 0 Doall
      for i = k × s until  (k + 1) × s − 1 do
            for j = 0 until  n − 1 do
                  P(lock(pixel(i, j)));
                  histog(pixel(i, j)) = histog(pixel(i, j)) + 1;
                  V(lock(pixel(i, j)));
            Enddo
      Enddo
Endall
```

(b) The maximum speedup of the parallel program over the serial program is p, provided there is no conflict in accessing the *histog* vector and the overhead associated with synchronization is negligible. An alternative approach is to associate a local histog vector with each processor, which will obviate the use of critical sections. At the end of the algorithm, the values in local histog vectors are added to obtain the final result.

Proble 12.15
(a) If a conservative policy is used, at most $20/4 = 5$ processes can be active simulta-

neously. Since one of the drives allocated to each process can be idle most of the time, at most 5 drives will be idle at a time. In the best case, none of the drives will be idle. (b) To improve the drive utilization, each process can be initially allocated three tape drives. The fourth one will be allocated on demand. In this policy, at most $\lfloor 20/3 \rfloor = 6$ processes can be active simultaneously. The minimum number of idle drives is 0 and the maximum number is 2.